Long-Term Ecological Research

Felix Müller · Cornelia Baessler ·
Hendrik Schubert · Stefan Klotz
Editors

Long-Term Ecological Research

Between Theory and Application

Springer

Editors
Felix Müller
University of Kiel
Ecology Centre
Olshausenstraße 75
D-24118 Kiel
Germany
fmueller@ecology.uni-kiel.de

Hendrik Schubert
University of Rostock,
Institute of Biosciences
Albert-Einstein-Straße 3
D-18055 Rostock
Germany
hendrik.schubert@uni-rostock.de

Cornelia Baessler
Helmholtz Centre for Environmental
Research UFZ
Department of Community Ecology
Theodor-Lieser-Straße 4
06120 Halle
Germany
cornelia.baessler@ufz.de

Stefan Klotz
Helmholtz Centre for Environmental
Research UFZ
Department of Community Ecology
Theodor-Lieser-Straße 4
06120 Halle
Germany
stefan.klotz@ufz.de

ISBN 978-90-481-8781-2 e-ISBN 978-90-481-8782-9
DOI 10.1007/978-90-481-8782-9
Springer Dordrecht Heidelberg London New York

Library of Congress Control Number: 2010925420

Cover Images: Bavarian Forest National Park. Photos courtesy of Heinrich Rall.

Printed on acid-free paper

Springer is part of Springer Science+Business Media (www.springer.com)

Preface

The components of global change operate on different spatial and temporal scales. Scientific analyses of this issue, however, often deal with shorter time scales, due to the typical funding duration of research projects. In spite of this practice, long-term observation is indispensable for the detection of long-term processes and changes and thus is the foundation needed to develop sustainable strategies. Long-term observation and monitoring imply that data are saved and documented and that they stay accessible for a long time after individual research or projects have been completed. This is in line with the long-term horizon of large-scale strategies for environmental protection and the sustainable use of nature, such as the *EU Habitats Directive* and the *EU Water Framework Directive*, which consider time periods of over 20 years for planning and observation.

Long-term approaches are particularly important in investigations of environmental change, because the respective modifications usually occur only gradually, accompanied by larger temporal fluctuations garbling the trend. Under such circumstances, only an adequately long observation period can be a sound basis to secure significant results and to support predictions.

These long-dated phenomena are the subjects of the *Long-Term Ecosystem Research Initiative* (LTER). LTER is organized into networks ranging from the global to national scale. Networking is essential for the development of and tuning into common standards and research strategies. The national networks are especially important, as they provide the contacts to the scientific base, i.e., and the investigation sites. For example, the German network for long-term ecological research, LTER-D, is intended to be a platform for communication, documentation, and collaboration of scientists in long-term, system-oriented, and interdisciplinary environmental research in Germany. Actually it covers 17 sites and platforms all over Germany that are performing long-term ecological research in all relevant ecosystem types from the high mountains to the Wadden Sea. LTER-D is a member of the international LTER umbrella organization, ILTER, and the European regional network LTER-Europe. Both networks are exclusively defined by the commitment of the members to their shared goals.

Founded in 2004, the German LTER is still in its developing and growing phase. The further development of LTER is closely tuned to recommendations of and cooperation with LTER-Europe and ILTER in order to make the networks powerful. LTER-D enables and advocates the collaboration between the main long-term environmental research projects and infrastructures such as *The Terrestrial Environmental Observatories – TERENO*, funded by the Helmholtz Association of

German Research Centres and the *Biodiversity Exploratories* funded by the German Research Foundation. Additionally, universities and other research institutes, as well as the biosphere reserves and national parks of Germany, are the backbones of LTER-D.

This publication is an initiative of the German network but is not a book focusing on the activities of LTER-D only. We tried to bring together theoretical ecological questions of long-term processes, as well as an international dimension of long-term monitoring, observations, and research. By doing so we produced an overview on different aspects of long-term ecological research. Aquatic as well as terrestrial ecosystems are represented, as concepts and results of case studies in both are discussed. The different time dimensions, as well as scales from the community and ecosystems up to the landscape scale are included. Finally we tried to link research with application in different fields of ecology and to describe urgent infrastructural, methodological, and research demands and challenges for the future.

This work was only possible by the joint effort of all authors from different countries and networks worldwide. We are also grateful to the publisher for the encouragement and support to complete the book.

Kiel, Germany Felix Müller
Halle, Germany Cornelia Baessler
Rostock, Germany Hendrik Schubert
Halle, Germany Stefan Klotz

Contents

Contributors

Christian Aden Institute of Landscape Ecology, University of Vechta, 49364 Vechta, Germany, caden@iuw.uni-vechta.de

Jean Duclos Alegue Feugo Department of Ecosystems and Environmental Informatics, Brandenburg University of Technology, Konrad-Wachsmann-Allee 1, D-03046 Cottbus, Germany, jeduclo@yahoo.com

Sabine Augustin Eidgenössisches Departement für Umwelt, Verkehr, Energie und Kommunikation (UVEK) Bundesamt für Umwelt (BAFU) Abteilung Wald Worblentalstrasse 68, 3063 Ittigen, CH-3003 Bern, Switzerland, sabine.augustin@bafu.admin.ch

Cornelia Baessler Helmholtz Centre for Environmental Research – UFZ, Department of Community Ecology, Theodor-Lieser-Straße 4, 06120 Halle, Germany, cornelia.baessler@ufz.de

Dietmar Barkusky Leibniz Centre for Agricultural Landscape Research (ZALF), Eberswalder Str. 84, D-15374 Müncheberg, Germany, dbarkusky@zalf.de

Olaf Bastian Leibniz Institute of Ecological and Regional Development (IOER), Weberplatz, 01217 Dresden, Germany, o.bastian@ioer.de

Burkhard Beudert Bavarian Forest National Park, Freyunger Straße 2, 94481 Grafenau, Germany, bbeudert@npv-bw.de

Heye Bogena Agrosphere Institute (ICG-4), Forschungszentrum Jülich GmbH, 52425 Jülich, Germany, hbogena@fz-juelich.de

Michael Bredemeier University of Göttingen, Forschungszentrum Waldökosysteme, Büsgenweg 2, D-37077 Göttingen, Germany

Vera Breuer Ecology-Centre, Christian-Albrecht University, Kiel, Germany

Christian Buschbaum Alfred-Wegener-Institute for Polar and Marine Research, Wadden Sea Station Sylt, Hafenstrasse 43, 25992 List/Sylt, Germany, christian.buschbaum@awi.de

Ulrich Callies GKSS Research Centre, Geesthacht, Germany, Ulrich.Callies@gkss.de

Jonathan Chipman Environmental Remote Sensing Center, Space Science and Engineering Center, University of Wisconsin-Madison, 1225 West Dayton St, Madison, WI 53706, USA

Warren B. Cohen USDA Forest Service, Pacific Northwest Research Station, Corvallis, OR 97333, USA

Ralf Dannowski Leibniz Centre for Agricultural Landscape Research (ZALF), Eberswalder Str. 84, D-15374 Müncheberg, Germany, rdannowski@zalf.de

Britta Diederichs Schleswig-Holstein Agency for Coastal Defence, National Park and Marine Conservation, Dep. 3: National Park and Marine Conservation, Schlossgarten 1, 25832 Tönning, Germany, britta.diederichs@lkn.landsh.de

Oliver Dilly School of Integrated Climate System Sciences, Klima Campus, University of Hamburg, Grindelberg 5, 20144 Hamburg, Germany, oliver.dilly@uni-hamburg.de

Walter Durka Department of Community Ecology, Helmholtz Centre for Enironmental Research – UFZ, Theodor- Lieser-Straße 4, 06120 Halle, Germany, walter.durka@ufz.de

Brian D. Fath Biology Department, Towson University, Towson, MD 21252, USA; Dynamic Systems, International Institute for Applied Systems Analysis, Laxenburg, Austria, bfath@towson.edu

Markus Fischer Institute of Biochemistry and Biology, University of Potsdam, Maulbeerallee 1, D-14469 Potsdam, Germany; Institute of Plant Sciences, University of Bern, Altenbergrain 21, 3013 Bern, Switzerland, markus.fischer@ips.unibe.ch

Glyn Francis Crop and Food Research, Soil and Water Management, Gerald Street, Lincoln, New Zealand, glyn.francis@agresearch.co.nz

Otto Fränzle Ecology Centre, University of Kiel, Olshausenstrasse 40, 24098 Kiel, Germany

Dirk Freese Research Centre for Landscape Development and Mining Landscapes, Brandenburg University of Technology, Cottbus, Germany

Mark Frenzel Helmholtz Centre for Environmental Research – UFZ, Department of Community Ecology, Theodor-Lieser-Straße 4, 06120 Halle, Germany

Veronika Gaube Institute of Social Ecology, Klagenfurt University, Schottenfeldgasse 29, 1070 Vienna, Austria, veronika.gaube@uni-klu.ac.at

Werner Gerwin Research Centre for Landscape Development and Mining Landscapes, Brandenburg University of Technology, Cottbus, Germany

Albrecht Gnauck Department of Ecosystems and Environmental Informatics, Brandenburg University of Technology, Konrad-Wachsmann-Allee 1, D-03046 Cottbus, Germany, albrecht.gnauck@tu-cottbus.de

James R. Gosz Department of Biology, University of Idaho, 709 S Deakin Ave, Moscow, ID 83843, USA, jgosz@uidaho.edu

Clemens M. Grünbühel Sustainable Ecosystems Division, Commonwealth Scientific and Industrial Research Organisation (CSIRO), Canberra, Australia, Clemens.Grunbuhel@csiro.au

Helmut Haberl Institute of Social Ecology, Klagenfurt University, Schottenfeldgasse 29, 1070, Vienna, Austria, helmut.haberl@uni-klu.ac.at

Marco Heurich Bavarian Forest National Park, Freyunger Straße 2, 94481 Grafenau, Germany, marco.heurich@npv-bw.bayern.de

Wilfried Hierold Institute of Soil Landscape Research, Leibniz Centre for Agricultural Landscape Research, Eberswalder Straße 84, 15374 Müncheberg, Germany, whierold@zalf.de

Patrick Hostert Geography Department, Humboldt-Universität zu Berlin, Unter den Linden 6, 10099 Berlin, Germany, patrick.hostert@geo.hu-berlin.de

Reinhard F. Hüttl Soil Protection and Re-cultivation, Brandenburg University of Technology, Cottbus, Germany; Helmholtz Centre Potsdam, GFZ German Research Centre for Geosciences, Potsdam, Germany

K. Bruce Jones U.S. Geological Survey, 12201 Sunrise Valley Drive, Reston, VA, USA, kbjones@usgs.gov

Sven E. Jørgensen The Royal Danish School of Pharmacy, Section of Environmental Chemistry, Copenhagen, Denmark, sej@farma.ku.dk

Elisabeth K.V. Kalko Department of Experimental Ecology, University of Ulm, Albert-Einstein Allee 11, D-89069 Ulm, Germany, elisabeth.kalko@uni-ulm.de

Rolf Karez Landesamt für Natur und Umwelt Schleswig-Holstein, Hamburger Chaussee 25, 24220 Flintbek, Rolf.Karez@llur.landsh.de

Lukas Kleppin Institute of Landscape Ecology, University of Vechta, 49364 Vechta, Germany, lkleppin@iuw.uni-vechta.de

Stefan Klotz Department of Community Ecology, Helmholtz Centre for Environmental Research – UFZ, Theodor-Lieser-Straße 4, 06120 Halle, Germany

Hartmut Koehler Centre for Environmental Research and Sustainable Technology (UFT), University of Bremen, Leobener Strasse, D-28359 Bremen, Germany, hartmut.koehler@uni-bremen.de

Jan Köhler Leibniz-Institute of Freshwater Ecology and Inland Fisheries, Müggelseedamm 301, 12587 Berlin, Germany, koehler@igb-berlin.de

Alexander Kohler Institute of Landscape and Plant Ecology, Faculty of Agricultural Sciences, University of Hohenheim, D-70593 Stuttgart, Germany, alexander.kohler@uni-hohenheim.de

Martijn Kos Institute of Botany, Faculty of Biology and Preclinical Medicine, University Regensburg, D-93040, Regensburg, Germany, martijn.kos@biologie.uni-regensburg.de

Pavel Krám Department of Environmental Geochemistry and Biogeochemistry, Czech Geological Survey, Klarov 3, CZ-11821 Prague 1, Czech Republic, pavel.kram@geology.cz

Zdenka Křenová Šumavá National Park and Protected Landscape Area, Sušicka 399, 34192 Kašperske Hory, Czech Republic, gentiana1@centrum.cz

Bai-Lian Larry Li CAU-UCR International Center for Ecology and Sustainability, University of California, Riverside, USA, bai-lian.li@ucr.edu

K. Eduard Linsenmair Theodor-Boveri-Institute for Biosciences, University of Wuerzburg (Biocenter), Am Hubland, D-97074 Würzburg, Germany, ke_lins@biozentrum.uni-wuerzburg.de

Gunnar Lischeid Institute of Landscape Hydrology, Leibniz Centre for Agricultutral Landscape Research, Eberswalder Straße 84, 15374 Müncheberg, Germany, lischeid@zalf.de

Petru Lisivieveci Department of Systems Ecology and Sustainability (DSES), University of Bucharest, Spl. Independentei 91-95, Sector 5, Bucharest, Romania, lisip@spiruharet.ro

Martina Loebl Mount Allison University, Sackville, New Brunswick, Canada, martina.loebl@nioz.nl

Vera Luthardt Department of Landscape Management and Nature Conservation, Hochschule für nachhaltige Entwicklung (FH) Eberswalde, University of Applied Sciences Eberswalde, Friedrich- Ebertstr. 28, 16225 Eberswalde, Germany, vluthardt@hnee.de

Bernhard Luther Brandenburg University of Technology, Erich-Weinert-Straße 1, D-03046 Cottbus, Germany, luther@tu-cottbus.de

Julia Lutz Institute of Social Ecology, Klagenfurt University, Schottenfeldgasse 29, 1070 Vienna, Austria, julia.lutz@uni-klu.ac.at

John J. Magnuson Centre for Limnology, University of Wisconsin, 680 N Park St, Madison, WI 53706, USA, jjmagnus@wisc.edu

Bryan F.J. Manly Western Ecosystems Technology, Laramie, WY, USA, bmanly@west-inc.com

Peter Martens Alfred-Wegener-Institute for Polar and Marine Research, Wadden Sea Station Sylt, Hafenstrasse 43, 25992 List/Sylt, Germany, peter.martens@awi.de

Robin Matthews Integrated Land Use Systems Group, Macaulay Institute, Craigiebuckler, Aberdeen AB15 8QH, UK, R.Matthews@macaulay.ac.uk

Viesturs Melecis Institute of Biology, University of Latvia, Miera iela 3, LV-2169 Salaspils, Latvia, vmelecis@email.lubi.edu.lv

Michael Mirtl Federal Environment Agency Austria (Umweltbundesamt), Spittelauer Lände 5, 1090 Wien, Austria, michael.mirtl@umweltbundesamt.at

Felix Müller Ecology Centre, University of Kiel, Olshausenstraße 75, D-24118 Kiel, Germany, fmueller@ecology.uni-kiel.de

Lothar Müller Leibniz Centre for Agricultural Landscape Research (ZALF), Eberswalder Str. 84, D-15374 Müncheberg, Germany, lmueller@zalf.de

Seth Nii-Annang Soil Protection and Re-cultivation, Brandenburg University of Technology, Cottbus, Germany

Cornelia Ohl UFZ-Helmholtz Centre for Environmental Research, Permoserstraße 15, 04318 Leipzig, Germany; German Emissions Trading Authority at the Federal Environment Agency, Bismarckplatz 1, 14193 Berlin, Germany, Cornelia.Ohl@uba.de

Roland Pesch Institute of Landscape Ecology, University of Vechta, 49364 Vechta, Germany, rpesch@iuw.uni-vechta.de

Eva-Maria Pfeiffer Institute of Soil Sciences, University of Hamburg, Hamburg, Germany

Simone Pfeiffer Institute of Biochemistry and Biology, University of Potsdam, Maulbeerallee 1, D-14469 Potsdam, Germany, Simone.Pfeiffer@uni-potsdam.de

Peter Poschlod Institute of Botany, University Regensburg, D-93040, Regensburg, Germany, peter.poschlod@biologie.uni-regensburg.de

Daniel Prati Institute of Biochemistry and Biology, University of Potsdam, Maulbeerallee 1, D-14469 Potsdam, Germany; Institute of Plant Sciences, University of Bern, Altenbergrain 21, 3013 Bern, Switzerland, dani.prati@ips.unibe.ch

Heinrich Rall Bavarian Forest National Park, Freyunger Straße 2, 94481 Grafenau, Germany, heinrich.rall@npv-bw.bayern.de

Karsten Reise Alfred-Wegener-Institute for Polar and Marine Research, Wadden Sea Station Sylt, Hafenstrasse 43, 25992 List/Sylt, Germany, karsten.reise@awi.de

Stephanie Roauer Institute of Botany,University Regensburg, D-93040, Regensburg, Germany, roauer@gmx.de

Matthias Röder Landeshauptstadt Dresden, Postfach 12 00 20, D-01001 Dresden, Germany, mroeder@dresden.de

Wolfgang Schaaf Soil Protection and Re-cultivation, Brandenburg University of Technology, Cottbus, Germany

Claus-G. Schimming Ökologie-Zentrum der Christian-Albrechts-Universität zu Kiel, Olshausenstr. 40, 24098 Kiel, Germany, cschimming@ecology.uni-kiel.de

Uwe Schindler Leibniz Centre for Agricultural Landscape Research (ZALF), Eberswalder Str. 84, D-15374 Müncheberg, Germany, uschindler@zalf.de

Gunther Schmidt Institute of Landscape Ecology, University of Vechta, PO Box 1553, 49364 Vechta, Germany, gschmidt@iuw.uni-vechta.de

Joachim Schrautzer Ecology-Centre, Christian-Albrecht University, Kiel, Germany

Winfried Schröder Institute of Landscape Ecology, University of Vechta, 49364 Vechta, Germany, wschroeder@iuw.uni-vechta.de

Hendrik Schubert Institute of Biosciences, University of Rostock, Albert-Einsteinstr. 3, D-18055 Rostock, Germany, hendrik.schubert@uni-rostock.de

Ernst-Detlef Schulze Max Planck Institute for Biogeochemistry, Hans-Knoell-Str.10, D-07745 Jena, Germany, dschulze@bgc-jena.mpg.de

Peter Schwartze Biological Station of the District Steinfurt, 49545 Tecklenburg, Germany

Andreas Seemann Institute of Botany, University Regensburg, D-93040, Regensburg, Germany, andreasseemann@gmx.de

Kevin G. Sellner Chesapeake Research Consortium, Inc. , 645 Contees Wharf Road, Edgewater, MD 21037, USA, sellnerk@si.edu

Simron J. Singh Institute of Social Ecology, Faculty of Interdisciplinary Studies, Klagenfurt University, Schottenfeldgasse 29, 1070 Vienna, Austria, simron.singh@uni-klu.ac.at

Karina Stockmann GKSS Research Centre, Geesthacht, Germany, Karina.Stockmann@bsh.de

Frederick Swayne USDA Forest Service, Pacific Northwest Research Station, Corvallis, OR 97333, USA

Scott M. Swinton Department of Agricultural, Food and Resource Economics, Michigan State University, East Lansing, MI 48824-1039, USA, swintons@msu.edu

Ralf-Uwe Syrbe Centre for Landscape Research, Leibniz Institute of Ecological and Regional Development Weberplatz 1, 01217 Dresden, Germany, syrbe@iale.de

Filipa Tavares Rua da Saudade lte 614, 2840 Seixal, Portugal

Angheluta Vadineanu Department of Systems Ecology and Sustainability (DSES), University of Bucharest, Spl. Independentei 91-95, Sector 5, Bucharest, Romania, angheluta.vadineanu@cncsis.ro

Justus E.E. van Beusekom Alfred-Wegener-Institute for Polar and Marine Research, Wadden Sea Station Sylt, Hafenstrasse 43, 25992 List/Sylt, Germany, Justus.van.beusekom@awi.de

Harry Vereecken Agrosphere Institute (ICG-4), Forschungszentrum Jülich GmbH, 52425 Jülich, Germany, hvereecken@fz-juelich.de

Maik Veste Research Centre for Landscape Development and Mining Landscapes, Brandenburg University of Technology, Cottbus, Germany

Robert B. Waide US LTER Network Office, Department of Biology, University of New Mexico, Albuquerque, NM 87131, USA, rwaide@LTERnet.edu

Norbert Wasmund Leibniz Institut for Baltic Sea Research, Seestr. 15, D-18119 Warnemünde, Germany, norbert.wasmund@io-warnemuende.de

Wolfgang W. Weisser Institute of Ecology, University of Jena, Dornburger Str. 159, D-07743 Jena, Germany, wolfgang.weisser@uni-jena.de

Jake F. Weltzin U.S. Geological Survey and USA National Phenology Network, National Coordinating Office, 1955 East 6th Street, Tucson, AZ 85721, USA; jweltzin@usgs.gov

Karl-Otto Wenkel Leibniz Centre for Agricultural Landscape Research, Eberswalder Straße 84, D-15374 Müncheberg, Germany

Christina Weyer Ecological Modelling, University of Bayreuth, Dr.-Hans-Frisch-Str. 1-3, D-95440 Bayreuth, Germany, Christina.Mohr@bayceer.uni-bayreuth.de

Oliver Wiesmann Institute of Botany, University Regensburg, D-93040 Regensburg, Germany

Martin Wildenberg Institute of Social Ecology, Klagenfurt University, Schottenfeldgasse 29, 1070 Vienna, Austria, martin.wildenberg@global2000.at

Karen H. Wiltshire Biological Institute Helgoland, Helgoland, Germany, Karen.Wiltshire@awi.de

Georg H. Zeltner Institute of Landscape and Plant Ecology, University of Hohenheim, D-70593 Stuttgart, Germany, georg.zeltner@uni-hohenheim.de

Part I
Introduction

Chapter 1

Long-Term Ecosystem Research Between Theory and Application – An Introduction

Felix Müller, Cornelia Baessler, Mark Frenzel, Stefan Klotz, and Hendrik Schubert

Ecosystems are dynamic entities. Triggered by external and internal factors, ecosystems change on a multitude of spatial and temporal scales. Nevertheless, in the past, the analysis of ecosystem dynamics has had top priority with reference to short time spans. The exploration of ecological interrelations usually has been carried out within a time period of three years, due to the typical funding and qualification duration of environmental research projects. As a consequence, to a great extent, the impacts of changes in the long-term have been neglected. This tendency has provoked a lack of information and methodological know how in this area. That strategy of course is not consistent with the long-term, precautionary way of thinking and acting expressed in political manifestations like the 2010 biodiversity target of the CBD or the EU Habitats Directive, which is an essential component of the sustainability principle. Additionally, real life has demonstrated the general significance of long-term processes: global climate change with its multiple consequences has fostered the awareness that there is an essential lack of scientific knowledge to build the ground for answers to the urgent long-term problems of mankind and the biosphere per se arising from these issues.

Many environmental impacts, like the widespread forest damages, the consequences of anthropogenic eutrophication and pollution, and the potential effects of genetic engineering, have been enfolding with slow change rates. Therefore reference values collected over a long period have become indispensable for sound analyses, predictions, and well-elaborated decision-making processes. Furthermore, the rising global trade does not only provoke long-term land use modifications, but leads to local ecosystems being increasingly burdened by the import of new plants and animals (Neobiota). Long-term observations are essential for evaluating the consequences of these changes. Additionally, long-term investigations are urgently needed to answer basic theoretical questions of ecology, e.g. to register the combined action of processes on different scales, to reduce the uncertainties concerning the irreversibility of structural ecosystem change, the orientation of successions, the success of restoration measures, to evaluate and improve the efficiency of protection measures, and to derive clear definitions of emergent ecosystem properties like resilience, stability, and adaptability. Moreover, these investigations are valuable even in terms of a concrete economic background, as the change in ecosystems often affects ecosystem services like pollination, food provision, or water purification.

This respective *need for research on long-term ecological processes* has been noticed by several scientists who have become active in the worldwide Long-Term Ecological Research Network. Also political institutions like the United Nations Convention on Biological Diversity, UNCBD, and the United Nations Framework Convention on Climate Change, UNFCCC, have formulated suggestions to ascertain long-term ecological research and long-term observations. Moreover, the Ecosystem Approach of the CBD (Principle 8, CoP V, UNEP, 2000, see http://www.cbd.int/decisions/?dec=V/6) claims that ecosystem management has to take long-term processes into consideration. Furthermore, the temporal

F. Müller (✉)
Ecology Centre, University of Kiel, Olshausenstraße 75, D-24118 Kiel, Germany
e-mail: fmueller@ecology.uni-kiel.de

F. Müller et al. (eds.), *Long-Term Ecological Research*,
DOI 10.1007/978-90-481-8782-9_1, © Springer Science+Business Media B.V. 2010

Table 1.1 Exemplary targets of long-term ecosystem research initiatives (from the LTER-D web page – http://www.lter-d.ufz.de/)

Focal targets of LTER-D:

- To investigate complex ecosystem long-term processes under conditions of global change;
- To improve systems analysis methods, scale distinctions, and comprehensions of systems dynamics by integrating research on long-term and short dynamics;
- To enhance the cooperation between theory, empirics, and applications;
- To conceive early warning systems on the base of retrospective investigations;
- To elaborate scientific fundamentals for sustainable ecosystem management strategies;
- To enhance the knowledge about the effects of long-term processes by comparative analyses between research sites;
- To improve the collaboration between natural scientists and social scientists to better understand the dynamics of human–environmental systems;
- To integrate long-term environmental research and environmental monitoring;
- To support these objectives by joint data management structures, information exchange, and international cooperation.

goals of current environmental protection programs (for example: the EU Water Framework Directive or the Fauna-Flora-Habitat-Directive, Natura 2000) and the spans of climate scenarios are calculated for long-term time periods of 20 years and more. The implementation of suitable sustainable management strategies that argue in generations, as well as the interpretation of long-term scenario analyses requires a comprehensive and sound scientific basis which can only be derived from observations carried out during adequate time periods.

In addition to these time-related factors, the UNCBD's (CoP V, 2000) Ecosystem Approach essentially aims to preserve *ecosystem structures and functions* in order to ensure their long-term social, ecological, and economic benefits (ecosystem services). The seventh principle of Agenda 21 points in the same direction: *health and integrity of ecosystems* have to be preserved, protected, and, if necessary, re-established. A third aspect of relevant international arrangements calls for a continuing *integration of socio-economic approaches* in environmental management. The combination of socio-economics and ecosystem analysis is the basis for the realization of these holistic concepts, which should be used for both the ecosystem research and the development of principles for advanced decision making.

The application of these three basic principles (long-term investigations, ecosystem approaches, linkages between human and environmental systems) and the development of appropriate actions require specific investigations of long-term ecological development in which ecological and socio-economical aspects are combined by an interdisciplinary long-term ecosystem approach. These are basic motivations of several national and international research networks, which are cooperating within the ILTER programme (International Long-Term Ecosystem Research). To illustrate the targets of these groups, Table 1.1 demonstrates some basic objectives of the German LTER group.

The typical timescales of long-term investigations are demonstrated in Fig. 1.1. In this context, it has to be noted that the time span between a year and a century should be understood as an eligible expansion of the traditional research periods. To understand ecosystems, of course the integration of scales has to be put into the focus. Long-term processes and short-term dynamics are interlinked, and potentially the analysis of the respective temporal networks will provide new emergent properties, which arise due to the overall connectivity. Therefore the theory-based *integration* of scales is much more important than the distinction of spatio-temporal units.

Taking the described necessities as a basic requirement, the prime objective of this book is to focus on studies dealing with the investigation of complex, long-term ecological processes with regard to global change, the development of early warning systems, and the acquisition of a scientific basis for strategic conservation management and for the sustainable use of ecosystems. The central motivation of this book is to stimulate the international discussion to foster the cooperation within the worldwide ILTER organization. Additionally this book shall demonstrate the high significance of long-term-oriented research questions for the understanding of ecosystems, as well as a better application of scientific methodology in environmental practice. Following these targets, the book is structured into eight sections, each of them posing

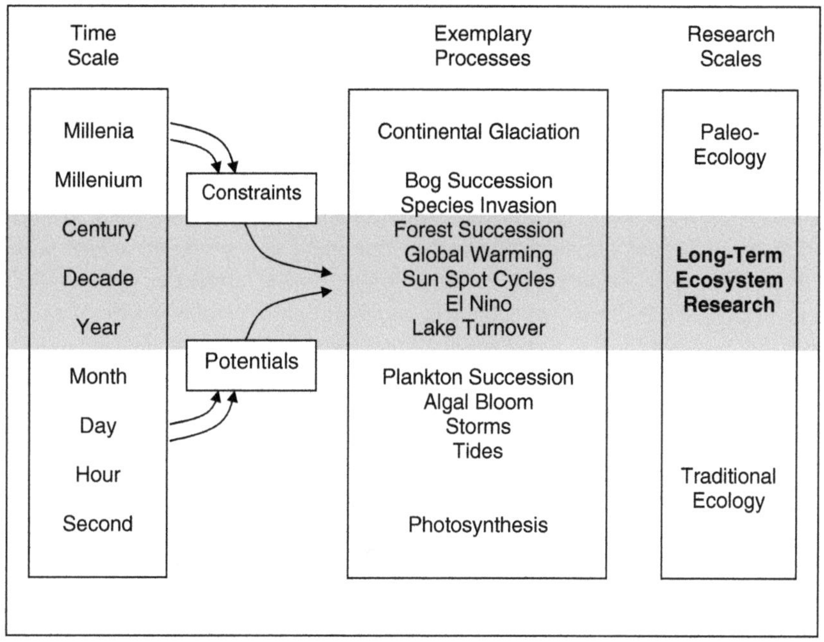

Fig. 1.1 Assigning different timescales to exemplary ecological processes and typical research scales, after Hobbie, Carpenter, Grimm, Gosz, and Seastedt (2003)

distinct questions (Fig. 1.2). The basic outline of this structure will be sketched in the following part of this Introduction.

Part II: The significance of ecological long-term processes: The focal questions of this first part of the book are the following: *Which are the most important processes calling for long-term-research? Which questions arise from these objects? and Why is it necessary to investigate their interrelationships?* To find arguments for a thorough discussion of these questions, Müller et al. discuss these items from the viewpoints of ecosystem theory, trying to list and describe recent theoretical problems which need long-term observations to be better understood or even solved. Besides the respective discussion of research demands that originate in theoretical considerations,

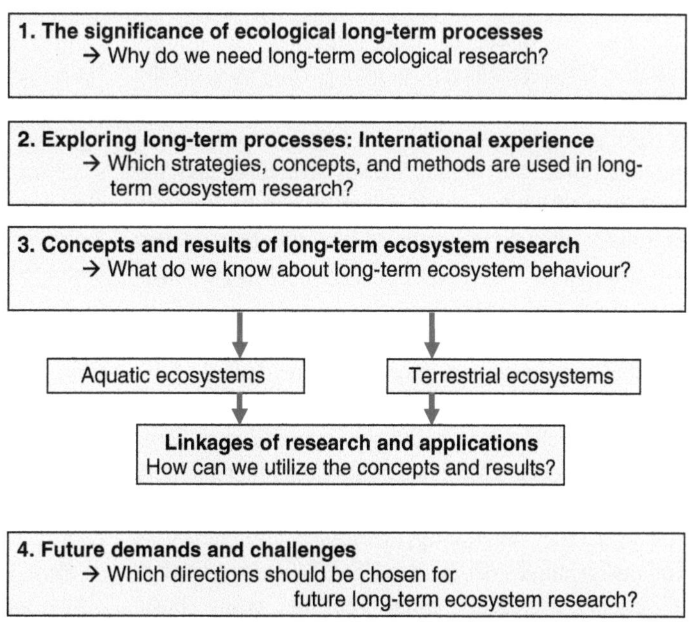

Fig. 1.2 Basic structure of the book

the potentials of modelling, data management, and statistics are demonstrated. One of the actual theoretical fields mentioned in the introductory chapter is described in more detail in the second chapter: Fath and Müller analyse environmental change as an explication of self-organized trajectories. Some basic ideas about ecosystem growth and development are reviewed, and the classical succession approach is linked with actual concepts of collapse, break down, and decay. While these two chapters are highlighting the necessity of long-term ecosystem research from aspects of systems theory and thermodynamics, Schimming et al. are illuminating long-term research conceptions in relation to environmental monitoring activities. Of course both branches of environmental analysis can mutually profit from each other: Scientists can be enthusiastic interpreters of monitoring data to better solve their long-term problems, and on the other hand, these solutions can be extremely valuable for any application in environmental management. As long-term management concepts should be developed although they are actually available only in a small portion, this symbiosis should be enhanced in the future.

Part III: Exploring long-term processes – International experience: The chapters of this section are dealing with the following questions: *How are LTER programmes developing? Which are their typical concepts, results, and research questions? Which experience has been made in the utilization of specific long-term related methodologies?* These questions are discussed from different viewpoints: Initially Gosz et al. report about 28 years of the US-LTER Program. The authors demonstrate the goals of their exemplary initiative, its history, motivations, and outcomes, the organizational structure, and the challenges which US-LTER scientists will be facing in the future. In many points those research demands are also realized by ecologists in Europe as demonstrated in the chapter of Mirtl.

In the two subsequent theoretical chapters, the potentials of ecosystem modelling are described by Gnauck et al. which concentrate on the powerful tools of statistics and their applications in long-term ecology. Hostert et al. additionally show the role of remote sensing techniques and their contributions for the mission of LTER, illustrating theoretical approaches as well as examples from the US LTER Programme. Of course, besides remote sensing, many basic methodologies are applied in long-term ecosystem research. Several examples can be found in the next section.

Part IV: Concepts and results: Presenting and interpreting long-term ecological processes: In this section, the following questions are discussed: *Which are the basic results of existing long-term research projects? Which processes have been observed in different ecosystem and landscape types? Which are the main results of these investigations? How can those results be used to link research and application?*

In this part (IV) case studies from *different aquatic ecosystem research* projects are presented. Starting with interesting experience from several coastal time series van Beusekom et al. demonstrate recent changes in the ecological structures and functions of the Wadden Sea ecosystems in Northern Germany. Their main questions are related to the consequences of temperature, salinity, and nutrient changes, the consequences for plankton and benthos communities, and the influence of alien species, which can be observed very realistic in the Wadden Sea. Stockmann et al. are expanding the aquatic research area, searching for externally forced signals in biological time series in the overall North Sea. The authors are using extensive simulations of varying external physical factors to explain the trends and shifts of biological parameters. A third type of aquatic ecosystems is described in the chapter of Schubert et al.: Based on several long-term data sets, typical developmental tendencies in brackish ecosystems are reported, including gradient-related comparisons of the Baltic Sea, coastal lagoons at the Baltic and Chesapeake Bay. The authors underline the necessity and the added value of long-term investigations to better understand these highly variable systems. Finally long-term developments in two freshwater ecosystem types are described: On the one hand, Köhler shows recent drivers, pressures, and impacts on lake ecosystems, taking into account the consequences of eutrophication, acidification, species invasions, and climate change, stressing the potentials and limitations of long-term research concepts. On the other hand, Poschlod et al. present the results of long-term monitoring investigations in rivers of South Germany since the 1970s. The authors are using macrophytes as focal indicators for the assessment of water quality.

Part V refers to the results of *terrestrial ecosystem research* investigations. Also in this part of the book, several different ecosystem types are illuminated, and different ecosystem components are analysed. Focussing, i.e. on soil mesofauna, Koehler and Melecis provide a methodological framework for respective faunistic, ecosystem-based analyses and show results from Collembola and Gamasina dynamics in different study sites. Highlighting the ecosystem water balance, Lischeid et al. show the results of tracing biogeochemical processes in small catchments using non-linear statistical methods. Schindler et al. concentrate on long-term measurements to quantify the impact of arable management practices on deep seepage and nitrate leaching in agricultural ecosystems. In the following chapter, the focus is changed to long-term processes in forest ecosystems: Tavares et al. are demonstrating the flux-based results of long-term ecosystem research in a beech forest in Northern Germany, concentrating on the flows and storages of elements through the compartments of the investigated beech forest ecosystem. Subsequently, Syrbe et al. provide a report about concepts, methods, and results of monitoring landscape change. They create a methodological framework, which is exemplified, i.e. with reference to hypothesis testing. Baessler et al. demonstrate the relative importance of historical to recent landscape structure and environmental conditions on plant species diversity and genetic variation. They demonstrate the importance of long-term studies to understand ecological processes and to consider this knowledge in the future. Finally Schmidt et al. develop a system for the integration of long-term environmental data by the example of the UNECE heavy metals in mosses survey in Germany. The authors are using a WebGIS-based metadata system, which is well suited for applications in LTER systems.

Part VI the *linkages between research and applications* are focal items of four chapters: While Luthardt refers to her experience with monitoring ecosystems in Biosphere Reserves, Heurich et al. show the potential of National Parks to be used as model regions for interdisciplinary long-term ecological research. The authors demonstrate the added value of long-term investigations for actual problems of environmental management in an exciting environment, and they demonstrate potentials for long-term collaboration between different nations. In contrast to Heurich's mountains of the Bavarian Forest, Diederichs refers to the experience of National Park management in the Wadden Sea of Schleswig-Holstein, describing how to turn long-term monitoring into policy. In the fourth chapter of this section, the conceptual framework and the design of ecological monitoring networks are described by Jones et al. from two sides: On the one hand, the design of the US NEON network is considered at the intersection point between research and application, and on the other hand, the new research network TERENO is described, which is planned to investigate ecosystem processes with a long-term conception. All of these chapters convincingly demonstrate the demands for long-term knowledge in theory and practice, and they illustrate the high potential of linking research and management.

Part VII: Future demands and challenges: The chapters of this section try to find answers to the following queries: *Which are the future steps of long-term ecological research? Which challenges have to be solved in the forthcoming years? How can we better integrate different research directions?* In this context, Singh et al. present a conceptualisation and a theoretical basis for long-term socio-ecological research (LTSER), trying to integrate the socio-economic dimension into long-term ecological research. The authors also provide some results and modelling studies from Austrian pilot test areas. Ohl and Swinton are concentrating on the linkages of environmental and societal dynamics, also proposing a future network of integrated human–environmental research sites. Thereafter Dilly et al. demonstrate the value of long-term research with reference to ecosystem manipulation and restoration on the basis of long-term conceptions. Here long-term approaches are used not only to control the success of restoration measures but also to develop intelligent directions of restoration. Following, Fischer et al. present a new integrated project in Germany, the exploratories for large-scale and long-term functional biodiversity research. Finally, there are some summarizing *conclusions (Part VIII)* of the book, concentrating on future long-term-related research questions and challenges.

References

Hobbie, J. E., Carpenter, S. R., Grimm, N. B., Gosz, J. R., & Seastedt, T. R. (2003). The US Long Term Ecological Research Program. *Bioscience, 53*, 21–32.

Part II
The Significance of Ecological Long-Term Processes

Chapter 2

Theoretical Demands for Long-Term Ecological Research and the Management of Long-Term Data Sets

Felix Müller, Albrecht Gnauck, Karl-Otto Wenkel, Hendrik Schubert, and Michael Bredemeier

Abstract Long-term ecological studies are required to understand ecosystem complexity, to develop integrated dynamic models, and to explore appropriate measures for the assessment and control of ecosystem behaviour. The knowledge derived from long-term ecological studies should be a prerequisite to formulate ecosystem management plans, i.e. with reference to recent environmental processes. Long-term problems are especially interesting elements of ecosystem theory. Many hypotheses only can be tested for validity from a long-term aspect, and many generalized ideas about ecosystem development could not be verified satisfactory up to now due to a lack of long-term data sets. In this introductory statement chapter, some theoretical aspects of ecosystem dynamics are sketched. These outlines are used to derive demands and questions from ecosystem theory towards long-term ecological studies. The queries are posed referring to six main objectives of long-term research: (i) understanding large-scale variabilities, (ii) understanding the interactions of short-term and long-term fluctuations, (iii) understanding self-organization, (iv) understanding rare events and disturbances, (v) better understanding the impacts of anthropogenic use of landscape resources on ecosystem functions, and (vi) generation of knowledge and data for the development and evaluation of new generations of ecosystem models for resource management. These items can be taken to demonstrate the enormous research demands referring to long-term environmental dynamics and to develop and apply techniques of ecological statistics, data management, and ecosystem modelling.

Keywords Ecosystem research · Long-term observation · Ecosystem theory · Data management

2.1 Introduction

Ecology is striving for new dimensions: The recent processes of global climate change, land use change, demographic change, and the consequences of globalization have yielded to slow and extremely complex environmental modifications. These processes are not detectable on the base of, e.g. annual investigations; it needs decades to observe their dynamics and to evaluate their consequences. Their potential significance for natural and human systems thus enhances the demand for widening the scales of ecological observations: The roles of long-term processes have become defining, and therefore they are turning into focal items of actual scientific as well as political debates. And although the importance of long-term processes has been obvious since the impacts of forest dieback and eutrophication or since the first analyses of ecotoxicological effects, our recent knowledge still has many lacks of understanding. Consequently, the management of the overwhelming global dynamics urgently depends on new knowledge, tools, and information about long-term ecological processes.

Besides these practical demands from environmental management and policy, long-term dynamics are also basic objects from theoretical viewpoints: Many theoretical conceptions have been tested with short-term data sets only, thus their factual significance

F. Müller (✉)
University of Kiel, Ecology Centre, Olshausenstraße 75, D-24118 Kiel, Germany
e-mail: fmueller@ecology.uni-kiel.de

F. Müller et al. (eds.), *Long-Term Ecological Research*,
DOI 10.1007/978-90-481-8782-9_2, © Springer Science+Business Media B.V. 2010

cannot really be tested or improved. And many of these approaches, such as succession theory or orientor theory, particularly refer to long-term dynamics. Thus, long-term data sets are extremely valuable for the comprehension, improvement, and testing of several ecological theories as well as the better comprehension of the ecological impacts of climate and land use changes.

Apart of these fundamental motivations, ecologists have to be aware that several short-term-based deductions are wrong if applied to longer periods. For instance, many environmental constraints and processes operate in cycles of several years or decades. Therefore an extrapolation of our typical 3 years results is always in danger of being incorrect if it is raised to an abstract – theoretical – level. The factual range of validity of ecological conclusions consequently can be determined only if longer time series are available.

It took quite a long time until this necessity became obvious: Since the introduction of the term 'ecology' by Haeckel (1866), the main focus of the discipline was to understand the interactions between organisms and their biotic and abiotic environments. Regarding the fact that even under extreme conditions a large number of species coexists on different trophic levels, influenced by a multitude of abiotic factors while utilizing several resources, it becomes understandable that most ecological research of the past has been short-term intense work, trying to gather data about as many organisms and factors as possible, simultaneously. Additionally, the main funding principles did (and partially do) not allow the execution of long-term investigations, i.e. because the expected success requires a high portion of patience and methodological repetitions: The experience of innovation is postponed too far into the future with reference to our 'modern' (short-term) scientific evaluation conceptions. As a result, the main bulk of ecological knowledge is based on 'snapshots'. However, it is obvious that the respective local and short-term data cannot easily be extrapolated to longer time periods and different landscape types because of a lack of knowledge about the underlying dynamics.

The respective long-term ecological research conceptions have consequently been designed to analyse, understand, and document the interrelationships between environmental changes and ecosystem development. The need for long-term ecological studies was formulated more than 20 years ago by Likens (1983), followed by encouraging concepts, e.g. from Likens (1989), Franklin (1989), or Risser (1991). These articles and books were oriented towards the understanding of slow phenomena, rare events, subtle processes, and complex processes (see Pickett, 1991). The scientific target activities of the arising LTER programmes were set on processes that operate on year to decade to century timescales (Magnuson, 1990). The respective US LTER investigations have produced promising preliminary results, which have been published recently in a coherent volume (e.g. Greenland et al., 2003; Hobbie, Carpenter, Grimm, Gosz, & Seastedt, 2003; Kratz, Deegan, Harmon, & Laurenroth, 2003; Symstad et al., 2003; Turner et al., 2003), demonstrating the high significance of long-term investigations.

Another central motivation for LTER has been arising from ecosystem research. While trying to identify and model the interactions between ecosystem elements and the impacts of land use changes on ecosystem functionality, it became more and more obvious that all relevant processes are operating on different spatio-temporal scales (Allen & Starr, 1982) and that the internal regulation schemes are based on scale-specific interactions and controls (O'Neill, DeAngelis, Waide, & Allen, 1986). Furthermore, the inherent complexity of ecosystems, the irreversibility of ecosystem dynamics, and the non-linear character of ecological relations must be taken into consideration, and the role of rare events has to be clarified. Therefore, ecosystem dynamics cannot be understood if only small time periods are taken into account and if ecological research is concentrated on one certain scale only.

Taking these general objectives as a basis, this introductory statement article aims at illuminating the demands for a better understanding of long-term environmental processes for the development of ecological theory, the development of modelling tools for ecological long-term impact assessment, and sustainable landscape resource management. The sequence of argumentation is arranged around the following propositions:

- Ecosystems have become focal units of research.
- Long-term research is strictly linked with the ecosystem approach.
- Theoretical ecosystem comprehension needs long-term investigations.

These points are discussed and illuminated in Sections 2.2 and 2.3, referring to different classes of theoretical approaches for ecosystem comprehension and to several questions posed to long-term data sets. Section 2.4 builds a relation to the requirements of ecosystem modelling and time series analysis, and in Section 2.5 the outcome of this essay will be summarized.

2.2 Ecosystems Have Become Focal Units of Environmental Research

Highly related to the growing consciousness about the significance of long-term processes, ecosystem science has performed a rapid development during the last decades and ecosystem complexity has become a focal object of investigation: The reductionistic methodology has been accomplished by systems approaches, and the linkages between the compartments of ecological and human–environmental systems have been analyzed, integrating structural, functional, and organisational entities. Long-term ecological studies have been carried out, e.g. to investigate relationships between biodiversity and ecosystem functioning (Schulze & Mooney, 1994) and to study focal relations within pelagic nutrient cycles (Andersen, 1997), shallow water bodies, marine ecosystems (Bruno, Boyer, Duffy, Lee, & Kertesz, 2005; Stachowicz, Fried, Osman, & Whitlach, 2002), or forest ecosystems (Bredemeier et al., 1998; Hauhs & Schmidt, 1999). In Germany, five Ecosystem Research Centres have been installed and supported within that time period (see e.g. Fränzle, 1998; Fritz, 1999; Gollan & Heindl, 1998; Hantschel, Kainz, & Filser, 1998; Widey, 1998; Wiggering, 2001) and additional research projects have been carried out in national parks (e.g. Kerner, Spandau, & Köppel, 1991), biosphere reservations (e.g. Schönthaler et al., 2001), and coastal districts (e.g. Dittmann et al., 1998; Kellermann, Gätje, & Schrey, 1998). The common features of these approaches have been documented by Schönthaler, Müller, and Barkmann (2003). These characteristics are

- considering indirect, chronic, and de-localized effects;
- integrating ecological processes and relations into management procedures;
- combining water balances, nutrient and matter budgets, energy budgets, and community dynamics;
- investigating the basic features of ecological complexity;
- considering self-organisation as a basic process;
- aggregating structural and functional items;
- realizing the irreversibility of changes in ecosystem structure and function;
- utilizing multiple extents and resolutions in terms of time, space, contents, disciplines, and analytical depth.

All of these items are correlated with the demand for long time series. For instance chronic effects for sure cannot be observed on a short time basis, indirect effects need long periods to become telling, the interactions between different element cycles operate on a multitude of scales, and self-organized processes do need long times to become detectable. Furthermore, the fact, that changes in ecosystems are mostly irreversible processes, illuminates the demand for historical attitudes to understand systems dynamics, and thus – once again – long-term investigations are essential components for the understanding of environmental systems. Only on the basis of the system's history, its actual state can be comprehended. Consistent future scenarios or even predictions related to new boundary conditions need to be founded on the historical situation and the system's development.

Ecosystemic attitudes have become favourable also in environmental practice: While in the past, environmental activities were restricted to specific resorts, today we can find resort spanning environmental politics and interdisciplinary cooperation is increasing continuously. The ecosystem concept has also been applied in political programmes: In Principle 7 of the Rio Convention from 1992 states are asked to 'cooperate in a spirit of global partnership to conserve, protect and restore the health and integrity of the Earth's ecosystem.' And in the Ecosystem Approach, the UN Commission on Biodiversity (Decision V/6 CBD, 2000, see http://www.iucn.org/ themes/cem/ea/) anticipates the consequences of these demands in various principles which include the following statements:

- 'Ecosystem managers should consider the effects . . . of their activities on adjacent and other ecosystems' (Principle 3).

- 'Conservation of ecosystem structure and functioning, in order to maintain ecosystem services, should be a priority target of the ecosystem approach' (Principle 5).
- 'Ecosystems must be managed within the limits of their functioning' (Principle 6).
- 'The ecosystem approach should be undertaken at the appropriate spatial and temporal scales' (Principle 7).
- 'Recognizing the varying temporal scales and lag-effects . . . of . . . ecosystem processes, objectives for ecosystem management should be set for the long term' (Principle 8).
- 'Management must recognize that change is inevitable' (Principle 9).

With reference to our subject, several long-term-oriented questions can be posed from that listing: How can we consider the chronic, indirect, and delocalized effects (Principle 3) without taking into consideration the lag-phases between a disturbance and the complex reaction? How can we consistently conserve structures and functions if we do not know how they interact (Principle 5)? How can we determine the limits of ecosystem functioning without the respective long-term observations (Principle 6)? How should we determine the correct scale if we only know the short-term parts of the hierarchy (Principle 7)? How can we cope with change without experience on its effects (Principle 8)? And how can we find the optimal compromise between the inevitable change and the conservation without basic knowledge on the dynamics of ecosystem change (Principle 9)?

The focal problem to answer these questions derives from the enormous complexity of ecosystems, which has been investigated on the basis of several conceptions (see Jørgensen & Müller, 2000). The most prominent theoretical basis is the theory of dynamic systems (Gnauck, 2000). This theory is used to search for dynamic laws underlying the changes of ecosystems and to abstract from real systems those particular characteristics which all systems or at least one particular class of systems have in common. Under the notion of complexity and with today's computing power, ecological, social, or economic systems (among many more) have become tractable units, at least in term of describing observed patterns. However, most of these applications use the physical notions of dynamic theory in a metaphorical sense only. For example, non-trivial predictions, showing that the underlying dynamics can indeed be extrapolated into realms not yet observed, the hallmark of successful physical theories, are still lacking in ecology. The reasons for this may lie in the complexity of ecosystems, in the technical limits of observation and computation, or in conceptual limits and unsuited abstractions, but also in the lack of a consistent, long-term database.

2.3 Ecosystem Comprehension Needs Long-Term Aspects

Not only ecosystem science and management provide motivations for the execution of long-term ecological studies. Tables 2.1 and 2.2 demonstrate a compilation of arguments from general ecological empiricism and environmental applications as well. It should be noted that there is a huge amount of processes which can be understood on the base of long-term data only. To illuminate the theory-based questions to long-term data sets, some relevant research targets and questions will be exemplified in the following text section.

2.3.1 Understanding Large-Scale Temporal Variabilities of Ecological Variables

Ecosystem dynamics is strongly influenced by the temporal development of external constraints. The environmental factors, such as precipitation or mean temperature, vary in geological timescales as well as in decadal, annual, and daily rhythms. The effects caused by such climatic changes are superimposed by short-term weather effects as well as medium-scale seasonal changes. The ecosystem itself responds to all of them, but in a non-linear and delayed mode. To extract the impact of the long-term changes in the abiotic external forcing functions, a simple repetition of snapshots is insufficient. For this purpose, time-resolved continuous monitoring is necessary to allow a proper discrimination between the superimposing responses. As an example, the effects of global climate change provide numerous questions with reference to this point:

- How do different species react on the ongoing long-term temperature rise?

Table 2.1 Utility of long-term investigations; process classes and examples, after Gruttke and Dröschmeister (1998) and Franklin (1989)

Process classes	Examples
Investigating slow processes	• Population dynamics and life-histories of long-living species • Successions • modifications of species areas • integration of invasive species • soil development • ecochemical degradation processes • ecotoxicological effects • chronic effects • eutrophication processes • consequences of global climate change • consequences of land use change • consequences of restoration measures • studies of natural history
Investigating rare events and episodic phenomena	• gradations • epidemics and diseases • consequences of predator abundances • consequences of extreme weather conditions (drought, floods, erosion, fire, wind, ...) • episodic reproduction phenomena
Investigating processes with high variability	• filtering noise from time series • determining trends of variable processes • deriving stability characteristics (resilience, elasticity, buffer capacity) • development of spatial patterns
Investigating complex interactions	• meta population dynamics • behavioural changes of organisms • genetic distinction • extinctions • consequences of fragmentation • mosaic dynamics • ecotone dynamics • studies across organizational levels

Table 2.2 Some recent research themes in long-term ecological research, based on the US-LTER Programme (Hobbie et al., 2003, see also http://www.lternet.edu/)

Research theme	Exemplary topics
Consequences of climate change	• role of invasive species • role of physical forcing factors
Role of disturbance	• interactions between long-term and short-term disturbances • disturbance regimes, disturbance legacies
Mechanisms of ecosystem self-organization	• interactions of structures and functions, providing mutual causes, and effects of ecological fluxes • successions • control of food web structure
Provision of ecosystem services	• consequences of alteration for ecosystems' capacities to provide services
Investigation of human–environmental systems	• consequences of changed services for human demands • role of environmental perceptions for human behaviour • motivation of human-induced press and pulse disturbances • human activity and land use change

- Which will be the resulting patterns of their spatial distributions?
- Which are the extinction risks for sensitive species?
- Which will be the consequences of the biotic reactions for the flows, losses, and storage capacities of energy, matter, water, and information in ecosystems?
- Will these physico-chemical reactions act as positive or negative feedback processes, enhancing or buffering the global dynamics?

- Which will be the consequences for the provision of ecosystem services?

2.3.2 Understanding the Interactions of Short-Term Fluctuations Versus Long-Term Trends

Different time horizons of observations and measurements can lead to different conclusions, and consequently temporal extrapolations can be rather dangerous. For instance, the growth of populations is influenced by short-term fluctuations of environmental variables (Nisbet & Gurney, 1982). But of course, also long-term effects – e.g. the climate change variables mentioned above – are influential. The interrelations between these timescales under steady-state conditions have been investigated by hierarchy theory (e.g. Allen & Starr, 1982; Holling, 1985; O'Neill et al., 1986; O'Neill, Johnson, & King, 1989; Callahan, 1991). It has been suggested that long-term processes (which are mostly operating on broad spatial scales) provide constraints for the rapid processes (which function on small spatial scales), limiting their degrees of freedom (see Fig. 2.1). Thus, the theory imposes a control mechanism into the temporal order of ecosystems. These ideas have been adopted by panarchy theory (Gundersson & Holling, 2003), where they are applied for the analysis of human–environmental systems. Although the hierarchy concepts are well elaborated from a theoretical viewpoint, their empirical evidence has not been shown comprehensively up to now, due to a lack of respective data sets. On the other hand, those concepts provide one of the fundamentals of the resilience approach which has been moved into very high degrees of environmental policy. Thus, there is an urgent necessity to prove the hypotheses of 'hierarchy', 'panarchy', or 'holarchy' theory (Kay, 2000) on the base of empirical data. Referring to the distinctions of different types of dynamics and taking into consideration the concepts of chaos and catastrophe theory, there is still a lack of understanding, and many questions, which could be answered with better long-term data sets, are still discussed. Some examples

- Will it be possible to prove and improve the basic ideas of hierarchy theory on the base of long-term data sets?
- Do we really find empirical examples for chaotic dynamics on different scales in nature, and if yes: how can they be characterized?
- Which are the characteristics of signal transfers in instable states and how can we assess the distance of a system from bifurcations?
- Which is the role of hysteresis effects in ecosystem dynamics?

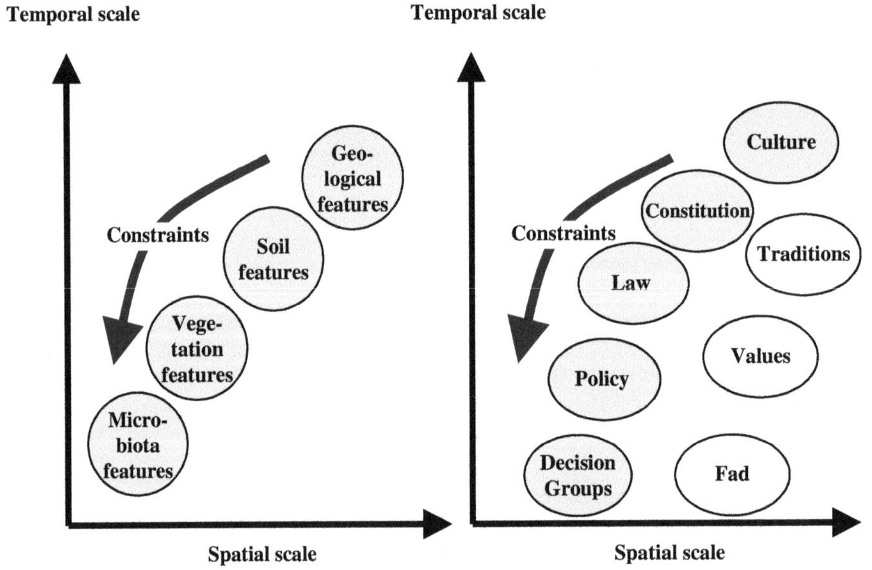

Fig. 2.1 Examples for hypothetical hierarchical structures of ecosystem components (*left side*) and elements of human–environmental systems (*right side*, after Gunderson & Holling, 2001)

- Which is the influence of stochastic dynamics in different states of ecosystem development?
- Can we develop new tools to distinguish different scale-based types of ecosystem dynamics?

2.3.3 Understanding Self-Organizing Ecological Mechanisms

Ecosystems are self-organized systems, creating macroscopic gradients which emerge from microscopic disorder and create a web of mutual interactions (Jørgensen, 1996). These interrelations and their structural implications change in time, and concerning the respective dynamics several hypotheses have been formulated by ecosystem theory. One of the corresponding approaches is orientor theory (Müller & Leupelt, 1998; Chapter 3) which states that throughout an undisturbed development, certain ecosystem variables – ecosystem orientors – are optimized up to certain, site-specific level. If we take a close look at the orientors which have been exemplarily presented in Fig. 2.2, it becomes obvious that many of them cannot be easily measured or even modelled under usual circumstances. Some orientors must be calculated on the base of very comprehensive data sets which are measured on a very small number of sites, only. Other orientors can only be quantified by model applications. In all cases, long-term validation data are basic necessities to empirically prove the significance of the following hypotheses.

The complexity of ecosystem structures increases during undisturbed developmental phases: While ecosystems are evolving, the number of abundant species is regularly increasing and the abiotic features are becoming more complex, too. This development can be indicated by a rising degree of information, heterogeneity, and complexity. Also specific life forms (e.g. symbiosis) and specific types of organisms (e.g. the ratio of r/k strategists, organisms with rising life spans and body masses) become predominant throughout the development.

The efficiency of ecosystem functions is enhanced during undisturbed developmental phases: Due to the increasing number of structural elements, the translocation processes of energy, water, and matter are becoming more and more complex, the significance of biological storages is growing as well as the degree of storage in general, and the residence times of the input fractions are increasing as a consequence. Due to the high degree of mutual adaptation throughout the long developmental time the efficiencies of the single transfer reactions are rising, cycling is optimized, and thus losses of matter are reduced. The respective ecosystem functions are usually investigated within three classes of processes which are interrelated to a very high degree:

- Ecosystem *energy balance*: The uptake of utilisable energy (exergy capture) is rising during the undisturbed development, the total system throughput is growing (maximum power principle, see

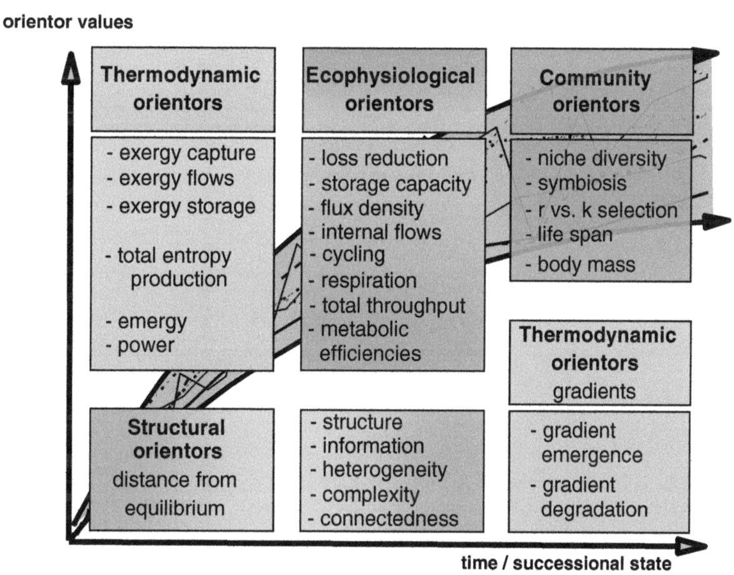

Fig. 2.2 Some ecological orientors. The respective theory claims that under undisturbed conditions ecosystems develop in a way that optimizes the values of the indicators (orientor values) in dependence of the specific site conditions (see also Chapter 3)

Odum, Brown, & Ulgiati, 2000) as well as the articulation of the flows (ascendancy, see Ulanowicz, 2000). Due to the high number of processors and the growing amount of biomass, the energetic demand for maintenance processes and respiration is also growing (entropy production, see Svirezhev & Steinborn, 2001).

- Ecosystem *water balance*: Throughout the undisturbed development of ecosystems and landscapes, more elements have to be provided with water. This means that especially the water flows through the vegetation show typical orientor behaviour. These fluxes are an important prerequisite for all cycling activities in terrestrial ecosystems.
- Ecosystem *matter balance*: Imported nutrients are transferred within the biotic community with a growing partition throughout undisturbed ecosystem development. Therefore the intra-biological nutrient fractions are rising as well as the abiotic carbon and nutrient storages, the cycling rate is growing, and the efficiencies are being improved. As a result, the loss of nutrients is reduced.

Although the described theoretical hypotheses have been tested with numerous sets of modelling results, empirical proofs are still missing. Therefore, the following questions are posed to long-term data sets:

- Can we find orientor behaviour in successional studies that include information on physiological and structural parameters?
- How strong are the deviations from the theoretically foreseen behaviour due to external inputs?
- Which orientors can be used as suitable indicators for ecosystem states?
- Which are the interrelationships between the orientors?

2.3.4 Understanding Rare Events and Disturbances

If the orientor hypotheses would be valid all the time, only growth and development processes would occur in nature. Realizing that this attitude is not correct, the theory of the 'adaptive cycle' has been developed. In Fig. 2.3, a respective sequence of ecosystem states has been illustrated as a function of the system's internal connectedness and the stored exergy. The respective approach refers to the conceptions and results of the resilience alliance (http://www.resalliance.org/) following the approach of Holling (1986): Starting with the exploitation function, there is a slow development which can be characterized by complexification procedures. The trajectory demonstrates a steady increase in mutual interactions as well as an increase in stored exergy, and the system follows an orientor dynamics. In spite of several variabilities (e.g. annual cycles), the long-term development shows a steady increase up to the mature state. Here the maximum connectivity can be found, which is a product of the system's

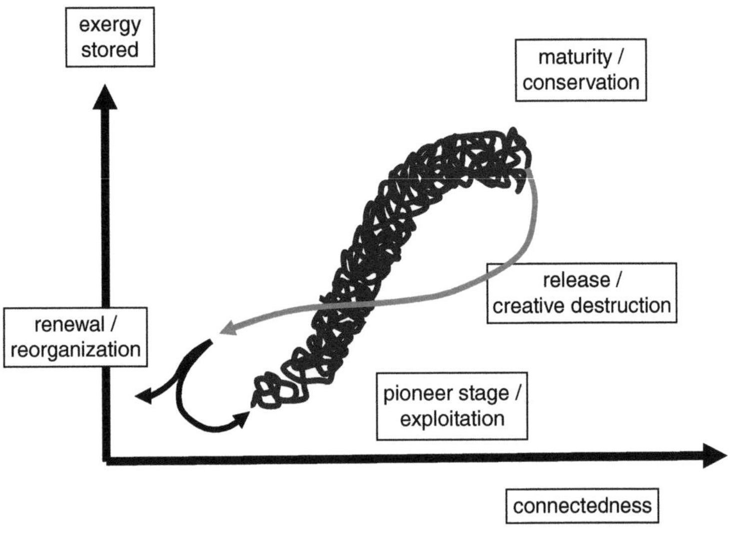

Fig. 2.3 A modified scheme of the adaptive cycle after Gunderson and Holling (2001)

orientation, but which also is correlated with the risk of missing adaptability that has been nominated as over connectedness by some authors. After the fast releasing event, the short-term conditions determine the further trajectory of the system, the controlling hierarchy is broken, and the system might turn into a similar trajectory or find a very different pathway.

All ecosystem components potentially develop within the described sequence of phases, and many ecological processes support these dynamics, theoretically. But of course, also this theoretical metaphor of ecosystem dynamics should be tested by empirical data sets. These can be used to solve, e.g. the following problems:

- When does an ecosystem lose its adaptability due to over connectedness?
- How do disturbances affect the future fate of the system and which are the sub-processes of the release function?
- Which parameters are responsible for the fate of the ecosystem after a creative destruction? Do the hypotheses of hierarchy theory function under these circumstances?
- Can a breakdown on a lower scale be understood as an adaptive process on a higher level of resolution?

The concept of the adaptive cycle is linked with investigations about the significance of external disturbances (Jørgensen et al., 2007), bifurcations (Schultz, 1995), and internal instabilities (Demazure, 2000) resulting in the irreversible development of ecosystems. For instance, for a given set of environmental conditions different ecosystem shapes can be found (Schultz, 1995). In the most extreme case, multiple alternative stable states reflect the theoretical ability of self-stabilisation of ecosystems, where the actual state becomes understandable only by knowledge about its history. The initializing bifurcations can be the result of single events which occur very seldom. To understand the underlying dynamics, long-term data provide a much higher probability that rare events are included, than conventional data sets. Thus, they can be extremely helpful to understand the interrelations between extremely short catastrophic events and continuous developmental traits, when disturbances are reduced or buffered. Furthermore, several theoretical constructions to describe systems dynamics would find much better explanations, quantifications, threshold criteria, and comparisons (e.g. the concepts of stability, resistance, elasticity, resilience, adaptability, or buffer capacity), if the respective data sets are available.

2.3.5 Understanding Anthropogenic Land Use Impacts on Ecosystem Functions

As stated in Table 2.2, the joint analysis of human and environmental (sub)systems has become a focal point of interest in science and management. This tendency can also be observed within the CBD ecosystem approach. In the operational guidance the following objectives are formulated: 'The Ecosystem Approach places human needs at the centre of biodiversity management. It aims to manage the ecosystem, based on the multiple functions that ecosystems perform and the multiple uses that are made of these functions. The ecosystem approach does not aim for short-term economic gains, but aims to optimize the use of an ecosystem without damaging it'. This program provides challenging demands for environmental management, which should be based on a better knowledge about the fundamental interrelations in human–environmental systems. These demands have been starting points for the LTSER program, which is described in Chapter 26, as well. Some respective demands are listed below:

- Which are the general interrelations within human–environmental systems?
- Can these interrelations be described on the base of hierarchical approaches (see Fig. 2.1) and which are the relevant scales of investigation?
- Which have been the general developmental pathways of ecosystem goods and services in the past? How do they develop from a long-term viewpoint?
- Which ecosystem services or functions of nature should be taken into account for environmental evaluation procedures and which is their historical background?
- What are the critical loads and levels for different landscape types?
- Which special requirements for long-term ecosystem management and monitoring can be drawn from long-term ecological research?
- How can ecosystem protection conceptions (e.g. ecosystem health and integrity) be implemented in such strategies?

- What can we learn to better solve actual long-term problems, such as the consequences of global climate change, demographic change, and land use change?

2.3.6 Development and Evaluation of New Ecosystem Models for Resource Management

Long-term ecological studies are required not only to understand ecosystems complexity but also to develop and evaluate integrated dynamic ecosystem and landscape models as a prerequisite for the sustainable resource management. Ecosystem and landscape models are important and necessary tools for studying and understanding ecological processes, for testing hypotheses of the functioning of ecosystems in a systematic manner, and for investigating environmental responses to human impacts (Seppelt, 2003). Simulation models can be used to augment data collection and data sets in a predictive mode. This step covers the use of models for the simulation of ecosystem dynamics at different spatio-temporal scales. The models are able to incorporate and combine different knowledge bases (data sets) and provide new results to higher level information. Therefore this approach can be understood as an analogue to empirical analyses and comparisons of long-term data sets of different observation sites. Simulation models can be used to test hypotheses for consistency with observations. In this modus models allow to formulate testable conjectures for further experiments and observations for ecosystem theories (Jørgensen & Müller, 2000). Furthermore, the results of model-based ecosystem simulations are important instruments to optimize the ecological long-term monitoring systems and give suggestions to more reliable experiments.

A new use of simulation models is their application as a communication tool. When the behaviour of an ecosystem cannot be reconstructed and predicted from known dynamics, it is still possible to represent the relevant observations in a simulation model. In this approach interactive simulation is regarded as a more powerful classification system of ecosystems than algorithmic computation (which is the basis of traditional simulation models).

Over recent decades marked progress has been achieved in ecosystem modelling. Traditionally, process and ecosystem models are developed for the one-dimensional case (point model), i.e. only for selected points within an area. Examples of such process models are soil–plant–atmosphere models such as CERES (Ritchie & Godwin, 1993), WOFOST (Supit, Hooijer, & van Diepen, 1994), HERMES (Kersebaum, 1995), APSIM (Keating et al., 2003), or AGROSIM (Mirschel, Schultz, & Wenkel, 2001; Wenkel & Mirschel, 1995). At the ecosystem or regional scale it is difficult to provide all this information for all spatial nodes. Here the available information is usually characterized by significant fuzziness and heterogeneity. Beyond it, there are many processes which are related to different spatial and temporal scales. For example, biotic processes described by habitat or population models usually are quite local. On the other hand, some abiotic processes such as matter flow driven by wind, water, or diffusion are strongly influenced by spatial features within catchments such as the heterogeneity of the land use types or elevation. To analyze, evaluate, and manage ecosystems, tools and models are required, which can represent and interpret the variety and complexity of the connections between biotic and abiotic landscape structures and functions. The recent concerns of climate change and the end of cheap oil will provide additional items to the mix of problems we face in the regional scale and provoke an even more urgent requirement for reliable decision-making instruments and policy support.

In spite of substantial progress in the development of mechanistic integrated ecosystem models in recent years, integrated space and time-related dynamic landscape models are exceptions to date. The development of such integrated regional models is a very ambitious task and requires its 'own inter-methodical approach' (Lausch, 2003). However, this is still a challenge as the diversity of knowledge from various scientific disciplines is increasing, requiring ever more creativity, versatility, and openness in research and communication. From the scientific point of view we are confronted with the following problems:

- How do we reduce the complexity of landscape processes within models?
- How much detail in process dynamics can we incorporate in regional models still capturing the most important feedback loops in regional landscape systems?

- How can we adequately capture the functional consequences of structural diversity and spatial heterogeneity in landscape models?
- How can we compromise between short-term and long-term developments adequately in landscape models? What do we mean by 'adequate?'
- Which model types and tools are best suited to landscape modelling, considering the fuzziness and uncertainties in ecological functions and data and the dynamic nature of processes?
- How can we bring together conceptually different ecological and economic models?
- Which spatial resolution is required to appropriately capture processes with regional significance?
- How can we adequately include spatial ecological heterogeneity and its influences on ecosystem functions?

To find answers to these problems we need new ideas, new thinking and a paradigm shift in integrated regional modelling. To cope with these problems we suggest an approach that promotes models of intermediate complexity (Wenkel et al., 2008). An important feature of Regional Models of Intermediate Complexity (REMICs) is that they are characterized by a lower degree of detail in the description of process dynamics, but a higher number of interacting components. The implementation of REMICs calls for specific modelling tools that can handle the spatial heterogeneity of GIS and can offer analytical capabilities of both process-based statistical modelling. One such tool, the Spatial Analysis and Modeling Toolbox (SAMT), is an open source package that can be used to develop decision support systems to help planners and decision makers better understand the complex reactions of the landscape to various forcing (Wieland, Voss, Holtmann, Mirschel, & Ajibefun, 2006).

A prerequisite for development and scientifically rigorous validation of complex integrated landscape simulation models is a network of long-term experimental study areas or, if feasible, specially designed landscape experiments. Back-casting and comparing model results to historical conditions offer a useful way to validate a model. The core research areas within the Long Term Ecological Research Network LTER (LTERNET, 2006) could be the basis for the much needed data sets and studies.

2.4 Data Management as a Pre-requisite of Long-Term Research

To understand the long-term dynamics of ecosystems and to formulate basic functional principles, besides the originating data, adapted tools for data management and interpretation are necessary. That is why ecological informatics as a field in the interface of ecology and ecosystem research needs to be further established and applied to long-term dynamics. For this reason, long-term monitored information bases as well as classical and modern mathematical procedures and informatic tools are needed to better portray ecosystem development.

The archetypical data sets from long-term research sites are time series based on different sampling or observation intervals. They may consist of single entries such as limnological, hydro-meteorological or terrestrial data, or repeated mappings, e.g. surveys of biota. A widespread appreciation of long-term data series is not only shared among ecologists but also among ecosystem managers what might be taken as indicative of favouring empiricism over a theory-based approach. Therefore, many management decisions in ecosystems appear to be predominately 'data-driven' and remain embedded in the local history of the respective ecosystems. Thus, long-term data provide the context in which short-term observational or experimental results can be interpreted. This analysis should be based upon mathematical tools. In the following text passages, some of these data management components are listed:

Missing data in long-term ecological data records often lead to some general problems for ecosystem research and simulation. They cause both difficulties in process identification and parameter estimation and misinterpretations of spatial and temporal variations of ecological variables. After Little and Rubin (1987) missing data can be classified as missing completely at random, missing at random, and as non-ignorable. Known procedures to compensate such gaps are replacing by smoothing, by interpolation, by application of trend functions, or by data from reference curves according to standards and conventional rules. The efficiency of all these procedures is quite different. Applying interpolation procedures on irregularly sampled raw data sets, time series with equidistant sampling intervals will be obtained. The application of approximation methods on such time series results in functional relationships. Another procedure is the so-called re-sampling method which requires data interpolation and, in the case of noisy information, data approximation to place sampled data on a regularly timescale. The goal of this method is to reconstruct time series with small sampling intervals. The following tasks are equivalent to re-sampling:

- filling the gaps of irregularly recorded data by interpolation,
- reconstruction of data of regularly recorded time series by means of analytical functions or signal estimations,
- filling the gaps of irregularly recorded data by means of measuring values of reference curves.

Mathematical equations describe either the time dependency (function of time t) or the frequency dependency (function of frequency ω or cycles per time unit) of environmental processes (Pollock, 1999). Some well-known linear and nonlinear interpolation methods exist while data estimation can be executed by static and dynamic approximation procedures. Regression type functions or – in the case of cycling time series – Fourier approximation are used. The re-sampling procedure can be extended by application of digital signal filter methods. On the base of equidistant data consistent time series based on major process frequencies will be obtained. Figure 2.4 shows the procedure to produce the respective, consistent data sets.

To obtain long-term ecological data series with equidistant sampling intervals from irregularly sampled data, *interpolation methods* should be used. Because of random, but not normal distributed long-term data and nonlinear effects of ecological processes observed, no R^2 statistics can be used. The suitability of the interpolation method is valuated by standard error $SE = s/\sqrt{n}$ between interpolated and original time series. Mostly, a linear interpolation method was found to be the best one in opposite of nearest neighbour, polynomial, and cubic spline functions. Another method for the re-sampling of long-term ecological records consists of estimations of data due to *approximating functions*. For parameter estimation mostly the least squares method or the maximum likelihood methods are applied. Useful procedures are multivariate regression models, polynomial models, or time-discrete transfer functions. Evaluations of the quality of the fit are given by linear or nonlinear coefficients of determination.

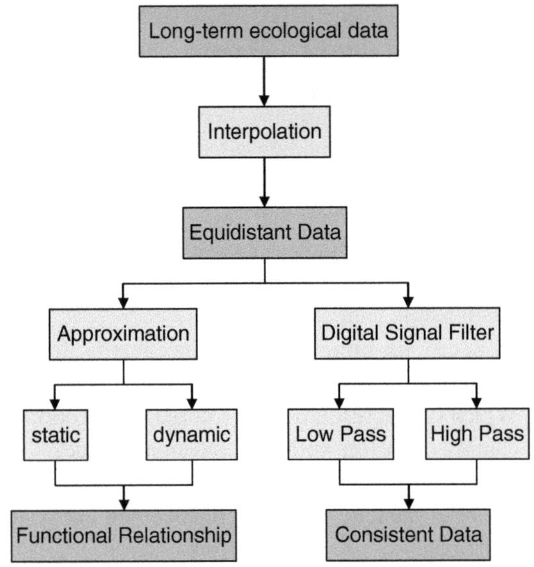

Fig. 2.4 Re-sampling procedures

2.5 The Outcome: Theory, Ecosystem Modelling, and Ecosystem Management Need Information About Long-Term Ecological Processes

In the foregoing text many theory-based questions have been asked and some mathematical concepts have been sketched which show that various hypotheses are waiting to be proved and that the instruments for the interpretation of long-term data sets are available.

Previous ecological research stated that the study of unusual events such as slow ecological processes, rare events, episodic phenomena, high frequent variable processes, subtle processes, and their combinations like complex phenomena have not been a commonly used object in the past (Weatherhead, 1986). Therefore, long-term ecological studies are extremely valuable for providing consistent data records for subtle processes and long-term changes in ecosystems, for providing early warning procedures due to expected rapid and hazardous changes, for the development of

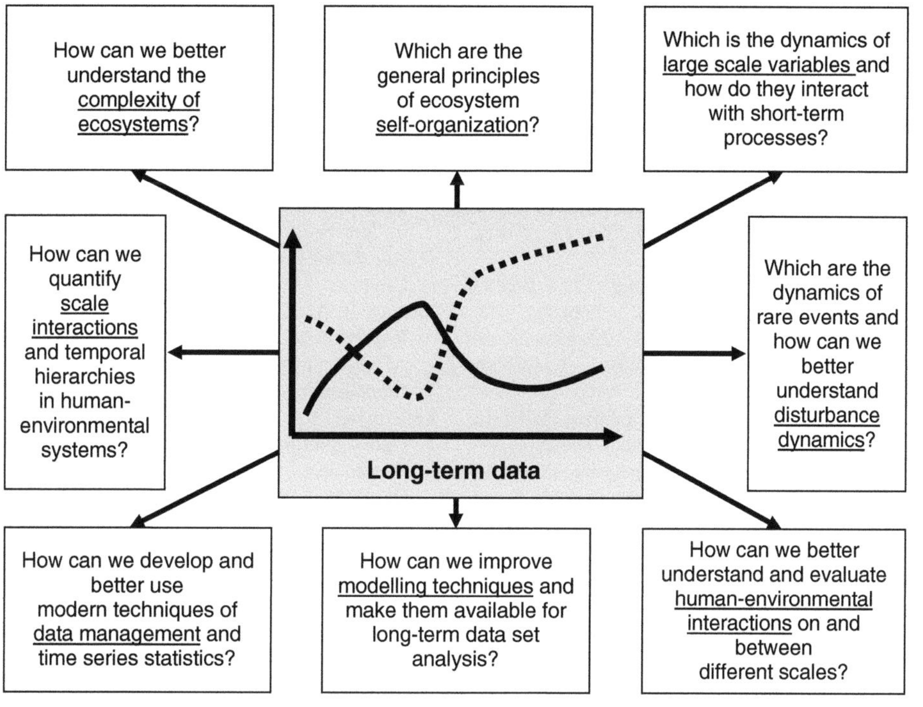

Fig. 2.5 Summary of important theoretical demands for long-term ecological research

integrated ecosystem models to a more reliable prediction of the spatio-temporal ecological impacts of land use and climate changes, and for suggesting and handling sustainable ecosystem management strategies.

The described demands have been focussed on a small number of approaches only. Further arguments can be taken from other theoretical concepts as well (Müller, 1997). For instance, network theory, information theory, or ecological thermodynamics could be taken to prolong the list of questions.

Summarizing, in fact there is a huge demand for long-term ecological analyses to fulfil these requirements (see Fig. 2.5). Therefore, also the scientific support policy should consider these demands more intensively and enhance the number of long-term studies on a worldwide scale.

Within the resulting conceptions, the collaboration between theory, statistics, modelling, and empirical ecology should be widened and deepened. Only an integrative interdisciplinary cooperation, based on the ecosystem approach, will be helpful to better understand the environmental interrelations as a holistic multiscale processual unit. Such knowledge will be suitable to find solutions for all the long-term problems our environment is facing recently. Therefore, a good theoretical background should not be attained as an academic exercise, but as a basis for better environmental applications of ecological knowledge.

Acknowledgements The authors wish to thank Michael Hauhs for important inputs and discussions. Ralf Seppelt and Sven-Erik Jørgensen have provided very valuable and constructive advices within their reviews, and Cornelia Baessler has been extremely helpful concerning the editorial procedures.

References

Allen, T. F. H., & Starr, T. B. (1982). *Hierarchy – perspectives for ecological complexity*. Chicago; London: The University of Chicago Press.

Andersen, T. (1997). *Pelagic nutrient cycles*. Berlin: Springer.

Bredemeier, M., Blanck, K., Dohrenbusch, A., Lamersdorf, N., Meyer, A. C., Murach, D., et al.(1998). The solling roof project – site characteristics, experiments and results. *Forest Ecology & Management, 101*, 281–293.

Bruno, J. F., Boyer, K. E., Duffy, J. E., Lee, S. C., & Kertesz, J. S. (2005). Effects of macroalgal species identity and richness on primary production in benthic marine communities. *Ecology Letters, 8*, 1165–1174.

Callahan, J. T. (1991). Long-term ecological research in the United States: A federal perspective. In: Risser, P. G. (Ed.), *Long-term ecological research – an international perspective* (pp. 9–21). Chichester: Wiley.

Demazure, M. (2000). *Bifurcations and catastrophes*. Berlin: Springer.

Dittmann, S., Schleier, U., Günther, S., Villbrandt P., Niesel, M. V., Hild, A., et al.(1998). *Elastizität des Ökosystems Wattenmeer. Projektsynthese. Forschungsbericht 03FO112 für den BMBF.* Wilhelmshaven: Terramare.

Franklin, J. F. (1989). Importance and justification of long-term studies in ecology. In: Likens, G. E. (Ed.), *Long-term studies in ecology* (pp. 3–19). New York, Berlin, Heidelberg: Springer.

Fränzle, O. (1998). Ökosystemforschung im Bereich der Bornhöveder Seenkette. In: Fränzle, O., Müller, F., & Schröder, W. (Eds.), *Handbuch der Ökosystemforschung* (Chapter V-4.3). Landsberg: Ecomed.

Fritz, P. (1999). UFZ-Forschungszentrum Leipzig-Halle. In: Fränzle, O., Müller, F., & Schröder, W. (Eds.), *Handbuch der Ökosystemforschung* (Chapter V-4.5), Landsberg: Ecomed.

Gnauck, A. (2000). Fundamentals of ecosystem theory from general systems analysis. In: Jørgensen, S. E. & Müller, F. (Eds.), *Handbook of ecosystem theories and management* (pp. 75–88). Boca Raton: Lewis Publishers.

Gollan, T., & Heindl, B. (1998). Bayreuther Institut für Terrestrische Ökosystemforschung. In: Fränzle, O., Müller, F., & Schröder, W. (Eds.), *Handbuch der Ökosystemforschung* (Chapter V-4.6). Landsberg: Ecomed.

Greenland, D., Hayden, B. P., Magnuson, J. J., Ollinger, S. V., Pielke, R. A. Sr., & Smith, R. C. (2003). Long-term research on biosphere–atmosphere interactions. *BioScience, 53*, 33–45.

Gruttke, H., & Dröschmeister, R. (1998). Ökologische Langzeitforschung für Naturschutz: Überblick, Abgrenzung und Einführung. *Schriftenreihe für Landschaftspflege und Naturschutz, 58*, 7–20.

Gundersson, L. H., & Holling, C. S. (Eds.). (2003). *Panarchy: Understanding transformations in human and natural systems*. Washington, DC: Island Press.

Haeckel, E. (1866). *Generelle Morphologie der Organismen. Bd. 2 Allgemeine Entwickelungsgeschichte der Organismen: Kritische Grundzüge der mechanischen Wissenschaft von den entstehenden Formen der Organismen.* Berlin: Reimer.

Hantschel, R., Kainz, M., & Filser, J. (1998). Forschungsverbund Agrarökosysteme München. In: Fränzle, O. , Müller, F., & Schröder, W. (Eds.), *Handbuch der Ökosystemforschung* (Chapter V-4.7). Landsberg: Ecomed.

Hauhs, M., & Schmidt, S. (Eds.). (1999). *Lange Bramke (Harz Mountains) – A Central European Forest Ecosystem (between Science and Management)*. Berlin: Springer.

Hobbie, J. E., Carpenter, S. R., Grimm, N. B., Gosz , J. R., & Seastedt, T. (2003). The US long-term ecological research program. *BioScience, 53*, 21–31.

Holling, C. S. (1985). Simplifying the complex: The paradigms of ecological function and structure. *European Journal of Operational Research, 30*, 139–146.

Holling, C. S. (1986). The resilience of terrestrial ecosystems: Local surprise and global change. In: Clark, W. C. & Munn, R. E. (Eds.), *Sustainable development of the biosphere* (pp. 292–317). Cambridge: Cambridge University Press.

Jørgensen, S. E. (1996). *Integration of ecosystem theories: A pattern*. Dordrecht: Kluwer.

Jørgensen, S. E., & Müller, F. (Eds.). (2000). *Handbook of ecosystem theories and management*. Boca Raton: CRC Press.

Jørgensen, S. E., Fath, B., Bastianoni, S., Marquez, J., Müller, F., Nielsen, S. N., et al.(2007). *A new ecology – the systems perspective*. Amsterdam: Elsevier Publishers.

Kay, J. J. (2000). Ecosystems as self-organised holarchic open systems: Narratives and the second law of thermodynamics. In: Jørgensen, S. E. & Müller, F. (Eds.), *Handbook of ecosystem theories and management* (pp. 135–160). Boca Raton: Lewis.

Keating, B. A., Carberry, P. S., Hammer, G. L., Probert, M. E., Robertson, M. J., Holzworth, D., et al.(2003). An overview of APSIM, a model designed for farming systems simulations. *European Journal of Agronomy, 18*, 267–288.

Kellermann, A., Gätje, C., & Schrey, E. (1998). Ökosystemforschung im Schleswig-Holsteinischen Wattenmeer. In: Fränzle, O., Müller, F., & Schröder, W. (Eds.) *Handbuch der Ökosystemforschung* (Chapter V-4.1.1). Landsberg: Ecomed.

Kerner, H. F., Spandau, L. & Köppel, J. (1991). *Methoden zur Angewandten Ökosystemforschung entwickelt im MAB 6 – Projekt Ökosystemforschung Berchtesgaden. MAB Mitteilungen 35.* Bonn: Deutsches National-Komitee Unesco Man and the Biosphere Programme.

Kersebaum, K. C. (1995). Application of a simple management model to simulate water and nitrogen dynamics. *Ecological Modelling, 81*, 145–156.

Kratz, T. K., Deegan, L., Harmon, M. E., & Laurenroth, W. K. (2003). Ecological variability in space and time: Insights gained from the US LTER program. *BioScience, 53*, 57–67.

Lausch, A. (2003). Integration of spatio-temporal landscape analysis in model approaches. In: Helming, K. & Wiggering, H. (Eds.), *Sustainable development of multifunctional landscapes* (pp. 111–123). Berlin, Heidelberg: Springer-Verlag.

Likens, G. E. (1983). A priority for ecological research. *Bulletin of the Ecological Society of America, 64*, 234–243.

Likens, G. E. (Ed.). (1989). *Long-term studies in ecology: Approaches and alternatives*. New York: Springer.

Little, R. J. A., & Rubin, D. B. (1987). *Statistical analysis with missing data*. Chichester: Wiley.

LTERNET (n.d.). Retrieved 2006, from http://www.lternet. edu.

Magnuson, J. J. (1990). Long-term ecological research and the invisible present. *BioScience, 40*, 495–502.

Mirschel, W., Schultz, A., & Wenkel, K.-O. (2001). Assessing the impact of land use intensity and climate change on ontogenesis, biomass production, and yield of Northeast German agro-landscapes. In: Tenhunen, J. D., Lenz, R., & Hantschel, R. (Eds.), *Ecosystem approaches to landscape management in Central Europe, Ecological Studies 147* (pp. 299–313). Berlin, Heidelberg, New York: Springer-Verlag.

Müller, F. (1997). State of the art in ecosystem theory. *Ecological Modelling, 100*, 161–165.

Müller, F., & Leupelt, M. (1998). *Eco targets, goal functions and orientors*. Berlin, Heidelberg, New York: Springer.

Nisbet, R. M., & Gurney, W. S. C. (1982). *Modelling fluctuating populations*. Chichester: Wiley.

Odum, H. T., Brown, M. T., & Ulgiati, S. (2000). Ecosystems as energetic systems. In: Jørgensen, S. E. & Müller, F. (Eds.),

Handbook of ecosystem theories and management (pp. 283–302). Boca Raton: CRC Press.

O'Neill, R. V., DeAngelis, D. L., Waide, J. B., & Allen, T. F. H. (1986). *A hierarchical concept of ecosystems.* Princeton: Princeton University Press.

O'Neill, R. V., Johnson, A. R., & King, A. W. (1989). A hierarchical framework for the analysis of scale. *Landscape Ecology, 3*, 193–205.

Pickett, S. T. A. (1991). Long-term studies: Past experience and recommendations for the future. In: Risser, P. G. (Ed.), *Long-term ecological research: An international perspective. SCOPE report 47* (pp. 71–88). Chichester: Wiley.

Pollock, D. S. G. (1999). *A handbook of time-series analysis, signal processing and dynamics.* San Diego: Academic Press.

Risser, P. G. (Eds.). (1991). *Long-term ecological research – An international perspective.* Chichester: Wiley.

Ritchie, J. T., & Godwin, D. C. (1993). Simulation of nitrogen dynamics in the soil plant system with the CERES models. *Agrarinformatik, 24*, 215–230.

Schönthaler, K., Meyer, U., Windhorst, W., Reichenbach, M., Pokorny, D., & Schuller, D. (2001). *Modellhafte Umsetzung und Konkretisierung der Konzeption für eine ökosystemare Umweltbeobachtung am Beispiel des länderübergreifenden Biosphärenreservats Rhön. Projektbericht zum F+E Vorhaben 109 02 076 des Umweltbundesamt Berlin.* München: Bosch & Partner GmbH.

Schönthaler, K., Müller, F. & Barkmann, J. (2003). Synopsis of systems approaches to environmental research – German contribution to ecosystem management. *UBA-Texte, 85/03*, 1–142.

Schulze, E.-D., & Mooney, H. A. (Eds.). (1994). *Biodiversity and ecosystem functioning.* Berlin: Springer.

Schultz, J. (1995). *The ecozones of the world.* Berlin: Springer.

Seppelt, R. (2003). *Computer-based environmental management.* Weinheim: Wiley-VCH Verlag.

Stachowicz, J. J., Fried, H., Osman R. W., & Whitlach, R. B. (2002). Biodiversity, invasion resistance and marine ecosystem function: Reconciling pattern and process. *Ecology, 83*, 2575–2590.

Supit, I., Hooijer, A. A., & van Diepen, C. A. (1994). *System description of the WOFOST 6.0 crop simulation model implemented.* Luxembourg: Office for Official Publications of the European Communities.

Svirezhev, Y. M., & Steinborn, W. (2001). Exergy of solar radiation: Thermodynamic approach. *Ecological Modelling, 145*, 101–110.

Symstad, A. J., Chapin, F.S., III., Wall, D. H., Gross, K. L., Huenneke, L. F., Mittelbach, G. G., et al.(2003). Long-term and long-range perspectives on the relationship between biodiversity and ecosystem functioning. *BioScience, 53*, 89–98.

Turner, M. G., Collins, S. L., Lugo, A. L., Magnuson, J. J., Rupp, T. S., & Swanson, R. J. (2003). Disturbance dynamics and ecological response: The contributions of long-term ecological research. *BioScience, 53*, 46–56.

Ulanowicz, R. E. (2000). Ascendancy: A measure of ecosystem performance. In: Jørgensen, S. E. & Müller, F. (Eds.), *Handbook of ecosystem theories and management* (pp. 303–316). Boca Raton: CRC Press.

Weatherhead, P. J. (1986). How unusual are unusual events? *American Naturalist, 128*, 150–154.

Wenkel, K.-O., & Mirschel, W. (1995). Agroökosystemmodellierung – Grundlage für die Abschätzung von Auswirkungen möglicher Landnutzungs- und Klimaänderungen. *ZALF-Bericht, 24*, 1–187.

Wenkel, K.-O., Wieland, R., Mirschel, W., Schultz, A., Kampichler, C., & Kirilenko, A. (2008). Regional models of intermediate complexity (REMICs)-a new direction in integrated landscape modelling. In: Jakeman, T., Voinov, A., & Rizzoli, A. E. (Eds.), *State of the art and futures in environmental modelling and software.* Amsterdam: Elsevier.

Widey, G. A. (1998). Forschungszentrum Waldökosysteme der Universität Göttingen. In: Fränzle, O., Müller, F., & Schröder, F. (Eds.), *Handbuch der Ökosystemforschung* (Chapter V-4.4). Landsberg: Ecomed.

Wieland, R., Voss, M., Holtmann, X., Mirschel, W., & Ajibefun, I. (2006). Spatial Analysis and Modeling Tool (SAMT): 1. Structure and possibilities. *Ecological Informatics, 1*, 67–76.

Wiggering, H. (2001). Zentrum für Agrarlandschafts- und Landnutzungsforschung – *Jahresbericht 2000/2001.* Müncheberg: ZALF.

Chapter 3

Long-Term Ecosystem Dynamics: Theoretical Concepts of Environmental Change

Brian D. Fath and Felix Müller

Abstract The question of ecosystem dynamics is important because when studying ecosystems, particularly over the long term, one must expect that natural endogenous changes will occur. In other words, observed changes may not be solely reflective of human-induced or other exogenous perturbations, but rather represent the natural long-term dynamic which the system experiences. Therefore, ecosystem management must account for these expectations, such that the goal might not be to preserve a system in its current state, but to allow the range of natural dynamics to occur, to allow the system to follow its self-organizing trajectory. The challenge for ecosystem theory and long-term ecological research is to identify this trajectory or direction in which ecosystems change. We can for instance look at some of the main system characteristics such as species composition, functional integrity, or biodiversity, and also look at changes in the thermodynamic patterns of organization in the ecosystem. In this chapter, we review some of the standard concepts on ecosystem growth and development and discuss the use of holistic orientors and indicators as a means to understand long-term ecosystem dynamics. We also try to demonstrate some respective linkages for the analysis of human–environmental systems and derive some suggestions for environmental management.

Keywords Ecological orientors · Goal functions · Ecological succession · Ecosystem growth · Ecosystem development, Ecosystem decay

B.D. Fath (✉)
Biology Department, Towson University, Towson, MD 21252, USA; Dynamic Systems, International Institute for Applied Systems Analysis, Laxenburg, Austria
e-mail: bfath@towson.edu

3.1 Introduction

There are many good reasons to support the long-term aspects in ecological research. Besides several urgent practical demands, theoretical considerations play a major role in long-term ecosystem analysis (see Müller et al., 2010). To illustrate these scientific motivations, we will discuss the state of the art in conceiving theoretical approaches for ecosystem growth and development. The following considerations therefore can be taken as a starting point for the derivation of new hypotheses on the long-term behavior of environmental systems.

We will start this short review with an outline of focal components from general succession theory, which has been focused on structural items and community features in the past. Therefore in a second step we will enhance the view and include functional processes throughout ecosystem growth and development, i.e., taking a thermodynamic perspective. Realizing that ecosystems do grow all the time, but death and decay also are parts of ecological dynamics, the adaptive cycle will be shortly described, and thereafter we will try to outline some principles from the long-term development of human–environmental systems. This will lead to a final short discussion about respective management options.

3.2 Components of Ecological Succession

The classical view of ecological succession is based on the idea that there are orderly changes in compositions and structures of biotic communities. The

F. Müller et al. (eds.), *Long-Term Ecological Research*,
DOI 10.1007/978-90-481-8782-9_3, © Springer Science+Business Media B.V. 2010

concept states that rapidly growing, short-lived, pioneer species with wide dispersal mechanisms colonize an area later to be replaced by slower growing, longer-lived, shade tolerant climax community species (Drury & Nisbet, 1973; Connell & Slatyer, 1977; Shugart, 1984; Pickett & White, 1985). This process is called primary succession if it originates on a landmass initially devoid of a biological community such as after glacial retreat, island formation, or volcanic activity (Walker & Del Moral, 2003; Glenn-Lewin, Peet, & Veblen, 1992); or secondary succession, if it occurs on an area which has the remnants of a biological community from which to rebuild, e.g., after a fire, hurricane, or human-induced clearing (Horn, 1974; Walker, Walker, & Hobbs, 2007).

This view of ecological succession theory describes a bi-modal development of species composition from r-selected species to K-selected species (MacArthur & Wilson, 1967). The biotic potential, r, for reproduction, expressed in exponential growth, characterizes the earlier pioneer stage:

$$\frac{dN}{dt} = rN \tag{3.1}$$

where N is the population size and r the rate of growth. During this pioneer period resources usually are not scarce relative to the biotic community so population grows unchecked, exponentially – except for density-independent factors not captured in the simple equation. Species that thrive during this stage are able to grow and reproduce quickly, colonizing new areas for establishment. In the second stage, crowding and other resource limitations express themselves as density-dependent factors such that the overall carrying capacity, K, limits the growth typically expressed using the logistic equation:

$$\frac{dN}{dt} = rN\left(1 - \frac{N}{K}\right) \tag{3.2}$$

In this view, the system grows quickly during the early stages when the population is low (when $N<<K$, this approximates Equation 3.1). Mathematically, there is an inflection point when $N=K/2$, after which the population growth slows as quickly as it rose during the first half until it reaches a steady state, also called a climax community (Whittaker, 1953; Roughgarden, May, & Levin, 1989). Under these conditions, when $N=K$ population growth is zero, but it does not mean that there is a static situation because it is not an equilibrium in the thermodynamic sense: Organisms are still growing and dying. Rather, the ecosystem level processes are equally balanced between growth and death. Of course nature is never so well behaved such that N asymptotically approaches K, nor remains poised there for an extended period of time. Consequently, there is little evidence that any empirical data sets will precisely fit this model (Hall, 1988).

In fact, a more common pattern is based on overshoot and collapse, where the growth momentum causes the population to exceed the carrying capacity, after which it cannot be supported by the environment, and collapses to a level below the carrying capacity (Taylor, Aarssen, & Loehle, 1990). If this pattern repeats, then there will be a series of overshoot and collapse. It is also possible that the overshoot and collapse degrades the environment in a way that permanently lowers the carrying capacity such that the rebounding population level will be lower. Furthermore, the carrying capacity itself is not fixed by environmental conditions, but is also influenced by the amount of resource extraction the species incur. More efficient use of resources allows a greater population sustenance that effectively raises the carrying capacity. These ecological concepts have been the fundamentals of several environmental approaches, such as carrying capacity or ecological footprint (e.g., Bossel, 1998, 1999, 2004; Wackernagel & Rees, 1996).

Summarizing the logistic model of succession from more exploitative to conservative life patterns provides a useful conceptual model to view ecosystem dynamics (Shugart, 1998). The emphasis in switch from growth in early stages to steady state is represented by the nomenclature for r- and K-selected species representing these two stages and captures an important transition in the overall dynamics.

While there have been several other variations of this standard succession model, one to call to attention has been developed by Grime (1979), specifically describing the dynamics of plant species. He expanded from the r and K classification to three ecological classes – ruderals, competitors, and stress tolerators – based on a population's response to the two categories disturbance and stress (Table 3.1). In this concept disturbance is any action that destroys the plant biomass in part or whole, and stress is any factor that prevents or inhibits growth. Grime saw no viable life strategy for

Table 3.1 Plant species succession model after Grime (1979)

	Stress	
Disturbance	Low	High
Low	Competitors	Stress tolerators
High	Ruderals	No viable strategy

a high stress, high disturbance scenario, and therefore left it blank.

A problem remains that both the bi-modal and tri-modal models downplay the significance of ecosystems as open, thermodynamic systems and do not sufficiently consider ecological functions. An open systems perspective, integrating ecosystem properties, was implicit in E.P. Odum's 1969 seminal paper describing trends to be expected in ecosystem growth and development because in it he focused on the thermodynamic development of ecosystems, which of course is dependent on a supply of energy. A sample of the type of dynamics he proposed as ecosystems go from early stages to mature stages is given in Table 3.2.

In this list it is clear that thermodynamically and organizationally the expected trend is toward a less growth dominated and more conservative stage in which the captured energy is utilized to maintain and support a large ecological community in terms of thermodynamics, which can be measured as biomass energy or exergy (Schneider & Kay, 1994a; Jørgensen, 2002; Kay, 2000; Odum, Brown, & Ulgiati, 2000). An efficient ecological community is one that supports large biomass storage for a relatively small, constant flow of energy. The ratio of biomass storage – with units of some amount – to flow (amount per time) gives units of time. In other words, the efficiency is a measure of how long the energy is retained in the system, which of course increases if there is a greater diversity of organisms forming a more complex network (Patten, 1992; Ulanowicz, 2000). The key to remember is that the amount of energy available to the system is mostly fixed by the solar constant. Within that constraint though, highly complex and diverse systems have arisen and are supported due to self-organized ecosystem dynamics (Müller & Nielsen, 2000; Jørgensen et al., 2007). The respective thermodynamic processes seem to constrain systems dynamics in a very general manner, whereby the increasing complexification can be seen as a key process (Weber et al., 1989; Müller et al., 1997).

Table 3.2 Some trends of ecosystem development after Odum (1969)

Ecosystem attribute	Developmental stage	Mature stage
Community energetics	–	–
Gross production/ community respiration (P/R ratio)	>1	~1
Biomass supported/unit energy flow (B/E ratio)	Low	High
Food chains	Linear	Weblike
Nutrients	–	–
Mineral cycles	Open	Closed
Nutrient exchange rate	Rapid	Slow
Nutrient conservation	Poor	Good
Overall homeostasis	–	–
Internal symbiosis	Undeveloped	Developed
Entropy	High	Low
Information	Low	High

3.3 Ecosystem Growth and Development

Another formulation of ecosystem growth and development by Jørgensen, Patten, and Straškraba (2000) and Fath, Jørgensen, Patten, and Straškraba (2004) proposed that there are four stages which ecosystems generally undergo. First, it is necessary to clarify that growth refers to the quantitative increase of some measure such as biomass or throughflow, and therefore has an upper limit based on physical constraints (Müller & Fath, 1998). Development, on the other hand, refers to qualitative change that transpires such as with the organization or information level (Ulanowicz, 1986). In principle, there is no upper limit to the process of development (Jørgensen et al., 2007).

The four stages of growth and development in this model are Boundary, Structural, Network, and Information Growth.

1. *Boundary Growth* brings the input of low-entropy material into the system. Ecosystems are open systems, dependent always on a constant inflow of low-entropy, high-quality energy and also on the

presence of sinks to absorb the discarded high-entropy, low-quality energy. The vast complexity of genetic, biological, and ecological diversity has emerged within these two constraints. The inflow for almost all natural ecosystems is solar radiation – the only known exception is the deep ocean sulfur vent community.
2. *Structural Growth* occurs when the physical quantity of biomass in the system increases, often as a result of the increase in the amount, number, and size of components in the ecosystem. Structural growth is a positive feedback mechanism because the additional biomass makes it possible for the system to capture more of the incoming solar energy, though also requiring more energy for maintenance.
3. *Network Development* is related to the internal organization of the connectivity in the system through additional or more effectively placed energy–matter transactions between system components, which results in more interconnected pathways increasing the cycling of matter and energy (Patten, 1998). Network development also raises the total system throughflow (since it is retained and utilized in the system longer, see Ulanowicz, 1997) and contributes to the role which indirectness plays in the system interactions (Fath & Patten, 1999).
4. *Information Development* is qualitative progression in system behavior from exploitative patterns to more conservative patterns, which are more energetically efficient. Information growth deals with the development of ecosystem compartments themselves, as they tend to increase their own performance within the system.

This four-stage model provides a general conceptual model for ecosystem growth and development. The typical sequence that occurs during primary and secondary succession corresponds roughly to these four forms, however, the expression of the four forms is not strictly sequential because the different functional components can operate in parallel (e.g., boundary input initiates growth, but continues on as long as the system organization remains; network and informational growth both can occur simultaneously). Of interest is its application to ecological processes. These stages help explain how the trends Odum (1969) observed come to pass. For example, initially, ecosystem biomass increases (physical structure), and while the overall system respiration increases, the respiration relative to biomass decreases as larger plants and animals become more prevalent. Total entropy production first increases and then stabilizes when further boundary flow importation is not possible. Next, feedback increases – including recycling of energy and matter – as the network structure evolves, which leads to decreases in biomass-specific entropy production. Lastly, information increases in terms of the biochemical and species diversity and genetic complexification.

Table 3.3 Several of the thermodynamic goal functions proposed to track ecosystem change

1. Maximize power (Lotka, 1922; Odum & Pinkerton, 1955) max (total system throughflow)
2. Maximize exergy storage (Jørgensen & Mejer, 1979) max (total system storage)
3. Maximize dissipation (Schneider & Kay, 1994a, 1994b) max (total system export)
4. Maximize cycling (Morowitz, 1968) max (total system cycling)
5. Minimize specific dissipation (Prigogine, 1955) min (total system export/total system storage)
6. Maximize retention time (Cheslak & Lamarra, 1981) max (total system storage/total system throughflow)

Several ecological 'goal functions' have been used in the literature to assess these growth and development trends (see, for example, Müller & Leupelt, 1998; Svirezhev, 1998, Müller et al., 2010). Table 3.3 lists some of the common ones based strictly on the thermodynamics of the system.

Fath, Patten, and Choi (2001) showed that the goal functions listed in Table 3.3 although derived to assess different aspects of ecosystem development are in fact not in conflict with each other, and often are complementary. For example, maximizing cycling leads to greater storage and greater throughflow. The property that seemed most at odds with the others was minimization of specific dissipation, but it became clear that this is possible first, because total system export is capped by the amount of inflow the system can capture, and second, the total system storage continues to increase due to the network and information development. One also sees from the maximization of residence time that over time the system storage outpaces the system throughflow, which is consistent with the idea that B/P is increasing in mature systems. In a more detailed analysis of these orientors Fath et al. (2004) showed that indeed the 'goal functions' while consistent in principle do behave differently during the

Table 3.4 Ecosystem growth and development during the four growth forms (↑ indicates an increase, ↓ a decrease, and ↔ no change in the condition or orientation)

	Boundary G&D	Structural G&D	Network G&D	Information G&D
Thermodynamic Indicator				
Specific entropy	↔	↔	↓	↓
Energy throughflow	↑	↑	↑	↑
Exergy degradation	↑	↑	↔	↔
Exergy storage	↑	↑	↑	↑
Retention Time	↔	↔	↑	↑

different phases (Table 3.4). We see that only system storage and throughflow continue along their intended trajectory during all growth and development stages.

At this point of the discussion, the question arises, whether ecosystems only grow or develop in a complexifying manner. They do not. The further the systems have been moved away from thermodynamic equilibrium, the higher seems to be the risk of getting moved back (Schneider & Kay, 1994b). The more time has been used for complexification, the higher is the risk of being seriously hit by disturbance (see Table 3.5), and the longer the elements of the system have increased their mutual connectedness, the stronger is the mutual interdependency and the total system's brittleness. Table 3.5 combines some features of mature ecosystems and lists some risk-related consequences of the orientor dynamics. In general it can be concluded that the adaptability after changes of

Table 3.5 Some characteristics of mature ecosystems and their potential consequences for the system's adaptability: Maturity is attained due to a long-term mutual adaptation process. In the end of the development the interrelations between the components are extremely strong, sometimes rigid. Reactivity and flexibility are reduced. If the constraints change this state runs the risk of being seriously disturbed (after Jørgensen et al., 2007)

Orientor functions	Related risks
High exergy capture	The system operates on the base of high energetic inputs → high vulnerability if the input pathways are reduced
High exergy flow density	Many elements of the flow webs are dependent on inputs which can be provided only if the functionality of the whole system is guaranteed → high risk of loosing mutually adapted components
High exergy storage and residence times	Exergy has been converted into biomass and information → high risk of internal eutrophication
High entropy production	Most of the captured exergy is used for the maintenance of the mature system → minor energetic reserves for structural adaptations
High information	High biotic and abiotic diversity → risk of accelerated structural breakdown if the elements are correlated
High degree of indirect effects	Many interactions between the components → increase of mutual dependency and risk of cascading chain effects
High complexity	Many components are interacting hierarchically → reduced flexibility
High ascendancy and trophic efficiency	Intensive flows and high flow diversities result in a loss reduction → changing one focal element can bring about high losses
High degree of symbiosis	Symbiosis is linked with mutual dependencies → risk of cascading chain effects
High intra-organismic storages	Energy and nutrients are processed and stored in the organismic phase (Kay, 1984) → no short-term availability for flexible reactions
Long life spans	Focal organisms have long life expectancies → no flexible reactivity
High niche specialization and K selection	Organisms are specialized to occupy very specific niche systems and often have a reduced fecundity → reduced flexibility

the constraints may decrease when a high degree of maturity is attained.

3.4 Adaptive Cycle

In this context, C.S. Holling (1986) made an important contribution to the standard ecosystem dynamics model by explicitly addressing the issue of disturbance of the steady-state, climax community. He proposed a four-stage model of ecosystem dynamics that includes growth (or exploitation), conservation, disturbance (which he called creative destruction), and reorganization (renewal, see Fig. 3.1). This figure is laid out as a sideways figure-eight to represent the continual growth and regenerative dynamics ecosystems undergo in a metaphoric sense. This conceptual model is further defined by identifying the axes as connectivity (abscissa) and the stored capital (ordinate). Other dimensional formulations have been proposed, such as Kay considered exergy storage and exergy consumed, while Hansell and Bass (1998) used carbon storage and nutrients. Jørgensen et al.(2007) re-oriented the model to show that the potential and dynamics follow an inverse pattern (Fig. 3.2; see also Müller et al., 2010). This is consistent with Holling's original concept in that the disturbance phase was termed creative destruction because the system becomes rigid

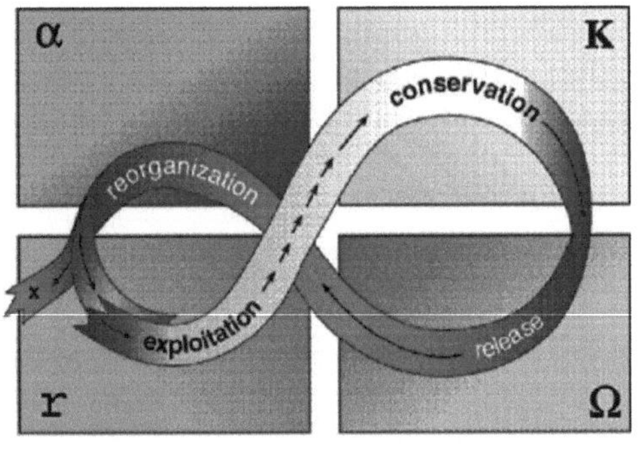

Fig. 3.1 Holling's 'figure-eight' curve. The concept of the adaptive cycle states that ecosystem development touches four different fundamental stages, which are passed with different velocities. Breakdown ('release') and 'reorganization' after disturbance are integrated processes. The slow growth phases of ecosystems are situated between the stages 'exploitation' and 'conservation.' From the resilience alliance web page, http://www.resalliance.org, see also Gundersson and Holling (2003)

or locked into a specific configuration in which whole-scale change although maybe warranted is not possible.

Schumpeter (1911) used a similar model to describe economic development and the cyclic nature of entrepreneurship. Without disturbance the system becomes rigid in ways that make emergence of new markets difficult, and the current players achieve a status of economic lock-in (Arthur, 1990). A good example of lock-in and the role of creative destruction in human society could be the housing stock which when constructed has a long characteristic time frame. In some cases though the initial designs are not in the long run most desirable, but once become established are too costly to replace particularly if it involves turnover of the entire existing stock. When a natural disaster strikes – the destruction – there is an open opportunity to rebuild with a new approach – allowing for creativity. Such was the case in the fast growing years of the city of Chicago in the mid-1800s. The population grew from a small remote town of 30,000 people in 1850 to almost 300,000 people by 1870 (on its way to 1,000,000 by 1890 and 2,000,000 by 1920) and due to the easy availability of lumber, balloon-frame, wooden housing was the dominant design style. A great fire in 1871 destroyed large portions of the city and the existing housing stock, and re-development in many central areas implemented masonry and concrete materials. While it would have been impractical to initiate a leveling of the city to replace the housing stock, the fire provided an opening for putting in place a new, more lasting design pattern. Some speculate that the current suburban development prominent in the United States is a similar investment trap in that it is unsustainable in the long run (due to long transportation requirements), but there is too much invested in it to change. Oil shocks may make these areas obsolete, providing another form of creative destruction. Other examples include the San Francisco earthquake of 1906, after which new building codes were implemented to guard against seismic shocks, but could not have been instituted in the existing structures, as well as the impact of Hurricane Katrina on New Orleans in 2005. The devastation caused by Katrina provides an opportunity for New Orleans not only to re-establish the natural flow of the Mississippi River but also to rebuild the lost communities in more sustainable patterns, but the verdict is out whether the re-development learns from or repeats the same mistakes.

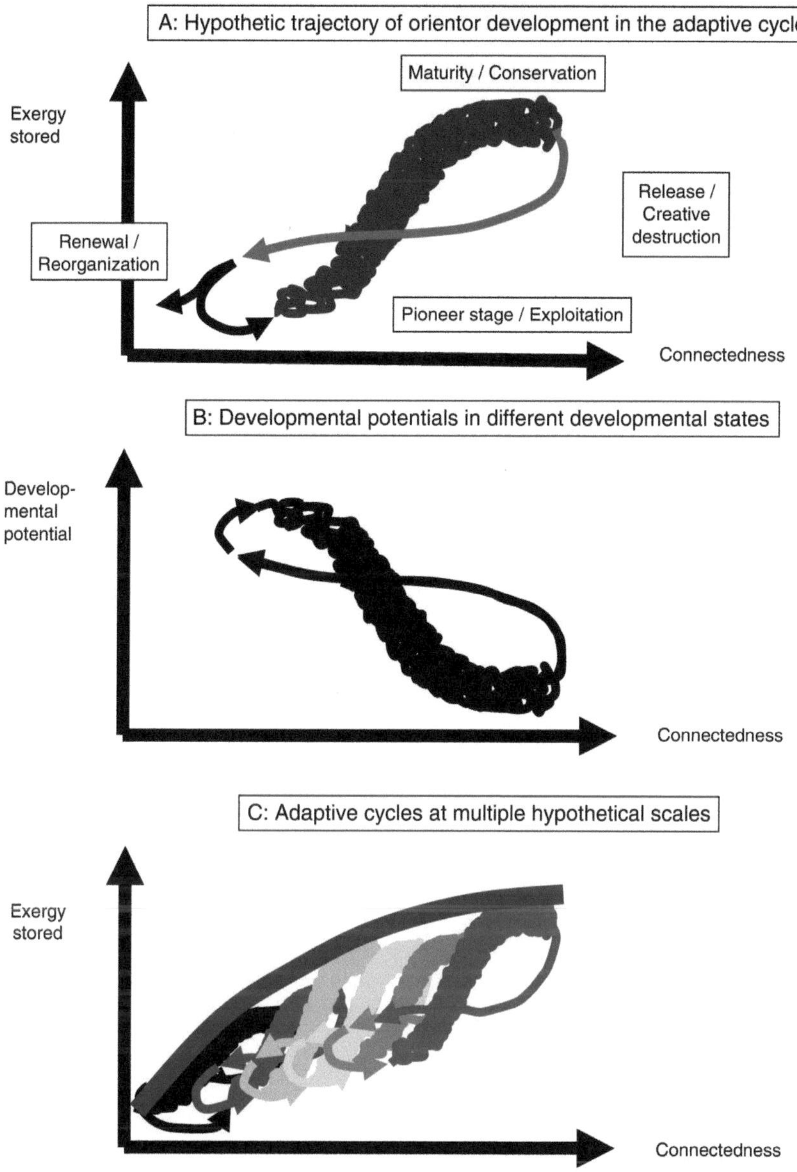

Fig. 3.2 Different interpretations of the adaptive cycle. (**a**) Hypothetic trajectory of orientor ('goal function') development in the adaptive cycle: In this case the function of connectedness (complexity) and stored exergy (biomass and information) is depicted. The system is operating in orientor dynamics for the longest time, while release and reorganization are passed rather quickly. During the renewal phase the stored exergy does not increase in contrast to the original 'figure-eight diagrams'. (**b**) In parallel with exergy increase the number of potential system states decreases throughout the orientor phase until around maturity the adaptability of the system is minimized. During the destruction the degrees of freedom of the system as well as the developmental uncertainty are rising strongly. (**c**) Depiction of a multiscale orientor hypothesis: Under certain continuously developing constraints breakdown events can provoke an optimization of long-term adaptability. Consequently destructions on a lower scale can support the orientor optimization on a higher spatio-temporal level

3.5 Complex Adaptive System Orientors

Ecosystems are a class of systems more generally referred to as complex adaptive systems. Characteristics of such systems are that they have many interacting parts operating at varying hierarchical scales (complexity), they maintain themselves in changing environments and create themselves in response to self-creativity in other systems (adaptivity), and they have emergent, irreducible properties (systemness). There have been proposals to describe holistically, the tendencies in which these systems change over time; one approach in particular by Hartmut Bossel (1998, 1999) worth mentioning identifies six fundamental orientors, which are meant to apply for all complex adaptive systems. These include

1. *Existence:* Attention to existential conditions is necessary to insure the basic compatibility and immediate survival of the system in the normal environmental state.
2. *Effectiveness:* In its efforts to secure scarce resources (exergy, matter, information) from, and to exert influence on its environment, the system should on balance be effective.
3. *Freedom of action:* The system must have the ability to cope in various ways with the challenges posed by environmental variety.
4. *Security:* The system must have the ability to protect itself from the detrimental effects of variable, fluctuating, unpredictable, and unreliable environmental conditions.
5. *Adaptability:* The system should be able to change its parameters and/or structure in order to generate more appropriate responses to challenges posed by changing environmental conditions.
6. *Coexistence:* The system must modify its behavior to account for behavior and interests (orientors) of other systems.

Orientors are defined as dimensions of concern not specific goals as they arise from the system interactions and are considered emergent system properties. They functions as attractors of systems development. The six orientors are responsive to the six general properties of the environment.

1. *Normal environmental state:* The actual environmental state can vary around this state in a certain range.
2. *Scarce resources:* Resources (exergy, matter, information) required for a system's survival are not immediately available when and where needed.
3. *Variety:* Many qualitatively very different processes and patterns occur in the environment constantly or intermittently.
4. *Reliability:* The normal environmental state fluctuates in random ways, and the fluctuations may occasionally take it far from the normal state.
5. *Change:* In the course of time, the normal environmental state may gradually or abruptly change to a permanently different normal environmental state.
6. *Other systems:* The behavior of other systems changes the environment of a given system.

Bossel proposes a one-to-one relationship between the properties of the environment and the basic orientors of systems. Therefore, the system equipped to secure better overall orientor satisfaction will have better fitness and will therefore have a better chance for long-term survival and sustainability. The point in including the orientor approach is to show the diversity of measures which have been used to identify the long-term trajectory and space in which systems adapt and evolve.

3.6 Long-Term Dynamics of Socio-ecological Systems

One would expect that the long-term dynamics observed in ecological systems could also provide insight into the development and dynamics of socio-ecological systems. Both systems are open, complex and far-from-equilibrium; they are ultimately dependent on the inflow of energy and the ability of the waste generated not to exceed the system's capacity for reuse or assimilation. While the orientors and goal functions can be seen as basic principles of the development of ecological systems there is the question if the related optimization functions are also of significance for human–environmental systems. At least three answers can be given to this question: (i) yes, because also human systems are self-organized systems; (ii) yes, because focal aspects of sustainability are founded on similar optimization strategies; (iii) no, because humans are optimizing their basic orientor expressions on different scales, thereby decreasing the potential development of the natural partner systems.

A relevant discussion in this context can be made about Jared Diamond's recent and popular book, 'Collapse: How societies choose to succeed or fail.' Diamond (2005) proposes 12 reasons for which human societies collapse related to their use of natural resources:

1. Deforestation and habitat destruction
2. Soil problems (erosion, salinization, loss of fertility)
3. Water management problems
4. Overhunting
5. Overfishing
6. Introduced species
7. Human population growth
8. Increased per capita impact
9. Human-induced climate change
10. Build up of toxic chemicals
11. Energy shortages
12. Full human utilization of earth's photosynthetic capacity.

The first eight reasons are affecting all epochs of human civilization up to the present; the last four have been emerging during the past few centuries since the industrial revolution (more appropriately referred to as the fossil fuel revolution). Diamond is clear in his statements that he believes societies make a choice to succeed or fail depending on how they deal with and manage their environmental resources. He provides numerous examples of societies that have collapsed as a result of environmental mismanagement, but the hypothesis lacks a guiding theory in that anecdotes exist of societies persisting or collapsing under similar activities.

In an earlier work on the same topic, Joseph Tainter (1988) takes a different, a more deterministic approach. First, he starts by critiquing many of the theories proposed for societies to collapse, including not only environmental ones, such as resource depletion, bountiful resources, catastrophes, intruders, and generally failure to adapt. The main problem, he contends, with these theories is that they are not causative in that in some cases societies experience such events (e.g., invaders, resource depletion), yet manage to thrive by winning in victory, finding new methods or resources, etc. In other words, it is not a good theory if sometimes it holds and others it does not. His thesis then rests more generally instead on the complexification of the society and its limitation of that as a problem-solving strategy.

'More complex societies are more costly to maintain than simpler ones. . .as societies increase in complexity, more networks are created among individuals, more hierarchical controls are created to regulate these networks, more information is processed, there is more centralization of information flow, there is increasing need to support specialists not directly involved in resource production, and the like' (Tainter 1988, p. 91). Note the similarity between these stages with those identified independently in ecosystem development, from structural, to network, to information growth. He goes on to say, 'All this complexity is dependent upon energy flow at a scale vastly greater than that characterizing small group of self-sufficient foragers or agriculturalists. The result is that as a society evolves toward greater complexity, the support cost levied on each individual will also rise, so that the population as a whole must allocate increasing portions of its energy budget to maintaining organizational institutions.' In other words, independent of the energy source, first solar, and now fossil, we rely on a greater allocation of this energy to maintain the complex structures created.

Ecosystems are very efficient at supporting a great amount of structure for a given amount of production (B/P high in mature ecosystems). Our current economic and political institutions are still in a phase that promulgates the creation of increased production – in addition to supporting all the existing built structure. This insistence on quantitative growth over qualitative development would be considered to be an early immature stage according to ecological succession. In fact, one could conclude that how production is treated in the long term for ecological versus societal systems is one of the main discrepancies and leads to much of the tension between resolving sustainable integration.

3.7 Ecosystem Dynamics and Environmental Management

It is clear that ecological systems change over time, so a policy that seeks only to maintain the system in its current state, i.e., preservation, counters that natural tendency. It takes energetic input to inhibit successional trajectories. For example, an agricultural field is maintained at an early stage of succession in order to benefit from the high productivity of fast-growing cereal crops, but at the expense of plowing, planting, and harvesting. In the case of agricultural production the motivation is obvious; however, there may be other cultural rationales for preserving socio-ecological systems in a state that opposes the natural tendency. For example, there are remnant bog ecosystems in the Midwestern United States that harbor species and habitats similar to those found thousands of years ago as the last glaciers retreated. Clearly, over time those ecosystems will fill in and give way to new ecosystems, but as a relict system and refuge for rare and endangered species environmental protection agencies may be encouraged or even required to intervene and preserve the habitats. While this could be an acceptable societal choice it should be acknowledged that this differs from protecting species within a current trajectory. Likewise, we at times protect social systems that are in conflict with the natural tendency. Venice is a cultural icon that will be protected at great expense in the face of sea level rise (and city sinking), however, as history has shown, over the long term it is likely that the costs will exceed the benefits to maintain and it will be abandoned or at least re-organized to better accommodate the current environmental conditions.

3.8 Conclusions

Understanding the long-term dynamics of natural ecosystems is an important goal both from a scientific and management question. It is clear that ecosystems undergo natural transitions from simpler, less diverse, less interconnected patterns, to ones that are more diverse, complex, and interconnected. Proposals to track the thermodynamic tendencies during this transition have produced encouraging results as there is an observable pattern that energy throughflow and storage both increase during growth and development. Ecosystems appear to naturally transition from a more growth-oriented phase to one in which conservative elements are dominant. But with the increasing connectedness, the risks of Holling's brittleness are growing as well, and destruction may take place due to a reduced buffer capacity. Nature conservation very often follows the basic principle to restore and maintain historical states, even if high inputs are necessary to fulfil that requirement. From a theoretical viewpoint, the question should be discussed in which cases this objective should be followed and in which situations self-organized dynamics might be allowed. It may in fact be easier to assess and protect the current (orientor) trajectory than the current state or a historical situation. Thus, environmental management might profit from learning from long-term ecological experience, which could be helpful to keep on developing the approaches of adaptive management strategies. The focal question should be: How to cope with change with respect to adaptability and future developmental capacity? To find constructive answers, on the other hand, the efforts to understand long-term ecological processes should be enhanced, theoretical propositions should be taken into account and should be tested, and the potentials and limitations to apply ecosystem principles for the analysis of human–environmental systems should be supported in the future.

References

Arthur, B. (1990). Positive feedbacks in the economy. *Scientific American, 262*, 92–99.

Bossel, H. (1998). Ecological orientors: Emergence of basic orientors in evolutionary self-organization. In F. Müller, & M. Leupelt (Eds.), *Eco targets, goal functions, and orientors* (pp. 19–33). New York: Springer.

Bossel, H. (1999). *Indicators for sustainable development: Theory, method, applications*. Winnipeg: International Institute for Sustainable Development.

Bossel, H. (2004). *Systeme, dynamik, simulation: Modellbildung, analyse und simulation.* Norderstedt: BoD-Books on Demand.

Cheslak, E. F., & Lamarra, V. A. (1981). The residence time of energy as a measure of ecological organization. In W. J. Mitsch, R. W. Bossermann, & J. M. Klopatek (Eds.), *Energy and ecological modelling* (pp. 591–600). Amsterdam: Elsevier.

Connell, J. H., & Slatyer, R. O. (1977). Mechanisms of succession in natural communities and their role in community stability and organization. *American Naturalist, 111*, 1119–1144.

Diamond, J. (2005). *Collapse: How societies choose to fail or succeed.* New York: Viking.

Drury, W. H., & Nisbet , I. C. T. (1973). Succession. *Journal of the Arnold Arboretum, 54*, 331–368.

Fath, B. D., & Patten, B. C. (1999). Network synergism: Emergence of positive relations in ecological models. *Ecological Modelling, 107*, 127–143.

Fath, B. D., Patten, B. C., & Choi, J. S. (2001). Complementarity of ecological goal functions. *Journal of Theoretical Biology, 208*, 493–506.

Fath, B. D., Jørgensen, S. E., Patten, B. C., & Straškraba, M. (2004). Ecosystem growth and development. *Biosystems, 77*, 213–228.

Glenn-Lewin, D. C., Peet, R. K., & Veblen, T. T. (1992). *Plant succession: Theory and prediction.* Berlin, Heidelberg, New York: Springer-Verlag.

Grime, J. P. (1979). *Plant strategies and vegetation strategies processes.* Chichester: John Wiley.

Gundersson, L. H., & Holling, C. S. (Eds.). (2003). *Panarchy: Understanding transformations in human and natural systems.* Washington, DC: Island Press.

Hall, C. A. S. (1988). An assessment of several of the historically most influential theoretical models used in ecology and of the data provided in their support. *Ecological Modelling, 43*, 5–31.

Hansell, R., & Bass, B. (1998). Holling's figure-eight model: A technical re-evaluation in relation to climate change and biodiversity. *Journal of Environmental Monitoring and Assessment, 49*, 157–168.

Holling, C. S. (1986). The resilience of terrestrial ecosystems: Local surprise and global change. In W. M. Clark, & R. E. Munn (Eds.), *Sustainable development of the biosphere* (pp. 292–317). Cambridge: Cambridge University Press.

Horn, H. S. (1974). The ecology of secondary succession. *Annual Review of Ecology and Systematics, 5*, 25–37.

Jørgensen, S. E. (2002). *Integration of ecosystem theories: A pattern.* Dordrecht: Kluwer.

Jørgensen, S. E., Fath, B., Bastianoni, S., Marquez, J., Müller, F., Nielsen, S. N., et al.(2007). *A new ecology – The systems perspective.* Amsterdam: Elsevier Publishers.

Jørgensen, S. E., & Mejer, H. F. (1979). A holistic approach to ecological modelling. *Ecological Modelling, 7*, 169–189.

Jørgensen, S. E., Patten, B. C., & Straškraba, M. (2000). Ecosystems emerging: 4. Growth. *Ecological Modelling, 126*, 249–284.

Kay, J. J. (1984). Self-organization in living systems. Dissertation, University of Waterloo.

Kay, J. J. (2000). Ecosystems as self-organised holarchic open systems: Narratives and the second law of thermodynamics. In S. E. Jørgensen, & F. Müller (Eds.), *Handbook of ecosystem theories and management* (pp. 135–160). Boca Raton: Lewis.

Lotka, A. J. (1922). Contribution to the energetics of evolution. *Proceedings of the National Academy of Sciences USA, 8*, 147–151.

MacArthur, R. H., & Wilson, E. O. (1967). *The theory of island biogeography.* Princeton: Princeton University Press.

Morowitz, H. J. (1968). *Energy flow in biology; Biological organization as a problem in thermal physics.* New York: Academic Press.

Müller, F., Breckling, B., Bredemeier, M., Grimm, V. Malchow, H., Nielsen, S. N., et al.(1997). Ökosystemare Selbstorganisation. In O. Fränzle, F. Müller, & W. Schröder (Eds.), *Handbuch der Ökosystemforschung* (Chapter III-2.4). Landsberg: Ecomed.

Müller, F., & Fath, B. D. (1998). The physical basis of ecological goal functions – An integrative discussion. In F. Müller & M. Leupelt (Eds.), *Eco targets, goal functions, and orientors* (pp. 269–285). Berlin, Heidelberg, New York: Springer.

Müller, F., Gnauck, A., Hauhs, M., Wenkel, K.-O., Schubert, H., & Bredemeier, M. (2010). Theoretical demands for long-term ecological research and the management of long-term data sets. In F. Müller, C. Baessler, H. Schubert, & S. Klotz (Eds.), *Long-term ecological research: Between theory and application* (pp. 11–25). Dordrecht, The Netherlands: Springer.

Müller, F., & Leupelt, M. (1998). *Eco targets, goal functions and orientors.* Berlin, Heidelberg, New York: Springer.

Müller, F., & Nielsen, S. N. (2000). Ecosystems as subjects of self-organising processes. In S. E. Jørgensen, & F. Müller (Eds.), *Handbook of ecosystem theories and management.* (pp. 177–194). Boca Raton: CRC Press.

Odum, E. P. (1969). The strategy of ecosystem development. *Science, 164*, 262–270.

Odum, H. T., Brown, M. T., & Ulgiati, S. (2000). Ecosystems as energetic systems. In S. E. Jørgensen, & F. Müller (Eds.), *Handbook of ecosystem theories and management* (pp. 283–302). Boca Raton: CRC Press.

Odum, H. T., & Pinkerton, R. C. (1955). Time's speed regulator: The optimum efficiency for maximum power output in physical and biological systems. *American Scientist, 43*, 331–343.

Patten, B. C. (1992). Energy, emergy and environs. *Ecological Modelling, 62*, 29–69.

Patten, B. C. (1998). Network orientors: Steps towards a cosmography of ecosystems: Orientors for directional development, self-organisation, and autoevolution. In F. Müller, & M. Leupelt (Eds.), *Eco-targets, goal functions, and orientors* (pp. 137–160). Berlin, Heidelberg, New York: Springer.

Pickett, S. T. A., & White, P. S. (1985). *The ecology of natural disturbance and patch dynamics.* Orlando: Academic Press.

Prigogine, I. (1955). *Thermodynamics of irreversible processes.* New York: Wiley.

Roughgarden, J., May, R. M., & Levin, S. A. (Eds.). (1989). *Perspectives in ecological theory.* Princeton: Princeton University Press.

Schneider, E. D., & Kay, J. J. (1994a). Complexity and thermodynamics: Towards a new ecology. *Futures, 24*, 626–647.

Schneider, E. D., & Kay, J. J. (1994b). Life as a manifestation of the second law of thermodynamics. *Mathematical and Computer Modelling, 19*, 25–48.

Schumpeter, J. A. (1911). *Theorie der wirtschaftlichen Entwicklung: Eine Untersuchung über Unternehmergewinn, Kapital, Kredit, Zins, und den Konjunkturzyklus.* Leipzig: Duncker & Humblot.

Shugart, H. H. (1984). *A theory of forest dynamics: The ecological implications of forest succession models.* New York: Springer.

Shugart, H. H. (1998). *Terrestrial ecosystems in changing environments.* Cambridge: Cambridge University Press.

Svirezhev, Y. (1998). Thermodynamic orientors: How to use thermodynamic concepts in ecology. In F. Müller & M. Lepeult (Eds.), *Eco targets, goal functions, and orientors* pp. 102–122. Berlin, Heidelberg, New York: Springer.

Tainter, J. A. (1988). *The collapse of complex societies.* Cambridge: Cambridge University Press.

Taylor, D. R., Aarssen, L. W., & Loehle, C. (1990). On the relationship between r/K selection and environmental carrying capacity: A new habitat template for plant life history strategies. *Oikos, 58*, 239–250.

Ulanowicz, R. E. (1986). *Growth and development: Ecosystems phenomenology.* New-York: Springer-Verlag.

Ulanowicz, R. E. (1997). *Ecology, the ascendant perspective.* Columbia: University Press.

Ulanowicz, R. E. (2000). Ascendancy: A measure of ecosystem performance. In S. E. Jørgensen, & F. Müller (Eds.), *Handbook of ecosystem theories and management* (pp. 303–316). Boca Raton: CRC Press.

Wackernagel, M., & Rees, W. E. (1996). *Our ecological footprint: Reducing human impact on the Earth.* Philadelphia: New Society Publishers.

Walker, L. R., & Del Moral, R. (2003). *Primary succession and ecosystem rehabilitation.* Cambridge: Cambridge University Press.

Walker, L. R., Walker, J., & Hobbs, R. (Eds.) (2007). *Linking restoration and ecological succession.* New York: Springer-Verlag.

Weber, B. H., Depew, D. J., Dyke, C., Salthe, S. N., Schneider, E. D., Ulanowicz, R. E., et al.(1989). Evolution in thermodynamic perspective: An ecological approach. *Biology and Philosophy, 4*, 373–405.

Whittaker, R. H. (1953). A consideration of climax theory: The climax as a population and pattern. *Ecological Monographs, 23*, 41–78.

Chapter 4

The Scientific Potential of Environmental Monitoring

Claus-G. Schimming, Sabine Augustin, and Rolf Karez

Abstract Over time environmental monitoring has produced an immense amount of data that was collected in order to assess environmental pollution impacts but also for ecological research. Although this research produced revelatory, valuable data and information on the structures and functions of ecosystems, amazingly few studies have focussed on the theoretical aspects of ecosystem evolution. This obviously is counter to the progress in the field of ecosystem theory and the provision of integrated concepts in relation to indicators of ecosystem performance. In this regard, the major intention and challenge of LTER (Long-Term Ecosystem Research) is to combine the theoretical background with the ecological knowledge already available in order to extrapolate monitoring information about environmental problems and to find solutions in line with recent progress at the global scale. With a view of ecosystem evolution in a long-term perspective and to support applied ecology with respective monitoring information, special methodical diligence has to be lenient towards temporal and spatial extrapolations. This focuses on problems of representativity and reference. The aim of this chapter is to inform the reader how information on ecosystems can be derived and the potentials and limitations of using available databases from monitoring in networks and results produced by integrated data evaluations in ecosystem research. Ideas on further optimisation of monitoring activities are also presented. Two environmental initiatives are presented as case studies including protection of forest ecosystems against the effects of atmospheric deposition (ICP-forests) and the definition of properties that define a good ecological state of aquatic ecosystems (European Water Directive); critical load indications and success control measures are respectively used and the benefits to environmental protection are discussed.

Keywords Monitoring · Ecological information · Data evaluation · Indicators · ICP-forests · Water Framework Directive · Reference state · Evaluation concepts

4.1 Introduction

Because of concern for environmental quality and human health issues, for a long time a great deal of monitoring data and long-term time series have been collected in many states exposed to pollution or environmental change problems. Data quality ranges from records of single values (e.g. air temperature, solar radiation, chemical concentrations, census) up to integrated data and information about ecological change (e.g. fluctuations in biocoenoses, productivity, element cycling). In many cases already integrated monitoring information is reported (Spellerberg, 2005; Vaughan et al., 2001). However, until recently, the collected data was sometimes incomplete with respect to environmental sciences and ecology (cf. Vaughan et al., 2001).

Unilaterally, the main focus has been emphasising the purity of ambient air, water and soil to determine safety issues regarding the quality of for example breathing air, drinking water, nutrient-production potentials in agriculture etc. (Larrsen & Hagen, 1996;

C.-G. Schimming (✉)
Ökologie-Zentrum der Christian-Albrechts-Universität zu Kiel, Olshausenstr. 40, 24098 Kiel, Germany
e-mail: cschimming@ecology.uni-kiel.de

F. Müller et al. (eds.), *Long-Term Ecological Research*,
DOI 10.1007/978-90-481-8782-9_4, © Springer Science+Business Media B.V. 2010

EEA, 2003, Buse et al., 2003, Fehér & Lázár, 2003, Carpenter et al., 2006, Álvarez-Robles, ÁngelLatre, Muro-Medrano, Zarazaga-Soria, & Béjar, 2007). The ascertained data was initially used for rectifying political actions or for juridical purposes in relation to critical value exceedances or environmental quality standards. Meanwhile monitoring activities by degree also became components of programmes to protect sensitive environments: The Rio Convention on Biological Diversity, the United Nations Framework Convention on Climate Change (UNFCCC), The Convention on Long-range Transboundary Transport of Air Pollution (CLRTAP) and the Kyoto-Protocol with the objective of reducing greenhouse gases that cause climate change. Because the global foci of programmes like the Kyoto-Protocol mostly require acquisition of data on socio-economic drivers and related environmental pressures, in topical juxtaposition numerous monitoring programs have been installed. Administrative bodies active in environmental protection mostly organise these, in order to record physical and biotic data, which describes environmental states and changes on regional or local scales.

Although bodies such as HELCOM (Governing body of the Helsinki Convention, Baltic Marine Environment Protection Commission) in the Baltic sphere install their own monitoring stations at the point and non-point sources of euthrophic elements or pollutant fluxes into water bodies and to study emissions, other international programmes like AMAP (Arctic Monitoring and Assessment Programme) refer to the existing stations and networks of national state authorities. The European Water Framework Directive (WFD) is based on a monitoring system which the EU member states are requested to set up individually, consisting of three intercalibrated types which grant for representativity of reporting in the individual EC states (EC, 2003a,b, Buffagni et al., 2005). Following the CLRTAP (Convention on Long-range Transboundary Air Pollution), International Cooperative Programmes (ICPs) received mandates to organise monitoring on the effects of air pollutants. Owing to the well organized and targeted framework, the Level I and Level II programmes of ICP-forests with focus on the state of the European forests and atmospheric deposition loads may be one of the most experienced systems in harmonization of methods worldwide especially qualified by the special manual and the recent data sets already recorded. Owing to the small-scale compliance with ecosystems and landscapes, especially data with local or regional reference is of great interest and relevance in ecology (Alcamo et al., 2003) and has been produced in a couple of major ecosystem research projects comprising ecosystems in a landscape context (cf. Fränzle et al., 2008).

Looking ahead, more integrated evaluation in particular of long-term observations in relation to holistic aspects of ecosystem research may produce sensitive indicators of ecological change in the global context. In this respect the purpose of this chapter is to rivet the reader's interest on the potentials of mentoring data in ecology and how to comply with the different data qualities in order to meet the challenges of using and integrating the data to infer possible changes using any kind of comprehensive information. The intention is neither to document the dimension of monitoring data and metadata information nor to document and evaluate the qualities of individual data sources but to stimulate ideas on how existing data can be combined in LTER (Long-Term Ecosystem Research network). The challenge is to proceed in regard to ecosystem evolution not only in the regional or landscape context but also to integrate information on ecosystem functions over the long-term perspective at the global scale.

Through two example case studies the recent state and pragmatism of monitoring data use is demonstrated. The first will enlarge on the problem of the European Water Framework Directive to find a reference for the "good state" of aquatic systems (EC, 2003). Secondly, evaluation of monitoring on forest ecosystems in the frame of ICP-forests will be demonstrated by relating tree vitality with atmospheric deposition load and exceedances.

4.2 Scientific Potentials of Monitoring Data in Ecosystem Research

Already in the early 1970s Ellenberg, Fränzle and Müller (1978), spearheading ecological principals, suggested a comprehensive ecological surveillance system for Germany, which reflects the basic ideas of UNESCO's Man and Biosphere Programme (MAB) (Kasstenholz & Erdmann, 1994). The concept was later tested for integrated environmental monitoring using existing network ecosystem research (cf. Schönthaler et al., 2003). In compliance with general aims of ecosystem research on the dynamics of structures and processes, the MAB concept in particular encompasses the important long-term aspects which allow for the identification of trends or periodicities in ecosystem evolution, for efficiency control of protection measures and for the scientific underpinning of spatiotemporal model extrapolations. The approach also provides data for early-warning systems (Ellenberg et al., 1978). As suggested by these primordial ideas and as a simple example, long-term monitoring of carbon transfers and atmospheric CO_2-concentration data on the global scale may have contributed a great deal of information to evaluating potential effects of global warming on the scale of ecosystems (several authors in Schulze et al., 2003; Piao et al., 2008). Although much revelatory and valuable knowledge about environmental impacts has been produced in the past with emphasis on structures and functions of ecosystems and landscapes in order to demonstrate the need for environmental protection and to sustain the attention of policy decision-makers via a scientific basis for such decisions, pitiably few studies have looked at the theoretical aspects of ecosystem evolution till recently.

However, on the basis of conceptual ideas about ecosystem functioning more advanced ecosystem analyses of data on climate, soil, the abundance of vegetation, animals, micro-organisms, and nutrient and energy fluxes could link these components together at the landscape scale. This for example was realised by distinguished projects in the framework of the International Biological Programme (IBG) such as in the Solling mountains in Germany (Ellenberg, Mayer, & Schauermann, 1986) or others like the Hubbard Brook Project in the White Mountains (NH, USA) (Likens et al., 1977) and the Bornhöved Lakes District in northwest Germany (Fränzle et al., 2008). The intention of LTER, also born from the idea that ecosystem evolution does not coincide with the average run of research projects, is to approach ecosystem analysis with an adequate time scale so that components can be linked together in the long-term perspective. Thus, the special aspects of LTER challenge databases produced in monitoring networks and ecosystem research to not only trace environmental change but to also test hypotheses on ecosystem evolution.

One of the major aims of applied ecology is to derive ecological information which describes ecosystems under human pressure (Nelson et al., 2006) and to use this information as an indicator for the need for environmental protection, and for the assessment of stressed ecosystems; modern concepts favour health or integrity indicators. These must be ideally integrated measures for example in context with primary production, aspects of biodiversity and other suitable principles (see Jorgensen et al., 2007; Udy et al., 2001). Rapport (1989) suggested more sophisticated indicator concepts on the bases of properties like vigour, organisation and resilience, whilst integrity refers to the ability of ecosystems to maintain organisation (Kay 2000), which both comprises and integrates structural and functional aspects of ecosystems (see De Leo & Levin, 1997). Because a range of disturbances during a preceding period is integrated these types of indicators provide expediently direct measures of ecological conditions whereas the measured values or indices they are integrated from only work indirectly. Although there is an immense amount of data collected by various environmental monitoring activities and research, the potential of these data and information resources to produce the integrative concepts implied above has not been realised. The reasons may be found in the amount of data itself but also in missing coordination which is especially needed especially to solve the environmental problems proceeding from the various issues on local or regional scales where they are assessable by ecosystem research in a global context.

In the first instance, evaluating census, climate and chemical concentration data comprising structural properties, and assessments of matter and energy interconnections in ecosystems can produce information about primary production and nutrient accumulation, and consequently proceed to functional indicators of euthrophication or pollution effects as well as to evaluating source and sink functions in ecosystems. Since the interconnections between physical and biotic environments make up ecosystems, beyond structural, biocoenitic and climate aspects, monitoring activities concern chemical concentrations in the different abiotic and biotic media not only for environmental purity and human health issues, but to monitor changes in ecosystems as a consequence of pollution and euthrophication. Element activities and matter-flux ratios are parameters of biogeochemical models that create couples of fluxes between mobile element species in the environment and biomass production. Thus, they refer to the aspect of connectivity as a measure of linkage in ecosystems and exchange with the environment as a functional indicator (Gardner & Ashby, 1970). Chemical concentration data are easily available from surface water monitoring activities comprising scales as of lakes, rivers, coastal and shelf seas, and the oceans as well as carried out at stations of atmospheric deposition measurements, of soil chemistry and groundwater pollution assessments. Solid matrices for chemical analyses are sampled from abiotic and biotic compartments of ecosystems and in respect of ecotoxical problems in relation to populations, organisms, or types of organs. Useful databases include element cadastres (see Lieth & Markert, 1990 and other publications of the authors), and chemical analyses carried out in soil inventories in any direction. Valuable databases to access and identify regional or local chemical loads may be acquired from environmental monitoring of sensitive acceptor plants possibly affected, typically mosses and lichens (Pott & Turpin, 2004, Kleppin, Schroeder, Schmidt, & Pesch, 2008). Surveys monitoring matter-transfer balances in ecosystems like in the IBG and related projects (see above), or carried out by soil scientists in experiments on the chemical equilibrium between mobile element activities and solid-phase species, make provision for more advanced knowledge about biogeochemical principles (see Nasidu, 2008).

Apparent dissimilarities in the chemical element concentrations of a source and a sink lead to the acquisition of useful information about imbalances in element cycling and of related chemistry in abiotic media. Considering ecological stochiometry (Sterner & Elser, 2002) which focuses on the definite proportion of chemical elements affected by organisms and their interactions in ecosystems, in a first step, element cadastres provide valuable databases. The basic approach considers definite proportions of elements that are affected by organisms and the evolution of ecosystems in the context of interactions with the environment. Assuming the relation between a medium in the abiotic environment and an autotrophic organism in a first step followed by consumers using food web resources, differences in the element contents between two steps provide useful information on imbalance, on trophical states and eutrophication issues. Structural aspects are explicitly respected and changes in simple element ratios from the uptake to the accumulation in biomass may serve as ecological strain indications (see Fränzle & Schimming, 2008). Matter flux estimates, more or less onerous to carry out, showing sampling and related scale problems are not essential.

Nevertheless, estimates of matter fluxes are essential for calculating rate-dependent loads in ecosystems since these strongly affect chemical change in ecosystems and control the overtopping of critical concentrations. For instance, a biogeochemical approach based on combined mass-charge balances of element fluxes, comprising atmospheric deposition, acidification and consequent change of soil-solution chemistry leads to greater understanding of forest decline in an ecological context (Ulrich, 1985). The degree of imbalance referenced by long-term budgets of positive and negative ionic charge correlates with the acid load as well as soil chemistry indicated by free acidity and other critical concentration values (Ulrich, 1983a,b). Finally, the evaluation of the results from the Solling project recently contributed to some theoretical aspects of ecosystem evolution for instance in regard to issues of resilience in forests (Ulrich, 1987). Beyond the aspects of ecosystem functions, reporting of chemical loads critical for ecosystems depend on matter flux data. Reporting of exceedances on the basis of mostly pragmatic concepts balances atmospheric deposition rates of acidifying air pollutants with the critical load of forest ecosystems by means of biogeochemical approaches (Bashkin, 2006). Steady state results are used as an indicator for air pollution control. The databases consist of air pollutant concentrations and information provided by landcover maps for the parametrization of a deposition model. Relevant soil properties are informative to calculate critical acid or eutrophic nitrogen loads and to balance the capacities with the deposition rates which proceeds to the degrees of critical loads exceedances. (see Gauger et al., 2002 and Chapter 18). The concept is also available for aquatic ecosystems (Nillson & Grennfelt, 1988; Kernan, 1995) and has been developed for assessing vegetation with respect to their sustainability against atmospheric deposition loads. Large amounts of census data are applied to calculate sink functions for vegetation cover with respect to the protection of rare species (see Schlutow & Kraft, 2008).

Following projects like that in Solling shows the onerous efforts over long periods needed for assembling adequate values and the combing of consistent data sets for useful evaluations. In the nature of things data sets produced in monitoring activities worldwide on the various media (air, soil, water, biocoenoses etc.) as well as spatial (micro-scales, local, regional. global) and temporal scale references are very heterogeneous in regard to eligibility for both ecosystem research and in particular LTER evaluations. Owing to scale inconsistency, the conflation of monitoring data obtained in various environments and different media but also related insufficiencies of data quality typically involves conformity problems with scale extrapolations which limits the scientific potentials of integrating ecological information from different sources.

It is clearly evident that use of the concepts mentioned above depends on appropriate databases being available that cover the important temporal aspect. Insufficient data sets that do not have the adequate completeness, duration or continuousness may be eventually supplemented or completed from other available data sources, where sampling, inventories, recording, mapping and surveying can contribute, but also authority-run environmental pollution control networks provide valuable data potentials eligible to LTER. In this regard, weather-service information that is available worldwide in certain temporal and spatial density is of great importance because of the available networks, the number of stations and the harmonised methods. Programmes especially focussing on ecosystem research on landscapes or on special kinds of integrated indicators (e.g. Ecosystem Health Monitoring Program, EHMP, www.ehmp.org/index.html) typically organise specific monitoring programmes on the local or regional scales under observation. Also the methodical compatibility, data quality, spatial and temporal representativity may be widely unknown, which is a concern for the evaluation of monitoring data. Evaluations of data by combination of the various sources and extrapolations to the ecosystem or landscape scale are mostly complex and counter intuitive. For instance, amounts of precipitation sampled by standard rain gauges do not necessarily comply with the amounts at the landscape scale. The general problem was exposed by Kluge and Fränzle (2008) for the evaluation of water and element relations in the Bornhöved Lakes District. The problem can also be identified during the calculation of element-flux budgets from soil chemistry data sampled on the basis of ceramic cup samplers of small volumes, and for atmospheric deposition rates, with typical model results presented at the scale of crown area projections or for certain grids. As a problem of theoretical physics, these difficulties are also implied by Petter (1956) and to be referred to as scale-, bottle-, edge-, and similar effects (Petter 1956). The edge effect has been adopted to ecology by Odum (1956) in his ecotone concept which describes the transition between two or more ecosystems and corresponding diversity of structure. This complex problem can be generalised by stating that data and related model extrapolations have to be carefully tested for connectivity regarding the desired information as elucidated in Fig. 4.1. Thus, in the sense of valid information on for example landscape scales data from measurements on process scales (e.g. leaves) may be questionable. The quality of data has to be reconciled with the accuracy of information needed and the validity of modelling with respect to the aspired scale (cf. Haraldsson & Sverdrup, 2004).

Therefore, potentials for ecology and particularly LTER purposes have to be carefully tested. Methods are suggested by Schröder, Schmidt, Pesch and Eckstein (2002) that are based on geostatistical considerations. In the scientific respect, mostly the obligatory manuals and data-quality control schemes contribute to and inform about the special quality of monitoring in regard to reproducible evaluations, valid modelling, and for mapping. However, the degree of elaboration and the applicableness of methods are very discriminative. Hitherto much more than in other monitoring activities, the competence of the expert panels of ICP-forests and ICP-mapping and the program manuals are most advanced in these fields (UNECE, 2006). Considering the objectives of LTER to investigate ecosystems in a long-term perspective and broad spatial scales in order to produce scientific explanations of e.g. cause-effect relationships, to find connectivity between driving forces and ecosystems, to describe their state and evolution, the use of monitoring data decisively depends on standards under two aspects: Methodically, which means accuracy and reproducibility related to the scale problem and the reference of standard or of baseline states (Paulus et al., 2005). From a more practical point of view, this means to choose a system or any object like a tree in crown condition inventories of known properties, a standard (healthy) tree respectively in time and space and then evaluate the relative performance of the observation in reproducible (calibrated) manner (cf. Rempel & Kushneriuk, 2003; Nillson et al., 2003; Redfern & Boswell, 2004, cf. EC, 2003). In more complex surveys on environmental problems, the properties of

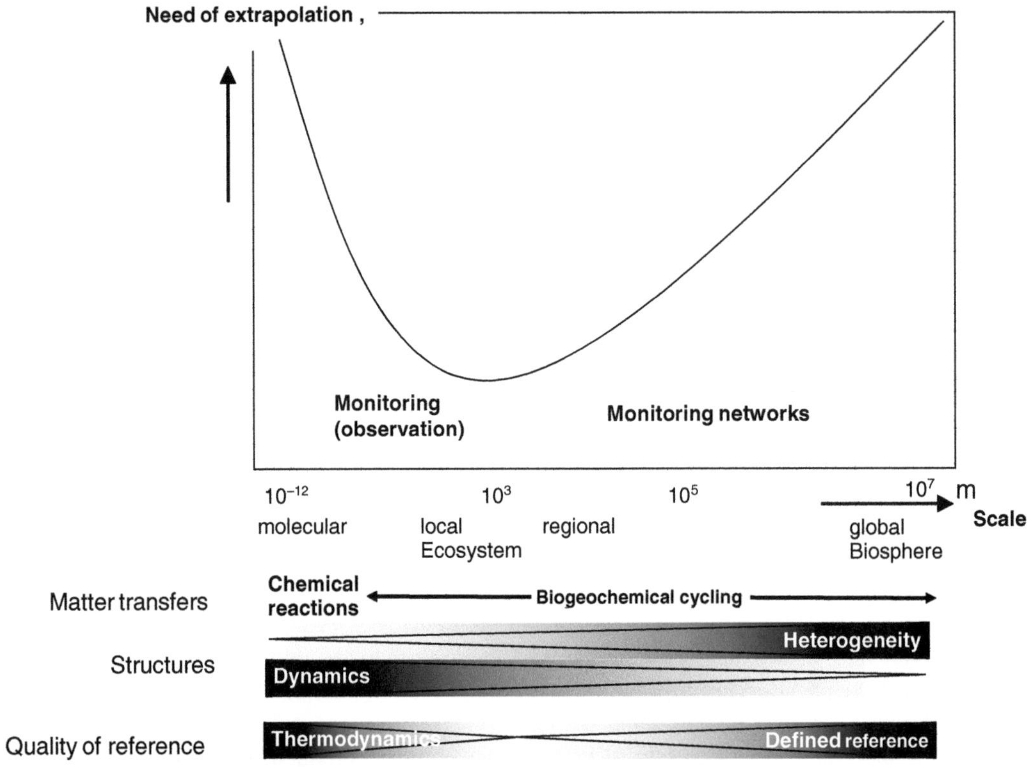

Fig. 4.1 Connectivity of monitoring data and significance of model extrapolations with respect to uncertainty and the quality of reference for environmental quality aims

an affected system can be compared with another supposedly unaffected system that has maintained sustainability or ecosystem health. The compared system has to be representative of the respective conditions which may be a problem particularly in LTER as the respective conditions change in the course of evolution independently from the compared system. With respect to information at higher scales, the specific local or regional situations can also be compared to those in representative ones as a reference. Methods for testing spatial representativity have been developed by Schröder et al. (2002) and suggested for harmonising environmental observation systems in Germany.

Other approaches are to objectify monitoring and environmental reporting by the integration of scale-conformed data on structures and functions of ecosystems using case studies or via a geostatistical approach using randomly selected landscape units in regard to representativeness (Hoffmann-Kroll, Schäfer, & Seibel, 1998; Dröschmeister, 2001; Statistisches Bundesamt, 2000). Last but not least, in Germany, an integrated and cross-functional approach tested harmonising of data from environmental monitoring in order to designate the reference to so-called representative Eco-regions. These were designated on the basis of geostatistical approaches as well as model applications on element transfers with spatial reference (Schönthaler et al., 2003). The project highlighted the potentials and limitations of environmental observations using existing databases and the consequent substantial reporting on diverse environmental issues.

4.3 Case Studies on Monitoring Activities in Two Environmental Protection Initiatives

4.3.1 *Water Framework Directive*

In 2000 the parliament of the European Union passed the Water Framework Directive (Directive 2000/60/EC of the European Parliament and of the Council establishing a framework for the Community action in the

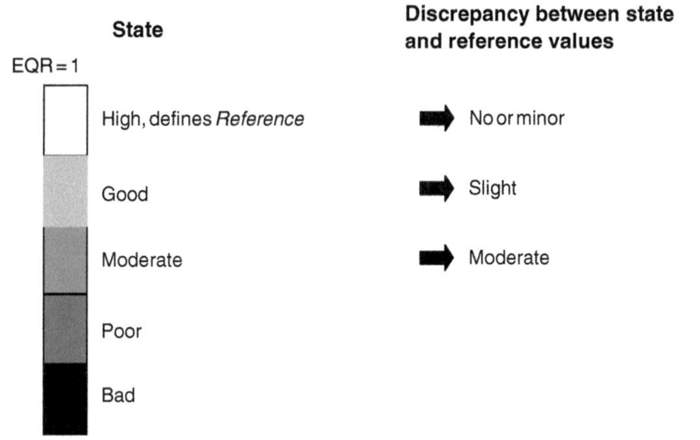

$$EQR \text{ (Ecological Quality Ratio)} = \frac{\text{Observed state value}}{\text{Reference state value}}$$

Fig. 4.2 Principle of the ecological state quality classification by use of a high-quality reference according to the guidance documents of the European Water Framework Directive (modified after EC, 2003a)

field of water policy – WFD), which was implemented in the member states' laws in the following years. The WFD requests that all aquatic habitats within the EC have to have good ecological status by 2015, and it concerns groundwater, rivers, lakes, estuaries and coastal areas. Since the WFD is an EC-directive the EU can amerce its member states if the directive is not properly implemented, i.e. if good status is not achieved in the agreed periods. However, upon proof of serious efforts by member states, extensions until 2021 or even 2027 are possible. For each of the water categories, various types were discriminated (e.g. alpine lakes, Mediterranean Sea) and for each of these types a five-step scale for the assessment of the type-specific ecological status had to be defined (EC, 2003a). This scale is based on so-called reference conditions (the "very good" status), which can be defined by reference sites of that type (if available) (Fig. 4.2), by historical data, or by modelling and expert judgement. After many discussions, the reference condition is more or less defined as the condition that would arise, if human impact on an aquatic system would be minimised without unacceptable retrenchments, although the latter is not explicitly stated in EC-guidance documents. It is, thus, not simply a pristine condition; however, there are various interpretations of feasible definitions. From here by certain deviations from the reference condition four more steps of degradation had to be defined (high, good, moderate, poor, bad), and only a good or very good status will be acceptable in 2015.

The WFD and related guidance (EC, 2003a,b, 2006) recently stimulated a lot of research to find scientific substantiation in order to calibrate this kind of assessment scheme and to define the so-called "Ecological Quality Ratios" (EQR) which range between 0 and 1 (Fig. 4.2) (EC, 2003a). The EQR values of status class limits can be variable among classification systems; however, an intercalibration process among member states sharing a certain water type shall assure a common understanding of status class limits (EC, 2003b). Theoretically reference properties are based on information on the pristine states of ecosystems which does not exist owing to the more or less ubiquitous anthropogenic influence. However, recently references have been defined based on information from historical research, modelling, or even qualified expertise and sometimes has been derived from a combination of these methods. The WFD recommends the methodical approaches necessary to establish the site-specific reference conditions (EC, 2003a). For each water category certain biological quality elements are mandatory to be used for defining the assessment tools for the ecological status, for example for estuaries this is phytoplankton, macro algae and angiosperms, the invertebrate benthic fauna and fish. Another important feature of the WFD is that the point of view has to be on river basin districts, which demands the cooperation of many agencies inside and among member states. Because of the legal bindingness and the demanding aims of the WFD, the directive had an enormous impact on the water agencies of the member states. In most cases, scientists in institutes and universities were entrusted with the definition of reference conditions and with the development of the assessment scales, thus the WFD also had an enormous impact on the scientific community and its publications and symposia.

In many cases the possibility to use contemporary reference sites was not available, since all water bodies (i.e. legally defined areas belonging to a certain type, e.g. bays, single lakes, parts of rivers) of that type today show degradation. Only in very few cases were historical data available that were of such comprehensiveness and scientific value that they could be used to define reference conditions with a certain reliability. One of these few data sets is the Danish data on the depth distribution of the Eelgrass *Zostera marina*.

Denmark has unique historical data sets on the distribution of eelgrass and benthic fauna in coastal waters around the year 1900. Even the interplay between the biology and the physico-chemical environment was described – decades before the term "ecology" was invented. The work was conducted by scientists connected to the Danish Biological Station led by C. G. Johannes Petersen with the main aim to investigate matters relevant to the fishery. The work included country-wide studies of the distribution (Petersen, 1901) and depth penetration of eelgrass as well as analyses of seasonal growth patterns and eelgrass growth in relation to light attenuation, bottom type, exposure and salinity (Ostenfeld, 1908) and assessment of the annual production of Danish eelgrass communities (Petersen, 1913). It was supplemented by detailed studies of eelgrass in specific areas (e.g. Petersen, 1892, 1893; Ostenfeld, 1918). The investigations documented that eelgrass penetrated to maximum depths of 11 m in the Kattegat, covered 6,726 km^2 and produced about 8 million tons dry matter per year.

These early studies could have been the foundation of a long-term-monitoring program, but only sporadic observations of the distribution and abundance of marine flora were made from the 1920s to the 1970s. In 1933, when eelgrass of the north Atlantic was hit by the eelgrass wasting disease, a questionnaire survey was conducted among fishery officers in order to map the extent of the disease (Blegvad, 1935). The decline of Danish eelgrass communities continued, and by 1941, the eelgrass area was reduced to only 1/13 of that around year 1900 (Lund, 1941). A few years later, investigations of the Isefjord showed that marked recolonisation of eelgrass had taken place (Steeman & Nielsen, 1951), and further vegetation surveys were conducted in a few estuaries (Mathiesen & Nielsen, 1956; Grøntved, 1958).

In the 1970s, Rasmussen (1977) reviewed the long-term changes in Danish eelgrass communities and its environmental effects. In the same period, a new generation of Danish marine biologists followed in the footsteps of the pioneer ecologists of the early 19th century and started a new era of Danish research in eelgrass ecology (Sand-Jensen, 1975; Wium-Andersen & Borum, 1980; Bak, 1980). During the 1970s and 1980s, eutrophication effects became more and more apparent and, as a consequence, the "Action Plan on the Aquatic Environment," which aimed at reducing phosphorus load by 80% and nitrogen load by 50%, was passed by the Danish Government in 1987, and a Danish national monitoring programme on marine and fresh waters was initiated in 1989. This monitoring programme now approaches a 20-year data series and is thus slowly becoming a long-term monitoring programme – with observations from the early 1900s as a reference scenario.

Comparisons between historical information and recent monitoring data have allowed assessments of changes in eelgrass distribution across the 20th century (Olesen, 1993; Rask et al., 1999; Boström, Baden, & Krause-Jensen, 2003). Changes in eelgrass area distribution over the period 1940/1950s–1990s have also been analysed based on information "hidden" in aerial photos found in the archives of the National Survey and Cadastre (KMS) (Frederiksen, Harris, Daunt, Rothery, & Wanless, 2004; Frederiksen, Wanless, Harris, Rothery, & Wilson, 2004). Moreover, the historical data have been used to define good ecological status with regard to Danish eelgrass communities as required by the WFD (Krause-Jensen, Greve, & Nielsen, 2005).

Quantitative analyses of changes in eelgrass communities across the century have been limited by the fact that no systematic, continuous sampling programme occurred before 1989. Sampling sites, metrics and methods have varied between investigations and the data sets have large time-gaps. Still, the historical data has proved to be very valuable for documenting and understanding the changes in Danish eelgrass communities and thereby underline the importance of long-term data sets.

The paucity of sufficient data sets for defining reference[1] conditions in most water types throughout Europe underlines the necessity for long-term ecological data sets. In many cases, expert judgement had to be used as the poorest of the proposed possibilities. There is an implicit danger in this option termed the "shifting baseline," meaning that a contemporary scientist only very rarely actually saw a pristine condition in his life-time and is apt to confuse what he saw as a young person with reference conditions. Only high-quality long-term data can really avoid this problem. Especially under the pressure of anthropogenic global change there is an urgent need to start many of these long-term surveys today for coming assessments in the future. However, on the other hand the WFD will use an enormous (though of course limited) amount of effort in the near and medium-term future for conservative ecological surveys, since the monitoring of the biological quality elements is mandatory with a type and element specific frequency, and every six years the results of this assessment-driven monitoring have to be reported to the EC. There are three different types of monitoring in the WFD. In cases where water bodies fail to reach the good ecological status, an "operative monitoring" has to indicate the success of measures that have to be taken to improve the ecological status (e.g. reduction of nutrient input). If water bodies are in a good or very good status, "surveillance monitoring" has to indicate any changes in the system. That means it has to be indicative for many potential pressures (cf. Müller, 2004). If there are such changes, but for unknown reasons, a temporally limited "investigative monitoring" is meant to find the causes of the observed changes. For all these monitoring types, it is essential that they are implemented in a scientifically sound way, since observed negative changes in ecological status (or the failure to improve it) have to be detected with a statistically sound certainty and since often economic interests are the ultimate reasons for pressures (e.g. fishery, dumping, fertilisers), monitoring data must be ready to endure court trials.

[1] According to EC (2003b) is the reference condition "a description of the biological quality elements that exist, or would exist, at high status. That is with no, or very minor disturbance from human activities. The objective of setting reference condition standards is to enable the assessment of ecological quality against these standards."

Therefore, there is an urgent need for help from the scientific community, but on the other hand there is an enormous chance for the scientific community to formulate sound monitoring standards that can also be used for "pure" science – at least as supporting and explanatory background for other changes and surveys. For the marine context this could, for example, be the controls for certain manipulative experiments concerning the communities of phytoplankton, benthic macro-vegetation, or macrozoobenthos. The data may contribute to science in autecology and physiology, community ecology, global change, invasion ecology to name only some of the most obvious.

4.3.2 Long-Term Forest Monitoring – Current Examples of the ICP Forests Level I and II Programs

In the following section two examples for the potential of the integrative evaluation of long-term and large-scale monitoring of the ICP Forests are given.

As a consequence of unexpected and wide-spread forest damages, that differ from those formerly observed in polluted areas (see also Chapter 18), in 1985 ICP Forests was launched under the CLRTAP. The mandate of ICP Forests is the scientific support of political decisions by monitoring of the effects of natural and anthropogenic stress on the condition and development of forests, and to contribute to a better understanding of cause-effect relationships in forest ecosystem functioning. This is obtained by a two-stage monitoring system: The extensive monitoring on Level I (Fig. 4.3), aims at the detection of the spatial and temporal pattern of crown condition. This survey is completed by a soil condition survey (Vanmechelen et al., 1997) and the assessment of the forest foliar element concentration (Stefan, Fürst, Hacker, & Bartels, 1997). The intensive monitoring on the 860 plots of the Level II system helps to detect key processes of ecosystem functioning by means of analysing for example element fluxes, tree growth and ground vegetation assessments (UNECE & EC, 2005).

The overall goal of the ICP Forests monitoring is the survey of forest vitality. Vitality of a tree is a theoretical concept (Dobbertin, 2005), which is defined as

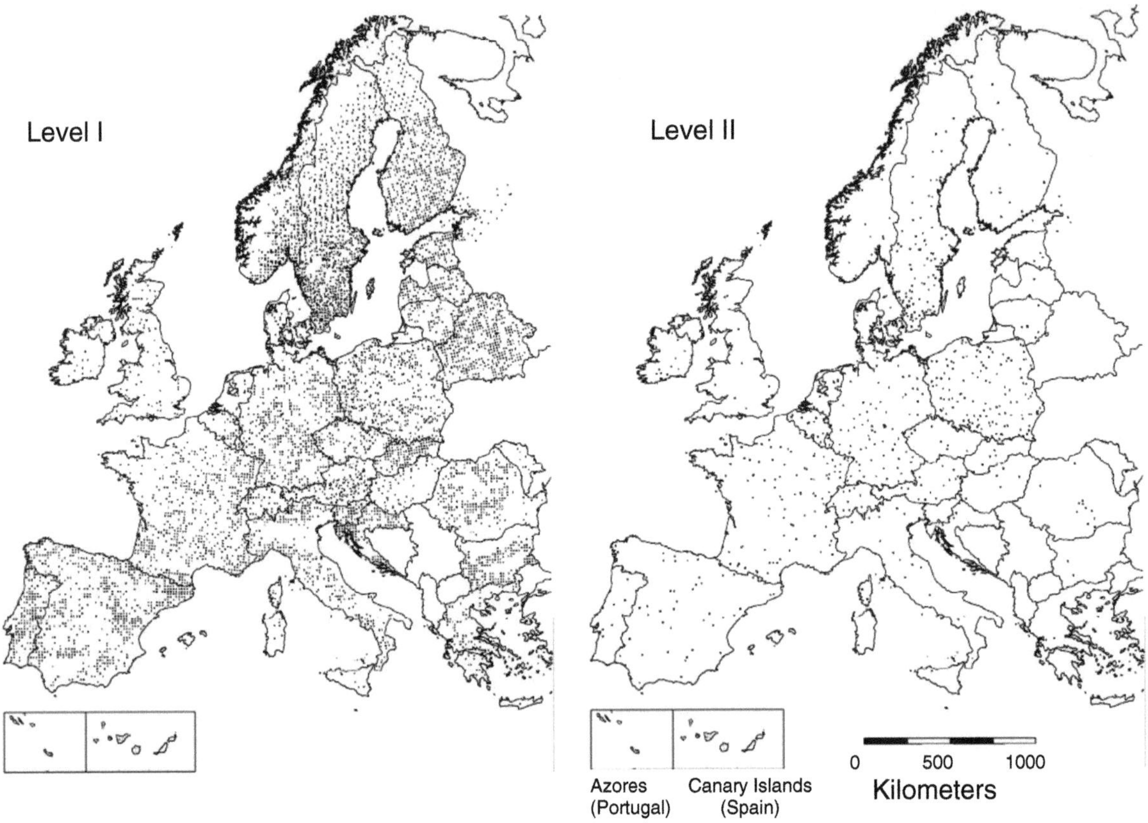

Fig. 4.3 Setting of the two intensity levels of forest monitoring of ICP-forests (modified after Lorenz et al., 2004)

a combination of the ability to resist stress, to grow and reproduce and longevity. In this context vitality can not be measured directly, so that various indicators serve to describe it (Gehrig, 2004). In the frame-work of the ICP Forests monitoring of crown condition is assessed as leaf/needle loss and discoloration (UNECE & EC, 2005). Other frequently used terms are crown transparency, crown density, foliage percentage, or defoliation. Crown condition is an unspecific indicator, reflecting all natural and anthropogenic factors acting together at a site.

The results of many studies on forest damage led to the conclusion that the damages are the effect of a complex forest disease and an expression of impaired functioning of the whole forest ecosystem (Ellenberg et al., 1986, Schulze, Lange, & Oren, 1989, Ulrich, 1989, Augustin & Andreae, 1998; De Vries et al., 2000). Possible causes were seen in both non-biotic and biotic factors, acting sequentially, concurrently, synergistically, or cumulatively on the forest ecosystem. However, to distinguish between triggering, accompanying and/or inciting factors is very difficult, because of the complexity of forest ecosystems and regionally varying affecting factors. The mean defoliation on the European scale is fluctuating. The amount of damaged trees (>25% defoliation) is relatively constant at ca. one fifth of the trees assessed (UNECE & EC, 2007). Evaluations indicate, that forests respond to all stress factors, but that their contribution differ depending on the dominance of certain stress factors in a given geographic region or site.

The tree growth reflects all beneficial and unfavourable environmental factors acting together at a forest stand. Forest growth can therefore act as another indicator for forest condition. Natural growth factors are nutrient and water availability and climate, and in the last decades the anthropogenically triggered input of acidifying and eutrophying depositions (sulphur, nitrogen), increased CO_2 concentrations in the atmosphere as well as higher temperatures and changes in forest management are additional factors influencing the forest growth. Overall, the forest growth has increased during the past several decades in many regions in Europe (Spiecker, Mielikäinen, Köhl, & Skovsgaard, 1996), which is mainly explained by the anthropogenic changes of growth conditions.

One of the aims of the Level II monitoring of the ICP Forests is the assessment of the response of forests and tree growth to environmental factors. Forest growth measurements have been carried out on Level II plots since 1994/1995. Despite a high variability of tree growth, there is a relation between high defoliation and reduced basal area growth, especially for Norway spruce. Trees with high defoliation have a higher risk to die in the following years (Dobbertin & Brang, 2001). Recent findings from 363 Level II plots confirm a positive growth response to nitrogen deposition (De Vries, Dobbertin, Reinds, & Solberg, 2007). On plots with high nitrogen inputs, growth of Scots pine, Norway spruce and common beech trees was consistently higher than expected for given site conditions and stand age. The magnitude of growth response depends highly on the nitrogen status of the site: trees on sites with limited nitrogen availability respond more strongly to the atmospheric input. On sites that were already nitrogen saturated this effect was smaller. Overall, an increase of 1 kg nitrogen per ha and year accounted for an increase of 1% in stem growth. Temperatures above the long-term annual mean during the growing season correlate with higher relative growth for the main tree species Scots pine, Norway spruce and common beech (De Vries et al., 2007). The extreme drought in Europe in 2003 revealed (in Switzerland and south Germany) that trees growing below 1,200 m a.s.l. showed up to 50% reduced growth, compared to the year 2002, whereas trees on sites above 1,200 m a.s.l. reacted with increased growth. Obviously water availability in 2003 had been the limiting factor at low altitude, while higher temperature at high altitude had favoured growth in 2003.

Independent of the possible causes for the increased tree growth the question arises of what this increase means in terms of tree vitality (Dobbertin, 2005)? Increased stem growth does not indicate automatically higher vitality. Results from fertiliser experiments show that increased above-ground growth can result in decreased root growth (Clemenssonlindell & Persson, 1995). The growth effect in fertiliser studies is often followed by a decline. A higher nitrogen nutrition state can lead to nutrient imbalances, lower frost resistance and more severe insect attacks. Higher trees can be more frequently subjected to wind throw. In some studies a higher sensitivity of the growth reaction to natural growth conditions was reported, starting in the 1970s, indicating a higher vulnerability to environmental stress (Beck, Müller, & Eichhorn, 2007, see also Dittmar and Zech (1994). Therefore, growth changes due to changes in environmental conditions must be interpreted with great caution and with a long-term perspective (Dobbertin, 2005).

In 1988 the International Co-operative Programme on Modelling and Mapping of Critical Loads and Levels and Air Pollution Effects, Risks and Trends (ICP Modelling and Mapping) was established. The aim is to develop an effects-based approach for air pollution abatement policy for the long-term maintenance of ecosystem sustainability and the minimising of risks due to the deposition of acidifying and eutrophying substances or heavy metals. This is realised with the modelling and mapping of critical loads. Critical loads were used as the basis of the CLRTAP protocols *Protocol on Further Reduction of Sulphur Emissions* (Oslo Protocol of the CLRTAP, 1999), the *Protocol on Abate Acidification, Eutrophication and Ground-level Ozone* (Gothenburg Protocol of CLRTAP, 1999) as well as the *National Emission Ceilings directive* of the EC (EC, 2001). The ICP Modelling and Mapping conducts research jointly with other ICPs to support the CLRTAP goals (UBA, 2004).

Critical Loads are defined as "the highest deposition of acidifying compounds that will not cause chemical changes leading to long-term harmful effects on ecosystem structure and function" (Nilsson & Grennfelt, 1988). Critical loads can be derived either on an empirical basis using observed sensitivities of plant species or communities (empirical critical loads for N: Bobbink & Roelofs, 2003) or by the site-specific modelling of all relevant acid and nitrogen consuming and producing processes and substances (steady state mass balance – SMB, biogeochemical models, see UNECE & EC, 2005). The critical load is an ecosystem property resulting from soil chemistry and stand characteristics (e.g. forest type). The exceedances of critical loads by depositions indicate a long-term potential risk for the ecosystem. This is not necessarily combined with the development of damage symptoms, since soils may buffer acids or ecosystems may accumulate nitrogen for a certain time span. This is the main reason for the so called "damage delay time" between the exceedance of critical loads and effects (Posch & Hettelingh, 2001). In contrast, the attainment of the threshold value of the critical limit (for acidity) reflects more or less the starting point of an "acid stress situation," with the possible onset of decline symptoms.

For the critical loads (acid) calculation the mostly used critical limit is ca. pH 4.2, which is the thermodynamically derived threshold of aluminium oxide dissolution. Between pH 5 and 4.2, manganese (Mn) oxides were dissoluted in soils and taken up by plants. High Mn concentrations in soils and plants are therefore an indicator for a considerable change in soil quality in the rooting zone (Guyette & Cutter, 1994) and the by-and-by reaching of the critical limit. Effects of pollution and changes in the availability of elements can be detected in tree rings and assigned to a distinct time in the past (Guyette & Cutter, 1994; Ferretti et al., 2002; Poszwa et al., 2003).

Within the framework of a German interdisciplinary study on the integrated evaluation of forest monitoring data (Augustin, Evers, et al., 2005), it was possible to link data on soil chemistry with tree chemical data in order to identify the attainment of the critical limit in nature (Augustin, Stephanowitz, Wolff, Schröder, & Hoffmann, 2005). Tree rings of 60-year-old Norway spruces from a plot in the Thuringian Forest show a distinct decline in Mn concentration, beginning in the late 1960s and ending in the late 1970s. This reflects the dissolution and subsequent depletion of Mn oxides in soils preceding the onset of an acid-stress situation for trees. These conditions are characterised by low availability of Mg, low pH and increasing concentrations of Al. First studies on the observed "novel forest damages" for this region in 1984 reveal for example those very low Mg contents in spruce needles and soils, leading to the deficiency symptoms. Consistently with this, often elevated Mn concentrations in plant organs discern damaged from undamaged tree collectives, and indicate the beginning of "novel forest damage" (multivariate evaluations of Level I data in German Bundesländer (Becher, 1986; Gärtner, Urfer, Eichhorn, Grabowski, & Huss, 1990).

The example shows that in principle it is possible to link information of the environmental monitoring in forests with indicators from tree ring analysis to reconstruct soil chemical history on the medium timescale. It could be demonstrated, that the thermodynamic and empirical derived value of the critical limit is linked to the onset of damages. The effect-based approach of the critical loads calculation is approved by the evaluation of ICP Forest data.

4.4 Conclusions

Considering the huge amounts of monitoring databases and environmental information produced, seemingly few have contributed towards unifying concepts that comprise advanced research results on abiotic and biotic environment interactions which make up ecosystems together with the theoretical aspects of ecosystem evolution. The integration of both aspects is not only a major emphasis of ecosystem research in general but is also subjected to LTER since solutions need to be found for environmental problems that, in a long-term perspective, have recently developed from local and regional to the global scale. The main tasks are determination of indicators and measures for sustainable environmental management and efficient protection of sensitive ecosystems and nature.

Beyond the value of monitoring information evaluations of information must face recent developments in ecosystem theories and various indicator concepts that are derived from holistic views as to how an ecosystem functions. By describing ecological states in terms of health and integrity they work more directly inclusive of the history of environmental impacts than the measured indices they are integrated from. In practice to date probably both the potential of monitoring data evaluations and that of scientific principles have not been combined to an extent that is useful for the assessment of ecological health and integrity as well as for applied ecology issues. Available data and information are not only produced by monitoring activities on various media but also research projects report evaluations for a better understanding of ecosystem functions. A great deal has been contributed to the understanding of biogeochemical cycling and the role of acidification in the functions of ecosystems and the succession of vegetation by forest decline research which later proceeded to critical load concepts as an indicator of exceedances that are calculated by means of information about atmospheric deposition rates provided directly from or calculated on monitoring data. Useful data sources are also provided by authority-run networks installed for environmental pollution control and sampling of solid matrices in biotic and abiotic environments under ecotoxicological considerations.

Data quality in environmental monitoring ranges widely covering categories of census, weather values, radiation, concentrations and information about ecological change but also integrated monitoring information is useful. However, until recently the collected data has sometimes been incomplete with respect to scientific evolution. Potentials and limitations are mostly determined by methodical compatibility and data quality performances which may be widely unknown. In this respect, evaluations of data by the combination of various sources and in particular different temporal and spatial scale references are typically counterintuitive. A generalisation of this complex problem is that data and related model extrapolations have to be carefully tested for valid representativity of values that are to be used for the extrapolation. Accordingly, it may be appropriate to organise specific monitoring programmes on the respective local or regional scales under observation.

As concluded from the presented examples of monitoring activities in the framework of the European Water Framework Directive and ICP-forests, evaluation of monitoring data can proceed to cause-effect relationships and help to find reference conditions to evaluate structural changes in ecosystems. In order to substantiate the scientific background of environmental policy and action plans, both initiatives undoubtedly stimulate a lot of useful contributions and have received considerable provision by their adjacent competent communities. Owing to the different emphases either on structural or vitality aspects and critical load exceedances as a functional aspect in ICP-forests, the values only to a limited extent are contributions to a unified position between ecology and solutions to solve environmental problems. Nevertheless, just for the specific aims a lot of valuable data and information has been produced. In particular the political mandate of the Convention on Long-range Transboundary Air Pollution has allowed for a long-term perspective for the adjacent monitoring conducted for specific data sets on forests dating back to more than ten years.

As demonstrated by the conceptual background suggested for a comprehensive surveillance system implemented by paradigmatic projects, conducting research on ecosystems and environmental interconnections can provide organisational structures by combining inter- and transdisciplinary cooperation with integral functions. The organisation of LTER structures may provide the expedient constraints as well as opportunities for panels to discuss integrated applications of ecosystem theoretical aspects. Conceivable are testing of hypotheses on ecosystem evolution using the potentials of monitoring data in order to develop knowledge about ecosystems functions, and their interactions proceeding from the local and regional toward the global scale in a long-term perspective.

Acknowledgements The authors thank Dorte Krause-Jensen (Danmarks Miljøundersøgelser, Aarhus Universitet) for her expertise especially in the context of the problem of a reference from long-term monitoring and in addition the Danish observations of eelgrass contributed a great deal of knowledge to the WFD in relation to how to identify possibly pristine states of these stands.

Financial support on the basis of "Forest Focus" (Regulation (EC) No 2152/2003 of the European Parliament and of the Council of 17 November 2003) is gratefully acknowledged for the results of the Level I and Level II programmes .

References

Alcamo, J., Döll, P., Henrichs, Th., Kaspar, F., Lehner, B., Rösch, Th., et al. (2003). Development and testing of the WaterGAP 2 global model of water use and availability. *Hydrological Sciences – Journal des Sciences Hydrologiques, 48*, 317–336.

Álvarez-Robles, J. A., ÁngelLatre M., Muro-Medrano, P. R., Zarazaga-Soria, F. J., & Béjar, R. (2007). Water quality monitoring to support the European Commission's Water Framework. Directive reporting requirements. *Transactions in GIS, 11*, 835–847.

Augustin, S., & Andreae, H. (Eds.). (1998). *Cause-Effect-Interrelations in Forest Condition – State of Knowledge. Study elaborated for the UN/ECE ICP Forests under the auspices of the Convention on Long-Range Transboundary Air Pollution.* Hamburg: BFH.

Augustin, S., Evers, J., Dietrich, H. P., Eichhorn, J., Haussmann, T., Icke, R., et al. (2005). Concept and feasibility study for the integrated evaluation of environmental monitoring data in forests. *European Journal of Forest Research, 124*, 289–300.

Augustin, S., Stephanowitz, H., Wolff, B, Schröder, J., & Hoffmann E. (2005). Manganese in tree rings of Norway spruce as an indicator for soil chemical changes in the past. *European Journal of Forest Research, 12*, 313–318.

Bashkin, V. N. (2006). *Modern biogeochemistry: Environmental risk assessment.* Dordrecht: Kluwer.

Bak, H. P., 1980. Age populations and biometrics in eelgrass, *Zostera marina* L. *Ophelia, 9*, 155–162.

Becher, G. (1986). Ergebnisse und methodisch-theoretische Überlegungen zur immissionsökologischen

Waldzustandserfassung (IWE) – dargestellt am Beispiel Hamburgs. *Forstarchiv, 57*, 167–174.

Beck, W., Müller, J., & Eichhorn, J. (2007). Impact of heat and drought on tree and stand vitality – dendroecological methods and first results from level 2-plots in southern Germany. In: Symposium: Forests in a changing environment – results of 20 years ICP Forests Monitoring. *Schriften aus der Forstlichen Fakultät der Universität Göttingen und der Niedersächsischen Forstlichen Versuchsanstalt Göttingen, 142*, 25–28.

Blegvad, H. (1935). An epidemic disease of the eelgrass (*Zostera marina* L.). *Danish Biological Station Report, 39*, 1–8.

Bobbink, R., & Roelofs J. G. M. (2003). Nitrogen critical loads for natural and semi-natural ecosystems. The empirical approach. *Water, Air, & Soil Pollution, 85*, 2413–2418.

Boström, C., Baden, S. P., & Krause-Jensen, D. (2003). The seagrasses of Scandinavia and the Baltic Sea. In E. P. Green, & F. T. Short (Eds.), *World atlas of seagrasses* (pp. 27–37). Berkeley: University Press.

Buffagni, A., Erba, S., Birk, S., Cazzola, M., Feld, C., Ofenböck, T., et al. (2005). Towards European Intercalibration for the Water Framework Directive: Procedures and examples for different river types from the E.C. project STAR. 11th STAR deliverable. STAR Contract No: EVK1-CT 2001-00089. *Quad Istituto di Ricerca sulle acque, 123*, 1–468.

Buse, A., Norris, D., Harmens, H., Büker, P., Ashenden, T., & Mills, G. (2003). *Heavy Metals in European Mosses: 2000/2001 Survey.* Bangor: UNECE ICP Vegetation Coordination Centre, Centre for Ecology and Hydrology.

Carpenter, S. R., DeFries, R., Dietz, T. Mooney, H. A., Polasky, S., Reid W. V., et al. (2006). Millennium ecosystem assessment: Research needs. *Science, 314*, 257–258.

Clemenssonlindell, A., & Persson, H. (1995). Fine-root vitality in a Norway spruce stand subjected to various nutrient supplies. *Plant Soil, 169*, 167–172.

De Leo, G. A., & Levin, S. (1997). The multifaceted aspects of ecosystem integrity. *Conservation Ecology [online] 1*(1), 3. Available from http://www.consecol.org/vol1/iss1/art3/

De Vries, W., Dobbertin, M., Reinds, G. J., & Solberg, S. (2007). *Nitrogen deposition and high temperatures accelerate forest tree growth. The condition of Forests in Europe. 2007 Executive Report.* Geneva: UNECE.

De Vries, W., Reinds, G. J., Van Kerkvoorde, M. S., Hendriks, C. M. A., Leeters, E. E. J. M., Gross, C. P., et al. (2000). *Intensive Monitoring of Forest Ecosystems in Europe, Technical Report.* Brussels and Geneva: EC and UNECE.

Dittmar, C., & Zech, W. (1994). Dendroökologische und ernährungskundliche Untersuchungen an gesunden und kranken Buchen (*Fagus sylvatica* L.) der Nordböhmischen Mittelgebirge. *Dendrochronologia, 12*, 45–58.

Dobbertin, M. (2005). Forest growth in Europe – A review. In: Naito, K. (Ed.), *International Symposium on The Role of Forests for Coming Generations – Philosophy and Technology for Forest Resource Management. October 17–22, 2004* pp. 21–38. Utsunomiya, Tochigi: Japan Society of forest Planning Press.

Dobbertin, M., & Brang, P. (2001). Crown defoliation improves tree mortality models. *Forest Ecology and Management, 141*, 271–284.

Dröschmeister, R. (2001). Bundesweites Naturschutzmonitoring in der "Normallandschaft" mit der Ökologischen Flächenstichprobe. *Natur und Landschaft, 76*, 58–69.

EC. (2001). Directive 2001 2001/81/EC of the European Parliament and of the council of 23 October 2001 on national emission ceilings for certain atmospheric pollutants. *Official Journal, L 309*, 22–30.

EC. (2003a). *Guidance on the Intercalibration Process 2004–2006* – Guidance document No, 10 – *Common implementation for the Water Framework Directive (2000/60/RC).* Luxembourg: Office for Official Publications of the European Communities.

EC. (2003b). *Transitional and Coastal Waters: Typology, Reference Conditions and Classification Systems* – Guidance document No. 5 – *Common implementation for the Water Framework Directive (2000/60/RC).* Luxembourg: Office for Official Publications of the European Communities.

EC. (2006). *Guidance on the Intercalibration Process 2004–2006 – Guidance Document No. 14 –Common implementation for the Water Framework Directive (2000/60/RC).* Luxembourg: Office for Official Publications of the European Communities.

EEA. (2003). *Review of existing national soil monitoring systems – an update.* Report of the European Topic Centre on Terrestrial Environment, Copenhagen. Copenhagen: European Environmental Agency.

Ellenberg, H., Fränzle, O., & Müller, P. (1978). Ökosystemforschung im Hinblick auf Umweltpolitik und Entwicklungsplanung. In: *Umweltforschungsplan des Bundesministers des Innern. Ökologie-Forschungsbericht 78-101 04 005* (pp. 1–144). Bonn: Bundesinnenministerium.

Ellenberg, H., Mayer, R., & Schauermann, J. (1986). *Ökosystemforschung. Ergebnisse des Sollingprojektes 1966–1986.* Stuttgart: Ulmer.

Fehér, J., & Lázár, A. (2003). European scale monitoring system to assess water quality of multi-national river basins in the accession countries. *Water Science and Technology, 48*, 103–109.

Ferretti, M., Innes, J. L., Jalkanen, R., Saurer, M., Schäffer, J., Spiecker, H., et al. (2002). Air pollution and environmental chemistry – what role for tree-ring studies? *Dendrochronologia, 20*, 159–174.

Fränzle, O., Kappen, L., Blume, H. P., & Dierssen, K. (Eds.). (2008). *Ecosystem organization of a complex landscape: Long-term research in the Bornhöved Lake District, Germany. Ecological Studies 202.* Berlin, Heidelberg: Springer.

Fränzle, O., & Schimming, C. G. (2008). Element fluxes in atmosphere, vegetation and soil. In O., Fränzle, L., Kappen, H. P., Blume, & K. Dierssen, (Eds.), *Ecosystem organization of a complex landscape: Long-term research in the Bornhöved Lake District, Germany. Ecological Studies 202* (pp. 189–206). Berlin, Heidelberg: Springer.

Frederiksen, M., Harris, M. P., Daunt, F., Rothery, P., & Wanless, S. (2004). Scale-dependent climate signals drive breeding phenology of three seabird species. *Global Change Biology, 10*, 12–14.

Frederiksen, M., Wanless, S., Harris, M. P., Rothery, P., & Wilson, L. (2004). The role of industrial fisheries and oceanographic change in the decline of North Sea black-legged kittiwakes. *Journal of Applied Ecology, 41*, 1129–1139.

Gardner, M. R., & Ashby, W. R. (1970). Connectance of large, dynamical (cybernetic) systems: Critical values for stability. *Nature, 228*, 784.

Gärtner, E. J., Urfer, W., Eichhorn, J., Grabowski, H., & Huss, H. (1990). Mangan-ein Bioindikator für den derzeitigen Schadzustand mittelalter Fichten in Hessen. *Forstarchiv, 61*, 229–233.

Gauger, T., Anshelm, F. Schuster, H. Erisman, J. W., Vermeulen, A. T., Draaisjers, G. P. J., et al. (2002). *Mapping of ecosystem specific long-term trends in deposition loads and concentrations of air pollutants in Germany and their comparison with critical loads and critical levels – Part 1 Deposition loads, part 2 Mapping critical levels exceedances, report 299.* Berlin: Umweltbundesamt.

Gehrig, M. (2004). *Methoden zur Vitalitätsbeurteilung von Bäumen*. Vergleichende Untersuchungen mit visuellen, nadelanalytischen und bioelektrischen Verfahren. Dissertation, ETH Zürich.

Grøntved, J. (1958). Underwater Macrovegetation in Shallow Coastal Waters. *Journal du Conseil Permanent International pour l'Exploration de la Mer, 24*, 32–42.

Guyette, R. P., & Cutter, B. E. (1994). Barium and manganese trends in tree-rings as monitors of sulfur deposition. *Water Air & Soil Pollution, 73*, 213–223.

Haraldsson, H., & Sverdrup, H. (2004). Finding simplicity in complexity in biogeochemical modelling. In: Wainwright, J. & Mulligan, M. (Eds.), *Environmental Modelling – Finding Simplicity in Complexity* (pp. 211–223). Chichester: Wiley.

Hoffmann-Kroll, R., Schäfer, D., & Seibel, S. (1998). Biodiversität und Statistik – Ergebnisse des Pilotprojekts zur Ökologischen Flächenstichprobe. *Wirtschaft und Statistik, 1*, 60–75.

Jørgensen, S. E., Fath, B., Bastianoni, S., Marques, J., Müller, F., Nielsen, S. N., et al. (2007). *A new ecology – The systems perspective*. Amsterdam: Elsevier Publishing.

Kastenholz, H., & Erdmann, K. H. (1994). Biosphere reserves in Germany: A contribution to support sustainable development. *The Environmentalist, 14*(2), 139–146.

Kay, J. J. (2000). Ecosystems as self-organised holarchic open systems: Narratives and the second law of thermodynamics. In S. E. Jørgensen, & F. Müller, (Eds.), *Handbook of ecosystem theories and management* (pp. 135–160). Boca Raton: Lewis.

Kernan, M. (1995). The use of catchment attributes to predict surface water critical loads – A preliminary analysis. *Water, Air & Soil Pollution, 85*, 2479–2484.

Kleppin, L., Schroeder, W., Schmidt, G., & Pesch, R. (2008). Entwicklung und Erprobung einer Metadaten- und WebGIS-Applikation für das Expositionsmonitoring mit Moosen in Deutschland. Ein Beitrag zum LTER-Netzwerk. *Zeitschrift für Umweltchemie und Ökotoxikologie, 20*, 38–48.

Kluge, W., & Fränzle, O. (2008). Transport processes between Lake Belau and its drainage basin. In O. Fränzle, H.-P. Blume, L. Kappen, K. Dierssen (Eds.) *Ecosystem organisation of a complex landscape – Long-term research in the Bornhöved Lake district*. Ecol. Stud. 202. Berlin: Springer.

Krause-Jensen, D., Greve, T. M., & Nielsen, K. (2005). Eelgrass as a bioindicator under the water framework directive. *Water Resources Management, 19*, 63–75.

Larrsen, S., & Hagen, L. O. (1996). *Air pollution monitoring in Europe – Problems and trends*. Copenhagen: European Environment Agency.

Lieth, H. H. F., & Markert, B. A. (1990). *Element Concentration Cadasters in Ecosystems (ECCE). Methods of Assessment and Evaluation.* Weinheim: Wiley VCH.

Likens, G. E., Bormann, F. H., Pierce, R. S., Eaton, J. S., & Johnson, N. M. (1977). *Biogeochemistry of a forested ecosystem*. New York: Springer.

Lorenz, M., Becher, G., Mues, V., Fischer, R., Ulrich, E., Dobbertin, M., et al. (2004). *Forest Condition in Europe 2004 Technical Report*. Geneva: UNECE.

Lund, S. (1941). Tangforekomster i de danske farvande og mulighederne for deres udnyttelse. *Dansk Tidsskrift for Farmaci, 15*, 158–174.

Mathiesen, H., & Nielsen, J. (1956). Botanical investigations in Randers Fjord and Grund Fjord. *Botanisk Tidsskrift, 53*, 1–34.

Müller, F. (2004). Ecosystem Indicators for the Integrated Management of Landscape Health and Integrity. In S. E., Joergensen, R., Costanza, & F. L. Xu, (Eds.). *Ecological indicators for assessment of ecosystem health* (pp. 277–304). Boca Raton, FL: Taylor and Francis.

Müller, F., Schrautzer, J., Reiche, E. W., & Rinker, A. (2006). Ecosystem based indicators in retrogressive successions of an agricultural landscape. *Ecological Indicators, 6*, 63–82.

Nasidu, R. (Ed.). (2008). *Chemical bioavailability in terrestrial environments, developments in soil science* (Vol. 32, 1st ed.). Amsterdam: Elsevier Science.

Nelson, N. M., Wright, A., Lowry, R. G., & Mutrie, N. (2006). Where is the theoretical basis for understanding and measuring the environment for physical activity? *Environmental Health Insights, 2008*(2), 111–116.

Nillson, J., & Grennfelt, P. (1988). *Critical loads for sulphur and nitrogen*. Copenhagen: Nordic council of Ministers.

Nilsson, J., & Grennfelt, P. (Eds.). (1988). *Critical loads for sulphur and nitrogen.* Report from a workshop held at Skokloster, Sweden, 19–24 March 1988 NORD Miljørapport 15, pp. 225–268. Copenhagen: Nordic council of Ministers.

Odum, E. P. (1956). *Fundamentals of ecology*. Saunders, Philadelphia, Pennsylvania, USA.

Olesen, B. (1993). *Bestandsdynamik hos ålegræs*. Dissertation, University of Århus.

Ostenfeld, C. H. (1908). On the Ecology and Distribution of the Grass-Wrack (*Zostera marina*) in Danish Waters. *Danish Biological Station Report, 16*, 1–62.

Ostenfeld, C. H. (1918). *Sea-grasses*. In: *Report of the Danish Oceanographic Expeditions 1908–1910 to the Mediterranean and Adjacent Seas* (pp. 1–17). Volume II Biology 2.

Paulus, M., Bartel, M., Klein, R., Nentwich, K., Quack, M., Teubner, D., et al. (2005). Effect of monitoring strategies and reference data of the German Environmental Specimen Banking Program. *Nucleonika, 50*, 53–58.

Petersen, C. C. J. (1892). Fiskensbiologiske forhold i Holboek Fjord, 1890–1891. *Beretning fra de Danske Biologiske Station, 1*, 121–183.

Petersen, C. G. J. (1901). Fortegnelse over ålerusestader i Danmark. *Beretning fra den Biologiske Station, 10*, 3–29.

Petersen, C. G. J. (1913). Om baedeltangens (*Zostera marina*) aarsproduktion i de danske Farvande. In H. P. E. Jungersen & E. Warming (Eds.), *Mindeskr. Japetus Steenstrups Føds* (Vol. 9, pp. 1–20). København: Gad.

Petersen, M. R. (1983). Observations of emperor geese feeding at Nelson Lagoon, Alaska. *Condor, 85*, 367–368.

Petter, G. (1956). Nuove ricerche sperimentali sulla totalizzazione percettiva. *Rivista di Psicologia, 50*, 213–227.

Piao, S., Ciaisl, P., Friedlingstein, P., Peylin, P., Reichstein, M., Luyssaert, S., et al. (2008). Net carbon dioxide losses of northern ecosystems in response to autumn warming. *Nature, 451*, 49–52.

Posch, M., & Hettelingh, J.-P. (2001). From critical loads to dynamic modelling. In M., Posch, P. A. M., DeSmett, J.-P., Hettlingh, & R. J. Downing, (Eds.), *Modelling and mapping of critical thresholds in Europe. RIVM Report No. 259101010* (pp. 33–39). Bilthoven: Coordination Centre for Effects, National Institute for Public Health and the Environment.

Poszwa, A., Wickman, T., Dambrine, E., Ferry, B., Dupouey, J.-L., Helle, G., et al. (2003). A retrospective isotopic study of Spruce decline in the Vosges mountains (France). *Water, Air & Soil Pollution, 3*, 201–222.

Pott, U., & Turpin H. (2004). Assessment of atmospheric heavy metals by moss monitoring with *Isothecium stoloniferum* Brid. in the Fraser Valley, B. C., Canada. *Water, Air, & Soil Pollution, 101*, 25–44.

Rapport, D. J. (1989). What constitutes ecosystem health? *Perspectives in Biology and Medicine, 33*, 120–132.

Rask, M., Eela, R., Heikkerö, T., & Neuvonen, A. (1999). *Teknologian arviointi, arvot ja osallistuminen-kokemuksia kasvigeenitekniikka-arvioista.* Espoo: VTT Group for Technology Studies.

Rasmussen, E. (1977). The wasting disease of eelgrass (*Zostera marina*) and its effects on environmental factors and fauna. In C. P. McRoy, & C. Helfferich, (Eds.), *Seagrass ecosystems* (pp. 1–52). New York: Marcel Dekker.

Redfern, D. B., & Boswell, R. C. (2004). Assessment of crown condition in forest trees: comparison of methods, sources of variation and observer bias. *Forest Ecology and Management, 188*, 149–160.

Rempel, R. S., & Kushneriuk, R. S. (2003). The influence of sampling scheme and interpolation method on the power to detect spatial effects of forest birds in Ontario (Canada). *Landscape Ecology, 18*, 741–757.

Sand-Jensen, K. (1975). Biomass, net production and growth dynamics in an eelgrass (*Zostera marina* L.) population in Vellerup Vig, Denmark. *Ophelia, 14*, 185–201.

Schlutow, A., & Huebner, P. (2004). *The BERN model: Bioindication for ecosystem regeneration towards natural conditions* (Research Report 200 85 221). Environmental Research of the Federal Ministry of the Environment, Nature Conservation and Nuclear Safety. Berlin: Umweltbundesamt.

Schönthaler, K., Meyer, U., Pokorny., D., Reichenbach, M., Schuller, D., & Windhorst, W. (2003). *Ökosystemare Umweltbeobachtung - Vom Konzept zur Umsetzung.* Berlin: Schmidt.

Schröder, W., Schmidt, G., Pesch, R., & Eckstein, T. (2002). Harmonisierung der Umweltbeobachtung. Instrumente zur Prüfung methodischer Vergleichbarkeit und räumlicher Repräsentanz. In: Fränzle, O., Müller, F., & Schröder, W. (Eds.), *Handbuch der Umweltwissenschaften. Grundlagen und Anwendungen der Ökosystemforschung* (Kapitel V-1.3.). Landsberg: Ecomed.

Schulze, E. D., Lange, O. L., & Oren, R. (Eds.). (1989). *Forest decline and air pollution. Ecological Studies 77.* Berlin-Heidelberg: Springer.

Schulze, E.-D. (2002). Understanding global change: Lessons learnt from the European landscape. *Journal of Vegetation Science, 13*, 403–412.

Spellerberg, I. F. (2005). *Monitoring ecological change.* New York: Cambridge University Press.

Spiecker, H., Mielikäinen, K., Köhl, M., & Skovsgaard, J. (Eds.). (1996). *Growth trends in European forests. European Forest Institute Research Report No. 5.* Berlin, Heidelberg, New York: Springer.

Statistisches Bundesamt & Bundesamt für Naturschutz (Ed.). (2000). Konzepte und Methoden zur Ökologischen Flächenstichprobe - Ebene II: Monitoring von Pflanzen und Tieren. *Angewandte Landschaftsökologie, 33*, 1–262.

Steemann-Nielsen, E. (1951). The marine vegetation of the Isefjord – A study on ecology and production. *Meddelelser fra Kommissionen for Danmarks Fiskeri- og Havundersøgelser, 4*, 1–14.

Stefan, K., Fürst, A., Hacker, R., & Bartels, U. (1997). *Forest Foliar Condition in Europe. Results of the large-scale foliar chemistry surveys (survey 1995 and data from previous years).* Brussels, Geneva, Vienna: EC/UNECE.

Sterner, R. W., & Elser, J. J. (2002). *Ecological stochiometry – The biology of elements from molecules to the biosphere.* New Jersey, Princeton: Princeton University Press.

UBA. (2004). *Mapping Manual 2004: Manual on methodologies and criteria for modelling and mapping of Critical Loads & Levels and air pollution effects, risks and trends.* Berlin: Umweltbundesamt.

Udy, J. W., Clapcott, J. E., Fellows, C. S., Harch, B. D., Bunn, S. E., & Davies, P. M. (2001). *Measures of primary production as indicators of ecosystem health.* Brisbane: Center for Catchment and Instream Research, Griffith University and South East Regional Water Quality Management Strategy.

Ulrich, B. (1983a). A concept of forest ecosystem stability and of acid load as a driving force for destabilization. In B. Ulrich, & J. Pankrath, (Eds.), *Effects of accumulation of air pollutants in forest ecosystems* (pp. 1–29). Dordrecht: D. Reidel Publishing Company.

Ulrich, B. (1983b). Interaction of forest canopies with atmospheric constituents: SO_2, alkali and earth alkali cations and chloride. In B. Ulrich, & J. Pankrath, (Eds.), *Effects of accumulation of air pollutants in forest ecosystems* (pp. 33–45). Dordrecht: D. Reidel Publishing Company.

Ulrich, B. (1985). Interaction of indirect and direct effects of air pollutants in forest ecosystems. In: Troyanowski, C. (Ed.), *Air pollution and plants* (pp. 149–181). Weinheim: Wiley VCH.

Ulrich, B. (1987). Stability and Resilience of terrestrial ecosystems with respect to matter balance. In E. D. Schultze, & H. Zwölfer, (Eds.), *Potentials and limitations of ecosystem analysis, Ecological Studies 61* (pp. 11–49). Berlin: Springer.

Ulrich, B. (1989). Effects of acidic precipitation on forest ecosystems in Europe. In D. C. Adriano, & A. H. Johnson,

(Eds.), *Acidic precipitation, Biological and Ecological Effects* (pp. 189–272, Vol. 2). New York: Springer.

UNECE. (2008). *Manual on methods and criteria for harmonized sampling, assessment, monitoring and analysis of the effects of air pollution on forests.* Hamburg: Federal Research Centre for Forestry and Forest Products.

UNECE & EC. (2005). *Manual on methods and criteria for harmonized sampling, assessment, monitoring and analysis of the effects of air pollution on forests*. Hamburg, Geneva: BFH.

UNECE & EC. (2007). *The condition of forests in Europe-2007 Executive Report*. Hamburg and Brussels: BFH & Institute of World Forestry.

Vanmechelen, L., Groenemans, R., & Ranst, E. (1997). Forest soil conditions in Europe. In: *Results of a large scale soil survey.* Technical Report. Brussels: Ministry of the Flemish Community, Geneva: UNECE.

Vaughan, H., Brydges, T., Fenech, A., & Lumb, A. (2001). Monitoring long-term ecological changes through the ecological monitoring and assessment network: Science-based and policy relevant. *Environmental Monitoring Assessment, 67*, 3–28.

Wium-Andersen, S., & Borum, J. (1980). Biomass and production of eelgrass (*Zostera marina* L.) in the Øresund, Denmark. *Ophelia, 1*, 49–55.

Part III
Exploring Long-Term Processes: International Experience

Chapter 5

Twenty-Eight Years of the US-LTER Program: Experience, Results, and Research Questions

James R. Gosz, Robert B. Waide, and John J. Magnuson

Abstract The U.S. Long Term Ecological Research (US-LTER) program consists of 26 research sites involving a wide range of ecosystem types and concentrates on the interactions of multiple ecosystem processes that play out at time scales spanning decades to centuries. Long-term data sets from programs such as US-LTER provide a context to evaluate the pace of ecological change, to interpret its effects, and to forecast the range of future biological responses to change.

The primary challenges for LTER type research during its history involved sustaining funding, partnership development to sustain growth, maintaining continuity in objectives, and linking scientists and data through communication and cooperation. These challenges have been successfully addressed over the decades of the US-LTER program through close cooperation and coordination of the scientific community and the funding agencies for these programs.

The scientific community benefits much from working with colleagues around the world that have other experiences, social cultures, and knowledge bases. Ultimately, the mission of LTER and the International LTER Network is to incorporate understanding of the role of humans in the environment to inform policy makers and translate understanding into action.

As the US-LTER Network approached its fourth successful decade of scientific achievement in the ecological sciences, it developed a scientific plan for the future to provide a unifying framework that proposes to understand how humans perceive the critical services provided by ecosystems at multiple human scales, how these perceptions change behavior and institutions, and how these changes in turn feed back to affect ecosystem structure and function and the ability of ecosystems to continue to deliver services over the long term. This initiative called *Integrative Science for Society and the Environment* will allow increased collaboration, experimentation, and synthesis that take full advantage of the power of a Network approach.

Keywords Long-term research · Time and space scales · Information management · Cyberinfrastructure · Synthesis · Legacies · Cross-site experiments · Ecosystem · National science foundation · International LTER · Partnerships · Networking · Transdisciplinary science · Education · Social ecology

5.1 Introduction

The U.S. Long Term Ecological Research program (hereafter US-LTER) concentrates on ecological processes that play out at time scales spanning decades to centuries. This focuses US-LTER research between the most common time scales for ecological studies (1–3 years; Tilman, 1989; Fig. 5.1) and the much longer temporal foci of disciplines such as paleoecology. The importance of the decade-to-century time scale is particularly evident in light of the rapid changes in ecological forcing functions that are occurring at a broad range of spatial scales (Millennium Ecosystem Assessment, 2005; Intergovernmental Panel on Climate Change, 2007). Long-term data sets from programs such as US-LTER

R.B. Waide (✉)
US LTER Network Office, Department of Biology,
University of New Mexico, Albuquerque, NM 87131, USA
e-mail: rwaide@lternet.edu

F. Müller et al. (eds.), *Long-Term Ecological Research*,
DOI 10.1007/978-90-481-8782-9_5,

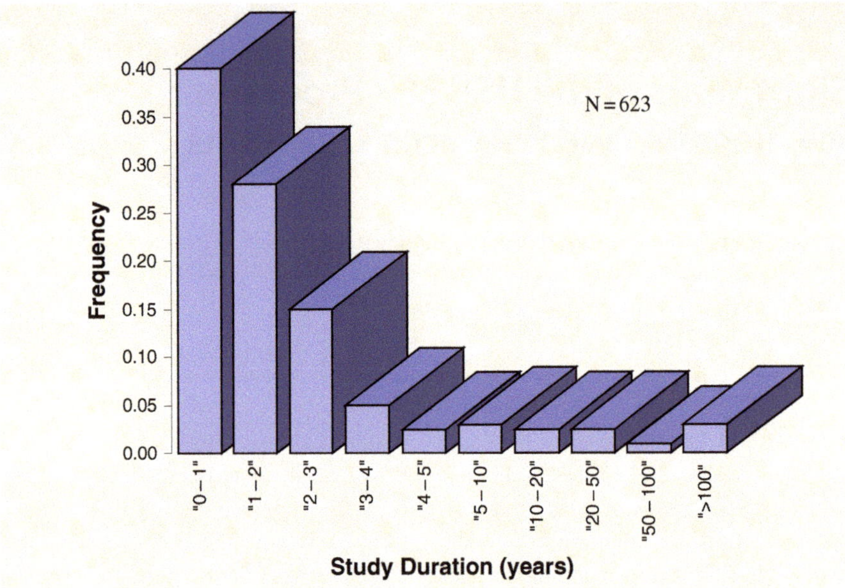

Fig. 5.1 Duration of observational and experimental studies randomly selected from Ecology between 1977 and 1987. From Tilman (1989)

provide a context to evaluate the pace of ecological change, to interpret its effects, and to forecast the range of future biological responses to change.

Although many US research sites and programs collect data over long time spans, the US-LTER is the only one that intensively addresses the interactions of multiple ecosystem processes acting over long time scales. Moreover, the 26 sites constituting the US-LTER function as a network to compare long-term trends and synthesize the results of those comparisons into general principles that underlie the behavior of ecosystems (LTER, 2007). Long-term data sets in the United States arise from many sources, including research programs sponsored by federal and state agencies (e.g., the Department of Energy's National Laboratories), monitoring programs instituted to support resource management and conservation (e.g., the Bird Breeding Survey), monitoring networks focused on environmentally significant factors (e.g., the National Atmospheric Deposition Program), and efforts to document trends in specific parameters undertaken by agriculturalists, academicians, conservation organizations, and private citizens. The existence of these other long-term data provides a wealth of information that complements results from the US-LTER and extends understanding beyond individual research sites.

5.2 Description of the US-LTER Program

Twenty-six research sites constitute the US-LTER at present (Fig. 5.2). These 26 sites include a wide range of ecosystem types spanning similarly broad ranges of environmental conditions and degrees of human domination of the landscape. The geographic distribution of sites ranges from Alaska to Antarctica and from the Caribbean to French Polynesia. The US-LTER Network includes agricultural lands, alpine tundra, barrier islands, coastal lagoons, cold and hot deserts, coral reefs, estuaries, forests, freshwater wetlands, grasslands, kelp forests, lakes, open ocean, savannas, streams, and urban landscapes. In addition to this wide range of ecosystem types, many sites encompass significant heterogeneity within their particular ecosystems. Each site develops individual research programs in five core areas: pattern and control of primary production; spatial and temporal distribution of populations selected to represent trophic structure; pattern and control of organic matter accumulation in surface layers and sediments; patterns of inorganic inputs and movements of nutrients through soils, groundwater, and surface waters; and patterns and frequency of site disturbances. The National Science Foundation (NSF) initially funded sites at $250,000/yr starting in 1980.

Fig. 5.2 Location of sites in the US-LTER Network. AND – H.J. Andrews Experimental Forest LTER, Oregon; ARC – Arctic Tundra LTER, Alaska; BES – Baltimore Ecosystem Study LTER, Maryland; BNZ – Bonanza Creek Experimental Forest LTER, Alaska; CAP – Central Arizona-Phoenix LTER, Arizona; CCE – California Current Ecosystem LTER, California; CDR – Cedar Creek Natural History Area LTER, Minnesota; CWT – Coweeta LTER, North Carolina; FCE – Florida Coastal Everglades LTER, Florida; GCE – Georgia Coastal Ecosystem LTER, Georgia; HBR – Hubbard Brook LTER, New Hampshire; HFR – Harvard Forest LTER, Massachusetts; JRN – Jornada Basin LTER, New Mexico; KBS – Kellogg Biological Station LTER, Michigan; KNZ – Konza Prairie LTER, Kansas; LUQ – Luquillo Experimental Forest LTER, Puerto Rico; MCM – McMurdo Dry Valleys LTER, Antarctica; MCR – Moorea Coral Reef LTER, French Polynesia; NWT – Niwot Ridge LTER, Colorado; NTL – North Temperate Lakes LTER, Wisconsin; PAL – Palmer Station LTER, Antarctica; PIE – Plum Island Ecosystem LTER, Massachusetts; SBC – Santa Barbara Coastal Ecosystem LTER, California; SEV – Sevilleta LTER, New Mexico; SGS – Shortgrass Steppe LTER, Colorado; VCR – Virginia Coast Reserve LTER, Virginia

Periodic increases have resulted in current funding levels of $820,000/yr through renewable, 6-year grants for base funding.

Sites in the US-LTER Network average 18 cooperating investigators and 20 graduate students per site (US-LTER Network Office, unpublished data). In addition, between 10 and 150 additional investigators make use of the research infrastructure provided at sites. Support for these additional researchers comes from many sources and augments the base funding available for research at sites by a factor of 2.9. Each of the 26 US-LTER sites integrates an education program with its research agenda. All sites incorporate graduate and undergraduate education in their programs, and most sites have kindergarten through 12th grade programs as well.

Network-scale research initiatives are developed through intensive planning activities (LTER, 2007) and fostered by annual meetings of a Science Council with representatives from each site and triennial meetings that involve most US-LTER investigators. The Network research agenda is supported by a coordinated program of information management that involves data managers from each site, common metadata standards, and a centralized information architecture that provides access to site data. The integration of data collection and data management activities provides a strong basis for cross-site and network synthesis.

The US-LTER is governed by an elected Chair and an Executive Board comprised of nine rotating site representatives and one member selected to provide expertise on information management. Eight standing committees (Climate, Education, Graduate Students, Information Management, International, Network Information System, Publications, and Social Science) support and inform the governance process. A Network Office, funded separately by the National Science Foundation, facilitates research, education, information management, and governance activities.

Goals – The mission of the US-LTER Network at this writing encompasses six goals:

Understanding: To understand a diverse array of ecosystems at multiple spatial and temporal scales.

Synthesis: To create general knowledge through long-term, interdisciplinary research, synthesis of information, and development of theory.

Information: To inform the US-LTER and broader scientific community by creating well-designed and well-documented databases.

Legacies: To create a legacy of well-designed and documented long-term observations, experiments, and archives of samples and specimens for future generations.

Education: To promote training, teaching, and learning about long-term ecological research and the Earth's ecosystems, and to educate a new generation of scientists.

Outreach: To reach out to the broader scientific community, natural resource managers, policymakers, and the general public by providing decision support, information, recommendations, and the knowledge and capability to address complex environmental challenges.

Five examples of the kinds of results that come from US-LTER research are provided in the following boxes. These examples demonstrate how LTER research can help uncover the causes of an outbreak of human disease, how LTER science can generate new concepts that involve both space and time, how an LTER site can contribute to evaluating competing needs for water, and results from intersite research.

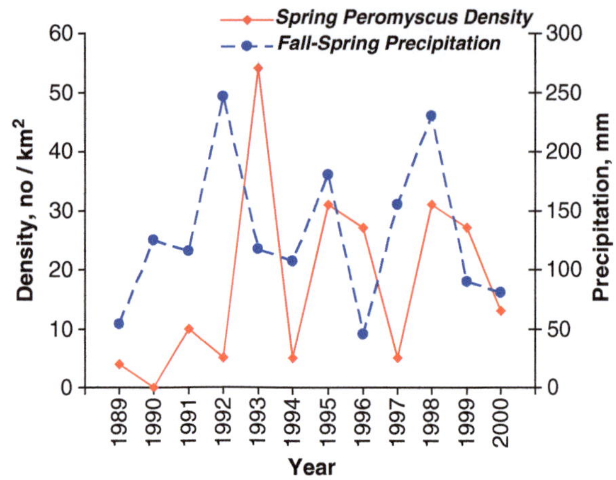

Fig. 5.3 Fall–spring precipitation is significantly correlated with spring *Peromyscus* densities a year later (r= 0.797, df = 9, $p < 0.01$) at the Sevilleta LTER site. High rodent densities in 1993 and 1998–1999 reflect post El Niño periods and coincide with the initial Four Corners Hantavirus outbreak in 1993 and increased prevalence of seropositive individuals sampled from the Four Corners area in 1998–1999. Figure by Michael T. Friggens from data published in Yates et al.(2002)

5.2.1 *Ecological Correlates to the Spread of Hantavirus*

Unexpected events (i.e., ecological surprises) driven by new phenomena or different combinations of variables that occur on an infrequent basis generate new questions and new discoveries. An example occurred in the North American Southwest in 1993 when a serious outbreak of a disease, subsequently identified as Hantavirus (Nichol et al., 1993), caused the deaths of nearly half of the people that contracted it. Research from the Sevilleta LTER project (Fig. 5.3) helped make possible the rapid discovery of the cause of the outbreak of the virus because the project was unique in the Southwest in having the necessary historical data and museum specimen collections to provide the critical information. Virus identification led to vector identification, the deer mouse (Parmenter, Brunt, Moore, & Morgan, 1993; Childs et al., 1994). Yates et al.(2002) showed that the outbreaks of Hantavirus in 1993 and in 1998–2000 were associated with the warm phase of the (ENSO) phenomenon that produced increased amounts of fall–spring precipitation in the arid and semi-arid regions of New Mexico and Arizona, and greater production of rodent food resources. These abundant food resources allowed greater reproduction in rodents that dispersed across the landscape and came into contact with humans in homes and businesses. The scientists also were able to confirm the existence of the deadly virus in tissues archived in the museum's collections prior to the original outbreak in 1993, proving the 'new' virus had been present in New Mexico probably for millions of years and was just now being discovered.

The value of the contribution has been identified by NSF as one of the 50 discoveries made with NSF funding that have had the most influence or biggest impact on the lives of Americans. Highlighted in an NSF publication titled 'Nifty 50' the discoveries were chosen from 10s of thousands of NSF-funded projects since its inception in 1950. This represents a classic case for the importance of long-term, basic research in helping to solve real societal problems.

5.2.2 *The Importance of Landscape Position*

The concept of the landscape position of lakes in the hydrologic flow system provides an important context for the factors that determine the status and dynamics of lakes (Kratz, Benson, Blood, Cunningham, & Dahlgren, 1991; Webster, Kratz, Bowser, Magnuson, & Rose, 1996; Kratz, Webster, Bowser, Magnuson,

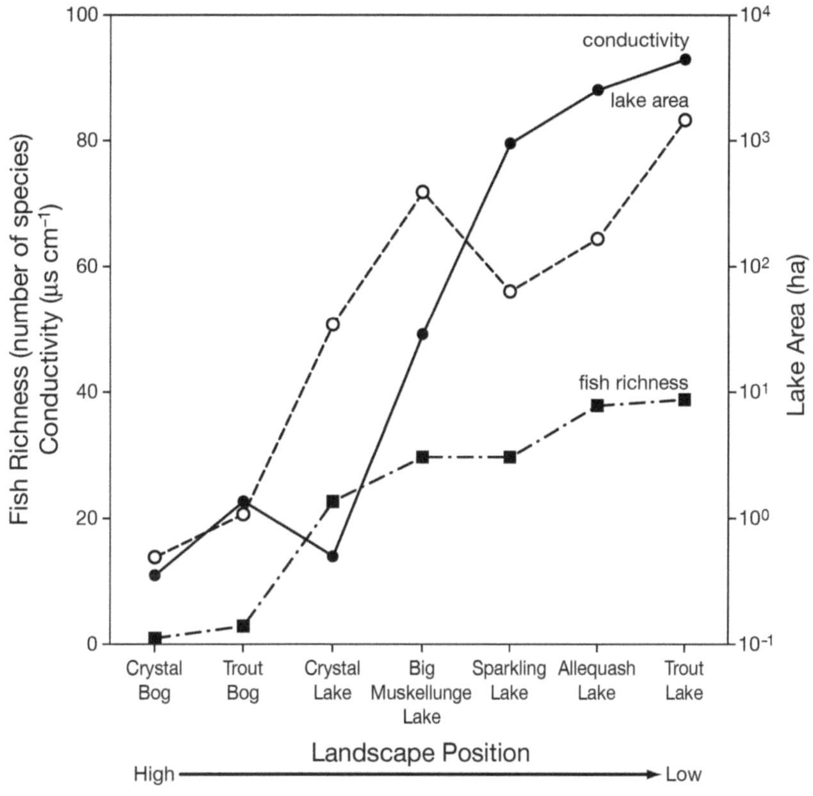

Fig. 5.4 Lake surface area, conductivity, and fish species richness as a function of landscape position for the seven primary LTER lakes in northern Wisconsin (Kratz et al., 1997, 2006). Modified from Fig. 3.3 on p. 54 from Magnuson et al.(2006). © Oxford University Press and used by permission of Oxford University Press, Inc.

& Benson, 1997; Magnuson & Kratz, 2000; Riera, Magnuson, Kratz, & Webster, 2000; Webster et al., 2000; Kratz, Webster, Riera, Lewis, & Pollard, 2006). It parallels, in some ways, stream continuum concept (Vannote, Minshall, Cummins, Sedell, & Cushing, 1980; Thorp & Delong, 1994).

In respect to status or condition, the position of lakes from high to low in the hydrologic flow paths at the North Temperate Lakes LTER determines, in part, the concentrations of phytoplankton and dissolved solids, and fish species richness (Fig. 5.4 from Kratz et al., 2006). A diverse set of other features, i.e., lake morphometry, optical properties, ion and nutrient concentrations, biology, and even human variables, are related to landscape position.

Landscape position is not a direct cause of these differences but a context related to more proximate factors that differ with landscape position. Concentrations of nutrients and other dissolved ions depend very much on the water source with precipitation (lower concentrations) dominating high in the landscape and surface and groundwater (higher concentrations) dominating low in the landscape. Lakes high in the landscape are more isolated from fish immigration because these higher lakes do not have surface streams. Lakes high in the landscape are less productive because they receive fewer nutrients in their water sources than do lakes lower in the landscape.

Landscape position also influences dynamics. During a drought over several years, lakes higher in the landscape (seepage lakes) had greater declines in water level than lakes low in the landscape, i.e., not only was precipitation lower, but seepage lakes received almost no groundwater during the drought (Webster et al., 1996). Lakes high in the landscape also had greater inter-annual variation in chemical concentrations (Kratz et al., 1991) owing to their greater responsiveness to drought.

The geomorphic template for landscape position of lakes is a legacy from geological events that patterned the landscape, in northern Wisconsin the retreat of the continental glaciers some 12,000 years ago. Thus, the influence of landscape position on lake status and dynamics is remarkably stable in ecological time; much more stable than the decade and century dynamics in ecological time.

5.2.3 Managing Water in Tropical Catchments

Increasing population, growth in tourism and industry, and agricultural needs combine in many parts of the tropics to create significant pressure on regional water supplies. This global issue has been addressed by the development of integrated water resources management initiatives, which are promoted by programs such as Hydrology for the Environment, Life and Policy (HELP), a joint effort of the United Nations Educational Science Organization (UNESCO) and the World Meteorological Organization (Ortiz-Zayas & Scatena, 2004). The problem is particularly acute on tropical islands, where groundwater resources are too limited to meet the demands of rapidly increasing populations (March, Benstead, Pringle, & Scatena, 2003). On these islands, dependence on surface water has led to a proliferation of small dams, the effects of which are poorly known. Because the majority of native macrofaunas in these islands migrate between rivers and coastal zones as part of their life cycle, the impact of dams is potentially great (March et al., 2003).

In Puerto Rico, long-term monitoring data and intensive studies of the biology of stream organisms conducted at the Luquillo LTER site provided the information necessary to evaluate the effects of dams and suggest alternative approaches. The effects of small dams on stream organisms are variable in the Luquillo Mountains depending on the structure of the dam, but in most cases dams serve as bottlenecks to migrating organisms (Benstead, March, Pringle, & Scatena, 1999, Fievet, Tito de Morais, Tito de Morais, Monti, & Tachet, 2001). For example, densities of juvenile fishes and shrimp migrating upstream increased below the dam, where the concentration of migrating organisms attracted a variety of predators (Benstead et al., 1999). Water withdrawal at the same dam led to high levels of mortality in shrimp larvae migrating downstream. In addition, the existence of dams alters physical habitat and the flow regime both upstream and downstream from the dam. The combined result of these effects can lead to changes in benthic community composition, increases in organic matter and sediment, and decreased rates of in-stream litter decay (Pringle, Hemphill, McDowell, Bednarek, & March, 1999; March, Benstead, Pringle, & Ruebel, 2001; March, Pringle, Townsend, & Wilson, 2002).

The effects of dams can be mitigated by maintaining existing fish and shrimp ladders and retrofitting dams without ladders (March et al., 2003). Knowledge of the biology and behavior of stream fauna can further mitigate the effects of water intakes through alterations in operating schedules. For example, Benstead et al. (1999) showed that larval shrimp mortality could be greatly reduced by limiting water extraction during the peak hours of downstream migration. A viable alternative to low-head dams is an in-channel withdrawal system combined with water storage outside of the stream channel. This type of system in the Río Mameyes in the Luquillo Mountains minimized the mortality of migrating organisms as well as the physical effects on the stream channel when compared to streams with low-head dams (March et al., 2003).

The importance of these measures is significant in Puerto Rico, where migrating stream organisms comprise a substantial portion of the diet of commercially important coastal fishes (Corujo, 1980; Pringle, 1997; March et al., 2001) as well as food and recreation for local human populations. The global impact may also be significant, since approximately 800,000 small dams exist worldwide (Rosenberg, McCully, & Pringle, 2000). The capacity to design low-impact water extraction systems depends on the availability of long-term records of flow rates and population fluctuations, such as those that exist in the Luquillo LTER program.

5.2.4 Network Level Studies

Two network science studies serve as examples of findings that could not be reached with site-level science: (1) a field experiment of forest litter decomposition and (2) variability in North American ecosystems.

(1) In the cross-site experiment of litter decomposition, standard sets of tree leaves were placed in bags at 21 terrestrial ecosystems (17 US-LTER sites) for 10 years. Even with the diversity among ecosystems types, differences in nitrogen release into the soil from litter bags were explained by the same two factors – the beginning concentrations of nitrogen in the litter and the amount of organic matter remaining. Such a robust result simplifies modeling of nitrogen release from litter decomposition as a non-ecosystem unique process.

(http://www.fsl.orst.edu/lter/research/intersite/lidet.htm)

(2) The analysis of variability in North American ecosystems used US-LTER site data to make general conclusions about spatial (position in the landscape) and temporal (inter-annual) variability within lake, stream, forest, grassland, alpine tundra, and desert ecosystems. The study used climatic, chemical, plant, and animal measures from each of 12 LTER sites. General and simple principles evolved from the analyses (Magnuson et al., 1991; Kratz et al., 1995). Relative variance increased from climatic, to edaphic, to plant, to animal variables. Additionally the proportion of variance explained by location in a landscape was greater than that explained by year for edaphic, plant, and animal measures. That such general emerging properties were robust across ecosystem types was initially surprising and demanded consideration of the causal basis of these general features of variability in North American ecosystems.

5.3 Origin and History of the US-LTER Program

Precursors – Long-term ecological or sustained ecological research became formalized in the United States as a National Science Foundation program originating from a natural evolution in the ecological sciences. Ecologists were well aware of the benefits of long-term observation and analyses from pioneering studies by Likens, Borman, Pierce, Eaton, and Johnson (1977) on Hubbard Brook, Edmondson (1991) on Lake Washington, and others at Windermere and elsewhere in Europe. They knew that extended time scales revealed new information, new phenomena, and new processes not apparent in short-term study. Ecologists were frustrated because funding sources were not available to more generally examine these important time scales except by piecing together a sequence of short-term projects or from monitoring by applied agencies or a diversity of groups that usually considered a limited number of parameters.

Long-term ecological research also grew out of a history of ecosystem studies from Forbes (1887), Tansley (1935), and Lindeman (1942) and from growing experience with large-scale interdisciplinary programs, especially the International Biological Program (Golley, 1993). Despite the advances made by these programs, ecologists were not able to routinely address the important ecological questions and ideas that played out over decades to centuries or that required comparisons among ecosystems and ecosystem types.

The focused development of ideas and possibilities for long-term research were stimulated in 1977 by a report by The Institute for Ecology (1977) called Experimental Ecological Reserves; a Proposed National Network (TIE, 1977). The rationales for setting up such a network were to provide sites for suitable ecological experiments in representative ecosystems, to support facilities for field sites, to ensure suitable baselines for responses of ecosystems, to encourage coordination so that researchers can complement each other in the study of complex ecosystem patterns, and to improve communication and cooperation between researchers and users of ecological knowledge. Subsequently, NSF sponsored three workshops from 1977 to 1979 to explore and provide the basis for long-term sustained, ecological research in the United States (Callahan, 1984; Franklin, Bledsoe, & Callahan, 1990). The leadership and the interaction of a number of NSF staff, especially John Langdon Brooks and James Thomas Callahan (Magnuson, Kratz, & Benson, 2006), were apparent.

Participants in these NSF workshops believed that the approach being proposed was of great value to advances in ecological sciences. They also knew that even though many came from programs that might compete for the new resources, not all worthwhile sites could be supported. Rather than select sites directly, they provided criteria for sites that could be used in the later selection process. Practical elements were dealt with such as guidelines for establishing long-term ecological measurements, the next steps toward implementation, enumeration of desirable long-term measurements, initial conditions, site characteristics, and intra- and inter-site coordination. The ideas were broadened to ensure a diverse suite of ecological sub-disciplines, i.e., not just ecosystem ecology, and provide opportunities for individual investigators in early efforts to reduce sub-disciplinary barriers among ecologists.

The reasons to initiate a Long-Term Ecological Research Program were enunciated by Marzolf (1982) at the beginning of the US-LTER program.

1. Investigations of ecological phenomena occurring at the time scales of decades and centuries were not normally supported by NSF funding in ecology.
2. Ecological experiments were conducted with little recognition of the high inter-annual variability in studied ecosystems.
3. Long-term trends were not being systematically monitored in ecological systems with the consequence that unidirectional changes could not be distinguished from more cyclic variation.
4. The absence of a coordinated network of ecological research sites inhibited comparative research and its benefits.
5. Natural ecosystems where research was being conducted were being lost to other uses.
6. Ecological research was often done on only a selected component of the system, and multilevel, integrated data were not available at intensive research sites.

Initiation – The call for proposals was made by NSF in 1979 for funding in October 1980. Of the many proposals received, six sites were funded for an initial 6 years at $250,000 per year per site. NSF based its funding choices among proposals on independent peer review by the ecological community. Choices were made on the basis of the individual proposals rather than an overarching plan designating the array of ecosystems needed in a national program. Funded sites in the first competition were Andrews Experimental Forest, Jornada Basin, Konza Prairie, Niwot Ridge, North Inlet Estuary, and North Temperate Lakes.

Leaders at these six sites were called to NSF's Washington DC office in October 1980 to discuss the program and establish a network for the inter-site aspects of the program. The NSF Program Officer for US-LTER convened the meeting. All the research that had been proposed in the original proposals was rather site specific with little thought being given to the network and inter-site research. This meeting reminded the participants of the network side of the equation and, with NSF's not so gentle urging, the lead principal investigators took the first steps in developing a network among the six sites and presumably any subsequent sites. A Principal Investigator from an LTER site was selected as chair, followed by the establishment of several inter-site committees including data management and several of the core areas specified in the call for proposals. Later chairs for the network continued the support and establishment of LTER committees to address specific needs and opportunities.

Site leaders began to meet as a network with support from network-level proposals funded by NSF. Attempts to conduct inter-site research and synthesis developed falteringly, but were stimulated by supplemental funding opportunities that were provided by NSF from the late 1980s until the mid-1990s. In 2006, the US-LTER established a Science Council, which has the role of planning and developing network-level science. The establishment of this body was provoked by several key considerations and changes in network governance including the continuing emphasis by NSF and decadal reviews of the US-LTER program on advancing network science, the re-location of the LTER Network Office to the University of New Mexico in 1997 with Robert B. Waide as Executive Director, and the formation of an Executive Board in 2006 to deal with the management issues of the network allowing the Science Council to focus on network science.

The developing science of the US-LTER is reviewed and synthesized in key papers by Tom Callahan (1984), in preparation for the 10-year review of the program (Magnuson, 1990; Swanson & Sparks, 1990; Franklin et al., 1990), in response to the 20-year review (Hobbie, Carpenter, Grimm, Gosz, & Seastedt, 2003), and in a growing number of site synthesis books (http://intranet.lternet.edu/committees/publications/oxford/).

Data management has been a key investment for the sites, the Network Office, and entire the US-LTER Network. The US-LTER program has been recognized as leading many aspects of information management because of its key role in collaboration among investigators and synthesis of results both within and across the Network of sites (Michener & Brunt, 2000; Hobbie et al., 2003). The history of developing this leadership started from the need to archive important long-term data sets, finding comparable archiving methods, facing the challenges of changing technologies for making data available, changing the culture among scientists to allow and promote data sharing, developing capabilities for tremendous increases in the volumes of data from new technologies, training new generations of data managers, and finally, providing the services needed for the entire Network through appropriate staffing and capabilities developed at the US-LTER Network Office (http://lno.lternet.edu/services/).

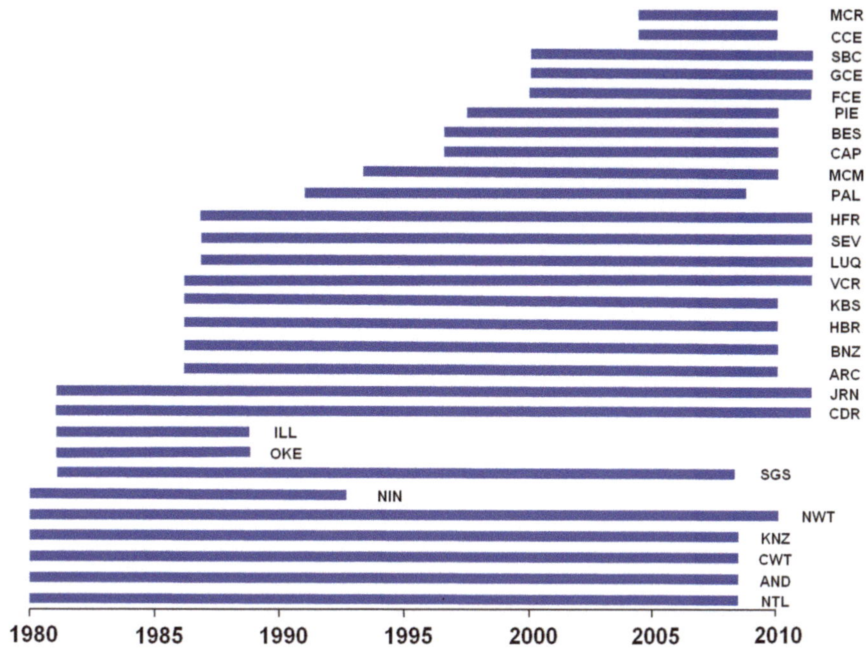

Fig. 5.5 Incremental addition of sites to the US-LTER Network. Initial cohorts were funded primarily by the Division of Environmental Biology of the National Science Foundation, but other Directorates added funding for specific sites

These services are based on the needs in Core Services, Cyberinfrastructure Support, Development and Outreach, and Synthesis Support. The types of services for these needs are collaboration support, consulting services, database support, data center services, email support, event support, outreach, training, and supporting various working groups for the Network.

Increases in sites and funding – The network of sites dedicated to long-term ecological research in the United States has grown steadily since it was formed in 1980 (Fig. 5.5). In addition to adding individual sites, the focus of the network has shifted from a concentration on individual site research to a broader synthetic view. As funding increased (Fig. 5.6), expectations of results have shifted to include not only individual researcher, single-site products but also cross-site, network-wide, and international collaborative studies. The goal of these latter studies is to search out general ecological principles that apply to many ecosystems at many different scales. Comparative and synthetic approaches have become the norm in the US-LTER Network.

At the same time that the Network was expanding its scope, the National Science Foundation initiated efforts to broaden participation in LTER research. The purpose of these efforts included a desire to involve additional investigators at satellite research sites, an interest in attracting scientists from different disciplines, and a need to apply results from US-LTER research to the solution of societal problems. Trial efforts to expand the studies conducted at individual sites were implemented at the Coweeta and North Temperate Lake sites by inviting the participation of social scientists and by expanding the geographical scale of the research. More recently, this trend has been carried further through an intensive planning effort (see below).

Governing an expanding network – The planning for an expanded research and education effort resulted in a new governance structure better suited to the coordination and management of the complexity inherent in network-level science (Fig. 5.7). The lead Principal Investigator (PI) of each site plus a rotating representative from that site and eight chairs of standing network committees make up a 60 member Science Council led by a chairperson elected for a 2-year term. The Science Council includes site scientists chosen to represent a diversity of research interests and sites. Science Council business is mostly limited to science-related reports and bylaw revisions, leaving operational decisions to a 12-member Executive Board, which is authorized to act on the Council's behalf. The Executive Board, acting on behalf of the Science

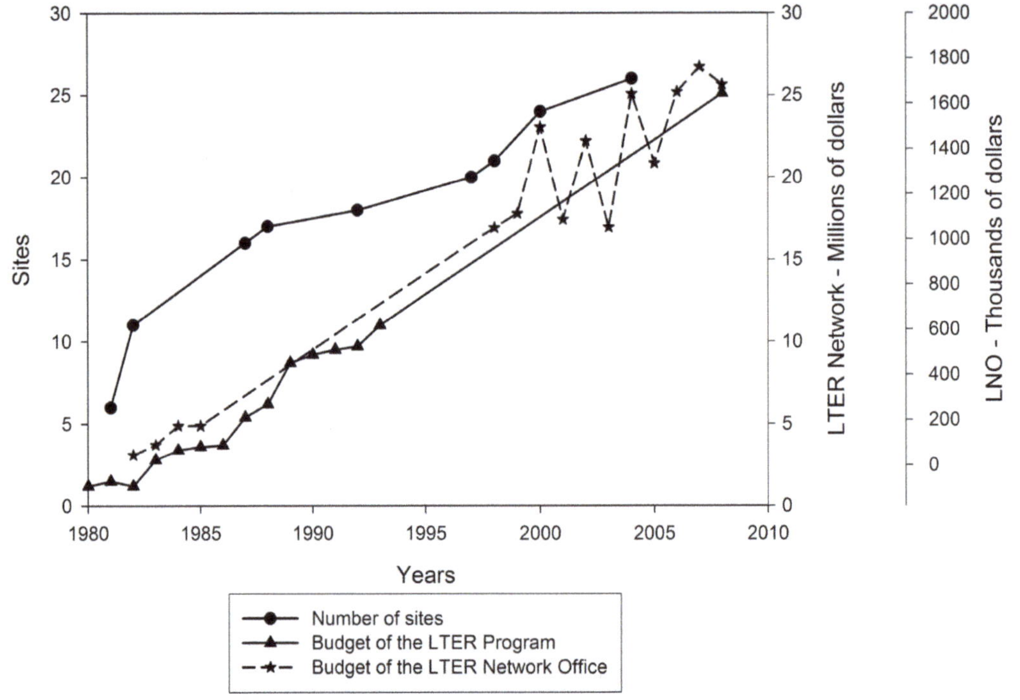

Fig. 5.6 Growth of the LTER Network since its inception. The overall increase in funding reflects a combination of increasing numbers of sites and increasing funding for each site. The budget of the LTER Network Office, which coordinates and supports Network activities, has grown at a pace with the Network budget

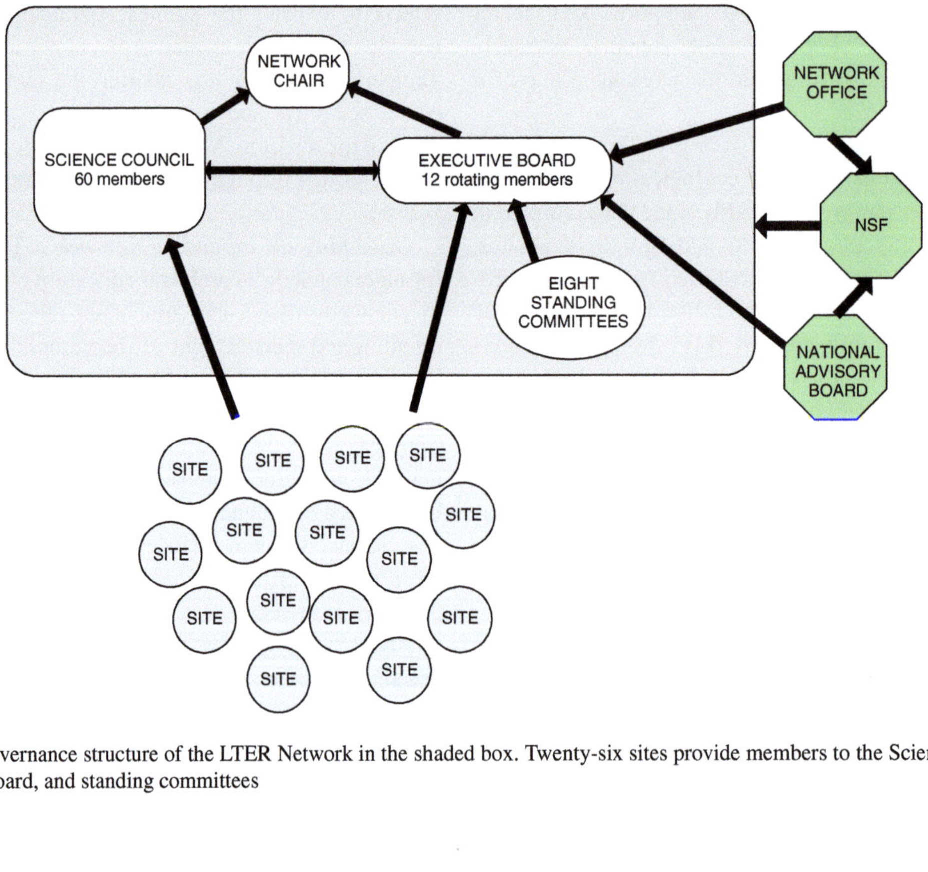

Fig. 5.7 Governance structure of the LTER Network in the shaded box. Twenty-six sites provide members to the Science Council, Executive Board, and standing committees

Council, is responsible for implementing the Network research plan. The Executive Director of the LTER Network Office is an ex officio member of the Science Council and Executive Board.

The global LTER network – Incorporating international awareness and participation is a critical aspect of the LTER Network initiative – from research to cyber infrastructure to education. The scientific community benefits much from working with colleagues around the world that have other experiences, social cultures, and knowledge bases. The US-LTER Network has and will continue to contribute to a significant effort at networking LTER programs in many countries (i.e., ILTER; International Long Term Ecological Network). Starting in 1993 at an All-Scientists meeting in the United States attended by scientists from 17 countries, the program has increased with 38 countries working together in the ILTER Network (http://www.ilternet.edu/). This international effort consists of networks of scientists engaged in long-term, site-based ecological and socioeconomic research. The mission is to improve understanding of global ecosystems and inform solutions to current and future environmental problems. ILTER's 10-year goals are to:

1. foster and promote collaboration and coordination among ecological researchers and research networks at local, regional, and global scales;
2. improve comparability of long-term ecological data from sites around the world and facilitate exchange and preservation of these data;
3. deliver scientific information to scientists, policymakers, and the public and develop best ecosystem management practices to meet the needs of decision makers at multiple levels;
4. facilitate education of the next generation of long-term scientists.

Ultimately, the ILTER mission incorporates understanding of the role of humans in the environment to inform policy makers and translate understanding into action. Ecological issues cross national boundaries, and with the appropriate infrastructure for collaboration and globally distributed sets of resources, data, and expertise, groups collaborating in the ILTER initiative can work at the scale of the ecological question. Also, just as students who are trained to work in interdisciplinary teams are better able to address the science of the future, students who can work in multi-cultural teams will be better able to compete in the global workforce and will be better in problem solving.

5.4 Challenges for LTER Type Research

The long-term sustainability of any research enterprise requires close coordination of goals among research teams and funding entities. Even with close coordination and excellent communication, many challenges face those who would organize a research program spanning decades. These challenges include the most fundamental question: How can a stable flow of funding be ensured to protect the long-term research investment? However, other key challenges include the building and maintenance of trust between funding agencies and researchers, the potential long-term effects of inflation, the development of appropriate funding partnerships that do not dilute the primary research goals, the development of mechanisms to assure smooth transitions in the research team across generations, and the many potential problems in coordinating a science program distributed among discrete sites and research teams. All of these challenges also present important opportunities to strengthen long-term research activities, but each problem must be identified and specifically addressed to ensure success. The experience of the US-LTER Network provides insights into mechanisms to address issues relating to inflation, partnerships, transitions, and communication.

Inflation and the unique nature of long-term research programs – Long-term research programs produce useful results even in the early stages of experiments and observations, but the ultimate values of increased understanding of lag effects, long-term change, and development of new concepts and principles from the research may be years or decades removed from the initiation of the program. This unique characteristic of long-term research has important implications for planning a long-term research program. Specifically, inflation in the cost of labor and analyses can be significant when compounded over the life of a long-term study.

In a constant funding environment with inflationary increases at best, long-term experiments can constitute a growing fraction of the total research budget, and this eventuality should be recognized and planned for from the beginning of the program. Some observations may need to be phased out over time to protect investments in long-term experiments. Alternatively, additional sources of funding may have to be sought to insure continuity of selected components of the program. The question of inflation is also important for funding agencies, who may find that their long-term programs create budgetary pressure on other research programs over time.

Partnerships as a method to sustain growth – A diversified funding portfolio provides protection from short-term changes in research support. One way to achieve diversification of funding is to establish partnerships among institutions and agencies that share common goals. Many of the US-LTER sites have entered into mutually beneficial partnerships with the owners or stewards of their research sites, whether these are government agencies, non-government organizations, or private individuals. These partnerships can produce direct funding for research, but more often they result in in-kind support that reduces the funding that must be raised from competitive sources and can buffer the research program against fluctuations in external funding. Such partnerships are also important in identifying new research directions and applications that in themselves may generate increased support. The US Department of Agriculture (USDA) has been an important partner of the US-LTER program in forest, rangeland, and agricultural sites, to the extent that the USDA has developed parallel efforts such as the Long Term Agricultural Research (LTAR; Robertson et al., 2008) and the Urban Long Term Research Area (ULTRA; http://www.nrs.fs.fed.us/urban/ultra/) programs.

Sustaining long-term research across generations – One unique challenge of long-term research is to maintain continuity in objectives in the face of changes in personnel and leadership that inevitably occur over the course of time. The US-LTER program has been in existence long enough (28 years at this writing) to experience considerable turnover in personnel, and there are numerous sites where LTER graduate students have, over time, moved into leadership positions in the research team (Magnuson et al., 2006). From the experience of the US-LTER, changes in leadership are particularly important events that must be planned for carefully. Many of the problems that have arisen with research programs at US-LTER sites have been related to changes in leadership, especially where individuals have exerted strong influence on research goals. Many sites address this problem by having inclusive leadership teams and well-defined plans for leadership succession.

Linking sites, scientists, and data sources – Networked long-term research programs require attention to communication and cooperation at two levels, within and among research teams. Individual scientists engaged in US-LTER research form relationships with colleagues that may last most of their research careers. The research teams so formed are dynamic and need to adapt to changes in their composition and to shifts in individual objectives over time. Team leaders need to foster communication and cooperation, especially when team members are from different institutions. The problem of communication is even more acute among sites, where the total number of researchers is greater and the geographic distribution greater. The US-LTER has addressed these challenges by focused efforts to bring site leaders together frequently and to involve the whole US-LTER community in a joint meeting every 3 years. One of the most important reasons to foster communication among sites is to encourage data sharing for comparison and synthesis. Data managers from each of the 26 sites have developed a well-integrated working group that provides for coordination of data access through a network information system. Other disciplinary groups (e.g., climate, social science) provide additional opportunities to foster communication among sites.

5.5 New Challenges for the US-LTER Program

As the US-LTER Network approached its fourth successful decade of scientific achievement in the ecological sciences, it undertook a significant self-analysis and planning strategy for future research and education. Starting in the late 1990s, the Network challenged itself to develop additional advancements in

Network science and a new kind of trans disciplinary science – one that ranges from local to global in scope, and that blends ecological and social science theories, methods, and interpretations in order to better understand and forecast environmental change in an era when no ecosystem on Earth is free from human influence. The results of over 7 years of planning are presented in a science plan submitted to the National Science Foundation on an Integrative Science for Society and the Environment (LTER, 2007; see also http://www.lternet.edu/decadalplan/). This effort describes a unifying framework that proposes to understand how humans perceive the critical services provided by ecosystems at multiple human scales, how these perceptions change behavior and institutions, and how these changes in turn feed back to affect ecosystem structure and function and the ability of ecosystems to continue to deliver services over the long term. The details of the plan involve a set of research themes pursued over the next decade and beyond. Land and water use change; climate change, variability, and extreme events; and nutrient mobilization and species introductions are considered as grand challenges in environmental science and all are important to society. They also affect every site in the Network – indeed, every part of the United States – and are intractable without full consideration of social–ecological interactions. These particular themes are also among those that best match the research strengths of the Network. They can be best addressed with new long-term data sets, cross-site experiments, and modeling activities.

At each of the Network's 26 sites there is an extraordinary amount of knowledge about the organisms and processes important at the site, about the way the site's ecosystems respond to disturbance, and about long-term environmental change. A growing number of cross-site observations and experiments also have revealed much about the way that key processes, organisms, and ecological attributes are organized and behave across major environmental gradients. The contributions of basic knowledge of ecological interactions, ability to forecast change, and testing ecological theory are well recognized by the scientific community.

The Science Plan involves a comprehensive summary of existing long-term data sets along with detailed blueprints for the education and cyberinfrastructure resources that will be crucial for its success. The cyberinfrastructure plans include building capacities to increase data acquisition, management, and curation at the site level; to increase data discovery, access, and integration at the network level; to increase modeling and analysis activities; and to integrate cyberinfrastructure into social ecological research, education, and training.

Linking research and education – Education plays a key role in US-LTER because of the need to inform and train the next generation of environmental scientists, inform the general public about ecosystem services to develop social–ecological change, and educate the public about the need for research. The Strategic Plan for Education in the US-LTER Network promotes a vision of an environmentally literate citizenry able to make informed choices about complex environmental issues and includes five parts: (1) a scientific endeavor that continues, builds on, and celebrates its rich history of basic scientific discovery; (2) a society with the environmental science literacy needed for sound environmental citizenship and thereby a society that makes best use of timely, accurate, and unbiased information in decision making, including the capacity to act proactively and with forethought; (3) engagement of the full spectrum of our diverse society in developing and applying understanding of environmental challenges; (4) a scientific community that is receptive and responsive to the knowledge needs of the public and is committed to the delivery of knowledge in a useful form; (5) an environmental research and education enterprise informed by an understanding of the science–society interface. Achieving this vision will require strategic initiatives to (1) develop leadership, organization, and cyberinfrastructure, (2) promote research and development around our goals of environmental science literacy and inclusion of diverse people and perspectives, and (3) develop programs for specific constituent groups: K-12 teachers and administrators, undergraduate and graduate students and professors, and active citizens.

Social ecology – Processes and technologies for using and maintaining ecosystem services depend on fundamental advances in scientific understanding of social–ecological systems. These systems are the basis of human well-being. Societies face a challenge of improving human well-being while maintaining current and future ecosystem services. The US-LTER Network is poised to contribute to that understanding through the social and ecological research described in its plan: (1) LTER sites will seek to understand social–ecological dynamics connecting human cognition, attitudes, behaviors, and institutions with ecological structures and processes. This will include understanding variations and similarities influencing human decision making among disparate social groups and under different social and ecological conditions (both in place and in time); (2) LTER sites will regionalize in order to understand these social–ecological dynamics in more diverse social–ecological contexts than they traditionally have pursued; (3) LTER sites will combine a long-term, spatially extensive, and multi-scale approach to understand these social–ecological dynamics in a regional context. This will expose temporal lags, spatial dependencies, and scale mismatches that generate decision-making surprises and sudden shifts; and (4) education will be both an activity and an object of study, engaging students and the public in the generation of ecological knowledge, sharing that knowledge broadly, and learning how to transmit ecological knowledge more effectively. The existing LTER connections between research and decision-making applications – public, non-profit, and private sectors – are likely to increase over time because of the enduring collaborations that are fostered by the long-term nature of US-LTER sites.

5.6 Summary

Long-term ecological research grew out of a history of ecosystem studies and from growing experience with large-scale interdisciplinary programs. The U.S. Long Term Ecological Research (US-LTER) program consists of 26 research sites involving a wide range of ecosystem types and concentrates on the interactions of multiple ecosystem processes that play out at time scales spanning decades to centuries. This focuses US-LTER research between the most common time scales for ecological studies (1–3 years) and the much longer temporal foci of disciplines such as paleoecology. Long-term data sets from programs such as US-LTER provide a context to evaluate the pace of ecological change, to interpret its effects, and to forecast the range of future biological responses to change.

Many issues and challenges are associated with maintaining a successful long-term research network. Governing an expanding network at both the United States and international levels requires management techniques that emphasize the coordination and management of the complexity inherent in network-level science. The US program involves a 60-member Science Council to address science-related efforts and bylaw revisions and a 12-member elected Executive Board authorized to act on the Council's behalf. Incorporating international awareness and participation is a critical aspect of the US-LTER Network initiative – from research to cyberinfrastructure to education. The scientific community benefits much from working with colleagues around the world that have other experiences, social cultures, and knowledge bases. The LTER Network contributes to a significant effort at networking LTER programs in many countries (i.e., ILTER; International Long Term Ecological Network). Ultimately, the ILTER mission incorporates understanding of the role of humans in the environment to inform policy makers and translate understanding into action.

The primary challenges for LTER type research during its history involved sustaining funding, partnership development to sustain growth, maintaining continuity in objectives, and linking scientists and data through communication and cooperation. These challenges have been successfully addressed over the decades of the US-LTER program through close cooperation and coordination of the scientific community and the funding agencies for these programs.

Integrative Science for Society and the Environment provides a framework to understand the interactions between critical ecosystem services, human perception of these services, and the behavior of individuals and institutions based on these perceptions. Education plays a key role in this plan, with the goal of an environmentally literate citizenry able to make informed choices about complex environmental issues.

References

Benstead, J. P., March, J. G., Pringle, C. M., & Scatena, F. N. (1999). Effects of a low-head dam and water abstraction on migratory tropical stream biota. *Ecological Applications, 9*, 656–668.

Callahan, J. T. (1984). Long-term ecological research. *BioScience, 34*, 363–367.

Childs, J. E., Ksiazek, T. G., Spiropoulou, C. F., Krebs, J. W., Morzunov, S., Maupin, G. O., et al.(1994). Serologic and genetic identification of *Peromyscus maniculatus* as the primary rodent reservoir for a new hantavirus in the southwestern United States. *Journal of Infectious Diseases, 169*, 1271–1280.

Corujo, I. N. (1980). A study of fish populations in the Espíritu Santo River Estuary. Master's thesis. University of Puerto Rico, Río Piedras.

Edmondson, W. T. (1991). *The uses of ecology: Lake Washington and beyond.* Seattle: University of Washington Press.

Fievet, E., Tito de Morais, L., Tito de Morais, A., Monti, D., & Tachet, H. (2001). Impacts of an irrigation and hydroelectric scheme in a stream with a high rate of diadromy (Guadeloupe, Lesser Antilles): Can downstream alterations affect upstream faunal assemblages? *Archiv für Hydrobiologie, 151*, 405–425.

Forbes, S. A. (1887). The lake as a microcosm. *Bulletin of the Peoria Science Association*. Reprinted in the *Illinois Natural History Survey Bulletin, 15*, 537–550.

Franklin, J. F., Bledsoe, C. S., & Callahan, J. T. (1990). Contributions of the long-term ecological research program. *BioScience, 40*, 509–523.

Golley, F. B. (1993). *A history of the ecosystem concept in ecology: More than the sum of the parts*. Connecticut: Yale University Press.

Hobbie, J. E., Carpenter, S. R., Grimm, N. B., Gosz, J. R., & Seastedt, T. R. (2003). The US long term ecological research program. *Bioscience, 53*, 21–32.

Institute of Ecology. (1977). *Experimental ecological reserves: A proposed national network*. Washington, DC: National Science Foundation.

Intergovernmental Panel on Climate Change. (2007). The Physical Science Basis: The Fourth Assessment Reports, Cambridge University Press.

Kratz, T. K., Benson, B. J., Blood, E. R., Cunningham, G. L., & Dahlgren, R. A. (1991). The influence of landscape position on temporal variability in four North American ecosystems. *American Naturalist, 138*, 355–378.

Kratz, T. K., Magnuson, J. J., Bayley, P., Benson, B. J., Berish, C. W., Bledsoe, C. S., et al.(1995). Temporal and spatial variability as neglected ecosystem properties: Lessons learned from 12 North American ecosystems. In: Rapport, D. & Calow, P. (Eds.), *Evaluating and monitoring the health of large-scale ecosystems* (pp. 359–383). New York: Springer.

Kratz, T. K., Webster, K. E., Bowser, C. J., Magnuson, J. J., & Benson, B. J. (1997). The influence of landscape position on lakes in northern Wisconsin. *Freshwater Biology, 37*, 209–217.

Kratz, T. K., Webster, K. E., Riera, J. L., Lewis, D. B., & Pollard, A. I. (2006). Making sense of the landscape: Geomorphic legacies and the landscape position of lakes. In: Magnuson, J., Kratz, T. K., & Benson, B. J. (Eds.), *Long-term dynamics of lakes in the landscape: Long-term ecological research on north temperate lakes* (pp. 49–66). Oxford University Press.

Likens, G. E., Borman, F. H., Pierce, R. S., Eaton, J. S., & Johnson, N. M. (1977). *Biogeochemistry of a forested ecosystem*. New York: Springer.

Lindeman, R. L. (1942). The trophic-dynamic aspect of ecology. *Ecology, 23*, 399–418.

LTER. (2007). *The Decadal Plan for LTER: Integrative Science for Society and the Environment.* LTER Network Office Publication Series No. 24, Albuquerque, New Mexico.

Magnuson, J. J. (1990). Long-term ecological research and the invisible present. *BioScience, 40*, 495–501.

Magnuson, J. J., & T. K. Kratz. (2000). Lakes in the landscape: Approaches to regional limnology. *Verhandlungen Internationale Vereinigung für Limnologie, 27*, 74–87.

Magnuson, J. J., Kratz, T. K., & Benson, B. J. (Eds.). (2006). *Long-term dynamics of lakes in the landscape: Long-term ecological research on north temperate lakes.* New York: Oxford University Press.

Magnuson, J. J., Kratz, T. K., Frost, T. M., Bowser, C. J., Benson, B. J., & Nero, R. (1991). Expanding the temporal and spatial scales of ecological research and comparison of divergent ecosystems: Roles for the LTER in the United States. In: Risser, P. G. (Ed.), *Long-term ecological research.* Sussex: John Wiley & Sons.

March J. G., Benstead, J. P., Pringle, C. M., & Ruebel, M. R. (2001). Linking shrimp assemblages with rates of detrital processing along an elevational gradient in a tropical stream. *Canadian Journal of Fisheries and Aquatic Sciences, 58*, 470–478.

March J. G., Pringle, C. M., Townsend, M. J., & Wilson, A. I. (2002). Effects of freshwater shrimp assemblages on benthic communities along an altitudinal gradient of a tropical island stream. *Freshwater Biology, 47*, 377–390.

March, J. G., Benstead, J. P., Pringle, C. M., & Scatena, F. N. (2003). Damming tropical island streams: Problems, solutions, and alternatives. *BioScience, 53*, 1069–1078.

Marzolf, G. R. (1982). Long-term ecological research in the United States: A network of research sites 1982. In: Halfpenny, J. C., Ingraham, K. & Hardesty, J. (Eds.), *Long-Term Ecological Research (LTER) Network.*

Michener, W. K., & Brunt, J. W. (2000). *Ecological data: Design, management and processing.* Oxford: Blackwell Science.

Millennium Ecosystem Assessment. (2005). *Ecosystems and human well-being: Current state and trends.* Washington, DC: Island Press.

Nichol, S. T., Spiropoulou, C. F., Morzunov, S., Rollin, P. E., Ksiazek, T. G., Feldmann, H., et al. (1993). Genetic identification of a hantavirus associated with an outbreak of acute respiratory illness. *Science, 262*, 914–917.

Ortiz-Zayas, J. R., & Scatena, F. N. (2004). Integrated water resources management in the Luquillo Mountains, Puerto Rico: An evolving process. *Water Resources Development, 20*, 387–398.

Parmenter, R. R., Brunt, J. W., Moore, D. I., & Morgan, E. S. (1993). The hantavirus epidemic in the Southwest: Rodent population dynamics and the implications for transmission of hantavirus-associated adult respiratory distress syndrome (HARDS) in the Four Corners region. *Sevilleta LTER Publication No. 41*, 1–45.

Pringle, C. M. (1997). Exploring how disturbance is transmitted upstream: Going against the flow. *Journal of the North American Benthological Society, 16*, 425–438.

Pringle, C. M., Hemphill, N., McDowell, W. H., Bednarek, A., & March, J. G. (1999). Linking species and ecosystems: Effects of different macrobiotic assemblages on interstream differences in benthic organic matter. *Ecology, 80*, 1860–1872.

Riera, J. L., Magnuson, J. J., Kratz, T. K., & Webster, K. E. (2000). A geomorphic template for the analysis of lake districts applied to the Northern Highland Lake District, Wisconsin, USA. *Freshwater Biology, 43*, 301–318.

Robertson, G. P., Allen, V. G., Boody, G., Boose, E. R., Creamer, N. G., Drinkwater, L. E., et al.(2008). Long-Term Agricultural Research (LTAR): A research, education, and extension imperative. *BioScience, 58*, 640–645.

Rosenberg, D. M., McCully, P. & Pringle, C. M. (2000). Global-scale environmental effects of hydrological alterations. *BioScience, 50*, 746–751.

Swanson, F. J., & Sparks, R. E. (1990). Long-term ecological research and the invisible place. *BioScience, 40*, 502–508.

Tansley, A. G. (1935). The use and abuse of vegetational concepts and terms. *Ecology, 16*, 284–307.

Thorp, J. H., & Delong, M. D. (1994). The Riverine Productivity model – An heuristic view of carbon-sources and organic-processing in large river ecosystems. *Oikos, 70*, 305–308.

Tilman, D. (1989). Ecological experimentation: Strengths and conceptual problems. In: Likens, G. E. (Ed.), *Long-term studies in ecology: Approaches and alternatives* (pp. 136–157). New York: Springer.

Vannote, R. L., Minshall, G. W., Cummins, K. W., Sedell, J. R., & Cushing, C. E. (1980). The river continuum concept. *Canadian Journal of Fisheries and Aquatic Sciences, 37*, 130–137.

Webster, K. E., Bowser, C. J., Anderson, M. P., & Lenters, J. D. (2006). Understanding the lake-groundwater system: Just follow the water. In: Magnuson, J. J., Kratz, T. K., & Benson, B. J. (Eds.) *Long-term dynamics of lakes in the landscape: Long-term ecological research on north temperate lakes* (pp. 19–48). Oxford: Oxford University Press.

Webster, K. E., Kratz, T. K., Bowser, C. J., Magnuson, J. J., & Rose, W. J. (1996). The influence of landscape position on lake chemical responses to drought in northern Wisconsin. *Limnology and Oceanography, 41*, 977–984.

Webster, K. E., Soranno, P. A., Baines, S. B., Kratz, T. K., Bowser, C. J., Dillon, P. J., et al.(2000). Structuring features of lake districts: Landscape controls on lake chemical responses to drought. *Freshwater Biology, 43*, 499–515.

Yates, T. L., Mills, J. N., Parmenter, C. A., Ksiazek, T. G., Parmenter, R. R., Castle, J. R., et al.(2002). The ecology and evolutionary history of an emergent disease: Hantavirus pulmonary syndrome. *Bioscience, 52*, 989–998.

Chapter 6

Introducing the Next Generation of Ecosystem Research in Europe: LTER-Europe's Multi-Functional and Multi-Scale Approach

Michael Mirtl

Abstract LTER-Europe is the umbrella network for Long-Term Ecosystem Research (LTER) and Long-Term Socio-Ecological Research (LTSER) in Europe. It forms part of the global LTER network (ILTER). Comprising 18 formal national member networks and five emerging ones LTER-Europe represents more than 400 LTER sites and 23 LTSER platforms. Besides this in-situ component LTER-Europe stands for a network of scientists, disciplines, institutions, data and metadata, and research projects. The Network of Excellence ALTER-Net (FP6) provided the frame for meeting the objective to integrate the highly fragmented European infrastructure with LTER potential across national and disciplinary boarders. LTER-Europe has become the terrestrial and aquatic component in the network of networks, currently organised by the ESFRI preparatory project LifeWatch. The interdisciplinary expertise represented by the ALTER-Net consortium allowed further development of the LTSER concept. LTSER platforms have been developed as multi-scale and multi-level infrastructure for investigating interactions of human and natural systems on the regional or sub-regional level. These hot spots of interdisciplinary research (IDR) and data sets are now complementing the first pillar of LTER-Europe's network, the network of LTER sites. The character and functional niche of LTER-Europe is best described by four core characteristics, namely "in-situ, long-term, system and process": LTER-Europe's research is generating or using data gathered together with a maximum of other sources of knowledge at concrete locations in the long term. This allows for the detection and quantification of processes of ecosystems and socio-ecological systems, which determine the sustainable provision of ecosystem services. Summarising, LTER-Europe is a multi-functional network, but also a process structuring and optimising a distributed research infrastructure, catalysing the development of research projects meeting societal needs and helping to streamline and harmonise the entire sector on the institutional, national, European and global level.

Keywords LTER-Europe network · Socio-ecological research · LTSER · Research infrastructure · Mode 2 research · European Research Area

6.1 Introduction

This chapter aims to provide an overview of Long-Term Ecosystem Research in Europe (LTER-Europe) as an umbrella framework and a process shaping ecosystem research and infrastructure in the continent. It outlines the history of LTER and describes the environment into which LTER-Europe has been moulded. We also describe the structure and governance of LTER-Europe with emphasis on two kinds of in-situ components – LTER sites and LTSER platforms – and the role it has played in establishing and further developing the concepts for socio-ecological research. Beyond that, LTER-Europe advanced to the regional implementation of LTSER proposing and testing a novel design of multi-functional research platforms or "LTSER platforms". Given the uniqueness of this proof of concept LTSER is indeed given disproportional space, not insinuating minor importance of the

M. Mirtl (✉)
Federal Environment Agency Austria (Umweltbundesamt), Spittelauer Lände 5, 1090 Wien, Austria
e-mail: michael.mirtl@umweltbundesamt.at

DOI 10.1007/978-90-481-8782-9_6, © Springer Science+Business Media B.V. 2010

traditional, site-based LTER, but anticipating that this component of LTER has broadly been accepted as state of the art. The representativeness of LTER-Europe's network for major environmental and socio-ecological zones will be illustrated.

An additional goal of this chapter is to highlight the multi-functional nature of LTER-Europe, which has started to serve a wide range of purposes across all levels from concrete research sites to institutions, national networks, the European Research Area, related sister networks and the global research environment. In that sense LTER-Europe can in parallel be perceived as

- A Network of LTER sites,
- A Network of LTSER platforms,
- A Network of national networks,
- A Network of institutions,
- A Network of scientists (a community),
- A Network of disciplines,
- A Network of data and metadata,
- Part of a network of European networks,
- A Network of site-based research and research projects,
- A process structuring and integrating all the above.

By creating a better understanding of the complex nature of LTER-Europe the chapter attempts to animate both the scientific community and those in charge of establishing a framework for environmental research. We highlight the wide range of added values of LTER-Europe that support a new generation of environmental science thereby contributing to a knowledge base about major ecosystem services at stake.

6.2 Definition, Terminology and History of LTER

The US National Science Foundation (NSF) established the US-LTER program in 1980 to support research on long-term ecological phenomena in the United States. The 26 US-LTER sites represent diverse ecosystems and research emphases (http://www.lternet.edu/). Supported by the powerful US-LTER and through the initiating efforts of NSF in the 1990s, 28 countries worldwide had established national LTER networks and were formally accepted as members of the global LTER network (International Long-Term Ecosystem Research, ILTER) by the end of the 20th century. "ILTER consists of networks of scientists engaged in long-term, site-based ecological and socioeconomic research. Our mission is to improve understanding of global ecosystems and inform solutions to current and future environmental problems" (http://www.ilternet.edu/about-ilter/mission). ILTER consists of continental or sub-continental regional groups. In 2001, the European Environment Agency (EEA) was formally represented at the global ILTER conference in London, advocating that Europe needed a powerful and unified continental research and monitoring network to cope with the complex challenge to better manage European ecosystems according to the "Late lessons from early warnings" report of the EEA (Gee, 2001). This report especially underpins the demand for a better link between ecosystem research and Long-Term Ecosystem Monitoring (LTEM). It openly addressed the inefficient fragmentation of European ecosystem research, including deficiencies in the analysis and synthesis of principally available information and the communication of results to better inform decision making and policy. To support the initiation of such a European LTER network, the EEA acted as a key stakeholder for the first European LTER conference, held in January 2003 in Copenhagen. Amongst other strategic efforts, the first call of the EU's 6th Framework Programme (FP6) considered LTER in combination with biodiversity as a potential Network of Excellence.

Concurrently US-LTER came under scrutiny. A review of two decades of LTER identified major challenges (http://intranet.lternet.edu/archives/documents/reports/20_yr_review/). The reviewers elaborated a list of 27 recommendations, including the necessity of efforts to establish proper interdisciplinary and cross-site projects and comparisons, focus on synthesis science, setting up a standardised and quality-assured monitoring and experimentation component, establishing informatics and cutting edge information management as core functions as well as starting to include the human dimension in LTER. Concerned with the restructuring of the US-LTER and the establishment of the high-tech sensor component NEON, the US National Science Foundation as key mentor targeted on making the global LTER network, ILTER, independent of continuous US support with American scientists holding key offices in the network. As a first step ILTER elaborated formal by-laws in 2004 (http://www.ilternet.edu/about-ilter/ILTER-bylaws-10-01-2004.pdf/view). International consultants and organisational developers were engaged in 2005 to develop a strategic and operational plan for ILTER, formally adopted by the ILTER Co-ordinating Committee in 2006 in Namibia (http://www.lter-europe.net/Structure/key-documents). ILTER had passed the first phase of establishing a permanent and globally acting institution representing the most powerful in-situ network of ecosystem research facilities. In 2007 ILTER became a legal body, registered as an international scientific association in Costa Rica. Responsibilities in the governing structures of ILTER are now evenly spread across the continental regional groups of ILTER. By the end of 2008 ILTER comprised 40 national member networks.

In the context of these positive international developments, LTER-Europe accelerated its efforts from 2004, mainly due to the activities in the Network of Excellence "ALTER-Net" funded by the European Union. Developing the concept for LTER-Europe required institutional integration and IT solutions alongside the biodiversity issue as a thematic trigger. ALTER-Net facilitated the creation of a database of 1,800 facilities with "LTER potential" across Europe. The extensive metadata collated for these facilities underpin the heterogeneity and fragmentation anticipated by earlier enquiries and reports (e.g. GTOS-TEMS). Moreover, the uneven distribution (across habitat types and environmental zones) of locations where long-term research and monitoring is carried out became apparent. Supported by the request of the European Commission for institutional integration the partner institutions of ALTER-Net started to act as stakeholders for the LTER processes in their respective countries. In parallel, ALTER-Net expanded its efforts beyond the ALTER-Net institutional consortium, succeeding in including most European countries into the LTER-Europe process. As a milestone, the former western and eastern LTER networks were merged in the course of the formal foundation of "LTER-Europe" in June 2007 in Balatonfüred, Hungary. LTER-Europe has become the most powerful regional group worldwide, since August 2008 consisting of 18 national networks (Fig. 6.1). In Europe LTER focuses on terrestrial and aquatic ecosystems. Besides being an umbrella term for activities in the field of ecosystem and environmental research, it is also the name for formal LTER networks of research sites and scientists on national and continental levels composing the global LTER Network (ILTER).

6.3 Multi-Functionality and Key Characteristics of LTER-Europe

Contrary to the setup of LTER in the United States of America a top-down approach for the infrastructural component of LTER-Europe was not feasible due to the diverse funding mechanisms of the European research area. Secondly, a research frame programme continuously supporting ecosystem and socio-ecological research projects at LTER sites and LTSER platforms cannot to be accomplished in the mid term. In contrast, US-LTER benefits from the NSF as key stakeholder and funding body securely providing resources for both, infrastructure and scientific LTER projects, for a network of 26 LTER master sites and with a time horizon of decades. Consequently, the design of LTER-Europe, its infrastructure, governance and related research projects had to take extremely heterogeneous framework conditions into account, including varying initial situations across countries in terms of involved stakeholder institutions, national research programmes and divergent possibilities in former East and West European countries. Another challenge consisted of the conceptual expansion of LTER to LTSER, dealing with entire socio-ecological systems and inducing a transformation of the scientific community and interest groups that had so far been involved in and carrying the traditional LTER. We discuss this later in more detail.

LTER-Europe was developed into an intricate landscape of European and national environmental monitoring schemes and nature conservation measures such as the UNECE International Co-operative Programmes, the UNESCO Biosphere Reserves and Natura2000 concerning their site networks, databases, responsible institutions and overlaps with environmental research. It has become a key element in the restructuring of the European research area in the field of life sciences (see below).

Consequently, LTER-Europe could not exclusively focus on single research topics, disciplines, administrational and economic levels, funding schemes, stakeholders, processes, or on infrastructure or research projects alone. It needed to serve multiple purposes

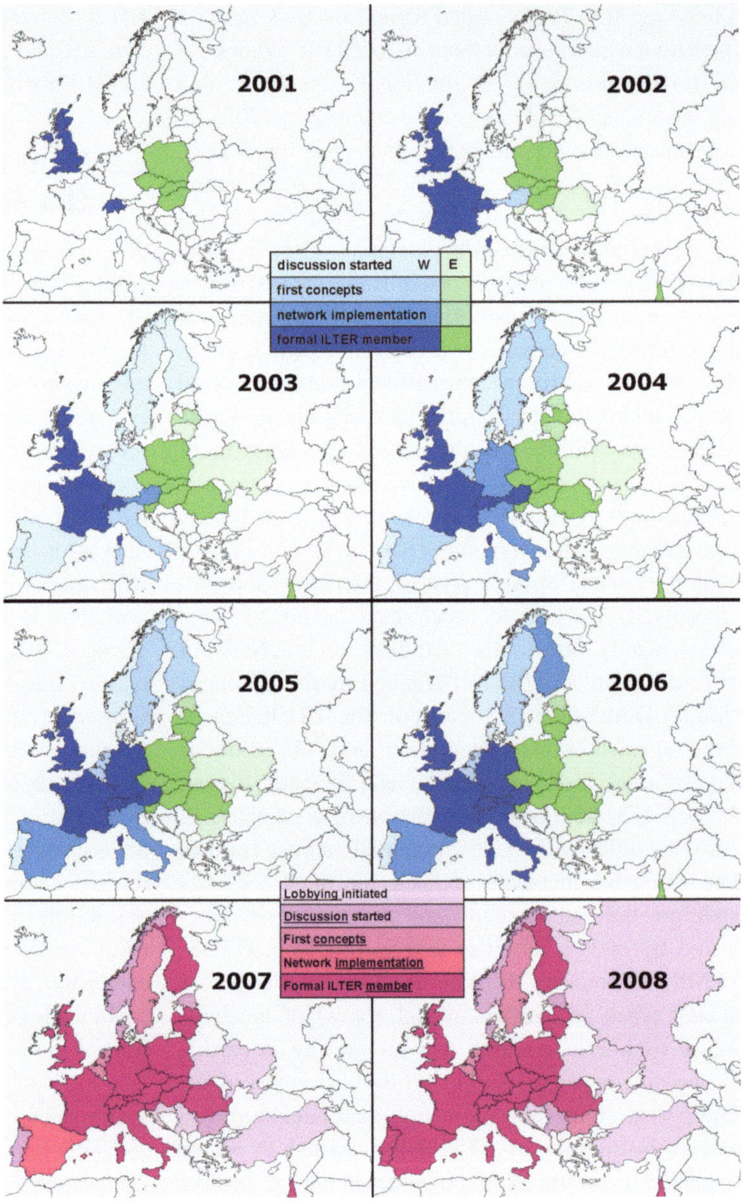

Fig. 6.1 Development of LTER-Europe 2001–2008; depths of colours indicate the level of formalisation with darkest colours for formal membership in the global LTER; in 2007 the former Eastern and Western networks were merged at the formal foundation of LTER-Europe (graph: M. Mirtl: http://www.lter-europe.net/Structure/key-documents)

and support a wide range of issues related to environmental research. Because of the general nature of a research infrastructure and research network like LTER-Europe, it was indispensable for LTER-Europe to identify the core characteristics and identity, thereby differentiating LTER-Europe from other networks. The identity of LTER-Europe is thus based on four characteristics:

In situ: LTER-Europe generates field data at different scales (up to the regional scale) and across ecosystem compartments. LTER projects make use of these interdisciplinary data hot spots, enhancing them with their own findings, but must not all necessarily gather raw data themselves.

Long-term: LTER-Europe dedicates itself to the provisioning, documentation and continuous use of long-term information and consistent data on ecosystems. The time horizon is decades in order to enable detection of trends. The long-term criterion accounts for the interdependency of LTER with the Long-Term Ecosystem Monitoring (LTEM), which is regarded as one of the most important but also most challenging aspects in establishing and securing sustainable LTER. It also requires complex information management to enable multiple and efficient use of data.

System: LTER-Europe contributes to a better understanding of the complexity of natural ecosystems and coupled socio-ecological systems, covering – as a network - the natural sphere of causation as well as the cultural sphere of causation (Fischer-Kowalski & Weisz, 1999). The typology of facilities of LTER-Europe (LTER sites and LTSER platforms) complies with the required disciplinary components, spatial and timescales.

Process: LTER-Europe's research aims at the identification, quantification and interaction of processes of ecosystems driven by internal and external drivers. The focal question is how these processes determine ecosystem services. This implies a scientifically sound combination of long-term monitoring (dynamics of state variables), research based on long-term data and short-term experimentation, covering all abiotic and biotic components of habitats and ecosystems across Europe's environmental and socio-ecological gradients.

As shown in Fig. 6.2 themes can be identified that benefit from being studied using a long-term process approach in an in-situ system (e.g. ecosystem services, climate and land use changes, biodiversity).

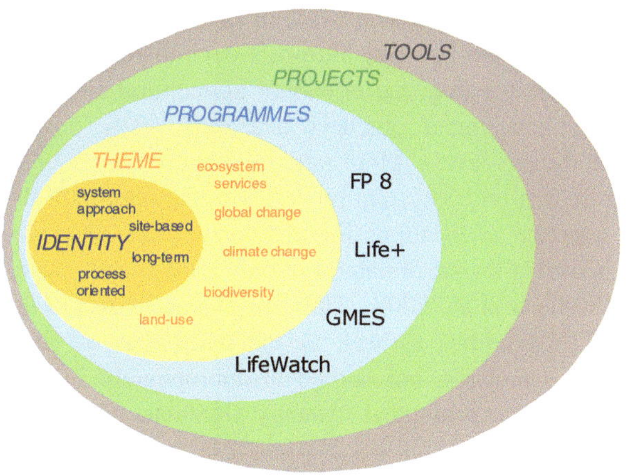

Fig. 6.2 The cascade spanning from the core characteristics of LTER-Europe to concrete projects at LTER sites and LTSER platforms and tools (graph: E. Groner: http://www.lter-europe.net/Structure/key-documents, modified with permission)

Specific programmes and projects on the continental (and global) scale, assigned to those themes and using a wide range of tools, have access to synergies enabled by LTER-Europe (comprising access to data, infrastructure in the field, laboratories and local teams of researchers etc).

6.4 The Integrated Network of LTER Sites and LTSER Platforms

LTER-Europe comprises two types of in-situ facilities, LTER sites and LTSER platforms, the primary difference being their spatial extension and structural complexity (Mirtl & Krauze, 2007). *LTER sites* extend over hundreds of hectares whereas *LTSER platforms* represent landscapes of thousands of square kilometres where natural, social and economic processes are intrinsically coupled. Therefore, LTER sites can be dominated by only one habitat type (e.g. grassland, forest) but the regions of LTSER platforms contain all elements of the landscape pattern, within the respective socio-environmental zone. Many of the LTER-sites were delineated to represent orographic and ideally also hydrological micro-catchments to allow for research related to matter and energy balances and hydrology; LTSER platform regions are or contain hydrological meso-catchments. As facilities for integrated system research both LTER sites and LTSER platforms represent a physical, logistical and research component, all of them being much more complex in the case of LTSER platforms (see below: The Design of LTSER Platforms). Activities at LTER sites concentrate on small-scale ecosystem processes and structures with core topics in the field of primary production, population ecology of selected taxonomic groups, biogeochemical cycles, organic matter dynamics, disturbances and – implicitly – biodiversity. LTSER platforms allow for investigating complex socio-ecological phenomena such as biodiversity conservation in the landscape context, including the influence of people's perception of biodiversity on decision making and the overall effects of biodiversity conservation measures in and outside protected areas. In Europe LTER has up to now focused on terrestrial and aquatic ecosystems. *LTER-facilities* are the umbrella term for wherever LTER might take place (location) and whatever might

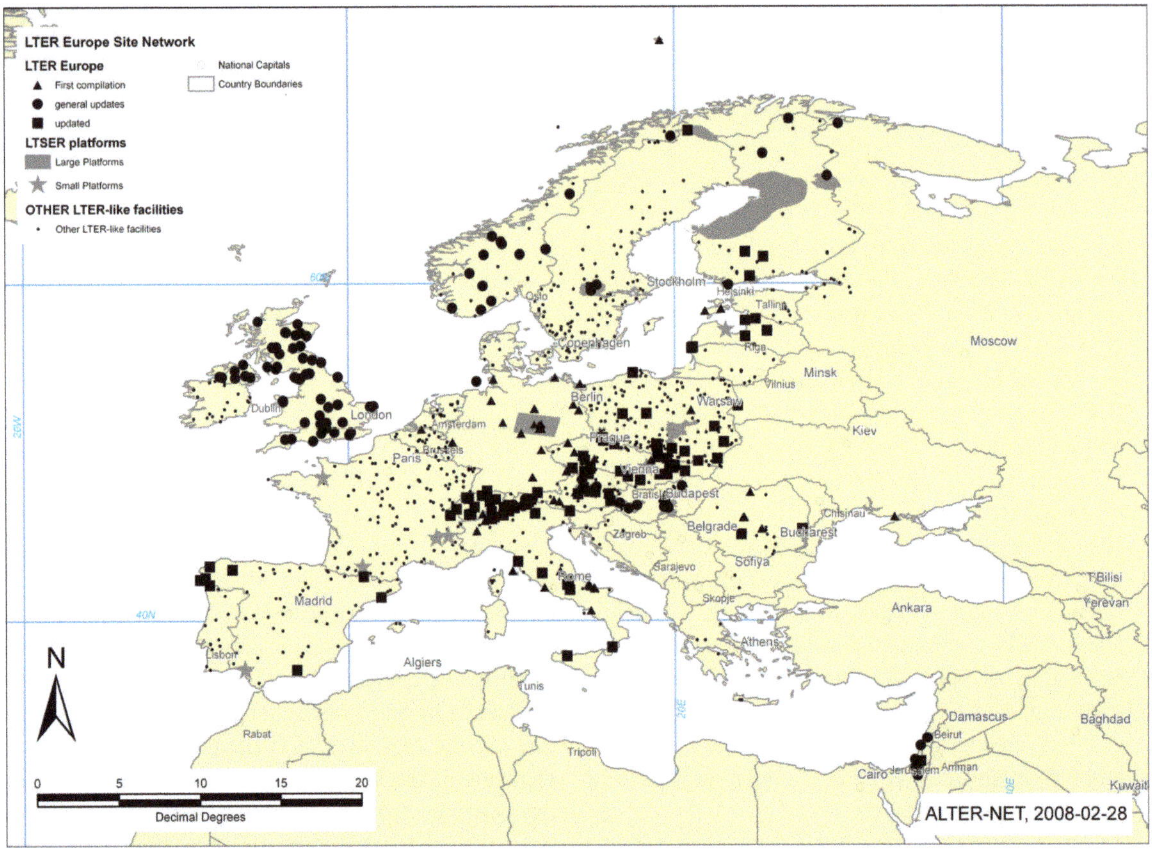

Fig. 6.3 Location of 406 LTER sites and 23 LTSER Platforms in Europe (graph: J. Peterseil: http://www.lter-europe.net/Structure/key-documents)

facilitate LTER-activities (e.g. logistics, laboratories, on-site supporting institutions). Figure 6.3 shows the location of 406 LTER sites and 23 LTSER Platforms in Europe.

6.4.1 Introducing the LTSER Component

Since the beginning of the 21st century existing national and continental networks for Long-term Ecosystem Research (LTER) underwent major reviews in a direction that emphasised more focus on their relevance for society, the efficiency of knowledge dissemination and adequacy of current designs to tackle urgent political questions many of which are related to the sustainable use of ecosystem services and the effects imposed on them by Global Change (Hobbie, Carpenter, Grimm, Gosz, & Seastedt, 2003).

The concept of LTSER platforms evolved in response to that demand implying the expansion of the traditional LTER approach. Important drivers and pressures and their impacts cannot be comprehensively investigated on the spatial scale of hundreds of hectares, even if a network of hundreds of sites of that scale covering Europe's environmental zones is available (e.g. the effects of agricultural subsidies on management practices and biodiversity on the landscape level). Moreover, to support fundamental research on ecosystem processes in the long-term the selection of locations for traditional LTER sites was biased in favour of natural or semi-natural ecosystems (Metzger et al., 2008). Thus, the characteristics of the LTER-facilities as well as the disciplines involved in research do not suffice to appropriately investigate socio-ecological systems (Redman, Grove, & Kuby, 2004). This has become a bromide since first stated in the course of evaluating and restructuring ecosystem research and ecosystem research infrastructure in the late 1990s (e.g. US-LTER review; http://intranet.lternet.edu/archives/documents/reports/

20_yr_review/). Since then even in studies designed to address complex interactions between society and natural resources mismatches between the observed spatial units and the related spatial scale of management as well as the level of political actions were detected (Dirnböck et al., 2008). This resulted in the request for a new-generation of LTER considering the human dimension in a scale- and level-explicit design and thus signalling the transition towards Long-Term Socio-Ecological Research or LTSER (Haberl et al., 2006; Chapter 26).

But *developing the framework for LTSER* needs to take a wide range of components into account, ranging from the underlying concepts to the challenge of developing the common language indispensable for proper interdisciplinary research (Kaljonen, Primmer, De Blust, Nijnik, & Külvik, 2007). Identifying appropriate regions and physically implementing LTSER has been revealed to be a major long-term effort, which strongly requires a shared vision and division of tasks on the European scale. The Network of Excellence ALTER-Net (FP 6) provided a frame for meeting the addressed challenges and integrates the strengths of the existing, but extremely fragmented LTER infrastructure on the site level. The interdisciplinary expertise represented by the ALTER-Net consortium allows further development of the main pillars of LTER-Europe's network, the now integrated networks of LTER sites and LTSER platforms.

At the network level the *strategic research intention of the LTSER component* is to establish a tool for building socio-ecological research capacity across Europe. The major socio-ecological systems of the continent shall be represented by at least one LTSER platform each, where exemplary research can efficiently take place, including the participation in assessments and forecasts of changes in structure, functions and dynamics of ecosystems and their services, and defining the socio-economic and socio-ecological implications of those changes. LTSER platforms are also to define and address key management issues according to complex local and regional settings – cultural and social values and expectations, economic conditions and constraints, inherent biophysical capacities, and impacts of internal and external factors. Finally, LTSER platforms will test and further develop tools and mechanisms for the communication and dissemination of knowledge across cultural and societal gradients.

LTSER is context-driven, problem-focused and interdisciplinary. It involves multi-disciplinary teams brought together for short periods of time to work on specific problems in the real world. Gibbons et al. (1994) labelled this *mode 2* knowledge production as opposed to traditional research (*mode 1*), which is academic, investigator-initiated and discipline-based knowledge production. So mode 1 knowledge production is investigator-initiated and discipline-based while mode 2 is problem-focused and interdisciplinary. *LTER-Europe provides an integrated framework for both types of knowledge production.*

6.4.2 The Design of LTSER Platforms

The required elements of LTSER platforms are firstly derived from the need to represent functionally and structurally relevant scales and levels. Secondly, the choice of elements of LTSER platforms depends on the characteristics of individual regions with regard to their landscape, occurring ecosystem types and administrational structures as well as economic, social and natural gradients within the target region. Nevertheless, the design of LTSER platforms in principle distinguishes *three functional layers* to which all elements can be assigned: (i) physical infrastructure comprising in-situ research sites, technical infrastructure like power supply and sensors, laboratories, sites of sectoral monitoring networks, collections, museums, visitor centres; (ii) research itself in the sense of research projects and a pro-active involvement of the research communities on the regional, national and international level and (iii) integrative management serving as the interface between all the above elements and providing services to enable efficient research and dissemination of results (Fig. 6.4).

Regarding the *physical infrastructure* LTSER platforms represent clusters of facilities supporting LTER activities and providing data. Ideally, we propose to distinguish between (i) site-level activities representing in-depth ecological research in quantitatively relevant habitat types, containing specific sampling plots, (ii) intermediate-scale elements such as national parks, biosphere reserves or investigated meso-catchments and finally (iii) the region as such (Fig. 6.5). Nested designs from the site- to the landscape levels and cascadedly harmonised sampling and parameter sets

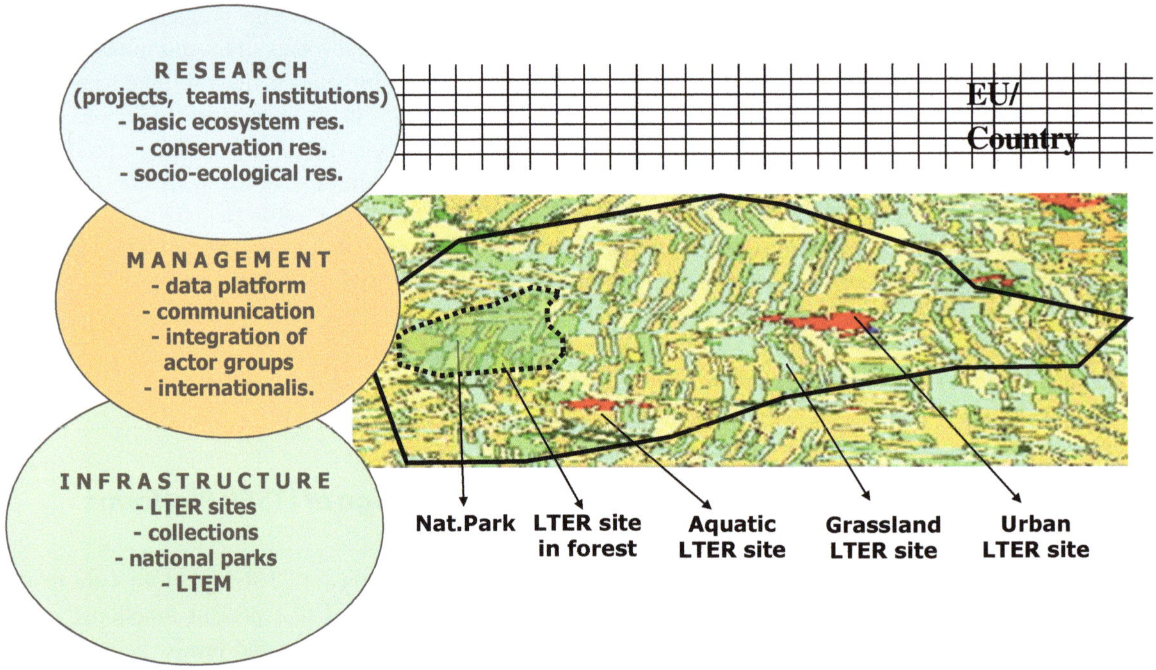

Fig. 6.4 The functional layers of LTSER platforms (*left*) and exemplary infrastructural elements according to landscape composition (graph: M. Mirtl: http://www.lter-europe.net/Structure/key-documents) (Mirtl et al., 2009)

enable the systematic assessment of the representativeness of individual elements for their vicinity. Elements belonging to bigger scale activities, including national and international monitoring schemes, are functionally linked for further up- and downscaling and crosswise validation (e.g. biodiversity indicators).

By means of land cover statistics, habitat and landscape type distributions and environmental parameter gradients the adequacy of existing research infrastructure is assessed (e.g. dominating land use sectors like agriculture ought to be covered by research on the effects of current and alternative management practices).

LTSER platform management: In order to achieve a scale and level explicit design the levels of administration, decision making and management impacting

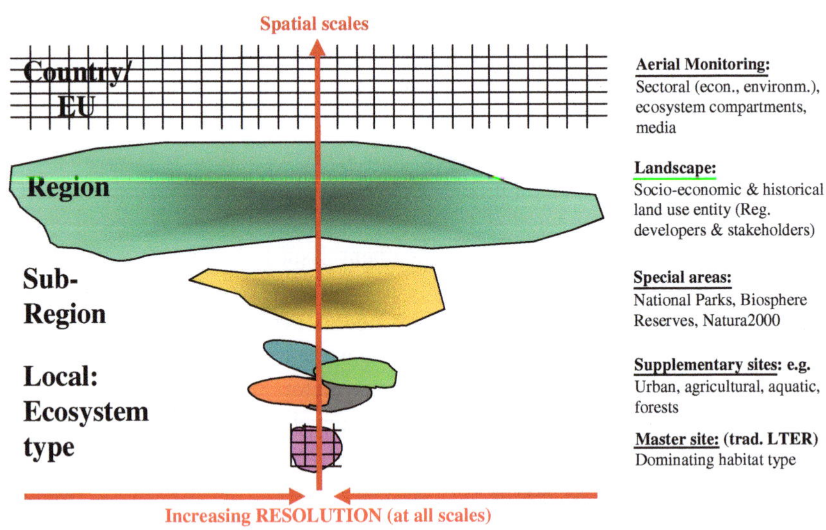

Fig. 6.5 Infrastructural elements of LTSER platforms across spatial scales within a LTSER platform region (graph: M. Mirtl: http://www.lter-europe.net/Structure/key-documents) (Mirtl et al., 2009)

the area at different scales are identified, differentiating between internal and external drivers. Socio-ecological profiling reveals key ecosystem services, environmental and economic compartments and societal factors driving the system. An actor analysis identifies the corresponding interest groups engaged in regional and local decision making, management, administration, regional development, education, monitoring and research itself, as well as stakeholders of dominating economic and land use sectors. Only interactive involvement of these key groups allows for the identification of research demands as regionally perceived. One of the processes LTSER platforms support is reconciling national and international top-down research strategies with bottom-up necessities of nature protection, regional development, environmental reporting, and the assessment of abatement strategies.

LTSER and LTSER projects in the proper sense mediate between strategies and requirements. Non-scientists should be involved in the definition phase of projects and the re-translation of scientific findings into guidelines for administration and management. Transdisciplinary and participatory approaches play an important role in the dissemination of knowledge and educational efforts to change behaviour where scientific findings recommend so. Accessory re-translation and implementation projects have access to other funding sources than research itself (e.g. LEADER, LIFE+, Interreg). All the above implies the necessity of establishing a multi-dimensional communication space considering a wide range of idioms spoken across actor groups of the same mother tongue. The same is true for science when it comes to the required data access and data flows. Without central facilitation the efforts for provisioning required data for complex LTSER projects alone would exhaust projects, even if these data were freely available.

It is obvious and has broadly been accepted that LTSER requires a *platform management* secured in the long term and providing a wide range of services deducible from the outlined work and communication flows. Amongst these services are:

- Conceptual work
- Project development
- Networking across interest groups, disciplines and stake holders
- Communication (inter- and transdisciplinary communication space, WEB site etc.)
- Data integration and policy
- IT-Tools
- Representation (nationally, internationally)
- Public relations
- Lobbying
- Fund raising

The management cares for an open communication space including the implementation of transdisciplinary and participatory approaches necessary to adopt research agendas to regional and local needs and for achieving access to and involvement of the regional population, key stakeholders and decision makers, all of whom can be seen as beneficiaries of the knowledge produced. Moreover, the management stands for a modern data policy and quick data exchange based on cutting-edge IT solutions for data integration (ontologies, tools for semantic mediation, disperse data sources). The research component of LTSER platforms has been described with the concept above.

6.4.3 Representativeness and Coverage

LTER projects scrutinise ecosystems in an exemplary way, assuming comparability of basic mechanisms in similar environments. The functional integration of LTER and sectoral monitoring schemes (water, air, biodiversity) set up with the ambition of probabilistic sampling in order to achieve reliable trend information, allows for empirical estimates of what LTER sites and LTSER platforms represent. Even though Long-Term Ecosystem Monitoring (LTEM) on the site- and platform level indispensably complements LTER-Europe's ecosystem research projects, the mission of LTER-Europe does not comprise providing representative information on ecosystem trends with full aerial coverage on the continental scale.

Hence, all European environmental zones must be represented by LTER sites covering the natural gradients within these zones. But the distribution of LTSER platforms needs to reflect the variation induced by humans, for example through land management, pollution and other disturbances. The extent and the ways in which humans influence the environment differ greatly, as do their impacts on ecosystems (Reid et al., 2005). Redman et al. (2004) give a list of socio-economic patterns and processes that influence ecosystems. Economic power and human population pressure form the basis of the majority of environmental pressures (Ehrlich & Holdren, 1971; Dietz & Rosa, 1994) whereas several stratifications of the European biogeophysical environment have been provided (Metzger, Bunce, Jongman, Mücher, & Watkins, 2005; Metzger, Leemans, & Schröter, 2005), LTER-Europe required a complementary stratification of socio-ecological regions of the continent in order to support the construction of its network of LTSER platforms. The ALTER-Net project elaborated the LTER Socio-ecological regions (LTER-SER) of Europe with a 1 km^2 resolution, delineating 48 European socio-ecological strata (Metzger & Mirtl, 2008). The data set is based on the Environmental Stratification of Europe (EnS) (Metzger, Bunce, et al., 2005; Jongman et al., 2006), a biogeophysical stratification developed using multi-variate clustering of largely climatic variables.

The EnS was combined with a newly developed socio-economic stratification based on an economic density indicator in order to overcome both the limitations in data availability at the 1 km^2 resolution across Europe and in distortions caused by using administrative regions (termed Nomenclature d'Unités Territoriales Statistiques (NUTS) regions). The economic density indicator, defined as the income generated per square kilometre (€ km^{-1}), can be mapped at a 1-km^2 spatial resolution. Economic density forms an integrative indicator that is based on the two key drivers that were identified above: economic power and human population pressure (Metzger, Bunce, van Eupen, & Mirtl, 2009). The indicator, which has been used to rank countries by their level of development (Gallup, Scach, & Mellinger, 1999; Sachs, Mellinger, & Gallup, 2001) can be considered a crude measure for impacts on the environment caused by economic activity (Radetzki, 2001).

As LTSER platforms are still scarce, each national decision on a new platform can close major gaps in the network (Fig. 6.6). In order to promote decisions in favour of an optimised division of tasks within the

Fig. 6.6 The detailed pattern of 48 LTER Socio-Ecological Regions of Europe with 23 LTSER platforms, including five preliminary platforms (Metzger et al., 2009; ALTER-Net; with permission)

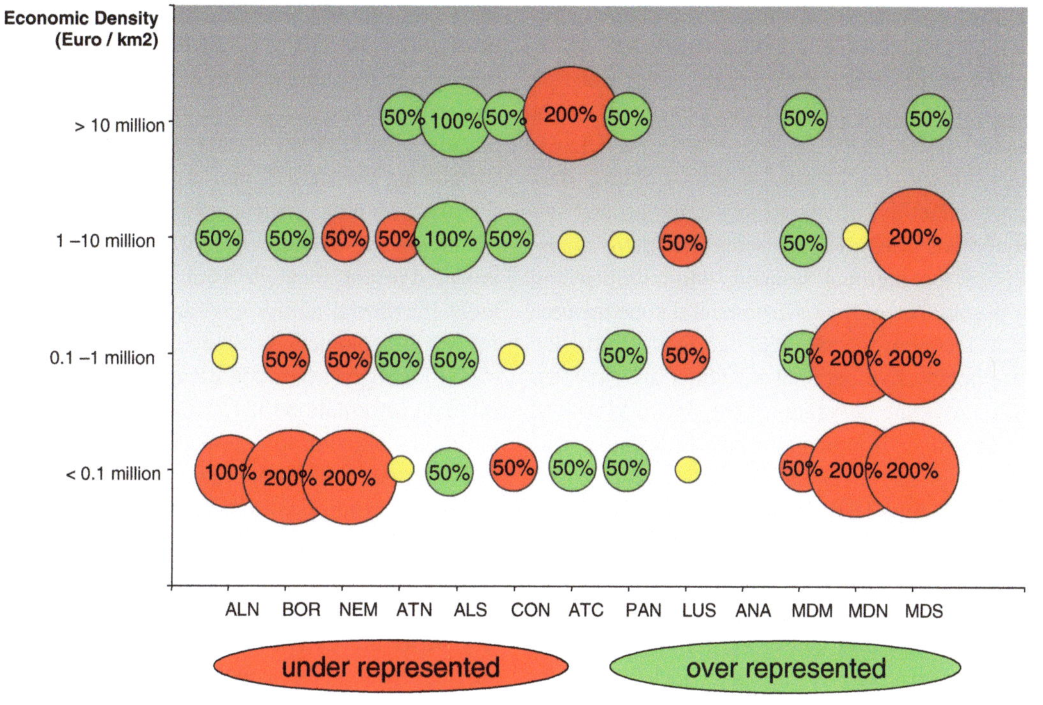

Fig. 6.7 Representativity analysis of Europe LTER facilities with European environmental zones (EnS) on the *x*-axis, from Atlantic north (*left*) to Mediterranean south (*right*); *dark/red* = under-represented; *light/green* = over-represented; *yellow/dots* = proportionally represented (Metzger & Mirtl, 2008; ALTER-Net; with permission)

European Research Area, a representativeness analysis was conducted disclosing environmental and socio-ecological zones over-, under- and proportionally represented by LTER facilities (Fig. 6.7).

The analysis shows two strong biases in the present LTER effort. Firstly, urban and disturbed regions, where humans interact most directly with nature, are consistently underrepresented, illustrating a bias for traditional ecological research away from human activity. Secondly, the Mediterranean, for which some of the most extreme global change impacts are projected, is receiving comparatively little attention. Both findings can help guide future investment in the European LTER network – and especially its LTSER component - to provide a more balanced coverage.

6.4.4 Implementing LTSER Platforms

The process outlined below has been developed and tested at a small number of LTSER platforms which were proposed by the FP 6 NoE ALTER-Net in 2006, especially the Austrian LTSER Eisenwurzen, the Finnish LTSER platforms, LTSER Donana in Spain and LTSER Braila Island in Romania. LTSER Eisenwurzen, where the implementation started in 2004, served as exemplar and basis for the training of LTSER platform managers.

According to the rules and governance of LTER-Europe the national LTER networks are responsible for choosing their LTER sites and LTSER platforms in the respective countries. LTER-Europe provides a framework to assist in national network building and decision making. Under the auspices of ALTER-Net a set of criteria for LTER networks, LTER sites and LTSER platforms have been developed since 2005 and was formally adopted in 2008 (Technical report on LTER-Europe Criteria: http://www.lter-europe.net/Structure/key-documents).

LTSER platforms contain clusters of LTER facilities located in the defined LTSER platform regions. The development of LTER-Europe was, on the request of the European Commission, to be based on existing infrastructure wherever possible. The first step in deciding potential areas for LTSER thus capitalises on inventories of existing infrastructure on the national level such as LTER sites, well-equipped sites of ecosystem monitoring schemes, protected areas, National Parks, Biosphere Reserves etc. Once, existing hot spots have been identified further decisions ought to consider (i) the LTER-Europe criteria for LTSER platforms (comprising aspects of infrastructure, data and data availability, access to key actor groups and streamlined activities), (ii) the scientific interests and strengths of the national research communities and (iii) the importance of the environmental zone the area represents (economy, ecosystem services). From the European perspective national networks are expected to help improve the coverage of the network as well as possible. All environmental zones (EnS) and socio-ecological regions (LTER-SER, see above) should be represented by LTER sites and LTSER platforms and the LTER facilities would ideally be evenly distributed over these zones. As mentioned, LTSER platforms are still scarce. Therefore, each national decision on a new platform can close major gaps in the network. The coverage of European LTER facilities across 48 socio-ecological strata was tested. LTER-Europe also provides country-specific analyses and maps to promote decisions in favour of an optimised division of tasks within the European Research Area.

Following the decision on the location of a new LTSER platform a range of analyses are done as a basis for further steps, aiming at the identification of

- dominant habitat and land use types to be dealt with by research facilities in the LTSER region,
- levels of decision making, administration and management affecting the region (provincial governments),
- actor groups to be considered in the LTSER platform consortium,
- existing data and other information on the region (environment, economy, administration, demography etc.),
- existing communication structures within the region, especially with respect to regional development and knowledge dissemination,
- prevailing human-caused pressures on ecosystems and conflicts on the use of natural resources,
- prevalent ecosystem services and their link with regional economy,
- environmental, economic and demographic gradients within the region.

In many of the LTSER platforms, consortia of those who control the major regional monitoring and research infrastructure form the core group promoting and implementing LTSER platforms. As their mission usually stretches over decades they offer ideal settings for hosting the LTSER platform management. Ideally the platform management is funded by the main beneficiaries of its services, namely the generation of better knowledge for local decision making and management (territorial authorities) and a unique framework for socio-ecological science (research funding agencies). Through promotion campaigns, workshops and bilateral information the LTSER concept and plans for its implementation in a new area are advertised in all relevant communities in order to build the LTSER platform consortium committed to by a Memorandum of Understanding (MoU), which should address the scientific and practical goals, governance structure and data policy of the platform. According to the respective MoU the LTSER platform management sets up services listed in the section above on LTSER platform design.

An alternative approach to the management of the research of a LTSER platform is to start from the bottom in a project-oriented way. Here various institutions develop the strategy for the research and plan the research activities jointly and – if possible – build the monitoring necessary for the research raised in the strategy. This approach is beneficial especially in the context of LTSER where innovations in carrying out research are required. The risk involved in the top-down approach is that the old traditions around ecological research will dominate the framing of the research which does not open space and build motivation for other disciplines to enter the LTSER research arena.

A crucial step in building LTSER platforms consists of the spatial delineation of the LTSER region. As empirical socio-ecological research capitalises on data and information from different realms these data need to refer to the same spatial units. In most cases the best available economic and census data are provided with a resolution on the level of municipalities. Municipal boundaries are self-evident borderlines for LTSER platform regions. Problems arise when ecological and social borders do not match. Preferably, the platforms provide a buffer zone to exclude this problem.

The LTSER platforms established so far vary considerably in composition, size and targets. Whereas some follow an integrated regional approach considering the entire policy cycle from user-oriented knowledge generation to management and political measures, others still are rather clusters of site-based research concentrated in a specific area. But there is clear evidence of a trend towards integrated approaches. As pointed out earlier, only structured access to key groups in the regions allows for the identification of research demands as regionally perceived and the dissemination and putting into practice of research findings.

6.5 Network of National LTER Networks

LTER-Europe consists of 18 national LTER networks in Austria, Czech Republic, Finland, France, Germany, Hungary, Israel, Italy, Latvia, Lithuania, Poland, Portugal, Rumania, Slovenia, Slovakia, Spain, Switzerland and the United Kingdom. In about eight more countries LTER networks are under development, ranging from the establishment of key contacts or stakeholders for a national LTER process to current final steps toward formal recognition. The current coverage of LTER-Europe is given in Fig. 6.8.

Each adopted national network has been established according to the national peculiarities regarding organisation and funding of research projects, institutions and infrastructure. Nonetheless LTER-Europe achieved comparable overviews of the respective national organisation in the form of "National LTER Mind Maps", mapping key network elements (stakeholders, institutions, LTSER platforms, LTER sites) and their relations (www.lter-europe.net). The mind maps contain references to the contact persons of each element. LTSER platforms and LTER sites are also linked with the LTER-Europe InfoBase. Key data on national LTER networks are annually reported through a LTER National Standard Report. The contents of these reports are mirrored in WEB pages interlinking the LTER-Europe WEB site with national LTER WEB sites, providing a comparable overview.

LTER-Europe and the national LTER processes have been key to issues of ecosystem research and related infrastructural requirements in almost all European countries. Since forest dieback research lost importance in the mid-1990s and environmental monitoring faced extensive economies hand-in-hand with declining public concern about environmental issues the strategic effort of the European institutions was to reorganise and integrate the distributed and fragmented LTER support avoiding an irrevocable loss of infrastructure, expertise and data.

6.6 Governance of LTER-Europe

A major goal of ALTER-Net was to create a permanent LTER network in Europe. This was achieved by establishing LTER-Europe as a regional group of the global LTER and adopting a governance structure in the LTER-Europe bylaws (http://www.lter-europe.net/Structure/key-documents). Each formal national LTER network is represented in the LTER-Europe Co-ordinating Committee (CC) as the central decision-making body. To actively include countries with LTER networks under development the National Networks Representatives Conference (NNRC) was set up as a platform for all countries with LTER.

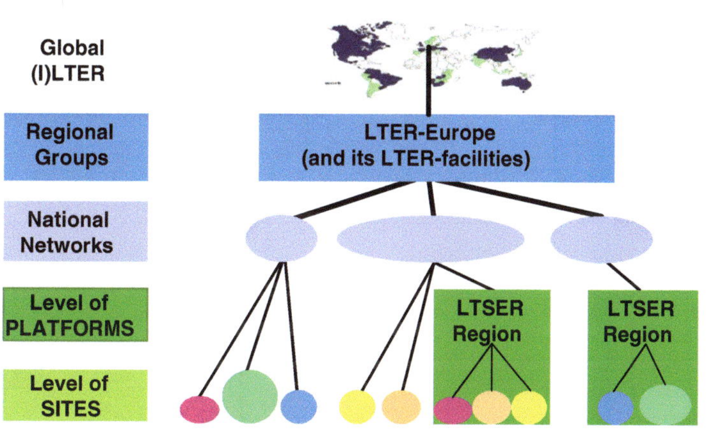

Fig. 6.8 Hierarchical organisation of LTER through national LTER networks. Not all national networks feature both LTER sites and LTSER platforms (graph: M. Mirtl: http://www.lter-europe.net/Structure/key-documents)

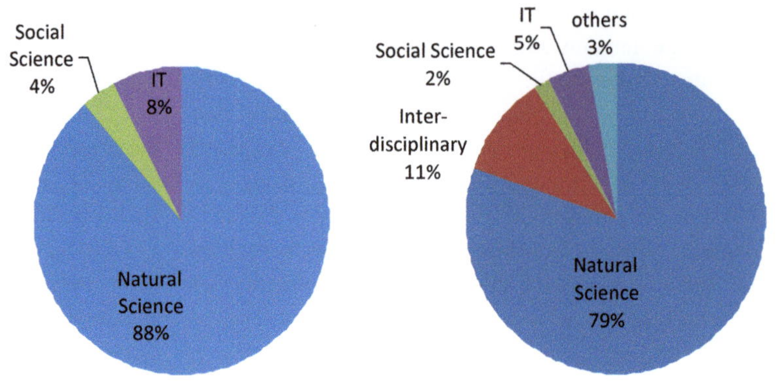

Fig. 6.9 Disciplinary background of LTER experts participating in the annual LTER conferences 2007 (*left*) and 2008 (*right*) (graph: M. Mirtl: http://www.lter-europe.net/Structure/key-documents) (Mirtl et al., 2009)

The continuity of LTER-Europe strongly depends on the group of experts co-ordinating LTER sites and managing LTSER platforms across the continent. This community is organised as the Scientific Site Coordinators Conference (SSCC). Annual LTER-Europe conferences are held to bring all bodies together. To secure key services and cover major scientific and technical aspects of the network, LTER-Europe Expert Panels (EP) were set up, comprising science strategy, communication, standardisation and harmonisation, technology, information management, LTSER and IDR as well as site management.

6.7 Network of Scientists and Disciplines

The modified scope of LTER implied changes in the composition of the involved scientific communities. In the initial phase of LTER-Europe only natural scientists carried the process. Because of the early agreement on improving information management and efforts to structure and collect metadata on LTER facilities and data IT experts joined the community. Despite a broad consensus on the necessity of a strong LTSER component in LTER-Europe it took several years until the disciplinary background of LTER experts started to reflect this fact. Figure 6.9 gives the share of disciplines of participants of major LTER conferences in 2007 and 2008.

6.8 A Network of Data and Metadata

The overarching goal "LTER" of the FP 6 NoE ALTER-Net has given information management high priority. The functionalities required to support the integrated and interdisciplinary approach outlined above were conceptualised and key elements were implemented (Fig. 6.10), comprising (i) a pilot version of an object-relational information system holding semantically structured data (MORIS), (ii) a core ontology for ecological and socio-ecological observations (SERONTO) and (iii) a metadata collection and entry system (LTER InfoBase).

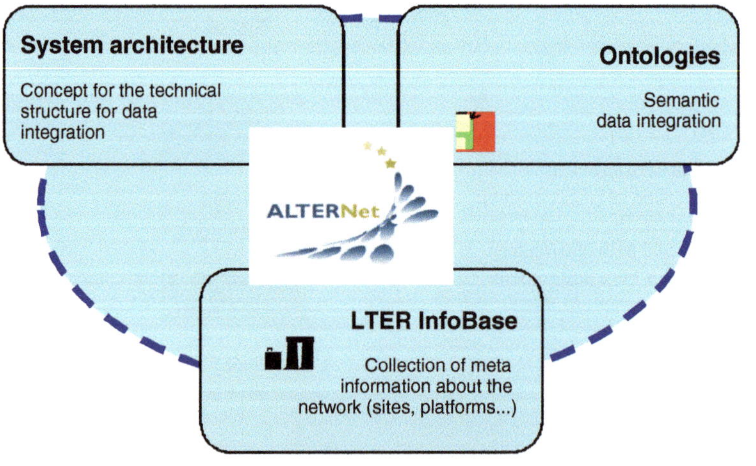

Fig. 6.10 Main pillars of LTER-Europe's information management (modified original of J. Peterseil, with permission: http://www.lter-europe.net/info_manage)

A major challenge for a network like LTER-Europe consists of the distributed nature of its infrastructure and data. Because of this heterogeneity and operational aspects it was decided to integrate data and databases through an ontology covering any kind of object investigated by LTER and all measured properties of these objects (parameters) across domains. Ontology formally represents a set of concepts within a domain and the relationships between those concepts. According to Gruber (1995) ontology is a formal specification of a shared conceptualisation of a domain of interest. Shared means a common understanding of the knowledge to be formalised as a basis for the domain experts, to which own specific views can be mapped. This opinion is supported also by Davenport (1997): "People can't share knowledge if they don't speak a common language". As LTER covers a wide range of domains core ontology was needed, determining in principle how observations are described. ALTER-Net set up a discussion and decision-making process to achieve agreement on the Socio-Ecological Research and Observation oNTOlogy (SERONTO; van der Werf et al., 2008). The mid-term goal of LTER-Europe is to achieve a seamless drill-down to primary data through metadata. As a first step a hierarchically structured metadata system was established to describe the distributed network of LTER sites and LTSER platforms (LTER Infobase based on eMORIS; http://www.lter-europe.net/sites-platforms). Information ranges from basic descriptors to comprehensive lists of measured parameters across domains, infrastructural and administrational data. The key purpose of the LTER Infobase is achieving an overview of ecosystem research facilities in the continent and enabling the search for LTER sites or LTSER platforms according to given criteria, for example grassland sites within a certain elevation range where temperature and nitrogen deposition have been measured.

6.9 Part of a Network of Networks

LTER-Europe was developed into an intricate landscape of European research networks and multi-site research projects (e.g. FP 6 IP ALARM, FP 6 IP SENSOR), national environmental monitoring schemes such as the UNECE International Co-operative Programmes and nature conservation measures like the UNESCO Biosphere Reserves (EuroMAB) and Natura2000. Even long-term research projects setting up networks of sites like the Field Site Network (FSN) of the IP ALARM cannot secure the maintenance of these networks beyond their runtime in the range of 3–5 years. On the other hand networks set up primarily for ecosystem monitoring purposes would benefit from a complementary research component which they cannot provide themselves (e.g. UNECE ICP Integrated Monitoring). Most of the site infrastructure both of European and country networks is funded on the national level. Nevertheless examples of efforts to systematically integrate different in-situ networks are scarce. Because of the necessity to build on existing infrastructure the process of establishing national LTER networks has initiated such network integration in many of the LTER-Europe member countries (e.g. Finland, Italy, Austria, Germany), where LTER sites and LTSER platforms tend to serve two or more purposes in other research, monitoring or nature conservation schemes (e.g. LTSER platforms containing National Parks, Biosphere Reserves, Nature2000 sites, sites of UNECE ICP Forests, UNECE EMEP stations etc.).

On the European level other sectors like the marine research communities, taxonomic collections and experimental ecology have launched networks and projects (MARBEF, EDIT, ANAEE). Each, again, with topical or infrastructural overlaps and a wide range of potential synergies from the point of view of a visionary restructuring of the European research area in the field of life sciences. As part of the effort to formalise this process across individual initiatives, the LifeWatch preparatory project (LifeWatch, 2007) was accepted in the ESFRI roadmap (ESFRI, 2006). Currently LifeWatch foresees the integration of in-situ networks on the European level with LTER-Europe as the terrestrial and aquatic component (Fig. 6.11). Information management, legal and financial aspects shall be covered as centralised services of LifeWatch across the topical networks.

6.10 Key Potential and Strengths of LTER

Nowak, Bowen and Cabot (2006) gave excellence evidence of the unique strength of LTER and LTSER. They found that land use and management had significantly changed in big parts of the Lake Mendota watershed, without prompting a decrease in phosphorous loads to the lake – with a resulting stagnation

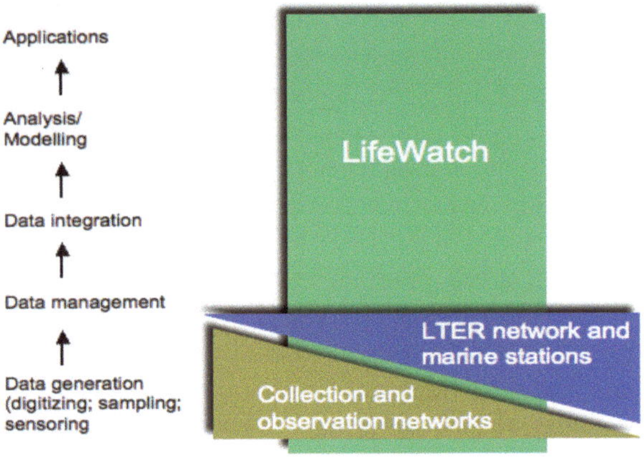

Fig. 6.11 LTER-Europe as the terrestrial and aquatic in-situ component of LifeWatch (from LifeWatch co-ordinator W. Los, with permission)

or even further decline of biodiversity. So, an interdisciplinary team of sociologists, agronomists and ecologists started searching for the reason for the detected phenomenon, applying the concept of disproportionality on the case. It is based on the fact that fundamental biological and social processes are characterised by positively skewed normal or even log-normal probability distributions. If, for example, the distribution between appropriate and inappropriate conservation behaviour of people in a specific watershed unfavourably overlap with the distribution of resilient and vulnerable parts of this watershed, easily overlooked malpractice may cause severe environmental impact. In the studied case dumping of manure by a few big farms in the Lake Mendota watershed had created hot spots of phosphorous in the landscape. These consequences of inappropriate social behaviour were frequently located close to farms or settlements and were thus unproportionally prone to erosion events (wind, storms). "Fast variables" such as (1) construction processes and (2) changes of vegetation cover influence such sensitivity, increasing or creating vulnerable environmental conditions. The consequence of a cross-wise overlap of two uneven probability distributions had disproportional influence on the overall system, because they accounted for a significant part of the overall P-load to the lake. As the P-Pools in the landscape are a so-called "slow variable" it will last decades before P-discharge into Lake Mendota will significantly decline in order to allow biodiversity to recover.

The lake Mendota study underpins key strengths and added values of LTER as a unique frame work for investigating the interaction of

- fast variables vs. slow variables (the long-term),
- processes at different spatial and timescales,
- processes in different realms (nature – society).

Alongside such examples the recommendations from the Millennium Ecosystem Assessment can be seen as a general framework towards policy-relevant ecosystem and biodiversity research. Carpenter et al. (2006) ask for tools capable of evaluating the success of human interventions in ecosystems, as well as the used indicators for monitoring biological, physical and social changes.

Within such a framework and the strong focus on in-situ research and monitoring, LTER and especially LTSER support or offer:

- *Evaluation and mechanic insight into main drivers* of ecosystem change, for example do the main drivers of biodiversity according to the Millennium Assessment (climate, N-input, land use change), have the expected effects in concrete regions and at concrete sites across Europe?
- Development and evaluation of *indicators of* the social and natural system across spatial and timescales.
- Research, monitoring and indicator development towards the assessment of the state and trend of *Ecosystem Services.*
- Possibility to *validate the applicability of socio-economic concepts and system-theories* in concrete human–natural systems.
- Evaluation of applicability of *socio-economic methods* to characterise systems, for example Socio-Economic Metabolism concept (Fischer-Kowalski & Haberl, 1997, 2007) and Human Appropriation of Net Primary Production (Haberl et al., 2002, 2004).
- Evaluating and monitoring interventions in socio-ecological systems (e.g. conservation measures).
- Recommendation of *measures for halting biodiversity loss* (e.g. Lake Mendota example).
- Application, validation and further development of *Integrated Assessments (IA)* on the regional scale.

Very importantly, LTER also represents an enormous potential for serendipitous science. Serendipity is the effect by which one accidentally discovers something fortunate, especially while looking for something else entirely (http://en.wikipedia.org/wiki/Serendipity) Firstly, sagacity is required to be able to link together apparently innocuous facts to come to a valuable conclusion. But – evenly important – one needs access to the facts to apply sagacity. Translated into environmental science and LTER, processes, cause–effect relationships and mechanisms eventually driving our socio-ecological systems and significantly affecting ecosystem services can only be identified on the basis of well documented long-term data and information. Creating such databases for a representative network of locations and securing the sustainable use of costly gathered legacy information belongs to the core of LTER-Europe's mission. Thus, while some scientists and inventors are reluctant about reporting accidental discoveries, others openly admit its role; in fact serendipity is a major component of scientific discoveries and inventions. According to Stoskopf (2005) it should be recognised that serendipitous discoveries are of significant value in the advancement of science and often present the foundation for important intellectual leaps of understanding. Bearing the importance of LTER's precautionary principle the 20 years review of US-LTER underpinned the importance of serendipitous science exploiting unexpected events as opposed to synthesis science looking forward and being hypothesis and theory driven.

Regarding biodiversity indicators LTER can deal with *two main questions*: (i) what are reliable indicators for biodiversity (what do they stand for in the context of the entire system) and (ii) biodiversity itself as an indicator for specific drivers (biotic, abiotic and anthropogenic), including the question of susceptibility and specificy, more specifically:

- Testing of biodiversity monitoring schemes and methods aiming at an integrated assessment across levels of biodiversity (e.g. in the frame of the FP 7 project EBONE, http://www.ebone.wur.nl);
- Comparison of identical indicators applied at different scales (e.g. bird inventories);
- Response time of indicators (e.g. eutrophication effects on ground-layer vegetation in forests);
- How do different indicators react to drivers relative to each other in time and space? How does the reliability, susceptibility and specific reaction (to drivers) vary across biogeographical regions?

Because of the limited usability of averaged or aggregated measures of processes and properties (social and environmental) to explain trends at lower scales the in-situ generation and provision of high-quality data at different spatial and timescales is indispensable. LTER facilitates synergies between long-term environmental monitoring providing reliable trend information and short-term measuring campaigns and experimentation data, which in return help with optimising and adapting monitoring designs. In many cases only interdisciplinary research (IDR) allows for the detection of complex phenomena.

The communication structure and governance established by LTER-Europe enables the efficient planning and execution of cross-site experiments and comparisons. Projects capitalising on LTER sites and LTSER platforms benefit from earlier investments in infrastructure and data by far exceeding the respective project budget, moreover each project adds to the value of LTER facilities, contribution data, knowledge and expertise. Thus, comparably low investments into the networking and central services of LTER have a significant financial leverage in the sense of a well-structured European Research Area.

6.11 Outlook

The NoE ALTER-Net ended in March 2009. It has fostered the LTER process at multiple levels from institutional commitments to the broad recognition of a distributed research infrastructure by European stakeholders and related networks. LTER-Europe forms the biggest regional group in the global LTER network and has set up permanent governance structures. Major strategic challenges ahead are the establishment of stable infrastructure funding mechanisms (nationally and on the European scale e.g. through ESFRI/LifeWatch) and the development of complementary research frame programs for LTER and LTSER projects. Data and metadata of LTER infrastructure have to be regularly updated to enable efficient selections of facilities that could be used by any kind of projects requiring an in-situ component. LTSER platforms provide a perfect basis for Integrated Assessments (IA) as well as the testing and further development of IA methods. The need for quality-assured long-term data series requires the integration of LTER with long-term environmental monitoring (LTEM), both in-situ and conceptually.

Acknowledgements We acknowledge financial support by the Network of Excellence ALTER-Net (FP6). We thank Simron Singh, Elli Groner, Eva Furman, Thomas Dirnböck, Kinga Krauze and Johannes Peterseil for valuable comments and discussions.

References

Carpenter, S. R., DeFries, R., Dietz, T., Mooney, H. A., Polasky, S., Reid, W. V., et al. (2006). Millenium ecosystem assessment: research needs. *Science, 314*, 257–258.

Davenport, T. (1997). *Knowledge management case study: Knowledge management at Microsoft.* From: http://www.itmweb.com/essay536.htm.

Dietz, T., & Rosa, E. A. (1994). Rethinking the environmental impacts of population, affluence and technology. *Human Ecology Review, 1*, 277–300.

Dirnböck, T., Bezák, P., Dullinger, S., Haberl, H., Lotze-Campen, H., Mirtl, M., et al. (2008). Scaling issues in long-term socio ecological biodiversity research. A review of European cases. Social Ecology Working Paper No. 100, Institute of Social Ecology, Vienna.

Ehrlich, P. R., & Holdren, J. P. (1971). Impact of population growth. *Science, 171*, 1212–1217.

ESFRI. (2006). European roadmap for research infrastructures – Report 2006. Office for Official Publications of the European Communities, 84 pp. ISBN 92-79-02694-1.

Fischer-Kowalski, M., & Haberl, H. (1997). Tons, joules, and money: Modes of production and their sustainability problems. *Society and Natural Resources, 10*, 61–85.

Fischer-Kowalski, M., & Haberl, H. (2007). *Socioecological transitions and global change: Trajectories of social metabolism and land use.* Cheltenham, UK, Northampton, USA: Edward Elgar.

Fischer-Kowalski, M., & Weisz, H. (1999). Society as hybrid between material and symbolic realms. Toward a theoretical framework of society-nature interaction. *Advances in Human Ecology, 8*, 215–251.

Gallup, J. L., Scach, J. D., & Mellinger, A. D. (1999). Geography and economic development. *International Regional Science Review, 22*, 179–232.

Gee, D. (2001). *Late lessons in early warning.* Copenhagen: European Environment Agency.

Gibbons, M., Limoges, C., Nowotny, H., Schwartzman, S., Scott, P., & Trow, M. (1994). *The new production of knowledge: The dynamics of science and research in contemporary societies.* London: Sage.

Gruber, T. (1995). Toward principles for the design of ontologies used for knowledge sharing. *International Journal Human-Computer Studies,* 43, 907–928.

Haberl, H., Krausmann, F., Erb, K.-H., Schulz, N. B., Rojstaczer, S., Sterling, S. M., et al. (2002). Human appropriation of net primary production. *Science, 296*, 1968–1969.

Haberl, H., Schulz, N. B., Plutzar, C., Erb, K.-H., Krausmann, F., Loibl, W., et al. (2004). Human appropriation of net primary production and species diversity in agricultural landscapes. *Agriculture, Ecosystems & Environment, 102*, 213–218.

Haberl, H., Winiwarter, V., Andersson, K., Ayres, R., Boone, C., Castillo, A., et al. (2006). From LTER to LTSER: Conceptualizing the socio-economic dimension of long-term socio-ecological research. *Ecology and Society, 11*, 13.

Hobbie, J. E., Carpenter, S. R., Grimm, N. B., Gosz, J. R., & Seastedt, T. R. (2003). The US Long Term Ecological Research program. *Bioscience, 53*, 21–32.

Jongman, R. H. G., Bunce, R. G. H., Metzger, M. J., Mücher, C. A., Howard, D. C., & Mateus, V. L. (2006). Objectives and applications of a statistical environmental stratification of Europe. *Landscape Ecology, 21*, 409–419.

Kaljonen, M., Primmer, E., De Blust, G., Nijnik, M., & Külvik, M. (2007). Multifunctionality and biodiversity conservation – Institutional challenges. In: Chmielewski, T. J. (Ed.), *Nature conservation management – From idea to practical results* (pp. 53–69). Lublin-Lodz-Helsinki-Aarhus: ALTERnet.

LifeWatch (2007). LifeWatch Preparatory Phase, supported by the European Commission with FP7 Grant Agreement No 211372-LIFE. Available from: http://www.lifewatch.eu/index.php?option=com_content&view=article&id=63-122008updateprep&Itemid=4.

Metzger, M. J., & Mirtl, M. (2008). LTER Socio-Ecological Regions (LTER-SER) – Representativeness of European LTER facilities – Current state of affairs. *AlterNet Technical Paper,* p. 5. Available from: http://www.alter-net.info.

Metzger, M. J., Bunce, R. G. H., Jongman, R. H. G., Mücher, C. A., & Watkins, J. W. (2005). A climatic stratification of the environment of Europe. *Global Ecology and Biogeography, 14*, 549–563.

Metzger, M. J., Bunce, R. G. H., van Eupen, M., & Mirtl, M. (2009). An assessment of long term ecological research activities across European socio-ecological regions. Journal of Environmental Management (accepted).

Metzger, M. J., Leemans, R., & Schröter, D. S. (2005). A multidisciplinary multi-scale framework for assessing vulnerability to global change. *International Journal of Applied Geo-information and Earth Observation, 7*, 253–267.

Mirtl, M., & Krauze, K. (2007). Developing a new strategy for environmental research, monitoring and management: The European Long-Term Ecological Research Network's (LTER-Europe) role and perspectives. In: Chmielewski, T. J. (Ed.), *Nature conservation management – From idea to practical results* (pp. 36–52). Lublin-Lodz-Helsinki-Aarhus: ALTERnet.

Mirtl, M., Boamrane, M., Braat, L., Furman, E., Krauze, K., Frenzel, M., et al. (2009). LTER-Europe Design and Implementation Report–Enabling "Next Generation Ecological Science": Report on the design and implementation phase of LTER-Europe under ALTER-Net & management plan 2009/2010. Umweltbundesamt (Federal Environment Agency Austria). Vienna. 220 pages. ISBN 978-3-99004-031-7. In press.

National Research Council (NRC). (2004). *NEON, addressing the nations environmental challenges.* Washington, DC: The National Academy Press.

Nowak, P., Bowen, S., & Cabot, P. E. (2006). Disproportionality as a framework for linking social and biophysical systems. *Society and Natural Resources, 19*, 53–173.

Radetzki, M. (2001). *The green myth – Economic growth and the quality of the environment.* Multi-Science Publishing Co. Ltd.

Redman, C. L., Grove, J. M., & Kuby, L. H. (2004). Integrating social science into the Long-Term Ecological Research (LTER) Network: Social dimensions of ecological change and ecological dimensions of social change. *Ecosystems, 7*, 161–171.

Reid, W. V., Mooney, H. A., Cropper, A., Capistrano, D., Carpenter, S. R., Chopra, K., et al. (2005). *Millennium Ecosystem Assessment Synthesis report.* Washington, DC: Island Press.

Sachs, J. D., Mellinger, A. D., & Gallup, J. L. (2001). The geography of poverty and wealth. *Scientific American, 284*, 62–67.

Stoskopf, M. K. (2005). Observation and cogitation: How serendipity provides the building blocks of scientific discovery. *ILAR Journal, 46*, 332–337.

Van der Werf, B., Adamescu, M., Ayromlou, M., Bertrand, N., Borovec, J., Boussard, H., et al. (2008). SERONTO: A Socio-Ecological Research and observation oNTOlogy. In: Weitzman, A. L. & Belbin L. (Eds.), *Proceedings of TDWG.* Fremantle, Australia.

Chapter 7

The Role of Ecosystem Modelling for Long-Term Ecological Research

Albrecht Gnauck, Sven E. Jørgensen, and Bernhard Luther

Abstract Analysis, control, and management of ecosystems are complex tasks which have to cover broad ranges of operating environmental states and decision making. Current approaches to ecosystems modelling and simulation are mostly based on information theory, thermodynamics, topology, or balances of biological and chemical components. Ecosystems are considered from the point of view of information theory as complex dynamic communication networks with living and non-living components and their interrelationships. State space approaches are used for ecosystem management. But these models give no answer to structural ecosystem changes. The relational ecology approach covers more the structural problems of ecosystem modelling. The relationships are given qualitatively but cannot be quantified easily. The network approach seems to be the most appropriate approach to include behavioural and structural changes of ecosystems in models.

Keywords Ecosystem theory · Ecosystem modelling · State space · Ecological network

7.1 Introduction

Ecosystems are dynamic, nonlinear, and non-stationary thermodynamically open large-scale systems. They are characterised by following features:

A. Gnauck (✉)
Department of Ecosystems and Environmental Informatics, Brandenburg University of Technology, Konrad-Wachsmann-Allee 1, D-03046 Cottbus, Germany
e-mail: albrecht.gnauck@tu-cottbus.de

1. Complexity: The number of ecosystem elements and the amount of relationships among the elements of an ecosystem as well as between the ecosystem and its environment are tending to large.
2. Integrity: The ecosystem possesses properties which emerge only as a result of interrelationships between ecosystem elements.
3. Multidimensional stability: Nonlinear and non-stationary systems may have several stable regions. Their number depends on bifurcation points of the system.
4. Controllability: The ecosystem is capable of changing from one state to another within a finite period of time. The state is called controllable if it can be influenced in a target-oriented way by man.
5. Observability: Previous states of ecosystem can be concluded from the present.
6. Buffer and storage capacity: A transition from one state to another in response to interference does not occur explosively but is characterised by developments of processes.
7. Information processing and storage: Ecosystems are capable of converting received information according to its mode of action and to interconnect it with other stored information to produce new information. Information is represented by single data or by time series.
8. Qualitative differences between ecosystem elements: Elements of ecosystems can be living (e.g. single individuals, species, or groups of species) or non-living (e.g. pools of chemical substances, energy resources, natural internal driving forces). The elements of ecosystems are coordinated to such an extent that ecosystems work as indivisible units. A study on ecosystem

F. Müller et al. (eds.), *Long-Term Ecological Research*,
DOI 10.1007/978-90-481-8782-9_7,

components level will never reveal the ecosystem properties.

9. Structures in ecosystems are qualified by physical conditions (e.g. space divisions, light, and energy conditions), by chemical conditions (e.g. quantities and distributions of organic and inorganic substances), by biological conditions (e.g. trophic levels, species distributions, biodiversity), and by time structures (e.g. biological succession, ecosystems evolution).
10. Functions in ecosystems are related to circulations of matter as well as degradations of energy, are related to interrelations between ecosystem elements as well as between the ecosystem and its environment. Special ecological functions (e.g. competition, predator–prey relations, synchronisation, and symbiosis) exist between associated populations at and among various trophic levels.

Generally, an ecosystem is a biotic and functional system or unit, which is able to sustain life and includes all biological and non-biological variables in that unit. Spatial and temporal scales are not specified a priori, but are entirely based upon the objectives of the ecosystem study. To model the manifold of ecosystems and ecological processes different scientific disciplines are related to this subject: Hydrology, hydrochemistry, hydrobiology, systems ecology, geography, physics, especially thermodynamics, and mathematics, especially statistics, systems theory, control theory, economics, information theory, and operations research. Ecosystem simulation models are well-known informatic tools to manage environmental knowledge (Bossel, 1992) and to use them for ecosystem modelling and management (Jørgensen & Müller, 2000; Barnsley, 2007). Approaches to ecosystems modelling are theoretically based on thermodynamics (Jørgensen, 1981; Anderson, 2005), on systems theory (Odum, 1983), on information theory (Ulanowicz, 1986), on topology (Casti & Karlquist, 1986), or on economics (Costanza, 1991; Costanza, Cumberland, Daly, Goodland, & Norgaard, 1998). Practically, the approaches are focused on different types of study. On the one hand, empirical studies collect bits of information (Straškraba & Gnauck, 1985). Then, attempts are made to integrate and assemble the studies into a complete picture of an ecosystem. Comparative studies are presented to compare some structural and functional elements for a range of ecosystem types. Experimental studies are carried out where manipulations of an ecosystem are used to identify and elucidate ecological mechanisms. On the other hand, modelling and computer simulation studies are worked out to derive ecosystem management plans and ecotechnological tools for goal-oriented man-made control actions. Information systems and decision support systems studies support industrial, agricultural, and administrative ecological decisions as well as medium- and long-term development plans for ecosystem management.

7.2 Ecosystems as Stochastic Transfer Systems

Ecosystems can be seen as stochastic transfer systems (Pahl-Wostl, 1995). They are characterised by measurable inputs, immeasurable (stochastic) disturbances, as well as by measurement errors (Jørgensen, 1994). The inputs $\boldsymbol{x}(t)$ are transformed into outputs $\boldsymbol{y}(t)$ by an operator which describes the transient behaviour of ecological processes $\boldsymbol{y}(t) = \boldsymbol{F} \cdot \boldsymbol{x}(t)$ (Fig. 7.1). During transition processes all inputs will be smoothed. There exists some redundancy between inputs and outputs.

Nonlinear dynamic processes within ecosystems are initiated by switching processes of input and state variables with different transfer time constants. They are overlaid and produce ecosystem responses (or output signals). But it is not possible to determine which part of ecosystems response and intensity is caused by the different ecological components and/or matter transfer processes. In Fig. 7.2 an example is given for a freshwater ecosystem with a certain algal species composition. After 5 days the external perturbation (BOD) has been diminished in form of a negative jump nearly to zero. The resulting amplitudes of pH and DO as well as Do net production are much smaller than before. Ecosystem responses return to former values after increase of BOD load but with different transfer times.

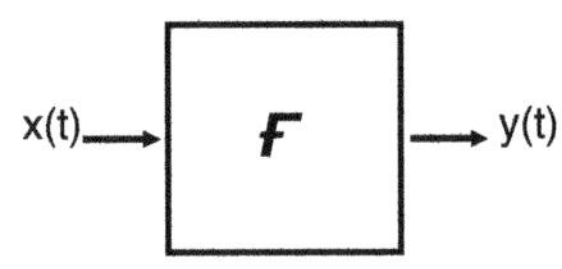

Fig. 7.1 Schematic representation of a transfer system

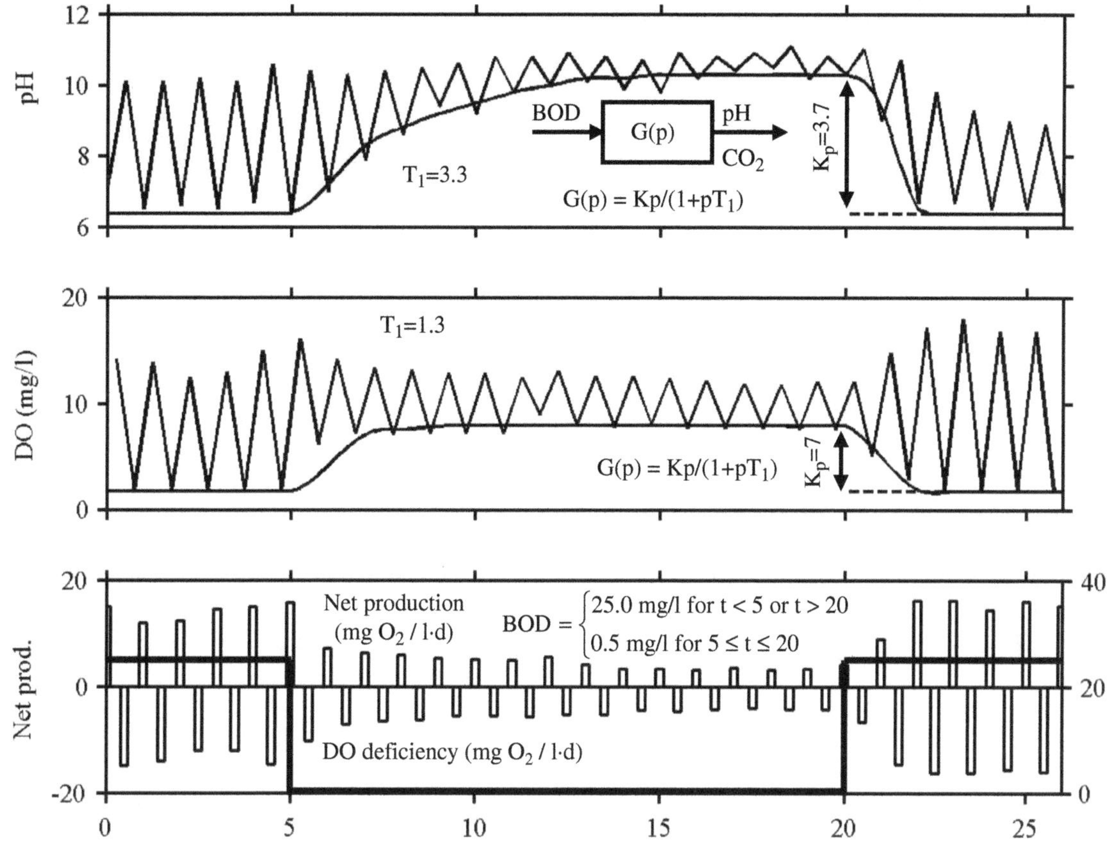

Fig. 7.2 Switching processes within an ecosystem influencing ecosystems responses

Linear and nonlinear transfer functions are represented by

1. *Pulse function* $x(t) = 0$ for $t < 0$ and $t > T, x(t) = x_0$ for $0 = t = T$,
2. *Jump function* $x(t) = x_0\sigma(t)$ with $\sigma(t) = 0$ for $t < 0$ and $\sigma(t) = 1$ for $t = 0$,
3. *Harmonic function* $x(t) = x_0 + \cos(\omega t + \varphi)$ for $-\infty < t < +\infty$ or $x(t) = x_0 e^{j(\omega t + \varphi)} = x_0^+ e^{j\omega t}$ with $x_0^+ = x_0 e^{j\varphi}$
4. *White noise function.*

Classical transfer functions known from control theory are given by

1. *Exponential function* $x(t) = x_0 e^{-t/T}$ for $0 = t < +\infty$ or $x(t) = x_0^+ e^{j\omega t} e^{\delta t}$ for $0 = t < +\infty$ and $\delta \neq 0$,
2. *Periodic function* $x(t) = a_0/2 + \Sigma_i a_i \cos(i\omega_0 t) + \Sigma_i b_i \sin(i\omega_0 t)$ or $x(t) = \Sigma_i c_i e^{j(j\omega_0 t)}$,
3. *Dirac impulse* (delta distribution) $x(t) = 0$ for $t < 0$ and $t > T, x(t) = \delta(t)$ with $\delta(t) = 0$ for $t \neq 0$ and $\int \delta(t)\mathrm{d}t = 1$,
4. *Ramp function* $x(t) = 0$ for $t < 0$ and $x(t) = at$ for $t = 0$
5. *Time discrete signal function* $x^{\sim}(t) = \sum_k x(\mathrm{kT})\,\delta(t - \mathrm{k}T)$ with $\mathrm{k} = 0, 1, 2, \ldots$ and $T = 1/(2\mathrm{f}_{\max})$ where $\mathrm{f}_{\max}$ is the maximum frequency contained in a signal.

7.3 Approaches to Ecosystem Modelling

Modelling of ecosystems is a complicated and complex task. Current modelling approaches are characterised by dynamic modelling and simulation procedures (Jørgensen, 1992), by methods of artificial intelligence (Chen, 2004) like neural networks (Recknagel,

2003) or knowledge based systems (Nielsen, 2000), by using genetic algorithms (Haefner, 1996), or by other approaches (Gaylord & Nishidate, 1996; Shields, 1997). All these approaches result in dynamic ecosystem models which are used to describe the time-varying behaviour of ecosystem outputs, to simulate the influence of changing environmental conditions on ecological processes, to produce forecasts of ecosystem developments, and to form an objectified base for ecosystem management and environmental decision making. Classifications of ecological models are often represented by state space characteristics (discrete or continuous), by the type of models used (linear or nonlinear), by the type of time behaviour of models (stationary or instationary), or by the type of parameters (lumped or distributed). Nonlinear feedbacks within an ecosystem cause changes of characteristics of signals and systems states during signal transfer processes by modulation of amplitudes, frequencies and phases, and/or by quantification (discretisation of time domain of amplitudes or sampling frequencies of signals) (Jørgensen, 1986). Another classification is given by the type of systems adaptability and stability (Straškraba & Gnauck, 1985). Looking on ecosystems as controllable transfer systems (Gnauck & Straškraba, 1980) notions of automata theory are of interest. Cellular automata procedures led to numerous models for environmental planning (Batty, Xie, & Sun, 1999; Bandini & Worsch, 2001), for urban planning (Liu & Phinn, 2003), for ecosystem description (Gronewold & Sonnenschein, 1998; Sonnenschein & Vogel, 2001), and for environmental management (White & Engelen, 1993; Thinh, 2003).

Mathematical approaches to ecosystem modelling may be distinguished by the type of mathematical generalisation (Gnauck, 2000). Set theory is well elaborated and widespread used. An ecosystem ***ES*** is defined by some weakly structured sets A, B, C, ..., and by some functions f, g, h, ... where A, B, C, ... are sets of signals, sets of biotic components, sets of states , sets of matter, energy and information storages, or others. The couplings between the elements of these sets are represented by linear or nonlinear functions. The system behaviour is described by mapping these sets (or their cross-products) into a set $\boldsymbol{ES} = (\mathrm{A, B, C}, \ldots, \mathrm{f, g, h}, \ldots)$. Another type of generalisation is given by mathematical algebra. An environmental system ***ES*** is represented by some general algebras A, B, C, ... and by morphisms $\phi, \gamma, \psi, \ldots$ which carry some algebras to other algebras without changes in their structural properties. The description is given by $ES = \{\mathrm{A, B, C}, \ldots, \phi, \gamma, \psi, \ldots\}$. This approach takes into account that the defining sets of an environmental system are structured by couplings of the elements between different system compartments. The third type of generalisation deals with mathematical theory of categories (Gnauck & Straškraba, 1980). An ecosystem ***ES*** is defined as a category ***C*** with sets as objects. Morphisms of ***C*** are functions with structure keeping behaviour. The description is given by $C = \langle \mathrm{S}_1, \mathrm{S}_2, \ldots, \mathrm{mor}(\mathrm{S}_1,\mathrm{S}_2), \mathrm{mor}(\mathrm{S}_1,\mathrm{S}_3), \ldots \rangle$, where S_i are subsystems with some common properties and $\mathrm{mor}(\mathrm{S}_i, \mathrm{S}_j), \ldots$ are sets of morphisms. Subsystems of ***C*** are also categories which are imbedded in ***C***. The interrelations between categories are realised by functors Φ. This type of generalisation forms a base for a linguistic (semiotic) analysis of environmental states. Interpreting ecosystems as discrete stochastic systems Markov chains and stochastic matrices will be helpful modelling tools. The transfers between states are characterised by jumps where $T = \{T_{ij}\}$ is a stochastic transfer matrix with $T_{ij} = \mathrm{Prob}\left(u_i, t_2 | u_j, t_1\right), t_2 \geq t_1, T_{ij} \geq 0$ and $\Sigma T_{ij} = 1$. Ecosystem states and transfer processes may be expressed by oriented graphs. Classical population models and modern workflow models are based on this approach.

Modelling the structure and functioning of ecosystems is characterised by representations and interpretations of relationships combining states and transfers. All modelling approaches have common roots but their applicability is quite different. The state space approach allows dynamic simulations of ecosystem changes due to changing input and environmental parameters. The relational ecology approach covers simulations as well as proofs of dynamic characteristics of ecosystems. The network approach results in simulations by modern workflow analysis. It allows interpretations of ecosystems as switching networks where binary relations play an important role. Modern signal analysis methods and their use for ecosystem management are included.

7.3.1 State Space Approach

Applying methods of systems theory to solve environmental problems two main tasks arise: Quantification of ecological processes and generalisation of simulation results to obtain a framework for ecosystem management. Bossel (1992) distinguished three basic types of systems which are important for ecosystem modelling and simulation: The general input–output system, the general state system, and the abstract automaton. Another classification given by five classes of dynamic systems can be defined (Table 7.1).

The application of models in ecology is almost obligatory for understanding the interrelationships between the structure and the functioning of complex systems as aquatic ecosystems. Mathematical models are useful instruments in the survey of complex systems. It is not possible to survey the many components and their reactions in an ecological system without the use of a mathematical model as synthesis tool.

In practice, modelling is started with a conceptual model of the system of interest (Thomann & Mueller, 1987). State variables are symbolised by boxes, while interrelationships are expressed by arrows. In general, the conceptual model may be interpreted as a directed graph showing these essential relationships. Figure 7.3 shows a model concept to simulate eutrophication processes in shallow water bodies. Model state variables are given by ecological state variables phytoplankton, zooplankton, orthophosphate phosphorus, ammonia nitrogen, nitrite nitrogen, and nitrate nitrogen as well as phosphorus remobilisation from sediment, dissolved oxygen, and biochemical oxygen demand. Q_{in} and Q_{out} denote discharges into and out of a river or lake segment. External driving forces are photoperiod (FOTOP), solar radiation (I), and water temperature (TEMP). To quantify the relationships some differential equations, site constants, and model-specific parameters have to be specified.

Table 7.1 State space classification of models

Type of system	State space characteristics
Space–time systems	Continuous, discrete
Automata	Discrete
Infinite-dimensional systems	Continuous, discrete
Linear time-dependent systems	Continuous, discrete
Nonlinear differential systems	Continuous

The relationships are quantified as follows:

Phytoplankton biomass – A (mg CHA/l)

$$dA/dt = Q/V \cdot (A_{\text{in}} - A) + G - UA \cdot A \cdot f$$
$$-FRZ \cdot Z \cdot A \cdot CR - RESP \cdot TEMP \cdot A$$

P-remobilisation from sediment – Psed (mg P/l)

$$dPsed/dt = \text{phi} \cdot \text{hs} \cdot (-Dsp/(1 - \log(\text{phi}^2))$$
$$\cdot(P - (\text{Psed}/(\text{hs} \cdot \text{phi})))/(\text{hs}/2)$$
$$+\Theta(cpcrit - cpEA)/cpcrit \cdot KFe$$
$$\cdot cp \cdot qp)$$

with $\Theta = 1$ if $cpEA \leq cpcrit$ and $\Theta = 0$ otherwise
Phosphate phosphorus – P (mg P/l)

$$dPO_4/dt = Q/V \cdot (PO_{4\text{in}} - PO_4) + FRZ \cdot A \cdot Z$$
$$\cdot/CR \cdot ((1 - AZP) \cdot KSA/(KSA + A))$$
$$+RESP \cdot TEMP \cdot A - G + (1/4)$$
$$\cdot dPsed/dt$$

Ammonia nitrogen – $NH_4 - N$(mg N/l)

$$dNH_4/dt = Q/V \cdot (NH_{4\text{in}} - NH_4) + B_3 \cdot NORG_{\text{in}}$$
$$-B_1 \cdot NH_4 - FA1 \cdot FUP \cdot G$$

Nitrite nitrogen NO_2-N (mg N/l)

$$dNO_2/dt = Q/V \cdot (NO_{2\text{in}} - NO_2) + B_1 \cdot NH_4$$
$$-B_2 \cdot NO_2$$

Nitrate nitrogen NO_3-N (mg N/l)

$$dNO_3/dt = Q/V \cdot (NO_{3\text{in}} - NO_3) - (1 - FUP)$$
$$\cdot FA1 \cdot G + B_2 \cdot NO_2$$

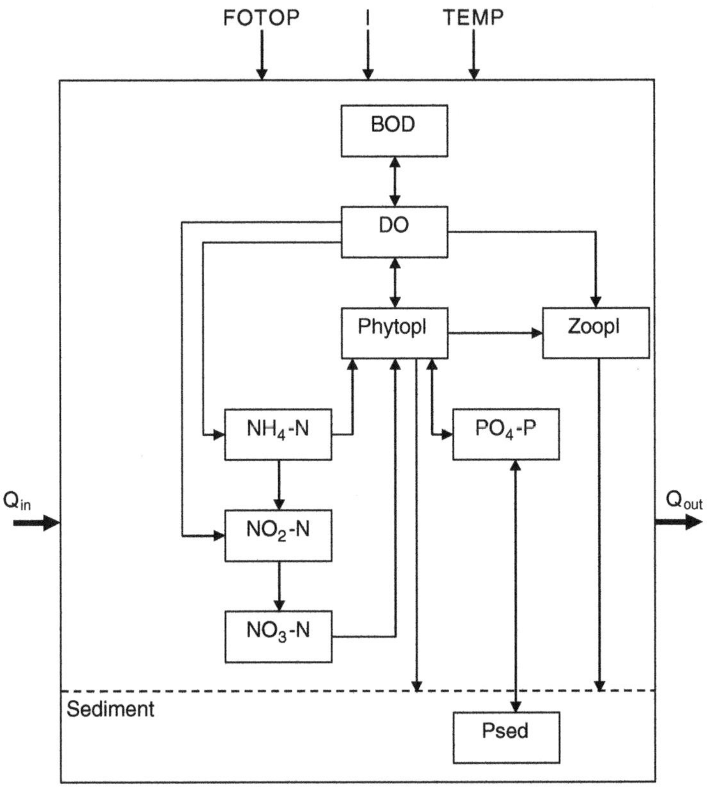

Fig. 7.3 Conceptual model of an eutrophication simulator

Zooplankton – Z (mg C/l)

$$dZ/dt = Q/V \cdot (Z_{in} - Z) + FRZ \cdot A \cdot Z \cdot CR$$
$$\cdot AZP \cdot C \cdot KSA/(KSA + A)$$
$$-MORT \cdot TEMP \cdot Z$$

Dissolved oxygen – DO (mg/l)

$$dDO/dt = Q/V \cdot (DO_{in} - DO) + K_2 \cdot (DO_{sat}$$
$$-DO) + (a_3 \cdot G/A - a_4 \cdot RESP \cdot TEMP)$$
$$\cdot A - K_1 \cdot BOD - K_4/zmix - a_5$$
$$\cdot B_1 \cdot NH_4 - a_6 \cdot B_2 \cdot NO_2 - a_7$$
$$\cdot MORT.TEMP \cdot Z$$

Biochemical oxygen demand – BOD (mg/l)

$$dBOD/dt = Q/V \cdot (BOD_{in} - BOD) + K_1$$
$$\cdot BOD - K_3 \cdot BOD$$

Temperature dependencies of physical water quality variables are modelled by sinusoidal functions according to Straškraba and Gnauck (1985) or Jørgensen (1986). DO saturation concentration is expressed by a third-order polynomial (Thomann, 1972). Other functional dependencies will be represented by regression type functions as well.

Results of this modelling approach are simulation runs. In Fig. 7.4 an example is presented of nitrate nitrogen variations at different locations along a chain of shallow lakes where the code 430 represents the measured input concentration and the codes Hv0110, Hv0120, and Hv0190 represent notation.

7.3.2 Relational Ecology Approach

The goal of application of algebraic methods to ecosystem modelling consists of a general mathematical description of the structural and functional components of an ecosystem. The interpretation of an ecosystem (an abstract automaton ***A***) as an algebraic structure ***S*** leads to an algebraic automata description and consequently to general formulations by category theory.

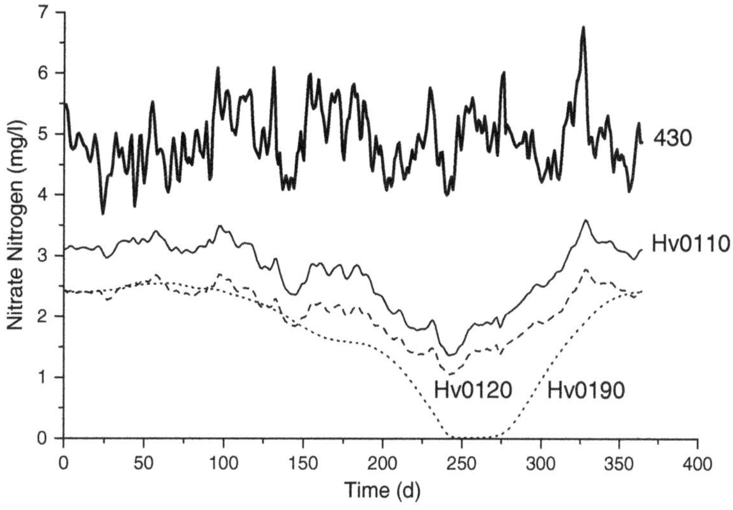

Fig. 7.4 Simulation runs of nitrate nitrogen concentration in shallow lakes

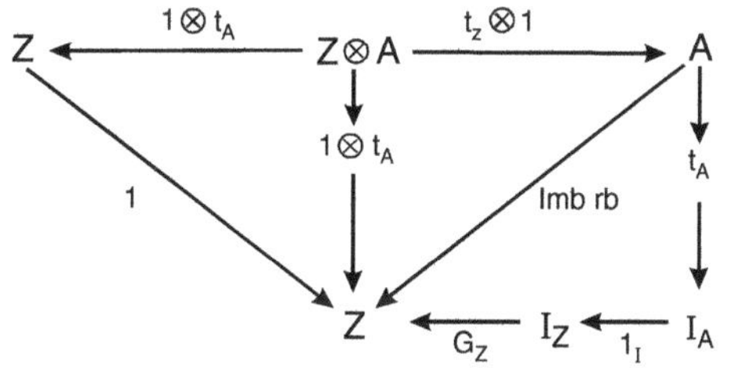

Fig. 7.5 Commutative diagram of a predator-prey relation

For this reason, it is necessary to use categories with multiplication. Representations and interpretations of relations will be proofed by means of commutative diagrams (see Fig. 7.5).

Between categories and stochastic automata exist the following relations: $M = \langle M, \otimes \rangle$ is a multiplicative category. A (Mealy-) automaton on $\boldsymbol{M}$ is given by a quintuple $\mathbf{M} = [X, Y, U, F, G]$ with $X, Y, U \in \text{ob}\, M, F, G \in \text{mor}\, \mathbf{M}$ with $G : U \otimes X \rightarrow Y$ (output morphism) and $F : U \otimes X \rightarrow U$ (state transition morphism). Ob $\boldsymbol{M}$ characterises the set of all objects of the category while mor $\boldsymbol{M}$ denotes a morphism belonging to the category $\boldsymbol{M}$. Because of a series of morphisms correspond to a multiplication of morphisms all these automata form a ***M-Aut*** category. Considering the stochastic relation $\forall u \in U, \forall x \in X$ and $U' \subseteq U, Y' \subseteq Y$: $\mathrm{H}[u,x](Y' \times U') = G[u,x](Y') \cdot F[u,x](U')$, then the category ***M(S)*** is performed with stochastic models as morphisms.

Then, an ecosystem system can be represented by a category ***C-Mat*** which results from ***C-Aut*** categories by means of an inclusion functor $Inc(C)$: **C-Aut** $\rightarrow$ **C-Mat**. The imbedding of subcategories into the category ***C-Mat*** will be realised by means of an imbedding functor *Imb(**C**)*. In ***C-Mat*** exists special morphisms like exchange morphism, diagonal morphism, terminal morphism, canonical projection (product) morphism.

On this way it is possible to model an ecosystem, trophic levels, or single elements by subcategories which are imbedded into other categories. Single elements are called elementary automata; their morphisms represent elementary switching circuits. Feedbacks are given by loop morphisms if the sets of objects may be connectable. Combining switching operations of elementary automata with internal and external couplings structural changes of an environmental system can described algebraically where parallel and serial couplings and combinations with other automata operations are included.

For ecological modelling the following semiotic expressions can be used, where U denotes a set of ecological states $u \in \{\mathrm{U}\}$ belonging to ob ***C-Mat***. *l(w)* denotes the length of a word, and **N** is the set of natural numbers:

1. *Growth of a state u:*

$$grow(u) : U \rightarrow U, U \in \text{ob}\, \boldsymbol{C\!-\!Mat}, \text{ where}$$

$$\forall u \in U, \forall w(u) \in W(U) : \exists t \in N \text{ with}$$

$$u \bullet w(u) \rightarrow \infty \text{ and } l(w(u)) = t > 0.$$

2. *Mortality of a state u:*

$$mort(u) : U \to U, U \in ob\ \mathbf{C\text{–}Mat}, \text{ where}$$
$$\forall u \in U, \forall w(u) \in W(U) : \exists t \in N \text{ with } u$$
$$\bullet w(u) = 0 \text{ and } l(w(u)) = t > 0;$$

3. *Lifetime of a state u:*

$$life(u) : life(u) = \min(t_i, t_j), \text{ where } U \in ob$$
$$\mathbf{C\text{−}Mat} \text{ and } \forall u \in U, \forall w(u) \in W(U) :$$
$$\exists t_i, t_j \in N \text{ with } u$$
$$\bullet w(u) > 0 \text{ and } l(w(u)) = \min(t_i, t_j) > 0$$

4. *Predator–prey relation between objects of C* (Fig. 7.5)

$$rb = (1_z \otimes t_A) : Z \otimes A \to Z \text{ with } F_z = (1_I \otimes t_A) :$$
$$I_A \otimes A \to I_z, G_z : I_z \to Z, A, A^* \in ob\ \mathbf{C_{ijk}\text{−}Mat},$$
$$Z \in ob\, \mathbf{C_{ij}\text{−}Mat},\ Imbrb\text{: } \mathbf{C_{ijk}\text{−}Mat}$$
$$\to \mathbf{C_{ij}\text{−}Mat}$$

5. *Mortality of an ecosystem element $\boldsymbol{E}_i$:*

$$morte\boldsymbol{E}_i = actlife\boldsymbol{E}_i \le t \in \mathbf{N}, \text{ where } \forall u \in U,$$
$$\forall w(u) \in W(U) : u \bullet w(u) = 0 \text{ and } l(w(u))$$
$$= \text{t} > 0.$$

Because each mortal state of $\boldsymbol{E}$ has a finite lifetime, the following is valid:

$$0 = morte_0\, \boldsymbol{E}_i \subseteq morte_1 \boldsymbol{E}_i \subseteq \ldots \subseteq morte_t$$
$$\boldsymbol{E}_i = morte\ \boldsymbol{E}_i.$$

6. *Actual lifetime of an ecosystem element $\boldsymbol{E}_i$:*

$$actlife\ \boldsymbol{E}_i = actlife(\boldsymbol{E}(u)) \text{ with } U \in ob\ \mathbf{C\text{−}Mat}$$
$$\text{and } \forall u \in U, \forall w(u) \in W(U)$$

7. *Stability of a compartment $\boldsymbol{E}_i$*

$$stab\boldsymbol{E}_{\text{i}} = \max(actlife(\text{u} - \text{uy})) = \text{n - 1 with n} \in \mathbf{N},$$
$$\forall u \in U, \forall y \in Y, U, Y \in ob\ \mathbf{C\text{−}Mat}.$$

8. *Process of accumulation in E* = [X, Y, U, F]:

$$acc\mathbf{E} : U' \otimes X \to Y \otimes U \in mor\ \mathbf{C\text{−}Mat}.$$

9. *Inter-specific concurrency of objects* A_i, $Nu \in ob$ **C-Mat**:

$$con\text{: } A_1 \otimes Nu \otimes A_2 \to A_1 \otimes A_2 \text{ with}$$
$$U = \{A_1, A_2, Nu\}, p = 1_{A1} \otimes t_{\text{Nu}}$$
$$\otimes\ 1_{A2} \text{ is an epimorphism, } p_1 - 1_{A1} \otimes t_{A2},$$
$$p_2 = t_{A1} \otimes 1_{A2}.$$

7.3.3 Network Approach

Ecosystems are considered as switching networks (Higashi & Burns, 1991) or information systems. This point of view allows not only a generalised mathematical description of ecosystems but also evaluations of signal transfers, statements on ecosystems stability, and detection of bifurcation points. Furthermore, a reliability analysis can be given based on Boolean variables of synchronous (natural) and asynchronous (man-controlled) system behaviour. Hazards of signals arise by time delays between internal sources and sinks of information.

Processes and changes of structural components are characterised by different time parameters like time delay τ_i, threshold values σ_j, altering of elements ϑ_k, physiological parameters φ_1, and others. Because of random changes of state variables and of fluctuations of environmental driving forces switching processes take place at different time strokes within intervals $(a_i(t), b_i(t))$ with probability densities $w_i(t)$ of time delays of ecosystem variables and probabilities $p_i(t)$ for each realisation of a state transition $p_i(t) = \int w_i(t)$ where $a_i(t)$ and $b_i(t)$ are the lower and the upper

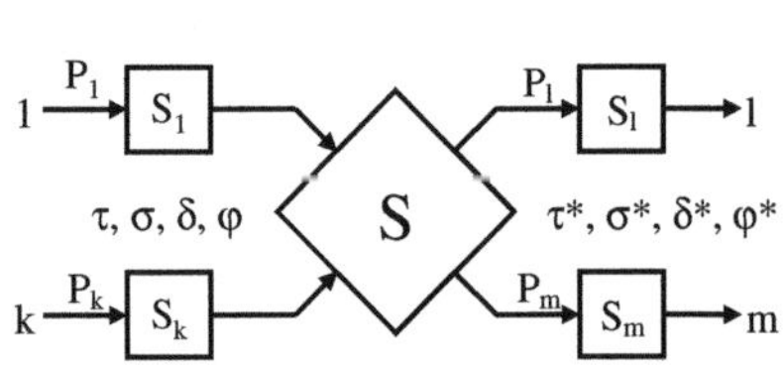

Fig. 7.6 Scheme of an ecosystem switching element

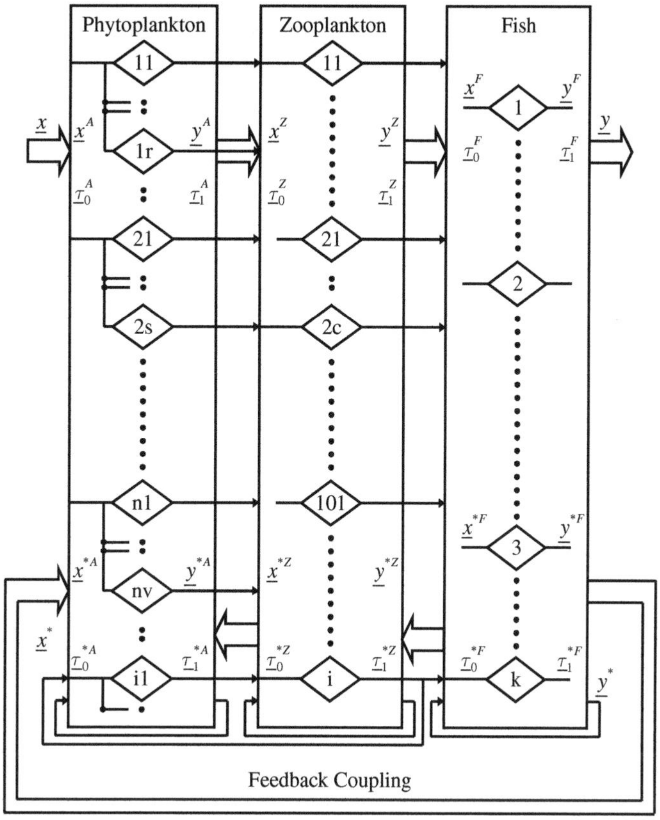

Fig. 7.7 A three-layer pelagic ecosystem as switching network

integration limits. Figure 7.6 shows such a switching element.

State transitions will be observed after certain time intervals only. Then, a state transition is characterised by a function $\Theta_i(t) = \{a_i(t), b_i(t), w_i(t), p_i(t)\}$. For $p_i(t) = 1$ a state transfer takes place absolutely (positive transfer). If $p_i(t) < 1$ (uncertain transfer), then there exists a probability $q_i(t) = 1 - p_i(t)$ that a transfer does not take place. Input and output of each system element is connected with a control element S on which the control parameter act. For any time events signal connections take place and cause changes of ecosystem states and responses to environmental perturbations. They can be observed in nature and denoted as positive ($0 \rightarrow 1$) or negative ($1 \rightarrow 0$) transfers. For a simple three-layer aquatic ecosystem a switching network is represented in Fig. 7.7. The trophic layers are interconnected and characterised by numerous feedbacks. The systems hierarchy is naturally given by the food chain. For each element the switch is visualised by a rhomb.

7.4 Ecosystem Models and Long-Term Research

Mankind has always used models. They are simplified pictures of reality or real world. Mathematical models are helpful informatic tools to solve research and management problems. Urbanisation and technological development have increasing environmental impacts. Energy, chemicals, and other pollutants are released into ecosystems, where they may cause more rapid growth of algae or bacteria, damage species, or alter the entire ecological structure. The quality of statements on ecosystem development depends on the flexibility of mathematical simulation models. Restricted information structures and aggregations of information lead to model errors.

The systems approach and ecosystems understanding are connected with the amount of data of ecological processes and compartments and their availability. Data sets (or time series) serve as information base for parameter estimations and for evaluation of mathematical models by verification and validation. Four cases may be distinguished (Grant, Pedersen, & Marín, 1997):

1. Many data and little understanding: Statistics have to be used for ecosystems modelling.
2. Many data and good understanding: Natural sciences (physics, mathematics, chemistry, biology) have to be used for modelling.
3. Few data and little understanding: Systems analysis methods and simulation procedures have to be used for ecosystems modelling.
4. Few data and good understanding: Systems analysis methods and simulation procedures have to be used for ecosystems modelling.

A model will never contain all the features of the real world. Then it would be the real system itself. A model contains the characteristic features, which are essential in the context of the problem to be solved or described. Therefore, an ecological model must contain the features which are of interest for ecological management or theoretical ecology. The reactions of the system might be not necessarily the sum of all individual reactions, but mathematical models can be used to reveal dynamic system properties. They reveal gaps in knowledge on ecological systems and can therefore be used to set up research priorities.

7.5 Conclusions

Continuous or discrete state space modelling is a well-known and widespread approach used for ecosystem management. But it expresses only the functioning of a system on the base of known physical, chemical, and biological behaviour. The subjective selection of state variables depends strongly on the basic knowledge on state variables and on the set of available data. Model parameter values are mostly unknown. They will be quantified by the assumption of stationary processes or special conditions. Environmental management options are derived on the base of a scenario analysis.

The qualitative relational ecology approach forms a theoretical base for ecosystems modelling. It combines structural elements and the systems functioning but without quantification. This approach is a connecting link to mathematical linguistics and graph theory. It describes the genesis of ecological states by chains of signs and the relations between structural elements. It is suited for theoretical environmental planning and decision making.

The network approach combines aspects of system functioning and structure. In opposite of the relational ecology approach, the relations will be quantified by signal transfers and the changes of ecological states are described by informational processes which occur in complex systems. The advantage of this approach is twofold. First, the hierarchic structure of an ecosystem and the couplings of subsystems are represented by the net itself. Second, in opposite of the state space approach, the cause–effect relation is replaced by a condition–event relation which allow the modelling of informational processes within a system.

References

Anderson, G. M. (2005). *Thermodynamics of natural systems*. Cambridge: Cambridge University Press.

Bandini, S., & Worsch, T. (Eds.). (2001). *Theoretical and practical issues on cellular automata*. Berlin: Springer.

Barnsley, M. (2007). *Environmental modeling*. Boca Raton: CRC Press.

Batty, M., Xie, Y., & Sun, Z. (1999). Modeling urban dynamics through GIS-based cellular automata. *Computers, Environment and Urban Systems, 23*, 205–233.

Bossel, H. (1992). *Modelling and simulation*. Braunschweig, Wiesbaden: Vieweg.

Casti, J. L., & Karlquist, A. (Eds.). (1986). *Complexity, language and life: Mathematical approaches*. Berlin: Springer.

Chen, Q. (2004). *Cellular automata and artificial intelligence in ecohydraulics modelling*. Delft: Balkema Publishers.

Costanza, R. (Ed.). (1991). *Ecological economics: The science and management of sustainability*. New York: Columbia University Press.

Costanza, R., Cumberland, J., Daly, H., Goodland, R., & Norgaard, R. (1998). *An introduction to ecological economics*. Boca Raton: Lewis.

Gaylord, R. J., & Nishidate, K. (1996). *Modeling nature*. New York: Springer.

Gnauck, A. (2000). Fundamentals of ecosystem theory from general systems analysis. In Jørgensen, S. E. & Müller, F. (Eds.), *Handbook of ecosystem theories and management*. (pp. 75–88). Boca Raton: Lewis.

Gnauck, A., & Straškraba, M. (1980). Theoretical concepts of ecosystem models. *ISEM Journal, 2*, 71–80.

Grant, W. E., Pedersen, E. K., & Marín, S. L. (1997). *Ecology and natural resource management*. New York: Wiley.

Gronewold, A., & Sonnenschein, M. (1998). Event-based modelling of ecological systems with asynchronous cellular automata. *Ecological Modelling, 108*, 37–52.

Haefner, J. W. (1996). *Modeling biological systems*. New York: Chapman & Hall.

Higashi, M., & Burns, T. P. (Eds.). (1991). *Theoretical studies of ecosystems: The network perspective*. Cambridge: Cambridge University Press.

Jørgensen, S. E. (1981). Application of exergy in ecological models. In D. Dubois (Ed.), *Progress in ecological modelling* (pp. 39–47). Liege: Cebedoc.

Jørgensen, S. E. (1986). *Fundamentals of ecological modelling*. Amsterdam: Elsevier.

Jørgensen, S. E. (1992). *Integration of ecosystem theories: A pattern*. Dordrecht: Kluwer.

Jørgensen, S. E. (1994). *Fundamentals of ecological modelling*. Amsterdam: Elsevier.

Jørgensen, S. E., & Müller, F. (Eds.). (2000). *Handbook of ecosystem theories and management*. Boca Raton: Lewis.

Liu, Y., & Phinn, S. R. (2003). Modelling urban development with cellular automata incorporating fuzzy-set approaches. *Computers, Environment and Urban Systems, 27*, 637–658.

Nielsen, S. N. (2000). Ecosystems as information systems. In S. E. Jørgensen, & F. Müller (Eds.), *Handbook of ecosystem theories and management* (pp. 217–247). Boca Raton: Lewis.

Odum, H. T. (1983). *Systems ecology*. New York: Wiley-Interscience.

Pahl-Wostl, C. (1995). *The dynamic nature of ecosystems*. Chichester: Wiley.

Recknagel, F. (2003). *Ecological informatics*. Berlin: Springer.

Shields, M. W. (1997). *Semantics of parallelism*. London: Springer.

Sonnenschein, M., & Vogel, U. (2001). Asymmetric cellular automata for the modelling of ecological systems. In I. M. Hilty, & P. W. Gilgen (Eds.), *Sustainability in the information society. Part 2* (pp. 631–636). Marburg: Metropolis.

Straškraba, M., & Gnauck, A. (1985). *Freshwater ecosystems – Modelling and simulation*. Stuttgart, Jena: Fischer.

Thinh, N. X. (2003). Contemporary spatial analysis and simulation of the settlement development of the Dresden city region. In A. Gnauck, & R. Heinrich (Eds.), *The information society and enlargement of the European Union. Part 1* (pp. 253–261). Marburg: Metropolis.

Thomann, R. V. (1972). *Systems analysis and water quality management*. New York: McGraw-Hill.

Thomann, R. V., & Mueller, J. A. (1987). *Principles of surface water quality modelling and control*. New York: Harper & Row.

Ulanowicz, R. E. (1986). *Growth and development*. New York: Springer.

White, R., & Engelen, G. (1993). Cellular automata and fractal urban form: A cellular automata approach to the evolution of urban land-use patterns. *Environment and Planning A, 25*, 1175–1199.

Chapter 8

The Role of Statistics for Long-Term Ecological Research

Albrecht Gnauck, Bai-Lian Larry Li, Jean Duclos Alegue Feugo, and Bernhard Luther

Abstract Sustainable management of natural resources requires a good understanding of ecosystems components and their interrelationships. Statistics is essential for understanding the structure and behaviour of ecological processes and provides the basis of predictive modelling. Mostly, physical, chemical, and biological variables are recorded across time and space. They serve as indicators, giving information concerning the state and changes of ecosystems. Most of monitored ecological indicators are non-stationary in time structure. The classical static statistical methods revealed the presence of trends and long memories in these data sets. On the other hand, modern dynamic statistical methods indicate the presence of long-term cycling processes. The Fourier polynomial is a technique for approximating periodic functions by sums of cosine and sine periodic functions, shifted and scaled. Therefore, it may be suitable for approximating cycling processes with a fixed frequency as portrayed by some ecological indicators. Wavelet analysis can be used to investigate the timescale behaviour of ecological processes. This analysis reveals the long-term evolution of an ecological indicator at different resolutions, the dominant scale of variability in the data set, and its correlation and cross-correlation with other ecological indicators on a scale by scale basis.

Keywords Statistical ecology · Statistical modeling · Descriptive statistics · Time series analysis

A. Gnauck (✉)
Department of Ecosystems and Environmental Informatics, Brandenburg University of Technology,
Konrad-Wachsmann-Allee 1, D-03046 Cottbus, Germany
e-mail: albrecht.gnauck@tu-cottbus.de

8.1 Introduction

In the past, ecology has been changed from a descriptive to a more quantitative science using statistical methods. This development was supported by a lot of famous books where univariate and multivariate statistical methods were presented (Bliss, 1970; Patil, Pielou, & Waters, 1971; Pielou, 1977; Eason, Coles, & Gettinby, 1980, Legendre & Legendre, 1983; Ludwig & Reynolds, 1988). Due consideration is given to temporal and spatial changes to the input and state variables as well as to the ecosystem parameters and their initial uncertainties. The use of statistical methods is necessary to understand the basic temporal–spatial changes of physical, chemical, and biological components of an ecosystem, their functions with another, and their interrelationships with the ecosystem environment. From this, the following questions should be answered: Are there relationships between ecosystems and their environments? Are there interrelationships between different ecological processes under consideration? Are there dependencies of one or several ecological variables (or indicators) on one or several others, and how can they be modelled? Are there dependencies between groups of ecosystems or ecological objects? Necessary are those approaches to solve ecosystem management problems which are compatible with the stochastic nature of ecosystem variables or indicators and ecological processes under consideration. Statistical procedures will be the adequate mathematical methods as long as the processes within an ecosystem and/or the ecosystem describing differential equations are unknown.

In order to assess the quality of ecosystem services as well as ecosystem management and environmental

F. Müller et al. (eds.), *Long-Term Ecological Research*,
DOI 10.1007/978-90-481-8782-9_8, © Springer Science+Business Media B.V. 2010

policies, ecological variables and indicators are observed or recorded over time and space by short-term or long-term monitoring programs (Mateu & Montes, 2001; Haining, 2003). These variables give information on the current states, the changes, and trends of ecological processes. The patterns displayed by these variables can be captured and used for forecasting by means of appropriate statistical modelling techniques. Modelling of ecological time series gives more insight to the time varying behaviour and the internal time-dependent correlation structure of an ecosystem compared with classical univariate and multivariate statistics which is helpful in comparing of samples, sites of investigation, intrinsic characteristics of ecosystems and environmental factors, environmental standards, and comparisons of habitats and landscapes. The extraction of process information from ecological signals usually requires statistical modelling by removal of the non-stationary events and then fitting a stationary stochastic model. For this reason advanced time series techniques are useful as trend analysis, ARMA or ARIMA modelling or Fourier analysis, or for long-term data sets by wavelet analysis. The information extracted may also be helpful in estimating the weight of input and disturbance relating to the output variables of the model concerned as well as serve as a practical tool when appropriate initial values for estimation procedures are to be chosen.

8.2 Statistical Approaches for Analysing Ecological Data

Statistical methods are based on theory of probability. From the point of view of systems theory a distinction is made between two groups, depending on whether the variable time is included into ecosystem analysis or not. Static statistical methods (without consideration of time as a variable) cover simple and multiple linear and nonlinear regression and correlation, analysis of variance, cluster, discriminant, principal component, and factor analyses and estimation procedures (techniques of direct estimation such as least squares estimation, Markovian estimation, maximum likelihood estimation, Bayes' estimation). A special position among the estimation techniques is held by the indirect techniques with which the time as a variable is included in evaluation by the sort of estimation procedure used (e.g. cusum technique, trend analysis). Another distinction can be made by recursive and non-recursive techniques (Aström, 1970). Dynamic statistical methods (with consideration of the variable time, also known as time series analysis) covers advanced procedures such as trend analysis, Fourier analysis, time correlation and spectral analysis, weighting and transfer functions, and wavelet analysis. These approaches have also become known under the notion of dynamic statistics. These techniques have usually become known in the literature by the name of time series analysis (Box, Jenkins, & Reinsel, 1994; Brockwell & Davis, 1998). They are standard techniques for statistical evaluation of results obtained also from sufficiently careful linearization of nonlinear systems.

8.2.1 Static Statistics

The question is asked if there is interdependence between two or more variables of an ecosystem. This question can be answered by a regression or correlation analysis depending on the purpose of the problem. The type of relationship between ecological variables or indicators is given by a regression equation, but the intensity of such a relationship is expressed by correlation measure.

8.2.1.1 Descriptive Statistics

Data series of ecosystem components collected by monitoring programs are extremely important for assessing the ecosystem state and developing an understanding of the interrelationships between the ecosystem components for ecosystem management. Data from monitoring programs contain errors that affect the results of statistical analysis and the quality of models developed (Pearson, 2005; Han & Kamber, 2006). This problem is more serious in high resolution data which usually contain missing data, outliers, and other impossible data as a result of sensor failure and external disturbances (Little & Rubin, 1987; Latini & Passerini, 2004). The errors cause not only difficulties in process identification and parameter estimation but also misinterpretations of spatial and temporal variations of ecological processes. Some classical techniques are aid in doing some preliminary checks and adjustments on ecological data series. Figure 8.1 shows the

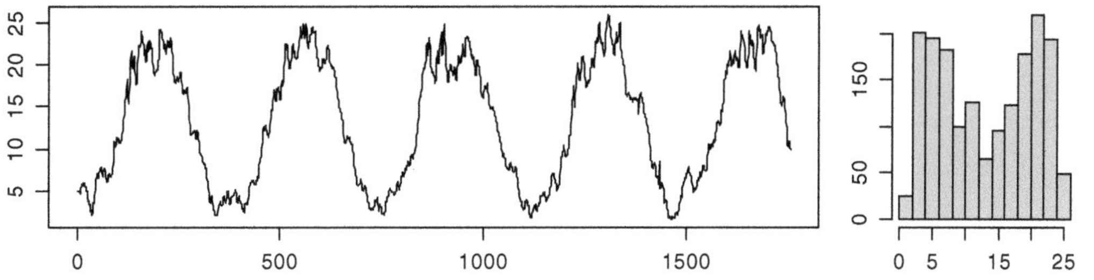

Fig. 8.1 Record (*left*) and probability density distribution (*right*) of water temperature data

basic statistical approach for ecological data analysis (Young & Young, 1992). The left-hand part contains the record of raw data (measurements of water temperature from a shallow lake) based on a regular time grid. Outliers cannot be detected from such a record. In the right-hand part, the probability density distribution of these measurements is presented. It shows a two-modal density probability distribution.

For classical statistical procedures such as calculations of averages, variances, and correlations as well as computation of regression functions or other multivariate statistics all outliers have to be removed from the data record. The data set has to be homogenous.

8.2.1.2 Regression and Correlation

A regression analysis is required for problems in which stochastic dependencies (stochastic cause–effect relations) are to be described by functions, with one or several variables being specified as independent (Ludwig & Reynolds, 1988). Simple regression with one independent variable x and one dependent variable y may be linear or nonlinear. Nonlinear regressions are usually obtained by linearization (transformation of variables) and short-cut methods of computation. This approach can have negative effects on the quality of approximation. Estimation techniques give unbiased estimates. When none of the variables can be considered as dependent, methods as Bartlett's best fit line or geometric mean are useful for statistical description. The method of geometric means is based on averaging the slopes of linear regressions $y = f(x)$ and $x = f(y)$, whereas Bartlett's best fit line is founded on subsample estimates.

Multiple regression is characterised by one dependent, but several independent variables. It is usually linear because the possibilities of linearization are limited (Chatfield & Collins, 1989; Krzanowski, 1990). There are many situations in which polynomials of nth order are used to describe ecological processes or ecosystems (polynomial regression) by the equation $y = a_0 + \Sigma a_i x^i$ for $i = 1, \ldots, n$. The highest occurring power n depicts the order of the polynomial. Whenever polynomials of higher order are used, there is a question of interpretability of variables in the ith power (see Table 8.1). There are meaningful physical interpretations of certain variables with powers up to the second-order maximum. In most cases, however, polynomials merely describe a relationship. For example, dissolved oxygen saturation $= f(T_W)$ can be depicted as polynomial of the third order (Thomann & Mueller, 1987).

Substitution of the variables x_i by sinus and cosinus functions will result in an equation of the form

Table 8.1 Interpretation of polynomials and exponential regression

Type of polynomial/ regression	Equation	Interpretation
Linear	$y(t) = a_0(t) + a_1(t)x(t)$	(a_0) – mean initial value, (a_1) – mean rate of change)
Squared	$y(t) = a_0(t) + a_1(t)x(t) + a_2(t)x^2(t)$	(a_0) – mean initial value, (a_1) – mean rate of change, (a_2) – mean process acceleration)
nth order	$y(t) = a_0(t) + a_1(t)x(t) + a_2(t)x^2(t) + \ldots.. + an(t)x^n(t)$	Interpretation of parameters is mostly impossible
Exponential	$y(t) = y(0)e^{-kt} + E$	*y(0)* – initial concentration value, k – rate of change, E – random quota (according to first-order kinetics)

$y = a + b_1 \sin x + b_2 \cos x$. This equation represents the simplest form of periodic regression or so-called Fourier polynomial. In an extended form this method is called Fourier analysis. This kind of regression is often used to determine a periodic trend, when time series of ecological indicators are analysed.

A correlation analysis is used to examine the tightness of correlations between two or more stochastic variable and to define the degree of stochastic interdependencies. This degree may be described by correlation coefficients. Bilateral and multilateral interdependencies are characterised by simple as well as partial and multiple correlation coefficients. A partial correlation coefficient is used also in the selection of those variables which influence an ecosystem. If multidimensional normal distribution is assumed, it is a measure of the linear dependence of two random variables, x_j and x_k, with elimination of the influence of all other random variables, $x_1, \ldots, x_{j-1}, x_{j+1}, \ldots, x_{k-1}, x_{k+1}, \ldots, x_n$. The square of correlation coefficient is called performance index $B = r^2$. Its value is a measure for the variance explicable by means of regression.

8.2.1.3 Variance and Covariance Analyses

The variance analysis is a statistical method for qualitative and quantitative studies into the effects of one or several variables on results or measurements. The basic idea is that the total sum of squares of deviations of all single measurements from the total mean can be split into two parts: A within-data series sum of squares and a between-data series sum of squares. In models with fixed effects merely mean values of several random samples are usually compared with one another. However, in models with random effects the factor of influence themselves are treated as random samples from the set of possible occurrences of these factors. This may be the case with data continuously monitored.

A covariance analysis may be used in a quantitative investigation of various degrees of effects of one or several variables on experimental results, with the action of additional random variables (covariables) being compulsorily taken into due consideration. This method actually unifies variance and regression analysis (each related to models with fixed effects).

8.2.1.4 Cluster Analysis and Discriminant Analysis

Cluster analysis is used in ecology for many purposes such as determination of interrelations between trophic state variables (Brezonik & Shannon, 1971), detection of pollution sources in streams (Einax, Zwanziger, & Geiß, 1997), in population ecology (Schulze & Mooney, 1994), or in landscape ecology (Webster & Oliver, 1990). Cluster analysis can be considered as a pattern recognition method (Massart & Kaufman, 1983). This set of methods is also known by the notions numerical taxonomy or automatic classification. It encompasses a family of methods which are useful for finding structures within a set of ecological data monitored. Cluster algorithms are divided into hierarchical and non-hierarchical (partitioning) techniques. Hierarchical clustering can be carried out in an agglomerative or a divisive way. The agglomerative techniques join similar objects into clusters and new objects will be added to clusters already found or to join similar clusters. Divisive techniques start with one cluster comprising all objects. Then, the most inhomogeneous objects will be stripped step by step. They form one or more new clusters with more homogeneous objects on a lower level of linkage. Outputs of hierarchical cluster algorithms are represented by dendrograms. Non-hierarchical methods allow rearrangements of objects. They need initial information on the number of clusters to be obtained. The centroids of each cluster form initial gravity centres where the objects are attached to these centres by Euclidean distance. Then, after computation of the new centroids the objects can be rearranged.

Discriminant analysis may be applied to the separation or classification of ecological objects and their association with two or more collectives (groups, populations) (Green, 1978; Legendre & Legendre, 1983; McLachlan, 1992). Separation is undertaken through analysis of quantitative characteristics and reference to a separating function by means of which a decision is made on classification. Illustrative examples of how useful discriminant analysis is for ecological data are given by Ciecka, Fabian, and Merilatt (1980) for trophic state classifications of lakes and by Reckhow and Chapra (1983) for construction of phosphorus loading criteria and others.

8.2.1.5 Factor Analysis and Principal Component Analysis

Factor analysis is used to examine correlations between random variables for common causes, that are factors, and to reduce these correlations. Emphasis is laid, in this context, on point estimations of parameters. Methods of factor analysis and principal component analysis have been worked out in general ecology and phytocoenology of higher plants for typifying space and time relations of different species and environmental variables. Detailed descriptions and examples are given by Legendre and Legendre (1983), Joliffe (1986), Einax et al. (1997), Lepš and Šmilauer (2003).

8.2.1.6 Estimation Techniques

Generally, it is quite difficult to derive ecological process parameters from a real ecosystem. Therefore, they have to be estimated from observations of various variables and from their sample functions. These may be point or interval estimates depending on whether the parameter proper is searched for or the interval in which the parameter is contained. Such estimates are conveniently obtained by statistical computer packages (cf. BMDP, SPSS, STATGRAPHICS, STATISTICA, and others (Einax et al., 1997)). For the nonlinear estimation of (simple) functions several procedures are known (for example, Gauss–Seidel algorithm, Newton–Raphson algorithm, Marquart procedure, and other techniques). Table 8.2 contains some estimation procedures used for ecosystem parameter estimations.

To obtain an unknown parameter a on the basis of measured data and to estimate it, it is necessary to calculate value g of a defined estimate function G with the latter's distribution depending on a. Value g is used for estimation of a and denoted $\hat{a}$. Function G is defined as estimate function of the unknown parameter a. Estimation techniques based on least squares methods have worked particularly well in ecological contexts since a long time (e.g. Shastry, Fan, & Erickson, 1973; Beck, 1979). The direct techniques (regression, Markovian estimate, Bayes' estimate, maximum likelihood estimate) estimate the parameters 'in one step' from a block of measured data at the input and output signals of a system. When these techniques are used, problems of application may rise up because of a priori information required on the system as a starting model. The indirect techniques are additionally subdivided into recursive and non-recursive procedures. Recursively organised procedures adjust a model of a given structure to the system in a stepwise manner, with any new set of measured data entailing another step of model adjustment. All parameters must be re-estimated throughout the procedure. The criterion by which to measure the quality of adjustment consists

Table 8.2 Parameter estimation procedures (after Straškraba & Gnauck, 1985)

Estimation procedure	Model structure	Numerical solution technique
Direct least squares t	$Ay - Bx = e$	Explicit matrix inversion, Gauss–Jordan technique, explicit pseudo-inversion
Recursive least squares	$Ay - Bx = e$	Orthogonal transformation, Cholesky technique, matrix expansion
Direct generalised least squares	$Ay - Bx = e/D$	Explicit matrix inversion, Gauss–Jordan technique, explicit pseudo-inversion
Recursive generalised least squares	$Ay - Bx = e/D$	Cholesky technique, matrix expansion, orthogonal transformation
Instrumental variable method	$Ay - Bx = e$	Explicit matrix inversion, Gauss–Jordan technique, Cholesky technique, explicit pseudo-inversion, matrix expansion
Maximum likelihood	$Ay - Bx = e/D$ or $Ay - Bx = eC$	Newton–Raphson technique
Stochastic approximation	$Ay - Bx = e$ or $Ay - Bx = eC$	Cholesky technique, matrix expansion, Newton method
Bayes estimation	*No special model structure*	Bayesian method
Prior-knowledge-fitting (PKF)	$Ay - Bx = Rx + V - E_R$	Orthogonal transformation

$\boldsymbol{A}, \boldsymbol{B}$ – z-transform polynomials of nth order (describing the deterministic part of an ecological process), $\boldsymbol{C}, \boldsymbol{D}$ – z-transform polynomials of mth order (describing the stochastic part of an ecological process), $\boldsymbol{E_R}$ – regression error matrix, $\boldsymbol{R}$ – z-transform of the discrete transfer function in the PKF-model, $\boldsymbol{V}$ – z-transform of drift polynomial, $\boldsymbol{e}$- error vector as non-parametrical procedures (e.g. Fourier analysis, correlation and spectral analyses), namely minimisation of a quality function Q by parameter vector $\hat{\boldsymbol{a}}$. The mean quadratic error, $Q = \|e\|^2$, and the simple quadratic error, $Q = {}^1/_2 \Sigma e_i^2$ with $e = (y_i - \tilde{y}_i)$ have been most commonly used as evaluation functions in statistical modelling.

The following two approaches to minimisation of Q have been proposed:

1. $\partial Q / \partial \hat{\boldsymbol{a}} = 0$. The estimation rule of normal regression $\hat{\boldsymbol{a}}_k = \left(\boldsymbol{x}_k^{\mathrm{T}} \cdot \boldsymbol{x}_k\right)^{-1} \boldsymbol{x}_k^{\mathrm{T}} y_k$ is obtained for k observations.
2. $\partial Q / \partial \hat{\boldsymbol{a}} \to 0$. The estimation rule of recursive regression $\hat{\boldsymbol{a}}_k = \hat{\boldsymbol{a}}_{k-1} + \boldsymbol{K}_{k-1}\left(y_k - \boldsymbol{x}_k^T \hat{a}_{k-1}\right)$ is obtained for k observations.

Good approximation and updating of the model to the system are achieved by adequate choice of weighting factor $\boldsymbol{K}$ (gain factor). Exponential weighting has proved to be particularly suitable in this context. The intensity of weighting will substantially depend on the rates of parameter variation and on the extent of disturbances. These two influences must be given different weightings. Hence, optimum weighting will be achievable only as a compromise.

8.2.2 Dynamic Statistics

Two classical approaches namely the time and frequency domain methods are essential for understanding ecological processes and in most cases, they provide the basis for predictive modelling (Koopmans, 1985). In the time domain approach, values of signals recorded as a function of time are analysed by means of techniques such as autocorrelation function, partial autocorrelation function. But, these time domain methods give no information concerning the frequency at which the changes in the signal occur. Given that many natural systems have a frequency-dependent variability, an understanding of this frequency dependence gives more information concerning the underlying physical mechanism that produced the signal. The Fourier transformation can be used to project a signal from the time domain into the frequency domain so as to reveal periodic components present in the signal, the active frequency bands in the signal, and their intensity or relative importance (Bloomfield, 1976; Percival, 1995). Fourier approximation of ecological processes is based on a certain but fixed frequency (Hipel & McLeod, 1994). However, the recorded observations of many ecological indicators are an amalgam of components or processes operating at different timescales (corresponding to different ranges of frequencies active at specific time intervals in the Fourier domain) but cannot be revealed by classical signal analysis methods. In opposite of that wavelet analysis considers variations in time and scale (Chui, 1992).

8.2.2.1 Statistics of Time Domain

A sequence of observations collected overtime on a particular ecological indicator is called a time series. This can be composed of a quantity observed at discrete times, averaged over a time interval, or recorded continuously with time. From the point of view of information theory, these data series represent full process information. Within this context they are also called signals. Most of the signals are non-stationary, where the statistical properties like the variance and mean are functions of time. A signal is strictly stationary if it is free of trends, shifts, or periodicity. This implies that the statistical parameters of the signal such as the mean and variance remain constant through time. Otherwise, the signal is non-stationary. Ecological signals are usually non-stationary as a result of internal and external processes operating in parallel manner not only at different timescales but as well with different frequencies (Powell & Steele, 1995).

Extracting process information from the signals by classical methods usually requires modelling by the removal of the non-stationary and then fitting a stationary stochastic model. Information can be extracted by means of distinguished time series techniques. To detect nonstationarities in the data set the cumulative sum and trend analysis methods can be used. The cumulative sum method enables the analyst to recognize changes in the general tendency of a signal. It specifically enables one to detect changes occurring in the mean value of the signal, the time at which the change appears and the mean value at homogenous intervals (Pollock, 1999). Given a signal $x(t)$ sampled at regular time intervals t, varying between 1 to N and a reference value r (for example the mean). This reference value is subtracted from all the estimations of the series, and then a cumulative sum of the successive values is calculated:

$$S_q = \sum_{i=1}^{q} X_i - qr$$

This cumulative sum is very sensitive to changes in the mean value of a signal. The advantage of the graphical plot of the values obtained is that all local mean is immediately deducted from the slope. Given two points X_i and X_j being the respective lower and upper limits of a relatively monotonous series, the slope p between these two values separated by K interval of time ($j - i = K$) will give $P = (X_j - X_i)/K$ where

$$p = \frac{X_j - X_i}{k}, \; p = \frac{\sum\limits_{i=i+1}^{j} X_i}{k} - r, \text{ and } \overline{X}_{ij} = P + r.$$

The local mean between the two distant points of K is equal to the slope of the graphic of the cumulative sum plus the chosen reference value r.

Figure 8.2 shows the time courses of a chemical water quality indicator of a freshwater ecosystem and

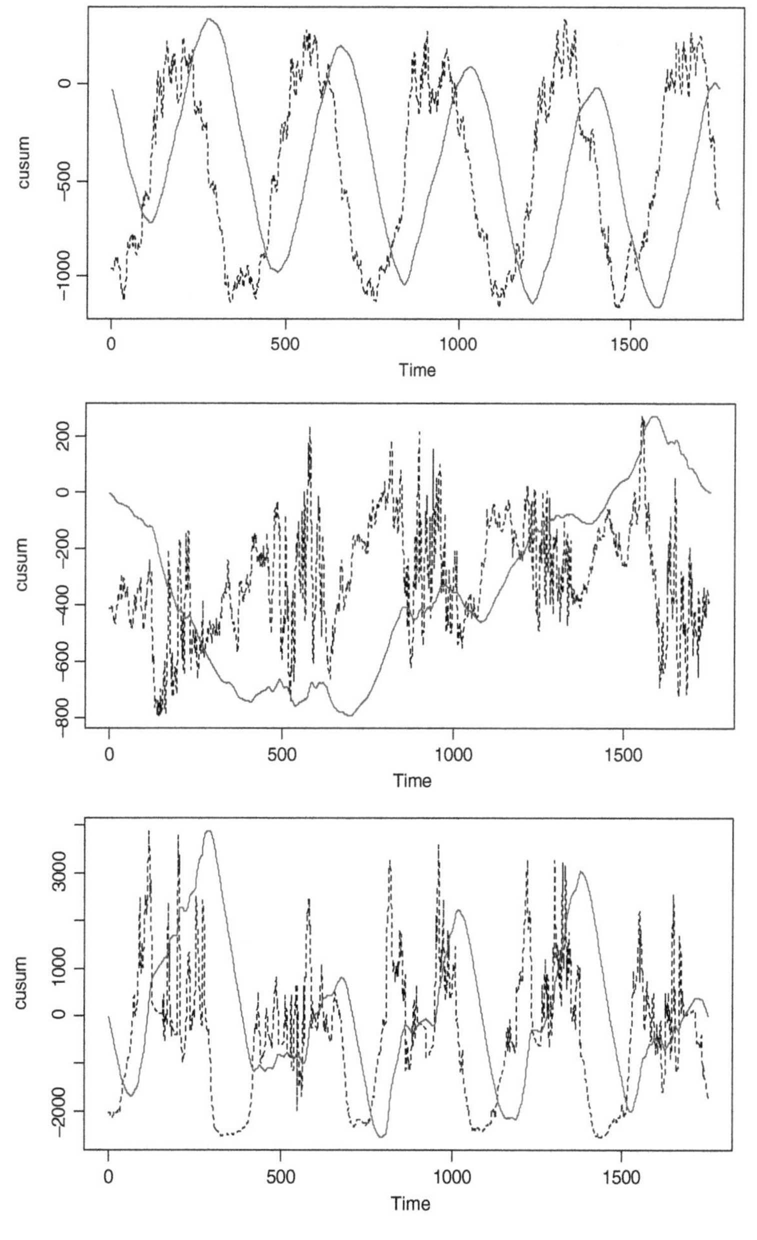

Fig. 8.2 Detection of local changes of mean for water temperature (*top*), dissolved oxygen (*middle*) and chlorophyll-a (*bottom*)

its cusum line. It can be seen that the mean is an incorrect statistical measure to describe time-dependent changes of the indicator.

The presence of trends in ecological data series can be analysed by linear or nonlinear regression functions (Rosenblatt, 2000). In their simplest way ecological data series of indicator values are related to time by an equation of the form $\mathbf{y} = b_0 + b_1(t)$ where $\mathbf{y}$ is the ecological indicator, t is the time, b_0 and b_1 are the least square estimates of the intercept and slope coefficients. The slope b_1 indicates the average rate of change in the indicator during each time instant of the time period. If the slope is significantly different from zero, the trend in the water quality indicator is equal to the magnitude of the slope and the direction of the trend is defined by the sign of the slope. The trend is increasing if the sign is positive and decreasing if the sign is negative. If the slope is not significantly different from zero, there is no trend in the water quality indicator. An advantage of this technique of trend analysis is that it is easy to apply to a long-term data set (Fig. 8.3).

The trend lines presented in Fig. 4.3 are computed as follows:

- Water temperature $y(t) = 11.8 + 0.00167 \cdot t$,
- Dissolved oxygen $y(t) = 8.935 + 0.00129 \cdot t$,
- Chlorophyll-a $y(t) = 50.37 - 0.00033 \cdot t$.

The method may fail to detect trends that are nonlinear but still monotonic (in one direction). Other approaches such as the Mann–Kendall test can equally be used for detecting monotonic and nonlinear trends, but it only indicates the direction and not the significance of the trends.

Another advanced statistical data modelling technique for stationary ecological processes is the so-called ARMA (autoregressive moving average) or ARIMA (autoregressive integrated moving average) modelling which is based on the theory of linear stochastic processes. Figure 8.4 shows examples of this procedure for a physical, a chemical, and a biological variable of a shallow lake ecosystem.

The stochastic signal $x(t)$ will be extracted from white noise $e(t)$. For equidistant data the following stochastic difference equation is valid: $a_0x(t) + a_1 x(t-1) + \ldots + a_r x(t-r) = b_0 e(t) + \ldots + b_1 e(t-l)$. Using the shift operator q^{-1} and the relation

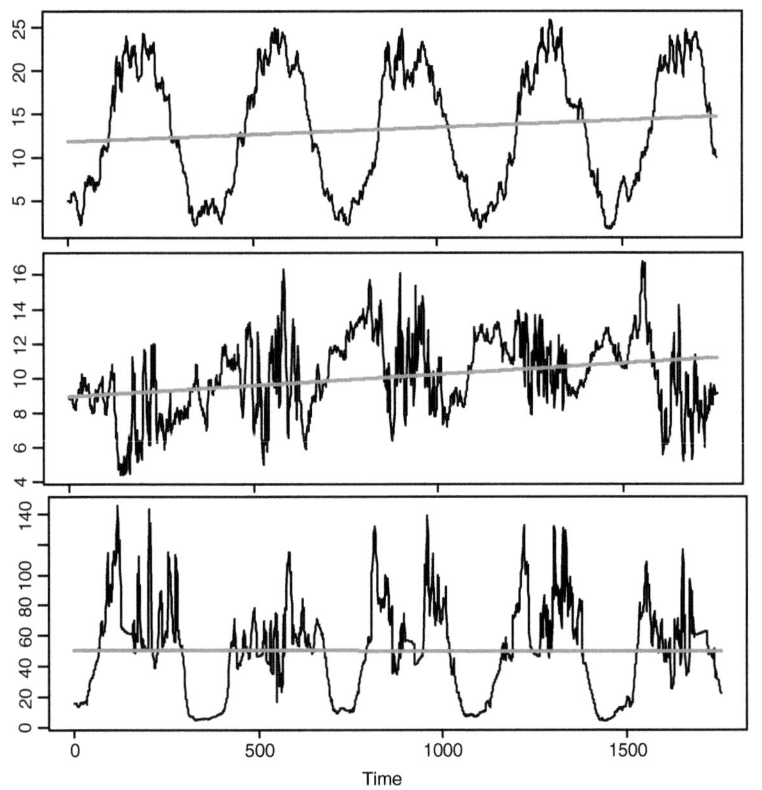

Fig. 8.3 Linear trend of water temperature (*top*), dissolved oxygen (*middle*), and chlorophyll-a (*bottom*)

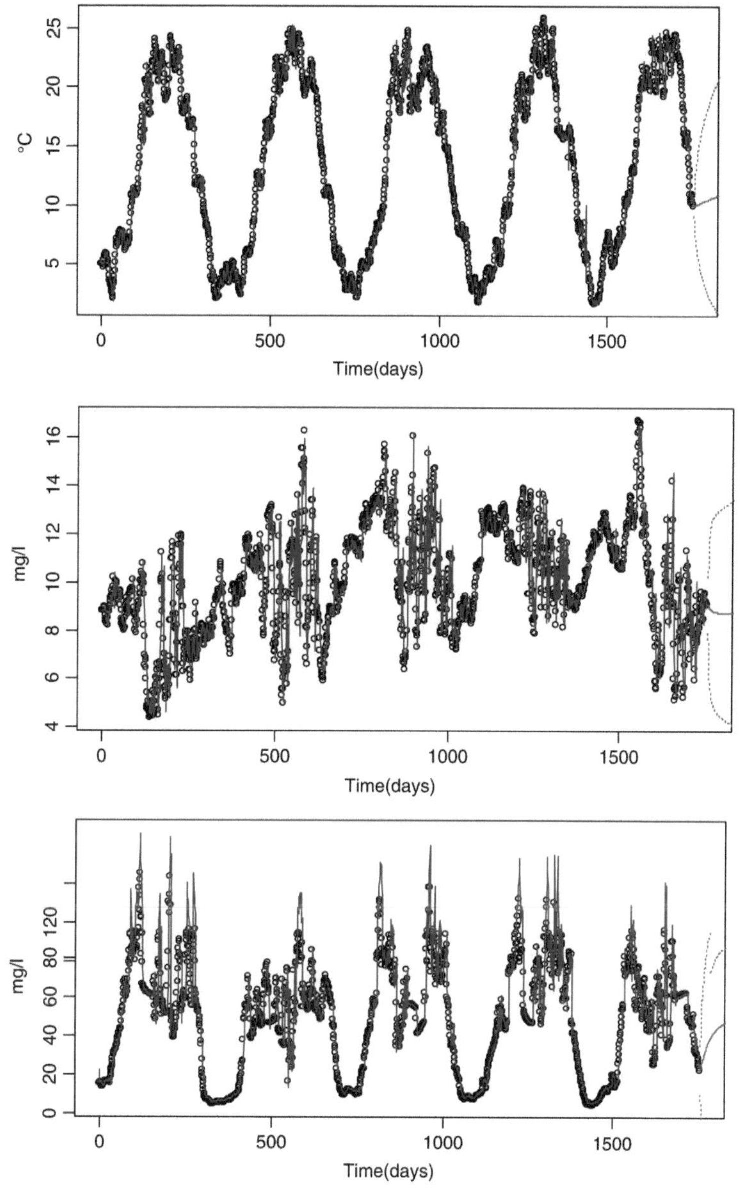

Fig. 8.4 ARMA (3,3) model for water temperature (*top*), ARIMA (1,1,2) model for dissolved oxygen (*middle*), ARIMA (3,1,3) model for chlorophyll-a

$x(t-n) \cong x(t)q^{-n}$ one gets $A(q^{-1})x(t) = B(q^{-1})e(t)$. From this equation the AR-model of order r and the MA-model of order l can be derived. Then, the ARMA-model of order (r, l) can be computed for $a_0 = b_0 = 1$, while the ARIMA-model is described by the equation $A(q^{-1})\Delta^d x(t) = B(q^{-1})e(t)$. The expression $\Delta^d x(t)$ means the difference of two neighboured data. These types of data models are valid for short-term and for long-term data sets. As can be seen at the right-hand site of the figure, the predicted values of indicators are sometimes far from reality.

8.2.3 Statistics of Frequency Domain

Any periodic signal of frequency f can be represented as the sum of properly chosen sinusoidal waves (Box et al., 1994; Brémaud, 2002). This resulted in the development of a family of mathematical techniques among which is the Fourier polynomial.

8.2.3.1 Fourier Analysis

Ecological processes which tend to undergo variation over time can be treated by means of orthogonal trigonometric (time-) function series which are called Fourier series. The elements of these series are conceived as scattered harmonic oscillations with the frequencies $\omega_k = k\omega_0 = k2\pi/T_0$, their amplitudes and phases being exposed to random variation. A basic oscillation with frequency ω_0 is obtained for $k = 1$ and corresponding distortions will be recordable if $k = 2, 3, \ldots$. With the coefficients being determined in the way described, the finite approximating trigonometric sum will have $y(t) = F(t)$ for any value of n:

$$F(t) = a_0 + \left[\sum_{k=1}^{n} (a_k \cos(k\omega_0 t))\right] + \left[\sum_{k=1}^{n} b_k \sin(k\omega_0 t)\right]$$

and a minimum mean quadratic deviation regarding $y(t)$. Relations $A_k = \sqrt{a_k^2 + b_k^2}$ and $\varphi_k = \text{arc}\tan(b_k/a_k)$ apply to the amplitude and phases of the approximated oscillations, with phases ϕ_k in the interval $[0, 2\pi]$ being uniformly distributed with constant probability density $w(\varphi_k) = 1/2\pi$, while the amplitudes satisfy the conditions of a Rayleigh distribution (Box et al., 1994).

In a system which consists of several dynamic sub-systems the output variable will be of an almost harmonic shape even with the presence of nonlinear intermediate elements, which enables the application of Fourier analysis to nonlinear systems. By using Fourier analysis for identification of linear or linearised systems, the periodic pattern of influence variable is given with sufficient accuracy by a function of the following form $y(t) = a_0 \Sigma A_k \sin(k\omega_0 t + \varphi_k)$. An estimation of the part of overall variance of the observed process explained by the harmonic function is possible by means of the following equation $var(\%) = 100 \cdot A_k^2/2 \cdot cov$ with $\text{cov} = 1/(n-1)\Sigma(y_i - \tilde{y})^2$. The percentual variance may be statistically secured by application of the F test to the ratio: variance of natural oscillation/variance of predicted oscillation.

The number of cosine and sine terms can be progressively increased in order to obtain an acceptable approximation of cycling behaviour. The advantage of Fourier approximation is the more or less

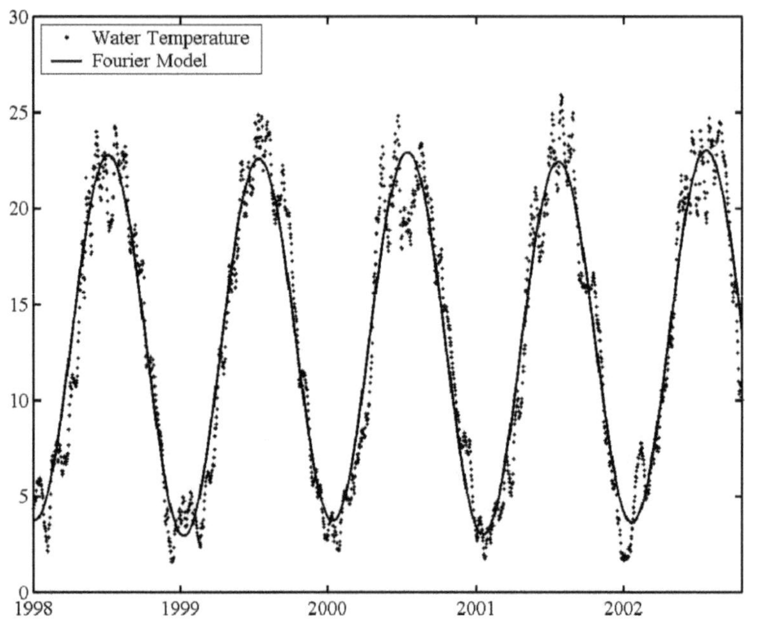

Fig. 8.5 Fourier approximation of water temperature, $R^2 = 0.95$

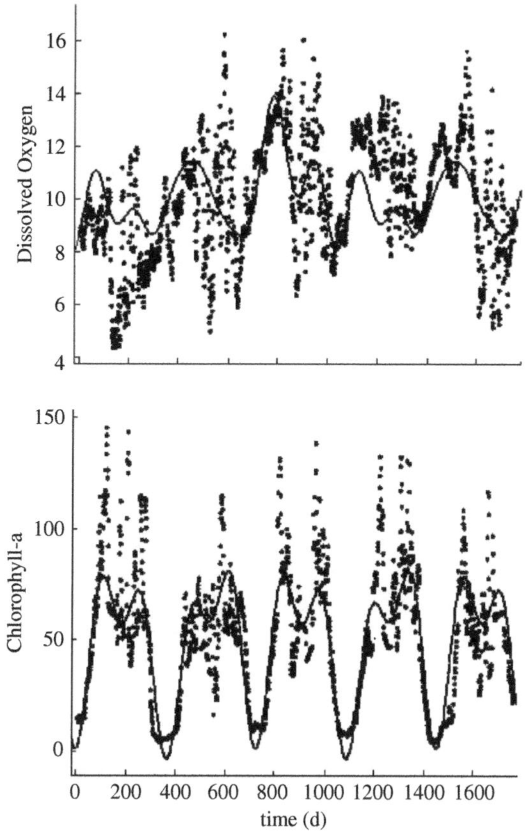

Fig. 8.6 Fourier approximation of dissolved oxygen, $R^2 = 0.30$ (*top*) and chlorophyll-a, $R^2 = 0.75$ (*down*)

very good description of physical cycling processes (cf. Fig. 8.5).

The model equation is given by $y(t) = 13.02 + 0.41 \cos(0.008t) + 0.029 \sin(0.008t) - 9.673 \cos(0.017t) - 0.47 \sin(0.017t)$.

For ecological processes with high levels of disturbances caused by internal and external driving forces this method can be used by restrictions only. Fourier polynomials are appropriate for processes with fixed frequencies, while natural processes especially those represented by chemical and biological indicators have varying frequencies. Figure 8.6 shows Fourier approximations which are not acceptable for predictions. This is because a physical indicator like water temperature is mainly influenced by external driving forces such as solar radiation, while high fluctuations observed in biological indicators like chlorophyll-a are dependent on external driving forces and internal ecological state variables such as light and nutrients availability. Chemical indicators such as dissolved oxygen are significantly influenced by natural and artificial external driving forces as well as by natural internal states. They fluctuate as these factors fluctuate.

8.2.3.2 Autocorrelation and Cross-Correlation

Steady-state ecological processes are characterised by their time-related linear and quadratic mean values. These, however, do not yet provide any information on mutual influences between the values of $x(t)$ due to the existence of energy storages and feedbacks in the ecosystem under review. Deeper insights into these correlations may be obtained with the aid of correlation functions (generalisation of quadratic mean values).

The autocorrelation function ACF gives the degree of relationship between the values of $x(t)$. It is defined by $\Phi_{xx}(\tau) = \lim 1/2T_0 \int x(t) \cdot x(t \pm \tau)\mathrm{d}t$ where the limit process runs from T to ∞, and integration is done from $-T$ to $+T$. The cross-correlation function CCF describes the correlation between two different variables $x(t)$ and $y(t)$: $\Phi_{xx}(\tau) = \lim 1/2T_0 \int x(t) \cdot x(t \pm \tau)dt$. The correlation time τ_Φ of a stochastic process $x(t)$ is defined as the time which has to pass before the amount of difference $\Phi_{xx}(\tau)$ and $x^2(t)$ drops below a specified ε bound (e.g. $\varepsilon \leq 10^{-5}$). If a periodic part is contained in process $x(t)$, a periodic part will be assigned also to ACF, all in the same period, whereas ACF of an accidental process, superimposed by noise, will decay.

The autocorrelation refers to the correlation of the signal with its past and future values. It is a method for characterizing the correlation within a signal over time. The autocorrelation function measures the correlation between two values of the same variable x_i and x_{i+k} and is used to detect non-randomness in the data (Rebecca, 1999). Correlograms are very practical for the determination of the dependence between successive observations of a time series (Fig. 8.7). If the correlogram indicates the existence of correlation between successive terms $x(t)$ and $x(t + k)$, the signal is assumed dependent or said to exhibit long memory (Beran, 1998).

The correlogram of the water temperature signal clearly shows the presence of a strong dependence of the future values on the present ones. The signal exhibits long memory with the absence of any form of randomness. The signal of the chemical indicator dissolved oxygen equally portrays persistence as shown in the correlogram, though not as strong as in the case of water temperature. This as well means that the future values of the signal are strongly influenced by the present values. The signal of the biological indicator chlorophyll-a also reveals the presence of dependence of the future values on the present ones. As is shown in Fig. 8.7, all the three signals have a long memory

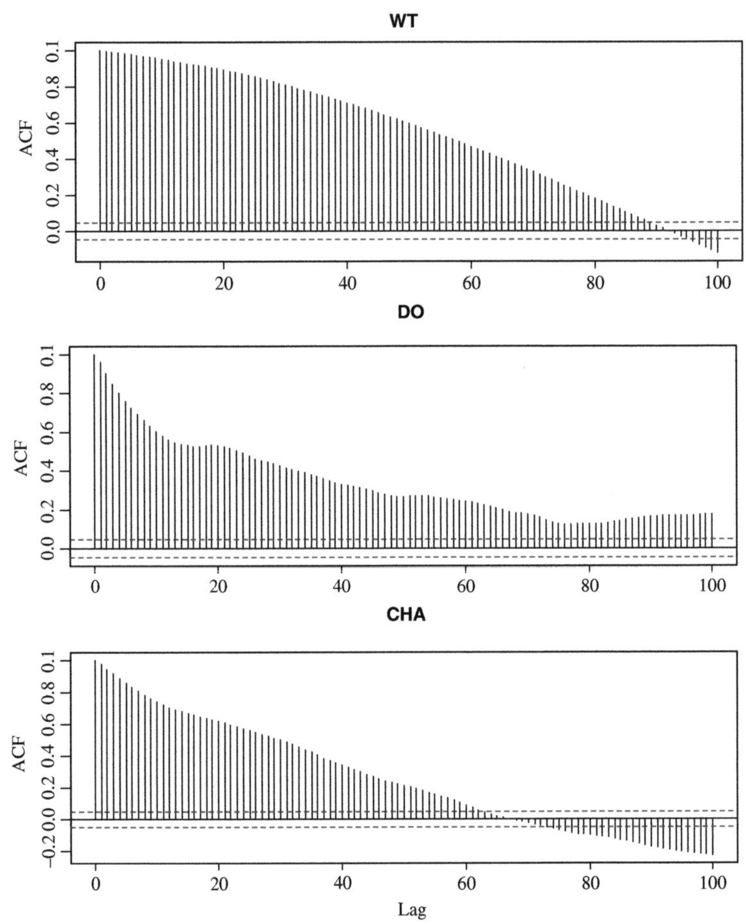

Fig. 8.7 Correlograms of dissolved oxygen (*top*), chlorophyll-a (*middle*), and water temperature (*bottom*)

with the signal of the physical indicator having the strongest autocorrelation compared to the other two. This has implications on monitoring in the sense that it will be more cost effective not to sample the signal at a high resolution so as to avoid a lot of redundant information.

8.2.3.3 Spectral Analysis

From time correlation functions the frequency correlation functions are derived by Fourier transformation. In a given frequency domain time functions can be depicted by their Fourier transformations, the spectral function, or power spectra. Such a transformation is not feasible unless $x(t)$ is absolutely integrable. This demand cannot be met, in the first place, by random signals, since convergent Fourier transforms do not exist in a stochastic process. The same condition, however, is satisfied by correlation functions.

The auto-power spectrum $S_{xx}(\omega)$ of $x(t)$ is the Fourier transform of ACF:

$$S_{xx}(\omega) = S_{xx}(-\omega) = 1/2\pi \cdot \int \Phi_{xx}(\tau)e^{-j\omega\tau}d\tau.$$

The variability of ecological processes are expressed by variations in signals (due to annual cycles, seasonal variations, diurnal rhythm, etc.) are clearly reflected in the auto-power spectrum by one or several peaks, though in processes with highly oscillation components, the associated spectral function tends to be more smoothed. It will be visualised as a function of frequency by means of the periodogram and cross-spectral analysis. Figure 8.8 shows such a periodogram for chlorophyll-a where two frequencies can be considered as dominant for algal development.

The cross-power spectrum $S_{xx}(\omega)$ of two stochastic processes $x(t)$ and $y(t)$ is the Fourier transform of CCF:

$$S_{xy}(\omega) = 1/2\pi \cdot \int \Phi_{xy}(\tau)e^{-j\omega\tau}d\tau.$$

$S_{xy}(\omega)$ is a complex function. The limitation of CCF is considered by what is called a window function $h(\tau) : S_{xy}(\omega) = 1/2\pi \cdot \int \Phi_{xy}(\tau)h(\tau) \cdot e^{-j\omega\tau} d\tau = S_{xy}(\alpha) \cdot H(\omega - \alpha)$ where $H(\omega)$ is the Fourier transform of $h(\tau)$ which distorts $S_{xy}(\omega)$ to $S^{-}_{xy}(\omega)$.

The coherency functions $CO_{xy}(\omega)$ is a measure of synchronicity of the ecological processes $x(t)$ and $y(t)$ and can be derived from spectral functions $S_{xx}(\omega)$, $S_{yy}(\omega)$ and $S_{xy}(\omega)$:

$$CO_{xy}(\omega) = \left|S_{xy}(\omega)\right|^2 / S_{xx}(\omega) \cdot S_{xy}(\omega) \text{ with } S_{xy}(\omega) = \sqrt{Re(S_{xy}(\omega))^2 + Im(S_{xy}(\omega))^2}.$$

The following relation applies to phase shifting between both signals:

$$\varphi(\omega) = \arctan\left(Im(S_{xy}(\omega))/Re(S_{xy}(\omega))\right).$$

Frequency resolution depends on both the length of the period of measurement and the time spacing between measured data. The highest recordable frequency ν_{max} can be calculated from the scanning step width, according to Shannon's theorem: $\nu_{max} = 1/2 \cdot \Delta\tau$ resp. $\omega_{max} = \pi/\Delta\tau$. If, due to scanning, the spectrum of a variable can be calculated only up to an upper frequency, ν_h, no higher frequencies must be contained in the signal itself. These will be undistinguishably 'folded into' the lower frequencies range, which will entail distortion of the spectrum (aliasing

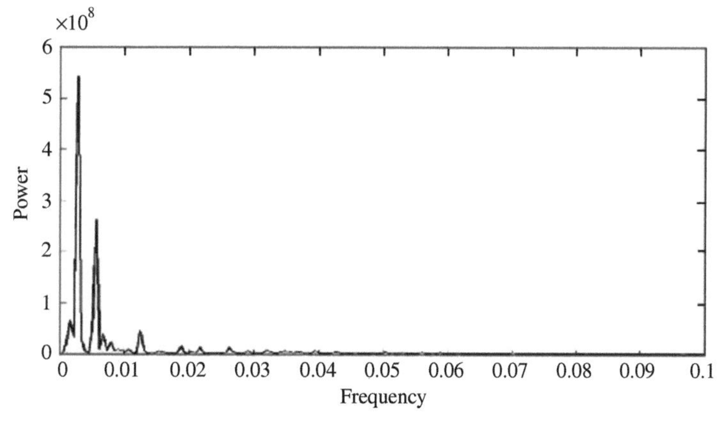

Fig. 8.8 Periodogram of chlorophyll-a

effect) (Jørgensen, 1994). Should such highly oscillation components be contained in the process, there are only two ways by which to cope with the problem:

1. Increase of the frequency of scanning (either by reduction of spacing between measured data or by lengthening of the data set by long-term observations).
2. To suppress these high oscillations by band-pass or low-pass filtering prior to computation of the spectrum.

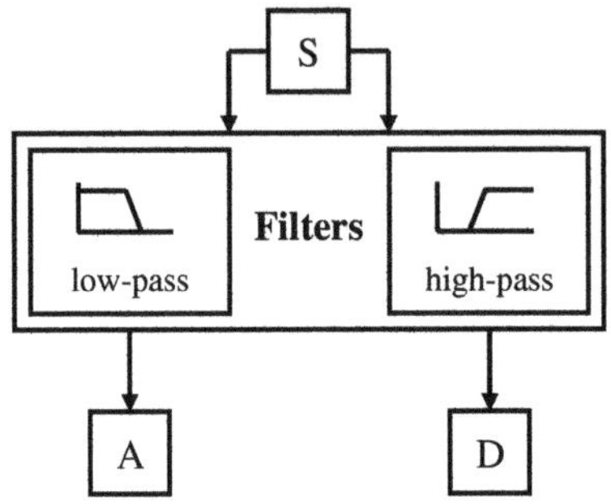

Fig. 8.9 Principle of wavelet decomposition of an ecological signal

8.3 Wavelet Analysis

Wavelets analysis is a mathematical tool that has been proven useful for timescale-based analysis of ecological processes. It is a solution to the timescale analysis problem because it offers an effective approach of extracting both the information on the time localization and the frequency content of the time series (Daubechies, 1990, 1992; Mallat, 1998). The main idea behind wavelet analysis is to imitate the windowed Fourier analysis, but using basis functions (mother wavelets) that are better suited to capture local behaviour of non-stationary ecological signals. They dilate and translate features which are local in time and frequency (Debnath, 2002; Shumway, 2005). Wavelet analysis makes use of the different wavelet basis functions in the wavelet transform to project a signal from the time domain into the timescale domain. It decomposes a signal into its constituents at different timescales (Shumway & Stoffer, 2000).

A wavelet is a small wave and a wave is a real valued function that is defined over the entire real axis and oscillates back and forth about zero with the amplitude of the oscillations remaining relatively constant everywhere (Percival & Walden, 2000) like the sine wave. Wavelets come in families generated by the father wavelet Φ and a mother wavelet ψ. A wavelet is a function of time that obeys the following wavelet admissibility conditions $\int \Psi(t)\mathrm{d}t = 0$ and $\int \Phi(t)\mathrm{d}t = 1$ (Gençay, Selcuk, & Whitcher, 2002). Father wavelets used to represent the long-scale smooth or low-frequency component of a signal integrates to one, while the mother wavelet used to capture the detailed and high-frequency components or deviations from the smooth components, integrates to zero. The father wavelet gives rise to the scaling coefficients, while the mother wavelet gives rise to the differencing coefficients. Hence, the father wavelet acts as a low-pass filter, while the mother wavelet acts as a high-pass filter (Fig. 8.9).

The wavelet transformation is a function W of two variables u and s obtained by projecting a signal $x(t)$ on to a particular mother wavelet Ψ which gives a translated and dilated version of the original wavelet function.

There are two essential wavelet transforms, namely the continuous (CWT) and the discrete wavelet transforms (DWT).

8.3.1 The Continuous Wavelet Transform

This is a function of two variables $W(u, s)$ obtained by projecting a signal $x(t)$ on to a particular wavelet Ψ and is given by

$$W(u,s) = \int_{-\infty}^{\infty} x(t)\Psi_{u,s}(t)dt,$$

and

$$\Psi_{u,s}(t) = \frac{1}{\sqrt{s}}\Psi\left(\frac{t-u}{s}\right)$$

where s is the scale parameter and u is the location parameter. Changing s produces dilating effects ($s > 1$) or contracting effects ($s < 1$) of the function $\psi(t)$. Changing u analyses the signal $x(t)$ around different points of u. The continuous wavelet transform is applied to functions $x(t)$ defined over the entire real axis. But, in ecological applications, there are only a finite number n of sampled values, rendering the CWT inadequate. Hence, there is a need for a discrete version.

8.3.2 Discrete Wavelet Transform

Implementing the wavelet transform on sampled data requires the discretisation of the scale and location parameters. Kumar and Foufoula-Georgiou (1997) demonstrated that in discretising the two parameters (s,u), one can choose $S = mS_0$, where m is an integer and S_0 is a fixed dilation step greater than one. Given $\sigma_s = S\sigma_1$, one can choose $t = nt_0\, mS_0$, where $t_0 > 0$ is dependent on $\psi(t)$ with n being an integer. By defining,

$$\psi_{m,n}(t) = \frac{1}{\sqrt{s_0^m}}\psi\left(\frac{t - nt_0\lambda_0^m}{s_0^m}\right) = s_0^{\frac{-m}{2}}\psi(s_0^{-m}t - nt_0)$$

then the discrete wavelet transform is given by

$$wf(m,n) = s_o^{-\frac{m}{2}}\int f(t)\psi(s_o^{-m}t - nt_o)dt.$$

When the DWT is applied to a time series or vector of observations x, it gives n wavelet coefficients

$$w = Wx$$

the coefficients can be organized into (J + 1) vectors $w = \left[w_1, \ldots, w_j, v_J\right]T$, with w_j being the length and $n/2J$ vector of scaling coefficients associated with averages on a scale of length $2J$ (Whitcher, 1998).

Applying shifted and scaled versions of a wavelet function $\psi(t)$ decomposes a signal $x(t)$ into simpler components. Wavelet analysis acts as a lens for inspecting the time varying structure of signals and relationships between signals. In multiresolution signal decomposition the time and frequency-dependent ecological process $x(t)$ will result in the wavelet transform results in a low-pass scaling filter (father wavelet) and in high-pass wavelet filter (mother wavelet) acting as a high-pass filter. The procedure can be seen from Fig. 8.10.

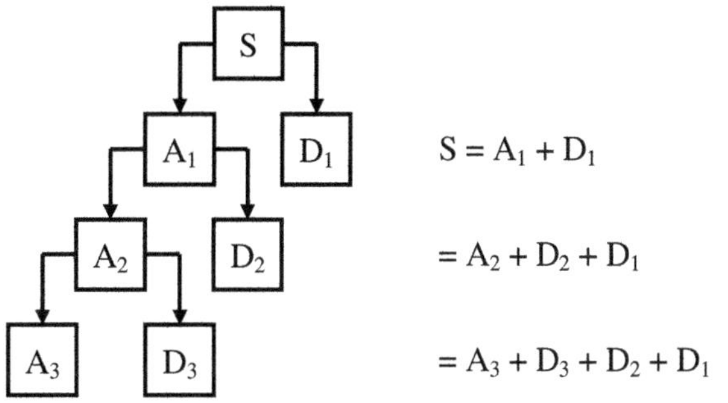

Fig. 8.10 Principle of signal approximation (A) and decomposition (D)

Table 8.3 Frequencies associated with level of decomposition

Frequency (d)	MRA scale	MRA scale
1	a_1	d_1
2	a_2	d_2
4	a_3	d_3
8 (appr. 1 week)	a_4	d_4
16 (appr. 2 weeks)	a_5	d_5
32 (appr. 1 month)	a_6	d_6
64 (appr. 2 months)	a_7	d_7
128 (appr. 4 months)	a_8	d_8

Given a signal of length $n = 2j$, a maximum of j filtering procedures can be performed creating j different resolution scales. The wavelet filter represents the details (d) or wavelet coefficients. It reveals the variations at different scales (multiresolution decomposition (MRD)). The scaling filter gives the smoothed version (a) of the original signal (cf. Fig. 8.11). This multiresolution approximation (MRA) filters information in the signal at different scales. If time records of daily data are available, Table 8.3 shows the frequencies associated with the different levels of decomposition or details.

Wavelet analysis was applied to dissolved oxygen time series of the River Havel, sampled at daily interval at the Potsdam monitoring station in the State of Brandenburg, Germany. The approximations shown in Fig. 8.11 reveal time-dependent variations and the long-term evolution of the dissolved oxygen record. The variations are not homogenous across time. The highest resolution a_7 indicates an upward movement with stable cycles from the third year. The decreasing effect of long-term DO changes after the fifth

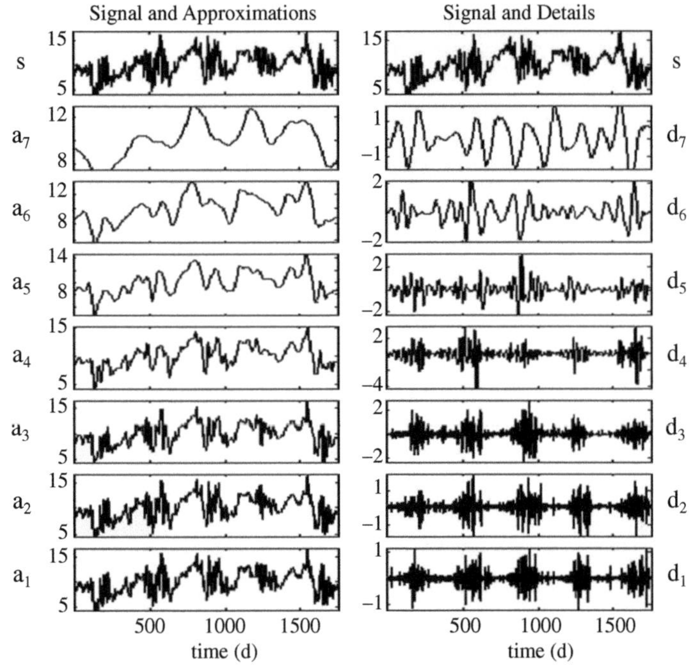

Fig. 8.11 Multiresolution decomposition of a dissolved oxygen time series using the Daubechies 8 mother wavelet

year indicates the changes of the underlying basic cycling behaviour. The details reveal that the daily variations (d_1) are less prominent than the weekly (d_4) and bi-weekly (d_5) variations. The variations become smaller at the bimonthly scale (d_7). This gives rise to the opportunity of checking the sampling strategy.

The variance of a signal can equally be decomposed using this technique. Gençay et al. (2002) showed that the time-varying variance for a signal $x(t)$ is the variance of the scale s_j wavelet coefficient $w_{j,t}$ using $\sigma^2_{x,t}(s_j) = (1/2s_j) \cdot var(w_{j,t})$.

The wavelet variance shown in Fig. 8.12 reveals the intensity of variation from one scale to the other of the dissolved oxygen time series.

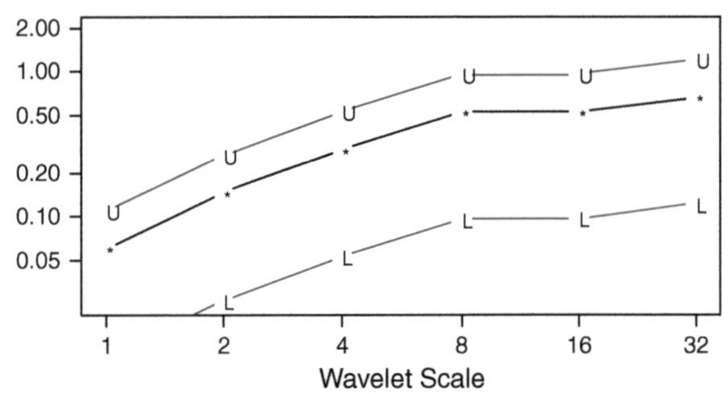

Fig. 8.12 Wavelet variance of dissolved oxygen using db8. U – upper 95% significance level, L – lower 95% significance level

The wavelet variance presented in Fig. 8.12 quantifies and indicates how much each scale contributes to the overall variability of the dissolved oxygen signal. It reveals that the variation in the time series increases progressively till scale 8 where a local maximum can be observed. Scales 8, 16, and 32 are more or less the same and contribute the most to the overall variability with the daily variations contributing the least. There is no significant increase in the variations that occur after scale 8. The dissolved oxygen curve is dominated by long-term changes or low frequencies. Rapid changes in algal productivity and respiration cannot be captured by dissolved oxygen measurements. Monitoring at scales lower than scale 8 gives redundant information. Hence, the optimal wavelet scale of monitoring is 8 for this indicator and the ecosystem under consideration.

Similar to the wavelet variance of a univariate signal, the wavelet covariance decomposes the covariance between two signals on a scale-by-scale basis (Whitcher, 1998). The wavelet covariance is the covariance between the scale s_j wavelet coefficients from a bivariate signal. After Gençay et al. (2002) the wavelet covariance for a bivariate stochastic signal of

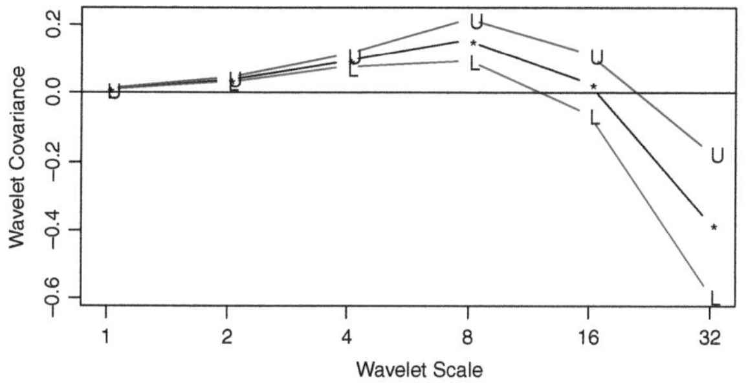

Fig. 8.13 Wavelet cross-covariance between dissolved oxygen and water temperature using db8

the scale s_j is given by

$$\gamma_x(s_j) = \frac{1}{2s_j}\text{cov}(w_{1,j,t}, w_{2,j,t}),$$

and a scale-by-scale decomposition of the covariance is given by

$$\sum_{j=1}^{\infty} \gamma_x(s_j) = \text{cov}(x_{1,t}, x_{2,t}).$$

Then, the wavelet cross-covariance is given by introducing a time lag τ between the signals:

$$\gamma_{x,\tau}(s_j) = \frac{1}{2s_j}\text{cov}(w_{1,j,t}, w_{2,j,t+\tau}).$$

It is known that the correlation between dissolved oxygen and water temperature is a negative one. The wavelet covariance between dissolved oxygen and water temperature in Fig. 8.13 shows a linear increase in the cross-covariance up to scale 8 (approximately 1 week) after which the co-variation starts dropping and increases negatively at scale 32. The changes of DO are determined mainly by the low-frequency behaviour of water temperature.

From the wavelet cross-covariance, the wavelet correlation (or normalised cross-covariance) is obtained:

$$\rho_x(s_j) = \frac{\gamma_x(s_j)}{\sigma_1(s_j)\sigma_2(s_j)}.$$

It must be noted that the wavelet correlation is not a test of cause–effect relationship. But comparisons of wavelet correlations enable a deeper insight in the long-term interrelationships of ecological variables. Results of wavelet correlations between dissolved oxygen and chlorophyll-a, and between dissolved oxygen and water temperature are presented in Fig. 8.14. The wavelet correlation of DO and chlorophyll-a (phytoplankton biomass) is quite different from those of DO and water temperature. At wavelet scale 8 (approximately 1 week) the highest influence of phytoplankton biomass on DO can be seen. In opposite of that the overall wavelet correlation of DO and water temperature indicates a more or less constant and positive relationship at scales 2, 4, and 8. There is a decrease in the strength of the positive relationship from scale 8 and becomes significantly negative at scale 32. It shows that the fluctuations of the DO time series are caused by the lower high-frequency scales (1–8) and the variations of the DO signal by the higher low-frequency scales (16 and 32).

Introducing a time lag τ to the wavelet correlation the wavelet cross-correlation is obtained:

$$\rho_{x,\tau}(s_j) = \frac{\gamma_{x,\tau}(s_j)}{\sigma_1(s_j)\sigma_2(s_j)}.$$

Figure 8.15 shows the lead–lag relationship between the indicators dissolved oxygen and water temperature. The behaviour has been decomposed into different timescales by means of wavelet cross-correlation. The correlation effects between the two

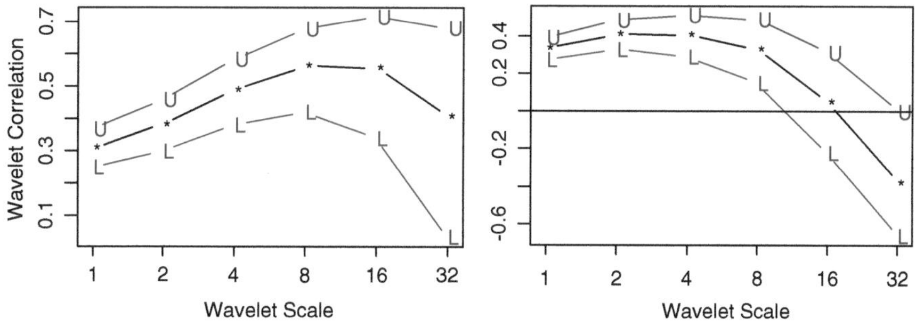

Fig. 8.14 Wavelet correlation of dissolved oxygen with chlorophyll-a (*left*) and water temperature (*right*)

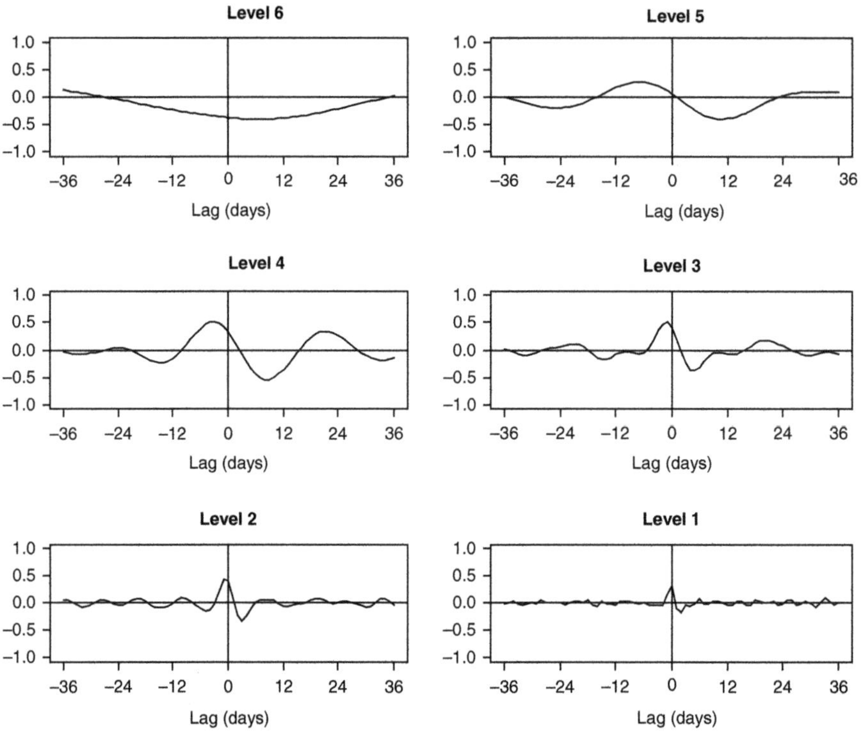

Fig. 8.15 Cross-correlation between dissolved oxygen and water temperature using db4

variables investigated is shown on each level. It can be seen that the high-frequency level (level 1) does not contribute to the expected shift between both indicators. This level characterises the noise contained in the original time series. Significant shifts are observed from level 2 to level 6 with increasing time lag. From level 2 to level 5 the values of wavelet cross-correlation at lag 0 are positive, while for level 6 this value will be negative. This confirms that water temperature effects on dissolved oxygen are stronger at low-frequency levels than at high-frequency levels.

8.4 Statistics and Long-Term Ecological Research

Ecosystems are complex dynamic nonlinear systems. Grant, Pedersen, and Marín (1997) compared methods of problem solving in ecology in terms of the relative level of understanding and the relative amount of data available on the ecosystem under consideration. From this point of view statistics plays an important role for long-term ecological research. While for short-term measurements (or observations) of ecological variables only classical static statistical methods should be applied to extract environmental information from the data, for the analysis of long-term ecological data records dynamic statistical procedures should be applied only. But, Fourier polynomials are based on fixed frequencies. From Fig. 8.16 can be seen that the Fourier polynomial is not able to follow a natural process like global radiation.

Therefore such polynomials are not well suited to analyse long-term records environmental time

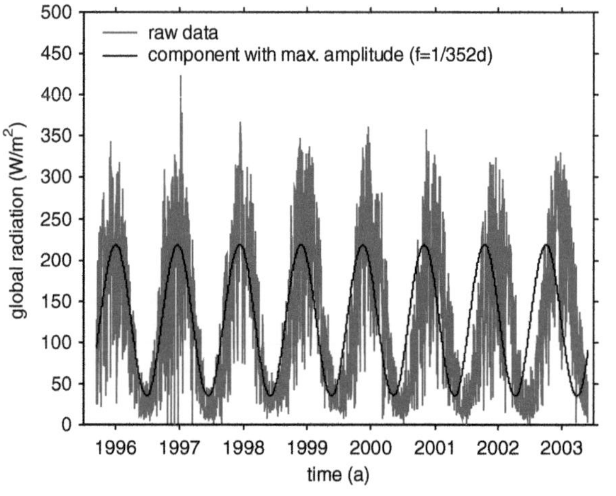

Fig. 8.16 Fourier polynomial of global radiation

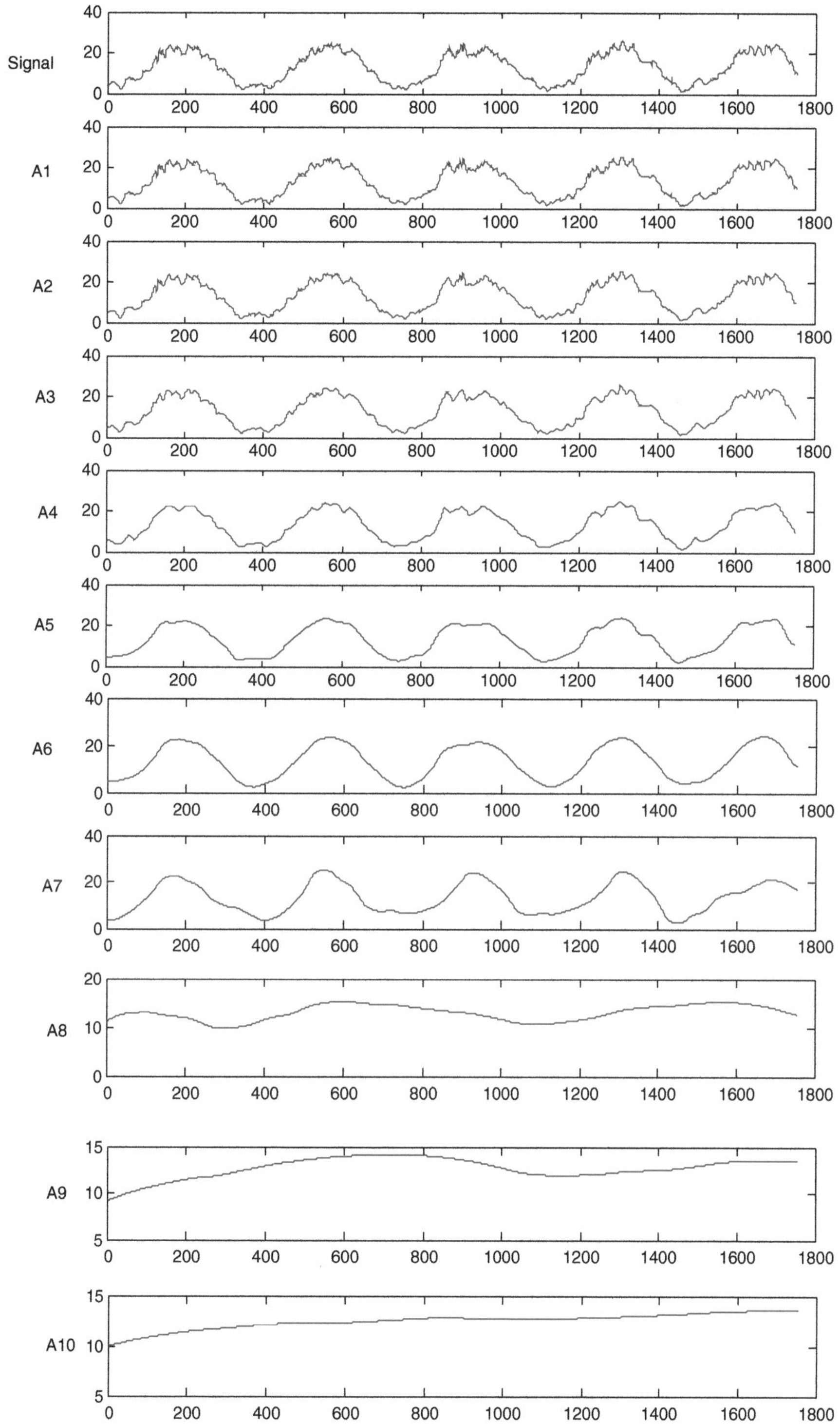

Fig. 8.17 Long-term approximations of water temperature

series. A wavelet analysis was applied to long-term ecological data series to study the underlying nature of ecological processes under consideration. The following Figs. 8.17, 8.18, and 8.19 present examples of ecological variables taken from a freshwater ecosystem for a physical, a chemical, and a biological variable. They were recorded from 1998 to 2003. Figure 8.17 gives an example for a physical variable.

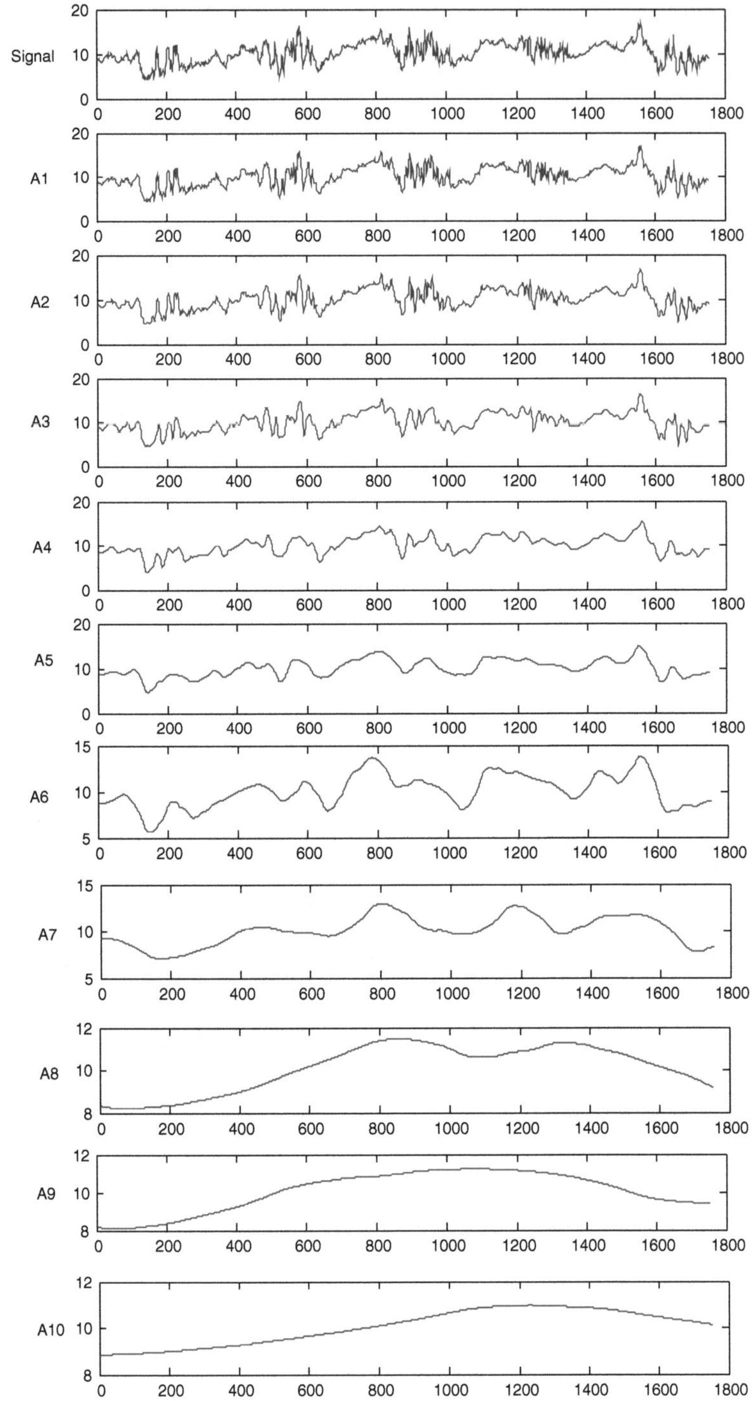

Fig. 8.18 Long-term approximations of dissolved oxygen

In each case, the upper panel shows the original time record.

When the wavelet approach was used to decompose the water temperature record till level 8, it was observed that a longer cycle was hidden within the yearly cycles. Extending it to level 10 a longer cycle can be seen. This shows the need for analyzing longer ecological records so as to unravel long cycles hidden

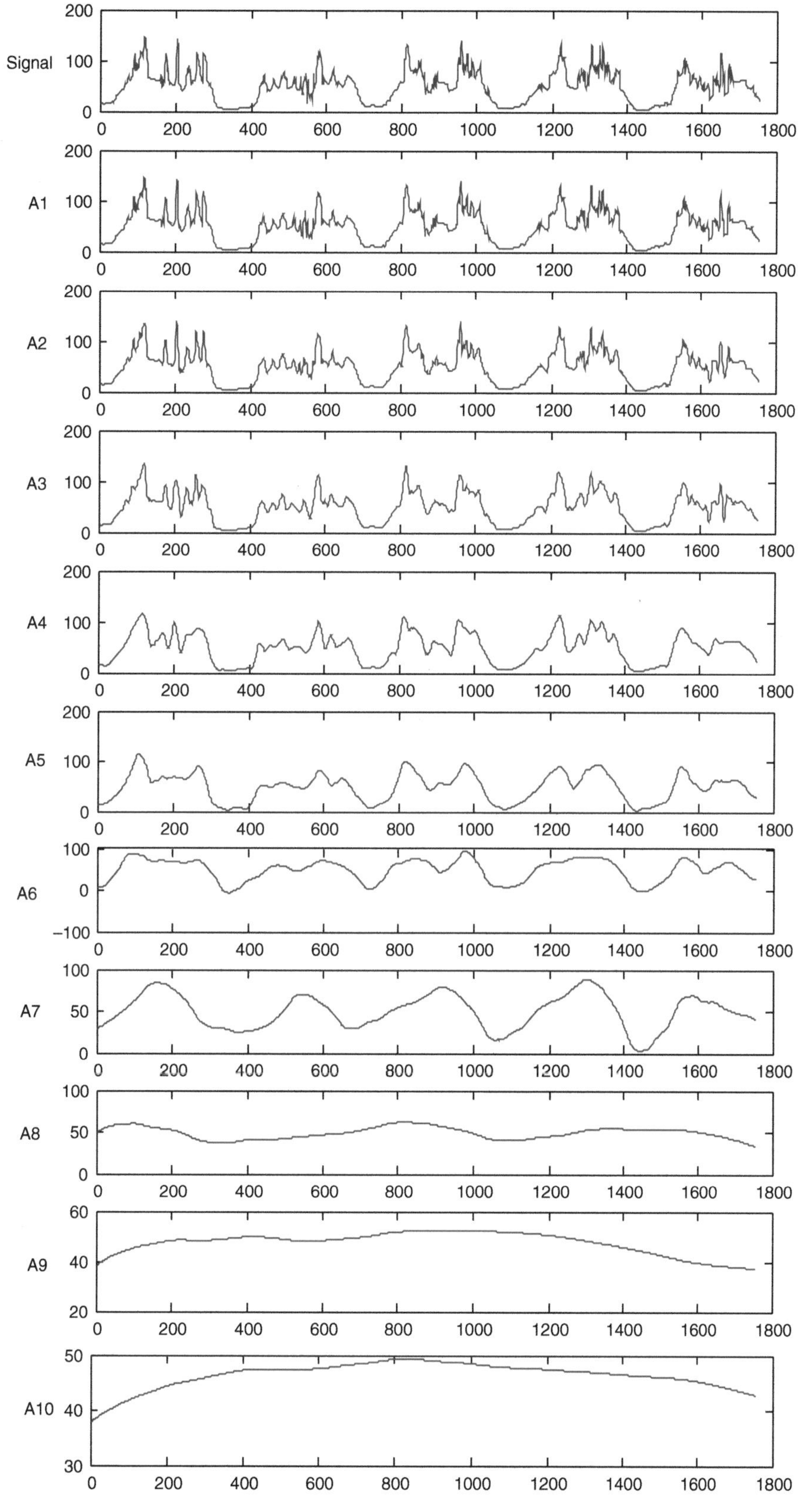

Fig. 8.19 Long-term approximations of chlorophyll-a

in signals. Investigating the reason for these cycles may be greatly relevant for a sustainable management of ecosystems.

The dissolved oxygen record (Fig. 8.18) contains a slow upward trend. By decomposing this record up to level 10 it was found that longer cycles which were not visible in the basic record are revealed.

The time record of chlorophyll-a (Fig. 8.19) shows a behaviour with two interesting areas of high algal biomass. They are caused by growth of diatoms in spring and of cyanobacteria in summer and fall. Taking of all environmental induced variations from the record a simple tendency underlying this record can be seen.

8.5 Conclusions

Statistical analysis of ecological data series by modern methods makes possible the investigation and quantification of the variation across different timescales as well as the dominant scale of variation and the identification of the relationships existing between complex ecological processes on a scale-by-scale basis. It allows the decomposition of signals according to their different frequency levels which characterise the intensity of natural and man-made disturbances. The analysis of long-term records of ecosystem signals is necessary for extracting information required for their sustainable management. This can be done using classical time series analysis methods with different levels of success. The structural characteristics of the signals such as the presence of trends, dependence, and long memory can be detected by techniques such as the cumulative sum, trend analysis, and the autocorrelation function. These techniques reveal that the ecological signals are non-stationary and have to be rendered stationary before applying time series models like the Box–Jenkins and the Fourier approximation modelling approaches. The non-stationary structure due to internal and external driving forces in ecosystems poses no problems to wavelet analysis which reveals the basic variation present in the signals. In so doing, it unravels any hidden long-term cycles which seem to be present in the ecological signals. It is of great importance to analyze the signals over longer periods of time so as to extract the underlying general tendency.

The following questions can be effectively answered by the help of wavelet analysis. What is the dominant scale of variation influencing the observed general tendency of the indicator? Are the variations from 1 day to the next more prominent than the variations from 1 week to the next? Are the statistical variations in the ecological indicator homogenous across time? What are the time-dependent variations such as the presence of trends? How are two indicators related on a scale-by-scale basis? How do they covary at different scales? These questions can be answered by modern statistical methods which consider not only the probability distribution of ecological data but also the inherent time structure of ecological processes.

References

Aström, K. (1970). *Introduction to stochastic control theory.* New York. Academic Press.

Beck, M. B. (1979). Model structure identification from experimental data. In E. Halfon (Ed.), *Theoretical systems ecology.* New York: Academic Press.

Beran, J. (1998). *Statistics for long-memory processes.* Boca Raton: Chapman & Hall.

Bliss, C. J. (1970). *Statistics in biology* (vol. I, II, III). New York: McGraw-Hill.

Bloomfield, P. (1976). *Fourier analysis of time series: An introduction.* New York: Wiley.

Box, G. E. P., Jenkins, G. M., & Reinsel, G. C. (1994). *Time series analysis.* Englewood Cliffs: Prentice Hall.

Brémaud, P. (2002). *Mathematical principles of signal processing.* New York: Springer.

Brezonik, P. L., & Shannon, E. E. (1971). Trophic state of lakes in North Central Florida. *Water Resources. Research Center, University of Florida, 13*, 1–102.

Brockwell, P. J., & Davis, R. A. (1998). *Introduction to time series and forecasting.* Berlin: Springer.

Chatfield, C., & Collins, A. J. (1989). *Introduction to multivariate analysis.* London: Chapman & Hall.

Chui, C. K. (1992). *An introduction to wavelets.* New York: Academic Press.

Ciecka, J., Fabian, R., & Merilatt, D. (1980). Eutrophication measures for small lake water quality management. *Water Resources Bulletin, 16*, 681–689.

Daubechies, I. (1990). The wavelet transform, time-frequency localization and signal analysis. *IEEE Transactions on Information Theory, 36*, 961–1005.

Daubechies, I. (1992). *Ten lectures on wavelets.* Philadelphia, PS: SIAM.

Debnath, L. (2002). *Wavelet transforms and their applications.* Basel: Birkhäuser.

Eason, G., Coles, C. W., & Gettinby, G. (1980). *Mathematics and statistics for the bio-sciences.* Chichester: Ellis Horwood.

Einax, J., Zwanziger, H. W., & Geiß, S. (1997). *Chemometrics in environmental analysis.* Weinheim: VCH.

Gençay, R., Selcuk, F., & Whitcher, B. (2002). *An introduction to wavelet and other filtering methods in finance and economics.* New York: Academic Press.

Grant, W. E., Pedersen, E. K., & Marín, S. L. (1997). *Ecology and natural resource management.* New York: Wiley.

Green, P. E. (1978). *Analyzing multivariate data.* Hinsdale: Dryden Press.

Haining, R. (2003). *Spatial data analysis.* Cambridge: Cambridge University Press.

Han, J., & Kamber, M. (2006). *Data mining – concepts and techniques.* New York: Morgan Kaufmann.

Hipel, K. W., & McLeod, A. I. (1994). *Time series modelling of water resources and ecological systems.* Amsterdam: Elsevier.

Joliffe, I. T. (1986). *Principal component analysis.* Berlin: Springer.

Jørgensen, S. E. (1994). *Fundamentals of ecological modelling.* Amsterdam: Elsevier.

Koopmans, L. H. (1985). *The spectral analysis of time series.* San Diego: Academic Press.

Krzanowski, W. J. (1990). *Principles of multivariate analysis.* Belfast: The University Press.

Kumar, P., & Foufoula-Georgiou, E. (1997). Wavelet analysis for geophysical applications. *Revues of Geophysics. 35,* 385–412.

Latini, G., & Passerini, G. (Eds.). (2004). *Handling missing data.* Southampton: WIT Press.

Legendre, L., & Legendre, P. (1983). *Numerical ecology.* Amsterdam: Elsevier.

Lepš, J., & Šmilauer, P. (2003). *Multivariate analysis of ecological data using CANOCO.* Cambridge: Cambridge University Press.

Little, R. J. A., & Rubin, D. B. (1987). *Statistical analysis with missing data.* Chichester: Wiley.

Ludwig, J. A., & Reynolds, J. F. (1988). *Statistical ecology.* New York: Wiley.

Mallat, S. (1998). *A wavelet tour of signal processing.* New York: Academic Press.

Massart, D. L., & Kaufman, L. (1983). *The interpretation of analytical chemical data by the use of cluster analysis.* New York: Wiley.

Mateu, J., & Montes, F. (2001). *Spatial statistics.* Southampton: WIT Press.

McLachlan, G. J. (1992). *Discriminant analysis and statistical pattern recognition.* Chichester: Wiley.

Patil, G. P., Pielou, E. C., & Waters, W. E. (1971). *Statistical ecology* (vol. I, II, III). London, Pennsylvania: State University Press.

Pearson, R. K. (2005). *Mining imperfect data.* Philadelphia: SIAM.

Percival, D. B. (1995). On estimation of the wavelet variance. *Biometrika, 82,* 619–631.

Percival, D. B., & Walden, A. T. (2000). *Wavelet methods for time series analysis.* Cambridge: Cambridge University Press.

Pielou, E. C. (1977). *Mathematical ecology.* New York: Wiley.

Pollock, D. S. G. (1999). *A handbook of time-series analysis, signal processing and dynamics.* San Diego: Academic Press.

Powell, T. M., & Steele, J. H. (Eds.). (1995). *Ecological time series.* New York: Chapman & Hall.

Rebecca, M. (1999). *Spectral analysis of time series data.* New York: Guilford Press.

Reckhow, K. H., & Chapra, S. C. (1983). *Engineering approaches for lake management. Vol. 1: Data analysis and empirical modelling.* Boston-London-Sydney-Wellington-Durban-Toronto: Butterworth.

Rosenblatt, M. (2000). *Gaussian and non-Gaussian linear time series and random fields.* Berlin: Springer.

Schulze, E.-D., & Mooney, H. A. (Eds.). (1994). *Biodiversity and ecosystem function.* Berlin: Springer.

Shastry, J. S., Fan, L. T., & Erickson, L. E. (1973). Nonlinear parameter estimation in water quality modelling. *Journal of the Environmental Engineering Division, 99,* 315–331.

Shumway, R. H. (2005). *Time series analysis and its applications.* New York: Springer.

Shumway, R. H., & Stoffer, D. S. (2000). *Time series analysis and its applications.* New York: Springer.

Straškraba, M., & Gnauck, A. (1985). *Freshwater ecosystems – modelling and simulation.* Amsterdam: Elsevier.

Thomann, R. V., & Mueller, J. A. (1987). *Principles of surface water quality modelling and control.* New York: Harper & Row.

Webster, R., & Oliver, M. A. (1990). *Statistical methods in soil and land resource survey.* Oxford: Oxford University Press.

Whitcher, B. J. (1998). Assessing nonstationary time series using wavelets. PhD Thesis, Department of Statistics, University of Washington, Seattle.

Young, P. C., & Young, T. (1992). Environmetric methods of nonstationary time-series analysis: Univariate methods. In C. N. Hewitt (Ed.), *Methods of environmental data analysis.* New York: Elsevier.

Chapter 9

The Role of Remote Sensing in LTER Projects

Patrick Hostert, Frederick Swayne, Warren B. Cohen, and Jonathan Chipman

Abstract All Long-Term Ecological Research (LTER) asks for spatially explicit information. Remote sensing-based data and related analysis products are major sources of such information. This chapter expands on general concepts of remote sensing and most important methodologies in relation to LTER. Examples from US-LTER sites exemplify opportunities related to remote sensing.

Keywords Remote sensing · Geomatics · Satellite imagery · Upscaling · Downscaling · Land cover change · Land use change · Landscape structure · GIS · Monitoring

9.1 Introduction

The first titles retrieved after entering the search phrase 'remote sensing' at the LTER All-site Bibliography (http://search.lternet.edu/biblio/) are

- Challenges in characterizing and mitigating urban heat islands – a role for integrated approaches including remote sensing,
- Canopy chlorophyll estimation with hyperspectral remote sensing,
- Accuracy assessment of vegetation cover estimated with remote sensing data,
- Remote sensing and climate from north to south in CAP LTER, etc.

While this is not truly representative for remote sensing-related research in LTER, it opens the eye for the breadth of topics in this context.

The mission of LTER is to document, analyze, and explain ecological patterns and processes operating over long time spans and broad ecological gradients; a specific objective is to detect signals of global environmental change and its impact on ecosystems across the world (Hobbie, Carpenter, Grimm, Gosz, & Seastedt, 2003). LTER on different ecosystems and in different ecoregions therefore relates to different data needs and corresponding information retrieval strategies. LTER relies on spatially explicit information, and one important source is remote sensing-derived information products. The purpose of this chapter is accordingly

(a) to explain the relevant background of terrestrial remote sensing data for LTER,
(b) to illustrate the most relevant methodological issues related to remote sensing data, and
(c) to provide two examples on how remote sensing is applied in US-LTER projects.

Accordingly, we first expand on general concepts of remote sensing. We will then describe the most important methodologies for qualitative and quantitative data analyses, remote sensing-based modeling, and concepts how to integrate remote sensing-derived information with information from other data sources. Moreover, examples from long-established and successful US-LTER sites give valuable insights in remote sensing-based LTER research. The chapter includes two examples from the Andrews Forest LTER and the North Temperate Lakes LTER sites, illustrating how remote sensing-based information plays a major role

P. Hostert (✉)
Geography Department, Humboldt-Universität zu Berlin, Unter den Linden 6, 10099 Berlin, Germany
e-mail: patrick.hostert@geo.hu-berlin.de

F. Müller et al. (eds.), *Long-Term Ecological Research*,
DOI 10.1007/978-90-481-8782-9_9,

for analysis and monitoring in different ecosystems. We close with most important conclusions on remote sensing in the context of LTER and an outlook on potential future developments.

9.2 Sensors, Resolution, Scale

LTER covers a broad spectrum both thematically and scale-wise: 'Ecosystems across the world' include forests, agricultural, marine, coastal and lake ecosystems, dry lands, wetlands, and also urban ecosystems. It is thus not possible to relate specific remote sensing approaches or sensor systems to LTER. We refer to general concepts here, with a focus on commonly utilized data.

It is often distinguished between ground-based, airborne, and satellite-based remote sensing, thereby implying the importance of scale, data accuracy, and information generalization (Aplin, 2006). In the broader context of remote sensing science, this spatial domain concept is extended to a spatio-spectral-temporal domain (e.g. Hostert, Röder, & Hill, 2003). This is important, as incorporating the revisiting time of a remote sensing system and the wavelength regions to be analyzed are crucial boundary conditions for information retrieval. Moreover, there is a kind of 'domain dilemma,' as spatial and spectral resolutions, spatial resolution and swath width (i.e., the area in the field of view of a sensor), as well as spatial and temporal resolutions are tightly connected. One dimension cannot be changed without restraints for the other – at least in the case of satellite-based systems.

When it comes to the question 'What can be detected?', a multitude of factors are playing a role, but any discussion on the appropriateness of a chosen approach will come back to the spatial and spectral resolutions (Fig. 9.1) of a sensor (e.g. Lee, Cohen, Kennedy, Maiersperger, & Gower, 2004; Underwood, Ustin, & Ramirez, 2007). Today's systems are usually designed as multispectral sensors (several bands), some also as hyperspectral systems (many bands, up to several hundreds). Such sensors capture the relevant spectral properties of the respective target variables either to separate different surface types or to characterize these surfaces quantitatively or qualitatively (see Section 9.3 – Methodological considerations). As many surfaces of interest (e.g., vegetation, soil, human-made materials) exhibit distinct spectral absorption bands, the spectral sensor characteristics very much determine what can be analyzed from a remote sensing perspective. The separability of different features thus largely depends on the position, number, and width of spectral bands for a given sensor (Bodechtel, 2001). Relevant spectral wavelengths in remote sensing range from the visible light (VIS), through near, shortwave and thermal infrared (nIR, SWIR, TIR, respectively), to the microwave wavelength region (radar). The most appropriate sensor for a given application depends on the remotely sensed surfaces, their spectral properties, and the chosen methods for image analysis. Most systems used for LTER are optical sensors, covering the

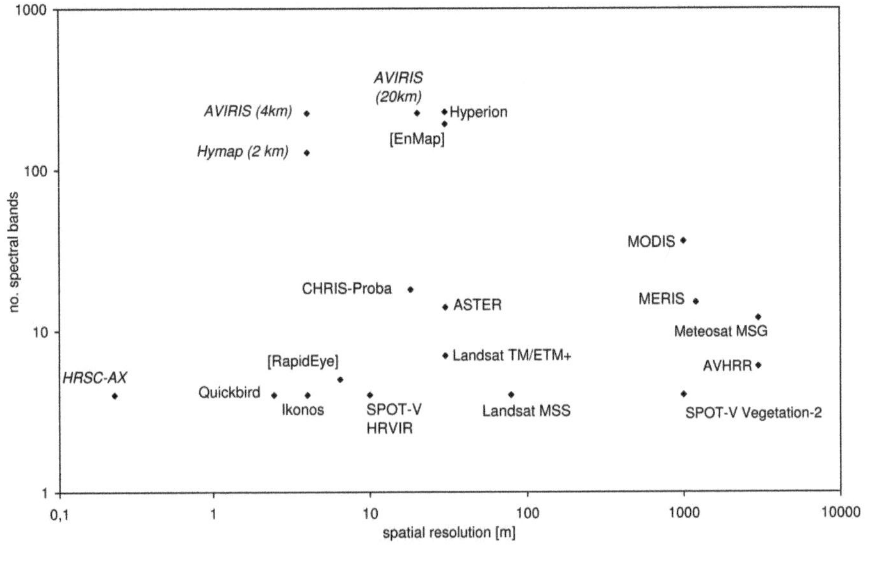

Fig. 9.1 Remote sensing systems with relevance for LTER-spatial versus spectral resolution

VIS-nIR-SWIR spectrum, i.e., wavelengths between 0.4 and 2.5 μm.

The imagery's spatial resolution is equally important when it comes to the detectability of objects. It directly refers to the scale of the objects or processes to be analyzed. Typical LTER-related scales range from local test sites, i.e., plots, to landscape and regional scale analyses. Accordingly, typical scales we want to observe vary from 1:1,000 to 1:250,000 – or even beyond (Hobbie, 2003; Hobbie et al., 2003; Kratz, Deegan, Harmon, & Lauenroth, 2003; Rastetter, Aber, Peters, Ojima, & Burke, 2003). It is thus legitimate to say that the use of airborne to satellite-based remote sensing with very high-resolution (VHR), high-resolution (HR), and medium-resolution (MR) data plays an important role in LTER-related approaches. It has, however, to be kept in mind that increasing spatial resolution will always decrease the spectral resolution of a sensor and vice versa. Also, if we want to monitor large areas (i.e., need a wide swath), the spatial resolution of a sensor will be limited (Fig. 9.2). In other words: depending on the problem at hand, we have to decide which compromise to make or how to combine different data sources to retrieve the optimum information.

When it comes to the question 'What can be monitored?', the temporal resolution gains importance. How often can a sensor revisit a site or a region, given the logistic constraints of airborne operations or the orbit characteristics (and the off-nadir viewing capabilities) of satellites? We may roughly differentiate between daily and monthly revisiting times, disregarding potential cloud cover or acquisition capacity limitations. The temporal resolution is essential for any monitoring program and questions focusing on change detection. This also includes phenological considerations in vegetation studies.

Airborne systems are most flexible and achieve the best spatial and spectral resolution. There is a multitude of sensors, from the classical aerial photograph to digital cameras and from VHR sensors with cm-resolution (e.g., Leica ADS 40, Digital Mapping Camera – DMC) to hyperspectral systems (e.g., AVIRIS, HyMap). However, this flexibility is connected to high costs and logistic constraints, relatively small coverage, and also a dedicated geometric preprocessing, as unsteady airborne platforms can result in severe image distortions. Satellite-based systems are often the better (or the only) choice, and the remote sensing community can employ data from VHR satellite systems such as Ikonos and Quickbird since the late 1990s (Dial, Bowen, Gerlach, Grodecki, & Oleszczuk, 2003; Wang, Sousa, Gong, & Biging, 2004).

Specifically in the case of landscape to global scale questions, using satellite data is the only way to cover large areas simultaneously and on a regular basis. The most prominent high-resolution system is Landsat, with multispectral scanner (MSS) data reaching back to 1972. Today's Landsat Enhanced Thematic Mapper Plus (ETM+) and its predecessor Thematic Mapper (TM) will probably be inherited by a Landsat follow-up mission, but are also augmented through

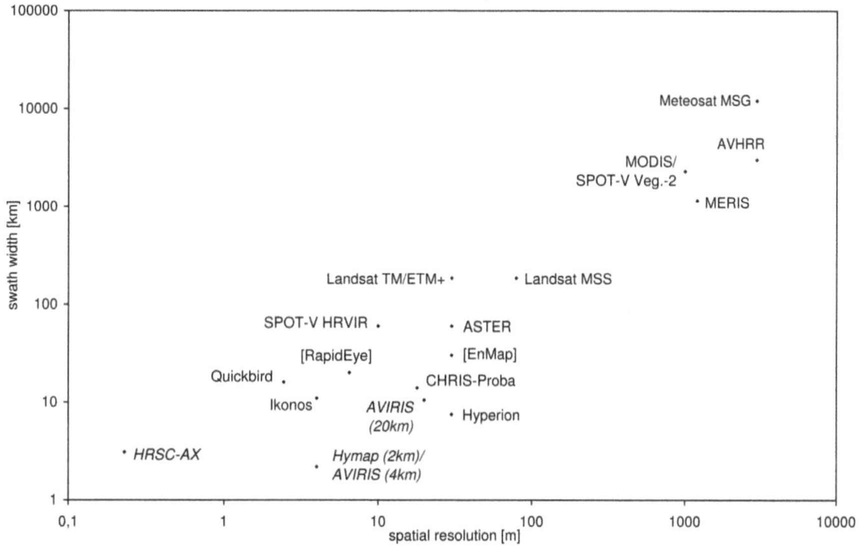

Fig. 9.2 Remote sensing systems with relevance for LTER-spatial resolution versus swath width

diverse missions worldwide (Mika, 1997). The Landsat satellite family can be regarded as the prototypical Earth observation remote sensing platform. It continuously acquires data almost worldwide and allows retrospective analyses for almost every region and ecosystem of the world. Similar sensors are, for example, the Indian Remote Sensing Satellite (IRS), the French Système Probatoire d'Observation de la Terre (also: Système Pour d'Observation de la Terre, SPOT), or the Advanced Spaceborne Thermal Emission and Reflection Radiometer (ASTER) (Capolsini, Andrefouet, Rion, & Payri, 2003; Gupta, Prasad, & Vijayan, 2000).

If spatial resolution is not of prime interest, but high re-visiting rates are mandatory (e.g., for phenological analyses), medium- or low-resolution satellites come into play: Geostationary satellites such as Meteosat may provide climatologic background information several times per hour. With Meteosat Second Generation (MSG, also Meteosat-8, -9), the first geostationary satellite with a 1 km-resolution in the panchromatic/SWIR band is available. Geostationary satellites of the future may therefore develop more impact than very coarse resolution satellites of the past. Until today, moderate resolution imagery is usually delivered from polar orbiting satellites. The Advanced Very High Resolution Radiometer (AVHRR) was the first polar orbiting mission to operationally deliver repetitive global data sets for environmental monitoring. Today, the Moderate-resolution Imaging Spectroradiometer (MODIS), the MEdium Resolution Imaging Spectrometer (MERIS), or SPOT-Vegetation provide high-quality global data sets with a temporal resolution of several days (Justice et al., 1998; Maisongrande, Duchemin, & Dedieu, 2004; Rast, Bezy, & Bruzzi, 1999; Tucker et al., 2005).

More specialized remote sensing techniques with potential relevance for different LTER are beyond the scope of this chapter and are only listed here:

- thermal remote sensing: temperature- and emissive-related techniques; especially important for characterizing the urban heat island, but also utilized in vegetation or geological studies (Torgersen, Faux, McIntosh, Poage, & Norton, 2001; Voogt & Oke, 2003)
- microwave remote sensing (radar): remote sensing based on rather long wavelengths (mm- to m-wavelengths); cloud penetrating; to a certain extent surface penetrating (Kimball, McDonald, Running, & Frolking, 2004; Treuhaft, Law, & Asner, 2004)
- LIght Detection And Ranging (LIDAR): Laser-based remote sensing, measuring the time delay between a sent laser pulse and its reflected detection signal; used for generating precision elevation models (see also Section 9.4 – Examples from US-LTER sites) (Hudak, Lefsky, Cohen, & Berterretche, 2002; Lim, Treitz, Wulder, St-Onge, & Flood, 2003; Means et al., 2000)

9.3 Data Analysis Strategies for LTER

While the mere image data may serve as a cartographic backdrop and for generic overviews, we are usually interested in problem-specific information derived from remote sensing imagery in LTER. Analyzed images become information layers that can be integrated in more complex data analysis approaches or modeling scenarios. Remote sensing scientists need to formulate a clear processing strategy and to define how LTER-related research can be supported with remote sensing-derived information products. Such a strategy always starts with identifying ecologically relevant indicators that can be tackled from a remote sensing point of view. These may be very diverse and equally diverse may be the appropriate processing strategies. Focusing, for example, on vegetation, we may map different vegetation types from standard remote sensing imagery or analyze photosynthetic activity in plants through modeling chlorophyll absorption from hyperspectral data. Indicators and processing strategies need to be valued against the background of relevant scientific questions, data constraints, accuracy needs, and scale dependencies. With respect to LTER, we may differentiate 'baseline products,' e.g., generic land use classifications, and problem-specific analyses like biomass estimates from vegetation canopies. We only concentrate on the most common and important approaches in the following.

9.3.1 Maps from Classified Image Data

Classifying data still builds the backbone of many LTER-related remote sensing tasks. Up-to-date land use and land cover (LULC) classifications are mandatory information layers for most LTER. Classifications do not necessary lead to generic LULC maps, but may also derive more specific results for any spectrally distinguishable surface. These may range from water pollution levels to urban surface sealing proportions.

One good example from US-LTER is the Agrarian Landscapes in Transition (Ag Trans) project under the lead of scientists from the Central Arizona – Phoenix Urban (CAP) LTER. This interdisciplinary project traces the effects of the introduction, spread, and abandonment of agriculture at six US-LTER sites, with cross comparisons in Mexico and France, using a variety of monitoring strategies, quantitative modeling, and comparative data. Major sources of information for this strategy are remote sensing-based land cover classifications, complemented by map information for the pre-remote sensing era (Stefanov, Netzband, Möller, Redman, & Mack, 2007).

Methodologically, classification approaches can be separated in unsupervised (with minimum user input) and supervised approaches (Richards & Jia, 2006). The latter will usually result in more precise information and are more often applied in LTER projects. Unsupervised approaches are preferred for exploratory and quick analyses or if training data is not available or of unknown quality. We often start unsupervised, gain knowledge about the spectral properties of an area, and then use these results for training supervised classifiers. Supervised classifications may be pixel- or object based, i.e., pixels are aggregated based on spectral homogeneity criteria or objects built from spectral gradients. Object-based analyses allow integrating image texture in the classification process, which can be an additional source of information to separate different surface components. Especially, urban, agricultural, or forest LTER may profit from such information. (Benz, Hofmann, Willhauck, Lingenfelder, & Heynen, 2004; Dorren, Maier, & Seijmonsbergen, 2003; Walker & Briggs, 2007).

Today, most unsupervised classifications are performed based on iterative clustering that defines optimum class boundaries based on image statistics and a user-defined number of classes. On the other hand, there are many established supervised methods, from simple box classifiers, over probabilistic approaches such as maximum-likelihood classifications, to distribution-free classifiers like neural networks or support vector machines (Atkinson & Tatnall, 1997; Huang, Davis, & Townshend, 2002; Pal & Mather, 2005; Richards & Jia, 2006). However, the most important point is always the quality of training statistics, i.e., how well we can define examples for a classification algorithm to be extrapolated to a larger area. LTER sites offer a great opportunity for high-quality supervised classifications, as there is a wealth of training areas well known from multidisciplinary research (Kratz et al., 2003). Classification-based results often build the basic input for more sophisticated thematic analyses or modeling approaches. Among others, classified maps are often used as input for structural analyses.

9.3.2 Analyzing Landscape Structure with Remote Sensing

Spatially explicit structural information on LTER sites delivers important input for solving numerous research questions related to landscape ecology and is easily deduced from remote sensing data. Information on landscape pattern can be derived directly from spectral data, e.g., based on texture (Kuemmerle, Hostert, St-Louis, & Radeloff, 2009; St-Louis, Pidgeon, Radeloff, Hawbaker, & Clayton, 2006), or from classified remote sensing data (Vogt, Riitters, Estreguil, Kozak, & Wade, 2007; Wickham, Oneill, Riitters, Wade, & Jones, 1997). We may, for example, prefer to derive pattern information from continuous remote sensing data when processes are determined along ecoclines and from classified data if ecotones prevail. In general, indicators on structural properties of a landscape relate to the geometric configuration of the analyzed surfaces in remote sensing data. Numerous studies in landscape ecological research have shown that relevant ecological indicators can be derived from remote sensing-based structural information across different scales and ecosystems. Most recently, urban ecological research in LTER has significantly benefited from remote sensing-based structural analysis (Cadenasso, Pickett, & Schwarz, 2007; Keys, Wentz, & Redman, 2007; Pickett et al., 2008).

Remote sensing approaches are directly congruent with several core concepts of landscape ecology: Remote sensing data offer input at multiple resolutions and for different points in time, i.e., allows for a spatio-temporal characterization of landscape heterogeneity across scales. Patch and edge statistics can be derived for large areas based on different processing levels of remote sensing data. These allow, for example, characterizing landscape fragmentation and related habitat attributes. This kind of analysis offers proxies for organism abundance at the landscape level and for understanding the behavior and functioning of the landscape itself (Malczewski, 1999; Sanderson & Harris, 2000; Turner, 1989, 2005; Turner & Gardner, 1991).

9.3.3 Quantitative Parameters and Remote Sensing-Based Models

Vegetation indices (as indices in general) can be regarded as the most basic, semi-quantitative measure from remote sensing data. Such approaches are simple, mostly relying on two or a few spectral bands. Indices may thereby disregard physical principles that are respected in the case of more sophisticated methods. However, indices are straightforward to calculate, reduce some remote sensing data inherent problems (such as topographically induced illumination differences), and are widely used in local to global scale studies (Lillesand & Kiefer, 2000). Numerous LTER studies are based on parameters derived from vegetation indices (Boelman et al., 2003; Duncan, Stow, Franklin, & Hope, 1993; Li, Ustin, & Lay, 2005; Stefanov, Ramsey, & Christensen, 2001). Indices as such offer a relative, physically uncalibrated view of 'more or less' for a targeted surface property. The most widely used Normalized Difference Vegetation Index (NDVI) simply relates the spectral reflectance in the visible red and the near infrared wavelength regions to derive a summarized measure of vegetation vitality and density. Such indices can again be empirically calibrated and related to physically meaningful parameters, for example, leaf area index (LAI) or biomass (Anderson, Hanson, & Haas, 1993; Baret & Guyot, 1991; Gupta et al., 2000).

More sophisticated and physically accurate remote sensing approaches rely on models to derive parameters quantitatively. Quantitative modeling asks for calibrated data, usually radiometrically corrected for atmospheric and topographic distortions. After careful radiometric pre-processing, we retrieve physically meaningful values, i.e., radiance or reflectance/emissive data. In other words, such data sets are regarded as remote sensing-based measurements, comparable to field or laboratory-based measurements, if we consider the accuracy constraints of the correction and measurement procedures (Song, Woodcock, Seto, Lenney, & Macomber, 2001). Such data exhibit characteristic spectral absorption features of surface components that – specifically in the case of spectral high-resolution data – allow for analysis methods based on such features. Moreover, such data serves as input for spectrally driven models and retrievable parameters can be diverse. These range from chlorophyll or water content and LAI or biomass in the case of vegetation, to soil organic carbon context, or water sediment freight. (Palacios-Orueta & Ustin, 1998; Sims & Gamon, 2003; Ustin, Roberts, Gamon, Asner, & Green, 2004)

9.3.4 Remote Sensing and GIS Integration – Knowledge-Based Image Analysis

Digital information layers beyond remote sensing derived information become more and more available – for different topics, for different regions of the world, and at multiple scales. For that reason, there is a tendency in remote sensing-related research to better integrate other geodata sources in the analysis process and to also integrate methods of digital image processing systems and geoinformation systems (GIS). Research teams in LTER sites collect substantial field-based data, compile statistical databases, and perform extensive field mapping. LTER is therefore a showcase of how heterogeneous information can be integrated to derive more detailed or complex indicators.

It is common to incorporate remote sensing derived information in GIS and further analyze heterogeneous data in an integrated way. It is less common – but equally logical – to incorporate this wealth of auxiliary geodata (from the remote sensing point of view) in the remote sensing data analysis process. A straightforward way to do so is to implement knowledge-based image analysis strategies, i.e., to develop rule sets that depend on additional information from non-remote sensing geodata. Usually, such rule sets will be implemented to stratify imagery for defining homogeneous regions with respect to certain features, e.g., geology, soils, plant communities, ground water levels, or slope.

It is obvious that such processing schemes allow to rule out otherwise confusing image characteristics and, for example, to decrease class confusion in a given stratum (Kontoes, Wilkinson, Burrill, Goffredo, & Megier, 1993; Stefanov et al., 2001; Wilkinson, 1996).

9.3.5 Multitemporal Data Analysis

LTER is, by definition, focused on monitoring ecosystems over long periods of time. Remote sensing based analysis must therefore be developed against the background of multitemporal comparability. Information products from remote sensing should be the result of multitemporal change analysis and reflect the modification of thematic classes or relevant parameters compared to their status at a given starting point in time.

Change analysis comprises several steps from an accurate pre-processing to change analysis procedures for the targeted variables, as well as their validation (Lu, Mausel, Brondizio, & Moran, 2004). Change detection algorithms need to work differently on different data sources, depending on the processing stage of the data source. Input can be raw imagery, a classified data set, or already quantitatively analyzed information. It can comprise two, several, or many steps in time. This mainly relates to the sensor system used (compare Section 9.2 – Sensors, Resolution, and Scale) and thus implicitly refers to the scale question and the temporal resolution problem.

Multitemporal images have to represent similar quantities to be compared or the change detection algorithm has to be robust against uncalibrated image input. Broad categories of change detection strategies comprise bi-temporal vs. multitemporal approaches, qualitative vs. quantitative approaches, and purely image-based vs. GIS-integrated approaches (Coppin, Jonckheere, Nackaerts, Muys, & Lambin, 2004; Lu et al., 2004).

9.4 Examples from US-LTER Sites

All US-LTER sites extensively utilize remote sensing-based data and derived information layers. Brief references have been given to selected activities in the previous section of this chapter. In the following, details are given for two examples from different ecosystems – the Andrews Forest LTER and the Northern Temperate Lakes LTER.

9.4.1 Remote Sensing for the Andrews Forest LTER

The H.J. Andrews Experimental Forest, a 6,400 ha USDA Forest Service research site established in 1948 in the Cascade Range of Oregon, has been a member of the US Long-Term Ecological Research (LTER) program since its inception in 1980 (AEF-LTER, 2002). This wet (2,500 mm annual precipitation), temperate conifer forest ecosystem ranges in age from plantation forests 35–55 yrs old (following clear cutting and burning) to 500-yr-old native forest established after wildfire. The structure, composition, and dynamics of forests and streams in this region have been the subject of intense science inquiry and social controversy concerning conservation vs. exploitation. This theme of science in its societal context appears in the central question of the Andrews Forest LTER program: *How do land use (forest cutting and roads), natural disturbances (mainly flood and wildfire), and climate variability affect key ecosystem properties, especially hydrology, carbon and nutrient dynamics, and biodiversity*? Several aspects of this question are well served by remote sensing science.

The Andrews Forest research program capitalizes on the complementary aspects of intensive, site-focused work of the LTER site and the much more extensive capabilities of remote sensing. For example:

- LTER studies have been concerned with how vertical and horizontal structure of forest of various age classes and origins influence the roles of the forest as habitat, sources of wood products, and sites of carbon sequestration. In this context, remote sensing scientists have explored applications of LIDAR to quickly and accurately assess forest structure and estimate various stand attributes and ecological functions (Lefsky et al., 1999; Means et al., 2000).
- Development of forest landscape patterns in response to wildfire and forest cutting has been a focus of landscape ecology and forest management and policy in the Andrews Forest LTER program

and across the Pacific Northwest for several decades (Franklin & Forman, 1987; Garman, Swanson, & Spies, 1999; Spies, Riple, & Bradshaw, 1994). Remote sensing of the history of stand-replacement disturbance has made it possible to examine effects of natural disturbance processes (e.g., wildfire) and land use (forest cutting) across millions of hectares spanning a wide range of environments and land ownerships and associated policies (Cohen et al., 2002).

- LTER scientists have been examining carbon stores and fluxes in forests of various structures, compositions, and disturbance histories. This information has been incorporated into models used in conjunction with remote assessment of disturbance history to map carbon sequestration/release in a 1 million hectare area of the Oregon Cascade Range (Cohen, Harmon, Wallin, & Fiorella, 1996). The broad geographic extent of remote sensing analysis makes it possible to carry the analysis across private and public ownerships, which have quite different land use practices.

The potential for future collaborations of LTER with remote sensing science is rich (LARSE, 2007). For example, new techniques for assessing partial stand disturbance (e.g., effects of low to moderate severity fire, cutting, and wind disturbance) now under development will be useful in examining effects of future fires and forestry practices. Further links of modeling with remote sensing will help extrapolate new findings from the LTER site to much larger areas. These are very appropriate ways to mutually benefit site-intensive work at sites like LTER's and geographically extensive programs, like those using remote sensing.

9.4.2 Remote Sensing for the North Temperate Lakes LTER

The North Temperate Lakes LTER site (NTL-LTER) was created to support research into the ecology of lakes over longer time periods and broader spatial scales than most traditional studies in limnology (Magnuson, Kratz, & Benson, 2006). Like the Andrews LTER site, NTL was among the six original sites in the US-LTER network at its inception. Today it focuses on two lake districts, one (the Northern Highlands Lake District) located in a forested landscape in northern Wisconsin and the other (the Madison Lakes) in an urban and agricultural setting in the southern part of the state. The site is operated by the University of Wisconsin-Madison, through the UW Centre for Limnology.

Research conducted at the site has been quite broad in scope, ranging from traditional studies of in-lake processes, to investigations of the connections between lakes and their watersheds, to studies of landscape-scale and regional-scale ecological processes (Carpenter et al., 2007). During the 1990s, the site's mission expanded to include social science alongside the natural science that represents the majority of research in the US-LTER network.

Remote sensing has played a significant role at the NTL site since the 1980s. The site has benefited from a close relationship with the University of Wisconsin's Environmental Remote Sensing Center. Examples of remote sensing-related projects at NTL-LTER include the following:

- NTL-LTER researchers have used Landsat imagery to map water clarity in approximately 8000 lakes state wide (Chipman, Lillesand, Schmaltz, Leale, & Nordheim, 2004). With support from NASA and the Wisconsin Department of Natural Resources, state wide lake clarity assessments were conducted for three periods (ca. 1980, 1992, and 2000). These assessments show general increases in water clarity across the region over this time period, a finding in accordance with field measurements and with similar studies in neighboring states (Peckham & Lillesand, 2006). The resulting water clarity database is now being used for a wide variety of studies to determine the relationships between lake water clarity and numerous other lake-level, landscape, and economic factors (e.g., Peckham, Chipman, Lillesand, & Dodson, 2006).
- Remote sensing of lake ice provides a robust indicator of regional-scale climate change. During the 1990s, scientists at NTL-LTER conducted a survey of trends in ice phenology in the northern US and Canada, using satellite imagery (Wynne, Lillesand, Clayton, & Magnuson, 1998). This eventually led to the formation of an international effort to examine long-term trends in lake ice across the northern hemisphere (Magnuson et al., 2000).

- The northern and southern lake districts in Wisconsin have experienced radically different patterns and trends in land use/land cover change. Researchers at NTL-LTER have used historical aerial photographs to interpret land use/land cover in the 1930s, 1960s, and 1990s for both study areas. In the Madison Lakes area, an increase in urbanization has led to dramatic changes in the extent of impervious surfaces and in the velocity and volume of runoff to the lakes following rainstorms (Wegner, 2001). In the northern lakes region, land use changes have included reductions in clear cutting along lakes' riparian buffer zones and increases in low-density residential development (Carpenter et al., 2007; Marburg, 2006).

Remotely sensed data and products derived from remote sensing are now among the 'core data sets' maintained by the NTL site. Over the coming years, it is expected that the use of remote sensing (and, more broadly, geographic information systems and spatial analysis) will continue to increase at the site. New focus areas for this activity include the use of field spectroradiometers for characterizing the bio-optical properties of lakes, and regional-scale monitoring of lakes and their watersheds at high temporal frequency using global monitoring satellite systems such as MODIS.

9.5 Conclusions and Outlook

It has been shown that remote sensing offers a wealth of applications related to LTER. It is, however, important to keep in mind the given limitations in order to realistically identify the potential of remote sensing. It also needs careful decisions of project managers and scientists to apply appropriate techniques and to choose the right degree of complexity in remote sensing-based analyses in the frame of LTER. In other words, there is an increasing necessity to incorporate remote sensing specialists in LTER studies to handle complex data and analysis tasks.

Future sensors and methods shall offer new opportunities, but will also add to the complexity of applications. Satellite-based hyperspectral data at Landsat-like geometric resolutions might be one of the forthcoming challenges in LTER-related remote sensing studies. Analyzing data with unprecedented spectral properties at high geometric resolutions based on operational acquisition intervals will certainly further promote quantitative analysis and remote sensing-based modeling approaches in LTER.

Finally, there is the need for future LTER to develop a closer integration at the cross-section of human–environmental systems (Haberl et al., 2006). Combining remote sensing and geoinformatics (also referred to as 'geomatics') is one way to integrate such interdisciplinary approaches through spatially explicit analysis and modeling. Research is needed to further develop methods to integrate 'soft' socioeconomic with 'hard' natural science information. LTER is one of the few arenas where the scientific breadth of involved disciplines facilitates substantial advances in integrating human–environment–systems analysis.

References

AEF-LTER (2002, available online). Andrews Experimental Forest LTER, http://www.fsl.orst.edu/lter/, [accessed June 29 2002].

Anderson, G. L., Hanson, J. D., & Haas, R. H. (1993). Evaluating Landsat Thematic Mapper derived vegetation indexes for estimating aboveground biomass on semiarid rangelands. *Remote Sensing of Environment, 45*, 165–175.

Aplin, P. (2006). On scales and dynamics in observing the environment. *International Journal of Remote Sensing, 27*, 2123–2140.

Atkinson, P. M., & Tatnall, A. R. L. (1997). Neural networks in remote sensing – Introduction. *International Journal of Remote Sensing, 18*, 699–709.

Baret, F., & Guyot, G. (1991). Potentials and limits of vegetation indexes for LAI and APAR assessment. *Remote Sensing of Environment, 35*, 161–173.

Benz, U. C., Hofmann, P., Willhauck, G., Lingenfelder, I., & Heynen, M. (2004). Multi-resolution, object-oriented fuzzy analysis of remote sensing data for GIS-ready information. *Isprs Journal of Photogrammetry and Remote Sensing, 58*, 239–258.

Bodechtel, J. (2001). Requirements on optical sensors for quantitative definition of surface parameters multispectral – Hyperspectral. *Advances in Space Research, 28*, 241–250.

Boelman, N. T., Stieglitz, M., Rueth, H. M., Sommerkorn, M., Griffin, K. L., Shaver, G. R., et al. (2003). Response of NDVI, biomass, and ecosystem gas exchange to long-term warming and fertilization in wet sedge tundra. *Oecologia, 135*, 414–421.

Cadenasso, M. L., Pickett, S. T. A., & Schwarz, K. (2007). Spatial heterogeneity in urban ecosystems: Reconceptualizing land cover and a framework for classification. *Frontiers in Ecology and the Environment, 5*, 80–88.

Capolsini, P., Andrefouet, S., Rion, C., & Payri, C. (2003). A comparison of Landsat ETM+, SPOT HRV, Ikonos, ASTER, and airborne MASTER data for coral reef habitat mapping in south pacific islands. *Canadian Journal of Remote Sensing, 29*, 187–200.

Carpenter, S. R., Benson, B. J., Biggs, R., Chipman, J. W., Foley, J. A., Golding, S. A., et al. (2007). Understanding regional change: A comparison of two lake districts. *Bioscience, 57*, 323–335.

Chipman, J. W., Lillesand, T. M., Schmaltz, J. E., Leale, J. E., & Nordheim, M. J. (2004). Mapping lake water clarity with Landsat images in Wisconsin, USA. *Canadian Journal of Remote Sensing, 30*, 1–7.

Cohen, W. B., Harmon, M. E., Wallin, D. O., & Fiorella, M. (1996). Two decades of carbon flux from forests of the Pacific northwest. *Bioscience, 46*, 836–844.

Cohen, W. B., Spies, T. A., Alig, R. J., Oetter, D. R., Maiersperger, T. K., & Fiorella, M. (2002). Characterizing 23 years (1972–1995) of stand replacement disturbance in western Oregon forests with Landsat imagery. *Ecosystems, 5*, 122–137.

Coppin, P., Jonckheere, I., Nackaerts, K., Muys, B., & Lambin, E. (2004). Digital change detection methods in ecosystem monitoring: A review. *International Journal of Remote Sensing, 25*, 1565–1596.

Dial, G., Bowen, H., Gerlach, F., Grodecki, J., & Oleszczuk, R. (2003). IKONOS satellite, imagery, and products. *Remote Sensing of Environment, 88*, 23–36.

Dorren, L. K. A., Maier, B., & Seijmonsbergen, A. C. (2003). Improved Landsat-based forest mapping in steep mountainous terrain using object-based classification. *Forest Ecology and Management, 183*, 31–46.

Duncan, J., Stow, D., Franklin, J., & Hope, A. (1993). Assessing the relationship between spectral vegetation indexes and shrub cover in the Jornada Basin, New-Mexico. *International Journal of Remote Sensing, 14*, 3395–3416.

Franklin, J. F., & Forman, R. T. T. (1987). Creating landscape patterns by forest cutting: Ecological consequences and principles. *Landscape Ecology, 1*, 5–18.

Garman, S. L., Swanson, F. J., & Spies, T. A. (1999). Past, present, and future landscape patterns in the Douglas-fir region of the Pacific Northwest. In J. A. Rochelle, Lehmann, L. A., & Wisniewski, J. (Eds.), *Forest fragmentation: Wildlife and management implication* (p. 322). Leiden, The Netherlands: Brill Academic Publishers.

Gupta, R. K., Prasad, T. S., & Vijayan, D. (2000). Relationship between LAI and NDVI for IRS LISS and LANDSAT TM bands. *Advances in Space Research, 26*, 1047–1050..

Haberl, H., Winiwarter, V., Andersson, K., Ayres, R., Boone, C., Castillo, A., et al. (2006). From LTER to LTSER: Conceptualizing the socio-economic dimension of long-term socio-ecological research. *Ecology and Society, 11*, 13.

Hobbie, J. E. (2003). Scientific accomplishments of the long term ecological research program: An introduction. *Bioscience, 53*, 17–20.

Hobbie, J. E., Carpenter, S. R:, Grimm, N. B., Gosz, J. R., & Seastedt, T. R. (2003). The US Long Term Ecological Research program. *Bioscience, 53*, 21–32.

Hostert, P., Röder, A., & Hill, J. (2003). Coupling spectral unmixing and trend analysis for monitoring of long-term vegetation dynamics in Mediterranean rangelands. *Remote Sensing of Environment, 87*, 183–197.

Huang, C., Davis, L. S., & Townshend, J. R. G. (2002). An assessment of support vector machines for land cover classification. *International Journal of Remote Sensing, 23*, 725–749.

Hudak, A. T., Lefsky, M. A., Cohen, W. B., & Berterretche, M. (2002). Integration of lidar and Landsat ETM plus data for estimating and mapping forest canopy height. *Remote Sensing of Environment, 82*, 397–416.

Justice, C. O., Vermote, E., Townshend, J. R. G., Defries, R., Roy, D. P., Hall, D. K., et al. (1998). The Moderate Resolution Imaging Spectroradiometer (MODIS): Land remote sensing for global change research. *IEEE Transactions on Geoscience and Remote Sensing, 36*, 1228–1249.

Keys, E., Wentz, E. A., & Redman, C. L. (2007). The spatial structure of land use from 1970–2000 in the Phoenix, Arizona, Metropolitan Area. *Professional Geographer, 59*, 131–147.

Kimball, J. S., McDonald, K. C., Running, S. W., & Frolking, S. E. (2004). Satellite radar remote sensing of seasonal growing seasons for boreal and subalpine evergreen forests. *Remote Sensing of Environment, 90*, 243–258.

Kontoes, C., Wilkinson, G. G., Burrill, A., Goffredo, S., & Megier, J. (1993). An experimental system for the integration of GIS data in knowledge-based image-analysis for remote-sensing of agriculture. *International Journal of Geographical Information Systems, 7*, 247–262.

Kratz, T. K., Deegan, L. A., Harmon, M. E., & Lauenroth, W. K. (2003). Ecological variability in space and time: Insights gained from the US LTER program. *Bioscience, 53*, 57–67.

Kuemmerle, T., Hostert, P., St-Louis, V., & Radeloff, V. C. (2009). Using image texture to map field size in Eastern Europe. *Journal of Landuse Science, 4*, 85–107.

LARSE (2007, available online). Laboratory for Applications of Remote Sensing in Ecology, http://www.fsl.orst.edu/larse/website/, [accessed June 29 2007].

Lee, K. S., Cohen, W. B., Kennedy, R. E., Maiersperger, T. K., & Gower, S. T. (2004). Hyperspectral versus multispectral data for estimating leaf area index in four different biomes. *Remote Sensing of Environment, 91*, 508–520.

Lefsky, M. A., Cohen, W. B., Acker, S. A., Parker, G. G., Spies, T. A., & Harding, D. (1999). Lidar remote sensing of the canopy structure and biophysical properties of Douglas-fir western hemlock forests. *Remote Sensing of Environment, 70*, 339–361.

Li, L., Ustin, S. L., & Lay, M. (2005). Application of AVIRIS data in detection of oil-induced vegetation stress and cover change at Jornada, New Mexico. *Remote Sensing of Environment, 94*, 1–16.

Lillesand, T. M., & Kiefer, R. W. (2000). *Remote sensing and image interpretation*. New York, USA: John Wiley & Sons.

Lim, K., Treitz, P., Wulder, M., St-Onge, B., & Flood, M. (2003). LiDAR remote sensing of forest structure. *Progress in Physical Geography, 27*, 88–106.

Lu, D., Mausel, P., Brondizio, E., & Moran, E. (2004). Change detection techniques. *International Journal of Remote Sensing, 25*, 2365–2407.

Magnuson, J. J., Kratz, T. K., & Benson, B. J. (Eds.) (2006). *Long-term dynamics of lakes in the landscape: Long-term ecological research on North Temperate Lakes*. New York: Oxford University Press.

Magnuson, J. J., Robertson, D. M., Benson, B. J., Wynne, R. H., Livingstone, D. M., Arai, T., et al. (2000). Historical trends in lake and river ice cover in the Northern Hemisphere. *Science, 289*, 1743–1746.

Maisongrande, P., Duchemin, B., & Dedieu, G. (2004). VEGETATION/SPOT: An operational mission for the Earth monitoring; presentation of new standard products. *International Journal of Remote Sensing, 25*, 9–14.

Malczewski, J. (1999). *GIS and multicriteria decision analysis*. New York: John Wiley & Sons.

Marburg, A. E. (2006). *Spatial and temporal patterns of riparian land cover, forests and littoral coarse wood in the Northern Highlands Lake District, Wisconsin, USA*. Wisconsin, USA, PhD dissertation, University of Wisconsin, Madison.

Means, J. E., Acker, S. A., Fitt, B. J., Renslow, M., Emerson, L., & Hendrix, C. J. (2000). Predicting forest stand characteristics with airborne scanning lidar. *Photogrammetric Engineering and Remote Sensing, 66*, 1367–1371.

Mika, A. M. (1997). Three decades of Landsat instruments. *Photogrammetric Engineering and Remote Sensing, 63*, 839–852.

Pal, M., & Mather, P. M. (2005). Support vector machines for classification in remote sensing. *International Journal of Remote Sensing, 26*, 1007–1011.

Palacios-Orueta, A., & Ustin, S. L. (1998). Remote sensing of soil properties in the Santa Monica Mountains I. Spectral analysis. *Remote Sensing of Environment, 65*, 170–183.

Peckham, S. D., & Lillesand, T. M. (2006). Detection of spatial and temporal trends in Wisconsin lake water clarity using landsat-derived estimates of Secchi depth. *Lake and Reservoir Management, 22*, 331–341.

Peckham, S. D., Chipman, J. W., Lillesand, T. M., & Dodson, S. I. (2006). Alternate stable states and the shape of the lake trophic distribution. *Hydrobiologia, 571*, 401–407.

Pickett, S. T. A., Cadenasso, M. L., Grove, J. M., Groffman, P. M., Band, L. E., Boone, C. G., et al. (2008). Beyond urban legends: An emerging framework of urban ecology, as illustrated by the Baltimore Ecosystem Study. *Bioscience, 58*, 139–150.

Rast, M., Bezy, J. L., & Bruzzi, S. (1999). The ESA Medium Resolution Imaging Spectrometer MERIS – a review of the instrument and its mission. *International Journal of Remote Sensing, 20*, 1681–1702.

Rastetter, E. B., Aber, J. D., Peters, D. P. C., Ojima, D. S., & Burke, I. C. (2003). Using mechanistic models to scale ecological processes across space and time. *Bioscience, 53*, 68–76.

Richards, J. A., & Jia, X. (2006). *Remote sensing digital image analysis. An introduction*. Berlin, Heidelberg, New York: Springer.

Sanderson, J., & Harris, L. D. (Eds.). (2000). *Landscape ecology: A top-down approach*. Boca Raton: Lewis Publishers.

Sims, D.A., & Gamon, J.A. (2003). Estimation of vegetation water content and photosynthetic tissue area from spectral reflectance: A comparison of indices based on liquid water and chlorophyll absorption features. *Remote Sensing of Environment, 84*, 526–537.

Song, C., Woodcock, C. E., Seto, K. C., Lenney, M. P., & Macomber, S. A. (2001). Classification and change detection using Landsat TM data: When and how to correct atmospheric effects? *Remote Sensing of Environment, 75*, 230–244.

Spies, T. A., Riple, W. J., & Bradshaw, G. A. (1994). Dynamics and pattern of a managed coniferous forest landscape in Oregon. *Ecological Applications, 4*, 555–568.

Stefanov, W. L., Netzband, M., Möller, M. S., Redman, C. L., & Mack, C. (2007). Phoenix, Arizona, USA: Applications of remote sensing in a rapidly urbanizing desert region. In M. Netzband, W. Stefanov, & C. L. Redman (Eds.), *Applied remote sensing for urban planning, governance and sustainability* (pp. 137–164). Berlin, Heidelberg: Springer.

Stefanov, W. L., Ramsey, M. S., & Christensen, P. R. (2001). Monitoring urban land cover change: An expert system approach to land cover classification of semiarid to arid urban centers. *Remote Sensing of Environment, 77*, 173–185.

St-Louis, V., Pidgeon, A. M., Radeloff, V. C., Hawbaker, T. J., & Clayton, M. K. (2006). High-resolution image texture as a predictor of bird species richness. *Remote Sensing of Environment, 105*, 299–312.

Torgersen, C. E., Faux, R. N., McIntosh, B. A., Poage, N. J., & Norton, D. J. (2001). Airborne thermal remote sensing for water temperature assessment in rivers and streams. *Remote Sensing of Environment, 76*, 386–398.

Treuhaft, R. N., Law, B. E., & Asner, G. P. (2004). Forest attributes from radar interferometric structure and its fusion with optical remote sensing. *Bioscience, 54*, 561–571.

Tucker, C. J., Pinzon, J. E., Brown, M. E., Slayback, D. A., Pak, E. W., Mahoney, R., et al. (2005). An extended AVHRR 8-km NDVI dataset compatible with MODIS and SPOT vegetation NDVI data. *International Journal of Remote Sensing, 26*, 4485–4498.

Turner, M. G., & Gardner, R. H. (1991). *Quantitative methods in landscape ecology*. New York: Springer.

Turner, M. G. (1989). Landscape ecology – the effect of pattern on process. *Annual Review of Ecology and Systematics, 20*, 171–197.

Turner, M. G. (2005). Landscape ecology in North America: Past, present, and future. *Ecology, 86*, 1967–1974.

Underwood, E. C., Ustin, S. L., & Ramirez, C. M. (2007). A comparison of spatial and spectral image resolution for mapping invasive plants in coastal California. *Environmental Management, 39*, 63–83.

Ustin, S. L., Roberts, D. A., Gamon, J. A., Asner, G. P., & Green, R. O. (2004). Using imaging spectroscopy to study ecosystem processes and properties. *Bioscience, 54*, 523–534.

Vogt, P., Riitters, K. H., Estreguil, C., Kozak, J., & Wade, T.G. (2007). Mapping spatial patterns with morphological image processing. *Landscape Ecology, 22*, 171–177.

Voogt, J. A., & Oke, T. R. (2003). Thermal remote sensing of urban climates. *Remote Sensing of Environment, 86*, 370–384.

Walker, J. S., & Briggs, J. M. (2007). An object-oriented approach to urban forest mapping in Phoenix. *Photogrammetric Engineering and Remote Sensing, 73*, 577–583.

Wang, L., Sousa, W. P., Gong, P., & Biging, G. S. (2004). Comparison of IKONOS and QuickBird images for mapping

mangrove species on the Caribbean coast of Panama. *Remote Sensing of Environment, 91*, 432–440.
Wegner, M. W. (2001). Long-term land use/land cover change patterns in the Yahara Lakes region and their impact on runoff volume to Lake Mendota. Master's thesis. University of Wisconsin, Madison.
Wickham, J. D., Oneill, R. V., Riitters, K. H., Wade, T. G., & Jones, K. B. (1997). Sensitivity of selected landscape pattern metrics to land-cover misclassification and differences in land-cover composition. *Photogrammetric Engineering and Remote Sensing, 63*, 397–402.
Wilkinson, G. G. (1996). A review of current issues in the integration of GIS and remote sensing data. *International Journal of Geographical Information Systems, 10*, 85–101.
Wynne, R. H., Lillesand, T. M., Clayton, M. K., & Magnuson, J. J. (1998). Satellite monitoring of lake ice breakup on the Laurentian Shield (1980–1994). *Photogrammetric Engineering and Remote Sensing, 64*, 607–617.

Part IV
Concepts and Results: Presenting and Interpreting Long-Term Ecological Processes: Aquatic Ecosystem Research

Chapter 10

Long-Term Ecological Change in the Northern Wadden Sea

Justus E. E. van Beusekom, Christian Buschbaum, Martina Loebl, Peter Martens, and Karsten Reise

Abstract The Wadden Sea is a shallow coastal region in the south eastern North Sea. Karl Möbius started ecological research in the northern Wadden Sea about 150 years ago studying the extensive oyster beds. With the foundation of a field station of the Biologische Anstalt Helgoland in List/Sylt in 1924 biological research in the Wadden Sea was continued to date. Several time series were initiated between the 1970s and the 1990s including a bi-weekly phytoplankton and zooplankton program and an observation program on macrobenthos. Three factors dominating the changes observed during the past decades are a rise in temperature, decreasing nutrients, and increasing invasions of non-native species. Phytoplankton blooms gradually decrease due to the combined effect of decreasing nutrient loads and increasing winter temperatures. Mean annual zooplankton abundance is stimulated by higher winter temperatures. Recently, invading species are increasingly dominating native mussel beds. For several invaders, a positive effect of temperature was shown. We expect that major pressures of change during the next years will be further species introductions, temperature increase, and reduced nutrient loads. On the long run (21st century), we expect sea level rise to be the key factor of coastal change through a loss of habitats with fine-grained sediments and intertidal sediments in general. A major challenge for coastal research will be to disentangle the interactive effects of these pressures on the long-term development of the Wadden Sea.

J.E.E. van Beusekom (✉)
Alfred-Wegener-Institute for Polar and Marine Research, Wadden Sea Station Sylt, Hafenstrasse 43, 25992 List/Sylt, Germany
e-mail: Justus.van.beusekom@awi.de

Keywords Long-Term Change · Wadden Sea · Sea level rise · Invaders · Eutrophication · Temperature

10.1 Introduction

The Wadden Sea is an area of outstanding beauty and has a high ecological relevance. The Dutch and German governments have jointly applied the status of a 'World Heritage Site' (CWSS, 2008). The Wadden Sea stretches from Den Helder in the Netherlands to the Skallingen peninsula in Denmark. It presently comprises a shoreline of about 900 km and an area of about 8000 km^2 (Reise, 2005). It is one of the largest coherent tidal flat areas in the world.

Ecological research in the Northern Wadden Sea has a long tradition. First investigations were carried out almost one and a half centuries ago by Karl Möbius (Möbius, 1877). He investigated the oyster beds of the European Oyster (*Ostrea edulis*) in the northern Wadden Sea (Fig. 10.1). This investigation was prompted by complaints of local oyster fishermen, who noticed a gradual reduction in catch and standing stock (e.g. Lotze, 2005). A 7-year fishery ban as proposed by Möbius did not bring about the expected recovery. During the 1920s Prof. A. Hagmeier from the Biologische Anstalt Helgoland (BAH) on Helgoland suggested to develop cultivation techniques to increase the yield of the Oyster beds. This resulted among others in the foundation of a field station of the Biologische Anstalt Helgoland in 1924 in List on Sylt. Since then numerous studies have been carried out on the biology and ecology of the northern Wadden Sea.

The early studies now are an indispensable tool to understand long-term coastal change in the Wadden

F. Müller et al. (eds.), *Long-Term Ecological Research*,
DOI 10.1007/978-90-481-8782-9_10, © Springer Science+Business Media B.V. 2010

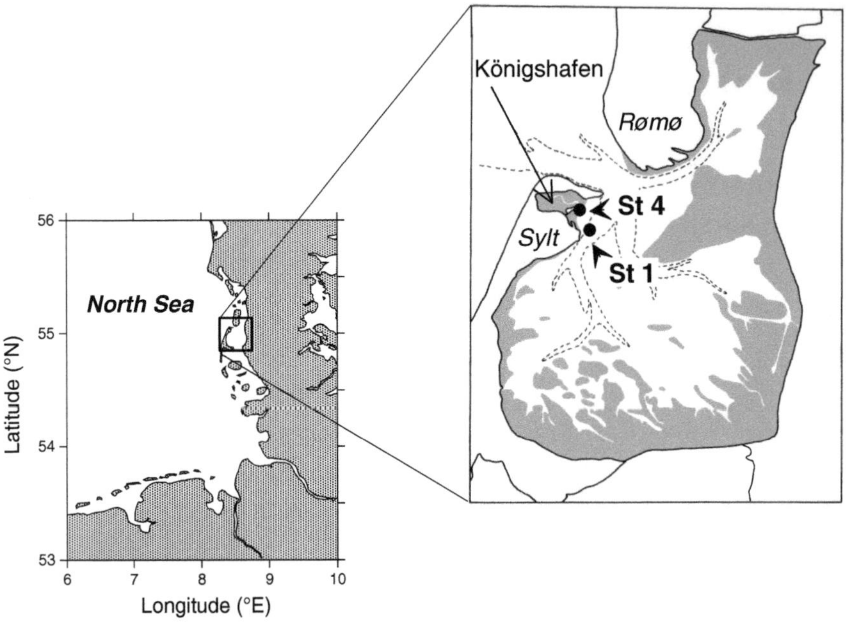

Fig. 10.1 Map showing the position of the Northern Wadden Sea and the List Tidal Basin

Sea. During recent decades, regular observations were initiated to study the temporal dynamics of this ecosystem. Several of these programs are now in place and they cover a wide range of aspects including hydrography, nutrients, phytoplankton, zooplankton, macrobenthos, macroalgae, mussel- and sea grass beds, and sedimentology (van Beusekom & Reise, 2008).

Focus of this contribution is to give an overview of ecological changes observed in the northern part of the Wadden Sea during the past century. We will show that at present temperature, eutrophication and species introduction are the main drivers of coastal change. First indications (Reise, Herre, & Sturm, 2008) point at sea level rise to be a prominent factor in coastal change during this century (Reise & van Beusekom, 2008). On longer time scales covering the time before the first scientific reports, other factors have induced ecological change in the Wadden Sea including fishing, hunting, and embankment (Lotze, 2005; Lotze et al., 2005; Reise, 2005). Fishing remains an important aspect of human-induced change in coastal ecology but is not explicitly dealt with in this study.

10.2 Area Description

Our data and investigations focus on the List Tidal Basin (54°50′–55°10′ N and 8°20′–8°40′ E), a 404 km^2semi-enclosed bight in the northern Wadden Sea (North Sea, Europe; Fig. 10.1). The basin is connected to the open North Sea by a single tidal inlet. To the north and to the south the basin is closed by two dams connecting the island of Rømø and the island of Sylt with the mainland. The southern dam (Hindenburgdamm) was built in 1927. The northern dam was finished in 1948. Water volume at mean tidal level is about 845 $\times$ 10^6 m^3. The mean water depth is 2.7 m (Loebl, Dolch, & van Beusekom, 2007) but reaches up to 40 m in the main tidal channel. The water column is mostly homogeneously mixed (Hickel, 1980). Tides are semidiurnal; the mean tidal range is about 2 m. During low tide, about 40% of the area is emerged. About 90% of the area is covered by sandy sediments. Salinity dynamics are dominated by the freshwater discharge of the rivers Weser and Elbe debouching about 150 km south of the Basin into the North Sea (Hickel, 1980; van Beusekom, Weigelt-Krenz, & Martens, 2008). Monthly mean salinity and temperature reach seasonal minimum values around February and maximum values around August (van Beusekom, Weigelt-Krenz, et al., 2008). Monthly mean salinity (1984–2005) ranges between 24.8 and 31.8 in February (mean, 28.1 $\pm$ 1.5) and between 28.6 and 32.4 in August (mean, 30.7 $\pm$ 1.1). Monthly mean temperature (1984–2005) ranges between –1.8 and +5.3 in February (mean, 2.3 $\pm$ 2.0) and between

16.3 and 21.9 in August (mean, 18.4 ± 1.4) (Martens & van Beusekom, 2008). Suspended matter ranges from ~20–60 mg/l in winter to ~5 mg/l in late summer. A detailed description of the area is given by Gätje and Reise (1998).

10.3 Historical Changes Since the Mid-19th Century

The earliest investigations in the northern Wadden Sea that allow at least a qualitative comparison with the present situation date back to the second half of the 19th century. Möbius (1877, 1893) and Hagmeier and Kändler (1927) observed a diverse epibenthos in tidal channels of the northern Wadden Sea. They described dense beds of subtidal eelgrass (*Zostera marina*) and mussels and reefs of oysters (*Ostrea edulis)* and *Sabellaria spinulosa* – a colonial and tube-building polychaete. These earliest investigations were repeated by Buhs and Reise (1997) using a similar gear, i.e., a traditional oyster dredge. Buhs and Reise (1997) and Reise and Buhs (1999) concluded that the epibenthic species richness had declined. Important communities like *Sabellaria* reefs, subtidal eelgrass beds, and oyster beds were diminished. The decrease in species richness was mainly attributed to fisheries. The disappearance of subtidal eelgrass beds is probably related to the so-called wasting disease by the slime mold *Labyrinthula* sp. that whipped out most of the eelgrass beds during the 1930s (Wohlenberg, 1935).

Earliest studies of the intertidal that allow a comparison with the recent situation date back to investigations in Königshafen during the 1920s and 1930s. Nienburg (1927) and Wohlenberg (1937) mainly focused on the benthic community of the intertidal of Königshafen, a small embayment in the northern part of the island of Sylt (Fig. 10.1). Reise et al. (2008) repeated Wohlenberg's investigation between 1988 and 2006 and detected major changes (Fig. 10.2). In the upper intertidal zone, muddy sediments originally inhabited by *Corophium volutator* are now replaced by a sandy berm with *Arenicola marina*. Below, former sea grass beds are now replaced by dense green algae mats (*Enteromorpha* spp., *Chaetomorpha* spp., *Ulva* ssp.). The lowest intertidal zone was dominated by extensive mussel beds and attached macroalgae (*Fucus vesiculosus* forma *mytili*) but nowadays these musselbeds are largely gone. Reise et al. (2008) suggest that the combination of four factors has lead to the observed changes during seven decades (1) sea level rise being responsible for increased sandiness at the expense of mud inhibiting *C. volutator* and facilitating *A. marina*, (2) eutrophication supported the development of massive green algae mats affecting most infauna and sea grass, (3) extreme weather events initiated the loss of mussel beds with attached fucoid algae, and (4) introduced exotic species (about nine) have added to the local species richness. A long-term

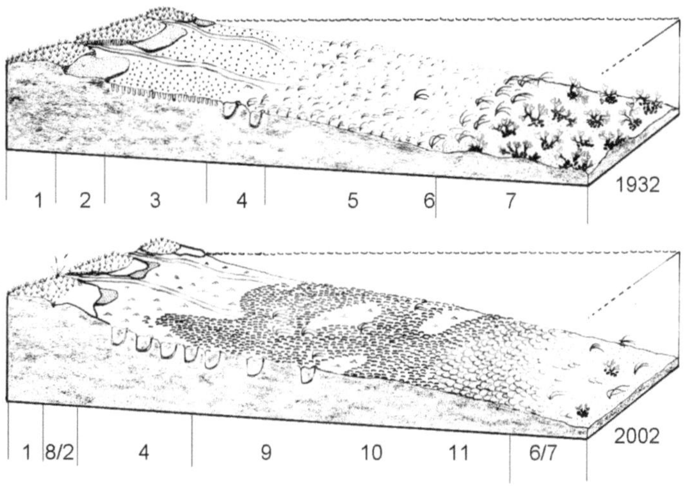

Fig. 10.2 Macrobenthic zonation in Gröning-Watt (Königshafen Sylt) from a *Puccinellia*-saltmarsh down to low tide level in 1932 (schematic after Wohlenberg, 1937; see Fig. 1.1) and seven decades later following a rise in high water level and eutrophication (see text). 1 – saltmarsh, 2 – Cyanobacteria-mats, 3 – *Corophium*-belt, 4 – *Arenicola*, 5 – *Zostera noltii*, 6 – *Z. marina*, 7 – *Fucus* anchored by mussels, 8 – sandy beach, 9 – *Enteromorpha*-mats, 10 – *Chaetomorpha*-mats, 11 – *Ulva*-mats. From Reise et al. (2008)

comparison of sediment composition in Königshafen by Dolch and Hass (2008) supports the general loss of fine sediment particles from Königshafen, presumably caused by increasing hydrodynamics due to sea level rise. According to the CPSL (2005) the annual mean high water level rose by 0.25 cm/year in the Wadden Sea from 1890 to 1989 and even by 0.67 cm/year for the period 1971–1989. A rising water level is associated with stronger currents that cause removal of fine-grained sediments (Dolch & Hass, 2008).

10.4 Recent Changes

The long-term comparisons described above give valuable information on changes on the decadal time scale. However, the resolution of the observation cannot adequately address short-term dynamics. Several observation programs are now in place revealing the annual dynamics of the ecosystem. These reveal that temperature and nutrients are the main drivers of ecological dynamics on annual time scales. In addition, the abundance of several invading species increased rapidly during the past decades.

10.4.1 Temperature Changes

The mean seasonal cycle of the water temperature shows on average minimum values of about 2°C in February and maximum temperatures of 18°C in August (Fig. 10.3a). We calculated the deviations of monthly water temperatures from the long-term monthly mean (Fig. 10.3b). Especially during the period 1984–1987 the winter temperatures were below the long-term mean. During recent years autumn temperatures are clearly above the long-term mean. A trend analysis for the period 1984–2005 (Martens & van Beusekom, 2008) revealed a significant increase in the mean annual temperature of about 1°C during the last 20 years, but interannual variability remains very high. On a monthly basis, a significant increase of more than 3°C with time was only found for September

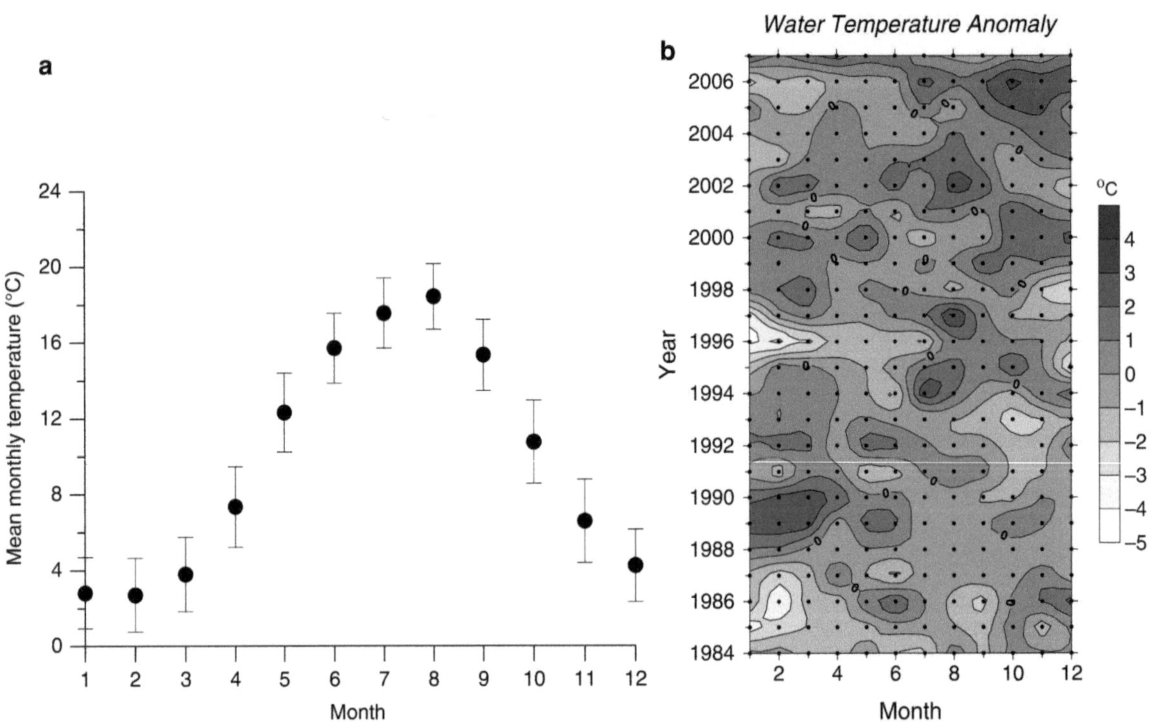

Fig. 10.3 (a) Mean seasonal cycle of mean monthly water temperature in the List Tidal Basin (1984–2007). The *error bars* are the standard deviation of the monthly means. (b)Temperature anomaly (monthly mean minus long-term monthly mean) in the List Tidal Basin, 1984–2007

indicating a gradual lengthening of the summer season (Martens & van Beusekom, 2008).

10.4.2 Salinity and Nutrients

The rivers Weser and Elbe are the major freshwater sources in the German Bight and adjacent northern Wadden Sea (e.g., Hickel, 1980). Regular data on the rivers Weser and Elbe are available since the late 1970s. Discharge shows large interannual differences and correlates with the mean annual salinity in the List Tidal Basin with highest salinities during years with a low riverine discharge (van Beusekom, Weigelt-Krenz, et al., 2008). Nitrogen is the main driver of coastal eutrophication in the northern Wadden Sea (van Beusekom, Loebl, & Marten, 2008). Riverine nitrogen loads attained a maximum during the mid 1980s and decreased continuously since then at a rate of about 2% per year (van Beusekom et al., 2005). Van Beusekom, Loebl, et al. (2008) found a similar decrease in winter nitrate concentrations in the List Tidal Basin of about 2% in a multiple regression including time and salinity as independent factors, but a large part of the variability remained unexplained. The authors suggested that denitrification or anammox (the anaerobic oxidation of NH_4 by NO_2 to N_2; e.g. Kuypers et al., 2003) is responsible for a variable amount of nitrate being removed during winter.

10.4.3 Plankton

Two time series focusing on phytoplankton and zooplankton are now being maintained in the list Tidal Basin (Martens & van Beusekom, 2008; van Beusekom, Loebl, et al., 2008). Inter annual phytoplankton dynamics are influenced by two factors: temperature and riverine nutrient loads. The diatom spring bloom which occurs between March and May is strongly influenced by temperature with higher biomass being attained after cold winters. A similar phenomenon is observed along the North American east coast and experimental evidence suggests that both enhanced pelagic grazing by zooplankton and enhanced filter feeding by benthic organisms are responsible for the low phytoplankton biomass attained after mild winters (Keller, Oviatt, Walker, & Hawk, 1999). Mean summer phytoplankton biomass on the other hand is significantly correlated with the riverine nitrogen loads by the rivers Weser and Elbe ($p = 0.0076$; Fig. 10.4). At present, much attention is given to background values as they may be used to assess the ecological state of coastal waters for the EU Water Framework Directive. No data are available, but we estimate that the pre-eutrophication levels were about five times lower than recent values (1984–2002) (van Beusekom, 2005). Van Beusekom (2006) estimated a summer phytoplankton biomass

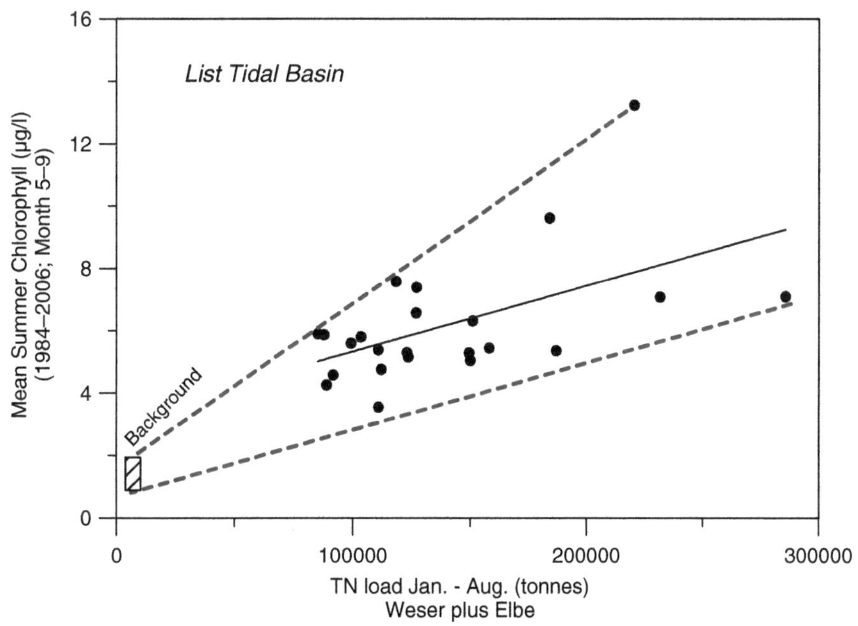

Fig. 10.4 Relation between summer phytoplankton biomass (as Chlorophyll *a*) and riverine Total Nitrogen loads (Weser plus Elbe, Month 1–8 only). The time-window from January until August was chosen because riverine nutrient discharge after August will not be able to influence the phytoplankton in List Tidal Basin. The *stippled line* denote the area of 50% uncertainty in the relation between riverine TN loads and Summer Phytoplankton biomass (as Chl a). The line is the regression line($Y = 2.1 \times 10^{-5}* X + 3.2$; $n=23$; $r^2 = 0.29$, significance $p = 0.0$)

(as Chl a) under pristine riverine nutrient loads of about 1–2 μg Chla L^{-1}. This value is also shown in Fig. 10.4. Tentatively, the following conclusions can be drawn from Fig. 10.4: A large uncertainty exists in the relation between riverine nitrogen loads and summer phytoplankton standing stock. The maximum and minimum summer phytoplankton that can be expected at a certain riverine nitrogen load lie between the two dashed lines that converge toward the estimated background chlorophyll values estimated for pristine nitrogen loads. This implies an uncertainty of about 50% at all levels of riverine nitrogen loads. Which factors are responsible for the observed variability remains unknown, both interannual differences in the magnitude pelagic grazing (Loebl & van Beusekom, 2008), benthic filter feeding and in the amount of organic matter imported into the Wadden Sea from the North Sea (van Beusekom & de Jonge, 2002) may contribute to the observed variability.

If nutrient availability drives the phytoplankton standing stock, we may expect an increase in nutrient limitation. This was indeed found by Loebl, Colijn, and van Beusekom (2008) who used a method developed by Cloern (1999) to identify potential limiting factors. Two trends were found: (1) between the 1980s and the 2000s potential nitrogen limitation increased from less than 1 to 2 months and (2) in September a shift from light limitation to a co-limitation of light and nitrogen occurred. These findings were corroborated by van Beusekom, Loebl, et al. (2008) who found a significant increase in the frequency of observations with low nitrate concentrations (<0.5 μM) during summer.

Zooplankton dynamics are mainly temperature controlled (Martens & van Beusekom, 2008). Mean annual adult copepod numbers and season length correlated positively with the preceding winter temperature. This makes sense as most of the copepod species develop from resting eggs and hatching rates are positively correlated with temperature (Holste & Peck, 2006).

10.4.4 Benthos

The long-term change of the benthos on the time scale of several decades has been addressed in a previous section. Here we will focus on recent changes (last two-three decades). The single-most factor changing the face of the Wadden Sea is the introduction of non-native species. Up to 52 exotic species are presently known for the Wadden Sea. Six of these species already do or are about to have strong effects on habitats and biota (Reise, Dankers, & Essink, 2005). In several cases, evidence exists that temperature plays an important role in the recent proliferation of these species.

The cord grass *Spartina anglica* – a fertile hybrid of S. marina and *S. alterniflora* – was introduced to the Wadden Sea coast in the 1920s for coastal protection. Recently, an enhanced spread is observed along the shoreline of Sylt and it is suggested that this plant may have taken advantage of a rise in spring temperatures (Loebl, van Beusekom, & Reise, 2006).

The Japanese sea weed *Sargassum muticum* was first observed near Sylt in 1993 (Schorries & Albrecht, 1995). The highly branched thalli offer a complex habitat for a rich associated community of native epibiontic species and motile fauna (Buschbaum, 2005; Buschbaum, Chapman, & Saier, 2006, Polte & Buschbaum, 2008).

The North American estuarine polychaete *Marenzellaria cf. wireni* is less important in the List Tidal Basin but can attain large densities in other parts of the Wadden Sea. It was first observed in the Ems estuary in 1983 (Essink, Eppinga, & Dekker, 1998).

The American razor clam *Ensis americanus* was introduced in the German Bight in 1978 and rapidly spread along the entire Wadden Sea (Armonies, 2001). Maximum abundances are reached in the shallow subtidal. Larvae reach very high abundances in the plankton (Strasser & Günther, 2001) and it may well be one of the most abundant large bivalves in the shallow subtidal reaching biomasses comparable to native cockles and mussels (Armonies & Reise, 1999; Reise, Dankers, et al., 2005).

The slipper limpet *Crepidula fornicata* was introduced to Europe with oyster cultures and was first observed in the list Tidal Basin in the 1920s. Cold winters mainly limit this filter-feeding snail (Thieltges, Strasser, van Beusekom, & Reise, 2004). Recently, a strong increase in abundances is observed (Thieltges, Reise, Prinz, & Jensen, 2009) which may be related to a series of relatively mild winters (compare Thieltges et al., 2004; Fig. 10.2).

The single-most event presently changing the face of the Wadden Sea (Fig. 10.5) is the recent spread of the pacific Oyster (*Crassostrea gigas*). In 1986

Fig. 10.5 Pictures of a mussel bed in 1995 without (*above*) and in 2006 with the pacific Oyster (photographs. C. Buschbaum, K. Reise)

an oyster culture was established in the List Tidal Basin that was stocked with spat mainly from British hatcheries and in 1991 the first free-living oyster was found (Reise, 1998). Recruitment of this bivalve occurs during July and August and is successful during years with above average temperatures (>18°C, Diederich, Nehls, van Beusekom, & Reise, 2005).

The strong increase in oyster densities from 10 individuals m^{-2} in 1995 to more than 1,000 individuals m^{-2} in 2006 (Reise, 2007) is accompanied by a loss of native blue mussel beds (*Mytilus edulis*) because oysters have been using mussels for attachment. Thus mussel beds changed to oyster reefs with many effects on the associated species community (Kochmann, Buschbaum, Volkenborn, & Reise, 2008). Apart from the Pacific Oyster, the abundance of other invasive species, most of them present in the area for several decades, increased during the past two decades (Büttger et al., 2008).

Seagrass is a quality indicator of the Wadden Sea (Reise, Jager, de Jong, van Katwijk, & Schanz, 2005). The wasting disease (an infection with the slime mold *Labyrinthula*) whipped out most seagrass beds in the 1930s. Whereas *Zostera noltii* re-established, the subtidal *Z. marina* beds did not return. From about the 1950s *Z. noltii* decreased (T. Dolch pers. comm.) but since the mid 1990s a recovery is observed (Reise & Kohlus, 2008). It remains to be tested whether the recovery is due to a change in storm frequency (Reise & Kohlus, 2008) or other factors like the decreasing eutrophication.

10.5 Outlook

We expect a further increase of invasive species in the Wadden Sea. Epibenthic suspension feeders such as Pacific Oysters (*Crassostrea gigas*), the razor clam (*Ensis americanus*) and the American slipper limpet (*Crepidula fornicata*) will expand their occurrence in the shallow subtidal zone providing additional hard substrate for further exotics like the Japanese seaweed (*Sargassum muticum*) and the suspension feeding tunicate *Styela clava*. The increase in suspension feeders in combination with decreasing nutrient loads will also affect the phyto- and zooplankton density in the water column with consequences for the nutrient cycles. On longer time scales, sea level rise will have a major impact on this coastal ecosystem (Reise & van Beusekom, 2008).

The Wadden Sea is currently rapidly changing due to a combination of rapid warming (IPCC, 2007), accelerated species introductions (Reise & van Beusekom, 2008) and nutrient reduction. Long-term observations in combination with appropriate experiments will be an indispensable tool for understanding the ecological change in the Wadden Sea and for providing scenarios of future coastal change.

Acknowledgments The authors thank all participants of the Workshop on Long-Term Ecological Change in the northern Wadden Sea (List, February 2007) for fruitful discussions. Comments by Franciscus Colijn considerably improved the manuscript.

References

Armonies, W. (2001). What an introduced species can tell us about the spatial extension of benthic populations. *Marine Ecology Progress Series*, *209*, 289–294.

Armonies, W., & Reise, K. (1999). On the population development of the introduced razor clam *Ensis americanus* near the island of Sylt (North Sea). *Helgoländer Meeresuntersuchungen*, *52*, 291–300.

Buhs, F., & Reise, K. (1997). Epibenthic fauna dredged from tidal channels in the Wadden Sea of Schleswig-Holstein: Spatial patterns and a long-term decline. *Helgoländer Meeresuntersuchungen, 51*, 343–359.

Buschbaum, C. (2005). Pest oder Bereicherung? Der eingeschleppte Japanische Beerentag an der deutschen Küste. *Natur und Museum, 135*, 216–221.

Buschbaum, C., Chapman, A. S., & Saier, B. (2006). How an introduced seaweed can effect epibiota diversity in different coastal systems. *Marine Biology, 158*, 743–754.

Büttger, H., Asmus, H., Asmus, R., Buschbaum, C., Dittman, S., & Nehls, G. (2008). Community dynamics of intertidal soft-bottom mussel beds over two decades. *Helgoland Marine Research, 62*, 23–36.

Cloern, J. E. (1999). The relative importance of light and nutrient limitation of phytoplankton growth: A simple index of coastal ecosystem sensitivity to nutrient enrichment. *Aquatic Ecology, 33*, 3–16.

CPSL. (2005). *Coastal Protection and Sea Level Rise – Solutions for sustainable coastal protection in the Wadden Sea region. Wadden Sea Ecosystem No. 21.* Wilhelmshaven: Common Wadden Sea Secretariat, Trilateral Working Group on Coastal Protection and Sea Level Rise.

CWSS. (2008). *Nomination of the Dutch-German Wadden Se as World Heritage Site –Volume One*. Wilhelmshaven: Common Wadden Sea Secretariat.

Diederich, S., Nehls, G., Van Beusekom, J. E. E., & Reise, K. (2005). Introduced Pacific oysters (*Crassostrea gigas*) in the northern Wadden Sea: Invasion accelerated by warm summers? *Helgoland Marine Research, 59*, 97–106.

Dolch, T., & Hass, H. C. (2008). long-term changes of intertidal and subtidal sediment composition in a tidal basin in the northern Wadden Sea (SE North Sea). *Helgoland Marine Research, 62*, 3–11.

Essink, K., Eppinga, J., & Dekker, R. (1998). Long-term changes (1977–1994) in intertidal macrobenthos of the Dollard (Ems estuary) and effects of introduction of the North American spionid polychaete *Marenzellaria* cf. *wireni. Senckenbergia Maritima, 28*, 21–25.

Gätje, C., & Reise, K. (1998). *Ökosystem Wattenmeer: Austausch, Transport und Stoffwandlungsprozesse.* Berlin: Springer Verlag.

Hagmeier, A., & Kändler, R. (1927). Neue Untersuchungen im nordfriesischen Wattenmeer und auf den fiskalischen Austernbänken. *Wissenschaftliche Meeresuntersuchungen (Helgoland), 16*, 1–99.

Hickel, W. (1980). The influence of Elbe River Water on the Wadden Sea of Sylt (German Bight, North Sea). *Deutsche Hydrographische Zeitschrift, 33*, 42–53.

Holste, L., & Peck, M. A. (2006). The effects of temperature and salinity on egg production and hatching success of Baltic *Acartia tonsa* (Copepoda: Calanoida): A laboratory investigation. *Marine Biology, 148*, 1061–1070.

IPCC. (2007). Summary for policymakers. In S. Solomon, D. Qin, M. Manning, Z. Chen, M. Marquis, K. B. Averyt, M. Tignor, & H. L. Miller (Eds.), *Climate change 2007: The Physical Science Basis. Contribution of Working Group I to the Fourth Assessment Report of the Intergovernmental Panel on Climate Change* (pp. 1–18). Cambridge: Cambridge University Press.

Keller, A. A., Oviatt, C. A., Walker, H. A., & Hawk, J. D. (1999). Predicted impacts of elevated temperature on the magnitude of the winter-spring phytoplankton bloom in temperate coastal waters: A mesocosm study. *Limnology & Oceanography, 44*, 344–356.

Kochmann, J., Buschbaum, C., Volkenborn, N., & Reise, K. (2008). Shift from native mussels to alien oysters: Differential effects of ecosystem engineers. *Journal Experimental Marine Biology and Ecology, 364*, 1–10.

Kuypers, M., Sliekers, A., Lavik, G., Schmid, M., Jørgensen, B. B., Kuenen, J., et al. (2003). Anaerobic ammonium oxidation by anammox bacteria in the Black Sea. *Nature, 422*, 608–611.

Loebl, M., & van Beusekom, J. E. E. (2008). Seasonality of microzooplankton grazing in the northern Wadden Sea. *Journal of Sea Research, 59*, 203–216.

Loebl, M., Colijn, F., & van Beusekom, J. E. E. (2008). Increasing nitrogen limitation during summer in the List Tidal Basin (Northern Wadden Sea). *Helgoland Marine Research, 62*, 59–65.

Loebl, M., Dolch, T., & van Beusekom, J. E. E. (2007). Annual dynamics of pelagic primary production and respiration in a shallow coastal basin. *Journal of Sea Research, 58*, 269–282.

Loebl, M., van Beusekom, J. E. E., & Reise, K. (2006). Spread of the neophyte *Spartina anglica* reently enhanced by increasing temperatures? *Aquatic Ecology, 40*, 315–324.

Lotze, H. (2005). Radical changes in the Wadden Sea fauna and flora over the last 2000 years. *Helgoland Marine Research, 59*, 71–83.

Lotze, H. K., Reise, K., Worm, B., van Beusekom, J. E. E., Busch, M., Ehlers, A., et al. (2005). Human transformation of the Wadden Sea ecosystem through time: A synthesis. *Helgoland Marine Research, 59*, 84–95.

Martens, P., & van Beusekom, J. E. E. (2008). Zooplankton response to a warmer northern Wadden Sea. *Helgoland Marine Research, 62*, 67–75.

Möbius, K. (1877). *Die Austern und die Austernwirthschaft.* Berlin: Hempel & Parey.

Möbius, K. (1893). Über die Tiere der schleswig-holsteinischen Austernbänke, ihre physikalischen und biologischen Lebensverhältnisse. *Sitzungsberichte der Königlich preussischen Akademie der Wissenschaften zu Berlin, 7*, 33–58.

Nienburg, W. (1927). Zur Ökologie der Flora des Wattenmeeres. I. *Der Königshafen bei List auf Sylt Wissenschaftliche Meeresuntersuchungen Kiel, 20*, 146–196.

Polte, P., & Buschbaum, C. (2008). Native pipefish (*Entelurus aequareus*) promoted by the introduced seaweed *Sargassum muticum* in the northern Wadden Sea. *Aquatic Biology, 3*, 11–18.

Reise, K. (1998). Pacific oysters invade mussel beds in the European Wadden Sea. *Senckenbergia Maritima, 28*, 167–175.

Reise, K. (2005). Coast of change: Habitat loss and transformations in the Wadden Sea. *Helgoland Marine Research, 59*, 9–21.

Reise, K. (2007). Oysters: Natives gone and aliens coming. *Neobiota, 7*, 259–263.

Reise, K., & Buhs, F. (1999). Reply to the comment of Damm and Neudecker (1999): Long-term decline in epibenthic fauna of tidal channels near the island of Sylt in the northern Wadden Sea. *Helgoland Marine Research, 53*, 143–145.

Reise, K., & Kohlus, J. (2008). Seagrass recovery in the Northern Wadden Sea? *Helgoland Marine Research, 62*, 77–84.

Reise, K., & van Beusekom, J. E. E. (2008). Interactive effects of global and regional change on a coastal ecosystem. *Helgoland Marine Research, 62*, 85–91.

Reise, K., Dankers, N. M. J. A., & Essink, K., (2005). Introduced species. In Essink, K., Dettmann, C., Farke, H., Laursen, K., Lüerßen, G., Marencic, H., & Wiersinga, W. (Eds.), *Wadden Sea Quality Status Report 2004. Wadden Sea ecosystem 19* (pp. 155–161). Wilhelmshaven: Common Wadden Sea Secretariat.

Reise, K., Herre, E., & Sturm, M. (2008). Mudflat biota since the 1930's: Change beyond return? *Helgoland Marine Research, 62*, 13–22.

Reise, K., Jager, Z., de Jong, D., VanKatwijk, M., & Schanz, A. (2005). Seagrass. In K. Essink, C. Dettmann, H. Farke, K. Laursen, G. Lüerßen, H. Marencic, & W. Wiersinga (Eds.), *Wadden Sea Quality Status Report 2004. Wadden Sea ecosystem 19* (pp. 201–207). Wilhelmshaven: Common Wadden Sea Secretariat.

Schorries, D., & Albrecht, H. (1995). *Sargassum muticum* der Beerentang im Deutschen Wattenmeer. *Natur und Museum, 125*, 92–98.

Strasser, M., & Günther, C.-P. (2001). Larval supply of predator and prey: Temporal mismatch between crabs and bivalves after a severe winter in the Wadden Sea. *Journal of Sea Research, 46*, 57–67.

Thieltges, D. W., Reise, K., Prinz, K., & Jensen, K. T. (2009). Invaders interfere with native parasite-host interactions. *Biological Invasions, 11*, 1421–1429. DOI 10.1007/s10530–008–9350-y.

Thieltges, D. W., Strasser, M., van Beusekom, J. E. E., & Reise, K. (2004). Too cold to prosper – Winter mortality prevents population increase of the introduced american slipper limpet *Crepidula fornicata* in northern Europe. *Journal of Experimental Marine Biology and Ecology, 311*, 375–391.

Van Beusekom, J. E. E. (2005). A historic perspective on Wadden Sea eutrophication. *Helgoland Marine Research, 59*, 45–54.

Van Beusekom, J. E. E. (2006). Eutrophication proxies in the Wadden Sea: Regional differences and background concentrations. In K. Laursen (Ed.), *Monitoring and Assessment in the Wadden Sea. Proceedings from the 11. Scientific Wadden Sea Symposium, Esbjerg, Denmark, 4–8. April 2005. NERI Technical Report No. 573* (pp. 45–51). Copenhagen: National Environmental Research Institute, Ministry of the Environment.

Van Beusekom, J. E. E., & de Jonge, V. N. (2002). Long-term changes in Wadden Sea nutrient cycles: Importance of organic matter import from the North Sea. *Hydrobiologia, 475/476*, 185–194.

Van Beusekom, J. E. E., & Reise, K. (2008). Editorial: Long-term ecological change in the northern Wadden Sea. *Helgoland Marine Research, 62*, 1–2.

Van Beusekom, J. E. E., Bot, P. V. M., Goebel, J. H. M., Lenhart, H., Paetsch, J., Peperzak, L., et al. (2005). Eutrophication. In K. Essink, C. Dettmann, H. Farke, K. Laursen, G. Lüerßen, H. Marencic, & W. Wiersinga (Eds.), *Wadden Sea Quality Status Report 2004. Wadden Sea ecosystem 19* (pp. 141–154). Wilhelmshaven: Common Wadden Sea Secretariat .

Van Beusekom, J. E. E., Loebl, M., & Martens, P. (2008). Distant riverine nutrient supply and local temperature drive the long-term phytoplankton development in a temperate coastal basin. *Journal of Sea Research, 61*, 26–33. DOI: 10.1016/j.seares.2008.06.005.

Van Beusekom, J. E. E., Weigelt-Krenz, S., & Martens, P. (2008). Long-term variability of winter nitrate concentrations in the Northern Wadden Sea driven by freshwater discharge, decreasing riverine loads and denitrification. *Helgoland Marine Research, 62*, 49–57.

Wohlenberg, E. (1935). Beobachtungen über das Seegras *Zostera marina* L. und seine Erkrankung im nordfriesischen Wattenmeer. *Nordelbingen, 11*, 1–19.

Wohlenberg, E. (1937). Die Wattenmeer-Lebensgemeinschaften im Königshafen von Sylt. *Helgoländer wissenschaftliche Meeresuntersuchungen, 1*, 1–92.

Chapter 11

Long-Term Model Simulation of Environmental Conditions to Identify Externally Forced Signals in Biological Time Series

Karina Stockmann, Ulrich Callies, Bryan F.J. Manly,and Karen H. Wiltshire

Abstract A case study is presented to demonstrate the added value which can be gained from combining long-term biological observations with model-based hind casts of physical environmental conditions. The study utilizes numerically simulated high-resolution fields of currents and water levels in the North Sea to investigate the relevance of hydrodynamic conditions for the occurrence of phytoplankton blooms, as observed at Helgoland Roads in the inner German Bight. Inter-annual variations as well as a possible regime shift are discussed with regard to the spring mean diatom day. The long-term high-resolution simulations of the North Sea circulation are taken from the data base 'coastDat'.

Keywords Hydrodynamics · Long-term time series · Phytoplankton spring bloom

11.1 Introduction

Trends and shifts of biological parameters in the North Sea and Wadden Sea are an important topic broadly discussed in the literature. Examples of interest are increased abundances of polychaetes in the western Wadden Sea (Beukema, Cadee, & Dekker, 2002), changes in the timing and abundance of zoo and phytoplankton species in the late 1970s (Edwards, Beaugrand, Reid, Rowden, & Jones, 2002) or the late 1980s (Dickson, Colebrook, & Svendsen, 1992; Edwards et al., 2002), changes in the abundance and distribution of North Sea fish stocks in the late 1970s (Corten, 1990) or 1980s (Reid, DeFatima-Borges, & Svendsen, 2001). The existence of discernible regime shifts has been proposed for the years around 1979, 1988, and possibly 1998 (Radach & Pätsch, 2007; Weijerman, Lindeboom, & Zuur, 2005).

The temporal evolution of ecosystems, via the growth of primary producers, is largely driven by variations of external physical forcing (e.g. Colebrook, 1986; Gröger & Rumohr, 2006; Wiltshire et al., 2008). The most obvious example is the seasonal cycle brought about mainly by changes in radiation and temperature. However, changes on many other time scales have a bearing on the behaviour of ecosystems and should therefore be taken into account for a meaningful interpretation of biological time series. Changes of the North Sea ecosystem have been assumed to be linked to the Atlantic inflow (Turrel, 1992; Reid, Edwards, Beaugrand, Skogen, & Stevens, 2003) or to local hydro-meteorological forcing parameters (Beaugrand, 2004; Reid et al., 2001). Therefore, long and homogeneous data on the physical environment are a prerequisite for the successful interpretation of time series of long-term ecological monitoring programs.

Ideally data on physical external forcing should be free of variation caused by changes in instrumentation, measurement techniques, or observational practices, which prevent reliable analyses of long-term trends and variability. Unfortunately for the oceans and the coastal zone such data are rare. Often observations have been taken only for a limited period and are spatially patchy or simply not available. Observed data almost never cover the whole domain of interest with a sufficient spatial or temporal resolution.

Numerical model simulations provide the option to at least partially overcome these difficulties. If models

K. Stockmann (✉)
GKSS Research Centre, Geesthacht, Germany
e-mail: Karina.Stockmann@bsh.de

F. Müller et al. (eds.), *Long-Term Ecological Research*,
DOI 10.1007/978-90-481-8782-9_11, © Springer Science+Business Media B.V. 2010

are run in a hindcast mode, methods of data assimilation may be applied to feed all available information into the numerical model system over a continuous time period (e.g. Salomon, Breton, & Gueguenіat, 1993; Weisse & Günther, 2007). Then the model's role is to transfer and spread this information in time and space according to known physical laws. Each time new data become available, these data will be merged with the model state which summarizes all past information up to the present point of time. The resulting best analysis may be stored on an hourly basis, for instance, to guarantee sufficient representation of short-term system variability. In meteorology such model-based weather re-analyses (hind casts) based on frozen state-of-the-art models have now become common tools for atmospheric analyses. The so-called global re-analyses projects (Kistler et al., 2001) provide gridded atmospheric data of best possible homogeneity for several decades, albeit at relatively coarse spatial and temporal resolutions.

For coastal and offshore purposes atmospheric data are the most essential boundary values. The global atmospheric re-analyses may be regionally refined by a nested modelling approach. The higher resolution of atmospheric data may then be employed as upper boundary conditions for numerical models dealing with storm surge and wave simulations. Similar to the global atmospheric model, all subsequent regional scale high-resolution models may be run for many decades to produce a consistent reconstruction of regional scale weather-related variability.

The main objective of this chapter is to illustrate the application of such detailed reconstructions of the North Sea circulation on long-term biological observations. We report a case study dealing with the relationship between diatom abundances observed in the vicinity of the island of Helgoland and reconstructed hydrodynamic circulation patterns in the inner German Bight.

11.2 The coastDat Data Set

The portal coastDat (www.coastdat.de) does not represent a hindcast initiative in itself. The idea of coastDat is to provide a unique platform which hosts, describes, and promotes such data sets (Feser, Weisse, & VonStorch, 2001) and their key applications (Weisse & Plüß, 2006). A unique combination of consistent atmospheric and oceanic parameters is offered at high spatial and temporal detail, even for locations and variables for which no measurements have been made. Coastal scenarios are provided for the near future to complement the numerical analyses of past conditions. All data sets meet the following minimum quality standards:

- Long and high-resolution reconstructions of recent offshore and coastal conditions
- Performed with state-of-the-art frozen model systems
- Performed with as homogeneous as possible boundary data
- Extensively validated with existing observational data (e.g. Plüß, 2004; Weisse & Plüß, 2006).

For the present study we used currents from the coastDat data set that was produced within the EU-funded project HIPOCAS (Hindcast of Dynamic Processes of the Ocean and the Coastal Areas). The data set consistently combines atmospheric simulations with simulations of currents, sea levels and waves. Figure 11.1 outlines how these partial data sets

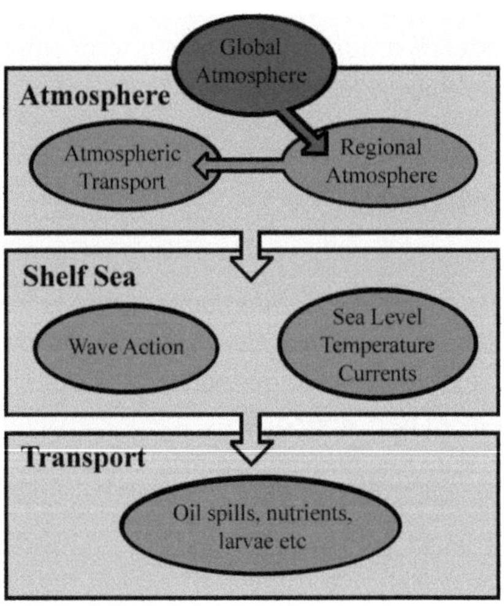

Fig. 11.1 Outline of the generation of consistent data sets stored in coastDat: Numerical models for the atmosphere, currents, or waves are coupled to or nested within each other to provide necessary boundary conditions. Lagrangian trajectory calculations (transports) including estimates of travel times, for instance, may be based on both shelf-sea currents and wind drift effects. Model results are stored with an hourly resolution for several decades; most data sets start in 1958

are related to each other. Current and sea level fields were produced by running the finite element model TELEMAC 2D (Hervouet & van Haren, 1996). For a more detailed simulation of the general model set up the reader is referred to Plüß (2004).

11.3 The Helgoland Roads Time Series

In 1962 a long-term pelagic monitoring program, including plankton species composition, was started at the island of Helgoland (54°11.3′N, 7°54.0′E) in the North Sea by the Biologische Anstalt Helgoland, on a work-daily basis (Hickel, Mangelsdorf, & Berg, 1993). This study focuses on the mean diatom day (MDD) – a variable, which represents the mean timing of the diatom bloom. Wiltshire and Manly (2004) computed it from the observed time series for the quarters of each year (January–March, April–June, July–September, and October–December). The MDD was defined as

$$MDD = \frac{\sum f_i d_i}{\sum f_i}$$

where f_i is the diatom count on day d_i of the quarter.

The MDD of the spring bloom is defined for the period from January to March. Observed annual values are shown in Fig. 11.2. Smoothing the MDD using a 5-year running mean Wiltshire and Manly (2004) show a distinct shift around the year 1978. The main motivation for the present study is to investigate whether this shift can be explained by corresponding changes of circulation patterns in the North Sea.

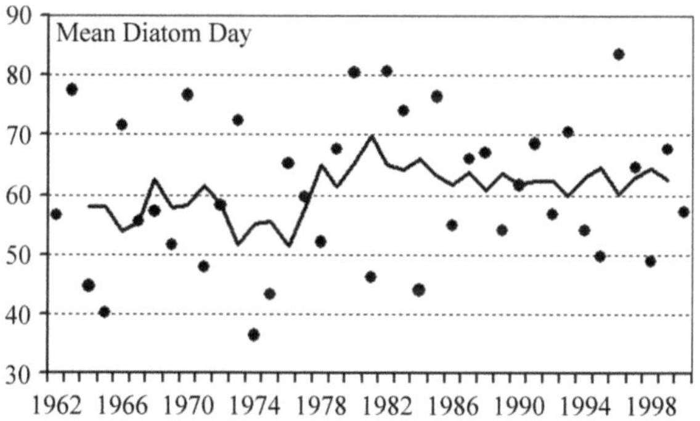

Fig. 11.2 Annual values of the spring bloom mean diatom day observed at station Helgoland Roads. The black line represents corresponding 5-years running means (after Wiltshire & Manly, 2004)

11.4 Methods

For the present study we chose water flow (product of velocity and total water depth) as our basic physical variable to be related to the biological observations. Using water flow instead of velocity puts less emphasis on near shore shallow water currents and is in line with a concept of mass balances. Since we were interested in anomalies, all externally forced periodic signals were removed from the current data as follows: A Fourier analysis was applied to extract all variations that are related to astronomically prescribed frequencies with time scales up to one year. The resulting filtered residual data contained neither tidal nor long-term seasonal cycles. The filtering has been applied to currents and water levels individually, prior to the calculation of water flows.

Next we subjected the time series of hourly residual water flow fields to empirical orthogonal function (EOF) analysis. This mathematical technique corresponds to principal component analysis (PCA), and a detailed description can be found in von Storch and Zwiers (1999). Spatial patterns of water transport anomalies in the inner German Bight are identified, the amplitudes of which (principal components, PCs) can be related to anomalies (i.e. inter-annual variations) of the MDD observed at Helgoland Roads.

11.5 Results

Figure 11.3 shows the mean residual flow pattern (left panel) together with the two most important anomaly patterns which explain 70 and 17%, respectively, of total residual variability. Time series of the annually averaged amplitudes of the residual transport patterns (i.e. the corresponding principal components, PCs) are shown in Fig. 11.4 (PC 1 for pattern 1, PC 2 for pattern 2). The mean residual transport not including tidal effects turns out to be very weak (close to zero). For this reason superimpositions of the two anomaly patterns shown in Fig. 11.3 multiplied with

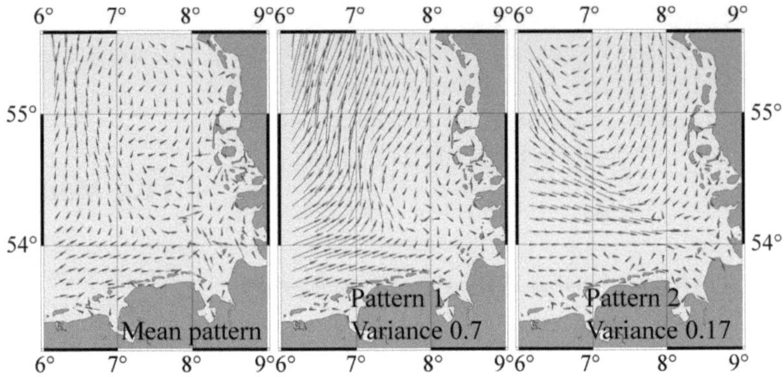

Fig. 11.3 Anomaly patterns of residual water flows as obtained from EOF analysis (see text). Amplitudes of the two anomaly patterns are shown in Fig. 11.4

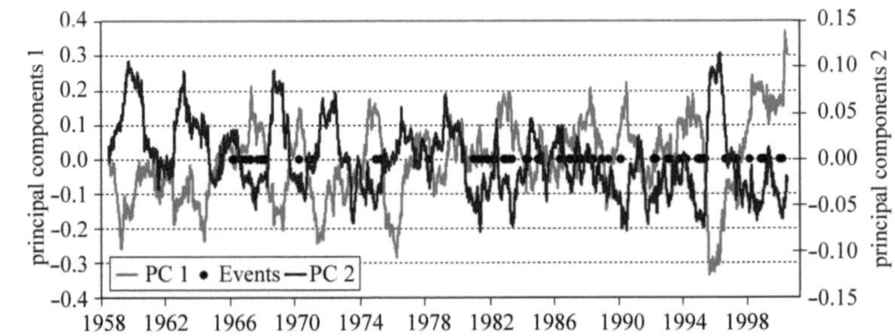

Fig. 11.4 Annually averaged time series of amplitudes (principal components, PCs) of the two anomaly patterns shown in Fig. 11.3. *Black dots* indicate observations of freshwater signals (low salinity in combination with high nutrient concentrations)

their corresponding amplitudes closely approximate actual residual flow patterns.

Due to the lack of sufficiently dense observations, validation of modelled water flow fields is a difficult task. It is even more difficult to validate water mass exchanges between different regions, as such calculations amount to an integration of changing flow fields over a certain span of time. Tracer experiments are the only way to check in detail whether modelled water mass transports are realistic (e.g. Maier-Reimer, 1977). For the present study we chose a related approach to check whether the model simulations can be regarded meaningful. From the time series of salinity and nutrient concentrations observed at Helgoland Roads, special freshwater inflow events were identified. Within six to eight hours salinity decreased by approximately 2‰ and simultaneously an increase of nitrate and silicate concentration was measured, which was mirroring the salinity curve. An automated procedure extracted about 150 of such events from the time series according to two pre-specified objective criteria. In Fig. 11.4 these events are marked as black dots.

The first observation from Fig. 11.4 is that the symbols which indicate freshwater signals are not evenly distributed but tend to be clustered during certain periods. The second very interesting observation is that there seems to be a relation between the occurrence of freshwater events and the annual running mean amplitudes of the water flow anomaly patterns. Freshwater inflow events never occur with high positive values of PC 2. According to the right panel of Fig. 11.3, a positive amplitude implies that the island of Helgoland will be hit by water from the northwest. It is consistent to assume that if this situation is dominant, freshwater inflow from the coastal area of the inner German Bight (or from the river Elbe) will be suppressed.

With regard to Fig. 11.4 the question of a possible conflict of time scales may be raised. The freshwater events are defined on a time scale of the order of two days. In the figure these events are combined with annual averages of the PCs representing instantaneous patterns of water transport. We suggest reading Fig. 11.4 in the sense that observed events are conditioned by means of hydrodynamic conditions. It should not be read in the sense that actual residual currents can be used as efficient predictors for the occurrence of freshwater events. Typically, freshwater from the river Elbe that arrives at Helgoland has been transported by changing currents for about two weeks, which by far exceeds the time span during which the resulting freshwater signal can be observed at Helgoland.

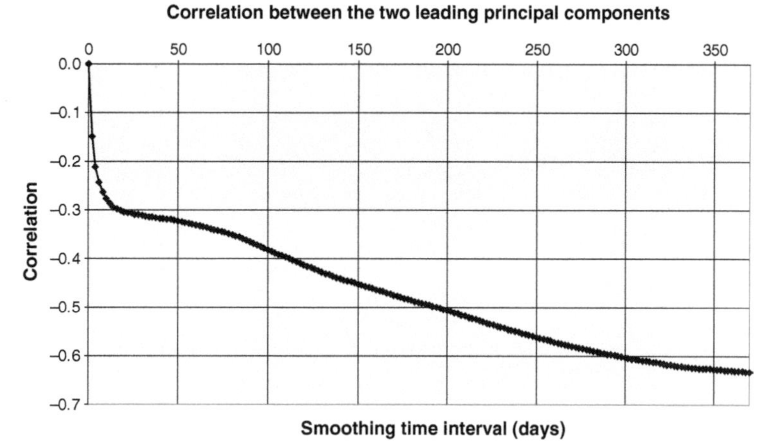

Fig. 11.5 Correlation between the two PCs, corresponding with the two anomaly residual flow patterns in Fig. 11.3. The correlation is shown as a function of the time interval used for moving averaging

By definition, the time coefficients (PCs) that belong to different EOFs (i.e. anomaly flow patterns) must be uncorrelated. This holds for time series of instantaneous values. According to Fig. 11.4, however, annual averaging produces a significant amount of negative correlation between the two PCs. Figure 11.5 shows correlation between the two PCs as a function of the smoothing interval. A negative correlation between the two time series originates rapidly when the smoothing interval is extended up to about two weeks. For longer time intervals the correlation increases at a lower rate. A spectral analysis of the data (not shown) reveals that the second PC has much more variability on short-term scales than the first PC. This short-term variability being present in only one PC seems to be the major reason for the lack of correlation of instantaneous values. It will be a topic of further research to investigate how the different behaviour of the two PCs can be linked to different modes of atmospheric forcing. Here, just two examples will be given to indicate which kind of links can be expected to exist.

It is reasonable to expect that inter-annual variations of the mean hydrodynamic conditions before spring algae blooms have a major impact on the variability of the MDD. Figure 11.6 compares the de-trended time series of observed MDDs with de-trended time series of the PCs of the second anomaly pattern and temperature, averaged over the 80 days prior to the MDD. We chose a time span of 80 days, because it produces the highest correlations. The MDD seems to be connected to the means of PC 2 (correlation coefficient 0.52) and temperature (correlation coefficient −0.40). It was possible for these correlations to reject the null hypothesis on a 1% significance level.

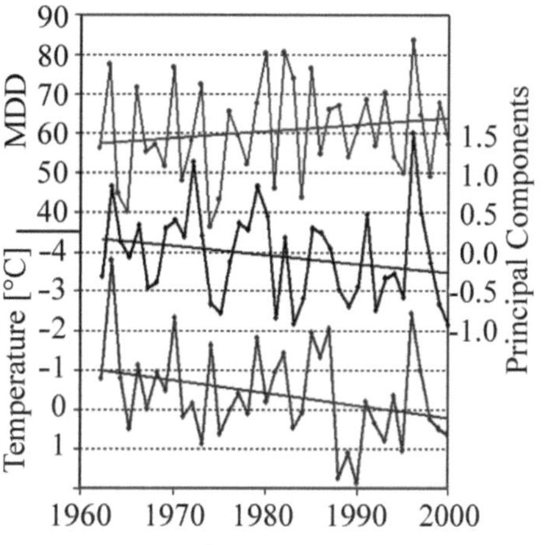

Fig. 11.6 Mean diatom days (MDD, spring diatom bloom), and temperature and time coefficients (PCs) of the second residual flow pattern, averaged over the last 80 days before the observed MDD. Correlation between de-trended data is 0.52

In agreement with the above correlations, the time series of the MDD, mean PC 2 and temperature show a similar inter-annual variability (Fig. 11.6). This implies that a higher PC 2 in autumn is followed by a delayed MDD and vice versa.

Notwithstanding its short-term correlation with the MDD, the second residual anomaly pattern, however, was found to be unable to explain the shift near year 1978 in the 5-year mean of spring bloom MDD (Fig. 11.2). It has already been mentioned that variability of the first PC is more concentrated in the long wave spectrum. Therefore the first PC may be anticipated to be a better predictor for long-term variations. Figure 11.7 compares the 5-year means of the time coefficient of the first PC with the MDD. Both

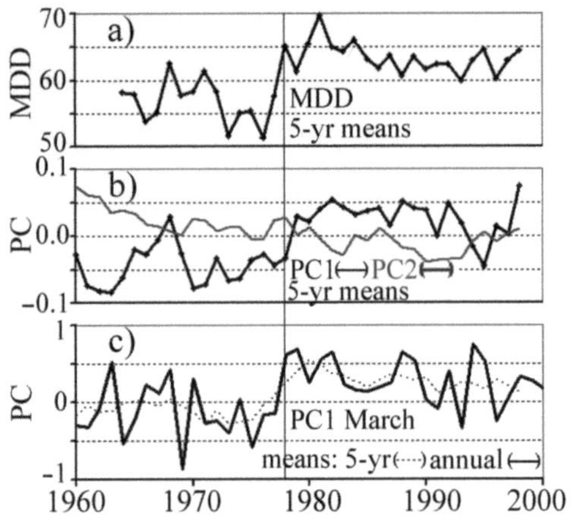

Fig. 11.7 5-yr means of (**a**) MDD, (**b**) PC 1, and PC 2 and (**c**) PC 1 in March

parameters show a simultaneous increase near the year 1978. Approximately at the same time (in 1977) the Great Salinity Anomaly reached its minimum in the western approaches to the English Channel (Dickson, Meincke, Malmberg, & Lee, 1988) and it is shown that changes in salinity in the northern North Sea during the 1970s were correlated with changes in fish stocks and plankton composition (Corten, 1990; Aebischer et al., 1990). Presumably the changes of the flow pattern in the German Bight with a stronger inflow from west, which change the composition of the water body (salinity, nutrients, temperature), also caused a delayed MDD.

The main changes of PC 1 occur in the month of March. This may also be the reason, why the step-like shift of the MDD can only be found for the spring bloom.

11.6 Discussion

A major difficulty with the interpretation of marine biological observations is the fact that local conditions are permanently affected by mostly insufficiently known advection. The objective of the present study was to demonstrate how this problem may be tackled by combining long-term observational evidence with detailed reconstructions of the physical environment.

The importance of long-term simulations for a proper interpretation of observed variability is illustrated by Fig. 11.4. In a previous paper, Kauker and von Storch (2000) investigated the short-term variability of a 15-yr general circulation simulation of the North Sea. Similar to our study the exposure of shelf-sea currents to re-analysed atmospheric forcing was taken into account. However, according to Fig. 11.4 the period 1979–1993 considered by Kauker and von Storch happens to seem qualitatively different from the times before and after, being characterized by the absence of negative mean amplitudes of the first anomaly pattern. At the same time positive mean amplitudes of the second pattern are less pronounced. This means (cf. Fig. 11.3) that during the 15-year period the mean inflow has turned counter clockwise from north-westerly to more westerly directions. If no simulations were available before 1978, the extension of the time series up to 1999 would probably suggest a recent change of in the system in terms of long-term variability. In the light of data before 1978, however, this interpretation becomes much less convincing.

It has been shown that a decomposition of simulated current fields in terms of different modes of circulation can be helpful for the proper interpretation of observational evidence on different time scales. For a more in depth analysis of long-term trends, however, the present study of the circulation in the inner German Bight should be extended by a similar analysis for the whole North Sea. Nevertheless, already the present analysis has allowed us to distinguish between the two time scales. A long-term shift and inter-annual variability, respectively, of the MDD were found to be related to different residual circulation patterns.

The type of data analysis we have described does not cover all the options possible based on a comprehensive data set like coastDat. To provide another example, Fig. 11.8 shows some result obtained from Lagrangian trajectory calculations based on the stored hourly current fields. For many applications in biology it is crucial to know transport rates between different areas of interest. This information may possibly be required in an aggregated version to properly represent the mean conditions during certain periods of time. For this purpose a large number of detailed transport simulations may be superimposed to get a composite picture. In many applications (the drift of larvae, for instance) it is reasonable to assume that the tracer material will not be passive but decreases according to certain pre-specified decay times. In Fig. 11.8 a decay time of 30 days has been assumed. Large numbers in

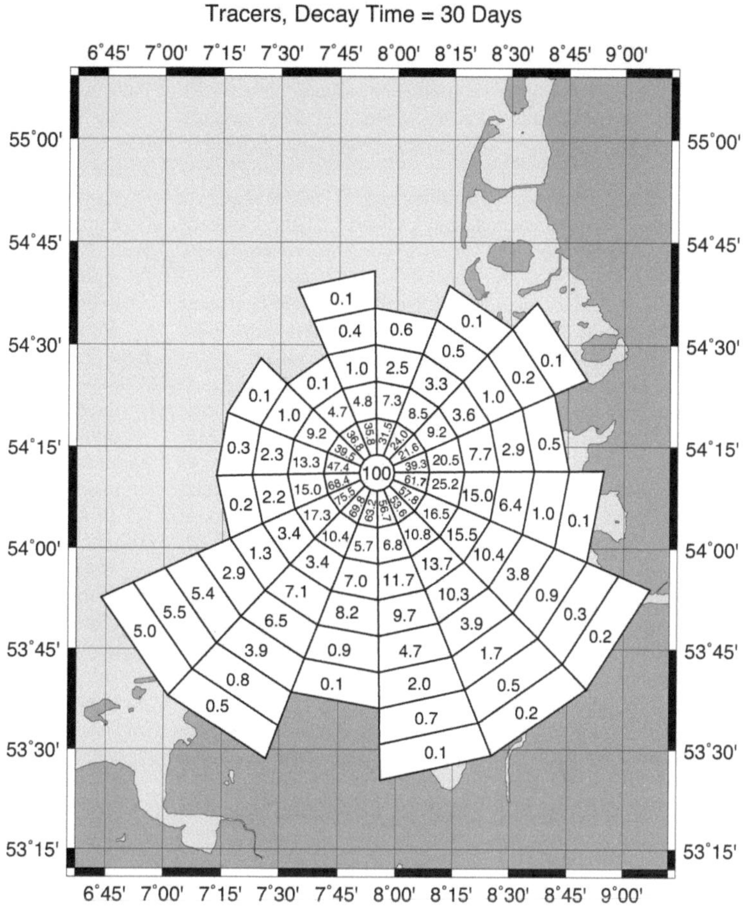

Fig. 11.8 Estimated Lagrangian particle transport rates from a given grid cell to Helgoland given the decay time of particle equals 30 days. The estimate is based on a composite of 200 daily simulations that are started within a period of 200 days (July 1982–January 1983). Each individual simulation considers 500-particle trajectories being started at Helgoland and integrated backward in time for 50 days. Numbers give the percentage of all 200 × 500 particles that reach Helgoland given the specified decay time

the figure represent regions from where large amounts of material were advected to Helgoland in relatively short times. Large values in the immediate surrounding of the island are due to tidal movements, while large values in more distant regions are due to the action of efficient residual currents.

In summary, we conclude that the strength of a database like coastDat lies in its provision of information on long-term trends and variability with sufficient detail in time and space. This type of model simulations efficiently complements long-term observations with short sampling intervals.

Acknowledgements Numerical simulation of North Sea currents were performed by Andreas Plüß from the Federal Waterways Engineering and Research Institute, Department Hamburg (BAW-AK). The atmospheric forcing has been produced within the project HIPOCAS (Hindcast of Dynamic Processes of the Ocean and the Coastal Areas) which was funded by the European Union under the contract EVK2-CT-1999-00038. The underlying global atmospheric data where provided by NCEP. We appreciate the support of Ralf Weisse with regard to data filtering and EOF analysis. We thank all those who helped obtain, analyse, and enumerate the Helgoland Roads Data over the years. In particular we would like to mention Peter Mangelsdorf and the Crew of the RV Aade.

References

Aebischer, N. J., Coulson, J. C., & Colebrook, J. M. (1990). Parallel long-term trends across four marine trophic levels and weather. *Nature, 347*, 753–755.

Beaugrand, G. (2004). The North Sea regime shift: Evidence, causes, mechanisms and consequences. *Progress in Oceanography, 60*, 245–262.

Beukema, J. J., Cadee, G. C., & Dekker, R. (2002). Zoobenthic biomass limited by phytoplankton abundance: Evidence from parallel changes in two long-term data-series in the Wadden Sea. *Journal of Sea Research, 48*, 111–125.

Colebrook, J. M. (1986). Environmental influences on long-term variability in marina plankton. *Hydrobiologia, 142*, 309–325.

Corten, A. (1990). Long-term trends in pelagic fish stocks of the North Sea and adjacent waters and their possible connection to hydrographic changes. *Netherlands Journal of Sea Research, 25*, 227–235.

Dickson, R. R., Meincke, J., Malmberg, S. A., & Lee, A. J. (1988). The "Great Salinity Anomaly" in the Northern North Atlantic 1968–1982. *Progress in Oceanography, 20*, 103–151.

Dickson, R. R., Colebrook, J. M., & Svendsen, E. (1992). Recent changes in the summer plankton of the North Sea. *ICES Marine Science Symposium, 195*, 232–242.

Edwards, M., Beaugrand, G., Reid, P. C., Rowden, A. A., & Jones, M. B. (2002). Ocean climate anomalies and the ecology of the North Sea. *Marine Ecology Progress Series, 239*, 1–10.

Feser, F., Weisse, R., & VonStorch, H. (2001). Multi-decadal atmospheric modeling for Europe yields multi-purpose data. *Eos, Transactions American Geophysical Union, 82*, 305.

Gröger, J., & Rumohr, H. (2006). Modelling and forecasting long-term dynamics of Western Baltic macrobenthic fauna in relation to climate signals and environmental change. *Journal of Sea Research, 55*, 266–277.

Hervouet, J. M., & Van Haren, L. (1996). *TELEMAC2D Version 3.0 Principle Note. Rapport EDF HE-4394052B, Electricité de France, Département Laboratoire National d'Hydraulique.* Chatou : Electricité de France.

Hickel, W., Mangelsdorf, P., & Berg, J. (1993). The human impact on the German Bight: Eutrophication during three decades (1962–1991). *Helgoländer Meeresuntersuchungen, 47*, 243–263.

Kauker, F., & Von Storch, H. (2000). Statistics of "synoptic circulation weather" in the North Sea as derived from a multiannual OGCM simulation. *Journal of Physical Oceanography, 30*, 3039–3049.

Kistler, R., Kalnay E., Collins, W., Saha, S., White, G., Wollen, J., et al. (2001). The NCEP/NCAR 50-year reanalysis: Monthly means CD-ROM and documentation. *Bulletin of the American Meteorological Society, 82*, 247–267.

Maier-Reimer, E. (1977). Residual circulation in the North Sea due to the M_2-tide and mean annual wind stress. *Deutsche Hydrographische Zeitung, 30*, 69–80.

Plüß, A. (2004). Das Nordseemodell der BAW zur Simulation der Tide in der Deutschen Bucht. *Die Küste, 67*, 83–127.

Radach, G., & Pätsch, J. (2007). Variability of continental riverine freshwater and nutrient inputs into the North Sea for the years 1977–2000 and its consequences for the assessment of eutrophication. *Estuaries and Coasts, 30*, 66–81.

Reid, P. C., DeFatima-Borges, M., & Svendsen, E. (2001). A regime shift in the North Sea circa 1988 linked to changes in the North Sea horse mackerel fishery. *Fisheries Research, 50*, 163–171.

Reid, P. C., Edwards, M., Beaugrand, G., Skogen, M. D., & Stevens, D. (2003). Periodic changes in the zooplankton of the North Sea during the twentieth century linked to oceanic inflow. *Fisheries Oceanography, 12*, 260–269.

Salomon, J.-C., Breton, M., & Gueguenìat, P. (1993). Computed residual flow through the Dover Strait. *Oceanologica Acta, 16*, 449–455.

Turrel, W. R. (1992). The East Shetland Atlantic Inflow. *ICES Marine Science Symposium, 195*, 127–143.

Von Storch, H., & Zwiers, F.W. (1999). *Statistical analysis in climate research.* Cambridge: Cambridge University Press.

Weijerman, M., Lindeboom, H., & Zuur, A. F. (2005). Regime shifts in marine ecosystems of the North Sea and Wadden Sea. *Marine Ecology Progress Series, 298*, 21–39.

Weisse, R., & Günther, H. (2007). Wave climate and long-term changes fort the Southern North Sea obtained from a high-resolution hindcast 1958–2002. *Ocean Dynamics, 57*, 161–172.

Weisse, R., & Plüß, A. (2006). Storm-related sea level variations along the North Sea coast as simulated by a high-resolution model 1958–2002. *Ocean Dynamics, 56*, 16–25.

Wiltshire, K. H., & Manly, B. F. J. (2004). The warming trend at Helgoland Roads, North Sea: Phytoplankton response. *Helgoland Marine Research, 58*, 269–273.

Wiltshire, K. H., Malzahn, A. M., Wirtz, K., Greve, W., Janisch, S., Mangelsdorf, P., et al. (2008). Resilience of North Sea phytoplankton spring bloom dynamics: An analysis of long-term data at Helgoland Roads. *Limnology and* Oceanography, *53*, 1294–1302.

Chapter 12

Long-Term Investigations in Brackish Ecosystems

Hendrik Schubert, Norbert Wasmund, and Kevin G. Sellner

Abstract Variability and complexity in brackish systems require long-term measurements in order to define base conditions, from which deviations can be ascertained. Long-term observations in three systems, lagoons, the Baltic Sea, and the Chesapeake Bay, are examined to identify system changes, unlikely detectable with sampling in single years or in temporally and spatially heterogeneous sampling. One basic condition in brackish systems is the gradient in salinity, which may be large-scale and rather stable stretching over the entire sea (marine gradient), or meso-scale and highly variable such as those in river plumes (estuarine gradient), and upwelling cells (upwelling gradient). For the first two gradients, and in some cases the third, distinct boundaries separate stenohaline taxa from more eurytopic taxa resulting in spatially explicit distributions of plankton, nutrients, and food web characteristics. The natural variability has to be ascertained through repeated long-term sampling in order to fix a baseline for shifts and trends in the ecosystem. A general trend during the last decades is cultural eutrophication, leading to increased phytoplankton biomass and sedimentation, and hypoxia in bottom water. In some areas, eutrophication was repressed in the 1990s, e.g., stabilization of chlorophyll concentrations in the Baltic Proper, recovery of macrophytes in the Darss-Zingst Bodden Chain (DZBC). In the coming years, the effects of declining nutrient loads are expected to cause a return to mesotrophic conditions in the DZBC, resulting in a return of nutrient limitation. Further monitoring will be performed to follow this unique event. It is therefore imperative that the community support long-term observations in these complex systems particularly as increasing human populations exacerbate impacts of global climate change that slowly warms waters, changes intensities and frequencies of meteorological events and responsive hydrologies, and shifts biogeographic ranges of many cosmopolitan taxa.

Keywords Abiotic limitation · Baltic Sea · Brackish Systems · Chesapeake Bay · Macrophytobenthos · Phytoplankton

12.1 Characteristics of Brackish Environments

Brackish environments are complex abiotic and biotic systems, more than just 'something between marine and limnetic'. As ecotones, they have unique flora and fauna resulting from transitions between freshwater stenohaline cellular functions to quite adaptable euryhalinic taxa. Chemistries also shift, reflecting saline effects on particle–organic matter–ion interactions, the most obvious detected as flocculation of clay minerals into large organic-rich aggregates and the sedimentation of particle-bound phosphorus for eventual release under increasing salinity and lower dissolved oxygen concentrations.

High variability especially can be seen with respect to salinity, dependent on the precedent marine system (tidal amplitude and salinity) and the freshwater inflow regime. Daily amplitudes of >10 psu at some locations have been described. Such changes are challenges for acclimation capabilities of organisms living there, as shown by initial investigations of Remane

H. Schubert (✉)
Institute of Biosciences, University of Rostock,
Albert-Einsteinstr. 3, D-18055 Rostock, Germany
e-mail: hendrik.schubert@uni-rostock.de

F. Müller et al. (eds.), *Long-Term Ecological Research*,
DOI 10.1007/978-90-481-8782-9_12, © Springer Science+Business Media B.V. 2010

(1940, 1955) who also noted that the mean salinity range influences the effects of this variability as well. A very pronounced sensitivity peak was observed around 5–8 psu, the so-called horohalinicum, where at least for animals a species minimum has been reported (Remane, 1940, 1955; Khlebovich, 1990). In general, species diversity sharply decreases down to the horohalinicum but species diversity can behave differently for individual groups of organisms: for benthic animals and phytoplankton, in general, the number of species increases again at salinities below the horohalinicum (because of increasing occurrence of freshwater species), while macrophytobenthic species numbers seem to decline further.

Salinity is only one factor that can be contrasted in freshwater or marine conditions – brackish systems exhibit large ion anomalies further restricting the range of species able to thrive and, especially in subarctic and boreal regions, brackish systems can be affected by ice scraping/ice cover phenomena, which act in two ways, mechanically as well as by diminishing light penetration.

Brackish systems (lagoons, river mouths, etc.) are mostly restricted to relatively small areas at the coast. Most of them have provided conditions favourable for human settlement and are of high economic importance (maritime industries, transport, fisheries, tourism). These human activities cause heavy deterioration of surrounding brackish ecosystems, with observed effects dependent on human impact *and* hydromorphology of the system. Examples of these cumulative impacts are obvious throughout the world, with dramatic examples in the Chesapeake and Baltic basins. For the former, the near highest water catchment area per water volume ratio of all large brackish systems insures maximum impact of land alterations on Chesapeake Bay responses while extremely high densities of human and animal activities on many Baltic and North Sea rivers guarantees excessive nutrient loads to these low salinity coastal systems. With increasing populations in catchment areas of these systems coupled with expected climatic changes in the coming 50 years, irreversible changes of the systems may occur, restricting the use of these systems for several services offered in the past.

The previously recorded deterioration of these and many other brackish systems as well as potential future stress likely from population growth and climate change support expanded observations over long time scales to assess ecological responses in these imperilled brackish systems in order to gain the knowledge needed for restoration measures and identification of long-term effectiveness of potential management actions. Long-term measurements in these nearshore systems are feasible as they are far more accessible for investigations than the open sea. Fortunately an extensive suite of investigations have been in place for some systems from all over the world over the last 30 years, resulting in several large long-term data sets.

As estuarine and other coastal waters are strongly influenced by highly variable freshwater input, upwelling events, and coastal jets, data gathered from these highly variable systems are difficult to evaluate. Quick changes and steep gradients are typical. Freshwater input from land and salt water input from the sea do create not only a horizontal but also a vertical salinity gradient in deeper waters as well. Most of the vertical gradients (light, temperature, and nutrients) are, however, not unique for brackish waters and therefore not addressed below. Of importance in the following discussion are salinity gradients, a 'marine gradient', an 'estuarine gradient', and an 'upwelling gradient'. The horizontal salinity gradient stretching from the inner parts of a brackish sea to the outlet to the ocean, rather long in some cases, is exemplified by the Baltic Sea and the Chesapeake Bay in this chapter. The meso-scale gradients in river mouths or lagoons are referred to as 'estuarine gradients'. The upwelling gradient refers to the intrusion of deep water of higher salinity into shallow waters, developing a meso-scale horizontal gradient.

The core of the chapter focuses on the value of long-term observations, such as those undertaken through multi-year field measurements, as critical observations to discerning patterns that would permit isolation of specific flora, fauna, and chemical signatures of these three salinity patterns. Long-term data sets for three brackish environments, i.e., lagoons, the Baltic proper, and a drowned river valley (the Chesapeake Bay), are considered for assessing unique system characteristics that otherwise would be difficult to discern with temporally brief observations.

12.2 System Analysis – Gradients

12.2.1 The Estuarine Gradient

The estuarine gradient not only is derived from a changing salinity but also shifts in nutrient concentration, light transmission, organic matter content, and consequently in species composition and

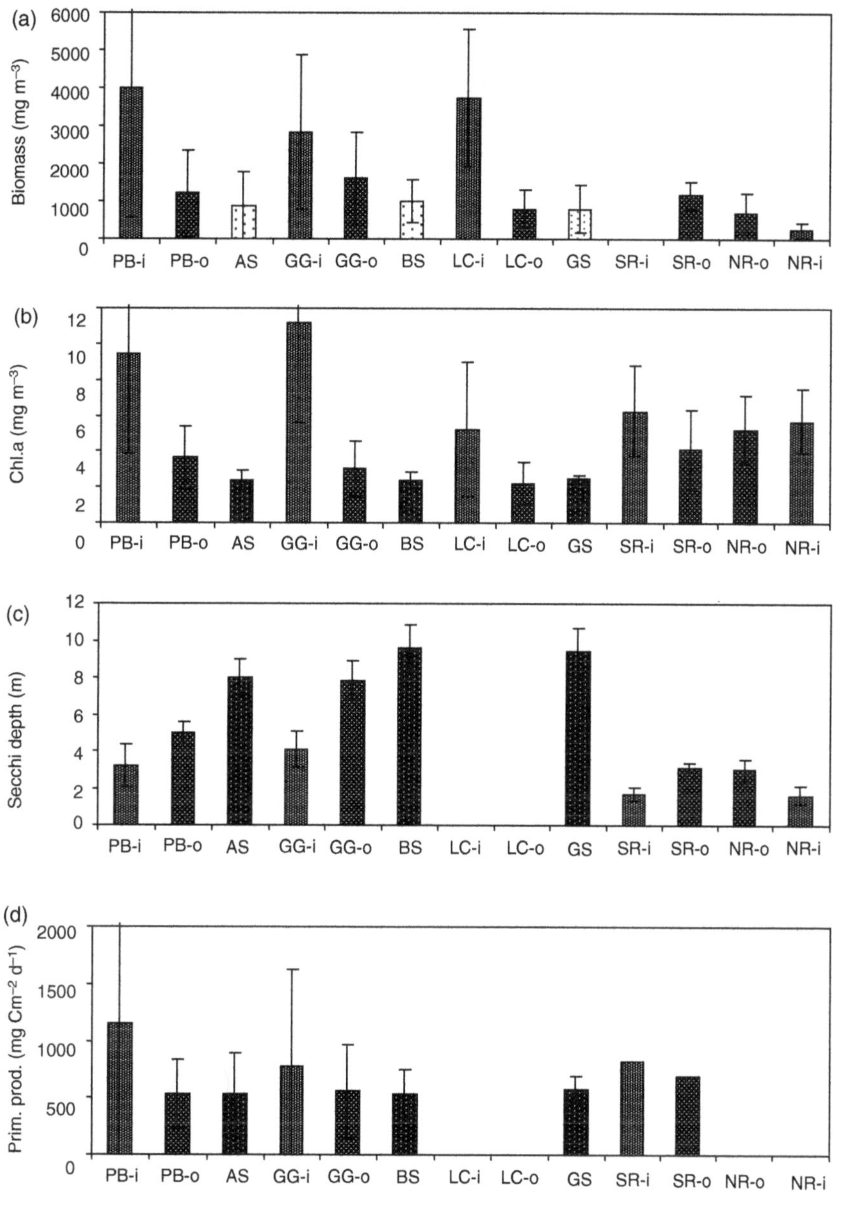

Fig. 12.1 Mean values (1993–1997) of (**a**) phytoplankton biomass, (**b**) chl *a* concentrations, (**c**) Secchi depth, and (**d**) in situ primary production in different Baltic Sea areas. The columns are represent results from short transects from the inner coastal regions (= i, columns densely stippled) to the outer coastal regions (= o) to the open sea (AS = Arkona Sea, BS = Bornholm Sea, GS = Eastern Gotland Sea, columns sparsely stippled); PB = Pomeranian Bay, GG = Gulf of Gdańsk, LC = Lithuanian coast. In the Gulf of Riga, the transect is arranged from the southern (SR) to the northern (NR) reach. The *bars* indicate confidence intervals ($p = 0.05$, $n = 5$). Primary production means from the southern Gulf of Riga calculated by pooling seasons of 1994–1997 first and estimating one combined annual average subsequently (after Wasmund et al., 2001)

trophic interactions (predators, food web structure). These different factors are interrelated and hamper causal analysis. The estuarine gradient along the Darss-Zingst Bodden Chain (DZBC) is well studied and provides an excellent opportunity to evaluate the importance of long-term measurements in system characterization. Wasmund (1990) compiled data mainly from the 1970s and found a reduction of phytoplankton biovolume from 27.9 to 3.6 $mm^3\ l^{-1}$ and phytoplankton primary production from 611 to 109 $g\ C\ m^{-2}a^{-1}$ from the inner to the outer regions of that lagoon-like water. Further, in large river plumes (Oder, Vistula, Klaipeda Strait, Daugava), phytoplankton biomass, chlorophyll *a* concentration, and primary production were higher in comparison with offshore waters (Fig. 12.1).

In the same areas, phytoplankton composition also shifted. In a transect from the Vistula River mouth to the Gdansk Deep (Fig. 12.2), it is obvious that phytoplankton biomass declined sharply, beyond what would be expected from simple dilution. Freshwater species disappear successively, first the green algae *Coelastrum microporum, Dictyosphaerium pulchellum, Pediastrum* spp., and *Scenedesmus* spp., followed by *D. ehrenbergianum. Microcystis aeruginosa* also disappears quickly but the

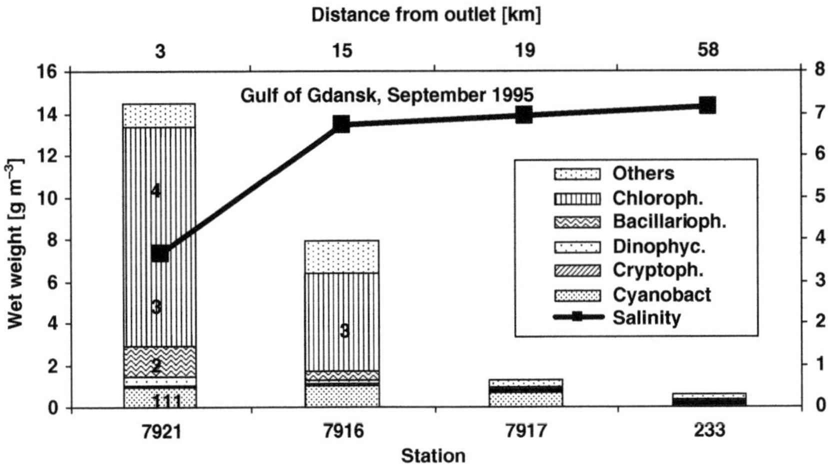

Fig. 12.2 Phytoplankton composition and salinity in surface water in a transect from the Vistula river mouth to the open Gulf of Gdansk on 14–15 September 1995. Important species: 1 = *Snowella septentrionalis*, 2 = *Aulacoseira granulata*, 3 = *Dictyosphaerium ehrenbergianum*, 4 = *Coelastrum microporum*, and *Scenedesmus* spp. (after Wasmund et al., 1999, modified)

cyanobacteria *Snowella septentrionalis, Aphanocapsa delicatissima, Pseudanabaena* sp., and *Merismopedia warmingiana* persisted until station 7917, at a salinity of 7.0 psu. Typical species of the open Baltic Proper were only found at station 233 (*Nodularia spumigena, Dinophysis norvegica, D. rotundata, Heterocapsa rotundata,* and *Chaetoceros densus*).

The Chesapeake Bay and its primary tributaries fall into this estuarine category as well. Similar changes in phytoplankton (loss of cyanobacteria, an increase in diatoms and flagellates) have been observed in the Chesapeake, exemplified in its second largest tributary, the Potomac River estuary (see below).

12.2.2 The Marine Gradient

The marine gradient originates from the inflow of oceanic water into a brackish sea. In contrast to estuarine gradients, the marine gradients are large-scale gradients occurring only in a few brackish seas, in particular the Baltic Sea. The marine gradient is less variable and mainly constitutes a salinity gradient with rather uniform light conditions, concentrations of organic matter, nutrients, etc. Therefore, it is well suited for studying the influence of salinity irrespective of the influences of other factors. Typical distribution patterns for phytoplankton are shown in Fig. 12.3, providing seasonal summaries. Some marine species like *Dactyliosolen fragilissimus, Proboscia alata,* and *Ceratium* spp. occur mainly in the western Baltic. Others, like *Peridiniella catenata, Nodularia spumigena, Aphanizomenon* sp., and *Coscinodiscus granii* are not observed in higher salinities. As expected, diversity is much higher in the western Baltic than in the central Baltic Sea, indicative of the horohalinicum discussed in the Introduction.

12.2.3 The Upwelling Gradient

Upwelling occurs in many marine waters, but the resulting gradients are rather weak. However, upwelling causes exceptionally strong horizontal gradients in stratified brackish seas and estuaries, as deep water is moved to the surface. In contrast to river water in estuarine gradients, the upwelled water has a higher salinity and lower temperature in comparison with the surrounding water. It is rather clear water containing only a few seeding cells (Gromisz & Szymelfenig, 2005) but substantial nutrients. Consequently, upwelling events lead initially to reduced biomass but may later initiate new blooms. At the upwelling frontal boundary, primary production and standing crop can be significantly enhanced (Nômmann, Sildam, Nôges, & Kahru, 1991). Kononen et al. (1996) and Vahtera, Laanemets, Pavelson, Huttunen, and Kononen (2005)

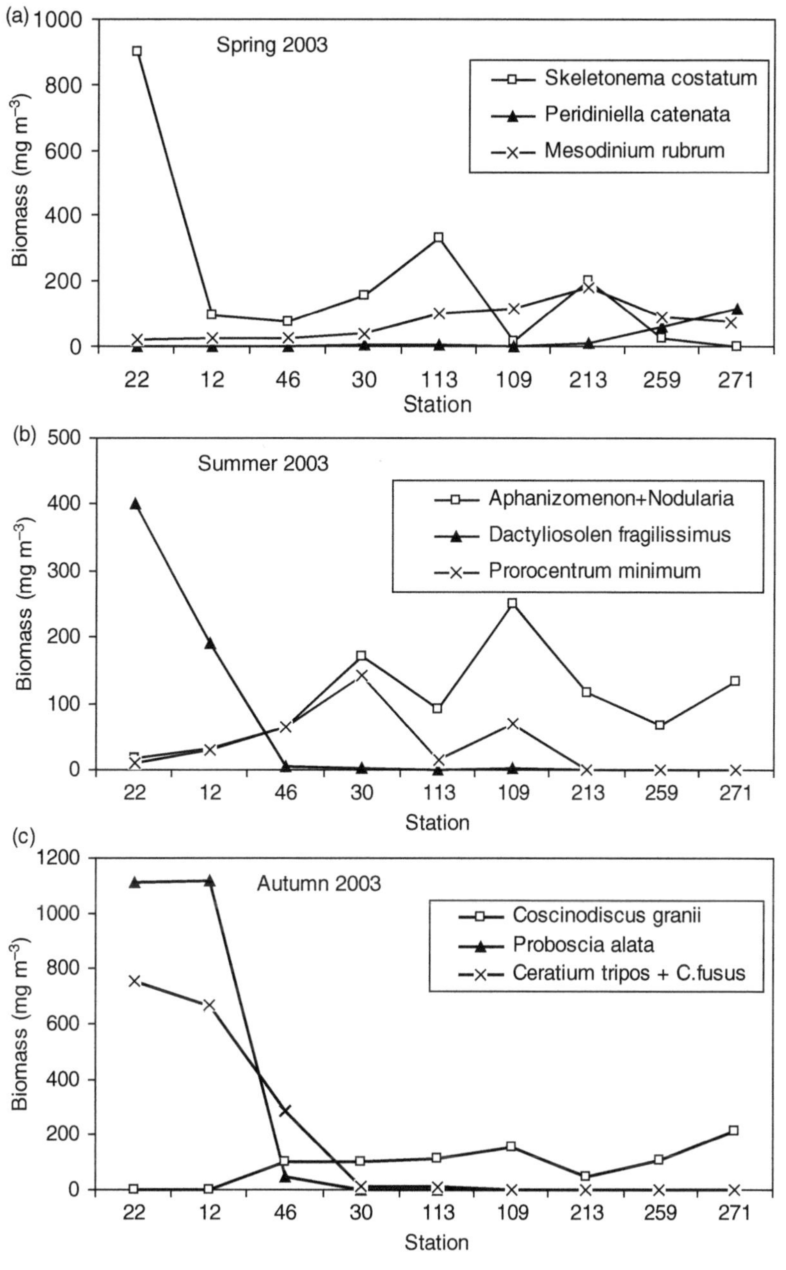

Fig. 12.3 Biomass of selected species along a transect through the Baltic Sea in (**a**) February–May, (**b**) June–September, and (**c**) October–December, 2003

described the initiation of cyanobacteria blooms in the Gulf of Finland by nutrient pulses due to vertical mixing, while several investigative teams in the Chesapeake have identified diatom and dinoflagellate (and associated chlorophyll levels) increases following summer 'tilting' of the pycnocline (see below), followed by dominance of microzooplankton grazers in response to the bloom (Sellner & Brownlee, 1990).

12.3 System Analysis – Long-Term Trends

12.3.1 Drowned River Valleys – Cheapeake Bay

The Chesapeake Bay is the brackish extension of the Susquehanna River along the mid-Atlantic coast of

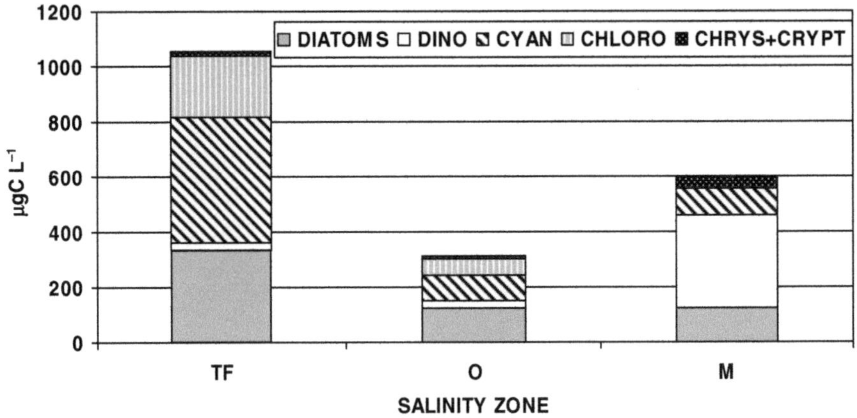

Fig. 12.4 Shift in biomass (μgC l^{-1}, averages from 1985 to 2006) of the most abundant phytoplankton groups in the Potomac River and estuary, a major tributary of the Chesapeake Bay; euglenophytes and prasinophytes not shown due to small contributions. TF = tidal fresh, 0–0.5 psu, O = oligohaline, 0.5–5 psu, and M = mesohaline, 5–18 psu

the USA. It formed after the last ice age when melting glaciers and retreating ice sheets raised sea level and flooded the lower Susquehanna River, yielding a partially stratified estuary, the current Chesapeake Bay.

Gravitational circulation insures mixing of fresh river water with oceanic water throughout the length of the Bay, yielding an extended estuarine gradient where large variability in chemistry and flora and fauna typify the system. Vertical and horizontal gradients in salinity lead to similar stratification in chemical processes and biotic signatures. The most marked gradient is at the freshwater–seawater interface, in the oligohaline zone of the system. In this zone, freshwater planktonic taxa experience severe osmotic stress, leading to rapid declines in abundances and dramatic shifts in taxonomic composition at salinities approximating 0.5–2 psu. For example, Sellner, Lacouture, and Parrish (1988) have documented salinity-induced aggregation and condensation of *Microcystis aeruginosa* typical of the tidal fresh Potomac River estuary, the second largest tributary to the brackish bay after the Susquehanna. The condensation leads to declining photosynthesis, loss of buoyancy control, and eventual rapid sedimentation of stressed, dense *M. aeruginosa* colonies. This loss of cyanobacteria is mimicked by similar declines in other freshwater taxa, particularly chlorophytes (Fig. 12.4).

As salinities increase to meso- and polyhaline conditions further down the estuary, a brackish flora emerges typified by cosmopolitan taxa including winter–spring diatoms *Skeletonema costatum*, *Cerataulina pelagica*, and *Rhizosolenia fragilissima* and the ubiquitous dinoflagellate *Prorocentrum minimum*. In summer, the flora shifts to euryhalinic small centric diatoms, dinoflagellates (*Karlodinium veneficum*, *Gymnodinium* spp.; see white bar in Fig. 12.4), and numerous unidentified microflagellates and cryptophytes. Interestingly, in the last decade, a filamentous cyanobacterium has also increased in the mesohaline and polyhaline bay (Hartsig, Lacouture, Sellner, & Imirie, 2007), leading to substantial cyanobacteria (picocyanobacteria and filament) contributions, 30–50%, to the summer autotrophic assemblage (hatched bar in mesohaline segment, Fig. 12.4).

These horizontal and seasonal shifts in phytoplankton vary year to year, reflecting annual variability in river discharge as well as river-delivered nutrient loads, particularly nitrogen (Malone, 1992; Harding, 1994). Long-term nitrate data collected in the upper Susquehanna River indicates a marked increase in concentration (Hagy, Boynton, Keefe, & Wood, 2004), reflecting population growth and accompanying 'cultural eutrophication', leading to the estuarine shifts in salinity and biomass of the phytoplankton taxa described above. Harding (1994) has described the effects of the elevated nitrate levels (Fig. 12.5) entering the upper brackish bay, largely governing the magnitude of the spring diatom bloom. This bloom thereafter governs the magnitude of summer anoxic volume via its decomposition as it settles to the bottom as largely ungrazed cells. The long-term pattern in the bay's

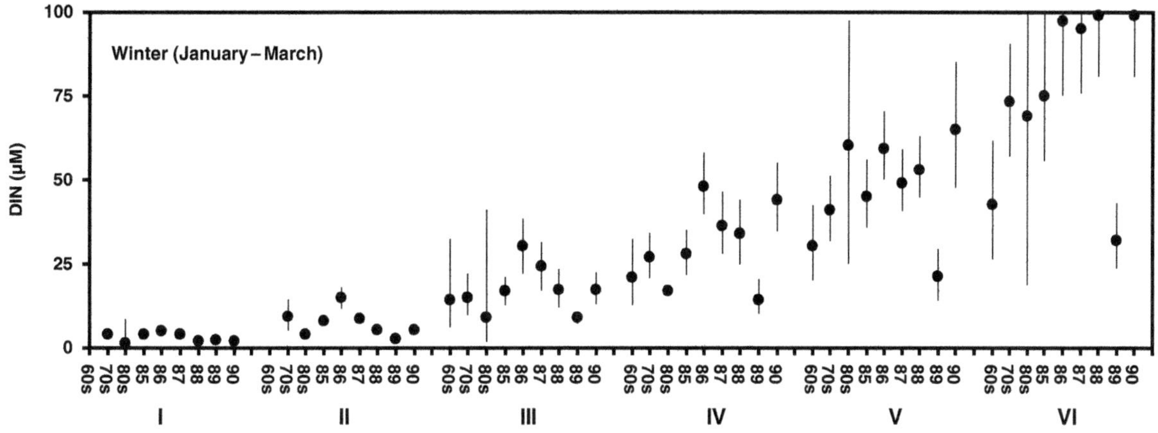

Fig. 12.5 Winter DIN concentrations (μM) by year and salinity regions of Chesapeake Bay (tidal fresh (I) to polyhaline (VI)) for 1960s, 1970s, 1980s, and individual years from 1985 to 1990. *Error bars* are 95% lower and upper confidence intervals (from Harding, 1994)

Fig. 12.6 Long-term (1950–2003) changes in bottom water hypoxia volumes in Chesapeake Bay as a function of Susquehanna River discharge (from Hagy et al., 2004)

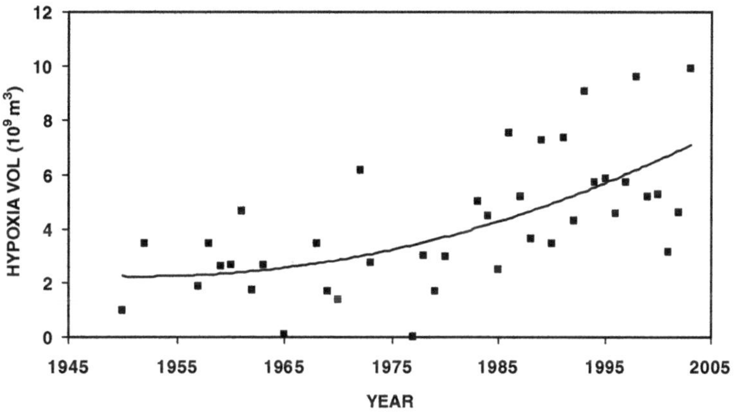

response to increasing load and algal accumulation is now obvious through the several decade increase in hypoxic volumes in the mesohaline bay, increasing at an exponential rate (Hagy et al., 2004; Fig. 12.6). The decomposition and elevated oxygen demand, in turn, lead to high benthic flux of ammonium and phosphate, fuelling elevated summer phytoplankton biomass over the last several decades (Fig. 12.7).

These recent trends suggest that eutrophication in the brackish bay is accelerating rapidly, perhaps having passed an assimilative capacity a decade or more ago. This increasing trend in system response to nutrient load could not be identified without the multi-year observations, thereby insuring acknowledgement of a rapidly worsening system state that would not be possible with short-term discrete measurements.

An interesting parallel in phytoplankton species selection is observed between the stratified brackish Chesapeake Bay and the open Baltic Sea. The seasonal shift from diatoms that occur in high flow, highly mixed surface waters of the winter–spring in the Chesapeake to small centrics, dinoflagellates, and cryptophytes in the summer can also be observed in the summer following wind-induced local 'upwelling'. In the Chesapeake, a shift from westerly to northerly or southerly winds >10 m s^{-1} leads to intrusion of sub-pycnocline, nutrient-rich water (derived from the same decomposition of diatoms noted above) into shallow depths along the shore. This in turn favours rapid growth of diatoms and a quick transition to bloom-forming dinoflagellates (Fig. 12.8), not unlike the spring-to-summer transition (Malone

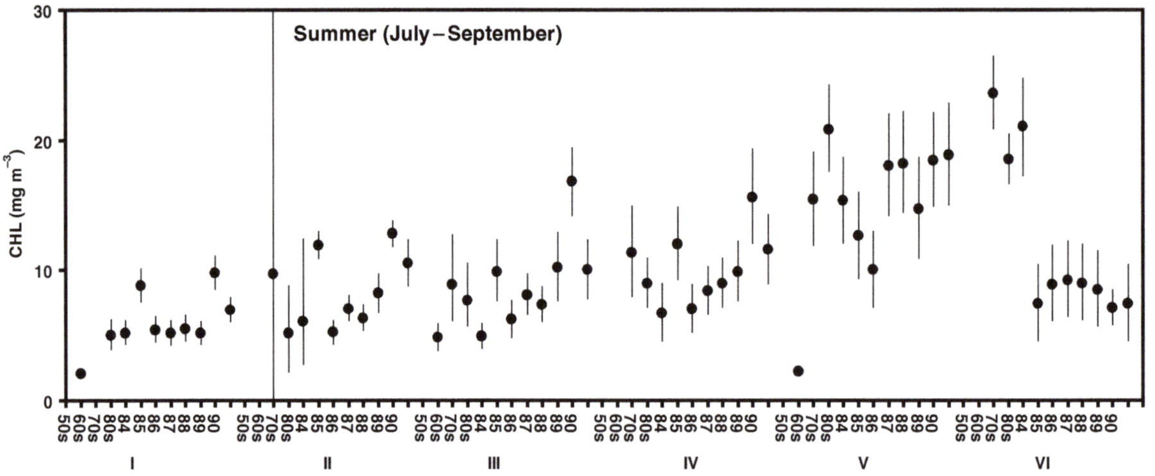

Fig. 12.7 Summer surface chlorophyll a concentrations (mg m^{-3}) in six Chesapeake Bay segments (I–VI) of increasing salinity (tidal fresh (I) to polyhaline (VI)) for the 1950s, 1960s, 1970s, 1980s, and individual years from 1984 to 1990 (*error bars* are 95% lower and upper confidence intervals) (from Harding, 1994)

et al., 1986; Sellner & Brownlee, 1990; Weiss et al., 2005).

In the open Baltic, a similar phenomenon occurs where the seasonal thermocline is displaced, enriching lighted surface waters with regenerated nitrogen (N) and phosphorus (P) with the latter giving rise to the thereafter dominant cyanobacteria *Aphanizomenon flos-aquae* or *Nodularia spumigena* (Kononen et al., 1996). In both systems, ambient flora are displaced to more opportunistic and thereafter dominant summer taxa through short-term, aperiodic wind events that alter surface mixing and nutrient availability. The importance of these event-induced shifts in phytoplankton and their subsequent dominant roles in summer production and trophic dynamics could be ascertained in the future through continuous records of nearshore trends in water temperature (satellite detection, observing systems) and subsequent species/chlorophyll responses.

This pattern for the Chesapeake likely reflects other drowned river valley system responses to increasing human population-derived land use alterations which mobilize nutrients into receiving waters and subsequent downstream advection. Once in the estuary, gravitational circulation insures substantial residence time in the brackish mixing zone, thereby favouring phytoplankton assimilation of elevated nutrient loads

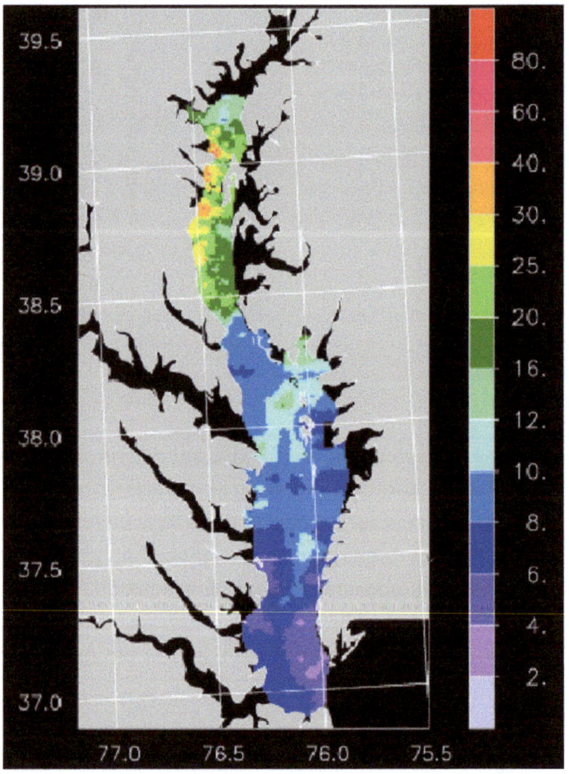

Fig. 12.8 Elevated chlorophyll a (µg l^{-1}) along Chesapeake Bay's western shore following summer pycnocline tilting, a local upwelling event (courtesy of L.W. Harding, Jr.)

and their subsequent utilization within the system prior to export to the coastal ocean.

12.3.2 Large Brackish Systems – Baltic Sea

The brackish character of the Baltic Sea (415,023 km^2, including the Kattegat; mean depth 52 m) is a function of basin geomorphology. It is connected with the North Sea only by narrow straits in the west but receives large riverine input mainly in the east. The resulting 'marine gradient' is discussed above. The sporadic inflow of North Sea water perpetuates this horizontal salinity gradient but also maintains the vertical salinity gradient because the heavier salt water flows into the deep central Baltic basins near the bottom. The resulting strong halocline is a barrier for vertical transport, leading to an oxygen deficit in deep water. Such a strong halocline is typical for deep brackish seas, e.g. also the Black Sea, and even in the Chesapeake (see above). As the halocline restricts the upward transport of nutrients, low human loads of the past centuries around the Baltic Sea likely assured that the system was largely oligotrophic. However, in the second half of the 20th century, high riverine nutrient loading from the large drainage area (1,729,000 km^2, Bergström & Carlsson, 1994) inhabited by more than 80 million people fertilized surface production and enhanced organic loading to the halocline and below, leading to the eutrophic conditions of the system now obvious in elevated summer cyanobacteria blooms.

The impacts of the increasing loads are dramatic. Elmgren (1989) described an increase in phytoplankton primary production by a factor of 1.3–1.7 from the beginning of the 20th century until the early 1980s. Kaiser, Renk, and Schulz (1981) estimated an average phytoplankton primary production of 84 g C m^{-2} yr^{-1}, which relates to 34.8×10^6 t C yr^{-1} over the entire Baltic Sea. A more recent compilation by Wasmund et al. (2001) estimated rates of 150 g C m^{-2} yr^{-1} and 62.1×10^6 t C yr^{-1}, almost a doubling within approximately two decades of the late 20th century. Similarly, chlorophyll *a* concentrations also increased in different areas of the Baltic Sea (e.g. Nakonieczny, Ochocki, & Renk, 1991; Suikkanen, Laamanen, & Huttunen, 2007). More recently, the increase seems to have levelled off in the middle of the 1990s as the former increasing trends are no longer apparent in recent time series. In the western Baltic Sea, chlorophyll *a* levels have been declining for more than 20 years (Wasmund & Uhlig, 2003).

Long-term changes were recorded not only in phytoplankton biomass and productivity but also in species composition. Wasmund, Nausch, and Matthäus (1998) described a strong decrease in the proportion of diatoms in spring blooms of the southern Baltic Proper in 1989, whereas dinoflagellates increased continuously. Despite of the high variability of the spring bloom data due to undersampling, Wasmund and Uhlig (2003) noted the same pattern for the central Baltic (Eastern Gotland Sea) using rigorous statistical analyses. The decrease in diatoms has direct implications on nutrient pools, e.g. the suddenly reduced silicate consumption in the spring bloom (see Fig. 12.9) in the southern Baltic Proper and the Gdańsk Basin (Trzosinska & Lysiak-Pastuszak, 1996) but not in Mecklenburg Bight. Despite decreasing winter silicate concentrations, diatoms are still not limited by silicate but by nitrogen, as shown by Brodherr (2006). Wasmund et al. (1998) attributed this pattern to a strong increase in winter temperatures since 1989. If the water temperature does not fall below the temperature of the highest density of the water (about 2.5°C in the brackish water of the Eastern Gotland Sea), the water column remains at least theoretically stable and therefore limited convective mixing ensues after warming in spring. Lack of mixing (stability) is disadvantageous for diatoms but beneficial for flagellates (see Harrison, Turpin, Bienfang, & Davis, 1986). The recovery of the diatoms since 2000 needs still to be investigated.

One interesting feature of the Baltic Proper is the low nitrogen: phosphorus ratio (N:P ratio), approximating eight, in the winter surface water. This deviation from the Redfield ratio (N:P = 16) is attributed to the oxycline, typical for deep brackish seas with stagnant water below the permanent halocline. In this suboxic region, denitrification and anaerobic ammonium oxidation occur, removing nitrogen from the water (Hannig et al., 2007). Moreover, phosphorus and silicate are liberated from anoxic sediments (Kuparinen & Tuominen, 2001, Conley, Humborg, Rahm, Savchuk,

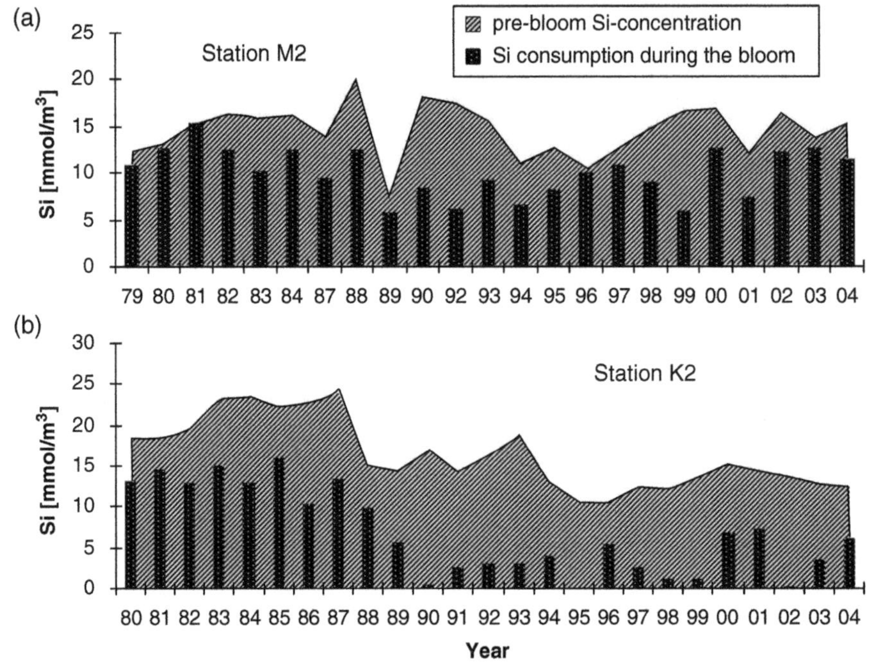

Fig. 12.9 Silicate concentration (mmol/m^3) in late winter (*line*) and difference in pre- and post-bloom silicate concentration (bars) from 1979/1980 to 2004 in (**a**) Mecklenburg Bight and (**b**) Bornholm Sea (updated from Wasmund et al., 1998)

& Wulff, 2002). The resulting decreased N:P ratios have far-reaching consequences. After the disappearance of the spring bloom, mainly by sedimentation, the inorganic nitrogen is exhausted, but a surplus of phosphorus remains in the water. Oligotrophic conditions would prevail but nitrogen-fixing cyanobacteria, mainly *Nodularia spumigena* and *Aphanizomenon* sp., grow in response to the phosphorus (see upwelling discussion above). Their growth leads to blooms, typical for the Baltic Proper in July or early August. These nuisance blooms receive high attention as they are normally toxic and restrict tourism in the bathing season. It is believed that the intensity of cyanobacteria blooms has increased, but statistical support is not yet available as data are highly variable due to high patchiness of blooms and inappropriate sampling. Satellite imagery is a useful tool especially for the monitoring of cyanobacteria blooms as they are buoyant and easily recognizable from space. From interpretations of satellite images, Kahru (1997) found large areas covered by cyanobacteria blooms in 1982–1984 and 1991–1994, which reflects an El Niño-like cycle. Trend analyses by Suikkanen et al. (2007) revealed a cyanobacteria increase in the Gulf of Finland and high singular cyanobacteria peaks in 1985, 1995, and 1996 in the Northern Baltic Proper. Mazur-Marzec, Krężel, Kobos, and Pliński (2006) stated that large-scale occurrences of *N. spumigena* were recorded for the first time in the Gulf of Gdańsk in 1994 and repeated in 2001, 2003, and 2004. Cyanobacteria blooms are not a new phenomenon in the Baltic Sea (Finni, Kononen, Olsonen, & Wallström, 2001), but they have probably intensified (Poutanen & Nikkilä, 2001).

As noted above, the importance of sub-pycnocline nutrients in Baltic summer phytoplankton and productivity is also important in summer phytoplankton production of the Chesapeake (see Fig. 12.7) with the difference, however, that the Baltic is dominated by N-fixing cyanobacteria and these are absent in the Chesapeake. As a result, the spring phytoplankton (diatom) maximum in biomass is followed by low ambient chlorophyll levels (although increasing through the last decade) but high productivity in the summer, the latter fuelled by regenerated N below the seasonal pycnocline in Chesapeake Bay and coastal areas of the Baltic. As this recycled N is introduced into surface waters through 'upwelling' (see above), short-term increases in biomass and productivity occur, visible as dinoflagellate blooms (see Fig. 12.8). It is reasonable to expect that similar disruptions of seasonal pycnoclines in the Bay's major tributaries result in temporary maxima in biomass and productivity, as sub-pycnocline N is delivered to shallow lighted depths of these systems. In the deep central Baltic areas, nitrogen fixation is the main N source for the pelagic system.

12.3.3 Coastal Lagoons – Darss-Zingst Bodden Chain

The Darss-Zingst Bodden Chain (DZBC) represents a typical element of the southern Baltic coast, a shallow lagoon with substantial freshwater inflow. This results in a strong salinity gradient, superimposed with an eutrophication gradient because of the heavy nutrient load from the catchment area (for details see Schiewer, 2007). Salinity ranges between 1 psu in the innermost areas and 8–12 psu at the opening to the Baltic Sea. With a total area of 197 km^2, DZBC is a rather small system and with a catchment area/water surface ratio of 8:1, the potential natural water quality is classified as mesotrophic/eutrophic (Schiewer, 2007).

In the 1970s, intense ecological investigations were undertaken which, through time, have revealed pronounced changes in ecosystem matter flux and species dominance. The first system analysis in the mid-1970s (Schnese, 1978; see also Schiewer, 2007) revealed that the main bulk of primary production was subject to sedimentation (see also C1 – Chesapeake) and not transferred via 'classical' food web to higher trophic levels (Fig. 12.10). Since then, nutrient load has substantially increased, leading to hypertrophic conditions (Schiewer, 1998a). A decade after the initial survey, Schiewer (1990, 1998b) noted that (A) primary production had decreased, caused to a large extent by the disappearance of macrophytes, (B) the importance of the planktonic food web further declined whereas, (C) the amount of detritus was almost stable (Fig. 12.10).

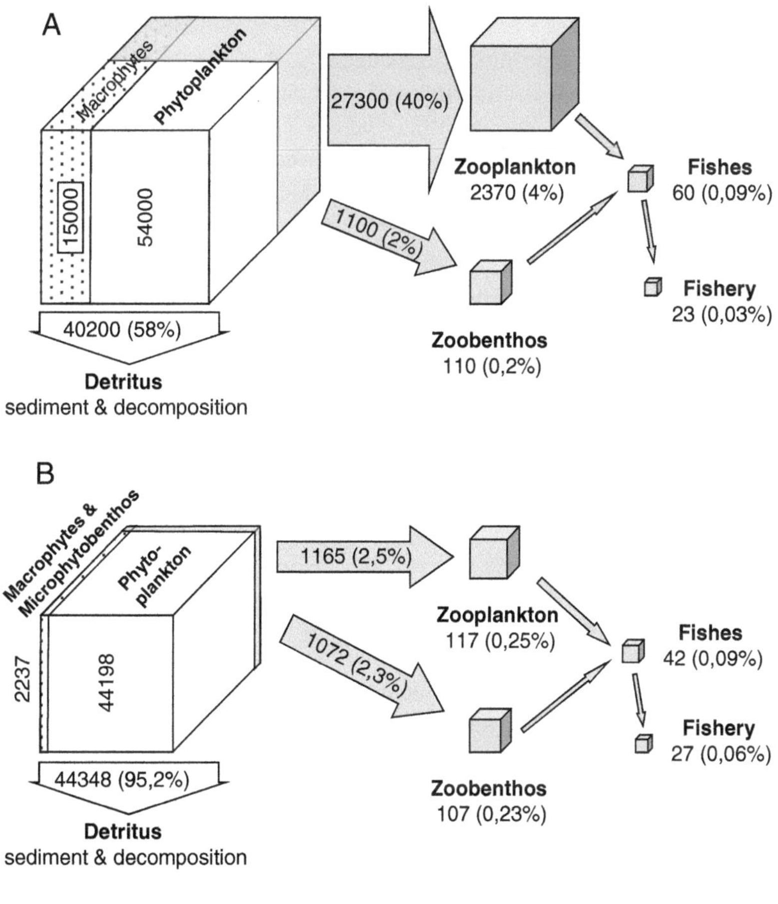

Fig. 12.10 Trophic relationships in the DZBK under (**a**) eutrophic and (**b**) hypertrophic conditions [re-drawn from Schiewer (2001); data from Schnese (1986, *upper panel*) and Schiewer (1985, *lower panel*) for the situation in the (**a**) mid-1970s and (**b**) 1981/1983]

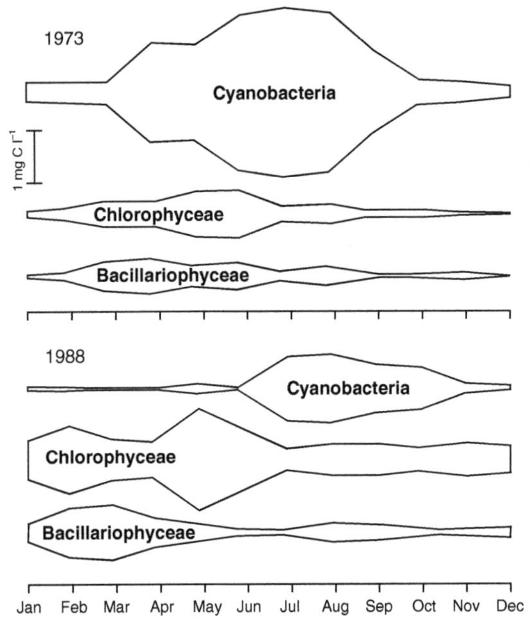

Fig. 12.11 Phytoplankton biomass (mgC l^{-1}) in the Darss-Zingst Bodden Chain under eutrophic (1973) and hypertrophic (1988) conditions (re-drawn after Wasmund & Schiewer, 1994)

Phytoplankton composition also changed. As shown in Fig. 12.11, cyanobacteria dominance retreated in the 1980s (see also Fig. 12.10), whereas chlorophytes benefited from the increased nutrient load. The declining cyanobacteria was associated with a declining pH (Wasmund & Heerkloss, 1993) and subsequent laboratory experiments with the abundant green alga *Tetrastrum* cf. *triangulare* and the cyanobacterium *Nodularia harveyana* revealed better green algae growth at low pH (7.5), while the cyanobacterium grew more rapidly at the elevated pH (9.0) (Wasmund, 1996).

The decrease of N-fixing species, however, can be interpreted as a sign of N-delimitation due to increased nutrient loads, i.e. with abundant N, there is no competitive advantage for diazotrophic cyanobacteria. The larger proportion of cyanobacteria as picocyanoplanktonic species seemed to be responsible for the marked decrease in trophic connectivity, shown already in Fig. 12.10. A regime shift in the same period has also been observed for the open sea (Wasmund et al., 1998, see also C2).

At the end of the 1980s, the common view, therefore, was that the DZBC system suffered strong light limitation as a result of elevated phytoplankton biomass and shading which in turn diminished macrophyte growth and led to a detritus-oriented food web. Reduced macrophytes further reduced productivity of the system, leading to sediment instability even further reducing irradiance availability for the macrophytes as well as phytoplankton.

This view was supported by the results of repeated macrophytobenthos inventories, irregularly investigated since 1895; the first thorough mapping was done in the early 1970s (for an overview see Blümel et al., 2002). In the late 1980s, macrophyte cover was reduced already by >70% (Lindner, 1978) and in the early 1990s, Schiewer (2001) reported that there were only a few remnants left of the former rich macrophytes. However, this statement, made on the basis of a 1993 mapping campaign, had to be revised in 1995, where a massive spread, particularly charophytes, were observed all over the DZBK (Yousef, 2000). Since then, macrophyte cover has increased to levels almost comparable to coverage noted in 1970s (Schubert et al., 2004). Interestingly, with respect to phytoplankton as well as dissolved nutrient concentrations, the situation has not changed much since 1990. Sediment phosphate pools still nourish massive phytoplankton development irrespective of the reduced nutrient loads from the catchment area (Schlungbaum, Baudler, & Krech, 2001).

Schumann (1993) analysed nutrient limitation of the DZBC by means of microcosm studies. Figure 12.12 shows the annual cycle of limitation events which can be described as follows:

After physical limitation by temperature/ice cover/light, rapid phytoplankton growth diminished both P and N, resulting initially in combined limitation by both macronutrients. However, the nutrient pool was sufficiently large to prevent effective zooplankton top-down control. After exceeding a N:P of 15, phosphate release from the sediment increased P-concentrations through onset of N-limitation of the still abiotic controlled (physical factors and nutrient – N) phytoplankton community, causing also a switch to diazotrophic cyanobacteria.

As can be seen from these kinetics, there is no general limitation scheme for these systems, but a complex succession of different limitation stages. These findings fit well with the 'PEG-model' derived from limnetic systems, which also describes a succession of multiple limitation stages occurring over the year in the planktonic system (Sommer, Gliwicz, Lampert,

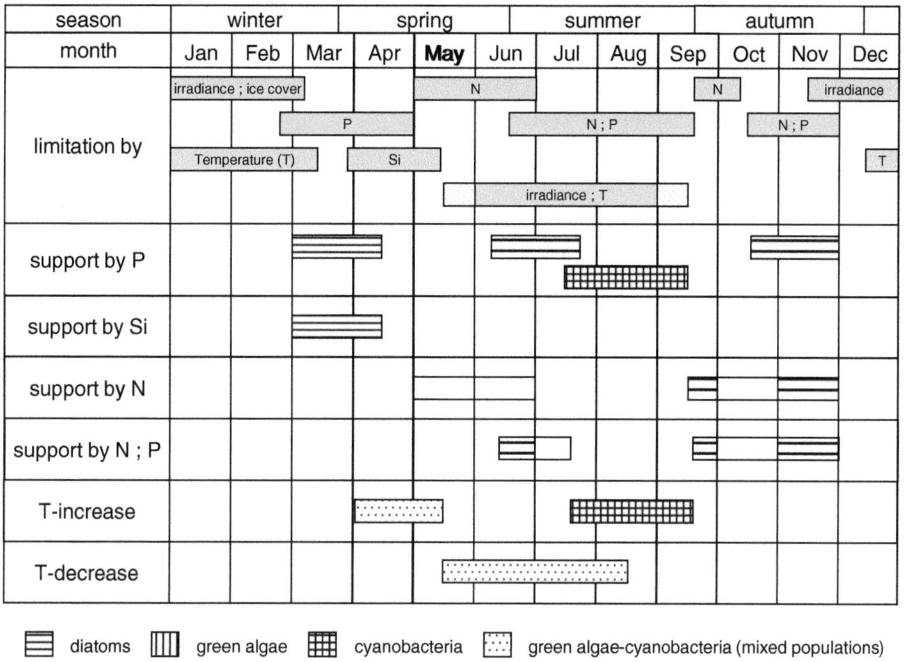

Fig. 12.12 Phytoplankton limitation pattern in the Darss-Zingst Bodden Chain. The limitation pattern was investigated by microcosm experiments with plankton samples from the Darss-Zingst Bodden Chain, in which the respective factors were increased/added for a 6-day period (for details see Schiewer, 1988)

& Duncan, 1986). However, the PEG model does not cover the hypertrophic conditions of the DZBC. Field investigations carried out in the mid-1990s showed that not only light limitation was responsible for the low trophic connectivity. In addition, an increased importance of the microbial loop caused a high turnover of nutrients and organic matter in the pelagic system. As a consequence, productivity of the system is no longer nutrient controlled; nutrient availability only controls species composition. It has been shown by several authors that under hypertrophic conditions irradiance availability and/or temperature follow nutrient availability as the controlling factor in such brackish systems (Schiewer, 2001; Schubert, 1996; Schubert & Wasmund, 2005). However, the mechanisms underlying (particularly) temperature dependency of planktonic biomass development are not yet understood and require further investigation (Schubert & Wasmund, 2005).

In the coming years, the effects of declining nutrient loads are expected to cause a return to mesotrophic conditions in the Darss-Zingst Bodden Chain, resulting in a re-onset of nutrient limitation. Further monitoring will be performed to follow this unique event.

12.4 Conclusions

Irrespective of hydromorphology-based differences and the large variability of almost all parameters typical of brackish waters, long-term studies of the three systems revealed some common features. First, the main driver for changes observed was anthropogenic in origin; irrespective of year-to-year or even decadal changes in weather conditions, eutrophication alone explains a large portion of the changes observed in ecosystem functioning of the systems. With respect to the specific eutrophication effects, decoupling of trophic interactions, leading to increased sedimentation and deposition, seems to be a general system response to increased nutrient availability. The specific changes in phytoplankton composition are more complex. At present, the switch from P- to N-limitation in the Baltic, indicated by an increase of diazotrophic cyanobacteria, can be interpreted as the first sign of hypertrophication. A further increase in nutrient loads may, as shown in C3, lead to light limitation even in shallow systems, conditions where diazotrophic Cyanobacteria are no longer favoured. However, there are several questions still to be answered: the effects of silica depletion in the Baltic Sea as well as the phenomenon of accelerated eutrophication in Chesapeake Bay are of general interest, but also specific details as the kinetics and the time lag in re-mesotrophication of the German coastal lagoon are ecological questions requiring long-term investigations.

Summarizing it can be stated that, beside the general vantage of long-term ecological studies to provide a robust backbone from whose output-specific investigations enable deeper understanding of system behaviour, such studies in brackish systems are requirements without alternatives. As shown in all three examples, the variability in brackish systems does not allow reliable detection of system shifts by point-to-point comparisons. Irregular mixing events can provoke development of dinoflagellates in Chesapeake Bay, and patchiness of cyanobacteria blooms in the Baltic in summer months, while year-to-year variation in weather conditions modulate the effects of nutrient limitation, etc. The latter effect probably will be of greater importance in future as global climate change gradually selects for new species and accelerates kinetics for most biogeochemical processes. Detection of the effects of global climate changes in highly variable systems requires not only profound knowledge about responses to 'non-climate drivers' but also robust data that bound system responses for 'pristine' weather-induced variability over several decades. Only long-term data can provide this analytical potential for detecting change.

References

Bergström, S., & Carlsson, B. (1994). River runoff to the Baltic Sea. 1950–1990. *Ambio, 23*, 280–287.

Blümel, C., Domin, A., Krause, J. C., Schubert, M., Schiewer, U., & Schubert, H. (2002). Der historische Makrophytenbewuchs der inneren Gewässer der deutschen Ostseeküste. *Rostocker Meeresbiologische Beiträge, 10*, 5–111.

Brodherr, B. (2006). Nutrient dependent growth dynamics of diatom spring populations in the southern Baltic Sea. Dissertation: Universität Rostock.

Conley, D., Humborg, C., Rahm, L., Savchuk, O. P., & Wulff, F. (2002). Hypoxia in the Baltic Sea and basin-scale changes in phosphorus biogeochemistry. *Environmental Science and Technology, 36*, 5315–5320.

Elmgren, R. (1989). Man's impact on the ecosystem of the Baltic Sea: Energy flows today and at the turn of the century. *Ambio, 18*, 326–332.

Finni, T., Kononen, K., Olsonen, R., & Wallström, K. (2001). The history of cyanobacterial blooms in the Baltic Sea. *Ambio, 30*, 172–178.

Gromisz, S., & Szymelfenig, M. (2005). Phytoplankton in the Hel upwelling region (the Baltic Sea). *Oceanological and Hydrobiological Studies, 34*, 115–135.

Hagy, J. D., Boynton, W. R., Keefe, C. W., & Wood, K. V. (2004). Hypoxia in Chesapeake Bay, 1950–2001: Long-term change in relation to nutrient loading and river flow. *Estuaries, 27*, 634–658.

Hannig, M., Lavik, G., Kuypers, M. M. M., Woebken, D., Martens-Habbena, W., & Jürgens, K. (2007). Shift from denitrification to anammox after inflow events in the central Baltic Sea. *Limnology and Oceanography, 52*, 1336–1345.

Harding, L.W., Jr. (1994). Long-term trends in the distribution of phytoplankton in Chesapeake Bay: Roles of light, nutrients, and streamflow. *Marine Ecology Progress Series, 104*, 267–291.

Harrison, P. J., Turpin, D. H., Bienfang, P. K., & Davis, C. O. (1986). Sinking as a factor affecting phytoplankton species succession: The use of selective loss semi-continuous cultures. *Journal of Experimental Marine Biology and Ecology, 99*, 19–30.

Hartsig, A. M., Lacouture, R. V., Sellner, S. G., & Imirie, A. L. (2007). *Increases in Cyanobacteria during Summer in Mesohaline Chesapeake Bay, 1985–2006.* Poster presented at the Estuarine Research Federation Meeting, Providence, RI, USA.

Kahru, M. (1997). Using satellites to monitor large-scale environmental change: A case study of cyanobacteria blooms in the Baltic. In M. Kahru, & C. W. Brown (Eds.), *Monitoring algal blooms* (pp. 43–61). Berlin: Springer.

Kaiser, W., Renk, H., & Schulz, S. (1981). Die Primärproduktion der Ostsee. *Geodätische und Geophysikalische Veröffentlichungen des Nationalkomitees für Geodäsie und Geophysik bei der AdW der DDR Reihe IV*, 33, 27–52.

Khlebovich, V. V. (1990). Some physico-chemical and biological phenomena in the salinity gradient. *Limnologica, 20*, 5–8.

Kononen, K., Kuparinen, J., Mäkelä, K., Laanemets, J., Pavelson, J., & Nômmann, S. (1996). Initiation of cyanobacterial blooms in a frontal region at the entrance to the Gulf of Finland, Baltic Sea. *Limnology & Oceanography, 41*, 98–112.

Kuparinen, J., & Tuominen, L. (2001). Eutrophication and self-purification: Counteractions forced by large-scale cycles and hydrodynamic processes. *Ambio, 30*, 190–194.

Lindner, A. (1978). Soziologisch-ökologische Untersuchungen an der submersen Vegetation in der Boddenkette südlich des Darß und des Zingst (südliche Ostsee). *Limnologica, 11*, 229–305.

Malone, T. C. (1992): Effects of water column processes on dissolved oxygen, nutrients, phytoplankton, and zooplankton. In D. E. Smith, M. Leffler, & G. Mackiernan (Eds.),

Oxygen dynamics in the Chesapeake Bay. A synthesis of recent research (pp. 61–112). College Park, MD: D Sea GrantCollege.

Malone, T. C., Kemp, W. M., Ducklow, H. W., Boynton, W. R., Tuttle, J. H., & Jonas, R. B. (1986). Lateral variation in the production and fate of phytoplankton in a partially stratified estuary. *Marine Ecology Progress Series, 32*, 149–160.

Mazur-Marzec, H., Krężel, A., Kobos, J., & Pliński, M. (2006). Toxic *Nodularia spumigena* blooms in the coastal waters of the Gulf of Gdańsk: A ten-year survey. *Oceanologia, 48*, 255–273.

Nakonieczny, J., Ochocki, S., & Renk, H. (1991). Long-term changes in primary production and chlorophyll concentrations in the southern Baltic. *Acta Ichthyologica et Piscatoria, 21*, 145–152.

Nômmann, S., Sildam, S., Nôges, T., & Kahru, M. (1991). Plankton distribution during a coastal upwelling event off Hiiumaa, Baltic Sea: Impact of short-term flow field variability. *Continental Shelf Research, 11*, 95–108.

Poutanen, E.-L., & Nikkilä, K. (2001). Carotenoid pigments as tracers of cyanobacterial blooms in recent and post-glacial sediments of the Baltic Sea. *Ambio, 30*, 179–183.

Remane, A. (1940). *Einführung in die zoologische Ökologie der Nord- und Ostsee. Lieferung 34 der Reihe: Tierwelt der Nord- und Ostsee.* Leipzig: Akademische Verlagsgesellschaft.

Remane, A. (1955). Die Brackwasser-Submergenz und die Umkomposition der Coenosen in Belt- und Ostsee. *Kieler Meeresforschung, 11*, 59–73.

Schiewer, U. (1985). *Analyse und Bewertung des im 5-Jahrplanzeitraume 1981/85 erreichten Kenntnisstandes und seiner volkswirtschaftlichen Verwertbarkeit auf der Grundlage der erbrachten Teilleistungen. Forschungsbericht.* Rostock: Wilhelm-Pieck-Universität Rostock.

Schiewer, U. (1988). Experimentelle Ökosystemanalyse – Ergebnisse und Probleme. *Wissenschaftliche Zeitschrift der Wilhelm-Pieck-Universität Rostock N-Reihe, 37*, 13–17.

Schiewer, U. (1990). Werner Schnese and the development of coastal waters ecology in Rostock, GDR. *International Review of Hydrobiology, 75*, 1–13.

Schiewer, U. (1998a). 30 years eutrophication in shallow brackish waters – lessons to be learned. *Hydrobiologia, 363*, 73–79.

Schiewer, U. (1998b). Hypertrophy of a Baltic estuary – changes in structure and function of the planktonic community. *Verhandlungen der Internationalen Vereinigung für Limnologie, 26*, 1503–1507.

Schiewer, U. (2001). Phytoplankton, Produktivität und Nahrungsnetze. *Meer und Museum, 16*, 39–45.

Schiewer, U. (2007). Darß-Zingst, Northern Rügener Bodden and Schlei. In U. Schiewer (Ed.), *Ecology of Baltic coastal waters* (pp. 35–86). Berlin: Springer.

Schlungbaum, G., Baudler, H., & Krech, M. (2001). Das Eutrophierungsproblem der Darß–Zingster Bodden – Nährstoffeinträge und Nährstoffbilanzen. *Meer und Museum, 16*, 25–34.

Schnese, W. (1978). *Produktionsbiologische Grundlagen für die Einbürgerung von Planktonfressern in der Darss-Zingster Boddenkette. Materialien der III. Wiss. Konferenz zu Fragen der Physiologie und Biologie von Nutzfischen.* Rostock: Wilhelm-Pieck-Universität Rostock.

Schubert, H. (1996) Ökophysiologie der Lichtanpassung des Phytoplanktons eutropher Flachgewässer. Habilitation, Universität Rostock.

Schubert, H., & Wasmund, N. (2005). Das Phytoplankton des Strelasundes und des Kubitzer Boddens. *Meer und Museum, 18*, 83–92.

Schubert, H., Blümel, C., Eggert, A., Rieling, T., Schubert, M., & Selig, U. (2004). *Entwicklung von leitbildorientierten Bewertungsgrundlagen für innere Küstengewässer der deutschen Ostseeküste nach der EU-WRRL. Report BMBF-Projekt 0330014.* Rostock: Universität Rostock.

Schumann, R. (1993). Zur Rolle des Pico– und Nanophytoplanktons im mikrobiellen Nahrungsgefüge der Darß–Zingster Boddenkette. Dissertation, Universität Rostock.

Sellner, K. G., & Brownlee, D. C. (1990). Dinoflagellate-microzooplankton interactions in Chesapeake Bay. In E. Granéli, B. Sundström, L. Edler, & D. M. Anderson (Eds.), *Toxic marine phytoplankton* (pp. 221-226). New York: Elsevier.

Sellner, K. G., Lacouture, R. V., & Parrish, C. R. (1988). Effects of increasing salinity on a cyanobacteria bloom in the Potomac River estuary. *Journal of Plankton Research, 10*, 49-61.

Sommer, U., Gliwicz, Z. M., Lampert, W., & Duncan, A. (1986). The PEG-model of seasonal succession of planktonic events in freshwaters. *Archiv für Hydrobiologie, 106*, 433–471.

Suikkanen, S., Laamanen, M., & Huttunen, M. (2007). Long-term changes in summer phytoplankton communities of the open northern Baltic Sea. *Estuarine, Coastal and Shelf Science, 71*, 580–592.

Trzosinska A., & Lysiak-Pastuszak E. (1996). Oxygen and nutrients in the southern Baltic Sea. *Oceanological Studies, 1*, 41–76.

Vahtera, E., Laanemets, J., Pavelson, J., Huttunen, M., & Kononen, K. (2005). Effect of upwelling on the pelagic environment and bloom-forming cyanobacteria in the western Gulf of Finland, Baltic Sea. *Journal of Marine Systems, 58*, 67–82.

Wasmund, N. (1990). Characteristics of phytoplankton in brackish waters of different trophic levels. *Limnologica, 20*, 47–51.

Wasmund, N. (1996). Periodicity and trends in the phytoplankton of a shallow coastal water. In A. Andrushaitis (Ed.), *Proceedings of the 13th Baltic Marine Biologists Symposium, Jurmala, Latvia, 31.8.1993–4.9.1993* (pp.63–66). Riga: Institute of Aquatic Ecology.

Wasmund, N., & Heerkloss, R. (1993). Seasonal and long-term succession of phytoplankton in shallow coastal water of the southern Baltic Sea. *Studia i materialy oceanologiczne, 64*, 203–213.

Wasmund, N., & Schiewer, U. (1994). Überblick zur Ökologie und Produktionsbiologie des Phytoplanktons der Darß-Zingster Boddenkette (südliche Ostsee). *Rostocker Meeresbiologische Beiträge, 2*, 41–60.

Wasmund, N., & Uhlig, S. (2003): Phytoplankton trends in the Baltic Sea. *ICES Journal of Marine Science, 60*, 177–186.

Wasmund, N., Andrushaitis, A., Łysiak-Pastuszak, E., Müller-Karulis, B., Nausch, G., Neumann, T., et al. (2001). Trophic status of the south-eastern Baltic Sea: A comparison of coastal and open areas. *Estuarine, Coastal and Shelf Science, 53*, 849–864.

Wasmund, N., Nausch, G., & Matthäus, W. (1998). Phytoplankton spring blooms in the southern Baltic Sea – spatio-temporal development and long-term trends. *Journal of Plankton Research, 20*, 1099–1117.

Wasmund, N., Zalewski, M., & Busch, S. (1999). Phytoplankton in large river plumes in the Baltic Sea. *ICES Journal of Marine Science, 56*, 23–32.

Weiss, G. M., Harding, L. W., Jr., Itsweire, E. C., & Campbell, J. W. (2005). Characterizing lateral variability of phytoplankton chlorophyll in Chesapeake Bay with aircraft ocean color data. *Marine Ecology Progress Series, 149*, 183–199.

Yousef, M. (2000) Ökophysiologie von Makrophyten und Epiphyten in Flachwasserökosystemen. Dissertation, Universität Rostock.

Chapter 13

Long-Term Ecological Research in Freshwater Ecosystems

Jan Köhler

Abstract Long-term changes of freshwater ecosystems are mainly caused by immissions from drainage basin and atmosphere (nutrients, acid substances, etc.) and by changing climatic conditions. Freshwater ecosystems often react in non-linear ways to these external forces. Beyond a certain threshold, gradual shifts may cause catastrophic switches to another state. The way back to the previous state rarely corresponds to the past changes because of memory effects of the system. Long-term studies are necessary, but they do not allow for a simple extrapolation of past observations into the future.

Freshwater systems are also influenced by rare events like invasion of new species, spates or droughts. Effects of perturbations should be studied until the system establishes a new equilibrium.

The analysis of long-term processes needs sound knowledge about natural oscillations or gradual changes of the baseline. Monitoring programmes of German lakes and reservoirs rarely last longer than 30 years. They usually started after serious environmental problems had emerged; they do not cover periods without human impacts (baseline conditions). Therefore, long-term monitoring should be accompanied and extended by palaeolimnological approaches.

Keywords Lake ecology · Long-term studies · Eutrophication · Acidification · Invasive species · Climate change · Feedback mechanisms · Hysteresis · Palaeolimnology

J. Köhler (✉)
Leibniz-Institute of Freshwater Ecology and Inland Fisheries, Müggelseedamm 301, 12587 Berlin, Germany
e-mail: koehler@igb-berlin.de

13.1 Introduction

German lakes were formed after the retreat of the last glaciation about 10,000–15,000 years before present. Since then, they underwent permanent developments caused by sediment accumulation, invasion by new and extinction of 'native' species, changes in vegetation, soil properties and hydrology of the drainage basin, or large-scale climatic shifts. Human activities influenced lake and river systems in many ways for at least the last millennium. First systematic studies of lakes were performed in the late 19th century. In the first decades of the 20th century, most limnological studies were focused on descriptions of several aspects of single lakes and attempts to compare lakes by different schemes of lake classification. Deep lakes were more frequently studied than shallow ones, and running waters were nearly neglected. Lakes were seen as microcosms or superorganisms; the effects of drainage basin and the atmosphere came into focus only in the 1970s (Kalff, 2002). Popularity of research topics and funding policy is always changing. Therefore, systematic studies of lakes or rivers rarely cover more than the last 30 years. This is a long time relative to the life span of the researcher but just a snapshot given the thousands of years' development time of the studied systems.

Man usually tries to extrapolate the own experience in order to understand past changes or to predict future developments. This strategy is of limited value, however, in case of rare events or slow shifts which eventually cause catastrophic switches. Obviously, even an extended study period will never cover all important timescales. But the reliability of extrapolations will increase with the duration of included

F. Müller et al. (eds.), *Long-Term Ecological Research*,
DOI 10.1007/978-90-481-8782-9_13, © Springer Science+Business Media B.V. 2010

observations. Even more important, long-term investigations are needed to understand the processes which shape ecosystems in the long run. Main driving forces for freshwater systems are immissions of nutrients and acid substances, invasion of new species and climatic conditions. In this chapter, I discuss the timescale of direct and indirect processes determining these driving forces and compare them with known monitoring programmes of German lakes and reservoirs.

13.2 Effects of Perturbations

Rare events or catastrophes may completely alter species composition or habitat structure of aquatic ecosystems. Running waters are especially prone to occasional floods or droughts. Obviously, long-term observations are necessary to estimate the frequency of rare events and to study their impact on the ecosystem.

We need to observe the effects even of frequent perturbations at least until the system reaches a new equilibrium. The response of an aquatic ecosystem to loading with nutrients or toxic substances depends on flushing rate or retention time. Rivers or flushed lakes will recover from perturbations much faster than isolated lakes. A new equilibrium appears only after some generations, i.e. it needs several years for fish or macrophytes but only few weeks for bacteria, algae or protozoa. Speed of equilibration also depends on reproductive strategies of the key species. On the other hand, environmental changes in temperate lakes or rivers follow an annual pattern, so that investigations even of quickly reproducing organisms should last whole years. Weather conditions (temperature, irradiance, precipitation and the related fluxes from the drainage basin) usually vary from year to year, i.e. observations from single years are not valid for longer periods.

Oscillations and long-term shifts are essential parts of dynamic systems. A serious analysis of any change in the system requires sound knowledge on temporal variations in the baseline from which the change occurred (see Elliott, 1990). In an ideal world, the system under investigation should be studied for some years before a perturbation starts.

Other processes cause gradual changes over long time periods, like nutrient enrichment (eutrophication), acidification and invasion of non-native species or climate change. These long-term changes may occur without any human activities so it is important to differentiate between natural dynamics and anthropogenic impacts. Apart from this baseline problem, the non-linear response to external forces requires long-term studies of aquatic ecosystems.

13.3 Eutrophication

Productivity and structure of most aquatic ecosystems depend on nutrient supply. In most freshwater systems, phosphorus availability limits growth of primary producers. Other potentially important nutrients are nitrogen and silicon. Nutrient immissions into lakes and rivers result from import to the drainage basin and release and retention processes within the basin. Release and retention of nutrients are influenced by properties of the catchment (relief, vegetation cover, geology, soil type, etc.), by hydrological and climatic conditions. Most of these factors are subject to permanent natural as well as anthropogenic changes. After the last glaciation, nutrient emissions from the catchment changed slowly due to soil formation, vegetation succession and development of stream systems. These long-term shifts were modified by climate changes. Human activities like deforestation, farming and stockbreeding, drainage of wetlands or construction of watermills influenced nutrient emissions in European catchments for at least the last millennium. Nutrient input increased dramatically after introduction of flushing toilets and sewer systems in urban areas and of artificial fertilisers in agriculture and horticulture. Phosphorus loading of European rivers and lakes was highest in the 1960s–1980s. It declined after the ban of P-containing washing powders and detergents and the construction of advanced sewage treatment plants which remove most phosphorus and nitrogen from wastewater. Nitrogen loading did not decline in many catchments due to still intensive use of fertilisers. Additionally, nitrate often travels some decades from application in agriculture via groundwater to the next river.

Aquatic ecosystems respond to changes in nutrient loading more or less delayed and often in non-linear ways. Lake sediments act as phosphorus sink. They accumulated large P amounts in decades of increased loading. This P pool delays the decline of in-lake P concentration despite reduced external loading. P remobilisation from sediments was enhanced by the positive feedback between high phytoplankton biomass and high oxygen demand at the sediment–water interface. In most lakes, a new equilibrium was attained only 10–15 years after reduction of external nutrient loading (Jeppesen et al., 2005). The length of this delay depends on flushing rate, P concentration in sediment and water and binding form of P. Often, lakes recovered initially at slow rates but improved once phosphorus concentration fell below a critical value (Moosmann, Gächter, Müller, & Wüest, 2005). Lakes respond nearly immediately to reductions in nitrogen loading because N accumulation in sediments is low and surplus N is removed from the system via denitrification (Jensen, Jeppesen, Kristensen, Christensen, & Sondergaard, 1992; Köhler et al., 2005).

Non-linear responses to changing nutrient loading are also caused by biotic interactions. Under eutrophic conditions, fish community was dominated by whitefish feeding on zooplankton and zoobenthos. A shift back to a higher proportion of piscivores is usually delayed and may be accelerated by artificial stocking. Shallow regions of lakes and rivers are often covered by aquatic vegetation. Dense stands of plants reduce resuspension of settled particles and provide shelter to planktonic and substrate to benthic filter-feeders. In this way, they reduce the turbidity of the water column and improve their own light supply. Increased nutrient loading caused higher biomass of planktonic and epiphytic algae which shaded the submersed macrophytes. As a consequence, higher plants disappeared and algal growth further accelerated. At reduced nutrient concentrations, submersed macrophytes re-appeared. This switch between macrophyte- and phytoplankton-dominated states may occur from one year to the next. The nutrient concentrations which trigger this switch are much lower for the re-colonisation by macrophytes than for their disappearance. Due to this hysteresis, two stable but completely different states of lake systems may occur at the same nutrient concentration (Scheffer, Hosper, Meijer, Moss, & Jeppesen, 1993). Hysteresis implies a delayed response of shallow lakes to reductions of nutrient loading.

13.4 Acidification

After the last glaciation, pH of soft-water lakes decreased slowly, caused by development of vegetation and soils releasing organic acids. During the last 2,000 years, man-made eutrophication produced a pH increase, followed by recent acidification (Psenner & Catalan, 1994). Acidification of rivers and lakes is caused by wet or dry deposition of acidic components like sulphur dioxide or nitrogen oxides in the drainage basin. Emissions of these gases to the atmosphere have increased since the industrial revolution, mainly due to burning of sulphur-containing coal in power plants or households and to oxidation of atmospheric nitrogen in combustion engines. Acid rain was first described by Smith (1872, cit. in Lenhart & Steinberg, 1992). Due to the long time lag between start of acidification (19th century) and public awareness of its consequences (second half of the 20th century), reconstruction of background conditions or a baseline is even more difficult than for eutrophication. Additionally, a large amount of sulphur dioxide stems from natural processes like volcano emissions, oxidation of biogenic dimethyl sulphide or sulphur-containing minerals. Acid substances are partly neutralised by dust which originates from natural (e.g. deserts) as well as anthropogenic sources (power plants, traffic, agriculture, etc.). Dust emission from power plants was earlier and more strongly diminished than emission of sulphur dioxide or nitrogen oxides, so a temporary more severe acidification resulted in regions affected by power plant emissions. During the last decades, sulphuric acid deposition declined drastically in Germany but was partly replaced by nitric acid. In many forest regions, the deposited NO_3- had been retained by soils and taken up by vegetation until the system became N-saturated. This process may cause a delayed, but sudden decline in pH. Decreases in acid deposition result in an incomplete or delayed recovery because large amounts of calcium and magnesium have been washed out from the soils so that their buffer capacity was reduced (Likens, Driscoll, & Buso, 1996).

Many aquatic organisms disappear if pH declines below 6.0–6.5 (Psenner & Catalan, 1994). Acidification affects certain stages of the life cycle, e.g. larval phases of fish or emergence of aquatic insects. Abundance of the affected species often declines only after some delay. Pulses of acid substances may reach the stream or lake only occasionally, e.g. after snow melt or heavy rain. Such pulse may or may not coincide with sensitive life stages.

In some lakes, a temporary recovery from acidification was produced by liming. The rate of recovery of aquatic organisms depends on their dispersal ability and reproductive strategy. Zooplankton recovery needed 10 years after liming in a moderately acidic Canadian lake, but more than 15 years in more acidic lakes (Yan, Keller, Somers, Pawson, & Girard, 1996). Molluscs and fish need an inoculum from refuges or artificial stocking. In any case, a time lag of several years occurs between chemical and biological recovery.

Global warming enhanced weathering of minerals and increased biological activity. Sommaruga-Wögrath et al. (1997) found a strong positive correlation between pH of alpine lakes and air temperature.

Sensitivity of a stream, reservoir or lake to import of acidic substances mainly depends on the geochemistry of its catchment. Most of northern Germany is covered by easily weathered carbonate-rich rocks and soils which neutralise incoming acids. Much more vulnerable to acidification are mountainous areas with insoluble surface material like granite, gneiss, basalt or sandstone. Streams, lakes and reservoirs with small drainage basin are most at risk. Therefore, acidified water bodies are concentrated in upper regions of the Ore Mountains, Bavarian Forest and Black Forest.

13.5 Invasion by Non-native Species

Compared to terrestrial or marine systems, freshwater ecosystems are more separated (less coherent) under natural conditions. Micro-organisms may spread by wind or birds over long distances. Many higher plants or animals, however, can migrate only within their river basin. Human activities (intended or accidental import, connection of river basins by canals, ship traffic, etc.) enabled or favoured invasion of new areas. Invasive species have large effects on diversity and structure of native communities, but also on pool sizes and main processes in aquatic ecosystems. Climate change favoured the northward spreading of warm-adapted species during the last decades.

Effects of many invaders are not constant over time. Processes like evolution, shifts in species composition, accumulation of materials and interactions with abiotic variables can increase, decrease or qualitatively change the impacts of an invader through time. Strayer, Eviner, Jeschke, and Pace (2006) differentiate between an acute phase just after arrival of a new species and a chronic phase after various ecological and evolutionary processes. The latter may last for decades or centuries. However, most studies of the effects of invasive species have been brief and lack a temporal context; 40% of recent studies did not even state the amount of time that had passed since the invasion (Strayer et al., 2006).

In the following, main processes requiring long-term studies of invasions are briefly discussed: Invasive species often have to acclimate to their new surrounding. This acclimatisation can occur quickly in the initial phase or as a response to changes developing later in the invasion process. The need to acclimate often causes a time lag between invasion and explosive population growth. On the other hand, the community that has been invaded also changes over time. Invaders often arrive in a new system without their associated enemies. Introduced populations host on average half the number of parasite species of native populations (Torchin, Lafferty, Dobson, McKenzie, & Kuris, 2003). Predators, parasites or competitors may follow from previously occupied regions or old established species may adapt to the invader. The time lag between invasion and arrival of the old or adaptation of new enemies often causes an initial mass development of the invasive species followed by a regulation to lower population densities. Additionally, composition of the invaded community can shift towards species resistant to effects of the invader or native species can adapt to reduce own losses caused by the invader.

The zebra mussel (*Dreissena polymorpha*) is a good example for many of the mentioned long-term processes. It expanded after construction of the Pripyat-Bug canal in 1780 from tributaries of the Black Sea to central Europe (Kinzelbach, 1995). Man-made changes favoured *Dreissena* during the last century: eutrophication increased the food concentration of this filter-feeder; hydraulic engineering provided hard substrates to settle. Zebra mussel filtration reduced the biomass (Makarewicz, Lewis, & Bertram, 1999; Caraco, Cole, & Strayer, 2006) and altered the species composition of phytoplankton (Vanderploeg et al., 2001). This species also creates new habitats by 'paving' soft sediments (Beekey, McCabe, & Marsden, 2004). The zebra mussel was followed by a specialised parasite, the trematode *Bucephalus polymorphus* (Kinzelbach, 1995). Some native waterfowls have adapted to feed on this mussel and may control its abundance (Petrie & Knapton, 1999). Ironically, water fowl foraging was facilitated by increased water clarity caused by filter-feeding mussels. Molloy, Karatayev, Burlakowa, Kurandina, and Laruelle (1997) reviewed the old and new predators, parasites and competitors of *Dreissena*. The zebra mussel was the most important filter-feeder in the Rhine during the 1980s but is now diminished by competition from new invaders, mainly the amphipod *Corophium curvispinum* (Kinzelbach, 1995).

13.6 Climate Change

'Climate change' means a significant change in mean temperature, precipitation or wind patterns during time periods of decades to millions of years. At the geological timescale, continental drift, uplift and erosion of mountains and CO_2 uptake by sedimentary rocks caused shifts in climatic conditions. At a shorter timescale, changes in the Earth's orbit and in solar energy output trigger glacial and interglacial phases. Volcano eruptions may cause short-term climatic changes. Human activities like deforestation, irrigation and fuel burning influenced the climate, at least locally, for many centuries. After the industrial revolution, emissions of greenhouse gases (CO_2, CH_4, etc.) increased dramatically, so that human activity is the main reason for the current rapid climate changes. Mean global surface temperature increased 0.74 K from 1906 to 2005 and will likely rise a further 1.1–6.4 K during the 21st century (IPCC, 2007). Several complex feedbacks may cause non-linear behaviour and rate-independent memory (hysteresis) of the climate system.

Higher temperature and lower discharge effect nutrient loading of rivers and lakes. Increasing evaporation and decreasing discharge may contribute to eutrophication of lakes which receive nutrients mainly from point sources. In contrast, lower discharge causes reduced P input from non-point sources. Lower precipitation reduces weathering and stream flow, resulting in lower silica concentrations in lakes (Schindler et al., 1996). Lakes in pristine catchments thus received less P and Si, resulting in a slight oligotrophication despite reduced water renewal (Schindler et al., 1996). Lower discharge and higher temperature favour nitrogen retention in the river system by denitrification.

In recent decades, water temperatures increased in many northern temperate lakes, especially in winter and spring (Gerten & Adrian, 2002). Duration of ice cover declined and the spring growth period of phytoplankton started earlier. Paradoxically, the shift of the spring growth period favoured phytoplankton species adapted to lower temperature, shorter day length, lower irradiance and, under high Si:P ratios, filamentous cyanobacteria (Shatwell, Köhler, & Nicklisch, 2008). Timing of zooplankton was relatively independent of the winter conditions (Shatwell et al., 2008), so that longer time was available for phytoplankton spring growth and less phytoplankton fuelled the trophic cascade from zooplankton to fish. Higher water temperatures also cause earlier and more stable stratification of the water column from late spring to autumn. In deep lakes, temperature of the epilimnion and gradients in the metalimnion increased. Moderately shallow lakes stratified more permanently and their hypolimnion tended to become colder (Gerten & Adrian, 2001), whereas the whole water column of very shallow, polymictic lakes became warmer. More stable stratification favours mobile phytoplankton species, enhances settling and reduces resuspension of particles. Higher temperatures at the sediments increase oxygen consumption and remobilisation of nutrients (P, Si) and heavy metals.

Lower silica input from the catchment, more stable stratification and higher temperatures favour some cyanobacteria and aggravate nuisance bloom formation (Schindler, 2006). In general, the aquatic organisms respond to climate changes in species-specific ways. The complex, often time-lagged, direct and indirect responses in biotic structure and trophic interactions are not fully understood yet.

13.7 Disadvantages and Limitations of Long-Term Investigations

Most limnologists agree that long-term investigations are necessary, but few of them want to spend time for repetitive monitoring. Less paper in high-ranking journals can be distilled per unit effort from long-term investigations than from studies on single processes. Once a long-term data series is accumulated, its utilisation is often profitable but its analyser stands on the shoulders of former colleagues which earned less for long efforts. Papers on popular topics are most cited. Researchers have to follow the fashion for funding. Of course, popularity of topics changes and so investigations are usually terminated after few years. Additionally, methods will advance but comparability within a data series often requires sticking on old methods. Therefore, data series without interruptions are very rare in research institutes.

Only few institutions can achieve to commit staff and facilities continuously for decades.

On the other hand, there are also theory-based arguments against long-term studies. Often, episodic, seasonal and interannual variations are higher than long-term directional changes. Stochastic disturbances can mask long-term shifts; in some freshwater systems frequent disturbances may prevent any steady state. On the community level, return intervals for the next disturbance are usually too short to attain a new equilibrium (Reice, 1994).

Even long-term studies do not allow for a simple extrapolation of past observations into the future. Freshwater ecosystems often react in non-linear ways to external forces. Beyond a certain threshold, gradual shifts may cause catastrophic switches to another state. The way back to the previous state rarely corresponds to the past changes because of memory effects of the system.

Research and corresponding management usually ignores timescales in the natural dynamics of lakes and rivers which exceed the human life span and experience. An assessment of the current state and of human influences should be based, however, on its deviation from natural reference conditions (European Union, 2000). Long-term monitoring usually started after serious environmental problems have emerged. They rarely cover the periods of nearly undisturbed conditions and even seldom the phase of deterioration. Therefore, long-term monitoring should be accompanied and extended by palaeolimnological approaches. Stratigraphical analyses of sediments or ice cores can cover much longer timescales than the most persistent limnological monitoring. On the other hand, the comparison between long-term observational data and recent sediment stratigraphy is very valuable for calibration of palaeolimnological methods (Battarbee, Anderson, Jeppesen, & Leavitt, 2005).

In some cases, we can change space for time. We can deduce the response of one system to temporal changes by comparing similar systems which experience the same impact at different degrees at the same time.

13.8 Long-Term Investigations of German Lakes and Reservoirs

In Germany, first freshwater laboratories were established in the late 19th century at Plön (1891) and Friedrichshagen (1893). These institutions were not intended to analyse certain lakes for longer time periods. In fact, comprehensive studies of lakes around Plön and Berlin-Friedrichshagen were occasionally organised by staff members, but these analyses lasted only a few years each, were not part of longer-lasting programmes and often used outdated or not well-documented methods. Lake Constance as the largest German lake and an important drinking water source of south-western Germany was regularly studied by the Institute of Lake Research at Langenargen since 1952, by the International Commission for the Protection of Lake Constance since 1961 and by the Limnological Institute of the University of Constance since 1979, although some programmes had been terminated or interrupted in the meantime [Arch Hydrobiol. Spec. Issues Advanc. Limnol. 53 (1998)]. Since the late 1950s, Lake Stechlin has been comprehensively and continuously monitored because of a small nuclear power plant which took lake water for cooling (Casper, 1985). Apart from these exceptions, monitoring programmes were not started before 1970. They were usually motivated by eutrophication problems and thus focused on growth and production of phytoplankton, nutrient budgets and trophic interactions in the pelagic zone. The benthic community was studied only occasionally and usually in qualitative terms. Long-term data on the fish stock of some lakes are (cautiously) extractable from catch statistics of commercial fishermen. Until now, consistent analyses comprising the major groups of organisms and most important processes at the same lake and during the same time are very rare.

Different institutions run long-term monitoring programmes for different purposes. Water authorities often monitor chemistry and some hydrological and biological parameters in large river systems. Reservoirs (and few lakes) serving as source of drinking water are analysed by the water suppliers. Many lakes are occasionally sampled by the local water authorities, often to test the suitability for bathing. Very few German lakes have been intensively investigated in high frequency for long time. Table 13.1 gives examples for monitoring programmes which have been

Table 13.1 Long-term monitoring of German lakes and reservoirs (selected sites): start, termination (if applicable) and frequency of regular measurements

Parameter	Lake Müggelsee	Lake Stechlin	Lake Constance	Saidenbach reservoir	Lake Haussee	Lake Luzin	Lake Tollense
Phytoplankton biovolume + species composition	1977-	1959-	1952-	1975-	Occasionally	–	–
Phytoplankton chlorophyll a	1994-	1980-	1976-	1994-	1978-	1981-	1981-
Zooplankton biovolume + species composition	–	1978-	1952-	1975-	1978-	1981-	1981-
Zooplankton abundance	1979-	1978-	1952-	1975- (without ciliates)	1978-	1981-	1981-
Fish stock	–	1995-	1909-	1981, 1998	Few observations	–	–
Macrozoobenthos abundance	1991–1992, 2001	Few observations	1968, 1972, 1978, 1985, 1993, 2005	–	–	–	–
Macrophytes coverage	1999, 2000, 2006, 2008	Few observations	1968, 1978, 1993, 2008	–	–	–	–
Periphyton biomass	–	–	–	–	–	–	–
Primary production (phytoplankton)	1978-	1970-	1979–1997	–	Few observations	–	–
Bacterial production (pelagic)	–	1976-	1980–1997, 2005	–	–	–	–
Water temperature	1976-	1958-	1961-	1975-	1978-	1981-	1981-
Ice coverage	1977-	1958-	875-	1975-	–	–	–
Secchi disc depth	1979-	1958-	1920–24, 1951-	1975-	1978-	1981-	1981-
Light attenuation	1976-	1970-	1979–1997	1975–1990, rarely	–	–	–
Discharge/retention time	1976-	–	1951-	1975-	–	–	–
Total phosphorus	1979-	1971-	1961-	1975-	1978-	1981-	1981-
Dissolved reactive phosphate	1979-	1971-	1961-	1975-	1978-	1981-	1981-
Total nitrogen	1976-	1993-	1961-	–	1993-	1993-	1993-
Dissolved inorganic nitrogen	1976-	1971-	1961-	1975- (only nitrate)	1978-	1981-	1981-
Dissolved reactive silica	1977-	1971-	1961-	1981-	1978-	1981-	1981-
Conductivity	1994-	1970-	1961-	1991-	1978-	1981-	1981-
pH	1991-	1970-	1961-	1991-	1978-	1981-	1981-
Oxygen content	1979-	1971-	1961-	–	1978-	1981-	1981-
Dissolved inorganic carbon	1978-	1971-	1976-	–	1978-	1981-	1981-
Sampling frequency summer (winter) in d	7 (14)	14 (28)	14	1975–1986: 7, 1987-: 14	14 (28)	14 (28)	14 (28)
Institution	IGB	IGB	IGKB, Uni Konstanz	TU Dresden	IGB	IGB	IGB

performed for more than 20 years at German lakes and reservoirs. This compilation is not complete, especially regarding reservoirs.

Due to the federal organisation of environmental protection, Germany lacks a central institution to coordinate monitoring programmes, to validate and standardise the used methods and to centralise data collation.

13.9 Conclusions

Natural systems are always changing. There is no constant baseline. Observed temporal changes are never exclusively caused by human activities. Long-term observations are necessary to understand natural fluctuations and shifts and to differentiate them from human impacts.

The temporal scale of investigations generally depends on the processes under study. Spatial scale and temporal scale determine tools and approaches which are effective at that level. The results and interpretations made at one scale may be wrong or inappropriate at another (Fisher, 1994). Long-term changes usually occur at the spatial scale of whole ecosystems. Long-term studies of lakes or rivers have to include the processes in its catchment and interactions with the atmosphere. On the other hand, it is often defensible to neglect the processes acting at lower scales of space or time. Statistical models with seasonal averaged responses of aggregated properties or proxies are appropriate for many long-term studies.

Long-term observations of lakes and rivers are not replaceable by palaeolimnological studies or by the comparison of systems at different development stages. Instead, these approaches should be combined to overcome limitations of the single concepts.

Acknowledgements Rainer Koschel, Reiner Kümmerlin, Hans Güde and Wofgang Horn provided information about long-term monitoring programmes at lakes Stechlin, Haussee, Luzin and Constance and at the Saidenbach reservoir.

References

Battarbee, R. W., Anderson, N. J., Jeppesen, E., & Leavitt, P. R. (2005). Combining palaeolimnological and limnological approaches in assessing lake ecosystem response to nutrient reduction. *Freshwater Biology, 50*, 1772–1780.

Beekey, M. A., McCabe, D. J., & Marsden, J. E. (2004). Zebra mussel colonization of soft sediments facilitates invertebrate communities. *Freshwater Biology, 49*, 535–545.

Caraco, N. F., Cole, J. J., & Strayer, D. L. (2006). Top down control from the bottom: Regulation of eutrophication in a large river by benthic grazing. *Limnology & Oceanography, 51*, 664–670.

Casper, S. J. (1985). *Lake Stechlin: A temperate oligotrophic lake*. Dordrecht: Junk Publishers.

Elliott, J. M. (1990). The need for long-term investigations in ecology and the contribution of the Freshwater Biological Association. *Freshwater Biology, 23*, 1–5.

European Union (2000). Directive 2000/60/EC of the European Parliament and of the Council of 23 October 2000 establishing a framework for Community action in the field of water policy. *Amtsblatt der Europäischen Gemeinschaften, L327*, 1–72.

Fisher, S. G. (1994). Pattern, process and scale in freshwater systems: Some unifying thoughts. In P. S. Giller, A. G. Hildrew, & D. G. Raffaelli (Eds.), *Aquatic ecology: Scale, pattern and process* (pp. 575–591). Oxford: Blackwell.

Gerten, D., & Adrian, R. (2001). Differences in the persistency of the North Atlantic Oscillation signal among lakes. *Limnology & Oceanography, 46*, 448–455.

Gerten, D., & Adrian, R. (2002). Effects of climate warming, North Atlantic Oscillation, and El Nino-Southern Oscillation on thermal conditions and plankton dynamics in northern hemispheric lakes. *The Scientific World Journal, 2*, 586–606.

IPCC (2007). Summary for policymakers. In S. Solomon, D. Qin, M. Manning, Z. Chen, M. Marquis, K. B. Averyt, M. Tignor, & H. L. Miller (Eds.), *Climate change 2007: The Physical Science Basis. Contribution of Working Group I to the Fourth Assessment Report of the Intergovernmental Panel on Climate Change* (pp. 1–18). Cambridge: Cambridge University Press.

Jensen, J. P., Jeppesen, E., Kristensen, P., Christensen, P. B., & Sondergaard, M. (1992). Nitrogen loss and denitrification as studied in relation to reductions in nitrogen loading in a shallow, hypertrophic lake (Lake Sobygard, Denmark). *Internationale Revue der gesamten Hydrobiologie und Hydrographie, 77*, 29–42.

Jeppesen, E., Sondergaard, M., Jensen, J. P., Havens, K. E., Anneville, O., Carvalho, L., et al. (2005). Lake responses to reduced nutrient loading – an analysis of contemporary long term data from 35 case studies. *Freshwater Biology, 50*, 1747–1770.

Kalff, J. (2002). *Limnology: Inland water ecosystems*. Upper Saddle River, NJ: Prentice Hall.

Kinzelbach, R. (1995). Neozoans in European waters – Exemplifying the worldwide process of invasion and species mixing. *Experientia, 51*, 526–538.

Köhler, J., Hilt, S., Adrian, R., Nicklisch, A., Kozerski, H. P., & Walz, N. (2005). Long-term response of a shallow, moderately flushed lake to reduced external phosphorus and nitrogen loading. *Freshwater Biology, 50*, 1639–1650.

Lenhart, B., & Steinberg, C. E. W. (1992). Gewässerversauerung. In W. K. Besch, A. Hamm, B. Lenhart, A. Melzer, B. Scharf, & C. E. W. Steinberg (Eds.), *Limnologie für die Praxis – Grundlagen des Gewässerschutzes* (pp. 347–404). Landsberg/Lech: ecomed.

Likens, G. E., Driscoll, C. T., & Buso, D.C. (1996). Long-term effects of acid rain: Response and recovery of a forest ecosystem. *Science, 272*, 244–246.

Makarewicz, J. C., Lewis, T. W., & Bertram, P. (1999). Phytoplankton composition and biomass in the offshore waters of Lake Erie: Pre- and post-Dreissena introduction (1983–1993). *Journal of Great Lakes Research, 25*, 135–148.

Molloy, D. P., Karatayev, A. Y., Burlakowa, L. E., Kurandina, D. P., & Laruelle, F. (1997). Natural enemies of zebra mussels: Predators, parasites, and ecological competitors. *Reviews in Fisheries Science, 5*, 27–97.

Moosmann, L., Gächter, R., Müller, B., & Wüest, A. (2005). Is phosphorus retention in lakes controlled by oxygen or phosphorus? *Limnology & Oceanography, 51*, 763–771.

Petrie, S. A., & Knapton, R. W. (1999). Rapid increase and subsequent decline of zebra and quagga mussels in Long Point Bay, Lake Erie: Possible influence of waterfowl predation. *Journal of Great Lakes Research, 25*, 772–782.

Psenner, R., & Catalan, J. (1994). Chemical composition of lakes in crystalline basins: A combination of atmospheric deposition, geologic background, biological activity and human action. In R. Margalef, (Ed.), Limnology now: Paradigms of planetary problems (pp. 255–314). Amsterdam: Elsevier.

Reice, S. R. (1994). Nonequilibrium determinants of biological community structure. *American Scientist, 82*, 424–435.

Scheffer, M., Hosper, S. H., Meijer, M. L., Moss, B., & Jeppesen, E. (1993). Alternative equilibria in shallow lakes. *Trends in Ecology & Evolution, 8*, 275–279.

Schindler, D. W. (2006). Recent advances in the understanding and management of eutrophication. *Limnology & Oceanography, 51*, 356–363.

Schindler, D. W., Bayley, S. E., Parker, B. R., Beaty, K. G., Cruikshank, D. R., Fee, E. J., et al. (1996). The effects of climatic warming on the properties of boreal lakes and streams at the Experimental Lakes Area, northwestern Ontario. *Limnology & Oceanography, 41*, 1004–1017.

Shatwell, T., Köhler, J., & Nicklisch, A. (2008). Warming promotes cold-adapted phytoplankton in temperate lakes and opens a loophole for Oscillatoriales in spring. *Global Change Biology, 21*, 2194–2200.

Sommaruga-Wögrath, S., Koinig, K.A., Schmidt, R., Sommaruga, R., Tessardi, R., & Psenner, R. (1997). Temperature effects on the acidity of remote alpine lakes. *Nature, 387*, 64–67.

Strayer, D. L., Eviner, V. T., Jeschke, J. M., & Pace, M. L. (2006). Understanding the long-term effects of species invasions. *Trends in Ecology & Evolution, 21*, 645–651.

Torchin, M. E., Lafferty, K. D., Dobson, A. P., McKenzie, V. J., & Kuris, A.M. (2003). Introduced species and their missing parasites. *Nature, 421*, 628–630.

Vanderploeg, H. A., Liebig, J. R., Carmichael, W. W., Agy, M. A., Johengen, T. H., Fahnenstiel, G. L., et al. (2001). Zebra mussel (*Dreissena polymorpha*) selective filtration promoted toxic *Microcystis* blooms in Saginaw Bay (Lake Huron) and Lake Erie. *Canadian Journal of Fisheries and Aquatic Sciences, 58*, 1208–1221.

Yan, N. D., Keller, W., Somers, K. M., Pawson, T. W., & Girard, R. E. (1996). Recovery of crustacean zooplankton from acid and metal contamination: Comparing manipulated and reference lakes. *Canadian Journal of Fisheries and Aquatic Sciences, 53*, 1301–1327.

Chapter 14

Long-Term Monitoring in Rivers of South Germany Since the 1970s – Macrophytes as Indicators for the Assessment of Water Quality and Its Implications for the Conservation of Rivers

Peter Poschlod, Martijn Kos, Stephanie Roauer, Andreas Seemann, Oliver Wiesmann, Georg H. Zeltner, and Alexander Kohler

Abstract Since 1970 long-term ecological research has been established in several running waters in Bavaria. An emphasis was laid on the record of the macrophytic vegetation. In this chapter we present long-term records from three rivers. The Moosach is situated in the calcareous-rich gravel plains north of Munich. The two soft-water rivers, the Pfreimd and the adjacent Naab, run out from the siliceous bedrock of the 'Upper Palatinate Forest.' Vegetation changes could be interpreted according to water quality changes and confirmed the suitability of macrophytes as water quality indicators, which recently obtained its official acceptance in the new Water Frame Directive of the European Union. The results show that there is a unification of the vegetation losing the extremes especially in the oligotrophic part. However, the regeneration potential of macrophytes is mainly low although in many species it is not yet understood. It is therefore concluded to lay an emphasis on the protection of still oligotrophic sections of a river to maintain the total species pool.

Keywords Long-term monitoring · Running waters · Macrophytes · Southern Germany · Hard water · Soft water

14.1 Introduction

Since a long time running waters were subjected to an increasing influence of chemical and organic load caused by sewage from agriculture, industry, and households with a peak during the 1960s and 1970s. Therefore, long-term ecological studies in running waters to assess water quality have a long tradition since the introduction of the basic principles of the saproby system (*Saprobiensystem*) by Kolkwitz and Marsson (1902, 1908, 1909; see also Kolkwitz, 1950) and the system on the so-called Güte-Klassen, meaning classes of water quality ranking from I to IV by Liebmann (1959, 1960, 1962) later taken over into official DIN standards (DIN 38410 Part 1, 1987; DIN 38410 Part 2, 1990). In these water quality assessment systems macrozoobenthic organisms played the major role as indicators. Kohler established in the 1970s another system based on experiments on the indicator quality of macrophytes that is emerged and submerged aquatic plants (Kohler, Zeltner, & Busse, 1972; Kohler, 1976). His classification system included three classes. At the same time he started the monitoring of rivers in the southern part of Germany, especially in Bavaria in regions of different geology. In this context he studied the river Moosach situated in the largely paludified calcareous gravel plain north of Munich (Kohler, Vollrath, & Beisl, 1971); the river Friedberger Au in the calcareous gravel plain near Augsburg (Kohler, Brinkmeier, & Vollrath, 1974); and the river Pfreimd in the siliceous bedrock of the Upper Palatinate Forest (Kohler & Zeltner, 1974). At this time macrophytes were not incorporated in the above-mentioned DIN standards despite the fact that the high value of macrophytes as bioindicators was widely acknowledged. However, with the implementation of the European Water Frame Directive to maintain running waters in good conditions macrophytes were now also 'officially' acknowledged as suitable indicators. Recently, the applicability of macrophytes as

P. Poschlod (✉)
Institute of Botany, University Regensburg, D-93040 Regensburg, Germany
e-mail: peter.poschlod@biologie.uni-regensburg.de

F. Müller et al. (eds.), *Long-Term Ecological Research*,
DOI 10.1007/978-90-481-8782-9_14,

indicators to assess water quality was confirmed again by Schaumburg et al. (2004, 2005) and Bayerisches Landesamt für Wasserwirtschaft (2005).

In this chapter, we present the monitoring scheme and results from the long-term monitoring of two river systems over 30–35 years to (1) show changes in macrophyte abundance and species composition and (2) its relation to environmental changes.

14.2 Study Area

The two systems studied are situated in two different geologic regions representing two extremes, the calcareous-rich river (Moosach) and the soft-water river (Pfreimd; Fig. 14.1).

The Moosach is a tributary of the Isar and runs through the largely paludified gravel plain in the north of Munich parallel to the southern edge of the tertiary hilly land. It drains the large peatlands 'Freisinger Moos' and 'Dachauer Moos' and crosses afterwards the unpaludified part of the gravel plain. However, throughout its whole length the Moosach is additionally fed by calcareous-rich groundwater. Therefore, it is characterized by the occurrence of indicators for oligotrophic calcareous-rich groundwater such as *Juncus subnodulosus*, *Potamogeton coloratus*, and *Groenlandia densa* (Fig. 14.2). The river Moosach

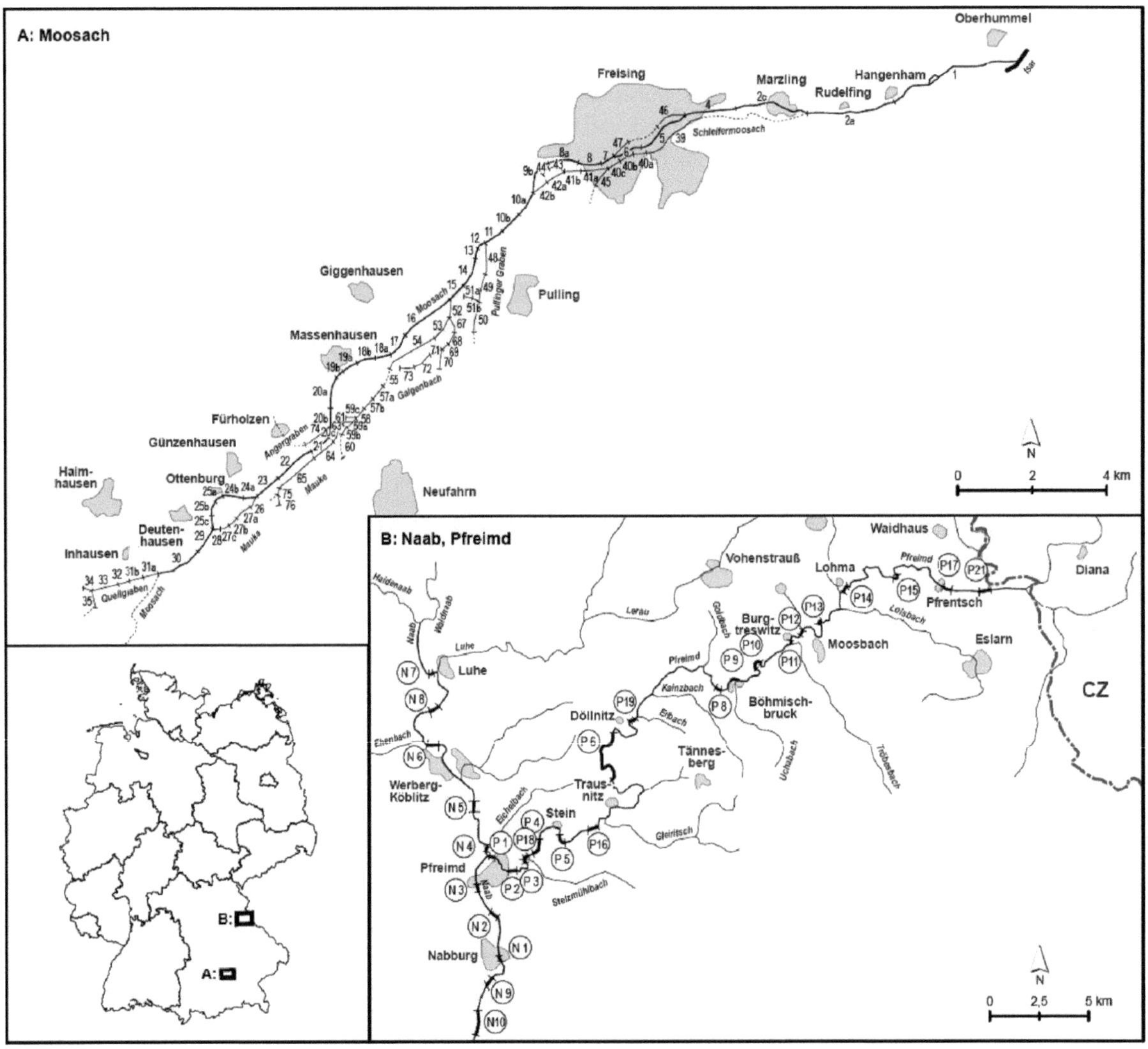

Fig. 14.1 Situation of the studied rivers and their division into monitoring sections

<table>
<tr><th>Species group</th><th>Zone A</th><th>Zone B</th><th>Zone C</th><th>Zone D</th></tr>
<tr><td colspan="5">Calcareous rich water rivers</td></tr>
<tr><td>I</td><td>Potamogeton coloratus, Chara hispida, Juncus subnodulosus, Mentha aquatica</td><td>→</td><td></td><td></td></tr>
<tr><td>II</td><td></td><td>←</td><td>Groenlandia densa, Potamogeton natans, Hippuris vulgaris</td><td></td></tr>
<tr><td>III</td><td></td><td></td><td colspan="2">Callitriche obtusangula, Ranunculus fluitans, Elodea canadensis, Myriophyllum verticillatum, Zannichellia palustris</td></tr>
<tr><td rowspan="2">Indiff. species</td><td colspan="4">Berula erecta, Ranunculus trichophyllus</td></tr>
<tr><td></td><td colspan="3">Agrostis stolonifera, Potamogeton crispus, P. pectinatus, Nasturtium officinale, Veronica anagallis-aquatica</td></tr>
<tr><td colspan="5">Calcareous poor soft water rivers</td></tr>
<tr><td>I</td><td colspan="2">Potamogeton alpinus, Myriophyllum alterniflorum, Ranunculus peltatus</td><td></td><td></td></tr>
<tr><td>II</td><td></td><td></td><td colspan="2">Potamogeton nodosus, P. perfoliatus, Ceratophyllum demersum, Myriophyllum spicatum, Ranunculus penicillatus</td></tr>
<tr><td>Indiff. species</td><td></td><td colspan="3">Hydrocharis morsus-ranae, Potamogeton crispus, Potamogeton natans, Sparganium emersum, Callitriche hamulata, Elodea canadensis, Ranunculus fluitans</td></tr>
</table>

Fig. 14.2 Indicative value of macrophytes of calcareous-rich and calcareous-poor running waters (see Kohler et al., 1971, 1974, 1992; Monschau-Dudenhausen, 1982; Veit & Kohler, 2003; Schaumburg et al., 2005)

is characterized by widths between 2 and 25 m and depths between 0.25 and 1.75 m. Flow velocity is low. Visible depth is relatively high except during floodwater and at the underflow behind Freising. The ground of the river is sandy and stony. Water chemistry measurements during the last monitoring in 2005 (Table 14.1) showed the highest calcium content and the fact that the Moosach was more polluted, especially by phosphate, than the other two rivers. This pollution is due to the localization of the river Moosach in an intensively used agricultural area, in which pasture farming prevails.

The river Naab and the river Pfreimd run partly through the region 'Upper Palatinate Forest' and drain parts of it. The underground of the 'Upper Palatinate Forest' consists of siliceous bedrock.

Table 14.1 Characterization of water chemistry of the studied rivers

	Moosach		Naab			Pfreimd		
Water chemical parameter	1970	2005	1968/1972	1982–1988	2004	1968/1972	1982–1988	2004
pH	7.9	7.0	6.9/6.9–7.5	7.1	7.3	6.9/7.0	7	7.3
Conductivity (μS/cm)	604	530	–	233	280	–	128	170
Ammonium (mg/l)	0.20	n.d.	0.66/0.40–0.92	0.51	n.d.	0.40/0.22	0.11	n.d.
Calcium (mg/l)	115	28	–	–	25	–	–	14
Carbonate (mg/l)	336.6	–	–	–	4.5	–	–	2.7
Nitrate (mg/l)	26.9	26.0	–/8.4–10.0	3.8	11.0	–	2.9	11
Phosphate (mg/l)	0.25	1.10	–/0.31–0.72	0.38	0.44	–	0.10/0.05	0.30

Moosach – 1970 – mean of 8–26 measurements at 14 sites; 2005 – mean of four measurements from June to August 2005 at 14 sites; Naab – 1968 – mean of eight measurements at one site; 1972 – mean of eight measurements at three sites; 1982–1988 – mean of 168 measurements at one site (Kohler et al., 1992); 2004 – mean of measurements in June 2004 at 10 sites; Pfreimd – 1968 and 1972 – mean of eight measurements at one site (Kohler & Zeltner, 1974); 1982–1988 – mean of 168 measurements at one site (Kohler et al., 1992); 2004 – mean of 20 measurements in June 2004 at 20 sites; n.d. = not detectable; – = not measured

Therefore, the rivers are characterized by soft water and by the occurrence of respective indicators such as *Callitriche hamulata*, *Myriophyllum alterniflorum*, *Ranunculus peltatus*, and *Ranunculus penicillatus* (Fig. 14.2).

The river Naab is characterized by widths between 20 and 40 m and depths between 0.25 and 1.5 m. Flow velocity and visible depth are low. The ground of the river is sandy and stony and that of the riverbanks muddy. Water chemistry was characterized by relatively high calcium contents affected by confluents from the Jurassic Franconian Alb and low pollution (Table 14.1).

The river Pfreimd is characterized by widths between 5 and 25 m and depths between 0.2 and 1.25 m. As in the river Naab, flow velocity of the river Pfreimd is low as well, but with frequent turbulences. In contrast to the river Naab, visible depth is high. The bottom of the river is stony and sandy, except for the upper reaches, which is muddy. Water chemistry showed the lowest degree of pollution within the studied rivers.

14.3 Monitoring Scheme and Methods

The monitoring scheme in all rivers followed the design described by Kohler (1978). Rivers were divided into sections. The lengths of the sections were defined by prominent and re-detectable features along the river such as houses, bridges, or larger trees. The river Moosach was mapped completely, whereas in the rivers Naab and Pfreimd only the macrophytes of selected sections were recorded. The monitoring in the river Moosach started in 1970 and was repeated in 1979, 1985, 1989, 1992, 1996, and 2005. In the rivers Naab and Pfreimd, it started in 1972 and was carried out again in 1980, 1988, and 2004.

The monitoring program included the mapping of the complete aquatic vegetation, i.e., all plants which had their roots below the water surface between June and August. The abundance of each macrophyte was estimated according to a five-step scale where 1 = very rare, 2 = rare, 3 = common, 4 = frequent, 5 = very frequent. Kohler (1978) explained the procedure in detail, which meanwhile is the standard method for the mapping of rivers according to the European Water Framework Directive (WRRL, 2000).

Water chemical parameters were either measured with an electronic instrument (WTW Multi 340i, pH, and conductivity) or an ion chromatograph was used (Dionex DX 120, calcium, ammonium, carbonate, nitrate, phosphate a.o.).

In this study, data were analyzed with respect to the number of occurrences of certain species with a strong indicator value for water quality. Four classes or zones, respectively, were differentiated following Schneider (2000) and Schorer, Schneider, and Melzer (2000): A = cleanest, unpolluted spring-fed brooks; B = low pollution; C = moderate pollution; D = high pollution. Indicator species were taken from Kohler et al. (1971, 1974) for the river Moosach (Veit & Kohler, 2003) and from Kohler, Lange, and Zeltner (1992) for the rivers Naab and Pfreimd (Fig. 14.2).

A detrended correspondence analysis (DCA) was carried out on the raw vegetation data (PC-ORD 4.0; McCune & Mefford, 1999) to assess the overall development. Sections and species abundances were ordinated, and species were correlated with this ordination to present a bi-plot. For all indicator species we calculated three indices for the first and last year of monitoring, which were the 'relative range length (L_T)' and the quantitative indices 'MMT' and 'MMO' (Kohler & Janauer, 1995).

The relative range length shows the occurrence of a species related to the total length of all mapped sections and is calculated according to the following formula:

$$L_T[\%] = \frac{\sum_{k=1}^{n} L_k}{L_{ges}} \cdot 100$$

where L_T = relative range length, L_k = length of section k, L_{ges} = total length of all mapped sections, k = running index, n = number of sections where the respective species is occurring.

The quantitative indices 'mean mass total (MMT)' and 'mean mass occurrence (MMO)' quote the dominance of a species related to the total length (MMT) or only to the sections where the species is occurring (MMO). The comparison of both values shows that in case of similar values of MMO and MMT a species is distributed over all monitored sections; in case of high values a species is very frequent and dominant; and in case of high MMO and low MMT values a species is only locally very frequent and dominant. MMT and MMO are calculated according to the following formulae:

$$\mathrm{MMT} = \sqrt[3]{\frac{\sum_{i=1}^{n} M_i^3 \cdot L_i}{L}}$$

$$\mathrm{MMO} = \sqrt[3]{\frac{\sum_{i=x}^{n} M_i^3 \cdot L_i}{\sum_{i=x}^{n} L_i}}$$

where M_i = dominance/frequency of a species in section i, L_i = length of section i, where the species is occurring, L = total length of all sections.

14.4 Results

The water analysis showed changes in all rivers within the long-term study, at least in some chemical parameters. In the river Moosach, calcium and ammonium decreased but phosphate increased. Pollution in the river Naab did not change except for ammonium which decreased and nitrate which decreased from 1972 to 1982/1988 but increased to the former level in 2004. In contrast to these rivers, both nitrate and phosphate increased strongly in the river Pfreimd.

Multivariate analysis of vegetation data shows continuous vegetation changes over time (Figs. 14.3, 14.4, and 14.5).

In the river Moosach, only the first and the last records were compared. Changes are correlated positively with *Callitriche obtusangula* and *Mentha aquatica*. These two species were increasing in contrast to *P. coloratus* and *J. subnodulosus* (Fig. 14.3), which is also shown by the comparison of the 'relative range length' (Table 14.2). This analysis clearly shows a decrease in indicators that point at unpolluted conditions or at low pollution. Even indicators for moderate pollution decline. With the exception of *M. aquatica*, only indicators for high-polluted areas increased such as *C. obtusangula* and *Zannichellia palustris*. Furthermore, a species group which indicates silting processes showed the largest increase of all (Tables 14.2 and 14.3). To this group belong typical reed plants (*Phalaris arundinacea*, *Phragmites australis*, *Iris pseudacorus*) and annual mud plants (*Veronica anagallis-aquatica*, *V. beccabunga*), which promote sedimentation. Some species of this group have established during the monitoring period only.

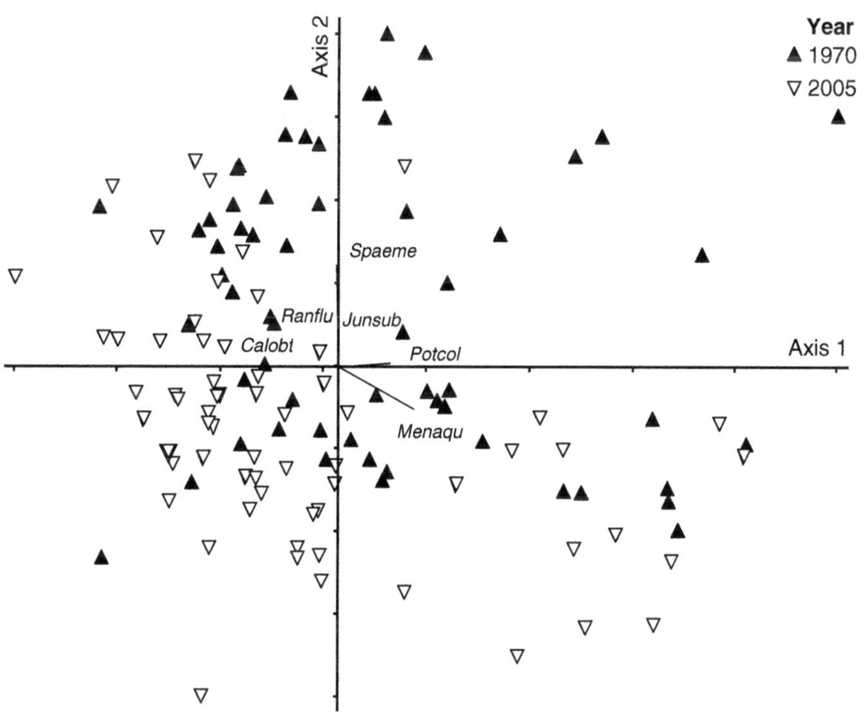

Fig. 14.3 DCA of vegetation data (abundance) of the river Moosach in 1970 and 2005. r^2-value = 0.36. Calobt – *C. obtusangula*, Junsub – *J. subnodulosus*, Menaqu – *M. aquatica*, Potcol – *P. coloratus*, Ranflu – *Ranunculus fluitans*, Spaeme – *Sparganium emersum et erectum*

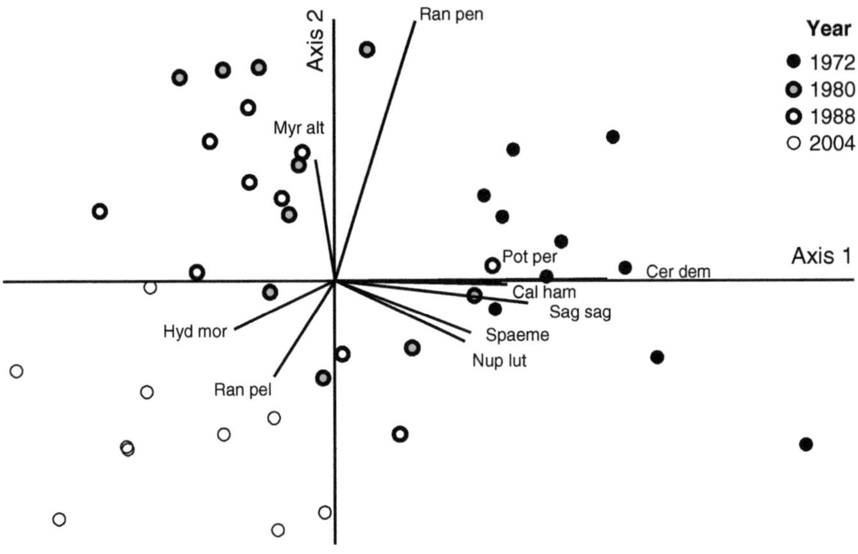

Fig. 14.4 DCA of vegetation data (abundance) of the river Naab from 1972 to 2004. r^2-value = 0.25. Calham – *C. hamulata*, Cerdem – *Ceratophyllum demersum*, Hydmor – *H. morsus-ranae*, Myralt – *M. alterniflorum*, Nuplut – *Nuphar lutea*, Potper – *Potamogeton perfoliatus*, Ranpel – *R. peltatus*, Ranpen – *R. penicillatus*, Sagsag – *Sagittaria sagittifolia*, Spaeme – *Sparganium emersum*

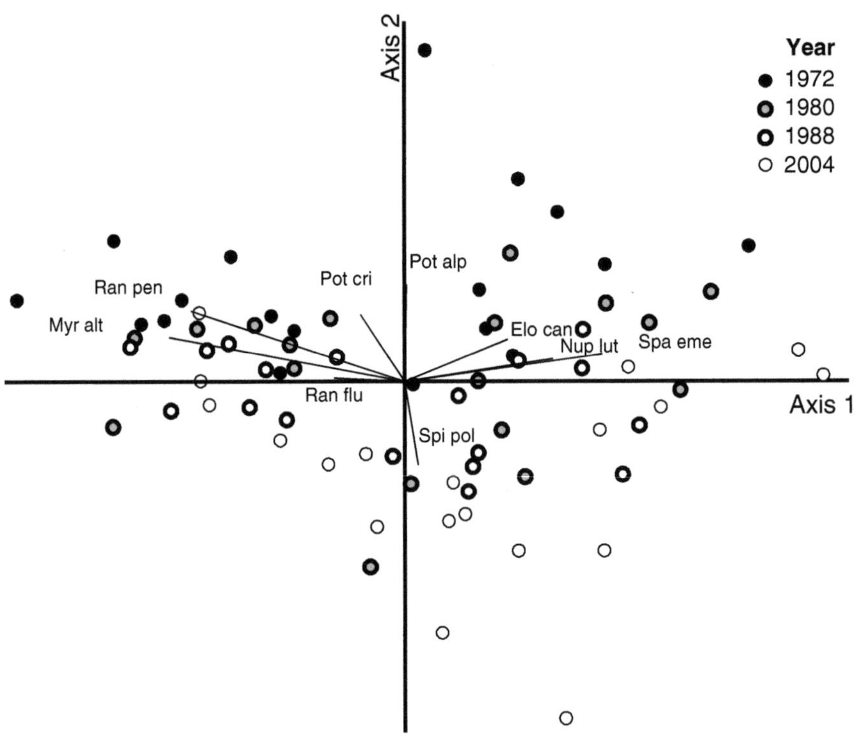

Fig. 14.5 DCA of vegetation data (abundance) of the river Pfreimd from 1972 to 2004. r^2-value = 0.15. Elocan – *Elodea canadensis*, Myralt – *M. alterniflorum*, Nuplut – *Nuphar lutea*, Potalp – *P. alpinus*, Potcri – *Potamogeton crispus*, Ranflu – *Ranunculus fluitans*, Ranpen – *R. penicillatus*, Spipol – *Spirodela polyrhiza*, Spaeme – *Sparganium emersum*

Table 14.2 'Relative range length' (L_T) of water quality (trophy) indicators and their changes in the Moosach from 1970 to 2005

Species	Zone	L_T 1970	L_T 2005	L_T gain/ loss (%)
Chara hispida	A	1.47	1.47	0.0
J. subnodulosus	A	4.78	3.10	−35.1
P. coloratus	A	3.10	3.10	0.0
M. aquatica	AB	33.91	70.87	109.0
G. densa	C	16.35	0.00	Extinct
Potamogeton natans	C	11.62	5.05	−56.5
Hippuris vulgaris	C	14.35	8.31	−42.1
Elodea canadensis	CD	47.21	57.26	21.3
M. verticillatum	CD	25.45	4.10	−83.9
C. obtusangula	CD	54.84	83.12	51.6
Ranunculus fluitans	CD	59.25	71.50	20.7
Z. palustris	CD	36.59	61.04	66.8
Agrostis stolonifera	Indifferent	42.01	12.93	−69.2
Potamogeton crispus	Indifferent	20.19	27.23	34.9
Potamogeton pectinatus	Indifferent	5.31	13.99	163.5
Berula erecta	Indifferent	62.09	91.96	48.1
Nasturtium officinale	Indifferent	54.15	33.54	−38.1
Ranunculus trichophyllus	Indifferent	48.42	19.19	−60.4
V. anagallis-aquatica	Indifferent, t	57.31	91.54	59.7
I. pseudacorus	t	0.00	22.40	New
P. arundinaceae	t	44.06	80.76	83.3
P. australis	t	3.63	57.05	1471.1
Typha latifolia	t	0.53	17.88	3273.6
V. beccabunga	t	0.00	28.13	New

Zone A to D – see text; t – indicators for terrestrialization and mud banks; new – species established only during the monitoring period

Table 14.3 Mean mass total (MMT) and mean mass occurrence (MMO) of water quality (trophy) indicators and their changes in the Moosach from 1970 to 2005

		1970		2005	
Species	Zone	MMT	MMO	MMT	MMO
Chara hispida	A	0.74	3.00	0.74	3.00
J. subnodulosus	A	1.58	4.36	1.05	2.90
P. coloratus	A	0.75	2.37	1.32	4.19
M. aquatica	AB	1.96	2.64	2.83	3.81
G. densa	C	1.23	2.25	0.00	0.00
Potamogeton natans	C	1.63	3.33	1.03	2.11
Hippuris vulgaris	C	1.70	3.24	1.26	2.40
Elodea canadensis	CD	1.57	1.93	2.58	3.16
M. verticillatum	CD	0.92	1.46	0.51	0.81
C. obtusangula	CD	3.21	3.77	4.19	4.92
Ranunculus fluitans	CD	3.57	4.09	3.45	3.95
Z. palustris	CD	1.92	2.54	2.15	2.84
Agrostis stolonifera	Indifferent	1.59	2.12	1.07	1.43
Potamogeton crispus	Indifferent	1.28	2.19	1.47	2.51
Potamogeton pectinatus	Indifferent	1.56	4.14	1.88	4.99
Berula erecta	Indifferent	3.58	4.04	4.16	4.70
Nasturtium officinale	Indifferent	1.79	2.11	1.70	2.00
Ranunculus trichophyllus	Indifferent	1.86	2.37	0.90	1.15
V. anagallis-aquatica	Indifferent, t	2.15	2.49	2.75	3.19
I. pseudacorus	t	0.00	–	1.04	–
P. arundinacea	t	1.88	2.35	2.32	2.90
P. australis	t	1.02	2.14	2.19	4.61
Typha latifolia	t	0.17	1.00	0.88	5.05
V. beccabunga	t	0.00	–	1.09	–

Zone A to D – see text; t – indicators for terrestrialization and mud banks

In the rivers Naab and Pfreimd, multivariate analyses showed a general trend correlated with a decrease of most species during the observation period (Figs. 14.4 and 14.5). Exceptions in the river Naab were *R. peltatus* and *Hydrocharis morsus-ranae* (Tables 14.4 and 14.5). *R. peltatus* was first recorded in 1978 and increased strongly until 2004. *H. morsus-ranae* appeared only in 2004, but already in half of the recorded sections. In the river Pfreimd, all indicator species decreased more or less continuously from 1972 to 2004 (Tables 14.6 and 14.7).

14.5 Discussion

Despite the general trend of water quality improvement in rivers throughout Germany (Länderarbeitsgemeinschaft Wasser, 2002) our study showed that only certain water quality parameters improved whereas other did not, especially in the rivers Moosach and Pfreimd. In all rivers ammonium was not anymore detectable indicating that the pollution by municipal sewages has stopped due to the establishment or improvement of sewage treatment plants during the 1970s and 1980s (Schmid, 2003). In contrast to that, phosphate concentrations in Moosach and Pfreimd have increased during the monitoring period, which is contributed by the more intense application of mineral fertilizer in agriculture.

Table 14.4 'Relative range length' of water quality (trophy) indicators and their changes in the Naab from 1972 to 2004

Species	Zone	L_T 1972	L_T 2004	L_T gain/ loss (%)
P. alpinus	AB	0.00	30.34	New
M. alterniflorum	AB	0.00	2.91	New
R. peltatus	AB	0.00	81.55	New
Potamogeton nodosus	CD	55.83	37.14	−33.5
Potamogeton perfoliatus	CD	36.65	0.00	Extinct
Ceratophyllum demersum	CD	92.72	0.00	Extinct
Lemna gibba	CD	21.12	11.89	−43.7
M. spicatum	CD	100.00	100.00	0.0
R. penicillatus	CD	100.00	41.26	−58.7
Spirodela polyrhiza	CD	25.97	50.00	92.5
H. morsus-ranae	Indifferent	0.00	66.44	New
Potamogeton crispus	Indifferent	27.43	22.33	−18.6
Potamogeton natans	Indifferent	30.10	0.00	Extinct
Sparganium emersum	Indifferent	100.00	97.09	−2.9
C. hamulata	Indifferent	94.31	95.68	1.5
Elodea canadensis	Indifferent	39.81	56.07	40.8
Ranunculus fluitans	Indifferent	0.00	7.28	New

Zone A to D – see text; new – species established only during the monitoring period; extinct – species extinct during the monitoring period

This fact has also triggered the higher nitrate concentrations in the river Pfreimd. In conclusion, water quality remained the same or slightly improved in the river Naab but worsened from nearly oligotrophic to mesotrophic in the river Pfreimd and from mesotrophic to eutrophic in the river Moosach. The number of river sections with unpolluted or low-polluted areas continuously decreased in these two rivers since the start of monitoring. This process is correlated with a decreasing number of oligotraphentic species. By contrast, the formerly high-polluted river Naab shows a significant decrease of eutraphentic species. Numerous species indicating lower pollution have re-established since the first record. Surprisingly, one of these species, *R. peltatus*, has established nearly in all sections. However, the contrary trends observed in the studied river systems support the assumption of a tendency toward unification. Currently, we lack more and more the extremes, not only high-polluted but also unpolluted or low-polluted river sections. The last German-wide overview on water quality in rivers assessed by the application of the saproby index showed this trend of unification as well, especially for

Table 14.5 Mean mass total (MMT) and mean mass occurrence (MMO) of water quality (trophy) indicators and their changes in the Naab from 1972 to 2004

		1972		2004	
Species	Zone	MMT	MMO	MMT	MMO
P. alpinus	AB	0.00	0.00	0.67	1.00
M. alterniflorum	AB	0.00	0.00	0.31	1.00
R. peltatus	AB	0.00	0.00	2.74	2.93
Potamogeton nodosus	CD	2.09	2.54	1.20	1.66
Potamogeton perfoliatus	CD	1.85	2.54	0.00	0.00
Ceratophyllum demersum	CD	1.95	2.00	0.00	0.00
Lemna gibba	CD	1.70	2.86	0.49	1.00
M. spicatum	CD	2.08	2.08	2.71	2.71
R. penicillatus	CD	4.56	4.56	1.36	1.83
Spirodela polyrhiza	CD	1.28	2.00	0.79	1.00
H. morsus-ranae	Indifferent	0.00	0.00	0.84	1.00
Potamogeton crispus	Indifferent	1.10	1.69	1.21	2.00
Potamogeton natans	Indifferent	0.00	0.00	0.00	0.00
Sparganium emersum	Indifferent	3.63	3.63	2.63	2.66
C. hamulata	Indifferent	2.61	2.61	1.12	1.35
Elodea canadensis	Indifferent	1.24	1.69	1.62	1.97
Ranunculus fluitans	Indifferent	0.00	0.00	0.42	1.00

Zone A to D – see text

Table 14.6 'Relative range length' of water quality (trophy) indicators and their changes in the Pfreimd from 1972 to 2004

Species	Zone	L_T 1972	L_T 2004	L_T gain/ loss (%)
P. alpinus	AB	13.31	0.64	−92.2
M. alterniflorum	AB	76.86	54.47	−29.1
R. peltatus	AB	87.42	78.38	−10.3
Potamogeton nodosus	CD	0.00	8.21	New
Lemna gibba	CD	0.00	7.30	New
R. penicillatus	CD	92.01	28.92	−68.6
Spirodela polyrhiza	CD	0.00	11.86	New
Potamogeton crispus	Indifferent	29.11	13.23	−54.6
Sparganium emersum	Indifferent	29.57	65.60	121.8
C. hamulata	Indifferent	94.31	79.93	−15.2
Elodea canadensis	Indifferent	13.31	13.87	4.2
Ranunculus fluitans	Indifferent	9.37	10.95	16.9

Zone A to D – see text; new – species established only during the monitoring period; extinct – species extinct during the monitoring period

Table 14.7 Mean mass total (MMT) and mean mass occurrence (MMO) of water quality (trophy) indicators and their changes in the Pfreimd from 1972 to 2004

		1972		2004	
Species	Zone	MMT	MMO	MMT	MMO
P. alpinus	AB	1.25	2.44	0.18	0.00
M. alterniflorum	AB	2.97	1.39	2.17	1.09
R. peltatus	AB	2.59	2.02	2.18	1.21
Potamogeton nodosum	CD	0.00	0.00	0.87	0.00
Lemna gibba	CD	0.00	0.00	0.42	0.79
R. penicillatus	CD	2.77	1.50	2.23	1.99
Spirodela polyrhiza	CD	0.00	0.00	0.49	0.73
Potamogeton crispus	Indifferent	1.66	1.72	1.02	1.45
Sparganium emersum	Indifferent	2.32	2.97	2.25	2.03
C. hamulata	Indifferent	2.41	2.14	2.17	1.71
Elodea canadensis	Indifferent	1.20	2.37	1.03	1.98
Ranunculus fluitans	Indifferent	0.72	1.47	0.67	0.00

Zone A to D – see text

lowland rivers (Länderarbeitsgemeinschaft Wasser, 2002).

Only single species behave opposite to the general trend of their indicator group such as *M. aquatica* in the river Moosach. This may be due to the fact that the occurrence of aquatic plants is affected by a large number of factors in which intensity and interaction are river specific (Schneider, Krumpholz, & Melzer, 2000). Furthermore, although the indicating value of most species is confirmed by other studies comparing water quality and species occurrence (e.g., Schneider, 2000), it was never validated, e.g., by transplantation experiments except for single species during the 1970s (*P. coloratus*, *G. densa*: Kohler et al., 1972).

Therefore, transplantation experiments of especially these indicator species with a different behavior in calcareous-rich and calcareous-poor waters (Fig. 14.2) are still one of the future tasks to better explain changes in certain species.

Today, the indicator group of oligotrophic waters exhibits the largest proportion of extinct and endangered species in Germany compared to any other habitat or ecosystem (Korneck et al., 1998). More than 80% of the 47 vascular plant species restricted to this habitat are listed in the Red Data Book; 4 are said to be extinct. In our systems, *Potamogeton alpinus, P. coloratus*, and *M. alterniflorum* belong to this group of endangered plants, *P. coloratus* and *Myriophyllum* being strongly endangered (Korneck, Schnittler, & Vollmer, 1996).

For this reason, there is an urgent need to know if these oligotraphentic indicator species are also able to re-establish in the case of local water quality improvements. Here we have the largest gaps of knowledge on aquatic plants. The regeneration potential of macrophytes is only poorly studied. Seed or spore (Characeae) bank and dispersal were only studied in more or less polluted rivers (Nilsson, Gardfjell, & Grelsson, 1991; Trottmann & Poschlod in Bonn & Poschlod 1998; Cellot, Mouillot, & Henry, 1998; Boedeltje, Bakker, & TerHeerdt, 2003; Boedeltje, Bakker, Bekker, & VanGroenendael, 2003). The *Chara* species have a high regeneration potential due to a long-term persistent oospore bank. In contrast, macrophytes like *Potamogeton* species seem to have a low regeneration potential taking into account their seed bank persistence which is either transient or short-term persistent (Thompson, Bakker, & Bekker, 1997). Concerning the above-mentioned endangered species no data are available except for *P. coloratus*. Despite the fact that this macrophyte often forms no seeds at all, a small seed bank could be found but in the vicinity of actual stands only (Trottmann and Poschlod, unpublished data). Therefore, it is not surprising that the re-colonization of macrophytes by vegetative propagules, e.g., by fragments, is considerably more relevant than by generative propagules (Capers, 2003). If we assume a transient or a short-term persistent seed bank for the above-mentioned strongly endangered species, at least a few sites of this species have to persist to assure its long-term viability in a river system. Probably this is the case with nearly all macrophytes. Therefore, the protection of oligotrophic conditions including the characteristic species pool should be a main issue in river management strategies to maintain the overall biodiversity for the future. In the case of the presented river systems the areas which should be protected are well outlined. On the one hand two valuable groundwater-fed ditches (Pullinger Graben, Mauke) feed the Moosach (Würzbach, Kohler, & Zeltner, 1997). On the other hand there is an area worth protecting in the upper reaches of the river Pfreimd just downstream the source. That spreading or re-colonization of a species even throughout a whole river system can occur fast is shown by the example of *R. peltatus* in the river Naab (Tables 14.4 and 14.5).

Although no individual of *R. peltatus* was found in 1972, this species had spread out to nearly all studied sections until 2004. For this reason, the study of life history of all indicator species with an emphasis on their regeneration potential is a further important task for the future with the objective of better understanding ecological changes in rivers systems.

Acknowledgments The authors thank all people and institutions who supported and contributed to the studies since the 1970s. In the most recent survey of the rivers Naab and Pfreimd, Julia Engels and Sabine Hartl collected distribution data. Uwe Veit has kindly commented an earlier draft of this chapter. An anonymous referee gave valuable comments that improved the chapter considerably. Sabine Fischer digitalized Fig. 14.1.

References

Bayerisches Landesamt für Wasserwirtschaft. (2005). Bewertungsverfahren Makrophyten & Phytobenthos. Fließgewässer- und Seen-Bewertung in Deutschland nach EG-WRRL. *Informationsbericht, 1/05*, 1–245.

Boedeltje, G., Bakker, J. P., & TerHeerdt, G. N. J. (2003). Potential role of propagule banks in the development of aquatic vegetation in backwaters along navigation canals. *Aquatic Botany, 77*, 53–69.

Boedeltje, G., Bakker, J. P., Bekker, R. M., & VanGroenendael, J. (2003). Plant dispersal in a lowland stream in relation to occurrence and three specific life-history traits of the species in the species pool. *Journal of Ecology, 91*, 855–866.

Bonn, S., & Poschlod, P. (1998). *Ausbreitungsbiologie der Pflanzen Mitteleuropas. Grundlagen und kulturhistorische Aspekte.* UTB. Wiesbaden: Quelle & Meyer, 404 pp.

Capers, R. S. (2003). Macrophyte colonization in a freshwater tidal wetland (Lyme, CT, USA). *Aquatic Botany, 77*, 325–338.

Cellot, B., Mouillot, F., & Henry, C. P. (1998). Flood drift and propagule bank of aquatic macrophytes in a riverine wetland. *Journal of Vegetation Science, 9*, 631–640.

DIN 38410 Part 1. (1987). *Biologisch-ökologische Gewässeruntersuchung*. Berlin: Beuth Verlag.

DIN 38410 Part 2. (1990). *Biologisch-ökologische Gewässeruntersuchung*. Berlin: Beuth Verlag.

Kohler, A. (1976). Makrophytische Wasserpflanzen als Bioindikatoren für Belastungen von Fließgewässer-Ökosystemen. *Verhandlungen Gesellschaft für Ökologie, Wien, 1975*, 255–276.

Kohler, A. (1978). Methoden der Kartierung von Flora und Vegetation von Süßwasserbiotopen. *Landschaft + Stadt, 10*, 73–85.

Kohler, A., & Janauer, A. (1995). Zur Methodik der Untersuchung von aquatischen Makrophyten in Fließgewässern. In: Steinberg, C., Bernhardt, H., & Klapper, H. (Eds.), *Handbuch Angewandte Limnologie* (pp. 3–22). Landsberg: Ecomed.

Kohler, A., & Zeltner, G. (1974). Verbreitung und Ökologie von Makrophyten in Weichwasserflüssen des Oberpfälzer Waldes. *Hoppea, Denkschriften der Regensburgischen Botanischen Gesellschaft, 33*, 171–232.

Kohler, A., Brinkmeier, R., & Vollrath, H. (1974). Verbreitung und Indikatorwert der submersen Makrophyten in den Fließgewässern der Friedberger Au. *Berichte der Bayerischen Botanischen Gesellschaft, 45*, 5–36.

Kohler, A., Lange, B., & Zeltner, G. (1992). Veränderung von Flora und Vegetation in den Fließgewässern Pfreimd und Naab (Oberpfälzer Wald) 1972–1988. *Berichte des Institutes für Landeskultur und Pflanzenökologie der Universität Hohenheim, 1*, 72–138.

Kohler, A., Vollrath, H., & Beisl, E. (1971). Zur Verbreitung, Vergesellschaftung und Ökologie der Gefäß-Makrophyten im Fließgewässersystem Moosach (Münchener Ebene). *Archiv Hydrobiologie, 69*, 333–365.

Kohler, A., Zeltner, G., & Busse, M. (1972). Wasserpflanzen und Bakterien als Verschmutzungszeiger von Fließgewässern. *Umschau, 72*, 158–159.

Kolkwitz, R. (1950). Ökologie der Saprobien. Über die Beziehungen der Wasserorganismen zur Umwelt. *Verein Wasser Boden Lufthygiene, 4*, 1–64.

Kolkwitz, R., & Marsson, M. (1902). Grundsätze für die biologische Beurteilung des Wassers nach seiner Flora und Fauna. *Mitteilungen aus der Königlichen Prüfungsanstalt für Wasserversorgung und Abwasserbeseitigung zu Berlin, 1*, 33–72.

Kolkwitz, R., & Marsson, M. (1908). Ökologie der pflanzlichen Saprobien. *Berichte der Deutschen Botanischen Gesellschaft, 22*, 505–519.

Kolkwitz, R., & Marsson, M. (1909). Ökologie der tierischen Saprobien. Beiträge zur Lehre von der biologischen Gewässerbeurteilung. *Internationale Revue der gesamten Hydrobiologie und Hydrographie, 2*, 126–152.

Korneck, D., Schnittler, M., & Vollmer, I. (1996). Rote Liste der Farn- und Blütenpflanzen (Pteridophyta et Spermatophyta) Deutschlands. *Schriftenreihe für Vegetationskunde, 28*, 21–187.

Korneck, D., Schnittler, M., Klingenstein, F., Ludwig, G., Takla, M. Bohn, U., et al. (1998). Warum verarmt unsere Flora? Auswertung der Roten Liste der Farn- und Blütenpflanzen Deutschlands. *Schriftenreihe für Vegetationskunde, 29*, 299–444.

Länderarbeitsgemeinschaft Wasser (LAWA). (2002). *Gewässergüteatlas der Bundesrepublik Deutschland. Biologische Gewässergütekarte 2000.* Berlin. Kulturbuch-Verlag.

Liebmann, H. (1959). Methodik und Auswertung der biologischen Wassergüte-Kartierung. *Münchner Beiträge zur Abwasser-, Fischerei - und Flussbiologie, 6*, 134.

Liebmann, H. (1960). *Handbuch der Frischwasser- und Abwasserbiologie, Band II.* München: Oldenbourg Verlag.

Liebmann, H. (1962). *Handbuch der Frischwasser- und Abwasserbiologie, Band I, 2. Auflage*. München: Oldenburg Verlag.

McCune, B., & Mefford, M. J. (1999). *PcOrd. Multivariate analysis of ecological data. User's Guide.* Oregon: MjM Software Design.

Nilsson, C., Gardfjell, M., & Grelsson, G. (1991). Importance of hydrochory in structuring plant communities along rivers. *Canadian Journal of Botany, 69*, 2631–2633.

Schaumburg, J., Schranz, C, Stelzer, D., Hofmann, G., Gutowski, A., & Foerster, J. (2005). *Bundesweiter Test: Bewertungsverfahren "Makrophyten & Phytobenthos" in Fließgewässern zur Umsetzung der WRRL.* Augsburg: Bayerisches Landesamt für Umwelt.

Schaumburg, J., Schranz, C., Foerster, J., Gutowski, A., Hofmann, G., Meilinger, P., et al. (2004). Ecological classification of macrophytes and phytobenthos for rivers in Germany according to the Water Framework Directive. *Limnologica, 34*, 283–301.

Schmid, A. (2003). *Handbuch der bayerischen Geschichte. Band IV, 1. Das Neue Bayern. Von 1800 bis zur Gegenwart. 2. Auflage.* München: C.H. Beck.

Schneider, S. (2000). *Entwicklung eines Makrophytenindex zur Trophieindikation in Fließgewässern.* Aachen: Shaker Verlag.

Schneider, S., Krumpholz, T., & Melzer, A. (2000). Trophieindikation in Fließgewässern mit Hilfe des TIM (Trophie-Index Makrophyten) – Erprobung eines neu entwickelten Index im Inninger Bach. *Acta Hydrochimica et Hydrobiologica, 28*, 241–249.

Schorer, A., Schneider, S., & Melzer, A. (2000). The importance of submerged macrophytes as indicators for the nutrient concentration in a small stream (Rotbach, Bavaria). *Limnologica, 30*, 351–358.

Thompson, K., Bakker, J. P., & Bekker, R. (1997). *The soil seed banks of North West Europe: Methodology, density and longevity.* Cambridge: Cambridge University Press.

Veit, U., & Kohler, A. (2003). Long-term study of the macrophytic vegetation in the running waters of the Friedberger Au (near Augsburg, Germany). *Large Rivers, 14*, 65–86.

WRRL. (2000). Richtlinie 2000/60EG des Europäischen Parlaments und des Rates vom 23. Oktober 2000 zur Schaffung eines Ordnungsrahmens für Maßnahmen der Gemeinschaft im Bereich der Wasserpolitik. *Amtsblatt der Europäischen Gemeinschaften, L327*, 1–72.

Würzbach, R., Kohler, A., & Zeltner, G.-H. (1997). Die Makrophyten-Vegetation des Fliessgewässersystems der Moosach (Münchener Ebene) – Ihre Entwicklung von 1970 bis 1996. *Beihefte Berichte Institut Landschafts- und Pflanzenökologie Universität Hohenheim, 4*, 243–312.

Part V
Concepts and Results: Presenting and Interpreting Long-Term Ecological Processes: Terrestrial Ecosystem Research

Chapter 15

Long-Term Observations of Soil Mesofauna

Hartmut Koehler and Viesturs Melecis

Abstract General problems connected with planning, sampling, and data processing of long-term research of soil mesofauna are discussed, based on two case studies: (i) the Bremen study of predatory mites (*Gamasina*) covering 20 years of secondary succession on a ruderal site in northern Germany and (ii) the Mazsalaca study of the effects of climate warming on Collembola of coniferous stands in the North Vidzeme Biosphere Reserve, Latvia, covering 11 years. The findings from both sites are embedded in an array of environmental data. The results from Bremen document the asynchrony of different biota in successional dynamics. The long-lasting increase of the species numbers of soil predatory mites (*Gamasina*) is contrasted by a decrease in plant species numbers. In the Baltic forests, climate change is indicated by the dynamics of collembolan community. Gradual decline in species richness has been observed from 1992 to 2002 attributed to global warming. The 'temporal window' or time unit to discern changes in soil mesofauna communities seems to span approx. 5 years, highlighting the necessity of long-term observations.

Keywords Climate change · Coniferous forest · Ruderal site · Soil mesofauna · Succession

15.1 Introduction

According to Reichle, O'Neill, and Harris (1975) ecosystem resilience depends on four attributes: energy base, energy reservoir, element recycling and rate regulation. Fluxes and flows are regulated to a large extent by soil mesofauna, being considered as ecosystem webmasters (Coleman & Hendrix, 2000). Changes in species composition of soil fauna may significantly affect litter decomposition and soil formation, making it vulnerable to degradation and loss of ecosystem services, such as soil fertility, water retention, and carbon sequestration. Below–above ground and multitrophic interactions are some of the complex relations soil mesofauna may trigger or at least play an important role in (e.g., Wardle, 2002; Poveda et al., 2003). Because of this intricate incorporation into the soil system, mesofauna groups are considered as useful bioindicators (e.g., Karg & Freier, 1995; Koehler, 1996; Van Straalen et al., 2005; Filser et al., 2008).

Long-term ecological research deserves high priority for the understanding of ecosystem functioning in response to environmental changes crucial for human existence, including that of climate. There is no doubt that climate change may alter species distributions (e.g., Hampe & Petit, 2005; Schroeter et al., 2005; Thuiller et al., 2005, Thomas et al., 2004).

Until recently most of the climate-related studies performed on soil fauna focused on experimentally simulated effects of drought (Kennedy, 1994; Briones et al., 1997; Lindberg, 2003; Lindberg & Bengtsson, 2006). However, the results of model experiments suffer from restricted spatial and temporal scales. They cannot reveal the long-term changes of soil ecosystems relevant for the consideration of effects of global warming. Also, space-for-time approaches as an alternative for long-term investigations do not reflect natural transitions and the role of year-specific conditions, e.g., in areas of glacial retreat or along

H. Koehler (✉)
Centre for Environmental Research and Sustainable Technology (UFT), University of Bremen, Leobener Strasse, D-28359 Bremen, Germany
e-mail: hartmut.koehler@uni-bremen.de

F. Müller et al. (eds.), *Long-Term Ecological Research*,
DOI 10.1007/978-90-481-8782-9_15, © Springer Science+Business Media B.V. 2010

precipitation gradients (as models for succession or climate change; Clements, 1916; Likens, 1988; Kaufmann, 2002; Araujo & Rahbek, 2006; Fleischer & Sternberg, 2006). These can only be traced by long-term research in natural environments.

Worldwide, comprehensive long-term research projects neglect soil organisms in spite of their importance sketched above (see also Melecis, 1999). The few studies seem to become even fewer in the future (Dunger, 1975; Dunger et al., 2001; Takeda, 1987; Hogervorst et al., 1993; Rusek, 1993; Christian, 1995; Wolters, 1998; Koehler, 2000; Koehler & Müller, 2003; Irmler, 2006). A major reason is that soil mesofauna research increasingly is facing a loss of taxonomic expertise and even reference collections. This is all the more disturbing since the soil harbors overwhelming species diversity (e.g., Anderson, 1975; André et al., 2001). Also, soil animals are less well known to the public compared to flowers or butterflies. Finally, soil organisms are not so attractive to researchers and also to stakeholders and policy makers because no quick results are to be expected since research on soil fauna is time consuming and tedious.

Nature is not static. Ecosystem change (succession) never stops (Connell & Slatyer, 1977) and has explicitly to be considered in long-term studies. Being embedded in the slow processes of the soil system, the development of the mesofauna community exhibits extended temporal dynamics, with the almost trivial consequence that studies have to be performed over longer periods of time.

General aspects of soil mesofauna methodology are summarized, e.g., by Edwards (1991), Dunger and Fiedler (1997), Schinner (1993). In this chapter, we highlight methodological peculiarities encountered in our long-term soil mesofauna research projects, dealing with ecological succession on a ruderal site (Siedenburg LTER site, Germany; Koehler, 1999, 2000; Koehler & Müller, 2003; Weidemann, 1985; Weidemann et al., 1982) and with the impact of global warming on Scots pine forest ecosystems (Mazsalaca LTER project, Latvia; Juceviča & Melecis, 2002, 2005, 2006). Some results are selected to discuss the importance of long-term observation. We hope that our findings will inspire the scientific community and funding agencies to sustain and intensify long-term research projects in this important branch of soil ecology.

15.2 Methodological Approach

15.2.1 Basic Requirements

In long-term research, foresighted planning is detrimental. The methodological protocol has to be followed over decades by successive researchers. It should be simple and robust enough to provide statistically representative and comparable data and to allow for adaptation or expansion according to methodological innovation.

Studies of soil mesofauna are imbedded in the general ecological context sketched in the introduction. With an eco-survey a basic set of environmental factors is monitored, such as climate, soil type, vegetation cover, successional stage, history. An interdisciplinary ecology team is advantageous, but not mandatory. Mesofauna studies include the following successive steps:

- Site selection
- Experimental design
- Arrangement of sampling plots
- Sampling
- Extracting animals from soil samples
- Sorting and counting on group level
- Preparation, species identification, documentation in collection
- Data processing.

Each of these steps has specific methodological problems relating to the peculiarities of soil mesofauna, such as distribution in space and time, scale, collecting, and taxonomy.

15.2.2 Site Selection and Arrangement of Sampling Plots

First of all, the availability of the study area to be selected has to be secured over decades by adequate property rights and exclusion of vandalism and unexpected interference (clear cutting, land use change, erection of buildings, etc.). The site should be accessible without major difficulties. Topographic and methodological constraints as well as the research questions determine the delimitation of the study area.

The site should be large enough not to be altered by the destructive sampling procedure necessary for studying soil mesofauna (taking soil cores over years) and be sufficiently homogeneous to allow for representative sampling. For an estimate of homogeneity of the sampling site, not only relief but also information on soil type and vegetation cover derived from an initial grid screening is important.

The experimental design of a long-term study depends on two general objectives. If the aim of the investigation is to reveal the long-term effects of different treatments (e.g., manuring, trampling, manipulating of plant cover) agricultural field research offers methods such as random block design (see Johnston & Powlson, 1994). A detailed discussion relating to long-term field experiments is found in Tilman (1989). When the factors under investigation cannot be controlled, as in global change research, a different situation arises. Correlations of several factors are likely and successional change is natural. To handle this situation, the experimental design has to allow for multivariate analyses. In addressing succession explicitly or in selecting comparable sites of different successional age, community change is included in the research question. The arrangement of sampling plots has to account for the destructive nature of sampling soil mesofauna.

15.2.3 Sampling

Soil sample units to study mesofauna are in the scale of centimeters and several units have to be taken per sampling campaign. Soil is taken with a corer, the design of which often depends on the methodological expertise of the researcher and the soil characteristics, e.g., the amount of gravel to be expected to challenge its mechanical stability. The type of augers may bias the results (Bruckner, 1998). Since the composition of soil mesofauna changes with depth, a vertical profile may be taken, usually to some 15 or 20 cm depth, divided into sub-cores of equal thickness of, e.g., 5 cm. Spatial limitations and time-consuming handling and processing of the samples restrict the number of cores and the amount of soil removed.

Soil mesofauna has a heterogeneous spatial distribution of individuals causing high data variation with a consistent dependence of variance on the mean and on the degree of taxonomic resolution (Ekschmitt, 1998). Theoretically, there are two ways to cope with high variances – either increasing sample unit size beyond the average patch size or reducing it to make it close to average distance between individuals (Greig-Smith, 1964). Since patch sizes depend on the site-specific environment and vary between taxonomic groups, no generalized approach can be given. Large sample units tend to result in very high numbers of individuals, which make analysis too laborious.

There exist two ways to facilitate the counting of soil mesofauna and at the same time allow for statistical analyses. First, small sampling units may be used with high sample size *n* (number of cores) to avoid underestimation of species richness. In the Latvian study, we compared estimates of Collembola community parameters obtained from large series of relatively small sample units (n = 100, surface area 1.0, 2.5, and 5.0 cm^2, depth 0–10 cm) with smaller series of larger samples (n = 10–20, surface area 25 and 75 cm^2, depth 0–10 cm; Melecis, 1988). Series of n = 100 considerably reduced standard errors of abundances in comparison with series of larger sample units with sample size of n = 10–20 (Fig. 15.1).

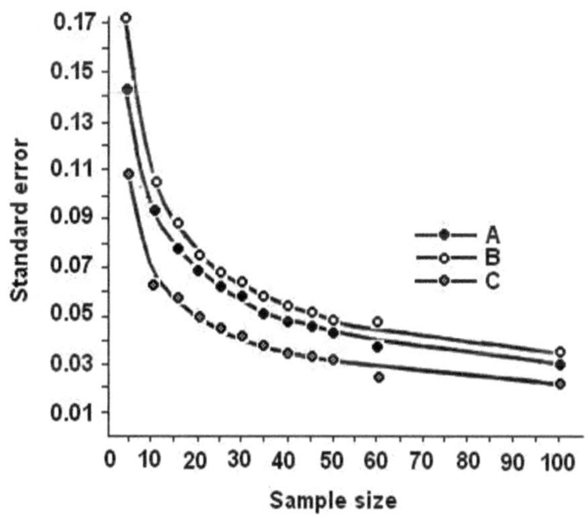

Fig. 15.1 Dependence of standard error of mean density calculated from logarithmically transformed data of Collembola from sample size by different sample unit sizes: A – 5 cm^2 × 4 cm, n = 100; B – 20 cm^2 × 4 cm, n = 100; C – 75 cm^2 × 4 cm, n = 100. Comparative study of sampling efficiency of parallel sample series with different unit size obtained by systematic sampling of birch forest plot. For small sample units high-gradient extractor was used, but for medium and large ones Tullgren's funnels were used. Standard errors were calculated by modeling random subsamples of different size (after Melecis, 1988)

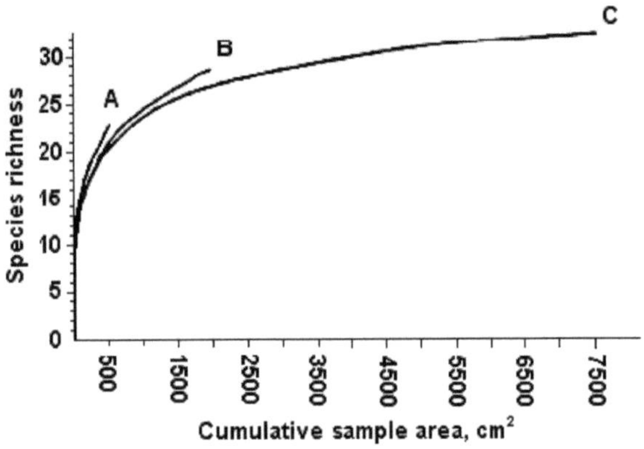

Fig. 15.2 Dependence of species richness of Collembola from cumulative sample area by different sample unit size: A – 5 cm^2 × 4 cm, $n = 100$; B – 20 cm^2 × 4 cm, $n = 100$; C – 75 cm^2 × 4 cm, $n = 100$. Comparative study of sampling efficiency of parallel sample series with different unit size obtained by systematic sampling of birch forest plot. For small sample units high-gradient extractor was used, but for medium and large ones Tullgren's funnels were used (after Melecis, 1988)

For one and the same total sampled area, highest estimates for species richness were obtained using the small sample units with correspondingly high n (Fig. 15.2). The use of large series of small sample units resulted also in a more representative coverage of the area of the sampling plot. However, this approach demands a large number of eclectors to provide simultaneous extraction for soil cores and is time consuming. Small cores may suffer for some groups and in some soil types from edge effects of the auger. Also, for taxa like the Gamasina with much lower abundances than Collembola, samples size of 20–25 cm^2 is appropriate.

Another way is the use of composite samples (Bruckner et al., 2000). This may increase representativity, but ecological information on the spatial distribution is lost.

An important problem in long-term studies of soil mesofauna is the placement of the sample units. Since most standard statistical tests depend on randomness (Jongman et al., 1987), random sampling rather than systematic or regular sampling is preferred. The most irreproachable method would be to choose sampling points by random coordinates. However, this was rejected in our studies because of the risk of impracticability in research over long periods of time. Instead, many researchers are using 'arbitrary but without preconceived bias' or 'haphazard' sampling schemes (McCune & Grace, 2002). To avoid effects of trampling and coring from previous campaigns, systematic sampling is prevalent in long-term studies of soil mesofauna. It may also provide more information per unit cost than simple random sampling because the sample is distributed uniformly over the entire population of study area (Ratti & Garton, 1994). Several studies (see review by Pawley, 2006) have demonstrated that systematic sampling has important advantages over alternative strategies in terms of statistical efficiency. Pawley (2006) also demonstrated that the negative effects of spatial autocorrelation cannot be validated when the aim of the study is the assessment of mean population numbers in a restricted space (such as a permanent plot in a long-term study). Therefore, systematic sampling might provide even better results than the random one.

For testing differences between mean population numbers an estimate of variance is necessary. Data from systematic sampling are not suitable to obtain an unbiased estimator of variance; it is underestimated when calculated by commonly used statistical methods (Pawley, 2006). Pawley (2006) compared 11 estimators derived from systematic sampling and identified 'Krige's additivity relationship' and variography to be most reliable.

Systematic sampling should not be used in case of latent periodicity within the sampled population. To avoid such a risk in the study of natural populations, stratified sampling schemes may be applied (Snedecor & Cochran, 1980).

Although phenology in the soil is by far not so important as above ground, the frequency and timing of the sampling campaigns over the year are an important issue. Frequent soil coring is laborious and may cause too much disturbance and destruction. Practicability may limit the sampling to one campaign per year. In the temperate zones, fall is preferred when high numbers of individuals and species tend to be observed (e.g., Takeda, 1987; Hogervorst et al., 1993). To avoid a bias caused by seasons, Irmler (2006) performed monthly sampling of mesofauna for 7 years of investigation with rather low sample size of $n = 4$ per campaign. He found relatively low yearly fluctuations and very high similarities between yearly species assemblages. The results of the Bremen study indicate that given a limited sample size for routine monitoring, more species can be secured when the samples are distributed over the year in seasonal sampling campaigns compared to an intensive effort in only one sampling campaign in late fall. Assessment of abundances is affected to a lesser extent (Koehler & Müller, 2003). These results underline the importance of standardized sampling intensities in long-term research.

Recently, inexpensive wireless sensor networks and advanced systems of continuous raw data collection have initiated significant improvements in measuring abiotic parameters particularly in long-term studies (Musaloiu-E et al., 2006).

15.2.4 Extracting Microarthropods from Soil Samples

The extraction process was studied systematically by several authors (e.g., MacFadyen, 1961; Usher, 1969; Edwards & Fletcher, 1971; Lasebikan et al., 1978; Edwards, 1991; Koehler, 1993). The original papers on the methodological development of mesofauna extraction in detail are strongly recommended to adapt the most efficient and robust method to the specific needs of the respective long-term research.

Extraction efficiency is affected by many factors, such as the extraction regime (gradient of heat and humidity in the soil sample, change of gradient during extraction process, duration) and whether the soil cores are intact or broken apart. In long-term studies the extraction efficiency has to be maintained at the same relative level over the whole study period and must neither be affected by environmental fluctuations of the seasons or between the years nor by the researcher's expertise. As was shown by Koehler (1984), field soil moisture affects the extraction process considerably: dry soil conditions reduce the efficiency, and very moist soil is instable so that coring clogs the pores, resulting in a reduced and selective catch. Very wet soils should not be sampled, dry soil cores can be humified before extraction with a water sprayer as used for pot plants. The extraction regime has to ensure gradual and finally efficient desiccation of the soil to drive the animals toward the collection vessel. As is compiled from various sources by Dunger and Fiedler (1997), pF values higher than pF 3.5 are critical for soil microarthropods and provoke direct movement. This threshold varies with different taxa (e.g., it is lower for Collembola and high for Oribatids; for details see also Vannier, 1970).

During and after extraction, collecting and preservation fluids are an important issue (Dunger & Fiedler, 1997). Since many of these are toxic or carcinogenic, safety precautions have to be observed.

15.2.5 Counting, Sorting, and Identification of Soil Animals

Counting and sorting the extracted material under the stereoscope is labor-intensive work. Automatization with image processing failed so far because of the large variation of the animals in shape and size and because of a background 'noise' from fallen-in organic and mineral matter. According to Bruckner et al. (2000), aliquots from composite samples may reduce labor and enhance representativity.

Identification of species often is seen as the most problematic step. However, given accessible collections and support by a taxonomic expert, learning to use a mite or Collembola key is just learning a method as any other. Self-teaching is not recommended and participation in an introductory course is strongly advised. Specimens have to be prepared for microscopic inspection. A collection of permanent slides is recommended as well as a data bank of digital photos. The use of very convenient polyvinyl-lactophenol is advised against: the specimens become defective after 5–10 years. Other media are found in the literature (e.g., Dunger & Fiedler, 1997). They often contain carcinogenic chemicals and safety precautions are obligatory.

Certain difficulties may arise with species identification of immature individuals of mites and Collembola, especially when the number of individuals collected is small. It is recommended to take several additional, bigger qualitative soil samples from the site and make extractions of soil mesofauna from them to obtain more individuals to assure identification and for reference collection. Some authors only give data for mature specimen; this is problematic particularly in mites, where immatures (deutonymphs) may contribute to a major proportion of the community resulting in a distorted community structure when omitted.

15.2.6 Data Processing

Frequently, in long-term research of soil mesofauna the following two questions are to be answered:

- What are the effects of temporal variation of certain environmental factors on soil animals?
- Are trends or cycles in population and/or community characteristics detectable?

It is important to note that commonly used methods of ecological data analysis (e.g., correlation/regression analysis, analysis of variance, and multidimensional ordination methods) should be used with precaution, considering the implications of the temporal aspect to long-term data series.

Long-term research is observational and natural variations in time are taken advantage of (succession, changes of environmental factors as in climate change). When in such 'natural experiments,' data are collected at certain intervals, they are called trajectory experiments (Gotelli & Ellison, 2004). The impact of factors in these experiments, commonly, is evaluated using analysis of variance (ANOVA). Specific experimental spatial designs are supported by corresponding models of ANOVA (see Sokal & Rohlf, 1987; Hairston, 1989). However, the application of commonly used models of ANOVA in trajectory experiments in which replications are temporal is not without problems: Each observation represents a different year, month, or season (Gotelli & Ellison, 2004), and factor levels are not independent. This is one of the most serious violations of the general assumptions of ANOVA models (Scheffe, 1959). Under such circumstances, ANOVA models based on repeated measures design or before–after control impact (BACI) are recommended (Gurevitch & Chester, 1986; Lindsey, 1993). Also it should be tested for homoskedasticity, e.g., the variances of the difference between any pair of times should be the same – the assumption which is rarely met in time series data (Gotelli & Ellison, 2004). Calculations can be made by software packages SPSS (Norusis, 2004) or SAS (SAS Institute Inc., 2004).

ANOVA design may be preferred when only few levels of predictor variables can be established. If the number of levels exceeds five and no replications are available, a regression approach is much more feasible. However, before using common regression analysis, time series data should be tested for the presence of autocorrelations of residuals ('error terms') of the dependent variable. Properties of time series data change over time, e.g., the time series are not stationary because the data are autocorrelated. The presence of autocorrelation violates the regression model assumption that the residuals are uncorrelated (Belsley et al., 1980). SPSS provides Durbin–Watson statistic as a traditional test for autocorrelations. If the test was not significant, the ordinary least square regression can be used (Bence, 1995). Otherwise, autoregressive correction procedures are to be applied (Belsley et al., 1980). SAS as well as SPSS program packages provide means to perform such analyses.

A new method called non-parametric multiplicative regression (NMR) is now available for the analysis of ecological data (McCune, 2006). NMR is not affected by the restrictions of the traditional least-square-based multiple regression. However, this method was designed to perform spatial data analysis and possibilities of its application to time series data evidently should be tested.

Properties of time series data mentioned above may negatively affect correlation analysis. Due to non-stationarity of the data, the Pearson correlation coefficient is not stable. Like in time series regression, the data should be tested for presence of autocorrelations, and correction procedures should be used to remove non-stationarity before the calculation of the correlation coefficient (Yang & Shahabi, 2005). Considering these problems, some authors (Wiwatwitaya & Takeda, 2004; Irmler, 2006) preferred non-parametric correlation coefficients instead of Pearson correlation coefficient in the analysis of soil mesofauna data. However, it should be noted that the use of non-parametric methods always is connected with loss of information contained in the data (Gotelli & Ellison, 2004). Use of time series analysis instead of simple regression allows to find out lag effects of environmental variables on soil mesofauna, provided frequency of sampling is high enough (see Irmler, 2006).

Similar problems are encountered in trend analyses of long-term data of soil mesofauna. The linear model does not consider effects of population growth through births and deaths, leading to autocorrelations between successive time series. Of course, the further apart in time the samples are separated from one another, the more they function as independent replicates. Based on information about life cycles of species, it has to be decided to which extent the linear regression is valid as a tool for trend analysis (Gotelli & Ellison, 2004). This can be tested by calculating partial autocorrelation function (PACF). Unless the test of presence of statistically not significant autocorrelation gives a negative result, autoregressive models are to be preferred in trend analysis of mesofauna abundance (Gotelli & Ellison, 2004).

Besides linear trends, the researcher might be interested in discovering nonlinearity or periodicity in time series separating them from random variation or 'white noise.' This could be important in the analysis of ecological successions. For this purpose spectral analysis can be used (see Box & Jenkins, 1970; Platt & Denman, 1975; Legendre & Legendre, 1998). However, existing time series on soil mesofauna (5–20 years) may appear too short to be analyzed by these methods. Programs of regression analysis and time series analysis are available in SPSS and SAS program packages.

Changes in community structure can be studied by calculating similarity/dissimilarity indices. In long-term studies of soil mesofauna, Morisita's index (Wiwatwitaya & Takeda, 2004) and squared Euclidian distance (Kampichler & Geissen, 2005) were successfully applied. Similarity of seasonal abundances between species over the period of study was examined by using Kendall's concordance (Wiwatwitaya & Takeda, 2004). Huhta (1979) tested 16 similarity indices as measures of succession in forest soil arthropod communities but only 6 of them seemed to express the changes satisfactorily (Kendall's rank correlation coefficient, Bray–Curtis measure, Renkonen's percentage similarity, Pearson correlation coefficient after logarithmic transformation of the data, Canberra metric, and diversity overlap index). Collins, Micheli, and Hartt (2000) proposed time-lag analysis for the investigation of changes of community composition over time. Some results using this method for the analysis of soil mesofauna data can be found in Kampichler and Geissen (2005).

When the aim of the study is to find out hidden factors responsible for temporal community changes, methods of multidimensional ordination should be used. It is possible to draw the trajectories of successive community states in ordination axes, which are interpreted as environmental factors (see Ponge, 1973). Catell (1951) called this kind of analysis the P-techniques, and it has been introduced and discussed first by psychologists (Anderson, 1963). There were also some objections against using multidimensional analysis of correlation matrices based on time series data (Überla, 1977): (i) dependence of observations leading to false correlations and (ii) possibility of lag effects reducing estimates of correlations. These also may refer to ecological data. Therefore, ordination methods of time series data based on correlation matrix should be used cautiously, e.g., principal components analysis (PCA), canonical correlation analysis (COR), and redundancy analysis (RDA; for a review of these methods see Jongman et al., 1987). Some methods have been worked out to overcome the problem of non-stationarity of multivariate time series data (Yang & Shahabi, 2005).

Evidently these problems are inferior to cluster analysis and distance-based ordination methods, e.g., detrended correspondence analysis (DCA) and non-metric multidimensional scaling (NMS). However, DCA which has been widely applied in ecology during the last decades has received serious criticism (see McCune & Grace, 2002). When the data matrix has nonlinear relationships it is recommended that nonmetric multidimensional scaling (NMS) is used. This method relies only on the rank order of similarities and is considered to be one of the most powerful ordination methods (McCune & Grace, 2002).

An interesting approach of multidimensional analysis of species abundance time series data was based on differences between successive abundance values, so-called before–after data (McCune & Grace, 2002). These data are distributed more or less normally and when distance measures providing non-negative before–after data values are used, the data can be analyzed by PCA (McCune & Grace, 2002). An example from aquatic ecology can be found in Allen and Koonce (1973).

For the analysis of aspects of density dependence in time series, partial rate correlation function is suitable (PRCF, Berryman & Turchin, 2001). Wolters (1998) used the method recommended by Pollard et al. (1978) for the analysis of time series of Collembola populations. However, the authors pointed out problems in obtaining statistically significant results even when 10–15 year-long time series were analyzed.

15.3 Case Studies

15.3.1 Community Succession of Predatory Gamasine Mites on a Ruderal Site (Siedenburg, Germany)

15.3.1.1 Experimental Design and Sampling

Secondary succession has been investigated in detail with synchronous and syntopic sampling of soil mesofauna, soil, and vegetation since 1980 on the plateau of the Siedenburg rubble and debris dump in the vicinity of Bremen, northern Germany. Various research questions concerning, e.g., recultivation, disturbance, and ecotoxicology were addressed in that framework for shorter periods of time. The site was covered with a layer of approx. 1 m of disturbed soils and did not exhibit any toxic influences. Prior to the start of the long-term experiment in 1980, 3–5 years of disturbed ruderal development took place. In early 1980, the whole plateau was thoroughly graded with heavy machinery. The majority of the almost 1 ha site was left for undisturbed succession (plot SUC), whereas an area of 1,000 m^2 was prepared by rotary till and seeding for recultivation (plot REC). A 10 by 10 m grid was laid out over the entire site and a screening of soil conditions was performed. With the help of this grid, plots for soil sampling, for permanent vegetation studies, and for specific experiments were assigned (details in Koehler & Müller, 2003).

For soil mesofauna, two quadrants 10 × 10 m were reserved on the SUC and REC plots, respectively. Soil cores of 25 cm^2 and 0–12 cm depth were taken stratified into three 4 cm layers. A seasonal sampling scheme with $n = 6$ cores in each of four seasonal campaigns per year was performed from 1980 to 1993, after that only one campaign in fall with $n = 12$ was undertaken. Interference with previous coring was avoided by a general documentation. The samples were assigned to the two 10 × 10 m subplots of SUC and REC, respectively, by 25% of total sample size each.

The change in sampling design violates the crucial aspect of maintaining a stable sample design over the years. This was inevitable because of unforeseeable cutbacks in research staff. The interpretation of results has to consider the two series 1980–1993 and 1994–2000.

In the evaluated period from 1980 to 2000, Gamasina were counted and identified to species from all samples. Modifications of high-gradient extractors were used for the extraction of microarthropods from the sample units. A humid regime (20–35°C) was followed by a drying regime in 5 K steps per 24 hours up 60°C surface of inverted intact cores (Koehler, 1993). Cooling from beneath was reduced toward the end of the extraction to make the animals leave even the bottom layers of the soil core. After drying in the extraction process, the cores were preserved for standard soil analyses.

Vegetation dynamics on the two experimental sites were studied in permanent quadrates and general surveys (Müller & Rosenthal, 1998). The influence of recultivation on soil ecological parameters could be assessed synchronously by comparison of the SUC and REC plots.

15.3.1.2 Results

Abundance of Gamasina

In the evaluated period from 1980 to 2000, a total of 2,750 samples were taken for microarthropods and 15,077 Gamasina were identified to species.

In most years, the mites (Acari) dominate the microarthropod community. The development of the abundances of mites and springtails (Collembola) is characterized by a very dynamic initial phase within the first 2–3 years, when maximal densities were found (pioneer maximum). In all groups, recultivation accelerates the buildup of initial population densities by approx. 1 year.

In Fig. 15.3, the y-axes are scaled according to the maxima of Gamasina abundances on the two plots, SUC and REC, respectively. The Gamasina abundances are considerably influenced by the recultivation measure, which stimulated abundances with maxima three times as high as on the successional plot SUC. Here, the pioneer maximum of 1982 is only surpassed in fall 1991 (11.2×10^3 ind./m^2), whereas on the recultivated plot REC maximal abundance in 1993 (33.4×10^3 ind./m^2) is up to 1.7 times as high as that of the pioneer maximum, which showed up very ephemerally in 1980.

Fig. 15.3 Abundances of Gamasina from SUC and REC plots from seasonal samplings 1980 to 1993, fall samplings from 1994 to 2001. Data are mean values for 0–12 cm depth, given as tsd. individuals/m^2, *y*1: SUC, *y*2: REC. Scale adjusted to the respective maxima

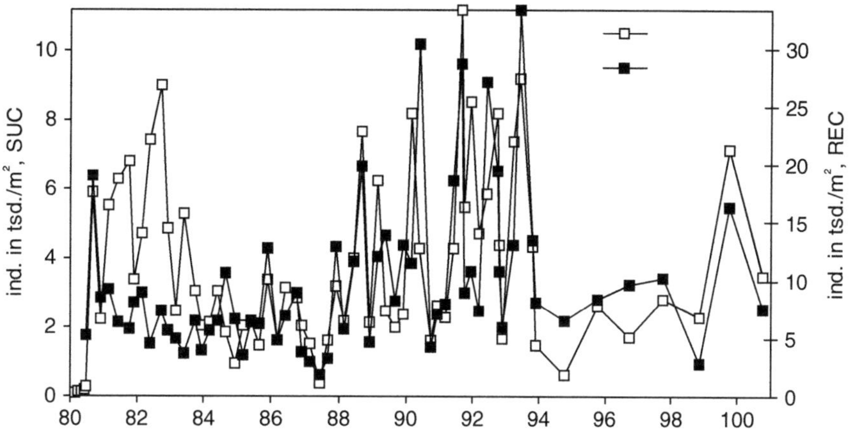

There is no good statistical correlation of the abundances of the two plots. However, this is to a large extent caused by irregular asynchronies of one season only, and the general dynamics after the pioneer phase until 1984 are surprisingly robust against plot differences.

Species Number of Gamasina

Over the years of the investigation, 86 species were identified from the two plots. Species numbers on SUC and REC are almost identical (68 and 65, respectively), with 50 common species (species identity Sörensen-index = 75%).

The assessment of the development of Gamasina diversity depends on data aggregation. When four seasonal campaigns are combined as year-wise means (1980–1993), species numbers are considerably higher than those gained from each of the seasonal campaigns (1994–2000; Fig. 15.4). Although performed with an increased number of sample units (n = 12), the fall samples from 1994 on cannot be taken as a comparable estimate with the aggregated year values, but only with the seasonal samples.

The most conspicuous result is the increase of species numbers irrespective of the dynamics of abundances for almost 20 years (seasonal samples). The interpretation of the last three data points would require evaluation of consecutive samples. When cumulative species numbers (not shown) are analyzed separately for the two sampling regimes, regression analysis reveals from 1980 to 1993 an increase of almost four species per year (linear models; R^2=0.95 [SUC], 0.98 [REC]). After 1993, the respective increase is even more (R^2=0.98 [SUC], 0.69 [REC]).

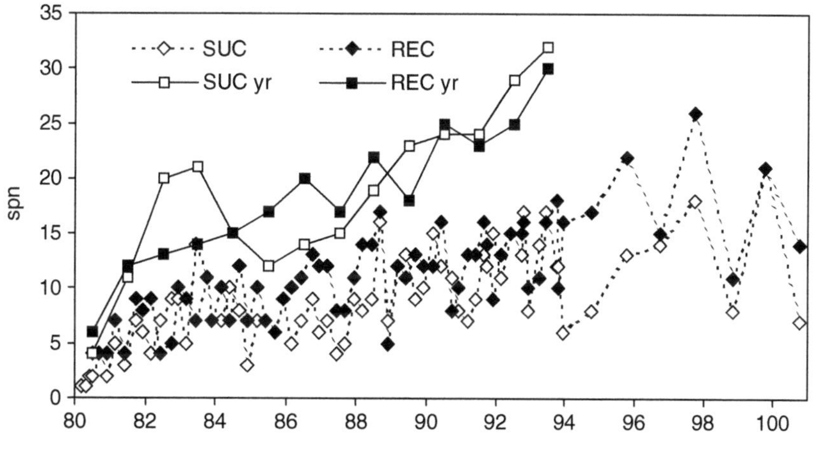

Fig. 15.4 Gamasina species numbers from SUC and REC plots. Higher values (*squares*, indexed yr): aggregation of four campaigns per year, lower values (*diamonds*): species per sample campaign (1980–1993: four per year; from fall 1994 onward: one per year)

Table 15.1 Asynchronous development of soil Gamasina and vascular plant communities. The Gamasina species are restricted to dominats, only species codes are given (for details refer to Koehler & Müller, 2003)

SUC

Gamasina
dominances, aggregated per year, 0-12cm depth

		80	81	82	83	84	85	86	87	88	89	90	91	92	93	94	95	96	97	98	99	00
1	*rsil*	46	40	74	51	18	18	30	39	33	32	22	25	13	27	50	46	23	6	56	61	52
2	*acet*	53	41	16																		
3	*cnec*		6																			
4	*pnor*		7																			
5	*lvag*				11	49	32	32		11	10									6		
6	*hang*					6	20	6	18	6												
7	*pcra*					6	6				7					6						
8	*ehor*							8		8									9			
9	*lbic*							7	20	6			7	12	15	6						
10	*vpla*									11	15	21	24	17	11	6	13	28	30		5	
11	*rrec*									7		10	12	12	6		24	7				
12	*pdod*										8	24	18	24	14			9	10			
13	*lcat*															13						
14	*rcla*															6				15	7	39
	+ 54 more species																					
	species number GA	4	11	20	21	15	12	14	15	19	23	24	24	29	32	8	13	14	18	8	21	7

Vegetation
% cover, aggregated per year

	80	81	82	83	84	85	86	87	88	89	90	91	92	93	94	95	96	97	98	99	00	01	02
cover %	18	93	97	70	97	95	95	100	99	85	97	99	90	99	100	100	100	100	100	99	85	100	100
mean height of growth	10	120	65	40	65	30	60	40	45	40	50	65	40	45	45	45	80	90	75	75	60	90	70
species number	26	17	17	15	16	10	9	8	8	7	9	8	8	7	7	7	9	9	10	9	11	11	11
Sinapis arvensis	5	1																					
Melilotus alba	3	30	15	1	10																		
Melilotus officinalis	3	30	10	1	10	5																	
Ranunculus repens	1	10	7	1	1	1																	
Agrostis stolonifera	3	20	35	20	1	1	1	1	3	1													
Tanac.-Artemis.-Holcus Arten	4	17	37	18	15	7	7	3	7	4													
Dactylis glomerata		1	15	10	7	7	25	30	40	25	20	15	7	15	20	10	10	10	30	25	10	15	15
Poa spp.	1	3	10	25	15	15	25	30	15	10	15	10	10	10	10	10	7	7	5	3	3	3	3
Agropyron repens		1	5	5	25	60	40	50	45	35	45	50	60	80	80	80	65	60	35	30	40	50	50
Urtica dioica			1	1	10	3	1	3	1	3	3	3	3	1	3	5	15	25	20	25	10	25	20
Solidago gigantea		1	3	1	1	1	3	3	5	15	25	40	25	15	15	35	35	40	40	40	10	20	10
Rubus armeniacus																					1	3	10
Cirsium arvense	1	1	1	15	25	10	20	7	3	3	10	5	3	3	3	15	25	15	10	10	10	5	3

increasing dominance of grasses → | high herbs | Rubus shrub

Succession

The comparison of the data sets from vascular plants and Gamasina reveal asynchronic 'stages' within the ecosystem process of succession. The deliminations shown in Table 15.1 are backed by cluster analysis for Gamasina.

The starting point of the secondary succession in 1980 is the grading of a young ruderal site, composed of disturbed soils of unknown origin. The soil mesofauna has suffered from the grading. The Gamasina pioneer community, after grading, is composed of two common species, the small deep dwelling survivor *Rhodacarellus silesiacus* and the phoretic colonizer *Arctoseius cetratus*. In contrast, the plants start from a diverse and vital seed bank, the emergence of which is governed by environmental factors.

In the course of succession, species numbers of Gamasina increase, as documented above (Fig. 15.4, seasonal samples), whereas diversity of vascular plants declines.

Additionally to the Gamasina, data were collected for Oribatids, Collembola, other minor microarthropod groups, and Enchytraeidae (separate samples and extraction). The abundances of the various mesofauna groups correlate only for some groups (no significances). Species numbers for different mesofauna groups, as far as documented, correlate only very poorly. These findings underline the necessity of a complex analysis for a process-oriented assessment of the biological diversity of a site. There is some indication of the influence of hotter and dryer summers on the species composition of Gamasina in the first 10 years of the study, which could not be substantiated in the following decade (Koehler, 1999).

Total C started from 2% DM to reach a plateau around 3.5 % DM within 8 years (1980–1988) on SUC, highly correlated with N (R^2 = 0.95; C/N in the range of 14.5). Because of seeding and grass sward development, values on REC are one unit higher. [1]

15.3.2 Long-Term Effects of Global Warming on Soil Collembola (Mazsalaca, Latvia)

15.3.2.1 Experimental Design and Sampling

Long-term effects of global warming on soil Collembola in the Scots pine forests of the North Vidzeme Biosphere Reserve near the small town of Mazsalaca (Latvia) have been investigated since 1992 (Jucevĭca & Melecis, 2002, 2005, 2006). The study has been carried out in three forest stands of different age, young (30–40 years), middle aged (50–70 years), and old (150–200 years) to partition the eventual effects of natural ecological succession from the effects of global warming. Selected sites were homogenous by relief and vegetation. Forest age structure of the region did not offer forest sites of equidistant age.

The sampling scheme in Mazsalaca was adapted from the Integrated Monitoring Programme (IMP; Anonymous, 1989). The systematic design provides spatially coherent sampling for soil microarthropods, enchytraeids, litter-dwelling arthropods, soil macrofauna, and soil for chemical analyses, being taken on each sampling occasion. A plot of 40 × 50 m was subdivided into 20 subplots of 10 × 10 m each. Four transect lines were defined, crossing five subplots each. These four transects are sampled for soil microarthropods annually with 25 samples each, traversing in total all 20 subplots (10 × 10 m; total n =100). The sample areas (0.5 × 0.5 m) within the transects are approx. 2 m apart. Within these areas, sampling is random. Each year, transects are moved for 1 m. So, repeated use of transects is only possible every 10 years (Fig. 15.5).

Once a year in August/ September, one hundred soil samples (5 cm^2, 0–10 cm depth) were collected from each of the three sites, yielding 300 soil samples

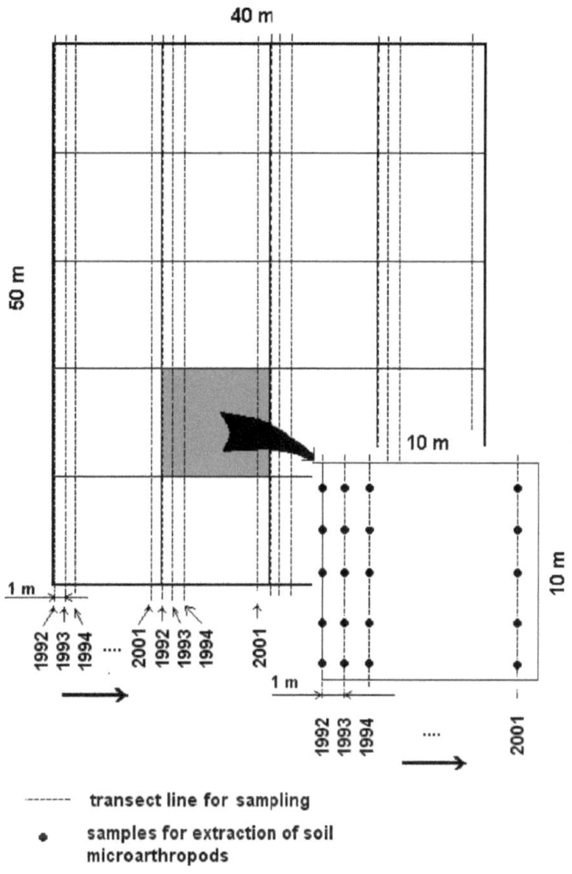

Fig. 15.5 Arrangement of soil sampling plots at Mazsalaca pine forest long-term ecological research site in the North Vidzeme Biosphere reserve, Latvia

in total. The cores were not stratified vertically. Soil samples were transported to the laboratory in special plastic containers. These were inserted top-down in a modified high-gradient extractor for simultaneous extraction of 150 samples (Jucevĭca & Melecis, 2002). So the 300 samples had to be divided into two sets. Each set included 50 random samples from each of the three sites, accounting for 150 samples. Before extraction, all samples were stored in the fridge at +4°C temperature for some days, the second portion for the time until the extraction of the first series was finished. To maintain temperature and moisture gradients within the soil cores, the samples were heated by infrared bulbs with thermo-elements. Efficient extraction was ensured by checking control samples every day. The process was suspended when no more animals emerged from the samples which generally was the case after approx. 2 weeks.

[1] C = soil carbon by incineration, N = total nitrogen content, DM= dry mass.

15.3.2.2 Results

Many results of the study were discussed in detail by Juceviča and Melecis (2006). Here we focus on methodical aspects.

The main objective of the study was to detect temporal trends in total density and species richness of Collembola. The time series were tested for autocorrelations by calculating PACF. As no significant autocorrelations were found, we applied linear regression analysis. The densities of Collembola showed year-to-year fluctuations, but for the old forest site a statistically significant decrease ($P = 0.05$) was observed. Species richness decreased during the period of study irrespective of forest age from 29–36 to 13–26 ($P = 0.05$). This similarity of these trends is interpreted to be caused by some regional level environmental influence and not by ecological succession.

Our environmental data support this interpretation. A statistically significant increase in the sums of positive air temperatures (> +4°C) was recorded during the period of investigation, while precipitation and thereby soil moisture showed considerable year-to-year fluctuations (Juceviča & Melecis, 2006). Linear regression/correlation analysis was performed by SPSS to evaluate eventual effects of positive temperature sums on species richness and density of Collembola. The validity of the method was tested by Durbin–Watson statistics. In all cases, d values ranged from 1.36 to 2.4 and exceeded critical value $d_{\text{upper},\alpha=0.05} = 1.32$ demonstrating no evidence of statistically significant positive or negative autocorrelation. Linear regression and correlation coefficients between species richness and temperature were statistically significant for young ($R = 0.787$, $P = 0.05$) and middle-aged stands ($R = 0.721$, $P = 0.05$), but not for old stands ($R = 0.482$).

For the investigation of trends in changes of community structure, nonmetric multidimensional scaling (NMS) by PC-ORD software was used. The data matrix contained 66 species and 33 sample-plots years. The options selected for NMS were the following: Sörensen's distance measure, number of runs with real data – 100, random starting configuration, number of randomized runs – 100, number of iterations for the final solution – 500. The input data were log transformed. NMS yielded a two-dimensional solution with the final stress value 15.9. Axes were statistically significant and explained 48.6 and 38.6% of the variation. Forest age showed no substantial effect on the community structure. Temporal changes of the community composition demonstrated pronounced trajectories with loops, which were similar for all three sampling sites (Juceviča & Melecis, 2006). So it was possible to calculate average scores and draw an average community trajectory within the two axes (Fig. 15.6). Axis 1 coincided with the trend of sums of positive air temperature ($r = 0.66$) and we interpreted it as temperature axis. Axis 2 was considered reflecting the effect of soil moisture fluctuations (correlation with moisture $r = -0.62$) on Collembola and it was interpreted as moisture axis. Correlation between axes ($r = 0.45$) indicated interaction effects between positive air temperatures and soil moisture.

Calculation of correlations of different collembolan species and life forms with NMS axes showed that euedaphic species were mostly affected by temperature increase, while the hemiedaphic ones

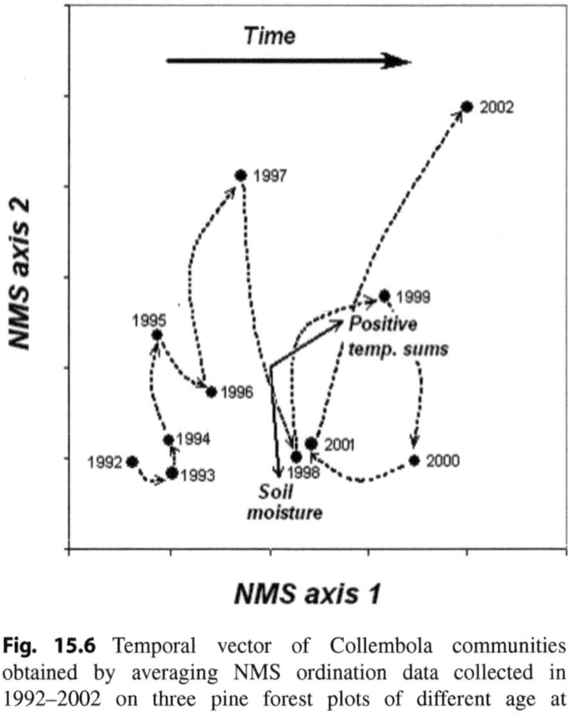

Fig. 15.6 Temporal vector of Collembola communities obtained by averaging NMS ordination data collected in 1992–2002 on three pine forest plots of different age at Mazsalaca long-term ecological research site in the North Vidzeme Biosphere Reserve, Latvia. NMS axis 1 was positively correlated with positive temperature sums ($r = 0.77$) because it was interpreted as temperature axis, NMS axis 2 was negatively correlated with soil moisture ($r = -0.62$) because it was interpreted as soil moisture axis (after Jucevica & Melecis, 2006)

appeared to be most sensitive to changes in moisture conditions (Juceviča & Melecis, 2006; here also details concerning the reaction of particular species can be found). In general, the results of our study were in concordance with results obtained by other investigations concerning the impact of climatic change on Collembola (Frampton et al., 2000; Pflug & Wolters, 2001; Lindberg, 2003; Irmler, 2006).

All these long-term studies explicitly demonstrated that Collembola significantly were affected by changes in positive temperature sums, soil moisture, and interaction of these factors, being sensitive indicators of climate change. It is important to note that the changes of these factors during our 11-year study were sufficient to cause a reduction in species richness. Considering the role of Collembola in northern coniferous forest ecosystems, this should be regarded as an important signal to expect certain changes in soil processes due to global warming. However, as can be seen from the results for Gamasina in the Bremen study, continued observation is needed to validate this interpretation.

15.4 Conclusions and Perspectives

Succession as an ecosystem process (Weidemann & Koehler, 2004) is at most partially synchronized within the system. The asynchronous development of elements of biodiversity documented in the Siedenburg study underlines that an assessment of biological diversity must not be restricted to one indicative group of organisms. A well-founded selection of a sufficient complex set of groups should reflect complexity and ecological importance or role within the ecosystem under study. Plants and soil mesofauna meet these demands largely, but other choices may be equally valid depending on the research question. They are major agents for three of the four attributes for ecosystem persistence proposed in the classical paper of Reichle et al. (1975) cited above: autotrophic energy base, nutrient cycling, and rate control (see also Coleman & Hendrix, 2000).

As summarized in Table 15.2, the two long-term case studies investigate two ecosystems differing in vegetation, soil type, successional stage, and history.

Table 15.2 Comparison of the two cases studies: Siedenburg study and Mazsalaca study

	Siedenburg (Germany)	Mazsalaca (Latvia)
Site	Ruderal, anthropogenic site, northern Germany two subplots: natural succession SUC, recultivation: REC	Scots Pine Forest, Three subplots, planted 30–40, 50–70, 150–200 years ago.
Research question	Succession, recultivation, ecotoxicology	Climate change
Focal taxa	Soil microarthropods (Gamasina), vascular plants	Soil microarthropods (Collembola)
Start	1980	1992
Unit: soil core	300 cm^3; 25 $cm^2 \times 12$ cm (stratified in three 100 cm^3 cylinders: 0–4, 4–8, 8–12 cm);	50 cm^3, 5 $cm^2 \times 10$ cm
Sample size/subplot	1980–1993: $n = 6$; from 1984 on: $n = 12$ Soil volume analyzed/yr: 7.2 L, 3.6 L	$n = 100$ Soil volume analyzed/yr: 5 L
Design	Two permanent plots (100 m^2 each)/subplot	Moving line transect (within permanent quadrat)
Sample frequency:	1980–1993: $m = 4$; from 1984 on: $m = 1$ (fall)	$m = 1$ (fall)
Extraction	Inverted intact cores, humid/dry regime, 5 K/d increment (10 d), 60°C final temp. core surface	Inverted intact cores, Up to 14 d (depending on emergence of animals)
Species determination	Gamasina (Collembola, Oribatei for shorter periods; Enchytraeidae)	Collembola Other selected taxa
Main statistical methods used	Descriptive statistics: arithmetic mean, variance Similarity: Renkonen-index, Sörensen-index Multivariate: DCA (CANOCO)	Descriptive statistics: statistical error calculations by using estimators derived for systematic sampling trend Analysis: linear regression, testing presence of autocorrelations prior to the analysis Multidimensional analysis: NMS
Measures of dominance/diversity:	Simpson, Shannon–Weaver, Evenness	Shannon–Weaver, Evenness
More details in	Koehler (1984, 1993, 1999, 2000); Koehler and Müller (2003)	Jucevica and Melecis (2002, 2005, 2006)

Also, the focus of research is different with some overlap in detail. Both studies consider succession and climate change, but with alternating emphasis: in the Siedenburg study, succession is central and climate change a secondary aspect, in contrast to the Mazsalaca study, where climate is central and succession an important, but secondary aspect. An objective of the Siedenburg study is whether the successional change of the ruderal system is influenced by climate change, whereas the more steady state situation in the Mazsalaca forest may adapt to climate change by allogenic successional change. Both studies investigate more than one site. From this comparative approach, important information is gained at the expense of reduced sample size per site.

The common focus on soil microarthropods generates similar problems that have been solved according to the research questions and the specific habitats. The predatory Gamasina have been chosen for reasons of bioindication, e.g., of successional change and the Collembola for their role in litter decomposition processes and also as bioindicators.

Both sites are protected from unwanted influences such as vandalism or building activities (efficient fencing, long-term contracting). Effects of destructive sampling are avoided by strict planning within strata or by moving line transects. Consequently, the experimental design has to be regular on the scale of approx. 0.25 m^2, but is random within this scale.

Practicability and prevention of too much disturbance due to frequent soil coring are the reasons for the restriction to one sampling campaign per year in the Mazsalaca study, similar to other long-term studies on soil mesofauna reported in the literature (see above). Additional sampling was done in the Siedenburg study in winter, spring, and summer from 1980 to 1993, confirming the findings of reduced information gained from only one sampling campaign. A seasonal or even more frequent sampling should be attempted whenever the working power can be assured over a long period of time. Considering the long time perspective, a design should be sought which is robust against eventual reductions in frequency and sample size due to changes in the supporting working power. Statistical evaluation evolves and assumptions may not always fully withstand critical review over decades, particularly when studying mesofauna with its heterogeneous distribution, risk of autocorrelations, insufficient true replications, and low sample size due to stratification, frequency of sampling, and comparative investigation of different sites.

Practically, only one sampling campaign per year seems to be possible to sustain. The preferred sampling time in both studies has been at the end of the vegetation period which is earlier in Latvia as compared to Bremen. Strictly, most of the results should be interpreted not as year-to-year changes but rather as year-to-year changes in fall.

Both authors have performed extensive methodological testing to optimize the extraction procedures to the needs of their research objects and objectives and came to profoundly different solutions for the apparatus and regimes. Interlaboratory tests would be necessary to prove the sturdiness of the methods. Such tests were started within the ad hoc Working Group of Soil Mesofauna in Germany but never were evaluated due to lack of funds. The approaches of the Siedenburg study focused on efficient sampling and on vertical stratification, those of the Mazsalaca study on statistical needs. The total amount of soil sampled is in the same range of 5 L per campaign in both studies.

The microarthropods were counted and Gamasina and Collembola identified to species level. Permanent slide collections allow lasting verification of all species (Collembola from Mazsalaca) or even all animals identified (Gamasina from Siedenburg). Electronic catalogs enable search and ensure access to the collections. As discussed above, mounting media may limit storage life.

Data analysis was quite similar in both studies. Descriptive statistics (arithmetic means, means for logarithmic data, variance, standard errors, etc.) were calculated and used as input data for certain types of analysis (graphical interpretation, trend analysis, correlation analysis). For analysis of Mazsalaca data, linear trend analysis was used, to make clear whether there were some trends in changes in population numbers and species richness. For Bremen data, however, linear trend analysis was not the appropriate method, because it did not reflect complex changes in species composition during ecological succession, which is rather a nonlinear than a linear process. For this reason, calculation of similarity indices between successive stages was the much more appropriate method. For both data, ordination techniques were used in order to discover the hidden factors responsible for changes in community structure. NMS of Mazsalaca data allowed not only to discover such factors but also to draw a temporal trajectory of community changes within the axes representing those factors.

Ecosystem development does not occur in months but rather in time units covering several years. From our studies it can be concluded that such a time unit extends over approx. 5 years in the middle to northern European region. Consequently, only long-term approaches can provide an understanding of successional development as an ecosystem process and to the related changes due to climate change (allogenic succession).

Both studies show that for an understanding of temporal changes in ecosystems the investigation of soil mesofauna is mandatory, both from an indicator and biodiversity point of view and from a functional perspective, and it has to be performed over the long term. Because of its role as promoter of ecosystem services (e.g., soil fertility, carbon sequestration, water infiltration), soil mesofauna should be considered in the assessment of risk associated with climate change and intensive use of the soil resource.

Acknowldgments Siedenburg Study: The first author is greatly indebted to Prof. G. Weidemann, the initiator and fervent advocate of long-term ecosystem-oriented research and to Dr. J. Müller, long-term vegetation researcher and friend. Many dedicated students contributed to our work and the technicians Ute Uebers and Annemarie Kissling strongly supported our investigations. The University of Bremen and the German Ministry of Education and Research (BMBF) have supported the studies temporarily. The evaluation of 20 years of data was supported within the BIOLOG framework (BMBF; 01LC0005 2000–2002).

The Mazsalaca study was supported by several successive grants from the Council of Science of Latvia (CSL). The CSL supported the development of the national long-term research network. We are grateful to the Environmental Office of US Embassy in Copenhagen and US National Science Foundation (NSF) for providing initial financial support for the implementation of the Latvia LTER in 2003–2004. We appreciate also further support coming from US LTER, NSF, European Environmental Agency as well as financial assistance from ALTER-net project making possible our participation in ILTER and European LTER meetings.

Many thanks to Dr. Klaus Birkhofer, University of Giessen, and an anonymous reviewer for their helpful comments on a previous manuscript.

References

Allen, T. F. H., & Koonce, J. F. (1973). Multivariate approaches to algal stratagems and tactics in systems analysis of phytoplankton. *Ecology, 54*, 1234–1246.

Anderson, T. W. (1963). The use of factor analysis in the statistical analysis of multiple time series. *Psychometrica, 28*, 1–25.

Anderson, J. M. (1975). The enigma of soil animal species diversity. In J. Vanek (Ed.), *Progress in soil zoology* (pp. 51–58). The Hague: Junk.

André, H. M., Ducarme, X., Anderson, J., Crossley, D., Jr., Koehler, H., Paoletti, M., et al. (2001). Skilled eyes are needed to go on studying the richness for the soil. *Nature, 409*, 761.

Anonymous (1989). *Field and Laboratory manual. International cooperative programme on integrated monitoring.* Helsinki: National Board of Waters and Environment.

Araujo, M. B., & Rahbek, C. (2006). How does climate change affect biodiversity? *Science, 313*, 1396–1397.

Belsley, D. A., Kuh, E., & Welsch, R. E. (1980). *Regression diagnostics*. New York: Wiley.

Bence, J. R. (1995). Analysis of short time series: Correcting for autocorrelation. *Ecology, 76*, 628–639.

Berryman, A., & Turchin, P. (2001). Identifying the density-dependent structure underlying ecological time series. *Oikos, 92*, 265–270.

Box, G., & Jenkins, G. (1970). *Time series analysis: Forecasting and control.* San Francisco: Holden-Day.

Briones, M. J. I., Ineson, P., & Piearce, T. G. (1997). Effects of climate change on soil fauna; responses of Enchytraeids, Diptera larvae and tardigrades in a transplant experiment. *Applied Soil Ecology, 6*, 117–134.

Bruckner, A. (1998). Augers may bias field samples of soil mesofauna. *Pedobiologia, 42*, 309–315.

Bruckner, G., Barth, G., & Scheibengraf, M. (2000). Composite sampling enhances the confidence of soil microarthropod abundance and species richness estimates. *Pedobiologia, 44*, 63–74.

Catell, R. B. (1951). P-techniques, a new method for analysing the structure of personal motivation. *Transactions of the New York Academy of Sciences Psychological Series, 14*, 29–34.

Christian, A. (1995). Succession of Gamasina in coal mined areas in Eastern Germany. *Acta Zoologica Fennica, 196*, 380–381.

Clements, F. E. (1916). Plant succession: An analysis of the development of vegetation. *Carnegie Institution* of *Washington publication, 242*, 1–51.

Coleman, D. C., & Hendrix, P. F. (2000). *Invertebrates as webmasters in ecosystems.* New York: CAB International.

Collins, S. L., Micheli, F., & Hartt, L. (2000). A method to determine rates and patterns of variability in ecological communities. *Oikos, 91*, 285–293.

Connell, J. H., & Slatyer, R. O. (1977). Mechanisms of succession in natural communities and their role in community stability and organization. *American Naturalist, 111*, 1119–1144.

Dunger, W. (1975). On the delimination of soil microarthropod cenosis in time and space. In J. Vanek (Ed.), *Progress in soil zoology* (pp. 43–49). The Hague: Junk.

Dunger, W., & Fiedler, H. J. (1997). *Methoden der Bodenbiologie*. Jena: Gustav Fischer.

Dunger, W., Wanner, M., Hauser, H., Hohberg, K., Schulz, H. J., Schwalbe, T., et al. (2001). Development of soil fauna at mine sites during 46 years after afforestation. *Pedobiologia, 45*, 243–271.

Edwards, C. A. (1991). The assessment of populations of soil-inhabiting invertebrates. *Agriculture, Ecosystem & Environment, 34*, 145–176.

Edwards, C. A., & Fletcher, K. E. (1971). A comparison of extraction methods for terrestrial arthropods. In J. Phillipson (Ed.), *Methods of study in quantitative soil ecology: Population, production, energy flow* (pp. 150–185). Oxford: Blackwell Scientific.

Ekschmitt, K. (1998). Population assessment of soil fauna: General criteria for planning of sampling schemes. *Applied Soil Ecology, 9*, 439–445.

Filser, J., Koehler, K., Ruf, A., Römbke, J., Prinzing, A., & Schaefer, M. (2008). Ecological theory meets soil ecotoxicology: A challenge and a chance. *Basic and Applied Ecology, 9*, 346–355.

Fleischer, A., & Sternberg, M. (2006). The economic impact of global climate change on Mediterranean rangeland ecosystems: A Space-for-Time approach. *Ecological Economics, 59*, 287–295.

Frampton, G. K., van den Brink, P. J., & Gould, P. J. L. (2000). Effects of spring drought and irrigation on farmland arthropods in southern Britain. *Journal of Applied Ecology, 37*, 865–883.

Gotelli, N. J., & Ellison, A. M. (2004). *A primer of ecological statistics*. Sunderland, MA: Sinauer Associates Inc. Publishers.

Greig-Smith, P. (1964). *Quantitative plant ecology*. London: Butterworths.

Gurevitch, J., & Chester, S. T., Jr. (1986). Analysis of repeated measures experiments. *Ecology, 67*, 251–255.

Hairston, N. G., Sr. (1989). *Ecological experiments: Purpose, design, and execution*. Cambridge: Cambridge University Press.

Hampe, A., & Petit, R. J. (2005). Conserving biodiversity under climate change: The rear edge matters. *Ecology Letters, 8*, 461–467.

Hogervorst, R. F., Verhoef, H. A., & VanStraalen, N. M. (1993). Five-year trends in soil arthropod densities in pine forests with various levels of vitality. *Biology and Fertility of Soils, 15*, 189–195.

Huhta, V. (1979). Evaluation of different similarity indices as measures of succession in arthropod communities of the forest floor after clear-cutting. *Oecologia (Berlin), 41*, 11–23.

Irmler, U. (2006). Climatic and litter fall effects on collembolan and oribatid mite species and communities in a beech wood based on a 7 years investigation. *European Journal of Soil Biology, 42*, 51–62.

Johnston, A. E., & Powlson, D. S. (1994). The setting-up, conduct and applicability of long-term, continuing field experiments in agricultural research. In D. J. Greenland, & I. Szabolcs (Eds.), *Soil resilience and sustainable land use* (pp. 395–421). Wallingford: CAB International.

Jongman, R. H., TerBraak, C. J. F., & VanTongren, O. F. R. (1987). *Data analysis in community and landscape ecology*. Wageningen: Pudoc.

Juceviča, E., & Melecis, V. (2002). Long-term dynamics of Collembola in a pine forest ecosystem. *Pedobiologia, 46*, 365–372.

Juceviča, E., & Melecis, V. (2005). Long-tern effects of climate warming on forest soil Collembola. *Acta Zoologica Lithuanica, 15*(2), 125–127.

Juceviča, E., & Melecis, V. (2006). Global warming affect Collembola community: A long-term study. *Pedobiologia, 50*, 2, 177–184.

Kampichler, C., & Geissen, V. (2005). Temporal predictability of soil microarthropod communities in temperate forests. *Pedobiologia, 49*, 41–50.

Karg, W., & Freier, B. (1995). Parasitiforme Milben als Indikatoren für den ökologischen Zustand von Ökosystemen. *Mitteilungen aus der Biologischen Bundesanstalt für Land- und Forstwirtschaft Berlin-Dahlem, 308*, 1–96.

Kaufmann, R. (2002). Glacier foreland colonisation: Distinguishing between short-term and long-term effects of climate change. *Oecologia, 130*, 470–475.

Kennedy, A. D. (1994). Simulated climate change: A field manipulation study of polar microarthropod community response to global warming. *Ecography, 17*, 131–140.

Koehler, H. (1984). Methodische, ökologische und experimentelle Untersuchungen zur Sukzession der Mesofauna der Abdeckschicht einer Bauschuttdeponie unter besonderer Berücksichtigung der Gamasina (Acari, Parasitiformes). Dissertation, University of Bremen.

Koehler, H. (1993). Extraktionsmethoden für Bodenmesofauna. *Informationen zu Naturschutz und Landschaftspflege in Nordwestdeutschland, 6*, 42–52.

Koehler, H. (1996). Soil animals and bioindication. In N. M. VanStraalen, & D. A. Krivolutsky (Eds.), *Bioindicator systems for soil pollution* (pp. 179–188). Dordrecht: Kluwer Academic.

Koehler, H. (1999). Gamasina in a succession of thirteen years. In J. Bruin, L. P. S. VanDerGeest, & M. Sabelis (Eds.), *Ecology and evolution of the Acari* (pp. 531–539). Dordrecht: Kluwer Academic.

Koehler, H. (2000). Natural regeneration and succession: Results from a 13 yrs study with reference to mesofauna and vegetation and implications for management. *Landscape and Urban Planning, 51*, 123–130.

Koehler, H., & Müller, J. (2003). *Entwicklung der Biodiversität während einer 20 jährigen Sukzession als Grundlage für Managementmaßnahmen. Report: BMBF, FKZ 01 LC 0005, 1–241*. Bremen: Universität Bremen, Arbeitsgruppe "Ökosystemforschung und Bodenökologie".

Lasebikan, B. A., Belfield, W., & Gibson, N. H. E. (1978). Comparison of relative efficiency of methods for the extraction of soil microarthropods. *Revue d'Ecologie et de Biologie du Sol, 15*, 39–65.

Legendre, P., & Legendre, L. (1998). *Numerical ecology*. Amsterdam: Elsevier.

Likens, G. E. (Ed.). (1988). *Long-term studies in ecology*. New York: Springer.

Lindberg, N. (2003). Soil Fauna and Global Change – Responses to Experimental Drought, Irrigation, Fertilisation and Soil Warming. Dissertation, Swedish University of Agricultural Sciences.

Lindberg, N., & Bengtsson, J. (2006). Recovery of forest soil fauna diversity and composition after repeated summer droughts. *Oikos*, *114*, 494–506.

Lindsey, J. K. (1993). *Models for repeated measurements*. Oxford: Oxford University Press.

MacFadyen, A. (1961). Improved funnel-type extractors for soil arthropods. *Journal of Animal Ecology*, *30*, 171–184.

McCune, B., & Grace, J. B. (2002). *Analysis of ecological communities*. Gleneden Beach, Oregon: MjM Software Design.

McCune, B. (2006). *Nonparametric Multiplicative Regression for Habitat Modelling*. Retrieved October 1, 2008, from http://www.pcord.com/NPMRintro.pdf.

Melecis, V. (1988). System approach in studies of soil Collembola as indicators of industrial pollution. Dissertation, Riga University of Latvia.

Melecis, V. (1999). Probleme des Boden-Biomonitorings. In H. Koehler, K. Mathes, & B. Breckling (Eds.), *Bodenökologie interdisziplinär* (pp. 133–147). Berlin: Springer.

Müller, J., & Rosenthal, G. (1998). Brachesukzessionen – Prozesse und Mechanismen. *Berichte des Institutes für Landschafts-Pflanzenökologie der Universität Hohenheim*, *5*, 103–132.

Musaloiu-E, R., Terzis, A., Szalay, A., Cogan, J., & Gray, J. (2006). Life Under your Feet: A Wireless Soil Ecology Sensor Network. In *Proceedings of 3rd Workshop on Embedded Network Sensors (EmNets 2006), May 30–31, 2006* (pp. 51–55). Cambridge, MA: Harvard University.

Norusis, M. (2004). *SPSS 13.0 statistical procedures companion*. Upper Saddle-River, NJ: Prentice Hall.

Ozer, S., Szalay, A., Szlavecz, K., Terzis, A., Musaloiu-E, R., & Cogan, J. (2006). *Using Data-Cubes in Science: An Example from Environmental Monitoring of the Soil Ecosystem. MSR-TR-2006–134*. Retrieved October 1, 2008, from ftp://ftp.research.microsoft.com/pub/tr/TR-2005–134.pdf.

Pawley, M. D. M. (2006). Systematic sampling in ecology. Dissertation, University of Auckland.

Pflug, A., & Wolters, V. (2001). Influence of drought and litter age on Collembola communities. *European Journal of Soil Biology*, *37*, 305–308.

Platt, T., & Denman, K. L. (1975). Spectral analysis in ecology. *Annual Review of Ecology and Systematics*, *6*, 189–210.

Pollard, E., Lakhani, K. H., & Rothery, P. (1978). The detection of density-dependence from a series of annual censuses. *Ecology*, *68*, 2046–2055.

Ponge, I. F. (1973). Application de l'analyse factorielle des correspondences a l'étude des variations annelles les populations de Microarthropodes. *Bulletin D'écologie*, *4*, 319–327.

Poveda, K., Stefan-Dewenter, I., Scheu, S., & Tscharntke, T. (2003). Effects of below- and above-ground herbivores on plant growth, flower visitation and seed set. *Oecologia*, *135*, 601–605.

Ratti, J. T., & Garton, E. O. (1994). Research and experimental design. In T. A. Bookout (Ed.), *Research and management techniques for wildlife and habitat* (pp. 1–23). Bethesda, MD: The Wildlife Society.

Reichle, D. E., O'Neill, R. V., & Harris, W. F. (1975). Principles of energy and material exchange in ecosystems. In W. H. VanDobben, & R. H. Lowe-McConnell (Eds.), *Unifying concepts in ecology* (pp. 27–43). The Hague: Dr. W. Junk Publishers.

Rusek, J. (1993). Air-pollution-mediated changes in alpine ecosystems and ecotones. *Ecological Applications*, *3*, 409–416.

SAS Institute Inc. (2004). *SAS/STAT® 9.1 User's Guide*. Cary, NC: SAS Institute Inc.

Scheffe, H. (1959). *The analysis of variance*. New York: John Wiley and Sons.

Schinner, F., Öhlinger, R., & Kandeler, E. (1993). *Bodenbiologische Arbeitsmethoden*. Berlin, Heidelberg: Springer.

Schroeter, D., Cramer, W., Leemans, R., Prentice, I. C., Araujo, M. B., Arnell, N. W., et al. (2005). Ecosystem service supply and vulnerability to global change in Europe. *Science*, *310*, 1333–1337.

Snedecor , G. W., & Cochran, W. G. (1980). *Statistical methods*. Ames: The Iowa State University Press.

Sokal, R. R., & Rohlf, F. J. (1987). *Introduction to biostatistics*. New Yorkf: W.H. Freeman & Co.

Takeda, H. (1987). Dynamics and maintenance of collembolan communities structure in a forest soil system. *Researches on Population Ecology*, *29*, 291–346.

Thomas, C. D., Cameron, A., Green, R. E., Bakkenes, M., Beaumont, L. J., Collingham, Y. C., et al. (2004). Extinction risk from climate change. *Nature*, *427*, 145–148.

Thuiller, W., Lavorel, S., Araujo, M. B., Sykes, M. T., & Prentice, I. C. (2005). Climate change threats to plant diversity in Europe. *Proceedings of the National Academy of Sciences of the United States of America*, *102*, 8245–8250.

Tilman, D. (1989). Ecological experimentation: Strengths and conceptual problems. In G. E. Likens (Ed.), *Long-term studies in ecology: Approaches and alternatives* (pp. 136–157). Berlin: Springer-Verlag.

Überla, K. (1977). *Faktorenanalyse*. Berlin: Springer Verlag.

Usher, M. B. (1969). The efficiency of counting samples of soil micro-arthropods. In G. O. Evans (Ed.), *Proceedings of the 2nd International Congress of Acarology, Sutton Bonington, England, 19–25 July, 1967* (pp. 57–60). Budapest: Akademiai Kiado.

Vannier, G. (1970). *Réactions des microarthropodes aux variations de l'état hydrique du sol: Techniques relatives à l'extraction des arthropodes du sol*. Paris: CNRS.

Van Straalen, N. M., Donker, M. H., Vijver, M. G., & VanGestel, C. A. M. (2005). Bioavailability of contaminants estimated from uptake rates into soil invertebrates. *Environmental Pollution*, *136*, 409–417.

Wardle, D. A. (2002). *Communities and Ecosystems: Linking the aboveground and belowground components*. Princeton: Princeton University Press.

Weidemann, G. (1985). Rekultivierung als ökologisches Problem: 1. Konzept und Probeflächen. *Verhandlungen der Gesellschaft für Ökologie*, *13*, 751–758.

Weidemann, G., & Koehler, H. (2004). Sukzession. In O. Fränzle, F. Müller, & W. Schröder (Eds.), *Handbuch der Umweltwissenschaften, 12. ergänzte Lieferung 6/04* (pp. 3–49). Landsberg: EoMed.

Weidemann, G., Koehler, H., & Schriefer, T. (1982). Recultivation: A problem of stabilization during ecosystem development. In R. Bornkamm, J. A. Lee, & M. R. D. Seaward (Eds.), *Urban ecology* (pp. 305–313). Oxford: Blackwell.

Wiwatwitaya, D., & Takeda, H. (2004). Seasonal changes in soil arthropod abundance in the dry evergreen forest of north-east Thailand, with special reference to collembolan communities. *Ecological Research*, *20*, 59–70.

Wolters, V. (1998). Long-term dynamics of collembolan community. *Applied Soil Ecology*, *9*, 221–227.

Yang, K., & Shahabi, C. (2005). *On the stationarity of multivariate time series for correlation-based data analysis*. Technical report, University of Southern California.

Chapter 16

Tracing Biogeochemical Processes in Small Catchments Using Non-linear Methods

Gunnar Lischeid, Pavel Krám, and Christina Weyer

Abstract Since the 1980s a variety of biogeochemical catchment studies have been set up to investigate the cycling of water and solutes. Groundwater and streams have been sampled to investigate the dominant processes of solute turnover in the subsoil and to monitor their long-term changes. Usually a variety of processes interact partly in a highly non-linear manner. Consequently, identifying the dominant processes is not an easy task. In this study, a non-linear variant of the principal component analysis was used to identify the dominant processes in groundwater and streamwater of two forested catchments in the East Bavarian–West Bohemian crystalline basement. The catchments are approximately 60 km apart, but exhibit similar bedrock, soils, climate, land use, and atmospheric deposition history. Both have been monitored since the end of the 1980s until today, that is, during a period of dramatic decrease of atmospheric deposition of sulfur and accompanying base cations. Time series of component scores at different sites were investigated. Non-linear long-term trends were determined using a low-pass filter based on a Lomb–Scargle spectrum analysis.

The first four components accounted for 94% of the variance of the data set. The component scores could be interpreted as quantitative measures of biogeochemical processes. Among these, redox processes played a dominant role even in apparently oxic parts of the aquifers. Low-pass filtered time series of the component scores showed consistent, although mostly, non-linear trends in both catchments.

Keywords Groundwater quality · Streamwater quality · Isometric feature mapping · Non-linearity · Time series analysis · Trend analysis

16.1 Introduction

Water resources management requires sound knowledge of the dominant hydrological and biogeochemical processes that affect water availability and water quality. Based on a multitude of process studies both in the laboratory and in the field, it is known that often many processes interact in a non-linear way and via feedback loops that make the determination of dominant processes at large scales difficult. Natural sources of spatial and temporal variability like heterogeneity of soil properties or interannual climatic variability have to be differentiated from anthropogenic impacts and effects of long-term climate change. In addition, small-scale effects have to be distinguished from more general patterns that are typical for large regions.

Environmental monitoring aims at assessing spatial or temporal patterns of single parameters, like concentration of solutes or abundance of species. In many cases of natural resources management, however, data about observed parameters are used to infer what and to what degree different processes had an impact on the system in order to evaluate single measures and to decide about further management strategies. To that end, environmental monitoring often focuses on single indicator parameters that are assumed to be very closely related to single processes. That approach

G. Lischeid (✉)
Institute of Landscape Hydrology, Leibniz Centre for Agricultutral Landscape Research, Eberswalder Straße 84, 15374 Müncheberg, Germany
e-mail: lischeid@zalf.de

F. Müller et al. (eds.), *Long-Term Ecological Research*,
DOI 10.1007/978-90-481-8782-9_16,

might work fine in some cases, but is clearly limited to certain boundary conditions.

In this study a different approach is followed for analyzing long-term water quality data. First, a joint analysis of groundwater and surface water is performed. Although a couple of processes are restricted to either groundwater or surface water systems, it is assumed that the most dominating processes are the same. However, it is taken into account that the extent to what these processes affect solute concentration differs between the two systems as well as they vary in space and time. Second, it is assumed that every single process alters solute concentration in a typical way. That can be used to determine these processes by a variant of a principal component analysis. If principal components can be ascribed to single biogeochemical processes, the scores of the components can be interpreted as quantitative measures of these processes. Then, time series of the component scores can be analyzed.

Third, relationships between different solutes in groundwater and streamwater systems are often non-linear. This fact is accounted for by using a non-linear version of the principal component analysis. Fourth, the long-term development of natural processes is very unlikely to follow a linear trend. On the other hand, the short-term variability of natural processes requires filtering the long-term components. In this study, a low-pass filter is applied based on a spectral analysis of the time series of components.

To summarize, the objective of this study was to identify the prevailing biogeochemical processes that affect groundwater and streamwater quality in two similar forested catchments at numerous sites, to investigate spatial patterns of processes, and to determine the long-term behavior with respect to these processes.

16.2 Data

The data set comprises groundwater and streamwater solute concentration data from the sub-mountainous region of the crystalline basement along the German-Czech border. Here, extensive biogeochemical monitoring programs have been run since the end of the 1980s in two catchments that are assumed to represent typical conditions in this region. In both catchments the bedrock consists of granite, overlain by dystric cambisols and podzols. Fibric histosols and dystric gleysols predominate in the extended riparian zones. A dense network of natural streams and artificial channels exists in both catchments. Land use is forestry exclusively consisting of Norway spruce (*Picea abies* (L.) Karst.) stands.

The catchments are located at about 60 km distance. The Lysina catchment is part of the Slavkovský les in the Czech Republic, at 50°03′N and 12°40′E. The Lehstenbach catchment belongs to the Fichtelgebirge region in Germany and is located at 50°08′N and 11°52′E. The catchment area is 0.273 km^2 at Lysina and 4.2 km^2 at Lehstenbach. Elevation ranges between 829 and 949 m a.s.l. at Lysina and between 695 and 877 m a.s.l. at Lehstenbach. The thickness of the regolith is about 2.5 m in the Lysina catchment and up to 40 m in the Lehstenbach catchment.

Annual mean precipitation of the 1991–2001 period was 933 mm at Lysina and 985 mm at Lehstenbach. Annual mean runoff was 432 and 470 mm, respectively. Annual mean air temperature is about 5.2°C at Lysina and 5–6.5°C in the Lehstenbach catchment, depending on altitude. Snowpack usually develops in December or January and final snowmelt occurs in March or April.

An extensive monitoring program has been performed in both catchments since the end of the 1980s (Krám, Hruška, Wenner, Driscoll, & Johnson, 1997; Matzner, 2004). Both sites have been severely impacted by nitrogen and sulfate deposition. Sulfate deposition peaked in the 1970s and decreased by more than 80% since then in both catchments (Hruška, Moldan, & Krám, 2002; Matzner, Zuber, Alewell, Lischeid, & Moritz, 2004; EMEP, 2007). In contrast, neither nitrogen deposition nor nitrogen concentration in the streams exhibited any clear trend at Lehstenbach. At Lysina, nitrogen deposition during 1998–2000 was only 73% of that in 1991–1993 (Hruška et al., 2002), and the nitrate peaks in the winter season started to decrease in the end of the 1990s. Besides, there was no change of land use or extended clearfelling. However, the upslope parts of the Lehstenbach catchment have been limed several times.

In total, the data set used for this analysis comprises 13 parameters of more than 4000 samples. Samples were taken from 1987 throughout 2007 at 27 different sites in the two catchments (Table 16.1). The location of the sampling sites is given in Fig. 16.1. The parameters comprise Al, Ca, Cl^-, DOC, Fe, K, Mg, Mn,

Table 16.1 Overview over the data sets used in this study

Component	Catchment	Number of sites	Period	Number of samples
Shallow groundwater	Lehstenbach	6	2001–2004	212
(0.9–4.5 m)	Lysina	7	2000	82
Deep groundwater	Lehstenbach	6	1988–2007	606
(1–24 m)	Lysina	–	–	–
Upslope springs	Lehstenbach	1	1988–2006	403
	Lysina	1	1989–2007	214
Streams	Lehstenbach	4	1987–2007	1157
	Lysina	2	1989–2007	1382
In total		**27**	**1987–2007**	**4056**

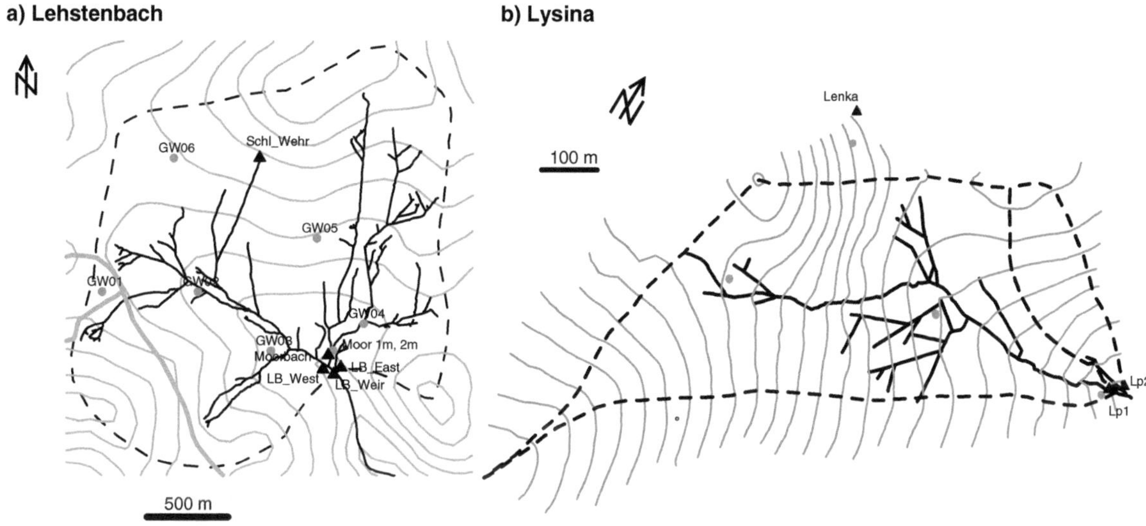

Fig. 16.1 Map of the Lysina and Lehstenbach catchments. *Solid black lines* denote streams, *gray lines* contour lines at 20 m (Lehstenbach) or 10 m interval (Lysina), *dashed black lines* the catchment boundaries, and *bold gray lines* public roads. Groundwater wells are given by *gray dots* and spring and streamwater sampling sites by *black triangles*

Na, NO_3^-, Si, and SO_4^{2-} and pH values. Streamwater samples had been taken as grab samples. Groundwater samples had been taken by a submersed pump and by a vacuum pump in the Lysina groundwater wells. Part of the Lehstenbach data was kindly provided by the Bavarian Environmental Agency which is highly appreciated. Quality assurance of the data was performed prior to this study by the Czech Geological Survey (Lysina data), the BayCEER central laboratory, and the Bavarian Environmental Agency (Lehstenbach data).

If any long-term trends were to be found in the water quality data, it should be tested whether they were related to trends of meteorological parameters. To that end, time series of daily precipitation and air temperature data were included into the analysis. At Lehstenbach precipitation and air temperature have been measured at an open field site in the catchment. For Lysina, precipitation data from Kladská (at 1 km distance) and Lazy (at 5 km distance) and temperature data from Mariánské Lázně (at 8 km distance) were available. In addition, data from the Braunersgrün meteorological station of the Bavarian Agency for Agriculture were used. Braunersgrün is located between the two catchments, 18 km east of Lehstenbach and 44 km west of Lysina, although at lower altitude (590 m a.s.l.). The data were downloaded at http://www.lfl.bayern.de. In this analysis, long-term shifts of meteorological parameters were investigated rather than absolute values. Thus, neither precipitation nor temperature data were corrected to account for different altitude or exposition.

16.3 Methods

Values less than the limit of quantization were replaced by 0.5 times that limit. Missing values were replaced by the mean of the respective parameter. The data were normalized to zero mean and unit variance for each parameter separately. All statistical analyses were performed using the R statistical software environment (R Development Core Team, 2006).

16.3.1 Principal Component Analysis

The basic idea of principal component analysis is that due to correlations between different parameters in a multivariate data set, a large fraction of the variance of the total data set can be represented by a small number of principal components. There are different ways how to determine these low-dimensional components.

16.3.1.1 Linear Principal Component Analysis

The linear principal component analysis (PCA) is based on an eigenvalue decomposition of the covariance matrix. In the case of z-normalized data, the covariance is equal to the Pearson correlation. The eigenvectors form the independent axes. These axes are usually interpreted based on their correlations with single parameters of the data set, i.e., the 'loadings'. The eigenvalues of the eigenvectors are proportional to the fraction of variance that is explained by the single axes or components and can be used to assess the importance of single 'processes.' Usually, only the axes with the highest eigenvalues are interpreted, e.g., components with an eigenvalue exceeding one.

Gaussian distribution of the data is a crucial prerequisite of the principal component analysis when the data set is rather small. In this study, however, the data set comprises some thousand data points where effects of skewed distributions are negligible. Furthermore, the Isomap approach does not require any specific distribution of the data. Thus, for the sake of comparability, the data set was not adjusted to a Gaussian distribution prior to the principal component analysis.

16.3.1.2 Isometric Feature Mapping (Isomap)

Principal component analysis is equal to the classical multidimensional scaling (CMDS) if Euclidean distances are used (Gámez, Zhou, Timmermann, & Kurths, 2004). The CMDS aims at preserving the interpoint distances of the high-dimensional data set while projecting the data into a low dimensional space. The solution is found by an eigenvalue decomposition of the distance matrix. Usually, only the first eigenvectors (sorted by decreasing eigenvalues) are considered for the subsequent analysis.

The isometric feature mapping (Isomap) method is a variant of the CMDS approach which differs with respect to the determination of the distance matrix (Tenenbaum, DeSilva, & Langford, 2000). In the first step, for every data point the k nearest neighbors are determined. Here, Euclidian distances were used. In the next step, the full distance matrix is set up, where distances are calculated as the smallest sum of the interpoint distances determined in the first step. These geodesic distances correspond to those of a road system where every town is directly linked only to the k nearest towns. These geodesic distances are equal to or larger than the corresponding direct distances used by planes. The approach aims at mapping non-linear manifolds by using a locally linear approach. The less linear the manifold in the data set, the more likely is the Isomap approach to outperform linear methods which has in fact been shown in some studies (Tenenbaum et al., 2000; Gámez et al., 2004; Mahecha, Martínez, Lischeid, & Beck, 2007). The impact of k on the Isomap performance was investigated by systematical variation of this parameter. In this study, the best performance was found for $k = 2000$.

The Isomap analysis gives components that can be interpreted correspondingly to the linear principal components. However, as they are not necessarily linear, it is not sufficient to use the Pearson correlation coefficient between single components and parameters. Instead, scatter plots were used in this study. In this study, the package 'vegan' by Oksanen, Kindt,

Legendre, O'Hara, and Stevens (2007) has been used for the Isomap analysis.

16.3.1.3 Measure of Performance

There are different measures of performance for the ordination methods. For the principal component analysis that is based on an eigenvalue decomposition of the covariance matrix the fraction of explained variance is proportional to the eigenvalues. However, that approach cannot be followed for the Isomap approach which is based on an eigenvalue decomposition of the distance matrix of the data. Here a more general approach is used for evaluating the performance of ordination methods. It is based on the Pearson correlation between the interpoint distances in the high-dimensional data space and in the low-dimensional projection. Both measures are related, and both give a value of one for a perfect fit, and close to zero for a very poor fit. However, they do not match exactly and thus cannot directly be compared.

16.3.2 Low-Pass Filtering of Time Series

One focus of this study was on investigating long-term shifts of the effect of different processes. Linear trend analysis was not considered to be an adequate approach for environmental data. Due to substantial short-term variability, the component scores had to be smoothed. A common way to do that is to calculate mean values of distinct or sliding windows of a certain length. Often annual mean values are selected as many environmental data exhibit a pronounced seasonal pattern. This approach was not used here as the sampling dates were not always equally spread over the year. Besides, considering only one mean value per year implies a considerable loss of information compared to weekly or monthly sampled data.

Instead, a low-pass filter was applied to the time series of component scores at single sites. The approach is based on a spectrum analysis. The samples had not always been taken at regular intervals. Thus, a Fourier analysis could not be performed. Instead, the approach by Lomb–Scargle was used. The approach and the package 'Lomb–Scargle' that was used for this study is described by Glynn, Chen, and Mushegian (2006). The package is available at http://research.stowers-institute.org/efg/2005/LombScargle/R/index.htm. Based on the results of the spectrum analysis, the low-pass filtering was performed by reconstructing the data using period lengths larger than a year only.

16.4 Results

16.4.1 Explained Variance

Figure 16.2 gives the Pearson correlation between the interpoint distances in the 13-dimensional data space and in the low-dimensional projection. The components are ordered according to the fraction of variance they explain. About 94% of the variance is represented by the first four Isomap components. These components are slightly more effective with respect to representing the variance of the data compared to the linear principal component approach, whereas there are only minor differences for the fifth and sixth components. In the following, only the first four components will be considered.

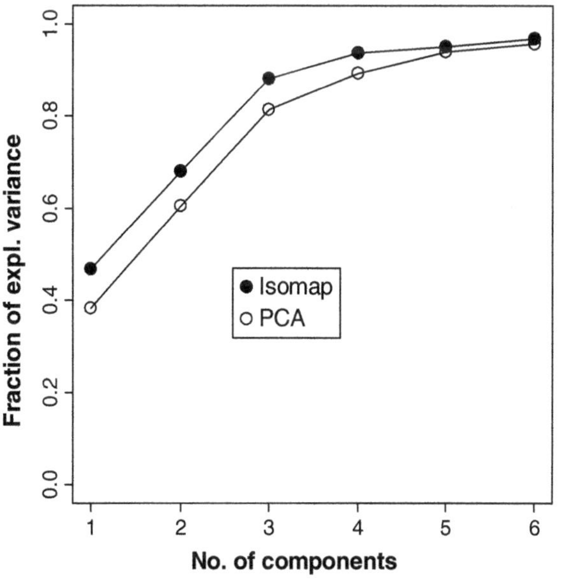

Fig. 16.2 Variance explained by the first six principal components and Isomap components

16.4.2 Identification of Processes

The next step aims at ascribing these components to biogeochemical processes. To that end, the relationships between single solutes and component scores are investigated as well as spatial and temporal patterns.

16.4.2.1 Component: Redox Status

The first component represents 46.8% of the variance of the data set. The component is positively correlated with NO_3^-, SO_4^{2-}, Al, Ca, K, and Mg and negatively correlated with Fe and DOC concentrations (Fig. 16.3). Appreciable NO_3^- concentrations are restricted to component scores > −2, whereas the opposite is true for Fe. The relationship between SO_4^{2-} and the first component can be split into two distinct groups with slightly different slopes and offsets. The point of intersection of both groups with the *y*-axis is roughly between −3 and −4, that is, slightly less compared to that of the NO_3^- data. A similar pattern emerges for Ca, which is presumably due to the same reasons. However, for the sake of brevity this is not presented in more detail.

The relationship between NO_3^-, SO_4^{2-}, and Fe concentrations and the first component resembles the well-known redox sequence: The lower the redox potential, the more pronounced is denitrification, desulfurication, and Fe reduction. The redox potential in the two catchments presumably is the lowest in the wetland soils, which exhibit the highest DOC concentration. In addition, NO_3^- and SO_4^{2-} are the prevailing anions in groundwater and streamwater. As the end products of denitrification and desulfurication are mostly released via degassing and precipitation, these processes result in a decrease of ionic strength which is reflected by a corresponding decrease of the concentration of most of the cations.

The lowest component scores, i.e., the most reduced water was found in the riparian wetlands of the Lehstenbach catchment, that is, for the deep groundwater wells GW02 and GW04, for shallow groundwater wells Moor_1m and Moor_2m and in the Moorbach stream that drains a wetland area close to the catchment outlet (Fig. 16.4, cf. Fig. 16.1). Please note that the redox component is not a measure of the redox potential of the water sample. In all streams including the Moorbach stream oxygen saturation usually was fairly high. In contrast, the redox component seems to be a measure of to what degree redox processes had an impact on water quality before it reached the stream.

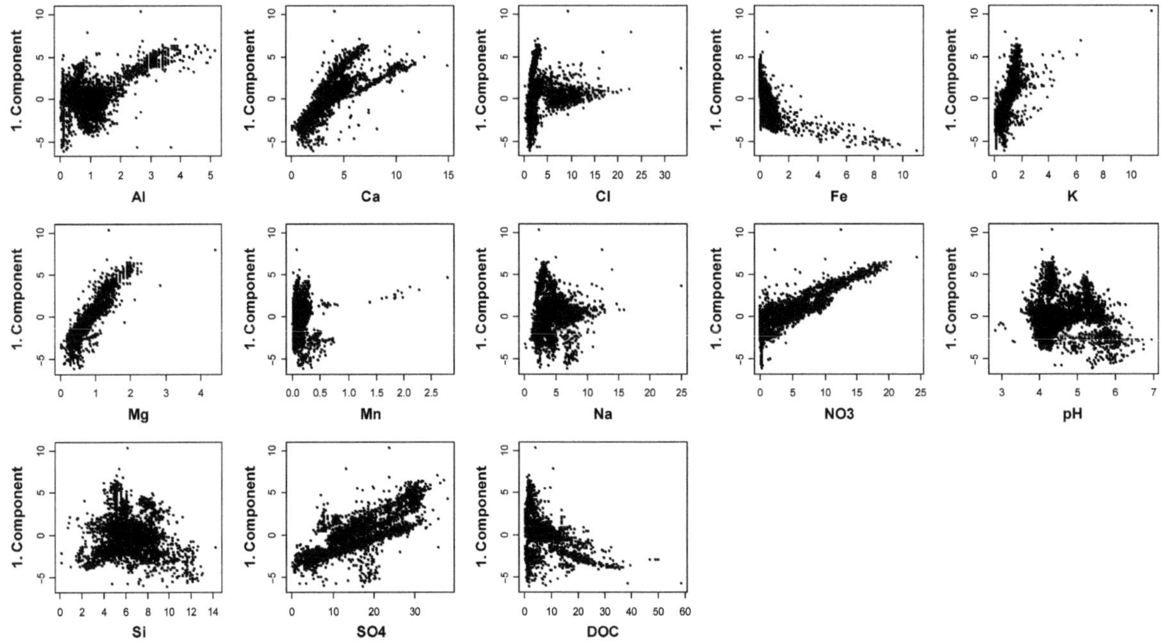

Fig. 16.3 Loadings of the first Isomap component. Concentrations are given in milligram per liter (mg/l)

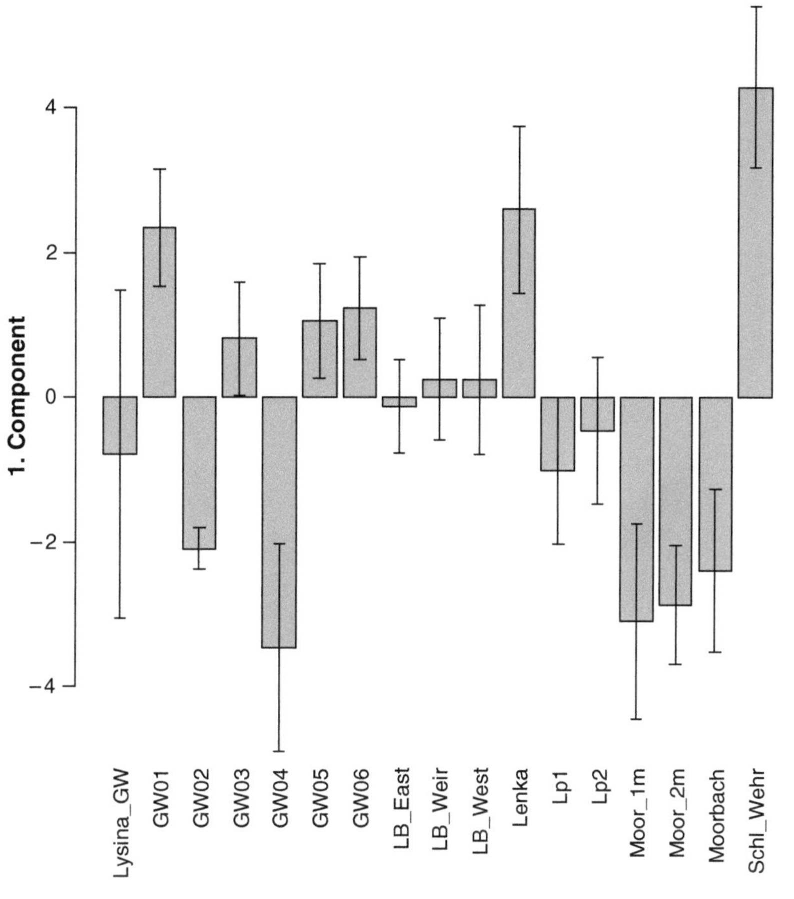

Fig. 16.4 Mean values and standard deviation of the first component scores for the different sampling sites

Consequently, this component can be interpreted corresponding to a shift of nitrate-^{15}N or sulfate-^{34}S values.

The upslope springs in both catchments (Lenka and Schl_Wehr) and the upslope groundwater well GW01 exhibit the highest values. It is interesting to note that the mean redox component score of the Lehstenbach catchment runoff (LB_Weir) is higher than those of the Lysina Lp2 stream, and that of Lysina stream Lp1 is even lower, which corresponds roughly to the fraction of riparian wetlands in the respective catchments.

16.4.2.2 Component: Topsoil

The second component adds another 21.3% of explained variance to that of the first component. High scores of this component are associated with high Al and DOC concentrations, low Si and Na concentrations, and low pH values (Fig. 16.5). This is typical for the soil solution in the uppermost horizons in both catchments. High Al concentrations and low pH are due to long-term deposition of sulfate and other acidifying solutes that leached base cations from the topsoil. However, due to the strong decline of acidifying deposition during the last two decades, there is no clear relationship between this topsoil component and $SO_4{}^{2-}$ concentration. Sulfate concentration is now the highest in groundwater of medium depth (Büttcher, 2001; Lischeid, Büttcher, & Hauck, 2003; Lischeid, Lange, Moritz, & Büttcher, 2004). There is a strong but non-linear relationship between this component and the pH values. Silica and Na originate mainly from weathering of feldspars of the granite bedrock (Weyer, Lischeid, Aquilina, Pierson-Wickmann, & Martin, 2008). Mobilization of Si has been shown to exhibit very slow kinetics in the Lehstenbach catchment, allowing using Si concentration as a proxy for groundwater residence time (Lischeid, Kolb, & Alewell, 2002).

Most of the scatter of the relationship between Fe concentration and the topsoil component can be

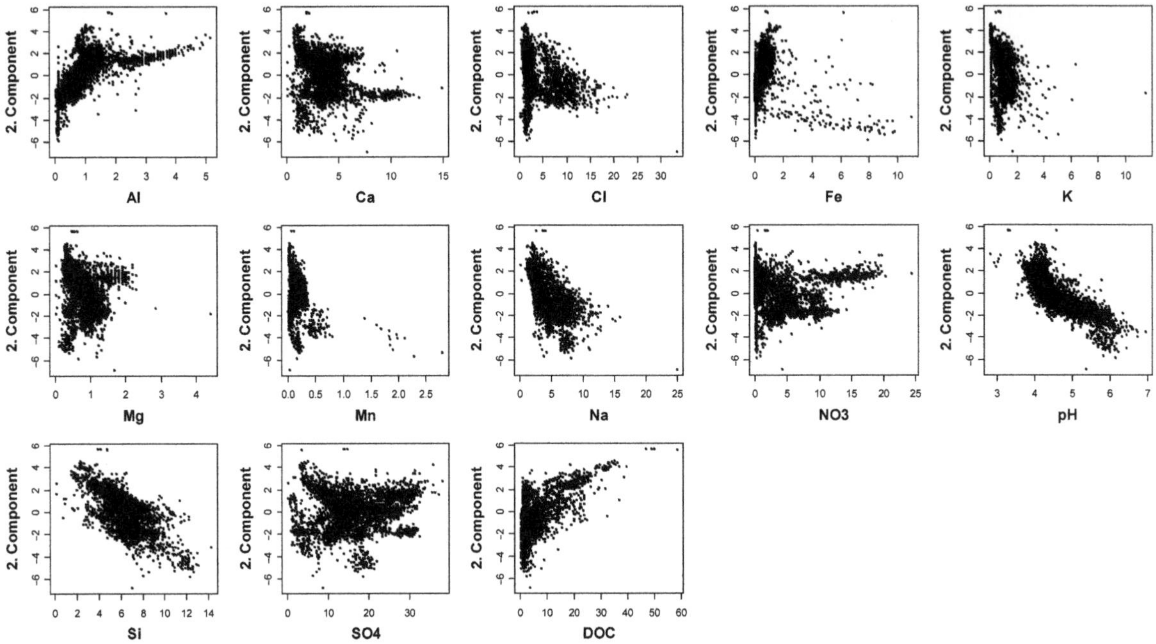

Fig. 16.5 Loadings of the second Isomap component. Concentrations are given in milligram per liter (mg/l)

ascribed to the redox component, especially for the high Fe concentration data. Putting aside this fraction of the data, there is an increase of Fe concentration with increasing component scores which corresponds to Fe(III) mobilization by low pH values around four.

The highest topsoil component scores were observed in the Lehstenbach streams Moorbach and Schl_Wehr that drain rather small subcatchments (Fig. 16.6). The lowest scores were found in the Lehstenbach groundwater wells GW04 and the Moor_2m wells. According to the results of a groundwater model, well GW04 samples groundwater with the longest residence time in the Lehstenbach catchment (Lischeid et al., 2003). The Moor_2m shallow groundwater wells are screened in a layer of very low hydraulic conductivity at 2 m depth which would correspond to a long residence time in spite of shallow depth. Comparing the three catchment outlet streams LB_Weir (Lehstenbach), Lp2 (Lysina), and Lp1 (Lysina), the topsoil component scores increase with the fraction of riparian wetlands in the respective catchments.

In the streams the topsoil component is closely related to discharge, although with different shapes (Fig. 16.7). Thus, this component could be used for chemical hydrograph separation similarly to the approaches used by Buzek, Hruška, and Krám (1995) and Lischeid et al. (2002).

16.4.2.3 Component: Road Salt

The third component is correlated with Cl^- and Na concentrations and inversely related to Mn concentration (Fig. 16.8). This component explains another 20.1% of the variance of the data. The highest scores were observed in the Lehstenbach groundwater well GW01, in the Lehstenbach runoff LB_Weir, and especially in its tributary LB_West (Fig. 16.9). These sites are affected by road salt application of a public road in the western part of the catchment (cf. Fig. 16.1). Because this effect is restricted to some sites only, it will not be considered in the following.

16.4.2.4 Component: Sulfate Contamination

The fourth component adds another 5.5% of explained variance to that given by the first three components. Correspondingly, the relationships between single solutes and the component scores are more difficult to interpret because they are masked by the effects of the first three components. Focusing on the bulk of

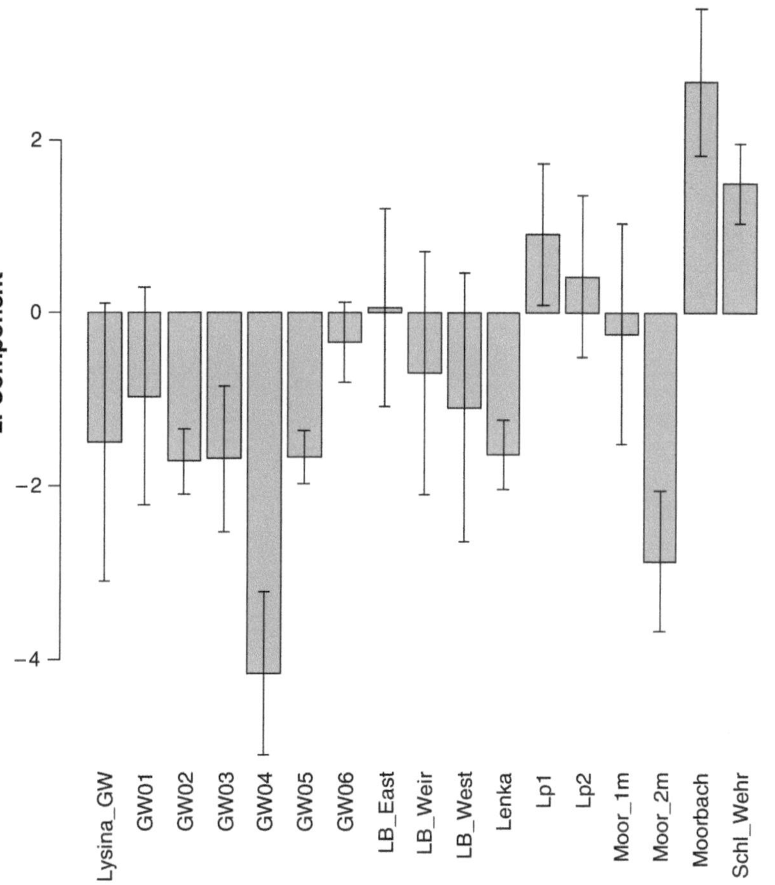

Fig. 16.6 Mean values and standard deviation of the second component scores for the different sampling sites

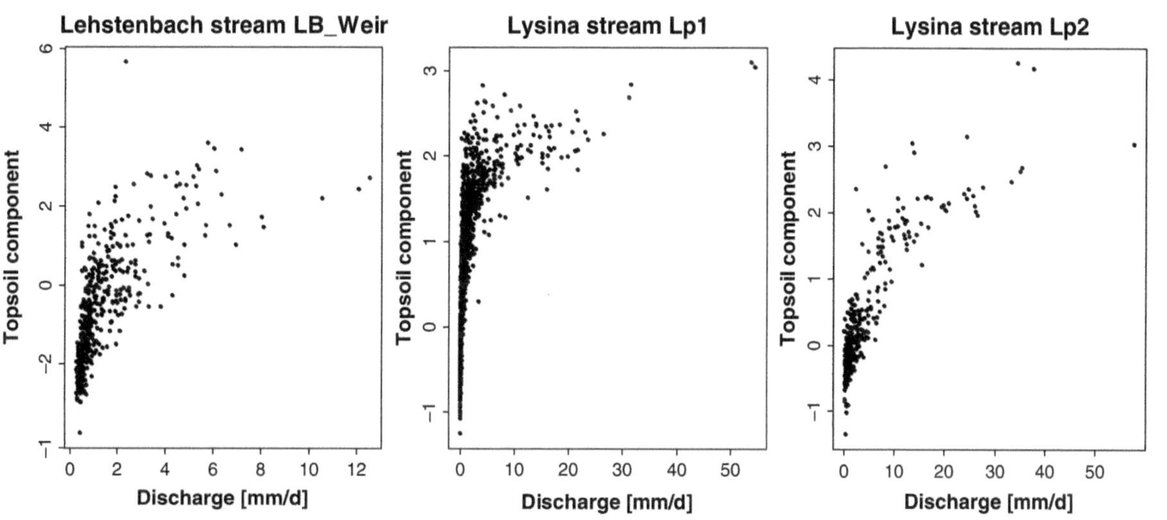

Fig. 16.7 Topsoil component scores versus discharge at the catchment outlets

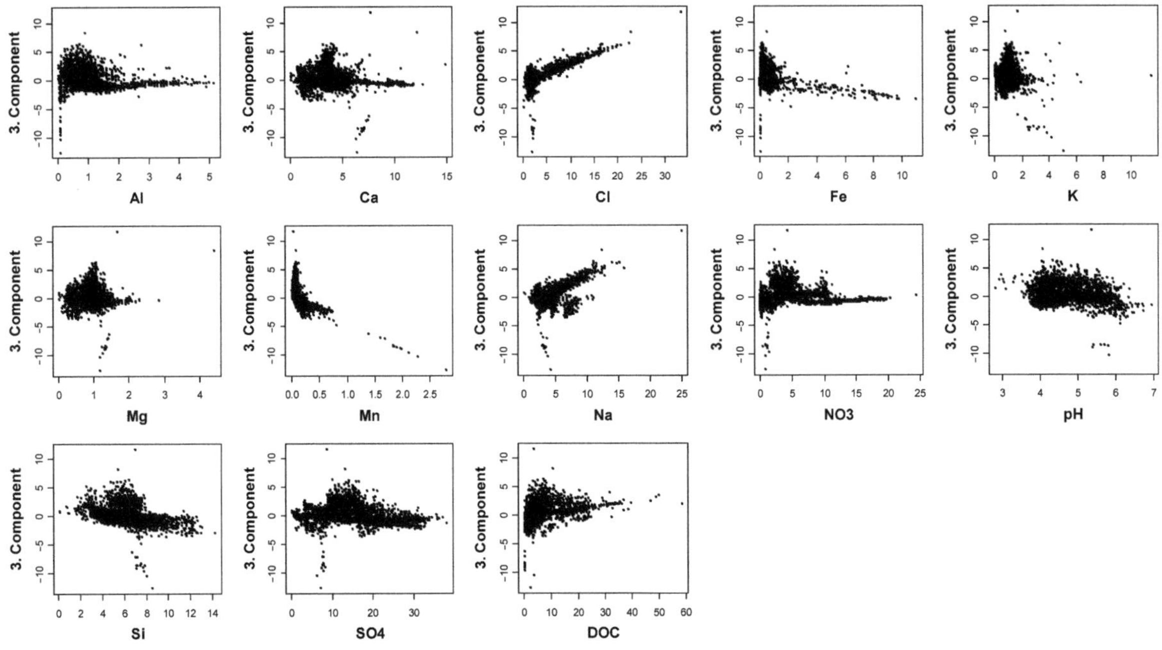

Fig. 16.8 Loadings of the third Isomap component. Concentrations are given in milligram per liter (mg/l)

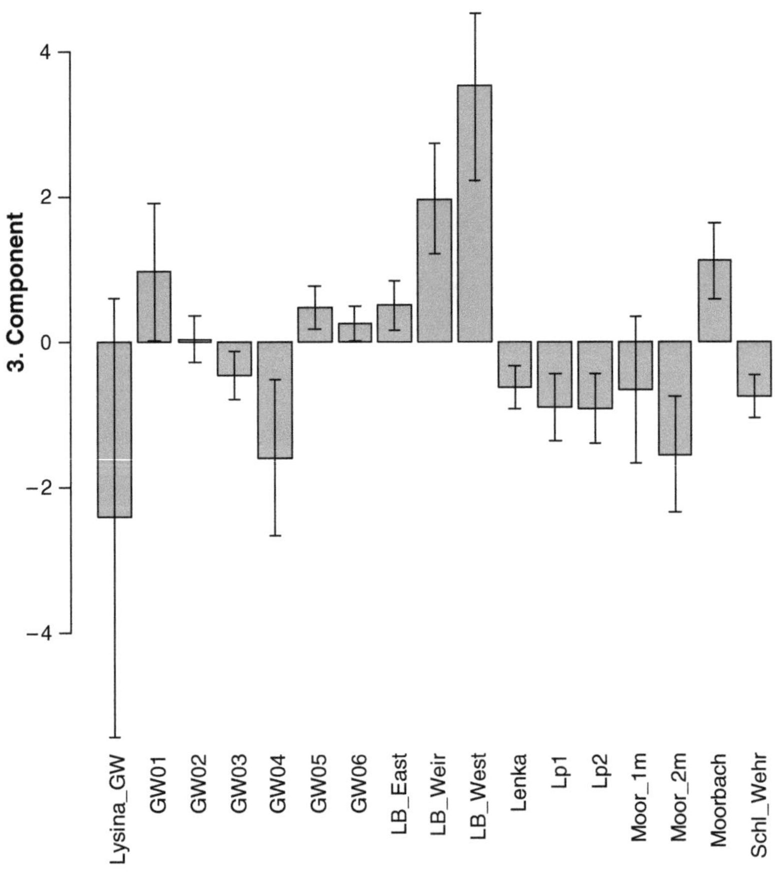

Fig. 16.9 Mean values and standard deviation of the third component scores for the different sampling sites

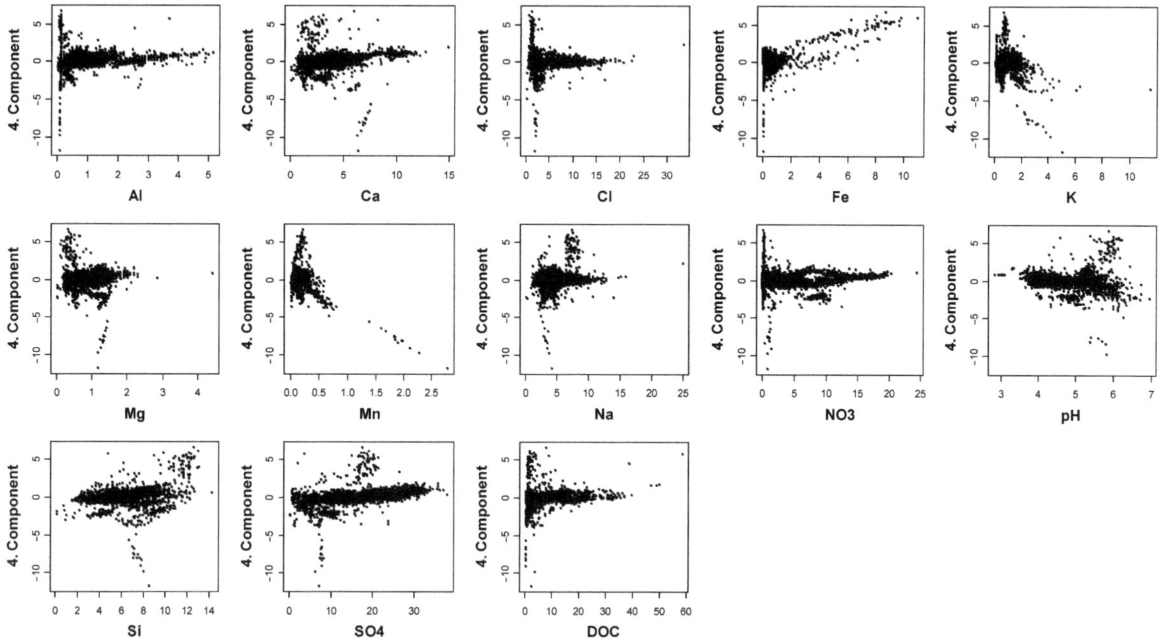

Fig. 16.10 Loadings of the fourth Isomap component. Concentrations are given in milligram per liter (mg/l)

the data, sulfate, Al, and Ca concentrations seem to increase and pH to decrease with increasing component scores (Fig. 16.10). Taking into account that most of the variance of these solutes is explained by the first three components, it is concluded that this component is associated with effects of sulfate contamination in addition to those described by the redox component.

A clear increase of Mn concentration with decreasing component scores is restricted to the lower quantile of component scores (Fig. 16.10). Component scores less than −5 are restricted to the deepest groundwater well of the Lysina catchment which is gauged between 4.2 and 4.4 m below surface. At this well K and Si concentrations show a similar relationship with that component, whereas the inverse pattern emerges for Ca and Mg (Fig. 16.10). This could point to a displacement of geogenic compounds by Ca and Mg that were associated with high sulfate deposition rates in the last decades. The samples with the highest scores of that component (exceeding +2) were taken at well GW04 and in the wetland piezometers Moor_1m at 1 m depth in the Lehstenbach catchment (Fig. 16.11). At GW04, oxygen concentration was almost, and at Moor_1m often close to zero. Only for these samples there was a positive correlation between component scores and Fe concentrations (Fig. 16.10).

Besides, the upslope springs Lenka (Lysina) and Schl_Wehr (Lehstenbach) show rather high values. These sites sample a high fraction of intermediate depth groundwater that now exhibits the highest SO_4^{2-} contamination (Büttcher, 2001; Lischeid et al., 2003).

16.4.3 Time Series

One focus of this study was to investigate if there are some common trends of single processes, including non-linear long-term patterns. To that end a low-pass filter was applied to the component scores. Please note that the low-pass filtered part of the fraction represents different fractions of the variance of the respective time series, due to different short-term variations at different sites. Besides, the beginning and end of the time series have to be interpreted with care, when only few samples were available from these periods. This holds especially true for the last years at the Lehstenbach catchment runoff (LB_Weir) and upslope spring (Schl_Wehr). Thus, the supporting points are given as black points in the graphs of components (Figs. 16.12, 16.13, 16.14, and 16.15).

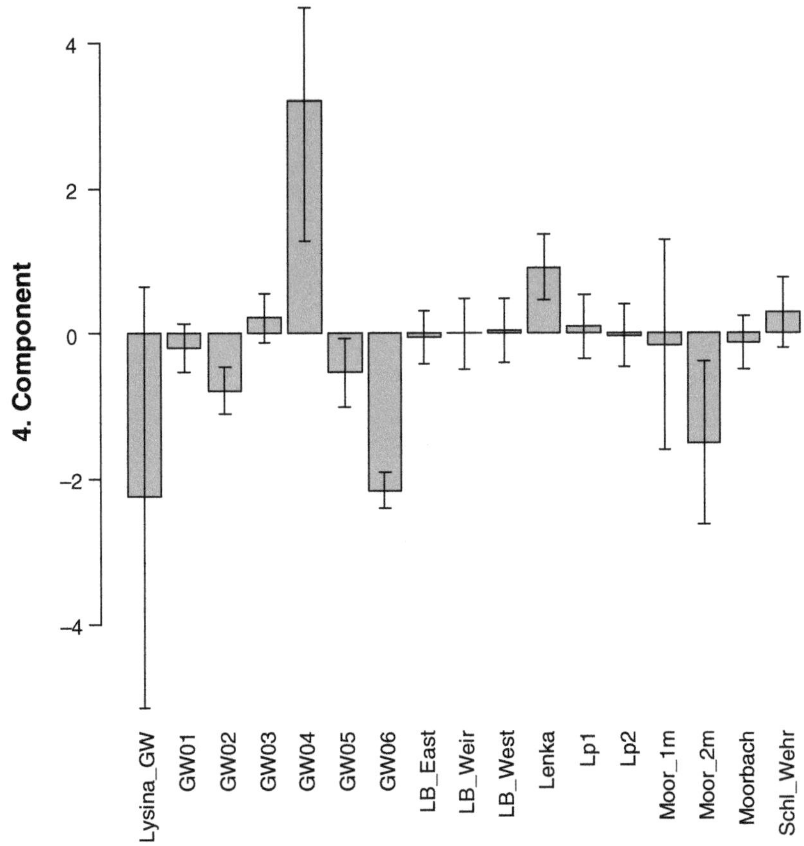

Fig. 16.11 Mean values and standard deviation of the fourth component scores for the different sampling sites

16.4.3.1 Meteorological Data

The meteorological data were low-pass filtered in the same way as the water quality data. Figure 16.12 shows the long-term trends during 1989–2007, normalized to the respective mean values in order to facilitate comparison between the two parameters. Please note that start and end of the single curves have to be interpreted with caution.

Both at Lysina and Braunersgrün, precipitation was fairly low during the early 1990s. At Lehstenbach and Braunersgrün precipitation exhibited a minor peak at 2001 and a minor decrease thereafter. The variance of the low-pass filtered temperature data was smaller compared to precipitation. Minor peaks were observed around 1993 and 2001, minor lows in 1997 and 2004.

16.4.3.2 Redox Component

Both of the Lysina streams (Lp1 and Lp2), the Lysina upslope spring (Lenka), as well as the Lehstenbach catchment runoff (LB_Weir) and upslope spring (Schl_Wehr), and one of the deep groundwater wells (GW03) exhibited a decrease during the last two decades (Fig. 16.13). However, these three Lehstenbach sites show a transient increase or a change of slope in the 1990s. In contrast, the redox component scores at the Lehstenbach groundwater wells GW01, GW04, GW05, and GW06 peak between 1995 and 2003. Around 1995 there is an intermediate or minor peak at many sites in the Lehstenbach catchment. In contrast, at the Lehstenbach well GW02 there is hardly any clear long-term shift of the redox component.

16.4.3.3 Topsoil Component

The low-pass filtered time series show two different patterns: either there is a minimum in the late 1990s and a steep increase thereafter (Lysina streams Lp1 and Lp2, Lysina upslope spring Lenka) or a pronounced decrease during the first years, a peak during 1995–2000 and decrease again thereafter (Lehstenbach groundwater wells). The latter behavior reflects that of low-pass temperature data, although

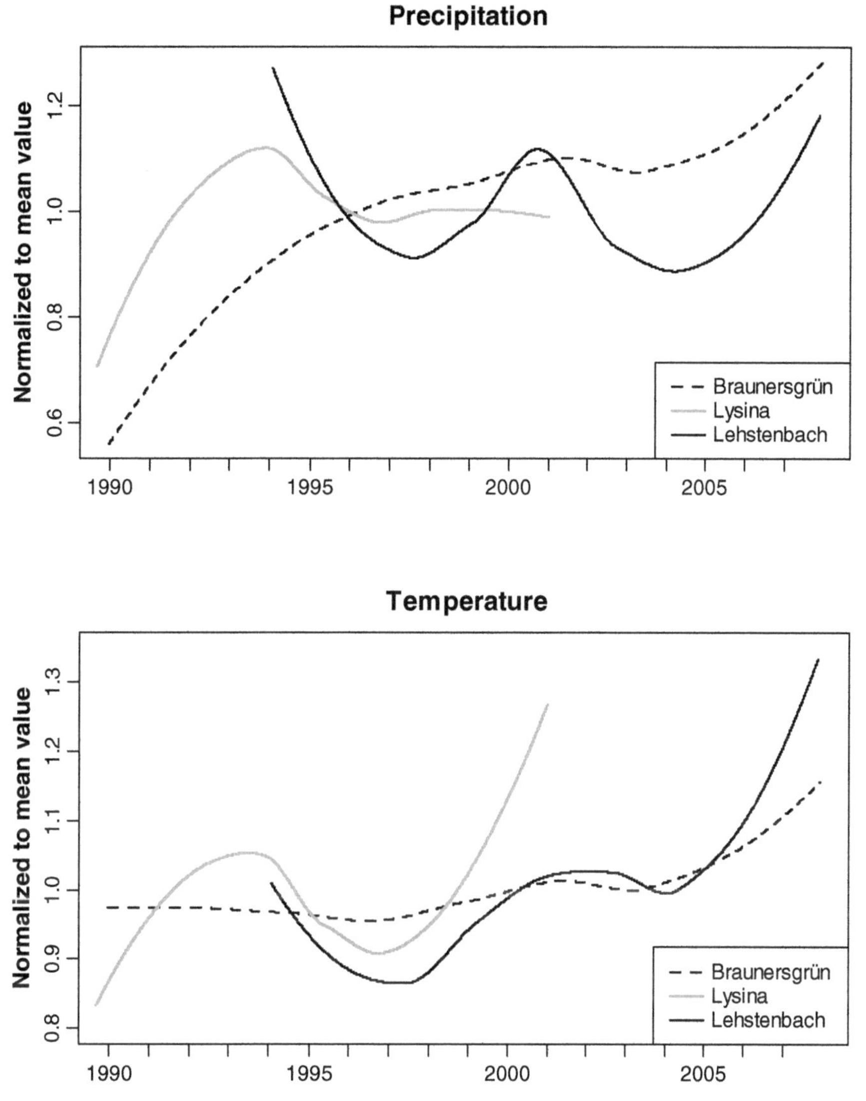

Fig. 16.12 Low-pass filtered time series of precipitation and air temperature at Braunersgrün, Lysina, and Lehstenbach 1989–2007, normalized to mean values for the whole period

inversely and slightly lagging behind temperature (cf. Fig. 16.12). The Lehstenbach stream (LB_Weir) and upslope spring (Schl_Wehr) exhibit features of both groups (Fig. 16.14). Only at well GW01 the long-term behavior of the topsoil component seems to parallel that of precipitation between 1995 and 2003. It has been known prior to this study that groundwater quality at this well often reflects preferential flow from the topsoil (Lischeid and Bittersohl 2008).

16.4.3.4 Sulfate Contamination Component

A clear decrease of the SO_4^{2-} contamination component was observed at the Lysina streams (LP1 and LP2) and the Lehstenbach upslope spring (Schl_Wehr). At the Lehstenbach well GW01 and the Lysina upslope spring Lenka this component peaked in 1995–1997 and clearly decreased thereafter. In contrast, most of the Lehstenbach groundwater wells and the Lehstenbach catchment runoff (LB_Weir) exhibit two peaks, the first one between 1990 and 1995 and the second one between 2003 and 2007 (Fig. 16.15). This behavior, especially at GW03, resembles that of low-pass filtered temperature in the same catchment (cf. Fig. 16.12).

16.4.4 Interplay of Different Processes

It has been stated prior to the study that single solutes were affected by different processes. This has been

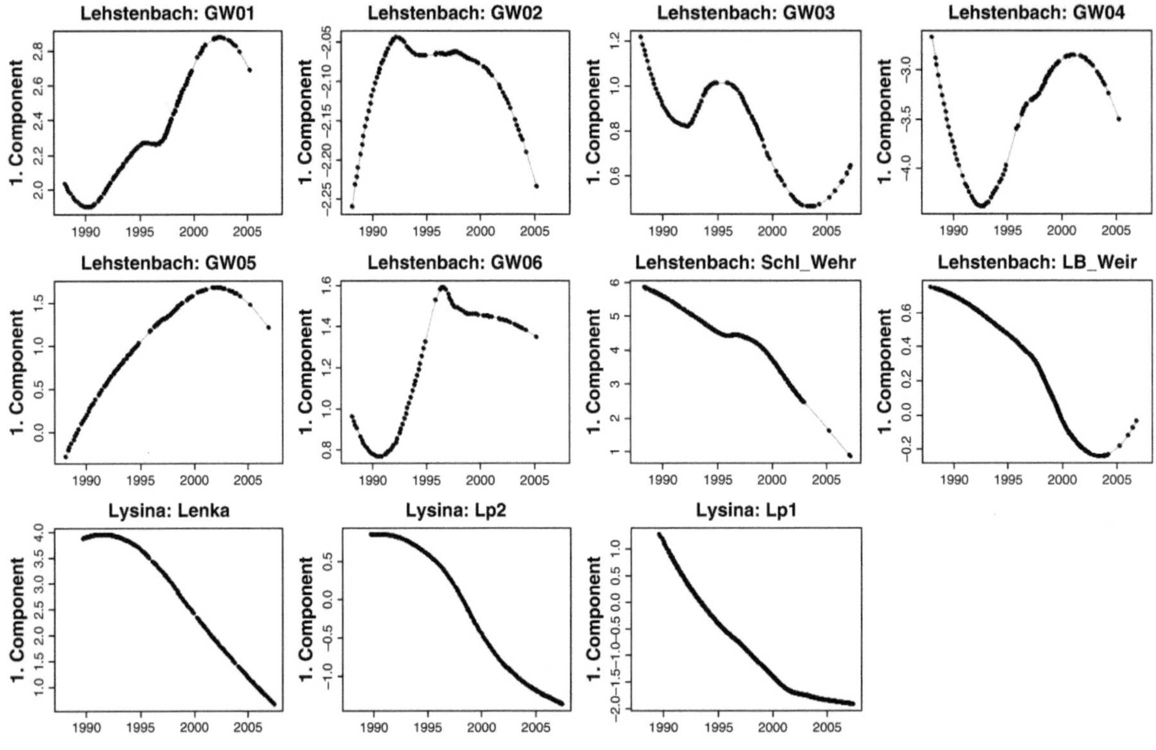

Fig. 16.13 Low-pass filtered time series of the first component at selected sampling sites

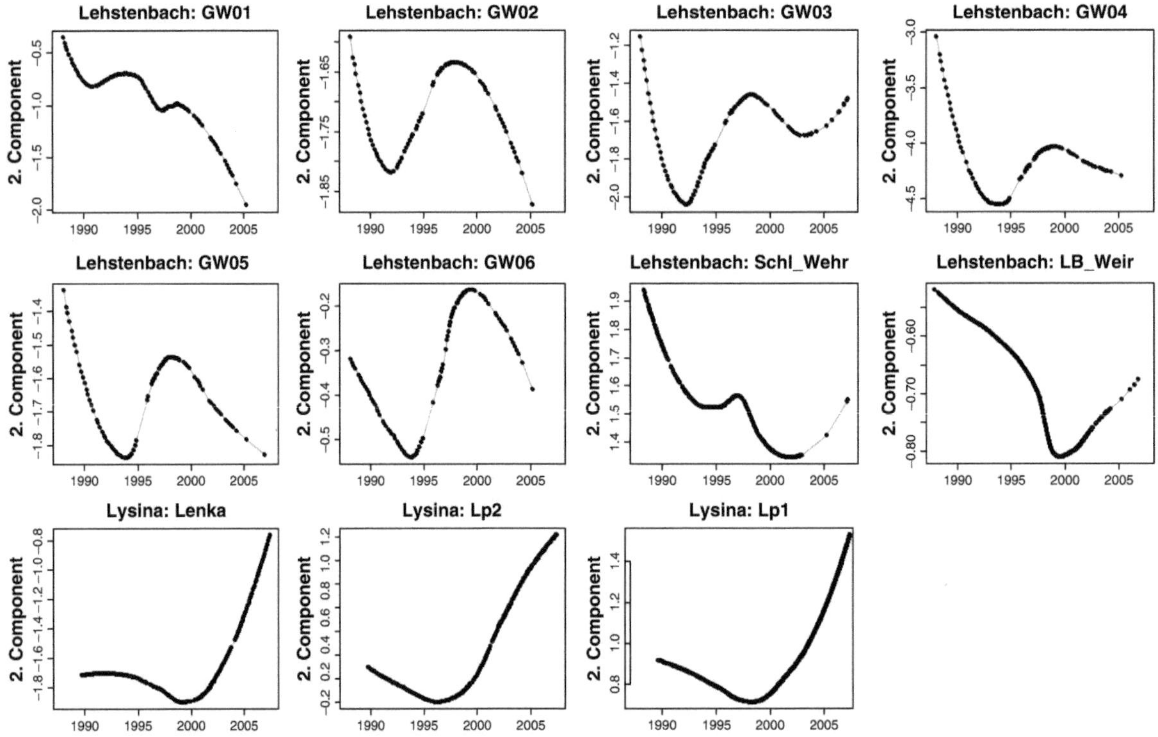

Fig. 16.14 Low-pass filtered time series of the second component at selected sampling sites

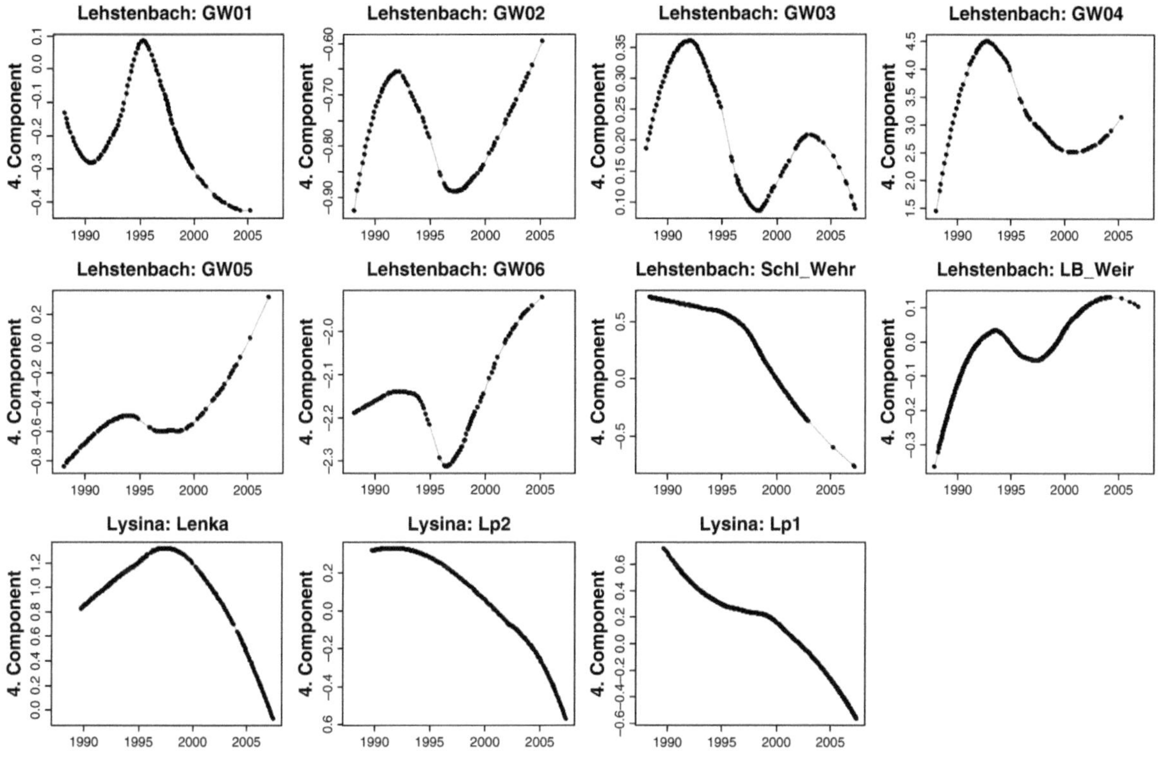

Fig. 16.15 Low-pass filtered time series of the fourth component at selected sampling sites

investigated in more detail using time series of the Lysina upslope spring Lenka and the Lehstenbach groundwater well GW01. To that end the prevailing anions NO_3^- and SO_4^{2-} have been selected.

16.4.4.1 Lysina Upslope Spring Lenka

The Lenka spring was unique among all sampling sites in the Lysina and Lehstenbach catchment as it exhibited a highly significant and nearly perfect linear decrease of NO_3^- concentration (Fig. 16.16). In contrast, SO_4^{2-} concentration slightly increased until 1995 and clearly decreased thereafter (Fig. 16.16). The redox component roughly paralleled the SO_4^{2-} time series and clearly decreased after 1994. In contrast, the SO_4^{2-} contamination component increased until 1998 and decreased thereafter. The topsoil component remained fairly stable until 2001 and started increasing during the last years of the study.

Sulfate deposition in this region decreased tremendously since the end of 1970s. Due to the effect of SO_4^{2-} sorption to the matrix, the SUNFLOW sulfate model predicted an increase of SO_4^{2-} contamination at this site until the 1990s (Büttcher, 2001) which is confirmed by the SO_4^{2-} contamination component. In addition, part of the SO_4^{2-} decrease seems to be due to increasingly less oxic conditions, as suggested by the decrease of the redox component after 1994 (Fig. 16.16). However, NO_3^- concentration started to decrease a few years before, coinciding with an initial increase of SO_4^{2-} concentration. This could point to an antagonism between the two anions.

Last, but not least the topsoil component did not start to increase before 2001/2002. In August 2002 the Lysina catchment was affected by the severe rainstorm that caused the Elbe flood further downstream, whereas there was no exceptional rainfall in the Lehstenbach catchment. It is likely that the increasing topsoil component amplified the decrease of SO_4^{2-} concentration in the spring water because the topsoil layers showed much more pronounced recovery of SO_4^{2-} deposition (Büttcher 2001, Shanley, Krám, Hruška, & Bullen, 2004).

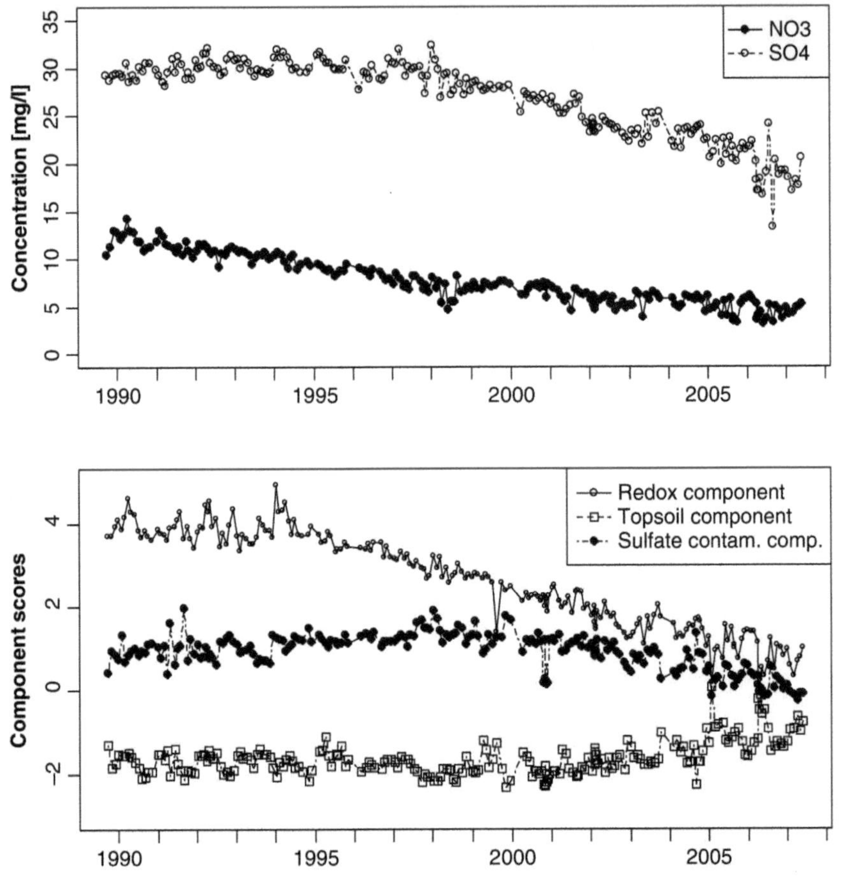

Fig. 16.16 Time series of NO_3 and SO_4 concentrations (*upper panel*) and component scores (*lower panel*) at the Lysina upslope spring Lenka

16.4.4.2 Lehstenbach Groundwater Well GW01

In contrast to the Lenka spring water, groundwater at the Lehstenbach well GW01 exhibits substantially higher short-term variability, especially for SO_4^{2-} concentration (Fig. 16.17). Sulfate concentration often increased by a factor of two during rainy or snowmelt periods in the dormant season. In fact, SO_4^{2-} concentration is closely related to groundwater level at this site (not shown). The SO_4^{2-} concentration peaks coincide with substantial increases of the topsoil component (Fig. 16.17), indicating preferential flow phenomena. This is especially true for fall and winter of 1998, a period of extensive rainfall (Lischeid et al., 2002). In addition, the redox component exhibits synchronous peaks, indicating preferential infiltration of oxic water from the topsoil. However, only a few of these peaks are reflected by a corresponding increase of NO_3^- concentration (Fig. 16.17). In the long term, the redox component showed an almost linear increase during 1990–2004, whereas the SO_4^{2-} contamination component peaked around 1997 and decreased thereafter. Nitrate in the groundwater slightly decreased until 1995 in spite of the decreasing redox component, but parallel to the increasing SO_4^{2-} contamination component. Like at Lysina, this could indicate an antagonism between NO_3^- and SO_4^{2-}. After 1995, the redox component increased and the SO_4^{2-} contamination component decreased at this well which seemed to have compensated each other, resulting in constant SO_4^{2-} and NO_3^- concentrations.

16.4.4.3 Antagonism Between NO_3^- and SO_4^{2-}

Both at Lenka and GW01 the data seem to indicate an antagonism between NO_3^- and SO_4^{2-} concentrations. Thus, the relationship between the redox component, the SO_4^{2-} contamination component, and NO_3^- concentration was investigated. This was rendered difficult due to some substantial outliers. Ignoring component scores less than the 0.05 and exceeding the 0.95 quantile, however, reveals a clear

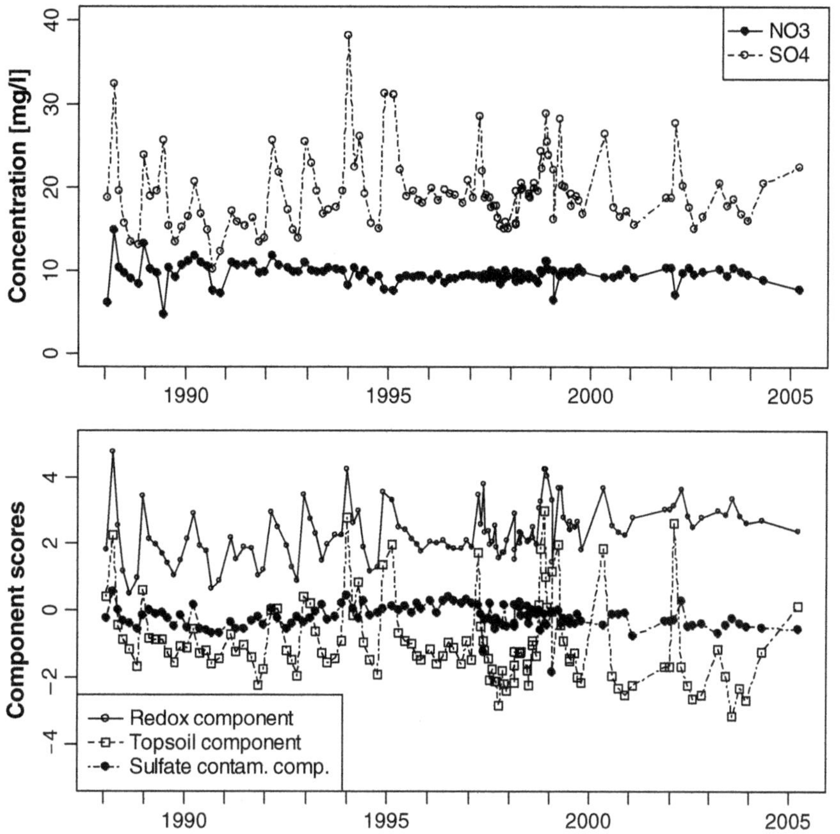

Fig. 16.17 Time series of NO_3 and SO_4 concentrations (*upper panel*) and component scores (*lower panel*) at the Lehstenbach well GW01

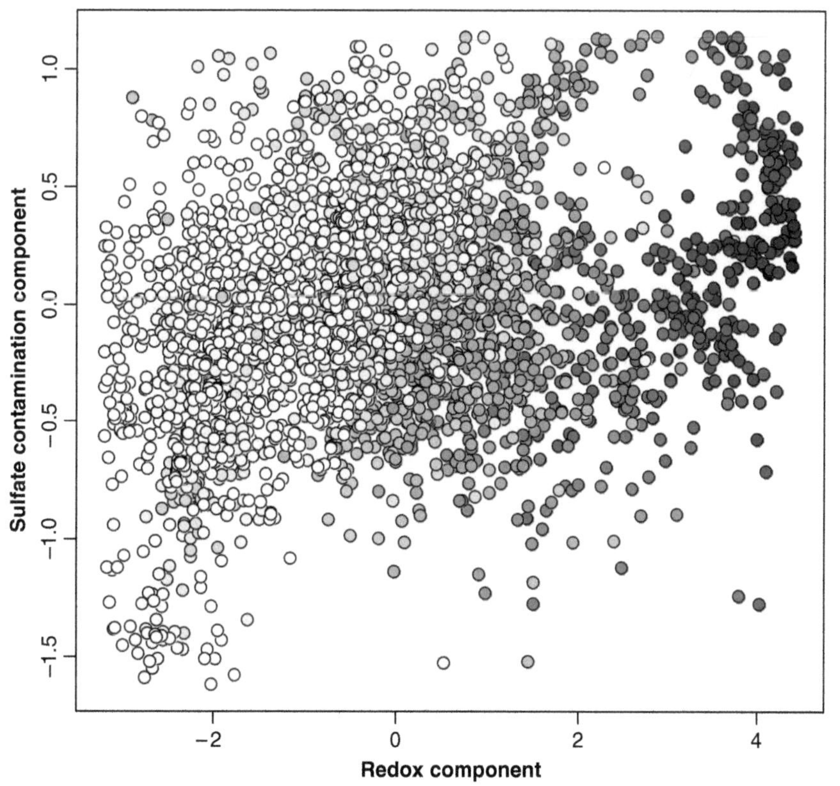

Fig. 16.18 Nitrate concentration (*white*: minimum and *black*: maximum) depending on scores of the first and fourth components. For the sake of clarity, extreme values of the components are not shown

Table 16.2 Coefficients of multivariate linear regression of NO_3^- concentration as a linear function of redox component and SO_4^{2-} contamination component scores, ignoring component scores less than the 0.05 and larger than the 0.95 quantile

	Coefficient of regression	Level of significance
Redox component	+0.421	$<2\times10^{-16}$
SO_4^{2-} contamination component	−0.359	$<2\times10^{-16}$

relationship (Fig. 16.18): Nitrate concentration does not only increase with increasing redox component scores but also decreases with increasing SO_4^{2-} contamination component scores. Thus, e.g., the transition between white symbols indicating very low NO_3^- concentration and slightly gray symbols indicating higher NO_3^- concentration occurs along a line that is inclined toward both axes. In fact, 74% of the variance of NO_3^- concentration in this data set can be explained by linear regression with the redox and the SO_4^{2-} contamination component, where the latter exhibits a negative regression coefficient (Table 16.2).

16.5 Discussion

16.5.1 Ad Methods

The objective of the study was to identify the key processes that affect groundwater and surface water quality in a region of the East Bavarian–West Bohemian crystalline basement and to investigate long-term trends of these processes. To that, non-linear statistical methods were used to account for the pronounced non-linearity of environmental data.

The non-linear isometric feature mapping proved to be slightly superior to the linear principal component analysis in terms of explained variance of the first components (Fig. 16.2). The difference is smaller compared to some other studies (Tenenbaum et al., 2000; Mahecha et al., 2007). Correspondingly, the first components of the linear and the non-linear principal component analysis were highly correlated. The larger the optimal *k* parameter for setting up the geodesic distance matrix for the Isomap analysis, the less the Isomap results differ from that of the linear principal component analysis. In this study, the optimal *k* was approximately equal to half the number of samples, pointing to rather weak non-linearities. However, the Isomap components were easier to interpret and to assign to biogeochemical processes. The same has been found by Lischeid and Bittersohl (2008).

Time series of environmental parameters usually exhibit substantial short-term variability, which is often traced back to a complex interplay of different processes. Water resources management, however, needs to consider long-term shifts. These do not necessarily show up as linear trends, and in fact rarely do so for longer periods. Thus, a low-pass filter was applied that was able to extract a rather smooth long-term pattern from the time series. As many hydrochemical time series exhibit pronounced seasonality, the cut-off for the considered periods was set equal to 1-year length.

Similarly to, e.g., linear regression, the fit of a low-pass filter to the data has to be considered with care for the beginning and for the end of the time series. The chosen approach is based on a spectrum analysis. Thus, it tends to fit low-frequent oscillations to the data. In this study, these structures have not been considered when they were restricted to the first or final 2 years of the time series.

16.5.2 Ad Results

Determination of the principal components does not require any a priori assumptions. In contrast, assigning components to biogeochemical processes is subject to expert's choice and can be questioned. It is primarily based on the relationships between component scores and measured parameters, but considering spatial and temporal patterns of the component scores as well.

16.5.2.1 Redox and Sulfate Contamination Components

The first component, which explained nearly half of the variance of the data, was interpreted as a redox component. This component provides a measure of the impact of redox processes on water quality on its way to the sampling site, but is not a measure of the actual redox potential at the time of sampling. The redox component was not clearly related to oxygen concentration, especially in the streams. For example, the wetland streams Moorbach and Lp1 exhibited low redox component scores in spite of oxygen saturation due to turbulent flow in the streams.

In a preceding study a systematic decrease of the redox component scores with residence time was found in the Lehstenbach deep groundwater, even at the upslope wells outside the wetland areas (Lischeid & Bittersohl, 2008). This decrease was associated with a corresponding decrease of NO_3^- concentration, pointing to denitrification. In spite of that, oxygen concentration usually exceeded 6 mg/l in the wells GW01, GW03, GW05, and GW06. In contrast, groundwater oxygen concentration at well GW02 usually was fairly low and close to zero at well GW04. Legout et al. (2007) could even show that denitrification in the vadose zone of a granite catchment in Brittany played a major role for the nitrogen dynamics. It was assumed that most of the denitrification occurred in anoxic microsites.

On the one hand, NO_3^- and SO_4^{2-} concentrations in anoxic groundwater decrease due to denitrification and desulfurication. On the other hand, enhanced input of NO_3^- and SO_4^{2-} due to atmospheric pollution increased the electron acceptor capacity in these microsites, thus increasing the redox potential of deep soil and groundwater. This effect plays a major role in oxygen-depleted lake sediments but is rarely considered in the vadose zone. The results of this and other studies, however, seem to indicate that the role of anoxic microsites has been substantially underestimated so far.

Deposition of inorganic nitrogen compounds (mainly dissolved NO_3^- and NH_4) showed substantially less pronounced trends compared to SO_4^{2-}. In deeper soil layers and in the groundwater, only NO_3^- was found in measurable quantities due to oxidation of NH_4 and NH_4 sorption to the soil matrix. During the last two decades nitrogen deposition exceeded the annual uptake by plants and microbes by far in botch catchments. In addition, NO_3^- sorption to the soil matrix is negligible. Thus, spatial and temporal patterns of NO_3^- in the streams and in the deep groundwater were primarily determined by denitrification, resulting in a rather close relationship to the redox component. In contrast, SO_4^{2-} deposition exhibited a tremendous decrease in both catchments and was subject to sorption to the regolith matrix. Consequently, SO_4^{2-} concentration in groundwater and streamwater was related to the flowpath length distribution in the respective subcatchments (Büttcher, 2001). This is reflected by the SO_4^{2-} contamination component that explained much of the scatter after subtracting the effect of the redox component (Figs. 16.3 and 16.10).

In addition, NO_3^- is slightly negatively correlated with the SO_4^{2-} contamination component. Please note that the relationship of the redox component with NO_3^- is much more pronounced, masking the NO_3^-–SO_4^{2-} antagonism. There is some anecdotal evidence in the literature about such an antagonism, e.g., by Nakagawa and Iwatsubo (2000). Nodvin, Driscoll, and Likens (1988) observed a decrease of streamwater SO_4^{2-} concentration after clear cut, which they trace back to increasing SO_4^{2-} sorption due to decreased pH as a consequence of enhanced NO_3^- release. Kaiser and Kaupenjohann (1998) found a corresponding relationship between pH and SO_4^{2-} sorption at two sites close to the Lehstenbach catchment.

16.5.2.2 Topsoil Component

The topsoil component was identified based on its relationship with Al, DOC, Si, and pH (Fig. 16.5). High Al and DOC concentrations, low Si concentration, and low pH were usually found in the soil solution of the uppermost horizons (Navrátil, Kurz, Krám, Hofmeister, & Hruška, 2007, Krám & Hruška, unpublished data, 2007, Lischeid, Kolb, Alewell, & Paul, 2007) that were not included in this analysis. In this data set the highest topsoil component scores were found in the streamwater samples that drain rather small subcatchments (Moorbach, Schl-Wehr, Lp1, Lp2; cf. Fig. 16.6). In addition, topsoil component scores clearly increased with discharge (Fig. 16.7). This is consistent with the observation that near-surface runoff constitutes a major fraction of stormflow runoff in both catchments (Buzek et al., 1995; Lischeid et al., 2002).

16.5.2.3 Long-Term Patterns

Time series of solute concentration and component scores of most of the sampling sites exhibited substantial short-term variation. Abstracting from that short-term dynamics the long-term pattern of component scores was remarkably similar at different sites. In general, the Lehstenbach and Lysina sampling sites can be split into three groups. The first group consists of the Lysina streams Lp1 and Lp2 and the Lysina upslope spring Lenka with nearly linear long-term patterns. The second group comprises the Lehstenbach deep groundwater wells with more cyclic long-term patterns. The Lehstenbach catchment runoff and the Lehstenbach upslope spring Schl_Wehr form a third group, exhibiting features of both groups. This grouping is likely to reflect the effect of increasing damping with depth which is the least pronounced for the first group and the most distinct for the second.

The long-term decrease of the first and fourth components at the Lysina sampling sites reflects in the first place the decrease of sulfur and nitrogen deposition. This effect can hardly be seen at the Lehstenbach deep groundwater wells. For the latter the interannual variability of climatic boundary conditions seems to be more decisive for the long-term redox conditions (Lischeid & Bittersohl, 2008).

Beyond that, there was no clear relationship between low-pass filtered component scores and meteorological data. Only the topsoil component scores at the Lehstenbach streams Schl_Wehr, LB_Weir, and the groundwater well GW04 seem to be related to the precipitation trend. However, a causal relationship is not very likely, as the latter exhibited the lowest topsoil component scores and the LB_Weir rather low component scores (Fig. 16.6).

16.5.2.4 Solute Concentration Versus Component Scores at the Lenka Site

Contrasting to all other Lehstenbach and Lysina sampling sites, there was a clear and almost linear decrease of NO_3^- concentration at the Lysina upslope spring Lenka since the end of the 1980s. This trend started almost 10 years before nitrogen deposition started to decrease at that site and a few years before SO_4^{2-} concentration started to decrease. In contrast, time series of the component scores at this site do not differ from those at other sites in the Lysina or Lenka catchment. Based on the results of this study, the apparently simple and intriguing trend of NO_3^- concentration seems to be the effect of the interplay of different processes that add to the observed long-term pattern (Fig. 16.16).

It is well accepted that the concentration of a single solute is usually due to a variety of different processes. On the other hand, simple long-term patterns are usually ascribed to a single dominant process. This study followed an alternative approach. It is concluded that investigating time series of components as representations of single processes instead of analyzing time series of concentrations of single solutes was successful in resolving some of the apparent inconsistencies of the latter.

16.6 Conclusions

The objective of this study was to identify the prevailing biogeochemical processes that affected groundwater and streamwater quality in two similar forested catchments at numerous sites, to investigate spatial patterns of processes, and to determine the long-term behavior with respect to these processes. A non-linear principal component analysis of long-term groundwater and streamwater solute concentration data was performed to identify the dominant processes and to investigate common long-term patterns in this region. The first four components could be ascribed to different processes and could be used as quantitative measures of the effect of these processes. Among these, only contamination by road salt was restricted to a few sites of the Lehstenbach catchment whereas the remaining three components and the associated processes had clear impacts in both catchments. The impact of topsoil solution infiltration could clearly be seen during major rainstorms and snowmelt. Redox processes seem to play a major role even in the oxic aquifer in both catchments which showed a seasonal pattern at various sites. The fourth component was interpreted as a sulfate contamination component which was less variable in the short term at most sites. The approach allowed investigating the different effects and their long-term behavior separately. Long-term behavior determined by low-pass filtering of the time series of component scores showed consistent trends. The dominant source of different behaviors at different sites seemed to be the varying thickness of the regolith and the fraction of wetland areas in the respective subcatchments. In the Lehstenbach deep groundwater wells long-term oscillations of single components were found whereas linear trends prevailed in the Lysina streams.

The analysis highlighted the relationship between different processes, namely an antagonism between NO_3^- and SO_4^{2-} concentrations that adds to their common dependence on redox conditions. This differentiation yielded a more clear and consistent picture than analysis of time series of single solutes. It is concluded that investigating spatial and temporal patterns of 'processes' (components) is more promising than looking at single solutes.

Acknowledgments Parts of the Lehstenbach data were kindly provided by Jochen Bittersohl, Klaus Moritz, and Uwe Wunderlich from the former Bavarian Water Resources Agency, now Bavarian Environmental Agency. Most of the Lehstenbach water samples were taken by Andreas Kolb and were analyzed in the Central Laboratory of BayCEER (former BITÖK) directed by Gunter Ilgen, which is highly appreciated. Most of the Lysina water samples were taken by Václav Kmínek from Kladská. Chemical analyses were performed in the Central Laboratories of the Czech Geological Survey, Prague.

A large fraction of the Lehstenbach monitoring program was financed by the German Ministry for Education and Research, Grant No. PT BEO 51 – 0339476 A-D, and by the German Research Foundation, Grant No. LI 802/2-2. Additional support for the Lysina data evaluation was provided by the 6th Framework EU project EUROLIMPACS (GOCE-CT-2003-505540) and by the research plan of the Czech Geological Survey (MZP0002579801).

References

Büttcher, H. (2001). Random variability or reproducible spatial patterns? Investigating sulphate dynamics in forested catchments with a coupled transport sorption model. Diploma thesis, University of Bayreuth.

Buzek, F., Hruška, J., & Krám, P. (1995). Three-component model of runoff generation, Lysina catchment, Czech Republic. *Water, Air, and Soil Pollution, 79*, 391–408.

EMEP modelled air concentrations, depositions. (n.d.). Retrieved December 21, 2007, from http://webdab.emep.int/Unified_Model_Results/.

Gámez, A. J., Zhou, C. S., Timmermann, A., Kurths, J. (2004). Nonlinear dimensionality reduction in climate data. *Nonlinear Processes in Geophysics, 11*, 393–398.

Glynn, E. F., Chen, J., & Mushegian, A. R. (2006). Detecting periodic patterns in unevenly spaced time series using Lomb-Scargle periodograms. *Bioinformatics, 22*, 310–316.

Hruška, J., Moldan, F., & Krám, P. (2002). Recovery from acidification in central Europe – observed and predicted changes of soil and streamwater chemistry in the Lysina catchment, Czech Republic. *Environmental Pollution, 120*, 261–274.

Kaiser, K., & Kaupenjohann, M. (1998). Influence of the soil solution composition on retention and release of sulfate in acid forest soils. *Water, Air, and Soil Pollution, 101*, 363–376.

Krám, P., Hruška, J., Wenner, B. S., Driscoll, C. T., & Johnson, C. E. (1997). The biogeochemistry of basic cations in two forest catchments with contrasting lithology in the Czech Republic. *Biogeochemistry, 37*, 173–202.

Legout, C., Molenat, J., Aquilina, L., Gascuel-Odoux, C., Faucheux, M., Fauvel Y., et al. (2007). Solute transfer in the unsaturated zone-groundwater continuum of a headwater catchment. *Journal of Hydrology, 332*, 427–441.

Lischeid, G., & Bittersohl, J. (2008). Tracing biogeochemical processes in stream water and groundwater using nonlinear statistics. *Journal of Hydrology, 357*, 11–28.

Lischeid, G., Büttcher, H., & Hauck, A. (2003). Combining data-based and process-based approaches to minimize the complexity of a reactive sulfate transport model. *IAHS-Publication, 277*, 402–408.

Lischeid, G., Kolb, A., & Alewell, C. (2002). Apparent translatory flow in groundwater recharge and runoff generation. *Journal of Hydrology, 265*, 195–211.

Lischeid, G., Kolb, A., Alewell, C., & Paul, S. (2007). Impact of redox and transport processes in a riparian wetland on stream water quality in the Fichtelgebirge Region, Southern Germany. *Hydrological Processes, 21*, 123–132.

Lischeid, G., Lange, H., Moritz, K., & Büttcher, H. (2004). Dynamics of runoff and runoff chemistry at the Lehstenbach and Steinkreuz catchment. In E. Matzner (Ed.), *Biogeochemistry of forested catchments in a changing environment. A German case study. Ecological Studies 172* (pp. 399–436). Berlin: Springer.

Mahecha, M., Martínez, A., Lischeid, G., & Beck, E. (2007). Nonlinear dimensionality reduction as a new ordination approach for extracting and visualizing biodiversity patterns in tropical montane forest vegetation data. *Ecological Informatics, 2*, 138–149.

Matzner, E. (Ed.). (2004). *Biogeochemistry of forested catchments in a changing environment. A German case study. Ecological Studies 172*. Berlin: Springer.

Matzner, E., Zuber, T., Alewell, C., Lischeid, G., & Moritz, K. (2004). Trends in deposition and canopy leaching of mineral elements as indicated by bulk deposition and throughfall measurements. In E. Matzner (Ed.), *Biogeochemistry of forested catchments in a changing environment. A German case study. Ecological Studies 172* (pp. 233–250). Berlin: Springer.

Nakagawa, Y., & Iwatsubo, G. (2000). Water chemistry in a number of mountainous streams of East Asia. *Journal of Hydrology, 240*, 118–130.

Navrátil, T., Kurz, D., Krám, P., Hofmeister, J., & Hruška, J. (2007). Acidification and recovery of soil at a heavily impacted forest catchment (Lysina, Czech Republic) – SAFE modeling and field results. *Ecological Modelling, 205*, 464–474.

Nodvin, S. C., Driscoll, C. T., & Likens, G. E. (1988). Soil processes and sulfate loss at the Hubbard Brook Experimental Forest. *Biogeochemistry, 5*, 185–199.

Oksanen, J., Kindt, R., Legendre, P., O'Hara, B., & Stevens, M. H. H. (2007). *The vegan package. Manual.* Retrieved from http://cran.r-project.org.

R Development Core Team. (2006). *R: a language and environment for statistical computing.* Vienna, Austria: R Foundation for Statistical Computing.

Shanley, J. B., Krám, P., Hruška, J., & Bullen, T. D. (2004). A biogeochemical comparison of two well-buffered catchments with contrasting histories of acid deposition. *Water, Air, and Soil Pollution: Focus, 4*, 325–342.

Tenenbaum, J. B., DeSilva, V., & Langford, J. C. (2000). A global geometric framework for nonlinear dimensionality reduction. *Science, 290*, 2319–2323.

Weyer, C. Lischeid, G., Aquilina, L., Pierson-Wickmann, A.-C., & Martin, C. (2008). Investigating mineralogical sources of the buffering capacity of a granite catchment using strontium isotopes. *Applied Geochemistry, 23*, 2888–2905. DOI:10.1016/j.apgeochem.2008.04.006.

Chapter 17

Long-Term Measurements to Quantify the Impact of Arable Management Practices on Deep Seepage and Nitrate Leaching

Uwe Schindler, Lothar Müller, Ralf Dannowski, Dietmar Barkusky, and Glyn Francis

Abstract Long-term soil hydrological measurements were used to quantify deep seepage and nitrate leaching in situ under undisturbed soil conditions. Deep seepage rates and nitrate losses from arable land managed under various farming regimes (integrated, integrated with irrigation, ecologic and low input) and tillage systems (plough and no-till) were quantified in the Pleistocene region of Northeast Germany from 1994 to 2005. Soil water content and tension measurements down to 3 m depth and soil water sampling were used to determine deep seepage dynamics and loss of nitrogen by leaching. As dependent on the management system, the nitrate concentration varied between 40 and 150 mg l^{-1}. In connection with annual deep seepage rates between 100 and 200 mm during the study period, the annual nitrogen loss varied between 14 and 41 kg ha^{-1}. Differences in nitrogen loss observed between the farming systems were low, but yields increased and nitrogen losses decreased as a result of irrigation throughout the variants. No-till treatment resulted in reduced nitrate leaching (18 kg ha^{-1}) as compared with the tillage system with plough and tooth cultivator (27 kg ha^{-1}). The method of quantifying deep seepage rates based on soil hydrological measurements was tested in comparison with lysimeter discharge measurements. Willmott's index of agreement was $d = 0.97$ revealing the validity of this simplified approach. Results underline the hypothesis of the well balanced, slow and continuous progression of the soil water content below the root zone. The suitability of long-term soil hydrological in situ measurements for quantifying arable management effects on ecological processes – deep seepage dynamics and solute leaching – was confirmed.

Keywords Long-term measurements · Soil hydrology · Field measurements · Deep seepage · Ground water discharge · Soil · Land use · Arable management · Nitrate leaching · Hydraulic conductivity · Water retention function

17.1 Introduction

Arable land is the main source of deep drainage and groundwater recharge in Northeast (NE) Germany (Müller, Dannowski, Schindler, Eulenstein, & Meißner, 1996). Land management practice is a decisive factor for the quantities of seepage flow and solute leaching (Benson, VanLeeuwen, Sanchez, Dohoo, & Somers, 2006; Köhler, Duijnisveld, & Böttcher, 2006), which constitutes two fundamental aspects of land use characterised by potentially conflictive ecologic implications. Efficient water use and intelligent water management are essential for NE Germany as a sub-humid region marked by an annual water balance deficit between 80 and 250 mm (Dyck & Peschke, 1989; Müller et al., 1996). Throughout that region, measures are in demand to support groundwater recharge. To meet this claim, knowledge is required about suitable land management systems providing seepage flow sustainable in quantity and quality.

Soil water content and tension are basic hydrologic variables that reflect effects of land surface processes. Soil moisture information, in spite of its importance to land use planning, agriculture and drought monitoring,

U. Schindler (✉)
Leibniz Centre for Agricultural Landscape Research (ZALF), Eberswalder Str. 84, D-15374 Müncheberg, Germany
e-mail: uschindler@zalf.de

F. Müller et al. (eds.), *Long-Term Ecological Research*,
DOI 10.1007/978-90-481-8782-9_17,

is not widely available. There is a lack of reliable methods and measured values for predicting effects of land use change on the soil water and solute status. Evaluating the impact of land use and management practices on soil water dynamics and solute leaching turns out to be difficult. This is because of few techniques which are capable of monitoring the quantity and quality of soil water flow below the root zone without disturbing the soil profile and affecting natural flow processes.

In general, underground lysimeter techniques are applied for measuring the seepage flow through the soil matrix by intercepting water and manually or automatically sampling the accumulated volume of water (Brye, Norman, Bundy, & Gower, 2001; Masarik, Norman, Brye, & Baker, 2004). Soil hydrological measuring installations, so-called virtual lysimeters (Kastanek, 1995), provide a good way for analysing the soil water and matter status in situ under undisturbed soil conditions. Compared with real lysimeters these methods are distinctly less invasive and just as much less expensive. In the first instance, their use is directed at obtaining knowledge about the processes including their dynamics of the soil water and solute status under different conditions of soil, land use, climate and agricultural management. Utilising this knowledge for transmitting quantitative results to the landscape level requires, in the next step, the development and application of models. For validating those models, in turn, soil hydrological measuring installations may provide the qualified basis (Mölders, Haferkorn, Döring, & Kramm, 2003; Miao et al., 2003; Bryant, Gburek, Veith, & Hively, 2006; Köhler et al., 2006). Potentially long residence times of water and solutes in the vadose zone, as well as the variability of weather conditions and management practices, dictate long-term experiments to be accomplished to assess solute transport processes (Narasimhan, Srinivasan, Arnold, & DiLuzio, 2005).

Starting in 1993, the impact of arable crop and soil management on seepage flow and solute leaching has been studied by the use of soil hydrological measuring installations under field experiments in the NE German lowlands. In the following, the method is detailed and results are presented to the reliability of those deep seepage estimations tested in comparison with real lysimeter measurements, the impact of different arable management practices on deep seepage dynamics and nitrogen leaching.

17.2 Materials and Methods

17.2.1 Site and Soil Characteristics

Agricultural investigations including the described soil hydrological measurements were carried out at the ZALF Müncheberg V4 experimental field on seepage-disposed sandy soils. Müncheberg is located in a Pleistocene end moraine landscape in NE Germany about 50 km east of Berlin and 40 km west of the Oder River, the German-Polish border. Long-term mean annual precipitation is 562 mm (323 mm in summer, 239 mm in winter time, measured values). Average air temperature of the last 50 years is 8.3°C with a tendency to increase. Soils are formed of Pleistocene parent material, classified as Haplic Albeluvisol (Eutric, Arenic) (WRB, 2006, Table 17.1) with a Bt horizon mostly beginning at about 70 cm depth. Thickness of the Bt horizon varies between 20 and 40 cm. With the exception of the Bt horizon, sandy substrates are dominating.

Measured annual precipitation largely varied over the study period (Table 17.2). It ranged from 418 mm in 1999 to 760 mm in 2002. Average annual precipitation was 543 mm and thus a little below the long-term average.

Table 17.1 Average soil parameters, Müncheberg V4 experimental field

Horizon	Depth cm	Clay %	Silt %	Sand %	DBD g cm^{-3}	Porosity cm^3 cm^{-3}	Θ_1 cm^3 cm^{-3}	Θ_2 cm^3 cm^{-3}
Ap	0–30	1	40	59	1.52	0.398	0.150	0.023
Bv	30–70	4	24	72	1.72	0.339	0.119	0.012
Bt	70–100	9	65	26	1.71	0.334	0.161	0.050
C	>100	1	22	77	1.70	0.344	0.073	0.009

DBD, dry bulk density; Θ_1, water content at pF2; Θ_2, water content at pF4.2

Table 17.2 Annual precipitation during 1994–2006 at the Müncheberg site, measured values

Year	1994	1995	1996	1997	1998	1999	2000	2001	2002	2003	2004	2005
***P* (mm)**	712	507	618	541	654	418	446	502	760	452	531	549

17.2.2 Arable Management Systems

In the measuring period (1994–2005) three management systems were practised and investigated at the Müncheberg V4 experimental field in three distinct stages.

From 1994 to 1998 different farming systems were tested (Table 17.3). Both at the integrated and the low input system mineral nitrogen fertilizer was applied. At the ecologic farming system exclusively organic nitrogen fertilizer (manure and slurry) was used. From 1999 to 2000 a break was inserted cultivating lucerne (alfalfa) grass at all plots. In 2001 tillage experiments started. The effect of different tillage systems (i – plough in combination with tooth cultivator and ii – no-till) on yield and on the soil water and solute balances was analysed. Table 17.3 provides an overview of the average annual nitrogen fertilizer input, annual irrigation rate and yield of the management systems as an average over the replications.

Crop rotations were:

1994–1998	Sugar beet, winter wheat, winter barley and winter rye with catch crop
1999–2000	Lucerne grass at all plots
2001–2005	Plough: winter wheat, winter barley, winter rape, triticale, winter rye + oil radish and potato
	No-till: winter wheat + oil radish, corn, winter rye + oil radish, pea, winter barley and winter rape

17.2.3 Soil Hydrological Measurements

Water and solute movement through the vadose zone should be compartmentalised into (i) movement through the root zone and (ii) movement below the root zone (Gee & Hillel, 1988). Fluxes within the root zone are spatially and temporally variable. Fluxes below the root zone may be transient or steady state (Gee & Hillel, 1988; Schindler & Müller, 1998).

The basic idea was to detect seepage flow preferably below the root zone where soil water movement is directed predominantly downwards, changes in the soil hydrologic status will proceed slowly and continuously, and rainfall events or fluctuations of evapotranspiration will not produce distinct changes in soil water content and tension.

For this reason, continuous measurement of tensions and water content values (16 soil hydrological plots, Fig. 17.1) was carried out down to 3 m depth. Based on soil tension measurement at two depths, the proof of downward water movement may be given directly from the hydraulic gradient, or where required, periods of capillary rise may be detached. The instruments were buried to enable agricultural equipment to pass over the measuring plots.

Precipitation was measured at two levels, 1 m above and next to the ground. Temperature, relative air humidity, global radiation and some other meteorological variables were measured directly at the

Table 17.3 Average annual management data, Müncheberg V4 experimental field, 1994–2006

Period	Management system	Plot number (according to Fig. 17.1)	N_{total} kg ha^{-1}	I mm a^{-1}	Yield[a] dt ha^{-1}
1994–1998	IO: integrated without irrigation	1, 4, 7,10	112	0	76
	IB: integrated with irrigation	1A, 4A, 7A, 10A	126	78	86
	OO: ecologic without irrigation	2, 5, 8, 11	45[b]	0	43
	EO: low input without irrigation	2, 6, 9, 12	86	0	73
1999–2000	Lucerne grass	All plots	–	–	–
2001–2005	P: plough with tooth cultivator	1, 3, 5, 7, 9, 11	165	–	59
	NT: no-till	2, 4, 6, 8, 10, 12	146	–	62

[a]Dry matter yield in dt ha^{-1}, [b]only manure and slurry

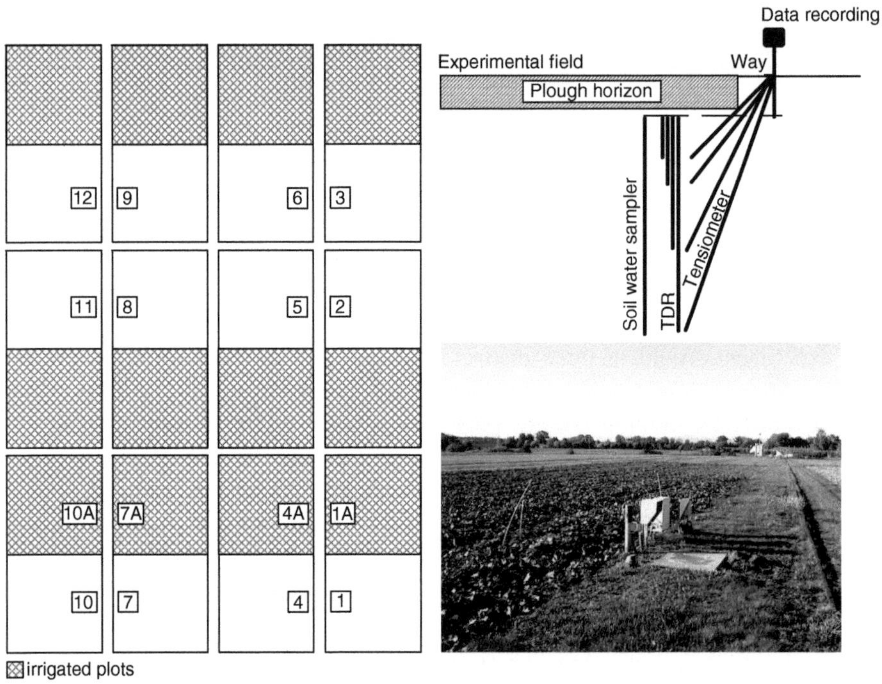

Fig. 17.1 Müncheberg V4 experimental field with measurement plots and schematic of the soil hydrological installation

Müncheberg weather station. Soil water samplers were used for analysing the nitrate concentration in seepage water. Taking deep seepage dynamics into consideration, nitrogen losses were estimated.

17.2.3.1 Description of the Method

Over time, the soil at measuring depth is considered as being a pipe of variable filling. Soil water content is taken as an indicator for the filling level (Schindler & Müller, 1998). Transformation of the filling level (soil water content) into flux (seepage flow rate v) is undertaken by the Darcy equation ($v = K(\Theta) \times i$) containing a non-linear scaling factor, the hydraulic conductivity function $K(\Theta)$ being dependent on water content Θ and the hydraulic gradient (i) as the driving force. Unit gradient is assumed to be valid. The method is feasible on sandy soils and deep water table. Daily seepage flow rates (v) are calculated based on water content measurements and the hydraulic conductivity function $K(\Theta)$ calibrated to the water balance.

Water Content and Tension Measurements

Water content (TDR technique: Malicki & Skierucha, 1989; Roth, Malicki, & Plagge, 1992) and tension are continuously measured below the root zone. For quantifying the hydraulic gradient tension is measured at two depths. From the water content and tension values at 3 m depth a field pF curve is constructed. This pF curve is valid only in the range of measurements and serves as the basis for calculating the unsaturated hydraulic conductivity function.

Prediction of the Relative Hydraulic Conductivity Function and Derivation of Relative Seepage Rates

The pF curve is fitted (van Genuchten, 1980) and the relative hydraulic conductivity function ($K_r(\Theta)$) is predicted (Mualem, 1976). Instead of tension, the water content is used because of no hysteresis effects appearing. Hydraulic gradients (i) are calculated from the tensions between the two depths and allow to conclude on flow direction (positive – downwards, seepage; negative – upwards, capillary rise). The relative seepage rates (v_r) are calculated by the simplified Darcy equation ($v_r = K_r(\Theta)$) based on the water content dynamics and the relative hydraulic conductivity function assuming the unit gradient 1 for periods with downward water movement.

Calibration of the Hydraulic Conductivity Function to the Water Balance

The water balance at the soil hydrological plot is calculated for a calibration period. Usually preference is given to an autumn/winter period with negligible error for the estimation of the evapotranspiration (ET). In the NE German region, the evapotranspiration during the winter period is estimated according to Ivanov (Wendling, Schellin, & Thomä, 1991). The precipitation (P) is measured and the soil water storage difference dΘ in the profile is calculated by means of water content measurements over the whole profile at the beginning and the end of this period. At the start and the end of the calibration period the soil profile should be free of frost. The total discharge (V) results from $V = P - ET - d\Theta$. V is divided by V_r (total relative seepage of the calibration period) to get the matching factor M ($M = V/V_r$) for transforming the relative hydraulic conductivity function into an effective level.

17.2.4 Deep Drillings

Additional deep drillings down to 4.2 m depth were used to analyse the nitrate concentration in soil water. These results were compared with nitrate concentrations of the soil water sampler to get an idea of the variability and accuracy of the latter values. Drillings (four drillings at each soil hydrological plot) were carried out at the end of 1995.

Water content and nitrate concentration were measured for 30 cm soil increments in the laboratory. Water content values were obtained by comparing soil sample weights before and after oven drying. Nitrate concentration was analysed photometrically after extraction with 0.02 N $CaCl_2$ solution.

17.2.5 Lysimeter Experiments

One gravitational lysimeter (1 m × 1 m; 2 m depth) from the Dedelow lysimeter station located in the Uckermark region about 100 km north from Müncheberg (Schindler, Wolff, & Kühn, 2001) was used for the comparison of deep seepage estimations based on the virtual lysimeter concept with the measured real lysimeter discharge. The soil stratification and soil properties of the lysimeter are given in Table 17.4.

Table 17.4 Stratification and soil properties of the Dedelow lysimeter

Horizon	Depth cm	C_t %	DBD g cm^{-3}	Clay %	Silt %	Sand
Ap	0–35	0.64	1.52	4	24	72
Bv	35–115	0.1	1.63	9	29	62
C	115–200	0.1	1.65	1	4	95

Two tensiometres (Ψ) in 1.6 and 1.85 m depth and one TDR-probe (Θ) in 1.85 m were installed. The water content dynamics in 1.85 m depth was taken as the basis for seepage calculations. Measuring values were recorded at 8-h intervals. Weather data were registered directly at the site.

The hydraulic conductivity function (effective function) was calibrated to the lysimeter discharge for the period from 1 November 2001 to 15 February 2002, whereas the total period of comparison lasted from November 2001 to February 2006.

Willmott's index of agreement, d (Equation 17.1) (Legates & McCabe, 1999), was used as criteria for assessing the performance of the measuring concept.

$$d = 1 - \sum (O_i - P_i)^2 / \sum (|P_i - O_m| + |O_i - O_m|)^2 \quad (17.1)$$

O – observed values, P – predicted values, O_m – mean observed value.

This index varies between 0.0 (poor agreement) and 1.0 (perfect agreement), similarly to the coefficient of determination r^2.

17.3 Results and Discussion

17.3.1 Comparison of Lysimeter Discharge and Calculated Seepage

The soil hydrological concept as introduced above was tested against lysimeter results. Tension and water content were measured in the lysimeter with high temporal resolution and accuracy. Both variables showed comparable dynamics with regard to drainage and saturation processes (Fig. 17.2).

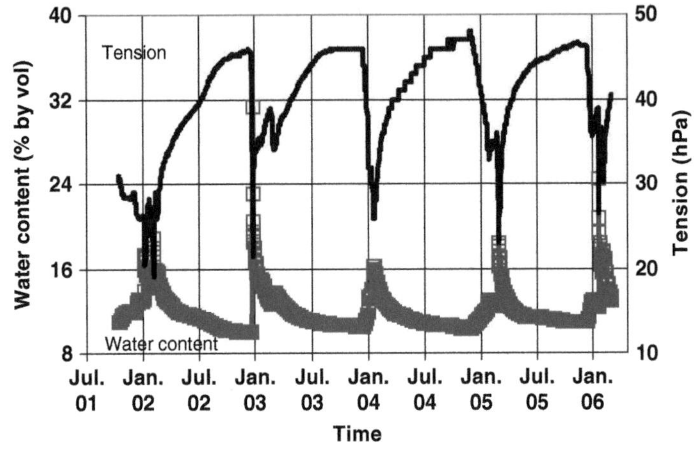

Fig. 17.2 Tension and water content dynamics in the Dedelow lysimeter, 1.85 m depth

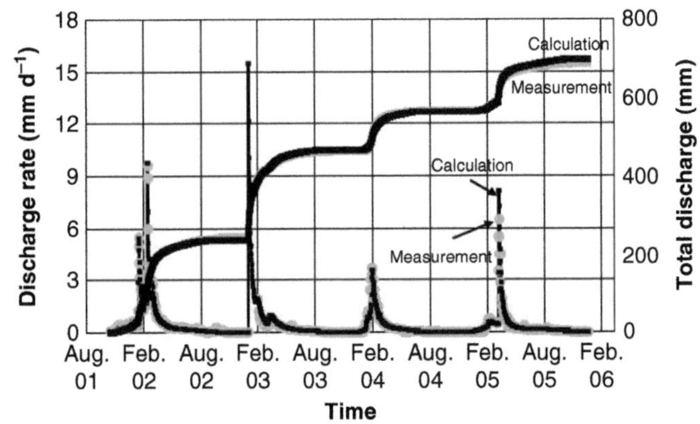

Fig. 17.3 Comparison of lysimeter discharge with discharge calculations

Seepage rates were calculated day by day on the basis of the water content dynamics and the calibrated hydraulic conductivity function (Fig. 17.3) and compared with daily lysimeter discharge values for the period 2001–2006. Willmott's index was $d = 0.97$, which confirmed the validity of this simplified approach for quantifying effective deep drainage flow rates.

17.3.2 Deep Seepage Dynamics and Arable Management Effects

In situ temporal soil water content and calculated seepage dynamics are presented in Fig. 17.4 exemplarily for one selected plot (period 1995–2005) of the Müncheberg V4 experimental field. Measurements of

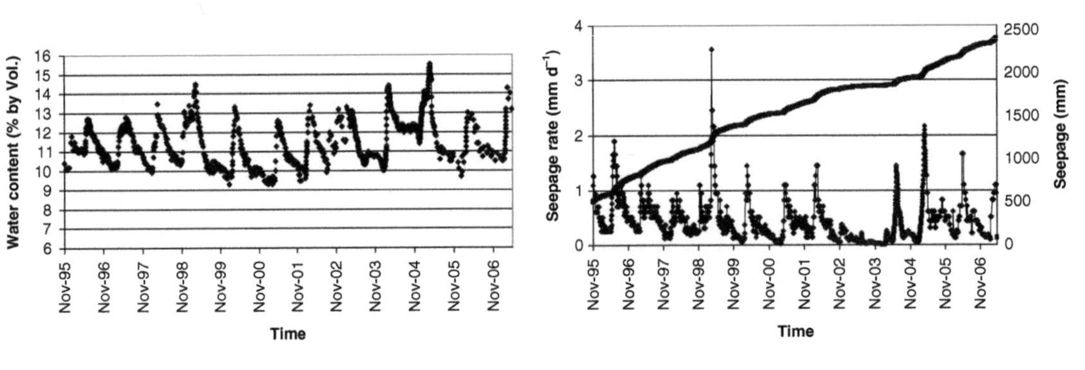

Fig. 17.4 Soil water content (*left*) and deep seepage dynamics (*right*) at 3 m depth, plot 2 at Müncheberg V4 experimental field

the soil water content allowed for a temporally highly disintegrated analysis of soil hydrologic conditions below the root zone. Periodic processes of seepage flow formation in the winter half-year characterised by rising soil water content or falling soil water tension and subsequent soil drainage during the vegetation period are clearly visible. Individual periods are well distinguished. Seepage periods started mostly between January and March followed by soil drainage during summer and autumn time.

Results confirm the hypothesis of the well balanced, slow and continuous progression of the soil water content in the deeper vadose zone below the root zone down to 3 m depth (Kutilek & Nielsen, 1994; Schindler & Müller, 1998). Daily changes in measured values performed small and allowed for the assumption of quasi-steady-state conditions within intervals of 1 day.

As the result of high variation in annual precipitation (see Table 17.1) and land management effects, the variation of annual deep seepage rates appeared high. Results differed as follows between the farming systems (stage 1994–1998). The integrated system with irrigation IB (191 mm a^{-1}, RMS 47 mm a^{-1}) produced highest seepage rates followed by the ecologic farming system OO (180 mm a^{-1}, RMS 33 mm a^{-1}), which in turn had the lowest yields (see Table 17.3). The integrated system without irrigation IO (161 mm a^{-1}, RMS 50 mm a^{-1}) and the low input system EO (154 mm a^{-1}, RMS 38 mm a^{-1}) produced nearly the same deep seepage rates and quite the same yields, too. The differences, however, were not significant. The lowest deep seepage rate was calculated with 130 mm a^{-1}, RMS 18 mm a^{-1} under lucerne grass (stage 1999–2000), but precipitation turned out to be remarkably low (436 mm a^{-1}) in this period. Results of tillage experiments are valid for the stage 2001–2005. Tillage with plough (191 mm a^{-1}, RMS 68 mm a^{-1}) produced higher deep seepage rates than the no-till system (158 mm a^{-1}, RMS 49 mm a^{-1}). Differences, however, were not significant again.

17.3.3 Nitrate Concentration and Nitrate Leaching

Figure 17.5 shows the dynamics of nitrate concentrations under the practised and tested arable management systems as an average of the replications from 1993 to 2005. There was an acceptable congruence between deep drilling results and the soil water samplers.

Nitrate concentrations at all Müncheberg measurement plots decreased until 1995, after which they levelled off. Obviously, the initial nitrate dynamics were a consequence of the preceding agricultural activities on that site. Since 1995 the nitrate concentrations

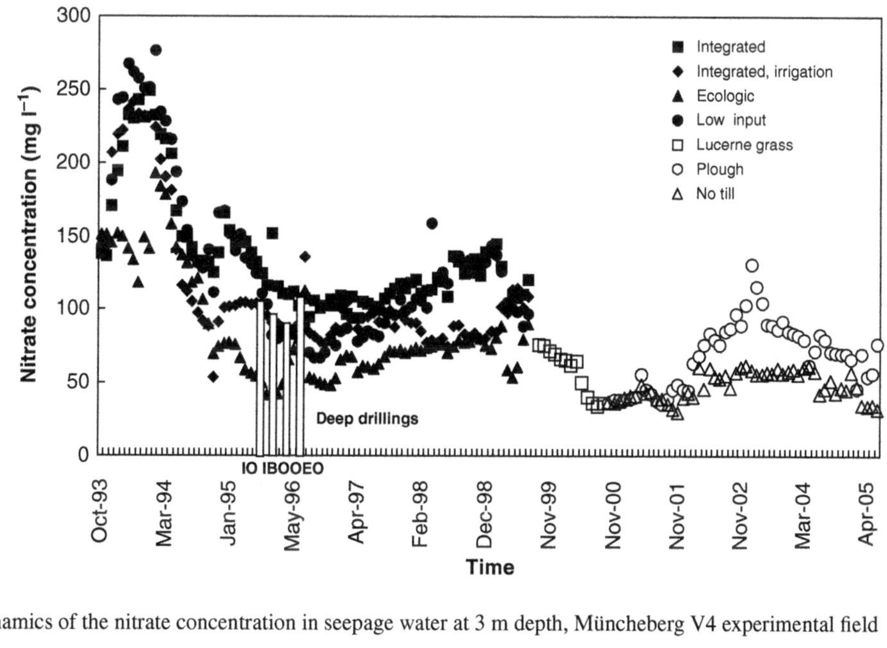

Fig. 17.5 Dynamics of the nitrate concentration in seepage water at 3 m depth, Müncheberg V4 experimental field

in seepage water have been reflecting the effect of the investigated farming systems. Lowest nitrate concentrations were observed at the ecologic farming system, where the nitrate concentration varied between approximately 50 and 80 mg l^{-1}. Highest nitrate concentrations (100–130 mg l^{-1}) were analysed under the integrated system without irrigation, where the nitrogen input was nearly twice as high as at the ecologic system. Most fertilizer was applied at the integrated system with irrigation. This treatment, however, displayed nitrate concentrations similar to the ecologic system between 80 and 100 mg l^{-1}. In all cases, nitrate concentrations consistently exceeded the 50 mg l^{-1} drinking water threshold (Bundesgesetzblatt, 1990). Starting in 1999, as soon as lucerne grass was growing, the nitrate concentration at all measurement plots decreased continually and levelled off at approximately 45 mg l^{-1} in 2000. With the beginning of the tillage experiments in 2001, nitrate concentrations in seepage water altered again. At ploughed plots concentration increased continually to reach its maximum of 140 mg l^{-1} in 2003. Afterwards the concentration decreased continually and approached the values of no-till plots in 2005. Reason for this dramatic increase after changing the management could be mineralisation processes of the organic matter. At no-till plots the nitrate concentration varied just a little over the observation period (2001–2005) and remained at the low level of 40–60 mg l^{-1}.

Nitrogen leaching rates were calculated based on nitrate concentration and deep seepage rates. The differences of annual nitrogen leaching rates between the farming variants were small and not significant. In the study period 1994–1998 the average annual nitrogen leaching rates varied between 41 kg N ha^{-1} (RSM 16 kg ha^{-1}) at the integrated plots IO (without irrigation) and 36 kg N ha^{-1} (RSM 12 kg ha^{-1}) at the ecologic variant (OO). Leaching rates of the low input variant EO (36 kg N ha^{-1}, RSM 16 kg ha^{-1}) and the integrated variant with irrigation IB (37 kg N ha^{-1}, RSM 9 kg ha^{-1}) did not differ much from the ecologic system. Irrigation, however, had a noticeable effect on yield and nitrogen intake. The lowest fertilizer input and the lowest yields occurred at the ecologic treatment. Nevertheless, this treatment caused nitrogen leaching rates similar to those of the other treatments. Nitrogen leaching under lucerne grass (1999–2000) was strongly reduced (14 kg N ha^{-1}, RSM 7 kg ha^{-1}) compared to the previous management. Tillage with plough in combination with tooth cultivator (27 kg N ha^{-1}, RSM 15 kg ha^{-1}) produced increased nitrogen leaching as compared with the no-till variant (18 kg N ha^{-1}, RSM 7 kg ha^{-1}). The differences, however, were not significant.

17.4 Conclusions

The effect of the investigated farming systems (integrated, integrated with irrigation, ecologic and low input) on nitrate dynamics was surprisingly small. The expected advantage of ecologic farming (Freyer, 2003; Pacini, Wossink, Giesen, Vazzana, & Huirne, 2003; Stark, Condron, Stewart, Di, & O'Callaghan, 2006) could not be confirmed. As ecologic treatment produced the lowest yields, it did not show noticeably lowered nitrogen leaching. Irrigation in turn provided a noticeable effect on yield and nitrogen intake, but nitrogen leaching did not significantly differ from the ecologic treatment. No-till treatment displayed lower nitrate leaching than the tillage system with plough and tooth cultivator, but mineralisation processes could have influenced this result. Also the differences of deep seepage and nitrate leaching rates between the tested farming systems were small and not significant. In order to obtain significance, a prolonged measuring time, more replications and additional sites are needed.

Over most of the time, nitrate concentration exceeded the 50 mg l^{-1} threshold value for drinking water at all management treatment sites. Due to the low deep seepage rates, nitrate surpluses of 20–30 kg ha^{-1} already led to nitrate concentrations of about 100 mg l^{-1}, suggesting that seepage water nitrate concentrations <50 mg l^{-1} as stipulated for sustainable agricultural practices are not realistic at the sandy and weak loamy arable sites in NE Germany.

The virtual lysimeter concept was tested against lysimeter discharge measurements and found to be valid. Preconditions are accurate water content and tension measurements in the deeper vadose zone with high temporal resolution and a hydraulic conductivity function thoroughly calibrated to the water balance. Test results confirm the hypothesis of the well balanced, slow and continuous progression of the soil water content below the root zone. The suitability of soil hydrological long-term measurements for quantifying arable management effects on ecological processes – deep seepage dynamics and solute leaching – was thus confirmed.

Acknowledgements The authors thank Brigitte Krüger and Gabriele Bähr for the preparation of the management data. We thank the persons Nico Zindler and Michael Bähr responsible for the instrumentation and measurements. Furthermore we thank Dr. R. Roth who was responsible for the agricultural experiments.

References

Benson, V. S., VanLeeuwen, J. A., Sanchez, J., Dohoo, I. R., & Somers, G. H. (2006). Spatial Analysis of land use impact on ground water nitrate concentrations. *Journal of Environmental Quality, 35*, 421–432.

Bryant, R. B., Gburek, W. J., Veith, T. L., & Hively, W. D. (2006). Perspectives on the potential for hydropedology to improve watershed modeling of phosphorus loss. *Geoderma, 131*, 299–307.

Brye, K. R., Norman, J. M., Bundy, L. G., & Gower, S. T. (2001). Nitrogen and carbon leaching in agroecosystems and their role in denitrification potential. *Journal of Environmental Quality, 30*, 58–70.

Bundesgesetzblatt. (1990). Verordnung über Trinkwasser und über Wasser für Lebensmittelbetriebe (Trinkwasserverordnung – TrinkwV) vom 5.12.1990. Bundesgesetzblatt, I, 2612.

Dyck, S., & Peschke, G. (1989). *Grundlagen der Hydrologie, 2. Auflage*. Berlin: Verlag für Bauwesen.

Freyer, B. (Ed.). (2003). *Beiträge zur 7. Wissenschaftstagung zum Ökologischen Landbau: Ökologischer Landbau der Zukunft, Contributions to the 7th Research Conference on Organic Agriculture: Organic Agriculture of the Future. Proceedings of 7. Wissenschaftstagung zum ökologischen Landbau: Ökologischer Landbau der Zukunft, Wien, 23–26.2.2003*. Wien: Universität für Bodenkultur Wien – Institut für ökologischen Landbau.

Gee, G. W., & Hillel, D. (1988). Groundwater recharge in arid regions: Review and critique of estimation methods. *Hydrological Processes, 2*, 255–266.

Kastanek, F. (1995). Kritische Bemerkungen zur Verwendung von Lysimetern. In: Bundesanstalt für alpenländische Landwirtschaft (Ed.), *BAL-Bericht 5. Gumpensteiner Lysimetertagung "Stofftransport und Stoffbilanz in der ungesättigten Zone": BAL Gumpenstein 25–26. April 1995* (pp. 93–102). Gumpenstein: Bundesanstalt für alpenländische Landwirtschaft.

Köhler, K., Duijnisveld, W. H. M., & Böttcher, J. (2006). Nitrogen fertilization and nitrate leaching into groundwater on arable sandy soils. *Journal of Plant Nutrition and Soil Science, 169*, 185–195.

Kutilek, M., & Nielsen, D. R. (1994). *Soil hydrology*. GeoEcology textbook. Reiskirchen: Catena Verlag.

Legates, D. R., & McCabe, G. J. (1999). Evaluating the use of "goodness-of-fit" measures in hydrologic and hydroclimatic model validation. *Water Resources Research, 35*, 233–242.

Malicki, M. A., & Skierucha W. M. (1989). A manually controlled TDR soil moisture meter operating with 300 ps rise-time needle pulse. *Irrigation Science, 10*, 153–163.

Masarik, K. C., Norman, J. M., Brye, K. R., & Baker, J. M. (2004). Improvements to measuring water flux in the vadose zone. *Journal of Environmental Quality, 33*, 1152–1158.

Miao, Z., Cheplick, M. J., Williams, M. W., Trevisan, M., Padovani, L., Gennari, M., et al. (2003). Simulating pesticide leaching and runoff in rice paddies with the RICEWQ–VADOFT Model. *Journal of Environmental Qualità, 32*, 2189–2199.

Mölders, N., Haferkorn, U., Döring, J., & Kramm, J. (2003). Long-term investigations on the water budget quantities predicted by the hydro-thermodynamic soil vegetation scheme (HTSVS) – Part I: Description of the model and impact of long-wave radiation, roots, snow, and soil frost. *Meteorology and Atmospheric Physics, 84*, 115–135.

Mualem, Y. (1976). A new model for predicting the hydraulic conductivity of unsaturated porous media. *Water Ressources Research, 12*, 513–522.

Müller, L., Dannowski, R., Schindler, U., Eulenstein, F., & Meißner, R. (1996). Gebietsabflüsse aus Agrarlandschaften Nordost- und Mitteldeutschlands. *Archiv Acker- und Pflanzenbau und Bodenkunde, 40*, 345–362.

Narasimhan, B., Srinivasan, R., Arnold, J. G., & DiLuzio, M. (2005). Estimation of long-term soil moisture using a distributed parameter hydrologic model and verification using remotely sensed data. *Transactions of the ASAE, 48*, 1101–1113.

Pacini, C., Wossink, A., Giesen, G., Vazzana, C., & Huirne, R. (2003). Evaluation of sustainability of organic, integrated and conventional farming systems. A farm and field-scale analysis. *Agriculture, Ecosystems & Environment, 95*, 273–288.

Roth, C. H., Malicki, M. A., & Plagge, R. (1992). Empirical evaluation of the relationship between soil dielectric constant and volumetric water content as the basis for calibrating soil moisture measurements by TDR. *Soil Sciences, 43*, 1–13.

Schindler, U., & Müller, L. (1998). Methods for the quantification of soil drainage rates in the north-east German lowlands. In American Society of Agricultural and Biological Engineers (Ed.), *Proceedings of the 7th International Drainage Symposium*, March 8–11, Orlando, FL (pp. 613–620). St. Joseph: ASABE.

Schindler, U., Wolff, M., & Kühn, G. (2001). Lysimeterstudie zum Einfluss von Düngung und Bewirtschaftung auf die Ertragsbildung, den Wasserhaushalt und die Nährstoffauswaschung im Trockengebiet der Uckermark. *Zeitschrift für Pflanzenernährung und Bodenkunde, 164*, 697–703.

Stark, C., Condron, L. M., Stewart, A., Di, H. J. & O'Callaghan, M. (2006). Effects of past and current management practices on crop yield and nitrogen leaching – a comparison of organic and conventional cropping systems. *New Zealand Journal of Crop and Horticultural Science, 34*, 207–215.

Van Genuchten, M. T. (1980). A closed-form equation for predicting the hydraulic conductivity of unsaturated soils. *Soil Science Society of America Journal, 44*, 892–898.

Wendling U., Schellin, H. G., & Thomä, M. (1991). Bereitstellung von täglichen Informationen zum Wasserhaushalt des Bodens für die Zwecke der agrarmeteorologischen Beratung. *Zeitschrift für Meteorologie, 41*, 1–16.

World Reference Base for Soil Resources. (2006). *World Soil Resources Reports No. 103.* Rome: FAO.

Chapter 18

Long-Term Ecosystem Research in a Beech Forest of Northern Germany

Filipa Tavares, Otto Fränzle, Felix Müller, and Claus-G. Schimming

Abstract In Central Europe, damage of forests has been clearly identified to be caused by atmospheric deposition of acidifying air pollutants and eutrophicating amounts of nitrogen. The rapidly growing public concern about this phenomenon induced numerous studies particularly in Germany already in the 1970s and 1980s, which produced an impressive number of pertinent botanical, parasitological, pedological and geographical information. In general, however, focus was on specific aspects of forest decline, which not infrequently led to monocausal explanations.

The concept of the interdisciplinary and transdisciplinary Bornhöved ecosystem research programme reflects basic principles of scientifically based programmes of environmental observation. The general aim of the Bornhöved Project was to study ecosystem structure and evolution in a highly complex landscape of the North German Plain. Twenty years of high-resolution time-series of pertinent meteorological, hydrological, biological and pedological data on the energy and material fluxes of the Bornhöved beech stands are available for both monitoring purposes and as a basis for ecosystem theory building. Atmospheric deposition and element behaviour in the soil solution exhibit distinct trends and periodicities in biogeochemical transfer processes in relation to the water use of the ecosystems. Efficiency measures such as the transpiration/evapotranspiration ratio, element fluxes or budgets and the stoichiometric definition of elemental imbalances have proved to be commendable indicators.

C.-G. Schimming (✉)
Ökologie-Zentrum der Christian-Albrechts-Universität zu Kiel, Olshausenstr. 40, 24098 Kiel, Germany
e-mail: cschimming@ecology.uni-kiel.de

Keywords Forest decline · Element fluxes · Atmospheric deposition · Soil-solution chemistry · Functional indicators · Element budgets · Soil acidification · Nitrogen load

18.1 Introduction

In Central Europe, damage of forests in the neighbourhood of industrial plants was attributed to flue gases, in particular sulphur dioxide, already several centuries ago (Agricola, 1556). But in the 1960s an apparently new type of forest injury became evident which was no longer related to local emitters but occurred to an increasing extent in presumably clean air areas, far away from conurbations or industrial centres. The rapidly growing public concern about this phenomenon induced numerous studies in Germany in the 1970s and 1980s, which produced an impressive number of pertinent botanical, parasitological, pedological and geographical information. In general, however, focus was on specific aspects of forest decline, which not infrequently led to monocausal explanations. On the international level, the Convention on Long-Range Transboundary Air Pollution (CLRTAP), stipulated under the auspices of the United Nations Economic Commission for Europe in 1985, led to the International Co-operative Programme on the Assessment and Monitoring of Air Pollution Effects on Forests (ICP Forests). In the following decades this programme developed into one of the major biomonitoring networks worldwide.

Although it had been noticed in early CLRTAP discussions already that a deeper understanding of the potentially harmful effects of air pollution on forests

F. Müller et al. (eds.), *Long-Term Ecological Research*,
DOI 10.1007/978-90-481-8782-9_18, © Springer Science+Business Media B.V. 2010

was a prerequisite for reaching agreements on effective pollution control, a comprehensive ecological analysis of the multifactorial phenomenon of forest growth and decline was realized only at a few centres, e.g. the Forest Ecosystem Research Centre at Göttingen, in the framework of the Solling Project (Ellenberg, Mayer, & Schauermann, 1986) or at the Ecosystem Research Centre of the Kiel University. As a consequence, comparative critical evaluations of the above forest decline studies with their mostly narrower scope and frequently controversial issues appeared only since the middle of the 1990s (cf., e.g. Ellenberg, 1995). Today, there is agreement that the decline of forest ecosystems is under the control of three groups of factors, namely, first a predisposing set of site factors such as soil and climate, second a set of interrelated triggering effects of both natural or human origin and third decline may prove liable to additional impacts which are not necessarily primary components of an injurious symptom complex but contribute to rapidly intensifying damage up to mortality (Elling, Heber, Polle, & Beese, 2007).

In the following, some results of the long-term inquiries into structure and functioning of representative beech ecosystems of the Bornhöved Lake district, situated some 30 km south of Kiel in Schleswig-Holstein, are presented as a contribution of the Kiel Ecology Centre to pertinent ICP-Forests studies which, in accordance with the CLRTAP philosophy, are to include effect-oriented analyses of airborne pollution fluxes. First an introductory summary of the major objectives of the ecosystem research in the Bornhöved Lake district is given, in which a beech forest stand belongs to the ICP network since 1995 and is part of LTER since 2006, then atmospheric deposition and element fluxes in the beech soil are analysed and finally typical seasonal variations of flux-related complex ecosystem parameters are illustrated.

18.2 Ecosystem Research in the Bornhöved Lake District

The concept of the interdisciplinary and transdisciplinary Bornhöved ecosystem research programme reflects the basic ideas of UNESCO's Man and the Biosphere (MAB) Programme as a follow-up action of the International Biological Programme (IBP). In accordance with the MAB project areas and in realization of the focal IBP proposition to study landscapes as ecosystems, which had remained unfulfilled, however, the general aim of the Bornhöved Project was to study ecosystem structure and evolution in a highly complex landscape of the North German Plain which, in the light of geostatistical analyses and regionalization procedures (Fränzle, 2000), is representative of the whole set of the structurally and functionally interrelated terrestrial and aquatic ecotope complexes along the margins of the Weichselian glaciation.

As a consequence of the complicated natural history and the human impact since the Early Neolithic, the ecosystems of the study area have faced long-term and short-term changes of their energy, water and nutrient cycles. Accordingly, the general aims of the transdisciplinary Bornhöved enquiries, which started in 1988 and covered a primary observation period of 12 years, were the analysis and modelling of ecosystem structures, dynamics and stability conditions and the assessment of environmental strains and related resilience mechanisms. Among the specific objectives of the research scheme, defined in correspondence with the distinctive structural features of the study area, the following have been of particular importance: First the determination and modelling of biotic, energetic and material exchange processes between terrestrial and aquatic ecosystems of different land use or fishery patterns with focus on the role of lentic ecotones, then modelling of agrarian and pasture ecosystems in the light of national and international production and marketing regulations and finally the development of novel instruments for landscape planning and agricultural management (Fränzle, Kappen, Blume, & Dierssen, 2008).

18.2.1 Selected Element Fluxes in the Beech Ecosystem

18.2.1.1 The Beech Reference System and Its Ecological Setting

The climate of the Bornhöved Lake district is humid mesothermal with a prevalence of moist maritime air masses and frequent cyclonic storms, particularly during the winter months. The soils originated from glacigenic, fluvioglacial, limnetic and organic deposits and form complicated spatial arrays of terrestrial,

Table 18.1 Site conditions of the beech reference system

Latitude	54°06′ 01″ N
Longitude	10°14′ 33″ E
Elevation	≤50 m
Vegetation type	*Galio odorati–Fagetum typicum*
Stand history	Afforestation of agricultural land in 1897
Mean height	34 m
Basal area	22 m^2 ha^{-1}
Bedrock	Meltwater sands covered by boulder clay
Soil type	Arenic Umbrisol
Humus type	Mould with raw humus characteristics
Precipitation[a]	720 mm
Mean temperature (1968–1998)	8.1°C

[a]Measured on the neighbouring field as a substitute of the above-canopy precipitation

semiterrestrial and subhydric soils. Among the terrestrial soils, Dystric and Stagnic Luvisols, Umbrisols and Stagnosols predominate under forest, Eutric Luvisols and Arenosols under tillage (Soil classification after World Reference Base for Soil Resources (2006)). The corresponding vegetation comprises communities of the *Alnetea glutinosae* and the *Querco-Fagetea.* Among the latter, the association *Galio-Fagetum* is of prime importance; its herb layer is often dominated by *Milium effusum.* Table 18.1 illustrates in greater detail the site conditions of this beech forest, where continuous deposition and flux measurements as a basis of the present contribution have been carried out since 1989.

A detailed insight into the soil conditions of the beech stand is provided by Table 18.2.

18.2.1.2 Atmospheric Deposition, Canopy Throughfall and Stemflow of Beech Stands

The comparative assessment of the deposition chemistry at both the country and the ecosystem scales, underlying the Bornhöved ecosystem studies, allows to reliably distinguish between local terrestrial sources of solute components from farther-distant marine ones. Thus, in comparison with the other federal states of Germany, Schleswig-Holstein is characterized by distinctly higher deposition rates of the airborne marine components, sodium, magnesium and chloride, while both the nitrate and sulphate rates are average. Owing to intensive emissions from agriculture, ammonium deposition is considerably increased, i.e. exceeding the federal average. Also, potassium and calcium deposition prove largely determined by local sources. With regard to increased ammonium and proportionately elevated sulphate concentrations, which are linked by gas-phase reactions (Fränzle, 1993), the seasonal pattern of agricultural activities with highly variable emissions from point and non-point sources is important. Wet deposition accounts for more than 85% of the inputs of sodium, copper, magnesium, chloride, ammonium, sulphate and nitrate into the ecosystems of the Bornhöved Lake district, while about 50% of the potassium and calcium inputs are due to dry deposition.

At the ecosystem scale, deposition rates were assessed by means of bulk samplers distributed in compliance with geostatistical requirements in order to obtain valid high-resolution areal data. Bulk samplers, however, can only provide estimates of the wet

Table 18.2 Physical and chemical characteristics of the Arenic Umbrisol under beech forest

			Texture					Exchangeable element pools					
Depth	Corg cm		Clay g kg^{-1}	Silt	Sand	pH $CaCl_2$	CEC $BaCl_2$	Al meq kg^{-1}	Fe	Mg	Ca	K	B.S.[1] %
L	−3.5	377	–	–	–	2.9	634	396	8.4	18	203	9.1	38
O	0	400	–	–	–	2.8	617	344	8	11	250	4.5	44
–	−3.5	472	–	–	–	–	–	–	–	–	–	–	–
Ah	7	340	70	150	780	3.0	980	819	14	24	114	9	16
RAp	38	90	90	180	730	4.0	480	336	8	8	127	1	30
Bw1	68	40	30	130	840	4.4	350	178	4	2	160	6	49
Bw2	91	40	20	170	810	4.5	240	121	4	2	106	7	50
RBg	112	<10	30	40	930	4.5	270	172	4	2	91	1	36
Bw3	125	<10	50	30	920	4.4	340	209	1	7	118	5	39
BdtC	148	<10	30	10	960	4.6	380	185	3	9	178	5	51
BwC	190	<10	30	10	960	4.4	290	180	1	4	102	3	38

[1]Base saturation

deposition and (an unknown part of) the dry deposition of gases and particles. Thus, apart from free acids, bulk deposition measurements yield only minimum values of the total deposition and, in combination with wet-only samplers, of the dry deposition, which assumes a particular importance in the framework of calibrating elemental flux models. Therefore, complementary estimates of total deposition were made on the basis of the Ulrich (1983) and van der Maas, Van Breemen, & Van Langenvelde (1991) approaches in combination with resistance models for dry deposition estimates (Wesely, 1989; Erisman, 1992). Figures 18.1 and 18.2 summarize the estimates of total nitrogen and sulphur deposition at the beech stand and, for comparison purposes, at the neighbouring agroecosystems.

Altogether, the figure shows that the long-term average deposition to the beech soil amounts to 1,911 mol N $ha^{-1}year^{-1}$ or 27 kg N ha^{-1} $year^{-1}$ with a minimum of 1,524 mol N ha^{-1} $year^{-1}$ (21.3 kg N ha^{-1} $year^{-1}$) in 1995 and a maximum of 2,385 mol N ha^{-1} $year^{-1}$ (33 kg N ha^{-1} $year^{-1}$) in 1989. Including dry deposition, a maximum deposition rate of about 40 kg N ha^{-1} $year^{-1}$ would be attained. Ammonia, preponderant among the NH_x species, is a result of increased emissions from decomposing animal manure which has been produced not only by cows and other ruminants grazing outdoors, but also by the increased number of cattle, pigs and poultry in highly intensive husbandry. Thus, integrating over the whole of the year, manured and fertilized plots of the Bornhöved Lake district are sources of ammonia emissions, while natural ecosystems in the area, such as the beech stands exhibit sink character (Branding, 1996). Dry deposition of ammonia to the canopy points to the importance

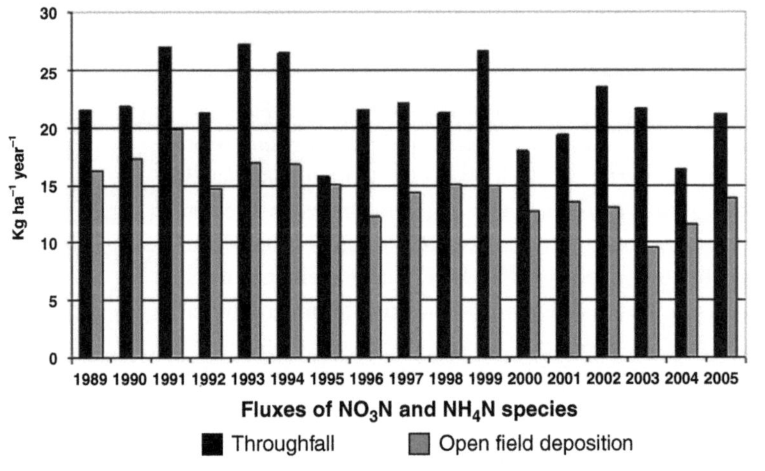

Fig. 18.1 Throughfall and open field deposition of NO_3N and NH_4N species to the beech Umbrisol during the 1989–2005 period

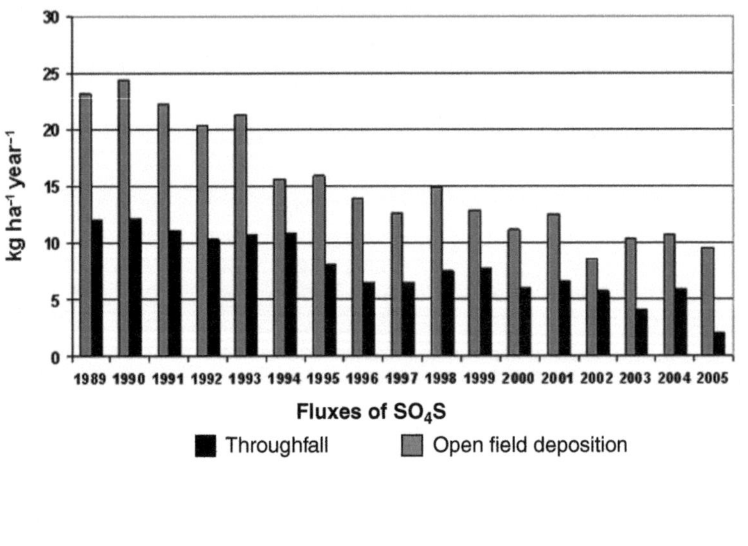

Fig. 18.2 Throughfall and open field deposition of SO_4S species to the beech Umbrisol during the 1989–2005 period

of regional or local sources; it seems to have been, in particular, low in 1995.

In comparison to Fig. 18.1, Fig. 18.2 shows on the basis of throughfall and stemflow analyses that the bulk deposition of sulphur species decreased distinctly during the whole observation period owing to legislative acts and technical innovations, while a concomitant reduction of canopy interaction processes proved less pronounced.

Comparing the atmospheric transport of sulphur species with those of the above nitrogen compounds, the faster conversion of NO_x to nitric acid and nitrate than the comparable conversion of SO_2 to sulphuric acid and sulphates and the higher deposition velocity of nitric acid involves that NO_x species contribute more to the acidification of rain nearby the source areas than at large distances. Weekly estimates of sulphur deposition exhibit strong seasonal flux dynamics with a pronounced winter maximum due to higher SO_2 emission rates as a consequence of elevated energy consumption on the one hand and much intensified sulphate transport from the North Sea by storms on the other.

18.2.1.3 Fluxes of Nitrogen and Sulphur Species in Soil

Soil is one of the major regulatory components of terrestrial ecosystem and therefore changes in soil-solution chemistry and the resultant element fluxes in seepage water reflect the complex influence of deposition, stand characteristics, climate and soil properties. The following description of element fluxes through the Arenic Umbrisol of the beech ecosystem, whose lower boundary is pragmatically equated with the maximum effective root zone, is based on the comparative analysis of solute concentrations and the corresponding seepage estimates modelled by means of the WASMOD (Reiche, 1991), VAMOS (Bornhöft, 1993) and CoupModel (Jansson & Karlberg, 2004) approaches. Soil solutes were obtained by continuous direct vacuum extraction in consecutive biweekly periods from litter (0 cm) and from 5, 12, 50, 150 and 400 cm depths.

Under beech the dominant nitrogen compound percolating into the soil is nitrate. The resultant concentrations in the litter attain maximum values in summer due to temperature-controlled nitrification processes of the organic matter and of coincident seasonally elevated ammonium inputs. In the deeper soil horizons both the concentrations and seasonal differences decrease rapidly, attaining the minimum below 150 cm with only very small intermonthly oscillations. Considering the low ammonium concentrations in the soil solutes of the Arenic Umbrisol (with the exception of the litter value), this decrease is indicative of intensive uptake and oxidation processes; the resultant NH_4^+ flow rates are consequently practically negligible. Figure 18.3 illustrates the considerable seasonal and interannual fluctuations of nitrogen input by throughfall and the corresponding nitrate output rates at 150 cm depth.

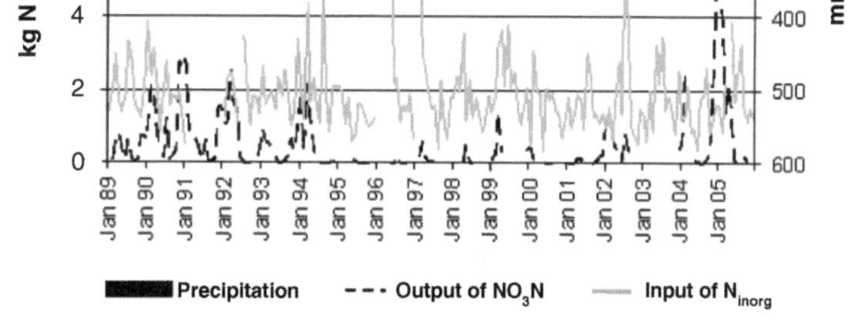

Fig. 18.3 Monthly precipitation, total inorganic nitrogen input by throughfall (kg N ha^{-1} $month^{-1}$) and nitrate export in seepage water (kg NO_3^-N ha^{-1} $month^{-1}$) at 150 cm depth during the 1989–2005 period

In comparison with the above depth function of nitrogen compound concentrations and the related flux dynamics, there is no clear seasonality of sulphate concentrations for any of the horizons of the forest soil, which seems to be due to the depth-related different influence of sulphate deposition, canopy interaction, throughfall, soil transport, uptake and adsorption–desorption processes (Bloem, Hopkins, & Benedetti, 2006). During the 1989–2003 period, the whole soil sulphur pool comprised 55% organic S (range: Ah 89%, Bw 21%), 26% adsorbed S (Ah 4%, AB 48%) and 19% in solution (Ah 7%, Bw 42%) (Mansfeldt, 1994); during the 1989–1999 period the average SO_4S-concentration of the soil solute has varied between 4 and 12 mg l^{-1} (Fränzle & Schimming, 2008). Figure 18.4 illustrates the relationships between precipitation and the sulphate input and output rates at the reference soil depth of 150 cm.

The diagram discloses a great deal of broad adjustment between rather high input and output rates until 1995; from 1996 to 2002. In Fig. 18.5 pH conditions, aluminium and nitrate ionic activities are summarized with the volumetric water content of the upper soil of 50 cm.

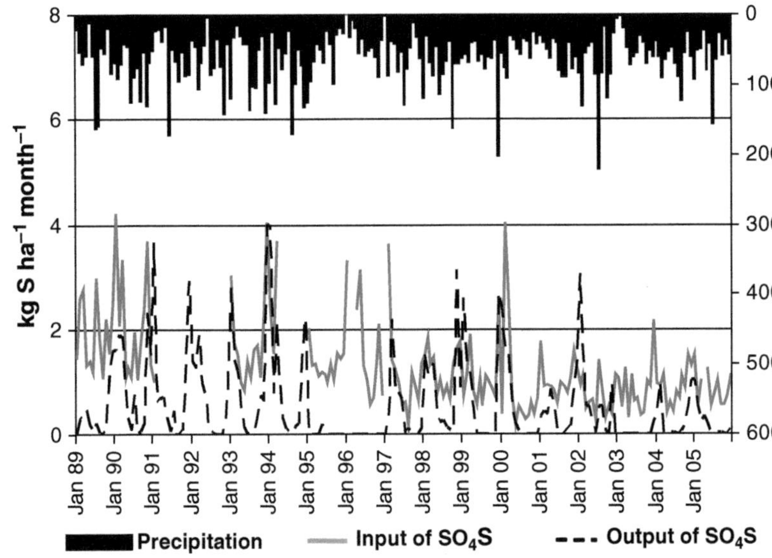

Fig. 18.4 Monthly precipitation and sulphate input by throughfall and output rates (kg SO_4S ha^{-1} month^{-1}) in seepage water at 150 cm depth during the 1989–2005 period

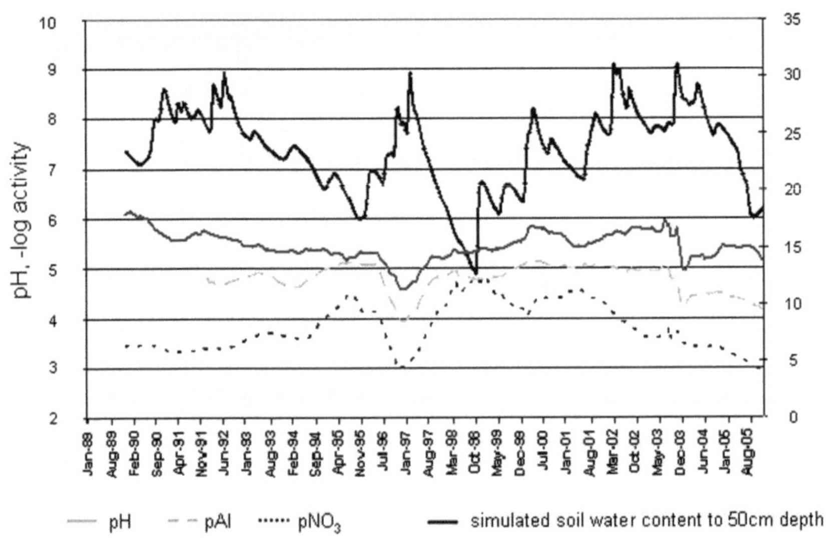

Fig. 18.5 Ionic activities of H^+, NO_3^- and aluminium species in the soil solution at 150 cm depth as related to modelled volumetric water content during the 1989–2005 period

The correlation of pH values and nitrate activity illustrates the importance of nitrate in soil acidification. Furthermore, it shows that the occurrence of the toxic aluminium species $Al(OH)_2^+$ (Marschner, 1995) is to be expected under the normally prevalent pH conditions, while the comparatively high concentrations of the most toxic species Al^{3+} in 1997 and 2005 appear related to the marked increase in acidity with pH values around four.

The juxtaposition of the nutrient supply from both atmospheric deposition and the potentially available element pool in the upper 70 cm of soil (N: 8,000; Ca: 2,300; K: 270 and Mg: 70 kg ha^{-1}) with the long-term element storage in the standing wood mass (N: 531; Ca: 280; K: 163; Mg: 56 kg ha^{-1}) is indicative of differential recycling and uptake mechanisms resulting in stoichiometric imbalances (Sterner & Elser, 2002) which can be interpreted as stability criteria of ecosystems (Fränzle & Schimming, 2008). These findings are also in agreement with Aber, Nadelhoffer, Steudler, and Melillo (1989) hypothesis that forest ecosystems respond to long-term elevated nitrogen inputs by an enhancement of internal N cycling processes, i.e. increased N fluxes by litterfall, higher microbial mineralization rates and a priming effect on nitrification (cf. also Dilly, Eschenbach, Kutsch, Kappen, & Munch, 2008; Irmler, Dilly, Schrautzer, & Dierssen, 2008). Furthermore, the present results confirm Prescott, Hope, and Blevins (2003) assumption that such effects can only be reliably assessed in the framework of longer-term studies, i.e. enquiries of (at least) 3–7 years duration.

18.2.2 Multivariate Characterization of Level II Stands

Long-term analyses of energetic and material flux rates in the soil–vegetation complex and the resultant element balances or the stoichiometric relationships between the different components of a plant and the

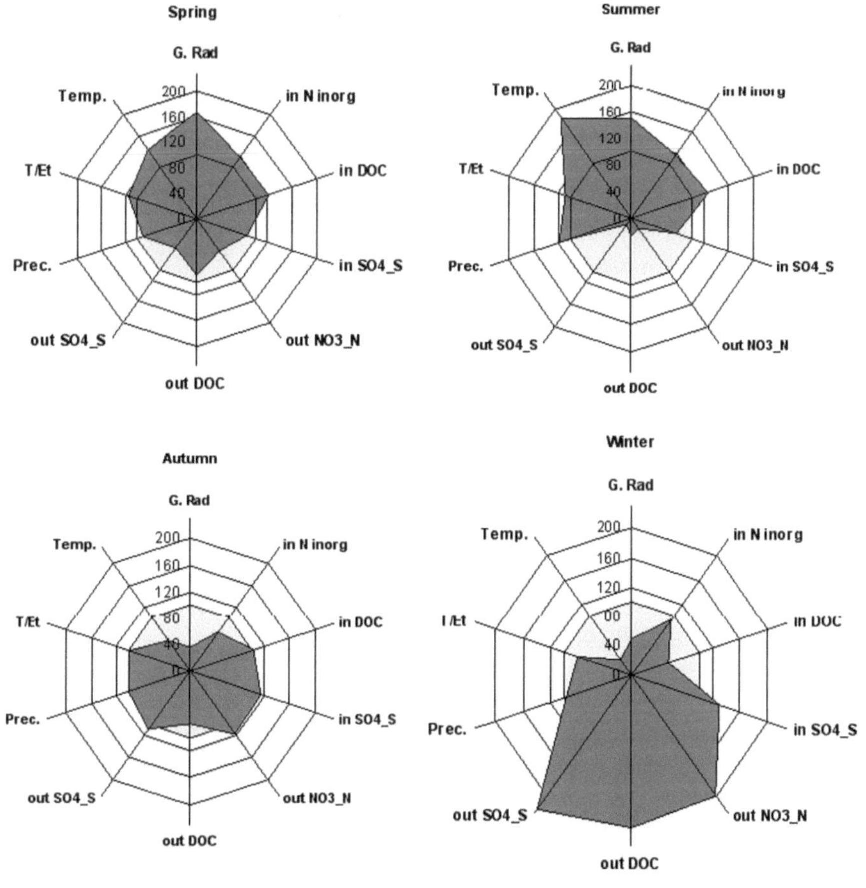

Fig. 18.6 Seasonal variations of selected ecosystem parameters of the Bornhöved beech stands. G Rad global radiation [Jcm^{-2}], in N_{inorg} = total input of inorganic nitrogen species in DOC = input of dissolved organic carbon compounds, in SO_4S = input of SO_4S, out NO_3N = output of NO_3N, out DOC = output of dissolved organic carbon compounds, out SO_4S = output of SO_4S [all values in kg ha^{-1} season^{-1}], Prec = precipitation [mm season^{-1}], T/Et = average transpiration-evapotranspiration ratio, Temp = average temperature [°C]. All values are normalized, 100% being the long-term average of the respective parameter and the reference period 1989–2005

chemistry of related soil solutes pave the way for a dynamic understanding of plant associations in accordance with the level II philosophy. The example of the Bornhöved studies shows, however, that data of the above type are commendably integrated into a more comprehensive characterization of ecosystem structure and dynamics by means of multivariate indicator sets, which is of particular importance in the framework of complex modelling approaches (Tavares, Hörmann, & Fohrer, 2007). In Fig. 18.6 the variables described above and some other physiologically relevant parameters are combined in the form of spoke-wheel or 'amoebae' diagrams in order to illustrate the state of the Bornhöved beech stands during the different seasons of the year.

Global radiation and *air temperature*, representing exergy capture, exhibit the normal seasonal pattern with maxima in spring and summer and minima in autumn and winter. The variability of these parameters in spring is a reflection of the typical macroclimatic situation of northern Germany. *Precipitation* has the normal summer maximum, while the *biotic water flow*, indicated by the T/Et ratio, attains its maximum in spring, when the biotic activity is optimum. The inputs of nitrogen and carbon compounds are mainly under the control of internal sequestration and release processes, e.g. leaching during autumnal leaf senescence, while sulphate input is highest in winter due to heating and increased seaborne import by storms from the North Sea. In contrast to the input situation, the *output rates of carbon, nitrogen* and *sulphur* compounds are strongly correlated with leaching processes and canopy interaction of the vegetation on the one hand and seepage chemistry on the other. The explicit definition of these relationships will be one of the focal tasks of process modelling approaches on the basis of the combined long-term data sets of both ecosystem research and level II monitoring programmes. Thus, they will serve to bring about a more detailed understanding of complex system state variables and functions like exergy capture, entropy production, storage capacity, loss reduction, hydrological and metabolic efficiencies, structural heterogeneity and biodiversity, which in combination yield a multidimensional characterization of terrestrial and aquatic ecosystems (Barkmann, Baumann, Meyer, Müller, & Windhorst, 2001; Müller, 2004, 2005; Müller, Schrautzer, Reiche, & Rinker, 2006, Jørgensen et al., 2007; Fränzle et al., 2008).

18.3 Conclusions

At Kiel University long-term transdisciplinary ecosystem research started in 1988 with the implementation of the Ecosystem Research Programme in the Bornhöved Lake district, which led to the foundation of the Ecology Centre that has also been in charge of the level II programme investigations. Thus, an almost 20-year, high-resolution time-series of pertinent meteorological, hydrological, biological and pedological data on the energy and material fluxes of the Bornhöved beech stands are available for both monitoring purposes and as a basis for ecosystem theory building.

Atmospheric deposition and element behaviour in the soil solution exhibit distinct trends and periodicities in biogeochemical transfer processes in relation to the water use of the ecosystems. Since terrestrial ecosystems depend on photosynthesis and autocatalysis, exergy capture and both negentropy production and export are expressions of site-specific thermodynamic boundary conditions. Their formulation in terms of complex ecophysiological or functional indicators (or orientors) permits to describe ecosystem state and development in relation to steady state or equilibrium situations. In the present context and also in consideration of practicability requirements, efficiency measures such as the transpiration/evapotranspiration ratio, element fluxes or budgets and the stoichiometric definition of elemental imbalances have proved to be commendable indicators. But their inherent high seasonal and interannual variability also points to the fact that the appropriate assessment of such indicators with a satisfactory degree of reliability and validity necessarily involves long-term measurements, or else premature, if not spurious, allegations and generalizations are inevitable.

References

Aber, J. D., Nadelhoffer, K. J., Steudler, P., & Melillo, J. M. (1989). Nitrogen Saturation in Northern Forest Ecosystems. *BioScience, 39*, 378–386.

Agricola, G. (1556). *De re metallica XII libri*. Basel: Froben.

Barkmann, J., Baumann, R., Meyer, U., Müller, F., & Windhorst, W. (2001). Ökologische Integrität: Risikovorsorge im nachhaltigen Landschaftsmanagement. *Gaia, 10*, 97–108.

Bloem, J., Hopkins, D., & Benedetti, A. (2006). *Microbiological methods for assessing soil quality*. Wallington: CABI.

Bornhöft, D. (1993). Untersuchungen zur Beschreibung und Modellierung des Bodenwasserhaushalts entlang einer Agrar- und einer Wald-Catena im Bereich der Bornhöveder Seenkette (Schleswig-Holstein). *EcoSys Suppl, 6*, 1–134.

Branding, A. (1996). Die Bedeutung der atmophärischen Deposition für die Forst-und Agrarökosysteme der Bornhöveder Seenkette. *EcoSys Suppl, 14*, 1–122.

Dilly, O., Eschenbach, C., Kutsch, W. L., Kappen, L., & Munch, J. C. (2008). Ecophysiological key processes in agricultural and forest ecosystems. In O. Fränzle, L. Kappen, H. P. Blume, & K. Dierssen (Eds.), *Ecosystem organization of a complex landscape: Long-term research in the Bornhöved Lake District, Germany. Ecological Studies 202* (pp. 119–167). Heidelberg: Springer.

Ellenberg, H. (1995). Allgemeines Waldsterben – ein Konstrukt? *Naturwissenschaftliche Rundschau, 48*, 93–96.

Ellenberg, H., Mayer, R., & Schauermann, J. (1986). *Ökosystemforschung. Ergebnisse des Solling-projektes 1966–1986*. Stuttgart: Ulmer.

Elling, W., Heber, U., Polle, A., & Beese, F. (2007). *Schädigung von Waldökosystemen. Auswirkungen anthropogener Umweltveränderungen und Schutzmaßnahmen*. Heidelberg: Elsevier-Spektrum Akademischer Verlag.

Erisman, J. W. (1992). Atmospheric deposition of acidifying compounds in the Netherlands. Dissertation, University of Utrecht.

Fränzle, O. (1993). *Contaminants in terrestrial environments*. Berlin: Springer.

Fränzle, O. (2000). Ecosystem research. In S. E. Jørgensen, & F. Müller (Eds.), *Handbook of ecosystem theories and management* (pp. 89–102). Baton Rouge: Lewis.

Fränzle, O., & Schimming, C. G. (2008). Element fluxes in atmosphere, vegetation and soil. In O. Fränzle, L. Kappen, H. P. Blume, & K. Dierssen (Eds.), *Ecosystem organization of a complex landscape: Long-term research in the Bornhöved Lake District, Germany. Ecological Studies 202* (pp.169–204). Heidelberg: Springer.

Fränzle, O., Kappen, L., Blume, H. P., & Dierssen, K. (Eds.) (2008). *Ecosystem organization of a complex landscape: Long-term research in the Bornhöved Lake District, Germany. Ecological Studies 202*. Heidelberg: Springer.

Irmler, U., Dilly, O., Schrautzer, J., & Dierssen, K. (2008). Site-related biocoenotic dynamics. In O. Fränzle, L. Kappen, H. P. Blume, & K. Dierssen (Eds.), *Ecosystem organization of a complex landscape: Long-term research in the Bornhöved Lake District, Germany. Ecological Studies 202* (pp. 119–145). Heidelberg: Springer.

Jansson, P. E., & Karlberg, L. (2004). *Coupled heat and mass transfer model for soil plant–atmosphere system*. Retrieved from Royal Institute of Technology, Dept of Civil and Environmental Engineering, Stockholm Web site: http://www.lwr.kth.se/CoupModel/CoupModel.pdf.

Jørgensen S. E., Fath, B., Bastianoni, S., Marques, J., Muller, F., Nielsen, S. N., et al. (2007). *A new ecology: The systems perspective*. Amsterdam: Elsevier Publishers.

Mansfeldt, T. (1994). Schwefeldynamik von Böden des Dithmarscher Speicherkoogs und der Bornhöveder Seenkette in Schleswig-Holstein. Dissertation, Universität Kiel.

Marschner, H. (1995). *Mineral nutrition of higher plants*. London: Academic Press.

Müller, F. (2004). Ecosystem indicators for the integrated management of landscape health and integrity. In S. E. Joergensen, R. Costanza, & F. L. Xu (Eds.), *Ecological indicators for assessment of ecosystem health* (pp. 277–304). Boca Raton: Taylor and Francis.

Müller, F. (2005). Indicating ecosystem and landscape organization. *Ecological Indicators, 5*, 280–294.

Müller, F. R., Schrautzer, J., Reiche, E. W., & Rinker, A. (2006). Ecosystem based indicators in retrogressive successions of an agricultural landscape. *Ecological Indicators, 6*, 63–82.

Prescott, E., Hope, G. D., & Blevins, L. (2003). Effect of gap size on litter decomposition and soil nitrate concentrations in a high-elevation spruce–fir forest. *Canadian Journal of Forest Research, 33*, 2210–2220.

Reiche, E. W. (1991). Entwicklung, Validierung und Anwendung eines Modellsystems zur Beschreibung und flächenhaften Bilanzierung der Wasser- und Stickstoffdynamik in Böden. *Kieler Geographische Schriften, 79*, 1–150.

Sterner, R. W., & Elser, J. J. (2002). *Ecological stoichiometry - The biology of elements from molecules to the biosphere*. Princeton: Princeton University Press.

Tavares, F., Hörmann, G., & Fohrer, F. (2007, October). *Modelling of soil water fluxes. Comparison of the soil-plant-atmosphere CoupModel with the SIMPEL soil water spreadsheets*. Paper presented at the 11. Workshop Modellierung und Simulation von Ökosystemen, Kölpinsee.

Ulrich, B. (1983). Interaction of forest canopies with atmospheric constituents: SO_2, alkali and earth alkali cations and chloride. In B. Ulrich, & J. Pankrath (Eds.), *Effects of accumulation of air pollutants in forest ecosystems* (pp. 33–45). Dordrecht: Reidel.

Van der Maas, M. P., Van Breemen, N., & Van Langenvelde, I. (1991). *Estimation of atmospheric deposition and canopy exchange in two Douglas fir stands in the Netherlands*. Internal publication. Wageningen: Department of soil science and geology, Agricultural University of Wageningen.

Wesely, M. L. (1989). Parametrization of surface resistances to gaseous dry deposition in regional-scale numerical models. *Atmospheric Environment, 23*, 1293–1304.

World Reference Base for Soil Resources (2006). *World Soil Resources Reports No. 103*. Rome: FAO.

Chapter 19

A Conceptual Framework for Integrated Functional Landscape Monitoring in the Wider Countryside of Central Europe

Ralf-Uwe Syrbe, Wilfried Hierold, Olaf Bastian, and Matthias Röder

Abstract Landscape monitoring is highly topical as we are faced again and again with challenges in the balance of nature and its utilisation. Knowing of this importance for the European Landscape Convention calls to support the landscape analyses for which suitable methods of international scale have to be developed. Landscape-related monitoring means, according to the authors, a trans-scale spatially nested and complex monitoring and evaluation of landscape change. Suitable methods were developed for the purpose and presented. The framework introduced has been tested successfully in three federal states of Germany since 1999 and may now be defined as being mature for application. As drawn up there are hypotheses, functions, indicators, and data to adapt to the spatial conception and the given conditions. The implementation of the various natural science and social science programmes which would require a high degree of compatibility for reason of their integrative properties represents a problem. They have to be implemented for entire project periods lasting for decades and implemented in continuous monitoring, the methodical consistency has to be assured, and a professional interpretation even of heterogeneous data and a reasonable, model-based forecast of the change of landscape have to be given.

Keywords Landscape change · Central Europe · Holistic concept · Monitoring programme · Indicators · Landscape functions · Metadata · Hypotheses · Acre · Sampling points · Dublin Core Set · Basis data

R.-U. Syrbe (✉)
Centre for Landscape Research, Leibniz Institute of Ecological and Regional Development Weberplatz 1, 01217 Dresden, Germany
e-mail: syrbe@iale.de

19.1 Introduction

Landscape changes are considered as qualitative landscape alterations concerning the balance of matter and energy, use of natural resources and landscape aesthetics (Bastian & Steinhardt, 2002). According to Neef (1983), the term landscape describes a concrete part of the earth's surface shaped by uniform structure and same process pattern. Consequently, landscape may include, e.g., the countryside, urban areas, mountainous and wetland areas. In this context, we set the focus on the countryside of Central Europe. Landscape is treated as system, as holistic concept comprising the interrelations between biotic and abiotic components, as well as the human impact upon them (Röder & Syrbe, 2000). Thus, landscape changes affect the functions of landscapes (cf. Section 19.4.2). An integrated approach would have advantages over methods based on single indicators, such as land cover, stock of landscape elements and biotopes, and biodiversity (Syrbe, Bastian, Röder, & Haase, 2002).

Landscape change is a complex process, encompassing ecological, socio-economic as well as cultural aspects. Natural landscape changes can take a very long time (hundreds and millions of years), proceed very slowly, and include a range of factors such as climatic fluctuations, origin and erosion of mountains, coastal and river dynamics. But there are also short-term events with duration of seconds through to years and which are often manifested as natural catastrophes, e.g. floods, hurricanes, droughts, earthquakes, volcanic eruptions, rock-slides, and avalanches. Landscapes, due to their unspoiled nature and physical complexity, were and are subject to steady change with regard to both space and time (Petit & Lambin, 2002). Change is often linked to burden and risk. Generally, landscape

F. Müller et al. (eds.), *Long-Term Ecological Research*,
DOI 10.1007/978-90-481-8782-9_19,

change can have both positive and negative effects. The loss of landscape functions is considered negative. The required sustainable land use calls for maintaining the range of functions in the landscape. However, there are also processes with undesired consequences. Degraded landscapes then develop, the functionality of which are considerably reduced compared to earlier conditions.

By all means, an alert observation is required (cf. Section 19.3). The term 'landscape monitoring' is defined as the regular long-term surveillance of a landscape. The monitoring of landscape changes has to be pragmatically aligned to man-made as well as natural changes because the causes are often too complex. Furthermore, at least the net consequences are relevant for society.

There are many specific reasons for monitoring. the identification of current changes and the forecast of further development, recognising the causes for undesirable developments, deriving suitable measures, controlling actions, and appraising success of environment policies and much more. The environment policy decision for more sustainability is to be supported by objective statements and quantitative data (Lütz, Bastian, & Weber, 2006).

Know-how gained about the change and about possible results is also suitable to sensitise the public and to counter possible undesirable developments. However, there are not only the current or expected ecological problems for the future that require well-aimed monitoring of the landscape. This task has, in the meantime, become politically pegged in many a way. Monitoring to fulfil landscape planning is ensured by an ever-increasing number of regulations. Provisions regarding environment monitoring and the respective reporting obligations are found, among others, in national and state environmental laws and planning directives such as in the European Union.

- Water Framework Directive (RL 2000/60/EG, Art. 9, 15),
- Habitats Directive (RL 92/42/EWG, Art. 17),
- Directive on the assessment of the effects on certain plans and programmes on the environment the so-called Strategic Environmental Assessment directive (SEA) (RL 2001/42/EG),
- Eco Management and Audit Scheme (761/2001 EG).

19.2 Historical Landscape Change

Increasingly, landscape changes are essentially influenced by the development of human society and its means of production. In a detailed survey for Central Europe, Bernhardt and Jäger (1985) and Bastian and Bernhardt (1993) distinguished four major stages of landscape development (see also Vos & Meekes, 1999; Bastian & Steinhardt, 2002).

Humans used nature for many 100,000s of years as a bran tub for hunting, harvesting, and cutting wood. The *agricultural acquisition and use* (c. 5,000–6,000 years) began with the crop cultivation and livestock rearing activities carried out by Neolithic peoples. Ploughs were applied while cereal cultivation spread widely.

Integrated development (c. 1,000 years) made wide use of all the other potentials and resources. Natural forests were increasingly disturbed and exploited for a range of resources, leading to forest management around the turn from the 18th to the 19th century. This utilisation increased the availability of habitats. So Central Europe reached its greatest biological diversity ever around the mid-19th century.

The industrial revolution (just over 150 years) led to marked agglomeration in the settlement structure and created areas for the large-scale exploitation of resources. About 10% of the earth's surface became completely transformed, or was sealed by residential construction, industrial development, transport routes, and mining. Agricultural land became more homogeneous because of fertilisation and land improvement.

The scientific and technological revolution (for the last 50 years) has drawn intensively on nearly all resources and potentials inherent in natural landscapes. Large machinery systems, chemicals, and automation have been used. Now all landscapes are exposed to human material and energy throughputs with an increase in diffuse deposition of a variety of pollutants. Over the long period of human-induced landscape change, there are some characteristic fundamental trends.

1. In the last few thousand years, landscape changes in Central Europe have been brought about almost exclusively by material and technological advances, and social developments.

2. Each principal stage in the process has been initiated and accompanied by a radical innovation in means of production.
3. The periods of time occupied by each of the main stages in Central Europe have become successively shorter (5,000–6,000 years, c. 1,000 years, c. 150 years, 50 years).
4. The acceleration in the pace of human intervention let doubtful for natural processes to stabilise landscape.
5. Human intervention and innovation, in the course of history, has diversified and spread to almost all elements of the landscape.
6. At first, environmental degradation was only local and limited. It spread to larger regions and has now reached global dimensions.
7. Changes in quality in the form of conspicuous landscape transformations are normally preceded by 'creeping' and invisible quantitative changes.
8. The intensity of land use and the ecological effect are continuously rising, with technogenic elements and largely homogenised farming areas.
9. There has been a rapid increase in the proportion of landscapes which have suffered from irreversible change.
10. Different landscapes with different assets react differently to the same type of human activity as reflected in various degrees of buffering and stress capacity.

This experience shows that land utilisation has to be planned for long-term viability with a view to consequences and risks. Status- and process-related monitoring is required for a sound management. Since the risks for the ecosystem integrity are increasing, the monitoring of landscapes is becoming an even more urgent task. It may not only be aligned to certain elements or partial systems of landscape, but must refer to its complexity. Besides, the interactions of landscape balance should be considered.

19.3 Examples of Landscape Monitoring

To date, many specific monitoring programmes have been developed by various institutions. Examples of important long-term international programmes are the Blue Plan (Antipolis, 1995), ROSELT/OSS (2005), AMAP (2005), and UNEP-WCMC (2005). Though these programmes are rather comprehensive, they are always focused on specific questions, such as

Table 19.1 Selected monitoring programmes in comparison

Programme	Affiliation	Covering	Scale	Analysis	Results
OEUB (Luthardt et al., 1999)	Federal state (BR, D)	Biosphere reserves	Multiple	System theoretical	Integrated assessed
SALMA (cf. this chapter)	Federal state (SN, D)	Area-wide	Multiple	System theoretical	Integrated assessed
ZALF (cf. this chapter)	2 Federal states (BR + MV, D)	Area-wide	Multiple	System theoretical	Integrated assessed
Niagara Escapement Plan (http.//www.escarpment.org)	Federal State (Ontario, CA)	Biosphere reserve	Multiple	Combined	Integrative
ROSELT/OSS. Réseau d'Observatoires de Surveillance Ecologique à Long Terme / Observatoire du Sahara et du Sahel	International network based on UN convention	25 observatories in 11 countries	One scale + extrapolations	Thematic selective	Assessed
Arctic Monitoring and Assessment Programme (AMAP)	8 Arctic littoral states	10 key areas	Multiple	Thematic selective	Integrated
Swedish Environmental Monitoring Program (NILS)	National (Sweden)	600 sample plots	5 × 5 km per plot	Thematic selective	Combined sectoral
Biological Monitoring of the Countryside Management Scheme	Northern Ireland	352 sample plots	Multiple	Thematic selective	Integrated assessed
Trilateral Monitoring and Assessment Program (TMAP)	NL, Germany, Denmark	Wadden Sea	Multiple	System theoretical	Integrated assessed

biodiversity and water protection. However, few initiatives are multifunctional – one exception is the British Countryside Survey (Haines-Young et al., 2000; Howard, Petit, & Bunce, 2000; McAdam, Flexen, & O'Mahony, 2008). Other initiatives are focused on specific land use systems (ADAS, 1997; Dramstad et al., 2002), nature reserves (Luthardt, Vahrson, & Dreger, 1999; Bastian, 2000), or special landscape types, e.g. the Wadden Sea (TMAP, 1997).

Table 19.1 gives some information about examples of landscape monitoring programmes with their administrative affiliations without intending to insinuate that they are complete. The following details are depicted in brief:

- What is the affiliation of the observers?
- Is the spatial strategy area-wide or selective?
- Do they use one special scale or multiple scales?
- How the analysis is theoretically based?
- Are the results integrative and are they described or assessed?

Our landscape monitoring framework has three main goals: early recognition and quantification of emerging landscape changes, assessment of their consequences, and predictions of future alterations. Additional benefits are

- differentiation between (for human society) critical and unproblematic landscape changes,
- data supply for modelling and for the assessment of landscape functions,
- problem-oriented analyses of landscape change, and
- continuous adjustment of recommendations for effective landscape management (Syrbe, Bastian, Röder, & James, 2007).

19.4 Developing an Integrated Functional Landscape Monitoring Framework

19.4.1 Research Background

This chapter has to reveal possibilities by which environment monitoring can be set up inter-disciplinarily and effectively. The authors introduce well-proven methods enabling possibilities to save resources and avoiding unnecessary piles of obsolete data.

The Leibniz-Centre for Agricultural Landscape Research (ZALF) and the Saxon Academy of Sciences (SAW) developed similar approaches for an integrated functional landscape monitoring framework as contribution to the German Long-Term Ecological Research (LTER-D) programme, which started in 2005 (http://www.lter-d.de). This framework is focused on the 'normal' landscape, i.e. it is not limited to protected areas or special landscape types. Our approach is also a response to the European Landscape Convention (CMCE, 2000), which requires standardisation of landscape-related investigations; therefore, the development of suitable and comparable monitoring methods is necessary. At present, the authors are not aware of another comparable holistic monitoring programme.

The monitoring framework is to be constructed from environmental problems that are formulated as hypotheses or assumptions (Section 19.4.2.1). To verify these hypotheses and assumptions, relatively simple indicators or the more complex landscape functions are selected (Section 19.4.2.2). The data set to be analysed (Section 19.4.3.1) depends on the assessment methods applied. The choice of adequate areas and representative test sites is essential for optimised monitoring programmes (Section 19.4.3.2). The results are interpreted to verify hypotheses, to construct metadata (Section 19.4.3.3), and to propose measures supporting a sustainable landscape development.

Once designed, the implementation of monitoring scheme starts from the basic data level and has to be executed in reverse order as depicted in Fig. 19.1.

The implementation starts with data collection (only once) to provide 'basic data' that can be used again and again. The monitoring procedure, which is repeated regularly, consists of the following steps (Bastian, Syrbe, & Röder, 2001):

- Collection of comparable monitoring data on landscape characteristics,
- Interpolation and partial integration of raw data through the derivation of quantitative indicators with fixed time and space references,
- Integration through the assessment of landscape functions, modelling, and interpretation,
- Validation of hypotheses on anticipated landscape changes, and
- Construction and updating of a metadata base to ensure full information about existing data, progress in work, and results.

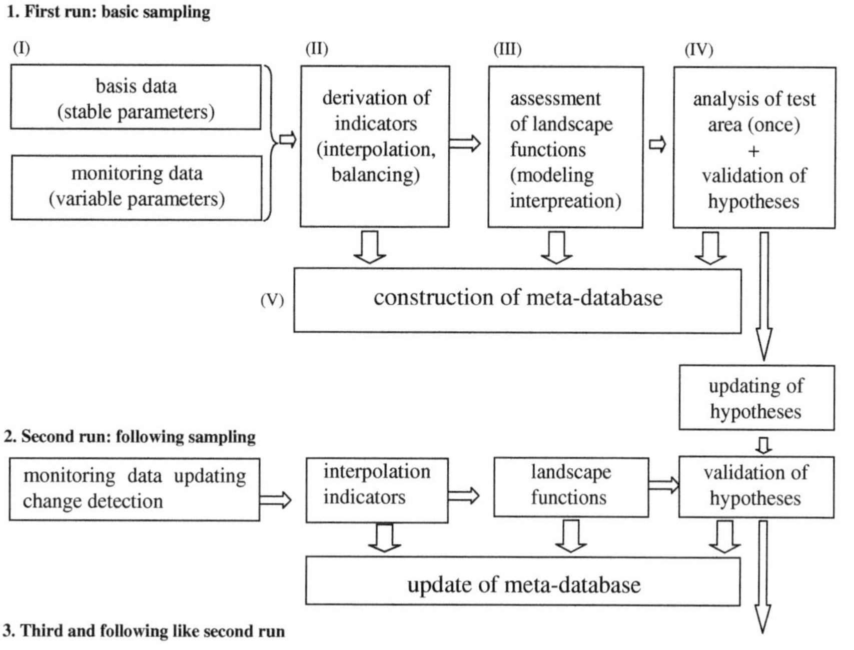

Fig. 19.1 Landscape monitoring – implementation (Syrbe et al., 2007; with kind permission of Elsevier)

The three characteristic features of this monitoring framework are briefly described as follows:

1. We understand landscape as a complex or an entity. Therefore, apart from common problems of data sampling, a main challenge facing this monitoring methodology is the integration of data to complex statements. For a versatile interpretation of monitoring results the clear documentation of the data is necessary as well as their purposeful aggregation, also in relation to social and political questions. As a key concept for the solution of the integration problem, we propose the assessment of landscape functions (cf. Section 19.4.2.1, Bastian & Schreiber, 1999; Bastian & Röder, 2002; Bastian & Steinhardt, 2002). The special methodical set of tools involved already triggers data collection (cf. Section 19.4.3.1).
2. The approach is problem-oriented. The selection of indicators (cf. Section 19.4.2.1) is referring to a set of hypotheses about current and anticipated landscape changes. By identifying modifications of landscape functions with regard to time, threats critical for the human society can be revealed.
3. The selection of test areas takes landscape classification into consideration. The character of basic reference units can influence the detection of landscape changes. Otherwise, essential landscape changes could fall into the limits of the spatial accuracy and remain undetected.

The identification and localisation of test areas, measuring stations and sampling sites requires consideration of spatial and scale factors to ensure that they are representative of the whole area. The identification of a reference landscape as a typical window of the area of interest is the first step of the landscape monitoring approach. The use of landscape classification systems and/or site types is essential. The choice of methods for landscape analysis and landscape assessments depends on scale (Syrbe, Röder, & Bastian, 2001). Furthermore the scale and size of test areas must be reconciled to financial resources.

19.4.2 Conceptual Considerations

19.4.2.1 Hypotheses

The presented landscape monitoring framework is focused on problems interrelated with landscape change and sustainable use. A hierarchical set of assumptions (hypotheses) meets this demand. The choice of all test areas and all measurements should be driven by this set. Hypotheses are derived from tendencies in landscape change observed already while emerging from scenarios and prognoses. Questions and assumptions about current trends in landscape development are identified. Monitoring tasks arise from practical decisions relevant to landscape ecological problems. They take into account undesired effects, ambiguous and critical development trends, and uncertainties regarding future problems. The success of landscape management, the impacts of human interventions, and the effects of environmental changes can be examined in this way. The assumptions are linked with the goals fixed in landscape plans in order to scrutinise, whether the plans contain appropriate strategies for foresight and prevention of impacts as well as for compensation.

Hypotheses describe possible future landscape changes, which can either support or counter these goals. The following examples show hypotheses, formulated exactly with respect to spatial validity. The hypotheses are differently tailored to the various scales (global, national, regional, local) or fitted to certain areas.

The following example relates to global level:

1. The global extinction of species is particularly caused by the loss of their habitats. The network of Natura-2000 areas (resp. Emerald Network) is suitable to conserve biodiversity (esp. endangered habitats and species).

Example hypothesis for the level of national states (Germany) are

2. International management plans for river catchment areas in connection with the technical progress in wastewater treatment lead to a distinct improvement of quality of the water courses.
3. The reduction of daily encroachment by sealed surfaces in Germany will be reduced to less than 30 ha by 2020 (Bundesregierung, 2002). Particularly measures in floodplains and areas with high runoff properties can minimise the increasing flood risk.

The hypotheses relate to landscape indicators and landscape functions.

19.4.2.2 Data Analysis and Interpretation up to Evaluation

In order to verify and to evaluate assumptions as listed above, we propose the assessment of landscape functions. Landscape functions are describing the performance of a landscape in the broadest sense. After de Groot (1992) landscape functions reflect the 'capacity of natural processes and components to provide goods and services that satisfy human needs (directly and/or indirectly)'. According to Haase (1978), the assessment of social functions of a landscape is a pre-condition of relating the actual landscape state to economic categories and processes. By the investigation of changes in landscape functions, fundamental landscape changes relevant to human society can be identified, even much better than comparing only single symptoms, such as land cover changes, loss of landscape elements, biotopes, and decrease in biodiversity. For most landscape functions (soil functions, habitat function, recreation function, etc.) a range of assessment methods at different scale levels was published in several textbooks (e.g. Marks, Müller, Leser, & Klink, 1992; Knospe, 1998; Bastian & Schreiber, 1999; Bastian & Steinhardt, 2002), see Table 19.2.

Monitoring indicators represent the medium stage of data integration. They abstract from raw data, they integrate different data, and they are related to geographical units. Also, indicators can be the output of an ecological model. They indicate a change in a landscape quantitatively and are value-free. Indicators can identify development trends and time balances. If new monitoring projects are designed, consideration should be given to incorporating indicators already used in national and international observation programmes, e.g. Environmental Aspects of Sustainable Development (DPCSD, 1999), EEA Environmental Indicators (EEA, 1999), OECD Environmental Indicators (OECD, 2001), or European Biodiversity and Indicator Framework (ECNC, 2002).

The degree of surface sealing is an example for a key indicator of landscape change. It has a decisive influence on the landscape water balance, the local climate, the recreation value and not least, on the habitat value. Therefore, the German Federal Government would like to reduce the current daily growth of sealed surfaces from over 100 ha to 30 ha by 2020. Recently, we have 120 ha/day and the hardly reachable objective of our government is 30 ha/day (the so-called

Table 19.2 Selected landscape functions and their determination as integration process for landscape monitoring

Item	Procedure by	Data and indicators used
Soil functions	–	–
Resistance to water erosion	USLE (Wischmeier & Smith, 1978)	Slope, land use, field core area, soil substrate
Resistance to wind erosion	RWEQ (Fryrear et al., 1998)	Soil roughness, soil substrate, vegetation cover, barriers silhouette, slope
Nitrate retention ability	AG Bodenkunde (1994)	Soil moisture cap., soil water balance
Physico-chemical filtering capacity	Syrbe (1994)	Soil substrate, humus content, groundwater table
Hydrological functions	–	–
Morphological stream quality	LAWA (2000)	Stream bed quality, stream edge quality, floodplain land use
Stream water quality	LAWA (2000)	Contents of nitrate, phosphate, oxygen/water biocoenosis
Water retention capacity	Zepp in Marks et al. (1992), Röder and Adolph (2006)	Soil texture, soil water balance, slope, geological structure, land use (sealing)
Habitat functions	–	–
Naturalness of vegetation	Schlüter (1987)	Land use, biotope types
Landscape habitat value	Bastian (1997)	Actual vegetation (plant communities, indicator species)
Production functions	–	–
Biotic yield potential	Klink and Glawion in Marks et al. (1992)	Soil type and texture, slope, temperature, erosion resistance

30-ha-goal). The sealing can be well estimated taking the land cover data. Corresponding to the differentiated hypotheses and evaluation processes, the indicator concept for every test area and every scale is to be compiled separately.

An increasing number of models are being developed and implemented to simulate future developments and thus to be able to adapt strategies already before changes take place. Models do not come up with forecasts, but scenario-based estimates. Thereby the ecological long-term monitoring is essential to develop such models and finally for their validation (Wegehenkel & Kersebaum, 2005; Wieland, Voß, Holtmann, Mirschel, & Ajibefun, 2006).

19.4.3 Methodical Steps

19.4.3.1 Data Collection

We define all information on the condition and the change of landscape and its elements as landscape data. These are collected directly from the terrain, are measured or analysed personally, are drawn from external sources or are derived by interpretation. The data concept is to be developed, differentiated by scale and analysed area. It is sensible to distinguish between basis data (more constant time attributes as evaluation background) and the variable values (condition properties). The latter are the actual target of long-term monitoring.

Basis data are, as a rule, only provided once. However, often considerable work has to be carried out in this respect, e.g. when no information or sources of varying suitability are available. Often the information is not available on a digital medium, or a resolution of adequate scale is missing. In some cases the necessary parameters first have to be derived.

When monitoring the landscape change we differentiate between basic programmes and special surveys. This is sensible when various specialists work in the same landscape area. The basic programmes supply data at representative locations (Section 19.4.3.2) which provide evaluation principles in the context of different questions. As an example we take the survey on management and cultivated plants in agricultural field cultures (inventory data, yield, etc.), which can, for instance, be interpreted with regard to erosion protection, floral diseases, habitat suitability, or economically for product profitability calculations. The basic programmes also provide a set of very well identified measuring locations. Every specialist is requested to check whether his own survey target – e.g. characteristics of soil biology (Joschko et al., 2006), grassland vegetation (Kaiser et al., 2005), or habitat distribution (Glemnitz & Wurbs, 2003) – can be sufficiently serviced. Often the specialists add further measuring locations which then also have to undergo basic inventory taking.

Depending on scale, level, and size of area various data sources are considered.

- For greater areas and large scales' primarily statistics, typified and aggregated geo-data, e.g. from satellite records, from sovereign measuring networks, databases, or environment information systems.
- Only for smaller areas and correspondingly smaller scales, the results of own field measurements and questionnaires are used apart from the derived values and (high-resolution) remote sensing data.

Which modifications of data prove real landscape changes, whereas many modifications result from another sensor, a new acquisition method, or the correction of former faults? Geodetic discrepancy, different scales, the enlargement or shift of roads, the growth of treetops, variable water levels, seasonal variations, and different interpretation, respectively, classification modes can cause a semblance of changes (Bastian & Röder, 1996; Neubert & Walz, 2000). Wegehenkel, Heinrich, Uhlemann, Dunger, and Matschullat (2006) discuss the importance of various sources on the current land coverage in landscape water balance modelling and thus confirm the above-mentioned methodical experiences.

Particularly for purposes of monitoring, nature protection, and landscape planning a comparability of different time cuts is essential (Bastian, 1999; Lausch, 2000; Syrbe, 2002). An update must consider the task of change detection already from the outset, like the CORINE project (Keil, Mohaupt-Jahr, Kiefl, & Strunz, 2002; Hölzl, 2003) and other primary remote sensing approaches (Werner, 2002) do it. But if new sources or new methods should be used, a proper method of data integration is needed, preserving the present information.

Table 19.3 Status codes indicating the cause and status of land use changes (Syrbe & Schulze, 2006)

Status code	Reason and action (concerning an individual polygon)
0	No landscape change
1	Change the geometry but not the attribute (remain part after secession)
2	Change the attribute but not the geometry
3	Change both geometry and attribute (separated part)
4	Insert new island polygons
5	Revision by field survey necessary
6	New data creation (without data up to now)
7	Correct an improper derivation from original source
8	Revision fault old data
9	Correct an improper geometry of old data

In order to design such integration, we recommend updating, for instance, geo-data with minimal alteration by help of an appropriate rule set. The original geometry should remain completely. Only new borders and attributes are added. One of the new attributes indicates the cause of each individual change (Table 19.3).

It should be possible to collect much monitoring data without having to revert to complex technical measuring processes. These include vegetation maps, fauna line taxation, trap and other random sample processes and the recording of relief, water and land use characteristics. To achieve this, standardised collection and recording provisions would be required. These are drawn up as compilation sheets or computer input masks and have to ensure that no important characteristic is left out locally. Parameters must be collected in a comparable manner. A good example for standardisation is the German soil-mapping manual (Ad-hoc-AG Boden, 2005). In cases of doubt reference material (reserve samples, photos, and other records) have to be available for clarification purposes.

For example, elements of the historical cultural landscape (memorial stones, barns, narrow passes, hedges, erosion ducts, and the similar) can be best mapped directly at their location. The essential recording criteria are the origin of the historic cultural landscape elements, their degree of conservation, their current utilisation and their importance for nature conservation and as landscape scenery. With a checklist, individual elements can be evaluated rationally and comparably (Syrbe & Palitzsch, 2002).

19.4.3.2 Choice of Reference Landscapes, Test Areas, and Sites

At this point the spatial concepts of the authors are introduced as two examples. Both approaches are similar. They go beyond the scales and are thus nested spatially (nested sampling). They are aligned at known and typified spatial structures. The one approach is more strongly oriented at the land use units (field structures); as a rule only aggregated data of the farmers per field or even per farm are available for the entire area. In the other study the measurement locations are derived directly from the natural structure of the landscape. These result in surveys oriented at representative locations. In the first example the types of surfaces provide the information. But there too, measurements take place as random point surveys in the context of local heterogeneity and thus point and areal results are generated. Other known statistical surface approaches of spatial random samples are not provided here. A good know-how of the geographic sample and the structures allows for stratified area and random point selection according to the authors.

In the first example, an integrative project of the agricultural landscape research in NE Germany, the steps of the choice of test areas, reference plots, and the selection of the measurement location are presented in brief. It is a cross scale and media project (Glemnitz & Wurbs, 2003; Wegehenkel et al., 2006; Sommer, 2006; Schindler, Steidl & Müller, 2007). It is developed further with a continuation of the hypothesis of the specialists. Basic programmes of monitoring remain constant, however, when required are supplemented by a further representative area types! The selection of the areas and measuring locations is presented in the example of the basic programme acre. According to the overall approach considering material balances the monitoring area is a sub-catchment of the River Ucker, flowing into the Baltic Sea. This landscape window of about 200 km^2 is typical in its geo-ecological character, dominated by ground moraines. Agricultural land use has dominated for over 750 years. Today about 50 farms work there, each mostly bigger than 1,000 ha. In this scale the sampling of management data is organised, on the example of 12 farms. The farmers told all agrarian information of each field every year. The rest of the whole area is subject to remote sensing by satellite data (Wegehenkel et al., 2006).

The target of the basic programme acre is to monitor the development of plant population of the most important crops over years under the influence of site, weather, and management. Finally 23 fields of average size (20–50 ha) are selected in various farms. The scientific criteria were soil and morpho-graphic characteristics (soil types, water conditions, and parent material). Others such as field size, the willingness of the farmer to co-operate or the accessibility were filters set later on. To be able to compare the fields with each other the information of the medium scale soil map 'MMK' (Schmidt, 1978; Schmidt & Diemann, 1981; Adler, Behrens, & Richter, 1995) was projected onto the field contours. Finally the identification and grouping of the fields according to relief and soil moisture is done by the help of index calculation which, for instance, took the surface ratio and intensity of water logging in the soils into account. So it was possible to sort the fields according to main characteristics. In line with the inventory of the overall area they could be selected according to area types.

Rare but interesting field types were included at the expense of more frequent locations. The fields observed in the monitoring area are thus placed in characteristic relation to the whole area. The continued monitoring of the floral population was drawn up in the next smaller scale on 250 m^2 lots of land, measured by GPS. Three lots of land each per field are distributed in such a manner that they reflect the internal field heterogeneity. Thereby the soil types, the substrate pattern, and/or relief characteristics were taken into account. A fourfold repletion of the inventory rating is affected in the lots of land. These points also have to be exactly located and reflect the variance in the soil at heterogeneous sites. Soil profile data were collected for the first time by mapping. We are currently examining whether the soil pattern can be identified better by geo-electric, remote sensing, and relief analysis. We hope to get better spatial interpolation of the measuring and laboratory results. As already mentioned, the basic programme acre, and also a grassland programme supply own monitoring data; however, they also provide basic information, e.g. for phyto-sanitary surveys, plant sociological, or microeconomic projects (Sattler, Kächele, & Verch, 2007). Despite being ready for integrative research, the involved nature, agriculture, and social scientists also need other spatial references.

The second approach, developed by the Saxon Academy of Sciences (SAW), is aligned at measurements and surveys which are not directed at cultivation fields. Therefore sampling points have to be determined. To be able to extrapolate the results to larger surface areas later, their representative validity should be considered when selecting these points. Instead of a random selection, the local conditions are drawn on for the design of the spatial sampling concept. The target is to capture the typical peculiarity and variance of an area with the fewest possible samples.

To explore the soil, for instance, the following representative analysis strategy is proposed (on the basis of spatial nature-type areas) which consists of the following work steps:

1. Determination of sampling density (depending on scale and hypotheses, e.g. ten samples/hectares)
2. Distribution of possible sample numbers according to frequency on existing soil types (maybe weighted distribution)
3. Thereby selection of above-average occurring soil–slope–utilisation combinations ('SSU', according to cross tabling)
4. Localisation of the sampling point in the mid of the largest polygon of every selected SSU type (to optimise the hit rate). Thus various SSU types are sampled for every type of soil. For statistic reasons, however, multiple samples of the same type are considered. Therefore the respectively largest polygons are selected.

The points thus evaluated are read into a GPS and with the latter found in the field. Should these positions prove to be non-accessible at site, replacement points are to be provided for alternative analyses, or points are to be re-measured at site.

19.4.3.3 Metadata

Metadata include information on the collected data and results. They should provide an overview of the data situation and selection situation and of the contact partners and of the availability. A suitable metadata concept is the prerequisite for successful collaboration, particularly for monitoring projects with cross-subject questions in the scope of larger associations. Metadata provide information on

- Goal and constraints of data collection
- Receipt, status, and degree of processing
- Reference to space and validity period
- Precision and statistic parameters
- Classification, interpretation rules
- Technical parameters (File format, software version)
- Availability and field of responsibility for all data, interim and final results of a monitoring project.

Nevertheless, the input to create the metadata must be minimal to ensure that the required updating is possible. A large amount also prevents users from accepting metadata. Often metadata are created in the form of XML files. The best known is the library information concerning inventory of literature for which the so-called Dublin Core Set (http.//www.dublincore.org/documents/dces/) was agreed upon as minimal database and for which there are a number of user interfaces (OPACS). To be able to store adequate information on spatial and ecological databases these description elements are partially extended by additional meta elements. Thereby, unfortunately, the above-named minimalist principle is often violated.

A two-step metadata construction has proven to be a suitable alternative to the dilemma described above:

1. The thematic arrangement of data according to the individual working steps. By insertion of hyperlinks, this working programme becomes a metadata bank. The hyperlinks show to what extent the working steps have been realised, and it opens the way to the (digital) results at least in intranet.
2. Documentation and putting together the 'technical information' in a tabular form according the Dublin-Core-Set (http.//www.bsz-bw.de/diglib/medserv/konvent/metadat/dcsyntax.html). This limited extent of information is sufficient for the documentation and the understanding of data (Table 19.4). This form is also common worldwide. Thus, comparability is guaranteed.

Table 19.4 Technical metadata according the Dublin Core Set

Element	Meaning	Example
DC.Title	Title of the data set	Soil map 'Kleine Spree'
DC.Creator	Main author (original data)	Syrbe
DC.Subject	Topic, due day	Soil profile point data 2000
DC.Description	Description (table columns, measurement units, etc.)	Profile, GPS, soil type, soil layer, depth, substratum, humus, munsell, admixture, macro-structure, moisture, groundwater, utilisation, relief, weather, observer, date
DC.Publisher	Published where	No
DC.Contributor	Co-authors + contact phone No.	Syrbe +49-351-81416805, Röder, Bauer
DC.Date	Date of last modification	10.05.2000
DC.Type	Type of document	Table
DC.Format	Data format	Excel workbook
DC.Identifier	Key	Spreebod
DC.Source	Data origin	Own elaboration
DC.Language	Coding	GER
DC.Relation	Hyperlink	\\Server\Benutzer\Monitoring\UGBiores\KleineSpree\Boden\SPREEBOD.XLS
DC.Coverage	Spatial coverage reference	5456903, 5467751, 5680994, 5691866
DC.Rights	Copyright	SAW

19.5 Monitoring Results and Experiences

Investigations are carried out in several scales. Due to this fact, some results concern to national or regional level, and other findings refer only to a special test area. The latter are much more precise, but they are representative of a special ecosystem type within a given area at most. Hence, monitoring results are comprehensive and manifold cf. Syrbe, Bastian, & Röder (2003). In all cases, they are ascribed to special monitoring purposes (for instance, success control of plans) or hypotheses (as explained in Section 19.4.2.1).

In this context, only some examples can be given related either to the whole covering area or to special nature regions.

19.5.1 Change Detection on the Data Level

Related to whole federal states, for instance, the growing of settlement and sealed area, the maintenance of nature-protected areas, of biotopes and species as well as climatic and hydrological characteristics are observed regularly.

The red lists of endangered species and habitats are an excellent example for hierarchically structured monitoring programmes. *Hypothesis 1* assumes that the global extinction of species and the loss of habitats can be halted by the international and national conservation area system. The inventories of the species are made at regional level by authorities, institutions, and environment organisations. Results are red lists issued in periodical cycles of administrations, for instance, by the federal states in Germany. At the next dimension level there is the aggregation of all available data at state level. Thus, according to the 'Red List of the endangered biotope types in Germany' 2006, 72.5% of all biotope types are identified as endangered (Riecken, Finck, Raths, Schröder, & Ssymank, 2006). The share of completely extinct habitats was 13.8%. Numerous species depend on these habitats and this fact is reflected in the Red Lists. The loss of habitats corresponds directly with the land use diversity and thus achieves economic and political dimensions. The agreement concluded at the Conference of Rio de Janeiro 1992 regarding the conservation of biodiversity also offers a global basis from species monitoring. These activities are, as is known, co-ordinated by IUCN (http.//www.iucnredlist.org) and by a large number of individuals and collaborating organisations.

The data provide proof that the number of endangered species is on the increase. The international network of protected areas was also of little effect in this respect. According to the authors, increasing efforts have to be made to conserve the species beyond the borders of the reservation (also in Flade, Plachter, Schmidt, & Werner, 2006).

The reasons for extinction of species and for declining of biodiversity could be found by means of detailed investigations of small sample areas. Some of these findings are surprising like the significant decrease of edge structures in connection with severer control of land grants by the European Union (EC European Commission, 2005). The consequence is a loss of marginal habitats particularly for reptiles and insects.

Considerable landscape changes are caused recently by urbanisation and sealing. Within the German federal state Saxony, for instance, the settled area has grown from 9.9% in 1992 up to 11.2% in 2000. Unfortunately, the sealed ratio increased due to the motorisation and subsequent upgrading of roads and spaces. The monitoring ought to clarify if at least the most damageable areas are omitted. While nature protection efforts led to success in the main part, particularly valleys, floodplains, and mountainous areas near the capital Dresden are affected. The sealing tendency within the valleys exceeds the average at 2.5%. Flood originating areas uphill the capital are sealed up to 150% of comparable landscapes outside.

Small-scale studies should cover catchments or floodplains because it is easy to compile balances of matter, energy, and water pollutants there. Such small areas also enable to scrutinise the land use intensity in more detail. One of these test sites is the Kleine Röder catchment which is characteristic for intensive used agricultural countryside at the edge of agglomeration areas. Within only 27 years, we found 9% changes of soil properties. But only one-third is caused by development. The largest change concerns changes in soil moisture. Due to the construction of a new highway and some commercial areas as well as the deterioration of drainage systems, the drain conditions changed considerably and almost 4% fell wet. Table 19.5 shows the changes found in detail.

Table 19.5 Changes of soil conditions due to land use intensity in the Kleine Röder catchment

Change	Area ratio (%)
Waterlogging	3.67
Erosion (truncated profiles)	2.63
Development	2.07
Accumulation/aggradation	0.45

Table 19.6 Compliance of the landscape plan of Moritzburg in 2004

Planned measure 1997	Compliance 2004
Maintain of biotopes	In essence
Renaturation of existing waters	None
Founding new ponds and sparse orchards	In less quantity
Maintain and emergence of edge biotopes	None
Enhencement of nature reserve	Yes

Another type of analysis deals with the realisation of definite planning objectives. Table 19.6 gives an example about the compliance of a landscape plan which is typical for such an implementation.

19.5.2 Complex Interpretations

The raw data of long-term and comprehensive monitoring programmes tend to piling up. Therefore, a sound cutback in terms of information grafting is recommended. Monitoring results are eminently meaningful if they are purposefully integrated. Since complex interpretations demand comprehensive well-matched data, they are predominantly used for smaller test areas, but also countrywide statements are available.

Hypothesis 2 postulates an increasing improvement of water quality as a result of technical innovation and political parameters. Contrary to the trend in numerous developing countries, Europe has been able to record an improvement of flowing water quality. With the European water framework directives (RL 2000/60/EG), a common instrument was created which bundles the monitoring activities in this respect. Physical, chemical, and biological parameters are the indicators. Quality monitoring that has been carried out over decades in some member states forms the basis for the evaluation of the actual state and the possibilities of the future development. Taking the example Saxony, the positive trend can be seen particularly clearly (Table 19.7) in the new federal states of Germany after 1990.

Mostly, the nutrient load of land is calculated by means of balances taking into account the application of fertilisers, harvest, turnovers as well as other in- and outputs. But the information must be taken by the

Table 19.7 Share of the quality grades seen in the waters network of Saxony (source: LfUG, 2003). I – not polluted, II – moderately polluted, III – severely polluted, IV – excessively polluted

Year	Share of the quality grades in %						
–	I	I–II	II	II–III	III	III–IV	IV
1994	1.2	4.1	26.4	38.8	21.0	6.2	2.2
1997	1.7	5.2	38.8	40.3	9.8	1.4	2.8
2000	2.2	8.1	60.9	23.8	4.4	0.5	0.1
2003	2.6	9.3	61.8	23.9	1.8	0.4	0.2

farmers interested in both achieving maximum yields and formal compliance of good agricultural practise. Only a systematic monitoring can clarify the accuracy of information. The investigations of several test sites resulted in a considerable difference between the nutrient balances and measured values in the receiving streams. For instance, a mathematical lowering of nitrogen load within the Dorbichtgraben catchment is accompanied by a constant high nitrate pollution of the creek (about twice the threshold) whereas adjacent catchments show a sufficient correlation.

The above mentioned Moritzburg monitoring site reveals a general lowering of high valuable biotopes and increasing low-value biotope types on about 4% within 7 years. The land use diversity (Shannon diversity) decreased in all investigated areas.

19.6 Methodical Experiences and Conclusions

Some measurements and observances must be done continuous (climate) or in close succession (water quality). But the most investigations should be done at a distance of several years. Depending on the rate of landscape change, suitable time intervals are between 5 and 7 years. At the border of Dresden, for instance, the repetition of land use and structure investigations revealed almost 100% changes due to the construction of a new highway and some developing areas. Another comparable study in the biosphere reserve showed only marginal changes within 7 years. But that interval should not be exceeded to prevent the lost of comparability due to new sampling methods, instruments, and landscape structures. In any case, the sample locations must be marked both by precise (GPS) coordinates and in terrain.

A common problem is the renewal of metadata. If the scheme of Section 19.4.3.3 is applied, the task could be shared. The technical metadata must be designed by the data deliverer, some software creates such metadata automatically (like ArcGis), which has just to complete. But the level I metadata have always to be maintained by the coordinator who gains an overview about the success of the programme as a whole. He makes a task schedule for all participants every year and supervises its realisation. Omitted investigations due to disease or technical problems are able to catch up within finite time.

References

ADAS. (1997). *Environmentally sensitive areas scheme, landscape monitoring in the clun ESA 1988–1995.*

AG Bodenkunde. (1994). *Bodenkundliche Kartieranleitung*. 4. Aufl., Bundesanstalt f. Geowiss. u. Rohstoffe & Niedersächs. Landesamt f. Bodenforsch. Hannover.

Ad-hoc-AG Boden. (2005). *Bodenkundliche Kartieranleitung*. 5.Aufl., Bundesanstalt f. Geowiss. u. Rohstoffe & Niedersächs. Landesamt f. Bodenforsch. Hannover.

Adler, G., Behrens, J., & Richter A. (1995). *Ergebnisse der Aufbereitung von Daten der Mittelmaßstäbigen Landwirtschaftlichen Standortkartierung (MMK) der ehem. DDR für länderübergreifenden Bodenschutz*. Bundesanst. f. Geowiss. u. Rohstoffe, Hannover/Berlin. Archiv-Nr. 2025093. Berlin.

AMAP. (2005). The AMAP Strategic Plan 2004+. AMAP Report 2004, 5 (http.//amap.no/documents/) 17p.

Antipolis, S. (1995). *Observation et évaluation de l'environment en Mediterranée (Phase preparatoire)*. PlanBlue, UNEP(OCA)/MED IG, 5/Inf. 5, Athens.

Bastian, O. (1997). Gedanken zur Bewertung von Landschaftsfunktionen – unter besonderer Berücksichtigung der Habitatfunktion. *NNA-Berichte, 10*, 106–125.

Bastian, O. (1999). Landschaftsbewertung und Leitbildentwicklung auf der Basis von Mikrogeochoren. In U. Steinhardt, & M. Volk (Eds.), *Regionalisierung in der Landschaftsökologie*. Fachtagung am UFZ Leipzig 31.3.1998–2.4.1998 (pp. 287–298).

Bastian, O. (2000). The assessment of landscape and vegetation changes. a case study – Upper Lusatian Heath and Pond Landscape. In A. Richling, J. Lechnio, & E. Malinowska (Eds.), *Landscape ecology. Theory and applications for practical purposes* (pp. 31–53). Pultusk School of Humanities, IALE Poland, Warszawa.

Bastian, O., & Bernhardt, A. (1993). Anthropogenetic landscape changes in Central Europe and the role of bioindication. *Landscape Ecology, 8*, 148–164.

Bastian, O., & Röder, M. (1996). Beurteilung von Landschaftsveränderungen anhand von Landschaftsfunktionen – am Beispiel zweier Testgebiete im sächsischen Hügelland. *Naturschutz u. Landschaftsplanung, 28*(10), 302–312.

Bastian, O., & Röder, M. (2002). Landscape functions and natural potentials. In O. Bastian, & U. Steinhardt (Eds.), *Development and perspectives of landscape ecology* (pp. 213–237). Dordrecht, Boston, London: Kluwer Academic Publishers.

Bastian, O., & Schreiber, K.-F. (1999). *Analyse und ökologische Bewertung der Landschaft* (2nd ed.). Heidelberg, Berlin: Spektrum.

Bastian, O., & Steinhardt U. (2002). *Development and perspectives in landscape ecology*. Dordrecht: Kluwer.

Bastian, O., Syrbe, R.-U., & Röder, M. (2001). Landscape-ecological monitoring in Germany. *Publicationes Instituti Geographici Universitatis Tartuensis, 92*, 252–257.

Bernhardt, A., & Jäger, K.-D. (1985). Zur gesellschaftlichen Einflussnahme auf den Landschaftswandel in Mitteleuropa in Vergangenheit und Gegenwart. In A. Bernhardt, K.-D. Jäger, K. Mannsfeld, & I. Hartsch (Eds.), *Beiträge zum Problemkreis des Landschaftswandels* (pp. 5–56). Sitz-Ber d. Sächs. Akad. Wiss. Leipzig, 117(4).

Bundesregierung. (2002). *Perspektiven für Deutschland. Unsere Strategie für eine nachhaltige Entwicklung*. Berlin.

CMCE Committee of Ministers of the Council of Europe. (2000). European Landscape Convention. Florence, 10-20-2000, from http.//www.nature.coe.int/english/main/ landscape/conv.htm.

de Groot, R. (1992). *Functions of Nature – Evaluation of nature in environmental planning, management and decision making*. Groningen: Wolters-Noordhoff.

DPCSD United Nations Department for Policy Coordination and Sustainable Development. (1999). Indicators for environmental aspects of sustainable development. In *Indicators of SD*. UN CSD Methodology Sheets from http.//esl.jrc.it/envind/un_meths/UN_ME.htm.

Dramstad, W. E., Fjellstad, W. J., Strand, G. H., Mathiesen, H. F., Engan, G., & Stokland, J. N. (2002). Development and implementation of the Norwegian monitoring programme for agricultural landscapes. *Journal of Environmental Management, 64*, 49–63.

EC European Commission. (2005). Council Regulation (EC) No 1698/2005 of 20 September 2005 on support for rural development by the European Agricultural Fund for Rural Development (EAFRD).

ECNC European Centre for Nature Conservation. (2002). European biodiversity monitoring and indicator framework. An inventory of biodiversity indicators in Europe. Prepared by B. Delbaere. from http.//nature.eionet.eu.int/activities/ products/indicators/index_html.

EEA European Environmental Agency. (1999). Environmental indicators. Typology and overview. Prepared by E. Smeets and R. Weterings, TNO Centre for Strategy, Technology and Policy, The Netherlands.

Flade, M., Plachter, H., Schmidt, R., & Werner, A. (2006). *Nature conservation in agricultural ecosystems. Results of the Schorfheide-Chorin research project*. Wiebelsheim, Quelle & Meyer.

Fryrear, D. W., Saleh, A., Bilbro, J. D., Schomberg, H. M., Stout, J. E., & Zobeck, T. M. (1998). Revised Wind Erosion Equation (RWEQ). Wind Erosion and Water Conservation Research Unit. USDA-ARS, Southern Plains Area Cropping Systems Research Laboratory. *Technical Bulletin No. 1*.

Glemnitz, M., & Wurbs, A. (2003). Analyzing spatial habitat distribution to improve the assessment of land use impacts on habitat functions. In K. Helming, & H. Wiggering (Eds.), *Sustainable development of multifunctional landscapes* (pp. 177–189). Springer: Berlin.

Haase, G. (1978). Zur Ableitung und Kennzeichnung von Naturraumpotentialen. *Petermanns Geogr. Mitt., 122*, 113–125.

Haines-Young, R. H., Barr, C. J., Black, H. I. J., Briggs, D. J., Bunce, R. G. H., Clarke, R. T., et al. (2000). *Accounting for nature. Assessing habitats in the UK countryside*. London: DETR.

Hölzl, M. (2003). Corine Landcover – Erfahrungen bei der Aktualisierung und Abgabe von Landbedeckungsdaten. In R. Grillmayer, & W. Schneider (Eds.), *Geodaten zur Landbedeckung in Österreich* (pp. 22–23).

Howard, D. C., Petit, S., & Bunce, R. G. H. (2000). Monitoring multi-functional landscapes at a national scale – guidelines drawn up from the Countryside Survey of Great Britain. In J. Brandt, B. Tress, & G. Tress (Eds.), *Multifunctional landscapes. Interdisciplinary approaches to landscape research and management* (pp. 51–62). Roskilde: Centre for Landscape Research.

Joschko, M., Fox, C. A., Lentzsch, P., Kiesel, J., Hierold, W., Krück, S., & et al. (2006). Spatial analysis of earthworm biodiversity at the regional scale. *Agriculture, Ecosystems & Environment, 112*(4), 367–380.

Kaiser, T., Käding, H., Kiesel, J., Müller, L., Hierold, W., & Behrendt, A. (2005). The derivation of grassland vegetation types on the basis of site and land use characteristics. *Archives of Agronomy and Soil Science, 51*(4), 405–416.

Keil, M., Mohaupt-Jahr, B., Kiefl, R., & Strunz, G. (2002). Das Projekt CORINE Land Cover 2000 in Deutschland. In S. Dech, et al. (Eds.), *Tagungsband 19. DFD-Nutzerseminar* (pp. 95–104), 15—16, Oktober 2002.

Knospe, F. (1998). *Handbuch zur argumentativen Bewertung. Methodischer Leitfaden für Planungsbeiträge zum Naturschutz und zur Landschaftsplanung*. Dortmunder Vertrieb für Bau- und Planungsliteratur, Dortmund.

Lausch, A. (2000). Raum-zeitliches Monitoring von Landschaftsstrukturen in der Tagebauregion Südraum Leipzig mit Methoden der Fernerkundung und Geoinformation. *UFZ-Bericht Nr. 12/2000*. Leipzig.

LAWA Länderarbeitsgemeinschaft Wasser (2000). *Gewässerstrukturgütekartierung in der Bundesrepublik Deutschland*. Verfahren für kleine und mittelgroße Fließgewässer..

LfUG Sächsisches Landesamt für Umwelt und Geologie (2003). *Gewässergütebericht 2003. Biologische Befunde der Gewässergüte sächsischer Fließgewässer mit Gewässergütekarte*. Materialien zur Wasserwirtschaft.

Luthardt, V., Vahrson, W.-G., & Dreger, F. (1999). Konzeption und Aufbau der Ökosystemaren Umweltbeobachtung für die Biosphärenreservate Brandenburgs. *Natur und Landschaft, 74*(4), 135–143.

Lütz, M., Bastian, O., & Weber, C. (2006). Rahmenmethodik zur Entwicklung lokaler Agrarumweltprogramme in Europa – Akzeptanz und Monitoring von Agrarumweltmaßnahmen. *Wasserwirtschaft 10/2006*, 34–40.

Marks, R., Müller, M.J., Leser, H., & Klink, H.-J. (1992). *Anleitung zur Bewertung des Leistungsvermögens des Landschaftshaushaltes (BALVL)*. (Forschungen zur deutschen Landeskunde, vol. 229, 2nd ed. Trier).

McAdam, J.H., Flexen, M., & O'Mahony, D. (2008). *Biological Monitoring of Countryside Management Scheme in Northern Ireland 2001/02–2006/07*. Report to DARD, Queen's University Belfast.

Neef, E. (1983). Die Landschaft als Integrationsebene gesellschaftlicher Raumgestaltung. *Geol. Mijnbouw, 62*, 531–534.

Neubert, M., & Walz, U. (2000). Der Landschaftswandel im Raum Pirna – eine Untersuchung auf der Grundlage des Amtlichen Topographisch-Kartographischen Informationssystems und des Vergleiches historischer topo-graphischer Karten. *Mitteilungen des Landesvereins Sächsischer Heimatschutz 1/2000*, 19–27, Dresden.

OECD Organisation for Economic Co-operation and Development. (2001). *Environmental indicators. Towards sustainable development*. Paris.

Petit, C.C., & Lambin, E.F. (2002). Impact of data integration technique on historical land-use/land-cover change. Comparing historical maps with remote sensing data in the Belgian Ardennes. *Landscape Ecology, 17*, 117–132.

Riecken, U., Finck, P., Raths, U., Schröder, E., & Ssymank, A. (2006). Rote Liste der gefährdeten Biotoptypen Deutschlands. Zweite fortgeschriebene Fassung 2006. *Natursch. Biol. Vielf., 34*.

Röder, M., & Adolph, B. (2006). Lokalisierung und Management von Hochwasserentstehungsgebieten. *Wasserwirtschaft, 96*(5), 22–26.

Röder, M., & Syrbe, R.-U. (2000). Relationship between land use changes, soil degradation and landscape functions. a case study within the, Upper Lusatian Health and Pond Landscape' biosphere reserve'. In A. Richling, J. Lechnio, & E. Malinowska (Eds.), *Landscape ecology – theory and applications for practical purposes* (pp. 235–246). The problems of landscape ecology, 4, Warszawa.

ROSELT/OSS (2005). Reseau d'Observatoires de Surveillance Ecologique á Long Terme. From http.//www.roselt-oss.org.

Sattler, C., Kächele, H., & Verch, G. (2007). Assessing the intensity of pesticide use in agriculture. *Agriculture, Ecosystems & Environment, 119*(3–4), 299–304.

Schindler, U., Steidl, J., & Müller, L. (2007). Influence of regional-scale soil heterogeneity and land use on water discharge formation in the "Stettiner Haff" catchment in North-East Germany. *Archives of Agronomy and Soil Science, 53*(1), 63–73.

Schlüter, H. (1987). Der Natürlichkeitsgrad der Vegetation als Kriterium der ökologischen Stabilität der Landschaft. *Ecology an creation of new environments* (pp. 93–102). Tokio.

Schmidt, R. (1978). Prinzipien der Standortgliederung der mittelmaßstäbigen Standortkartierung der DDR. *Arch. Acker-Pflanzenbau Bodenkd. Berlin, 19*(8), 533–543.

Schmidt, R., & Diemann R. (1981). *Erläuterungen zur Mittelmaßstäbigen landwirtschaftlichen Standortkartierung*. Akad. d. Landwirtschaftswiss. d. DDR, Inst. f. Bodenkunde, Eberswalde.

Sommer, M. (2006). Influence of soil pattern on matter transport in and from terrestrial biogeosystems. a new concept for landscape pedology. *Geoderma, 133*(1–2), 107–123.

Syrbe, R.-U. (1994). Spezielle Bewertungsverfahren im Oberspreewald. In H. Barsch, G. Saupe, et al. (Eds.), *Bewertung und Gestaltung der naturnahen Landschaft in Schutzgebieten, Erholungs- und Freizeitgebieten* (pp. 46–58). Potsdamer Geographische Forschungen 8.

Syrbe, R.-U. (2002). Aktualisierung generalisierter Landnutzungsdaten aus älteren CIR-Biotopinformationen mit Landsat 7. In S. Dech, et al. (Eds.), *Tagungsband 19. DFD-Nutzerseminar* (pp. 113–125).

Syrbe, R.-U., & Palitzsch, G. (2002). Historische Elemente der Kulturlandschaft am Südrand von Dresden. Beitrag zu einem repräsentativen Monitoring. *Mitteilungen des Landesvereins Sächsischer Heimatschutz 1/2002*, 43–51.

Syrbe, R.-U., & Schulze, T. (2006). A land use data update tool for rule-based query of remote sensing outputs. In P. Karrasch (Ed.), *NatureProtection GIS*. International Symposium on Geoinformatics in European Nature Protection Regions. Proceedings (pp. 121–132). Dresden.

Syrbe, R.-U., Bastian, O., & Röder, M. (2003). Landschaftsmonitoring als Grundlage für Bewertung und Modellierung. In O. Bastian, K. Grunewald, J. Schanze, R.-U. Syrbe, & U. Walz (Eds.), Bewertung und Entwicklung der Landschaft. *IÖR-Schriften, 40*, 225–236.

Syrbe, R.-U., Bastian, O., Röder, M., & Haase, G. (2002). Veränderungen der Landnutzung und Landschaftswandel. In W. Fritsche, & L. Zerling (Eds.), Umwelt und Mensch – Langzeitwirkungen und Schlußfolgerungen für die Zukunft. *Abhandlungen der Sächsischen Akademie der Wissenschaften zu Leipzig, Mathematisch-naturwissenschaftliche Klasse, 59*(5), 141–160.

Syrbe, R.-U., Bastian, O., Röder, M., & James, P. (2007). A framework for monitoring landscape functions. The Saxon Academy Landscape Monitoring Approach (SALMA), exemplified by soil investigations in the Kleine Spree floodplain (Saxony, Germany). *Landscape and Urban Planning, 79*, 190–199.

Syrbe, R.-U., Röder, M., & Bastian, O. (2001). Regionalisierungsansätze in der Landschaftsbewertung - dargestellt am Oberlausitzer Heide- und Teichgebiet. *Berichte zur deutschen Landeskunde, 75*(1), 67–89.

TMAP Trilateral Monitoring and Assessment Group. (1997). *TMAP Manual. The Trilateral Monitoring and Assessment Program (TMAP)*. Wilhelmshaven: Common Wadden Sea Secretariat.

UNEP-WCMW (2005). *Guidelines for biodiversity assessment and monitoring for protected areas*. Cambridge: UNEP-World Conservation Monitoring Centre.

Vos, V., & Meekes, H. (1999). Trends in European cultural landscape development. Perspectives for a sustainable future. *Landscape and Urban Planning, 46*, 3–14.

Wegehenkel, M., Heinrich, U., Uhlemann, S., Dunger, V., & Matschullat, J. (2006). The impact of different spatial land cover data sets on the outputs of a hydrological model. A modelling exercise in the Ucker catchment, North-East Germany. *Physics and Chemistry of the Earth, 31*(17), 1075–1088.

Wegehenkel, M., & Kersebaum, K.-C. (2005). The validation of a modelling system for calculating water balance and catchment discharge using simple techniques based on field data and remote sensing data. *Physics and Chemistry of the Earth, 30*(1–3), 171–179.

Werner, C. (2002). *Einsatzmöglichkeiten räumlich hochauflösender Satellitendaten in Landschaftplanung und Naturschutz. Beispiele der Auswertung von IRS-1-C-D-Daten.* Habilitationsschrift, Technische Universität Berlin.

Wieland, R., Voß, M., Holtmann, X., Mirschel, W., & Ajibefun, I. A. (2006). Spatial Analysis and Modeling Tool (SAMT). 1. Structure and possibilities. *Ecological Informatics, 1*(1), 67–76.

Wischmeier, W. H., & Smith, D. D. (1978). *Predicting rainfall erosion losses. A guide to conservation planning*. Aricultural Handbook 537, U.S. Department of Agiculture.

Chapter 20

Temporal Changes and Spatial Determinants of Plant Species Diversity and Genetic Variation

Cornelia Baessler, Stefan Klotz, and Walter Durka

Abstract Intensification of agricultural land use during the last century, combined with an increasing level of agrochemicals, has resulted in a decline of both habitat diversity and quality and to simplification and homogenization of Central Europe landscapes. For three agricultural landscapes in Central Germany we investigated (1) the influence of historical and current land-use and landscape structure on plant species diversity patterns in semi-natural habitats as well as on arable fields and (2) the extent to which genetic variation within populations (H_e) of the common forest herb *Geum urbanum* is related to population properties and to present landscape structure. Historical and present floristic field data were analysed in relation to land-use and landscape structure characteristics of the same periods (1950s, 1970s and 2000/2002). Changes in plant species richness and composition during the past 50 years varied among landscapes according to their land-use history and environmental characteristics, but were mostly in favour of ruderal species. Plant species richness for semi-natural habitats was negatively affected by increases in mean patch size of meadows and by increases in phosphorus application. Moreover, the application of mineral fertilizer, especially phosphorus, led to many habitat specialists being replaced by generalist species. Species richness of 'arable weeds' was significantly affected at both landscape and regional level by the proportion of semi-natural habitats, habitat diversity and habitat isolation due to landscape homogenization. More intensive land use, and particularly increased nitrogen application, was associated with decreased richness of 'arable weeds'.

The landscape genetic approach was extended to the same three landscapes and thus landscape-specific patterns could be disentangled from general relationships consistent across landscapes. Genetic variation of 70 populations was determined at eight microsatellite loci. Landscape structure was assessed in circular areas around populations and related to genetic variation within populations (H_e) by linear mixed-effects models. H_e was affected in an inverse manner by the size of *Geum* populations, by average patch size, by land-use diversity, by the area of woody habitat and by the area of roads. The study underlines the importance of habitat area and isolation as factors affecting genetic diversity, with both factors varying in a landscape-specific way.

These results suggest that regional and historical processes, as well as local environmental factors, influence local plant community and population structure. Therefore, long-term studies are of high importance to understand ecological processes. We conclude that for conserving biodiversity in agricultural landscapes it is as important to protect existing, historically developed habitat diversity as the protection of habitat quality.

Keywords Agricultural landscapes · Landscape structure · Land-use change · Land-use intensity · Biodiversity trends · Vegetation mosaic · Landscape genetics

C. Baessler (✉)
Helmholtz Centre for Environmental Research – UFZ,
Department of Community Ecology, Theodor-Lieser-Straße 4,
06120 Halle, Germany
e-mail: cornelia.baessler@ufz.de

F. Müller et al. (eds.), *Long-Term Ecological Research*,
DOI 10.1007/978-90-481-8782-9_20, © Springer Science+Business Media B.V. 2010

20.1 Introduction

The biodiversity of agricultural landscapes is strongly influenced by agricultural policies and management and production techniques (Luoto, Rekolainen, Aakkula, & Pykälä, 2003). In Europe, agriculture has been the primary land use for hundreds of years and played an important role in shaping landscapes. Traditional farming systems produced a small-scale patchwork of different semi-natural habitats and enhanced species diversity in agricultural landscapes (Piorr, 2003). These landscapes were rich in biodiversity in both semi-natural habitats and agriculturally used areas. Land-use changes, however, are predicted to have the largest global impact on species diversity in terrestrial ecosystems (Sala et al., 2000). Urbanization, but mainly intensification of agricultural land use during the second half of the 20th century supported by agricultural policy and advances in agricultural technology and management, combined with an increasing level of inputs of agrochemicals (fertilizers, pesticides and herbicides) reduced both diversity and habitat quality of existing agricultural landscapes all over Europe (Robinson & Sutherland, 2002). In East Germany, loss and degradation of semi-natural habitats have been especially marked since the 1960s when private farmlands were pooled to create large state farming units as a result of agricultural policy of the former German Democratic Republic. The structure and scale of the landscapes have changed significantly because of the trend to separate arable and livestock production systems (Baessler & Klotz, 2006). Since 1990, when all agricultural policy was adapted to meet E.U. norms and regulations there have been further changes in land use.

The application of previously limiting resources that influence ecosystem functioning (nutrients like nitrogen, phosphorus and water) and the application of pesticides and the conversion of natural ecosystems to agriculture cause several changes in agro-ecosystems (Tilman et al., 2001). Increased land-use intensity, in particular higher nutrient input, initiated species turnover and contributed to declines of species richness (Honnay, Hermy, & Coppin, 1999; Hilli, Kuitunen, & Suhonen, 2007), population size and genetic diversity of surviving populations in both managed areas and adjacent semi-natural habitats (Moilanen & Hanski, 1998; Endels, Jacquemyn, Brys, Hermy, & De Blust, 2002). As nitrogen and phosphorus released through fertilizers and animal waste enter surface and groundwater, these leakages from agricultural systems cause major environmental problems and changes (Stoate et al., 2001; Warren, Lawson, & Belcher, 2008). A large-scale increase in the nutrient input into more stable ecosystems causes eutrophication of different ecological systems and leads to an imbalanced trophic structure (Carpenter et al., 1998; Khan & Ansari, 2005). Hence, human impact on the environment has resulted not only in decline of species diversity, but also changes in the relative abundance of species in terrestrial and aquatic ecosystems, and thus to changes in ecosystem functioning (Chapin et al., 1997; Tilman, Reich, & Knops, 2006).

Landscapes changed significantly, both in structure and in spatial scale, as arable and livestock production systems became increasingly separated (Ihse, 1995; Hietala-Koivu, 2002). Linear elements were removed to enlarge fields and improve efficiency, leading to reduction and isolation of areas with high biodiversity, increasing uniformity of the spatial arrangement of habitat patches, and therefore to large-scale simplification and homogenization of agricultural landscapes. Habitat loss and fragmentation as well as habitat deterioration are the main causes of decline in biodiversity, connected with community simplification and loss of ecosystem services (Tilman et al., 2001; Green, Cornell, Scharlemann, & Balmford, 2005). Habitat destruction, especially fragmentation and reduction in habitat size, cause isolation, edge effects, invasion of non-native species and decline in population size (Luoto et al., 2003; Fahrig, 2003). Habitat fragmentation additionally reduces within-patch species and genetic diversity as a result of decreases in population and community size and reduced immigration rate (Frankham, Ballou, & Briscoe, 2002). In accordance to the species–area relationship, small habitat patches contain small communities and populations are therefore expected to lose species and genotypes via drift (Young, Boyle, Brown, 1996; Wilson & Chiarucci, 2000). These effects of drift may be counteracted by immigration, both at levels of the community and the population, as immigration provides new species or novel alleles. Drift and migration, therefore, influence species and genetic diversity in fairly similar and straightforward ways (Vellend & Geber, 2005). The processes of drift and migration determine the positive correlations of diversity with the area or connectivity of localities (Rosenzweig, 1995; Frankham et al., 2002). Additionally, these structural characteristics of the landscape can create positive correlations between species diversity and genetic diversity (Vellend, 2005). The extinction probability for populations is expected to decrease with increasing area, whereas immigration processes decrease with increasing isolation, both with strong variation among species (Bruun, 2000). The responses of plant species to the loss or deterioration of habitat are mainly determined by functional traits like reproduction, life history and dispersal mechanisms (e.g. McIntyre, Lavorel, Landsberg, & Forbes, 1999; Dupré & Ehrlén, 2002). Habitat specialists are often predicted to become extinct faster than generalist species like ruderals. Therefore, functional shifts within communities are expected as sets of species with particular traits are replaced by other sets with different traits (McCollin, Moore, & Sparks, 2000; Loreau et al., 2001). According to the particular traits of plant species, populations and metapopulations respond with shorter or longer time lag (Hanski, 1998; Hanski & Ovaskainen, 2003). Populations of larger species with a long life cycle and slow intrinsic dynamics may occur as remnant populations and communities in modern agricultural landscapes (Eriksson, 2000; Helm, Hanski, & Pärtel, 2006). Hence, pronounced decline in local and regional species richness as consequences of habitat loss and fragmentation may only become apparent after a substantial length of time, and therefore resulting in an 'extinction debt' (Tilman, May, Lehman, & Nowak, 1994). Recently it was shown that species richness and community composition are even better explained by the isolation of habitat fragments than by the fragment area (e.g. Petit et al., 2004; Bailey, 2007). In accordance to the rescue effect there is a higher probability for populations to persist in small habitat patches if they are well connected (Brown & Kodric-Brown, 1977; Piessens, Honnay, Nackaerts, & Hermy, 2004). Furthermore, landscapes with high historical connectivity are expected to be more resilient to disturbance and to maintain higher species diversity than more fragmented landscapes, as colonization is influenced by habitat connectivity (Lindborg & Eriksson, 2004; Lindborg, 2007). Past changes in environmental conditions as well as species composition may have large and often irreversible effects on the structure and dynamics of present-day ecosystems (de Blois, Domon, & Bouchard, 2001), thus in ecology, historical events play a major role. Additionally, remnant habitat patches may exist in non-equilibrium states due to a complex history of anthropogenic disturbance systems (Fischer & Stöcklin, 1997; Cousins & Eriksson, 2001). In dynamic landscapes, therefore, patterns of plant species occurrence in fragmented and isolated habitat patches are affected by both historical and present landscape configurations.

Landscape ecology previously mainly concentrated on spatial interactions across the landscape and the influences of spatial patterns on biotic and abiotic processes and has often been criticized for ignoring temporal changes (Naveh & Lieberman, 1984; Levin, 1992). Many studies focusing on fragmentation effects on species diversity and composition mostly reflect a short time frame, which cannot encompass the many indirect feedbacks that occur in anthropogenic disturbed landscapes (Debinski & Holt, 2000). But it is the dramatic changes that occur on varying timescales that are predominantly characterising agricultural landscapes. Therefore, effects of both short- and long-term processes are necessary to be detected. Recently, however, greater emphasis is placed on changes in functional aspects of remnant ecosystems in fragmented agricultural landscapes (e.g. Gustavsson, Lennartsson, & Emanuelsson, 2007; Liira et al., 2008).

Spatial heterogeneity is the main predictor of biodiversity in agricultural landscapes (Wiens, Stenseth, van Horne, & Ims, 1993). While several recent studies relate spatial heterogeneity and land-use intensity to species diversity of different taxonomic groups (e.g. Schweiger et al., 2005; Billeter et al., 2008), investigations including comparisons in space and time are rare. Additionally, recent studies have shown an effect of historical land-use management and landscape structure on present species diversity patterns (e.g. Honnay, Jacquemyn, Bossuyt, & Hermy, 2005; Johansson et al., 2008). Most of these studies have been performed on a local scale and focused on species diversity of one habitat type. Other recent studies on larger scales did not relate species diversity patterns to historical landscape characteristics (e.g. Dormann et al., 2007; Liira et al., 2008). Comparisons across space allow an evaluation of biodiversity and its predictors in agricultural landscapes on larger scales. However, to evaluate natural succession or environmental changes caused by anthropogenic or natural-induced effects comparisons over longer time periods are necessary (Duelli, 1997).

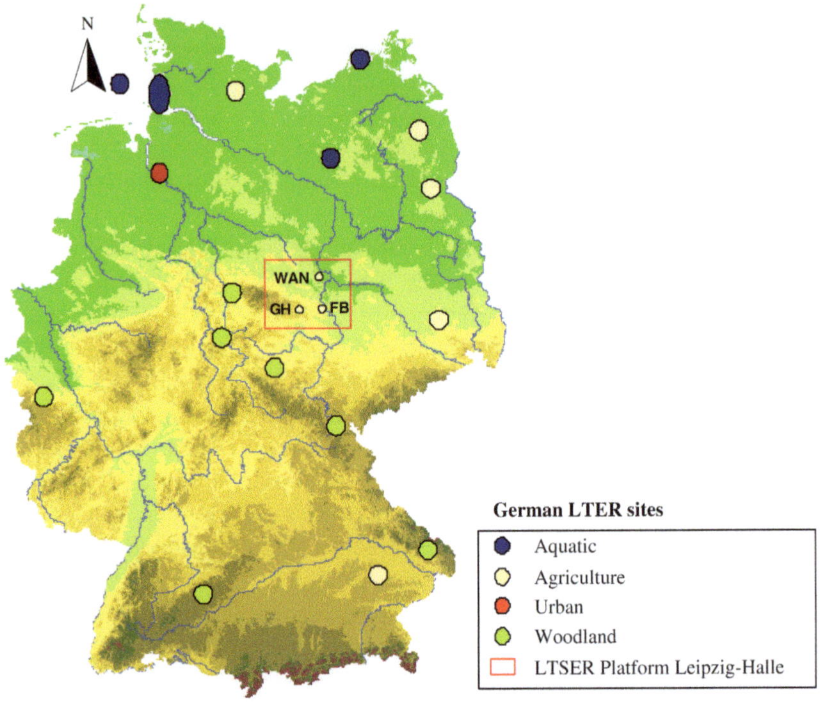

Fig. 20.1 Location of the three study sites: Wanzleben (WAN), Friedeburg (FB) and Greifenhagen (GH) as LTER sites within the LTSER platform Leipzig-Halle, in Germany; LTER sites of the German network, coloured according to the dominating habitat types

The complex interactions among historical, environmental and structural processes are reflected by the vegetation mosaic of a landscape (de Blois et al., 2001). Assessing the relative contribution of each of these variables and estimating relationships across landscapes are crucial steps to understand vegetation and landscape dynamics and to consider this knowledge in the future. In this study, different aspects of biodiversity were analysed to evaluate changes in ecological functions, characteristics and interrelations within agricultural landscapes. Changes in species diversity were linked to agricultural management and landscape structure. These relations were studied separately at the regional, landscape and species level. The relative importance of historical to recent landscape structure and environmental conditions on plant species diversity and genetic variation was assessed within three agricultural landscapes in Central Germany across three time periods (1950s, 1970s and 2000/2002). The study sites Wanzleben, Friedeburg and Greifenhagen (Fig. 20.1) were chosen according to historical vegetation inventories. These sites are dominated by agricultural land use and are representative for East Germany and other parts of Central Europe with respect to environmental conditions as well as temporal landscape changes. Each site is a LTER site within the LTSER platform Halle-Leipzig.

20.2 Effects of Changes in Land-Use Intensity and Landscape Structure on Plant Species Richness

20.2.1 Methods

20.2.1.1 Study Sites

The study sites are located in Saxony-Anhalt, Germany, near the villages Wanzleben (hereafter WAN; 11°26′E, 52°04′N), Friedeburg (FB; 10°34′E, 45°12′N) and Greifenhagen (GH; 11°25′E, 51°38′N) (Fig. 6.1). Each study site has subcontinental climatic conditions. The study sites were chosen according to historical vegetation inventories (Schubert & Mahn, 1959; Mahn & Schubert, 1961, 1962; Kiesel, 1980; Westhus, 1980) and cover areas of 9.6 km^2 (WAN), 6.7 km^2 (FB) and 7.9 km^2 (GH). The landscapes are representative of the region, being located within a much larger agricultural area with similar landscape

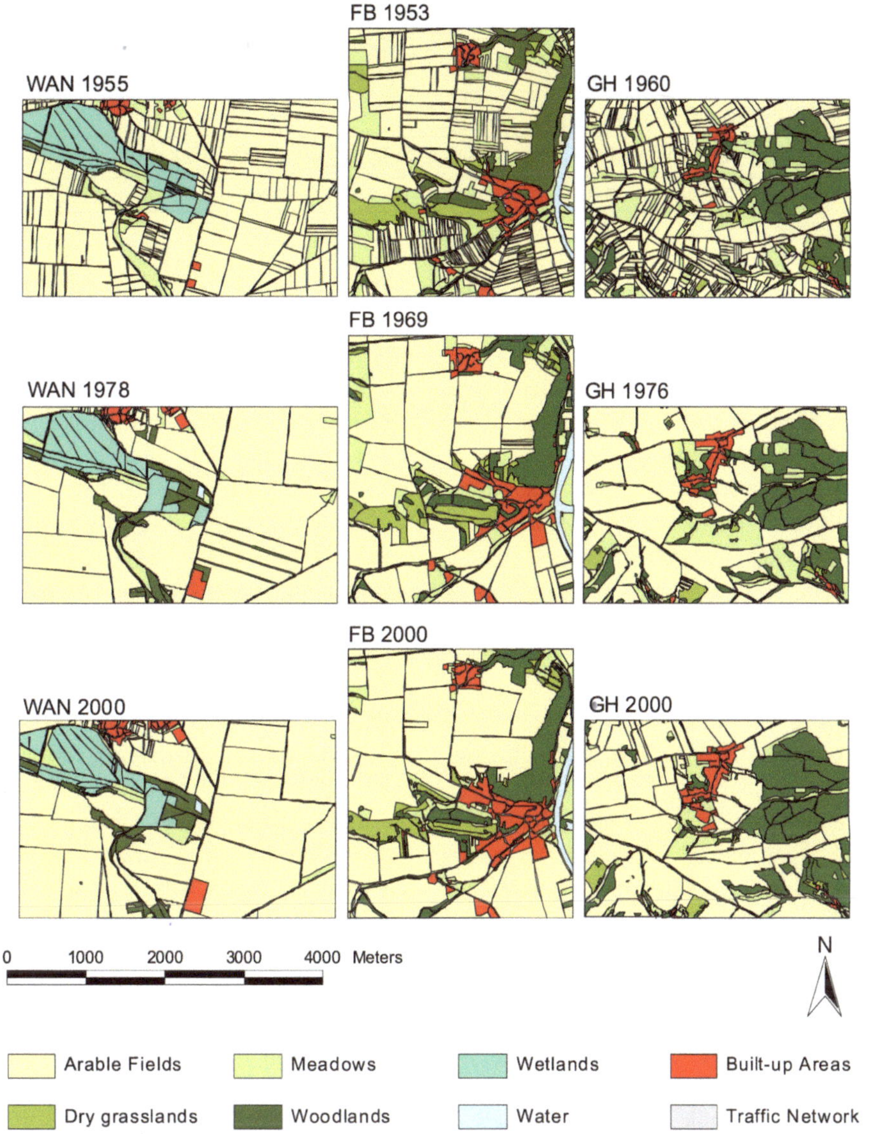

Fig. 20.2 Maps of study sites Wanzleben (WAN), Friedeburg, (FB) and Greifenhagen (GH) during three time periods

structure and land-use intensity. They are presently characterized by a mosaic of arable cultivation, grasslands and deciduous or mixed forest (Fig. 20.2).

20.2.1.2 Landscape Structure and Spatial Analyses

To calculate landscape and class metrics, landscape structure of each study site and period was mapped at 1-m resolution from aerial photographs (black and white orthophotos) using ArcView GIS 3.2 and ArcGIS 9.2 (ESRI). For the third period landscape structures reported by the aerial photos were updated supported by field validation in 2000/2002. Habitat types were classified into eight categories based on the landscape data in 2000: arable fields, meadows, dry grasslands, woodlands, wetlands, water, built-up areas and traffic networks. For each landscape a set of non-redundant landscape metrics was calculated at 1-m resolution using FRAGSTATS version 3.3 (McGarigal, Cushman, Neel, & Ene, 2002). Mean patch size (MPS), patch perimeter (PERIM), proximity index (PROX) and Euclidean nearest neighbour

Table 20.1 Land-use intensity indicators taken from statistical yearbooks and from annual abstracts of statistics of the former GDR (German Democratic Republic) and the states of East Germany (Statistisches Bundesamt, 2000; Statistisches Landesamt Sachsen-Anhalt, 2000; Staatliche Zentralverwaltung für Statistik, 1957, 1960, 1980, 1987). As data of mineral fertilizer application were only available on the level of administrative districts the same values had to be used for the study sites FB and GH, because both are situated in the same district

	Nitrogen (N, kg/ha)		Phosphorus (P_2O_5, kg/ha)		Potash (K_2O, kg/ha)		Lime (CaO, kg/ha)		Livestock density (1/ha)		
Period	WAN	FB/GH	WAN	FB/GH	WAN	FB/GH	WAN	FB/GH	WAN	FB	GH
1957	34	35	27	31	85	76	84	57	0.56	0.65	0.7
1979	117	124	51	61	69	68	145	143	1.10	0.98	1.1
2000	174	178	28	32	81	72	184	126	0.45	0.26	0.2

(ENN) were calculated as indicators of landscape fragmentation and isolation. Furthermore, landscape composition was measured by the Shannon's diversity index (SHDI) and the share of arable fields as well as semi-natural habitats on the total landscapes. Additionally, all metrics excluding SHDI were calculated for each habitat type individually.

20.2.1.3 Land-Use Intensity

The average amount of mineral fertilizers applied and livestock density were used as indicators of land-use intensity on the landscape scale (Table 20.1). Data for the three periods were taken for 1957, 1979 and 2000 from official statistical yearbooks (Statistisches Bundesamt, 2000, 1993; Statistisches Landesamt Sachsen-Anhalt, 2000, 2003; Staatliche Zentralverwaltung für Statistik, 1960, 1980, 1987). As data of mineral fertilizer application were only available at the level of administrative districts, the same values had to be used for the study sites FB and GH, both of which are situated in the same district. Data on pesticide use were not available for all time periods.

20.2.1.4 Plant Species Diversity Assessment

Species richness pattern on arable fields and in four semi-natural habitats – meadows, dry grasslands, woodlands and wetlands – were investigated during the three periods (1950s, 1970s and 2000–2003). Historical floristic composition was reconstructed from vegetation relevés that covered all land-use types (Schubert & Mahn, 1959; Mahn & Schubert, 1961, 1962; Kiesel, 1980; Westhus, 1980). The current floristic composition was documented by relevés made in 2000 (FB, 345 relevés), 2002 (GH, 183 relevés) and 2003 (WAN, 156 relevés). The relevés (100 m^2) were located using random coordinates generated in the GIS, but ensuring that each habitat type was represented by at least five relevés. All relevé plots on arable fields were placed at least 20 m from the border of the arable fields. Nomenclature of plant species followed Rothmaler (2002). The relevés of all floristic inventories were edited with the program SORT (Durka & Ackermann, 1993). To analyse changes in species composition all species were classified into six species groups based on their main habitat preference: species of woody habitats (comprise woods and shrubberies), dry grasslands (semi-dry and dry grasslands and dwarf-shrub heaths), meadows (meadows and pastures), wetlands (marsh areas), ruderal habitats (basically settlement areas) and 'arable weeds' sensu stricto (hereafter 'arable weeds'). Classification of species groups followed Haeupler (2002). Additionally biological traits (life span, pollen vector) and distributional characteristics (status) of each species were identified using the database BIOLFLOR (Klotz, Kühn, & Durka, 2002) to analyse changes in the composition of species traits during periods.

To consider the variation in sampling intensity of vegetation data and to facilitate comparisons, mean total numbers of plant species of 0.9 n relevés (n = lowest number of relevés of the appropriate level – period, study site and habitat type) were rarefied from the original data sets using the program package vegan in the free software R 2.6.0 (Venables, Smith, & R Development Core Team, 2006). Differences in species richness estimates in the three study sites among periods were tested by a percentile test (Slivka, 1970) with a two-sided significance level of 0.05. Finally, the share of species groups on the total species richness on arable fields as well as in each semi-natural habitat was estimated and differences among periods (within study sites) were tested using the binomial proportion test (Crawley, 2007).

Finally, all land-use indicators, landscape metrics and species richness estimates included in the analyses were log-transformed prior to statistical analyses.

20.2.1.5 Statistical Analyses

Changes in the composition of species traits within study sites among periods were tested by two-factorial analysis of variance (ANOVA, Crawley, 2007; species number∼species trait∗period), followed by a Tukey HSD test.

Relationships between species richness estimates and explanatory variables (landscape metrics and land-use intensity parameters) at different periods were tested using analysis of covariance (ANCOVA) with the explanatory variables as continuous variables and landscape (study site) as categorical variable. All two-way interactions were entered in the models to specify different relationships among landscapes. Model simplification was done by removing non-significant interactions ($p > 0.05$), until lastly the minimal adequate model with the highest explanatory power was obtained (Crawley, 2007). Models were calculated in each case, for total species richness in each habitat type and for species richness of each species group within each habitat type.

All tests were performed using the program package R 2.6.0 (Venables et al., 2006).

20.2.2 Results

20.2.2.1 Species Richness and Species Composition

Species richness of all semi-natural habitats and of individual habitat types varied among study sites and changed between time periods within sites (Fig. 20.3). With a mean of 255 plant species, site FB had the highest species richness of the three sites at all periods, followed by GH (203 species) and WAN (187 species). Total species richness decreased in WAN and GH between period one and two, followed by a large increase in period three, whereas in FB it showed a slight but nonsignificant increase.

Species composition of semi-natural habitats with respect to species groups differed between sites and mirrored habitat composition. In WAN wetland and ruderal species contributed most to the flora of semi-natural habitats at all periods supplemented by woodland species in the third period, whereas in FB ruderal species, dry grassland and woodland species contributed most (Fig. 20.3). In contrast, in GH the flora was composed predominantly of woodland species, followed by meadow and ruderal species. In two of the three sites (FB, GH), 'arable weeds' and ruderal species increased in almost all habitat types (woodlands, dry grasslands and meadows; results not shown).

The changes in the composition of biological species traits (life span and pollen vector) within all semi-natural habitats show different patterns across the landscapes according to the landscape history, landscape structure and land-use intensity (Fig. 20.4a, b). In FB and GH the number of annual and self-pollinated species increased significantly from the first to the third period. Perennial species contributed most to the flora of semi-natural habitats in all landscapes and at all periods, but decreased significantly in FB and GH from the first to the second period. In contrast, in WAN the number of perennial species increased during the periods.

In all landscapes species composition of semi-natural habitats is dominated by native species at all periods (Fig. 20.4c). However, in FB there is a

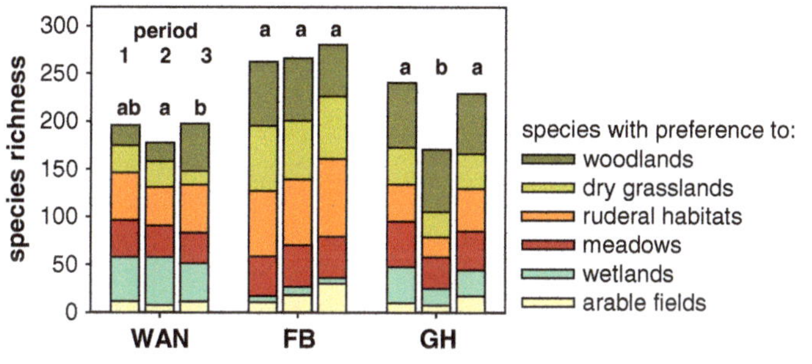

Fig. 20.3 Estimates of total species richness of semi-natural habitats for three sites in three time periods. Significant differences of total species richness between periods within study sites are indicated by *different letters* (results of percentile tests). *Colours* indicate species specialist groups (*dark green* – woodlands, *light green* – dry grasslands, *orange* – ruderal, *red* – meadows, *blue* – wetlands, *yellow* – arable)

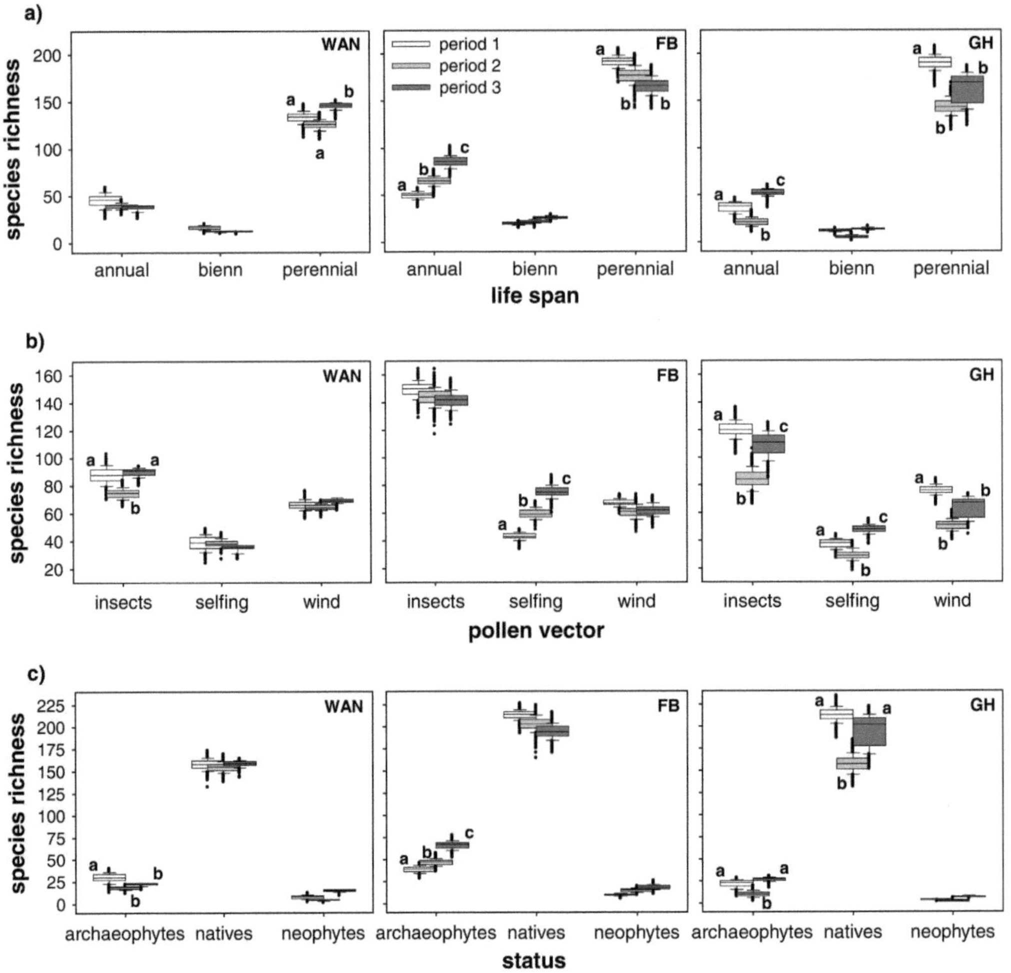

Fig. 20.4 Changes in the composition of plant species traits – biological traits (**a**: life span, **b**: pollen vector) and distributional characteristics (**c**: status). Significant differences within study sites among periods (results of ANOVA and Tukey HSD) are indicated by *different letters*

significant increase of archaeophytes and a decrease of native species during periods. In WAN and GH the pattern of changes in the number of archaeophytes, natives and neophytes is similar to the changes in species richness (Fig. 20.3).

Species richness on arable fields differed significantly between study sites and between periods within each study site. Overall, arable fields in FB sustained higher species diversity, except in the first period, than arable fields in WAN and GH (Fig. 20.5a). In WAN (104 species) and GH (147 species) highest numbers in plant species were recorded in period one, while in FB highest species richness (157 species) was recorded in the third period. There is a significant decrease of species richness on arable fields in FB and GH between period one and two, followed by a highly significant increase in FB and a slight but not significant increase in GH to period three. In WAN species richness did not change from period one to two, whereas it decreased by half in period three, and thus reached the smallest total species number (53 species) across landscapes and periods.

At the level of species specialist groups' significant changes were observed (Fig. 20.5b). 'Arable weeds' and ruderals accounted for the majority (60.5–90.9%) of species in all study sites across periods. In FB, significant changes in the share of 'arable weeds' and ruderals were estimated between period one and three, with a decrease of 'arable weeds' and an increase of ruderals. The same trend in the share of ruderals was recorded for WAN. However, within the weed flora of agricultural fields the share of meadow species in

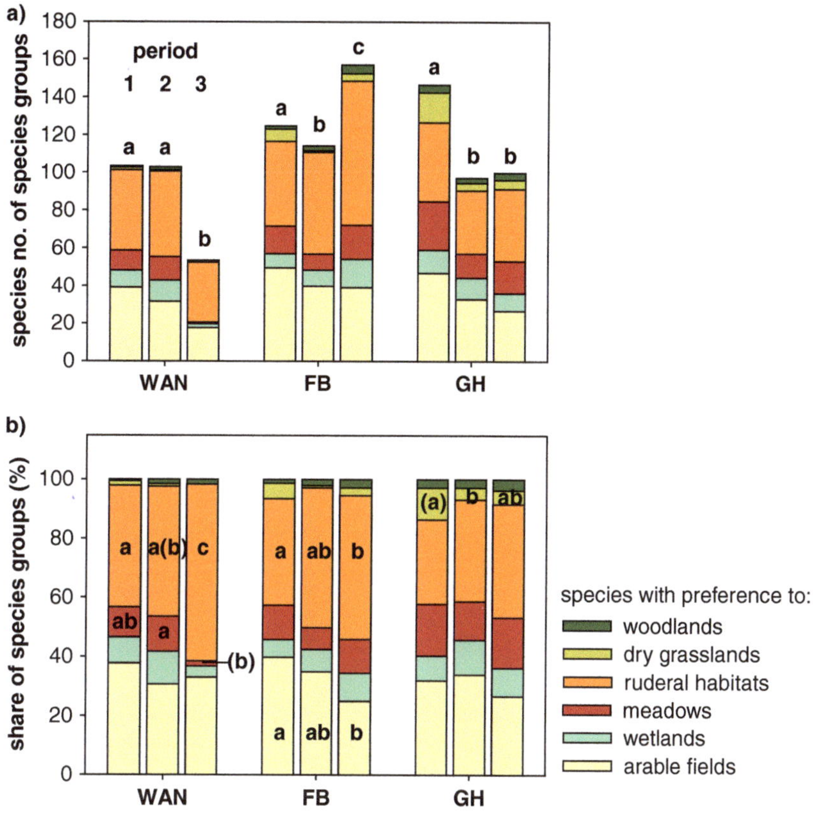

Fig. 20.5 (**a**) Species richness on arable fields and (**b**) percentage share of different species groups. Significant differences between total species numbers among periods (results of percentile tests) and between shares of a species group among periods (within study sites) are indicated by *different letters*. Marginal significant differences are in *brackets*. *Colours* indicate species specialist groups (*dark green* – woodlands, *light green* – dry grasslands, *orange* – ruderal, *red* – meadows, *blue* – wetlands, *yellow* – arable)

WAN and of dry grassland species in GH decreased significantly during periods.

The temporal changes in species richness of species groups varied among landscapes. At the landscape level, the mean numbers of 'arable weeds' were significantly higher at period one (WAN: 39.1, FB: 49.6, GH: 46.8 species) than in the following periods. The number of 'arable weeds' decreased significantly in WAN and in GH during studied periods, but remained constant in FB from period two to period three.

20.2.2.2 Plant Species Richness vs. Landscape Structure and Land-Use Intensity

ANCOVA models for species richness in the different habitat types as affected by landscape structure and land-use intensity showed that only few of the parameters had significant effects (Fig. 20.6). Species richness of semi-natural habitats was significantly related to mean patch size of meadows and to phosphorus application (Fig. 20.6a, c). Additionally, site-specific effects were indicated for phosphorus application by significant interaction terms with landscape (Fig. 20.6c).

Analysis of covariance for species richness estimates of species specialist groups on arable fields showed that only species richness of 'arable weeds' were significantly correlated with environmental variables. Species numbers of 'arable weeds' were significantly positive related to habitat diversity (SHDI; Fig. 20.6b). In contrast, species numbers of 'arable weeds' were negatively related to the amount of nitrogen applied per hectare (Fig. 20.6d).

20.3 Landscape Genetics of a Common Forest Herb

20.3.1 Methods

20.3.1.1 Study Sites

This study was carried out in the same three landscapes like in the first case study (Fig. 20.1) with one difference as each study site covered 16 km^2. Each study site represents an agricultural landscape with a network of potential habitats for *Geum urbanum* (Fig. 20.7).

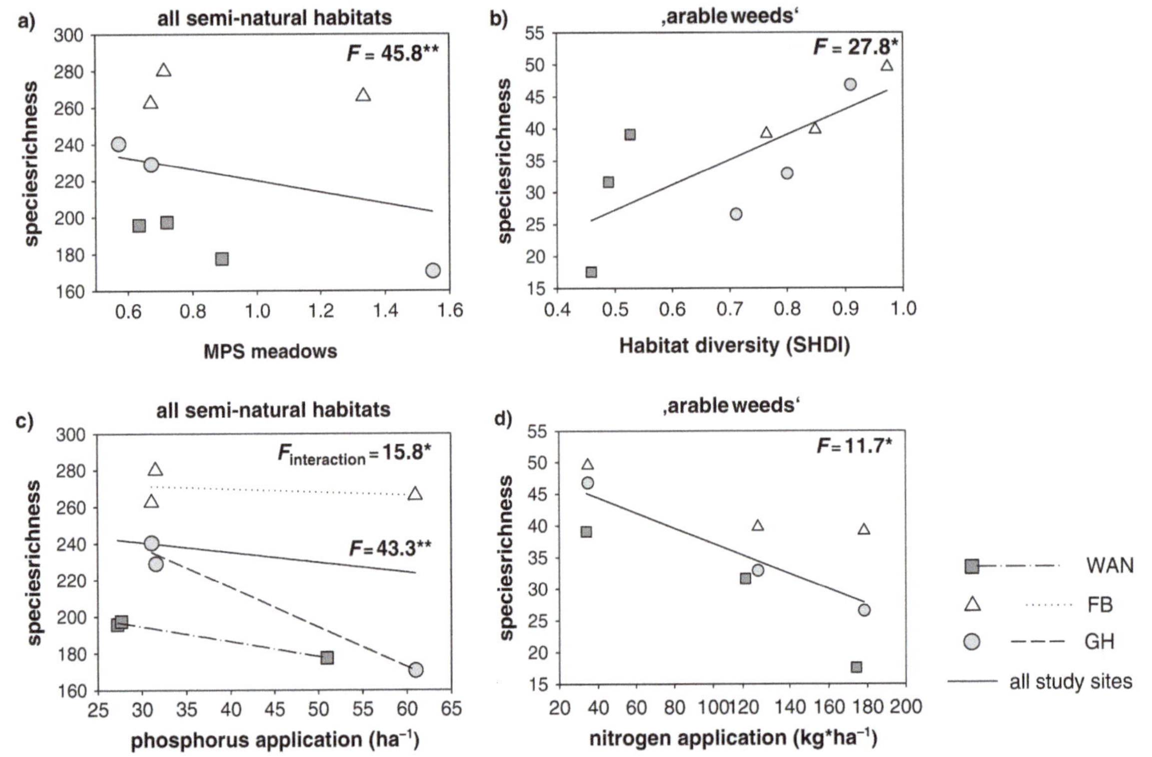

Fig. 20.6 Response of species richness estimates to landscape configuration as represented by mean patch size of meadows (**a**: all semi-natural habitats) and to landscape composition represented by habitat diversity (**b**: 'arable weeds') as well as to land-use intensity represented by phosphorus application (kg ha^{-1}) (**c**: all semi-natural habitats) and nitrogen application (kg ha^{-1}) (**d**: 'arable weeds')

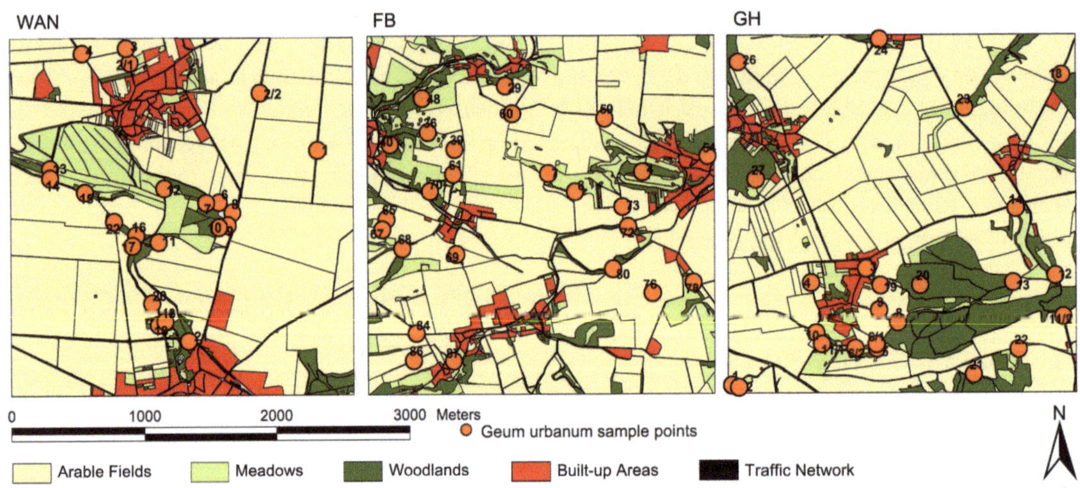

Fig. 20.7 Map of study sites and *Geum urbanum* sample points; WAN – Wanzleben, FB – Friedeburg, GH – Greifenhagen (Central Germany)

20.3.1.2 Study Species and Sampling Design

Wood Avens (*G. urbanum* L.) is a perennial herbaceous hexaploid ($2n = 6x = 42$) member of the Rosaceae, native to the Eurasian temperate zone. The species has a predominantly autogamous breeding system due to autodeposition of pollen (Taylor, 1997). *G. urbanum* occurs only in shaded habitats like deciduous forests, scrubs and hedgerows.

All the present-day populations of *G. urbanum* within the study areas were mapped and characterized in terms of size and area. Between 21 and 25 populations per study site (21 in WAN, 25 in FB, 24 in GH; Fig. 20.7) were randomly selected. Sampling of leaves (one leaf from 16 individuals per population) proceeded randomly within a circle of 10 m radius.

20.3.1.3 Genetic Analyses

In total, 1,105 individuals were sampled and genotyped at eight microsatellite loci according to Arens, Durka, Wernke-Lenting, and Smulders (2004). For each population we calculated mean expected heterozygosity (H_e) corrected for sample size (Nei, 1987). H_e was a convenient measure of genetic diversity for our purposes since it is less sample-size dependent than alternatives such as the number of alleles (Soule & Yang, 1973; Nei, 1986). H_e was computed using the software MSA (Dieringer & Schlötterer, 2003).

Tukey's honestly significant difference test (HSD) was used to test for differences of H_e among study sites using R 2.6.0 (Venables et al., 2006).

20.3.1.4 Landscape Structure and Spatial Analyses

To investigate the influence of landscape structure upon the genetic diversity of plant populations, circular areas with a radius of 200 m (12.56 ha) around each sampled population were defined. Landscape indices were determined for each circle based on the landscape elements of the habitat types and *G. urbanum* populations using FRAGSTATS, version 3.3 (McGarigal et al., 2002).

Within the circular areas, the following landscape indices were calculated representing, first, overall landscape structure and, second, the structure of woody patches as potential habitat of *G. urbanum*: number of land-use types (No. Land-use Types), number of patches (PATCHNUMBER), average patch size (PATCHSIZE), percentage area of the four most common land-use types (ARABLE, MEADOWS, WOODY, ROADS). Isolation of woody habitats (ISOL WOODY) was estimated by the Euclidian Nearest Neighbour (ENN) and density of woody habitats (DENS WOODY) was estimated by the PROXIMITY function of FRAGSTATS. To characterize the population structure of *G. urbanum* within the circular areas, the following parameters were used: population size (GEUM pop. size), percentage area of populations (GEUM AREA), number of population patches (No. GEUM pop.) and density of populations (DENS GEUM, estimated by the PROXIMITY function of FRAGSTATS). All landscape structure parameters and spatial population characteristics included in the analyses were log-transformed prior to statistical analyses.

20.3.1.5 Statistical Analyses

Linear mixed-effects models (Pinheiro & Bates, 2000) were used to test for the influences of landscape structure parameters and spatial population characteristics on genetic diversity of *G. urbanum*. One response variable was used in the models: expected heterozygosity (H_e). The mixed-effects models using maximum

Table 20.2 Results of best linear mixed-effects model for the response variable genetic diversity (H_e) with landscape as random factor. Fixed factors: no. land-use types – number of land-use types; roads – percentage area of traffic networks; patchsize – average patch size; woody – percentage area of woody habitats; Geum pop. size – population size of *Geum urbanum*

Fixed effects	DF	*F*-value	*p*-value
(Intercept)	1	968.96	**<0.0001**
Woody	1	0.10	0.7518
Roads	1	13.74	**0.0005**
No. land-use types	1	4.53	**0.0382**
Geum pop. size	1	8.62	**0.0050**
Patchsize	1	7.61	**0.0081**
Woody × landscape	2	3.64	**0.0335**
Roads × landscape	2	2.50	0.0932
No. land-use types × landscape	2	1.70	0.1934
Geum pop. size × landscape	2	0.43	0.6524
Patchsize × landscape	2	8.91	**0.0005**

likelihood were carried out using the routine 'lme' in the library 'nlme' (Pinheiro & Bates, 2000; Crawley, 2007) within the program package R 2.6.0 (Venables et al., 2006).

The explanatory variables of landscape structure and spatial population characteristics described above were included in the mixed-effects models as 'fixed factors'. The study sites (LANDSCAPE) were included as a 'random factor'.

20.3.2 Results

All microsatellite loci were polymorphic in almost all sampled populations within the three study sites. Expected heterozygosity (H_e) per population ranged from 0.05 to 0.60 in WAN, from 0.13 to 0.66 in FB and from 0.00 to 0.56 in GH. Mean values of H_e were 0.36, 0.38 and 0.34 in WAN, FB and GH, respectively, and did not differ significantly ($p > 0.2$).

20.3.2.1 Genetic Diversity vs. Landscape and Population Structure

Several fixed factors significantly explained the variance in genetic diversity within the *G. urbanum* populations (Table 20.2, Fig. 20.8a–e). Genetic diversity increased with *Geum* population size and in populations surrounded by a higher number of

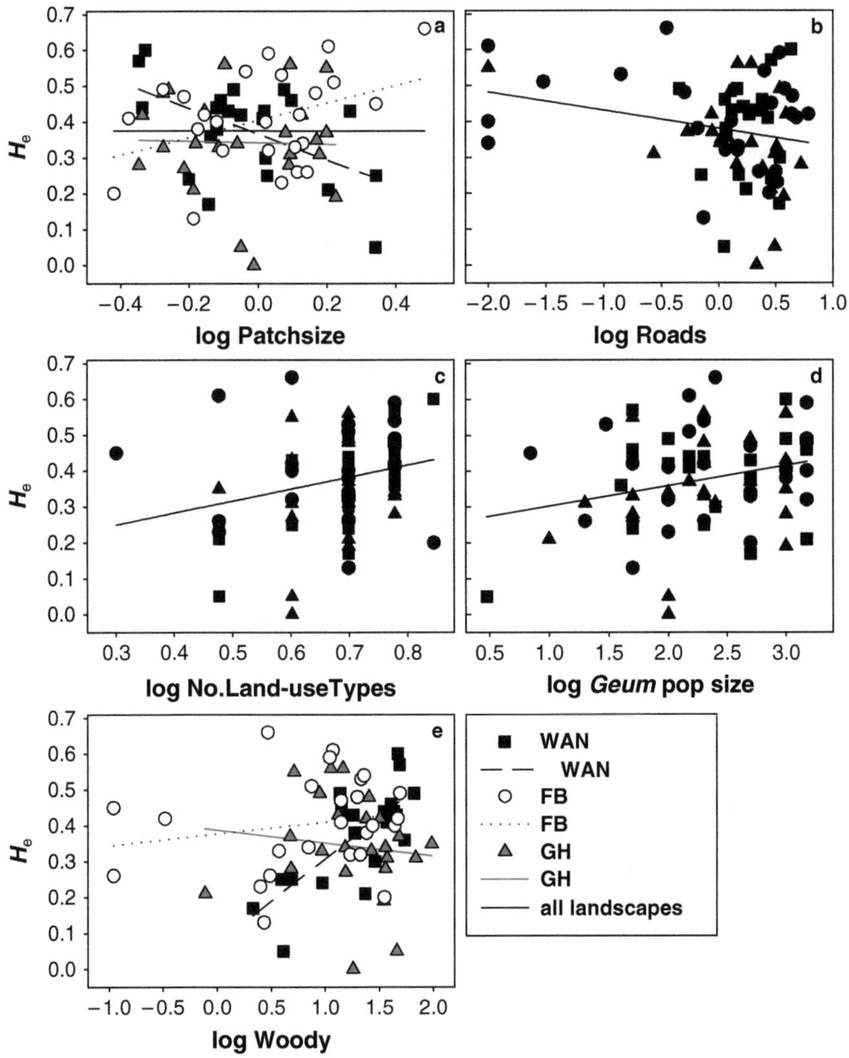

Fig. 20.8 Correlations between genetic diversity within populations (H_e) and landscape structure parameters within and across landscapes. *Hairline regression lines* indicate overall significant effects; *bold regression lines* indicate landscape-specific relationships (LANDSCAPE interactions of linear mixed-effects model). (**a**) Average patch size – Patchsize, (**b**) percentage area of traffic networks – Roads, (**c**) number of land-use types – No. Land-use Types, (**d**) Geum population size – Geum pop. size, (**e**) area of woody vegetation – Woody

different land-use types (Fig. 20.8c, d). In contrast, genetic diversity decreased with increasing percentage area of traffic networks (Fig. 20.8b). The interaction terms of these fixed factors and the random factor *LANDSCAPE* were not significant, suggesting that the relationships were consistent across the three study sites. However, study sites differed in the relationship between genetic diversity and average patch size and area of woody habitats (Table 20.2, Fig. 20.8a, e). There was a positive relationship between genetic diversity and the average patch size in FB, a negative relationship in WAN, and no relationship in GH. Similarly, genetic diversity increased with woody area in WAN but was less affected in FB and GH.

20.4 Discussion and Conclusions

20.4.1 Relationships Between Plant Species Richness and Landscape Structure and Land-Use Intensity

Recent studies of biodiversity in agricultural landscapes on the European scale have shown positive relationships between species numbers and area of semi-natural habitats (Dormann et al., 2007; Billeter et al., 2008; Liira et al., 2008). The share of semi-natural habitats and the number of semi-natural habitats within landscapes were the most important parameters influencing the current species richness of herbs. However, the (nonsignificant) increase in the percentage proportion of arable fields connected with loss of semi-natural habitats and decline of habitat diversity within the analysed landscapes did not have any significant effect on species richness and average species numbers of semi-natural habitats.

As observed by other authors (e.g. Liira et al., 2008), mean patch size of meadow habitats was negatively related to species richness and average species numbers per relevé of some species groups on landscape as well as on larger scales. In consideration of the species–area relationship higher species numbers would be expected with increasing patch size (Wilson & Chiarucci, 2000; Lennon, Kunin, & Hartley, 2002). However, the larger patch area of meadows and pastures may be indicators of more intensive use (as is also the share of meadows and pastures), which is related to livestock density (Herzog et al., 2006). Lowest species richness for both all semi-natural habitats and wetlands was recorded in the second period that was characterized by the largest patch size of meadows. Liira et al. (2008) have shown that, for species mainly confined to semi-natural habitats, species richness was negatively related to land-use intensity. In contrast, for species common in anthropogenic habitat types, such as ruderals and 'arable weeds' (especially annuals), species richness was positively related to land-use intensity. A high proportion of ruderals, including alien species, is one indicator of human impact (Hill, Roy, & Thompson, 2002). The human impact on biodiversity is mainly due to urbanization and agricultural land use, especially application of mineral fertilizer and pesticides, because of their devastating effects on habitat availability and habitat quality (Sala et al., 2000).

Mineral fertilizer is one of the major anthropogenic inputs of phosphorus into semi-natural habitats and has a positive impact on growth rates of plants causing eutrophication in wetlands and other semi-natural habitats (Khan & Ansari, 2005). Negative effects of soil phosphate contents on species richness patterns were recorded for woody habitats (Honnay et al., 1999), for meadows (Smith et al., 2003; Muller, Gusewell, & Edwards, 2003) and for wetlands (e.g. Aerts & Berendse, 1988; Colijn, Hesse, Ladwig, & Tillmann, 2002). Our results are consistent with these studies, as species richness of all semi-natural habitats was negatively related to the application amount of phosphorus. But we also could show that there was increasing species richness with decreasing input of phosphorus from the 1970s to 2000s combined with a high species turnover indicated by lower community similarity values. Phosphorus levels rather than nitrogen levels are generally limiting plant growth and thus may stimulate the establishment of dominant ruderal species like *Urtica dioica* in forests (Pigott, 1971), or *Phalaris arundinacea, Phragmites australis* and different *Typha* species in wetlands (Bruland et al., 2007; Khan & Ansari, 2005). These large, dominant plants reduce the chances for recolonization by non-ruderal specialist species of semi-natural habitats.

Similar relationships could be shown for species richness pattern on arable fields. Species richness of 'arable weeds' was significantly negatively affected by the average amount of nitrogen applications. This result is consistent with the observations of many other studies (e.g. Stoate et al., 2001; Gabriel,

Thies, & Tscharntke, 2005; Kleijn et al., 2006). The use of fertilizers, especially nitrogen, on arable fields increased substantially in the study sites during the second half of last century. Pysek and Leps (1991) found that species composition is affected by fertilizer application. Further, the same authors have shown that some ruderal species are supported by the addition of nitrogen (e.g. *Poa annua, Fallopia convolvulus, Veronica persica, Sonchus arvensis*). Similarly, Lososová, Chytrý, and Kühn (2008) pointed out that generalist species like ruderals are more demanding of nutrients than 'arable weeds'. Whereas the number of 'arable weeds' decreased with increasing nitrogen applications, the share of ruderal species increased in the studied landscapes.

The research indicates that richness of 'arable weeds' was additionally related to landscape composition and landscape configuration. The number of 'arable weeds' was positively affected by increasing habitat diversity. This result is consistent with the results of Gabriel et al. (2005), who suggest that both regional and local processes are important factors influencing diversity pattern of 'arable weeds'. Additionally, recent studies have shown that species richness including 'arable weeds' was higher in field edges than in the field centre (Walker et al., 2007; Cousins & Aggemyr, 2008). The contribution to local diversity could be related to source–sink relationships (Wagner & Edwards, 2001; Sosnoskie, Luschei, & Fanning, 2007) as well as neighbourhood effects (Dunning, Danielson, & Pulliam, 1992). Adjacent habitats provide the opportunity for both rare and common 'arable weeds' to retreat from agricultural fields and to immigrate again into the fields. In Friedeburg, for instance, some rare 'arable weeds' (*Sherardia arvensis, Adonis aestivalis, Caucalis platycarpos, Nigella arvensis*) were recorded at the border of arable fields to dry grasslands, meadows or along paths. Thus, 'arable weeds' do not used to be confined only to agricultural habitats, but they also occur in different semi-natural habitats. Thus, each habitat patch situated outside of the agricultural fields poses a potential for enhancement of biodiversity. Therefore, higher diversity of habitat types and higher share of semi-natural habitats in landscapes might be expected to enhance colonization probabilities and thus 'arable weed' diversity.

The studies about changes in plant species diversity in both semi-natural habitats and arable fields have shown that knowledge of current land-use intensity as well as of historical land-use and landscape structure helps to understand historical and present-day diversity patterns. In the landscapes studied, changes in diversity patterns of semi-natural habitats reflected the changes in the spatial pattern of landscape configuration and land-use intensity.

20.4.2 Landscape Genetics

Analyses of genetic variation of *G. urbanum* populations have shown that the population structure is affected by both population and landscape properties. The influence of several population and landscape structure variables on genetic diversity varied among landscapes.

In a subdivided population, genetic diversity within subpopulation is affected by the size of the local subpopulation, because large populations are less susceptible to genetic drift (Wright, 1943). Thus, within-population genetic variation is generally correlated positively to population size (Leimu, Mutikainen, Koricheva, & Fischer, 2006) and our results are consistent with this pattern. However, the rather weak influence of population size in our study may be due to the fact that levels of genetic diversity in present-day populations may reflect the populations' accumulated history of size fluctuations rather than present population size (Linhart & Premoli, 1994; Prentice, Lönn, Rosquist, Ihse, & Kindström, 2006). The weak influence of population size may also be due to predominant self-pollination; this is because the populations developing after the colonization of empty patches tend to be large but genetically depauperate (Pannell & Barrett, 1998).

The fact that the area of woody habitat was significant in the models in addition to *Geum* population structure indicates an important role of woody vegetation, e.g. for dispersal or in a metapopulation context. The positive association between genetic diversity and area of woody habitats in two of the landscapes is consistent with the expectation that close proximity to source populations enhances the variability of incoming gene flow (Slatkin & Voelm, 1991; Lönn & Prentice, 2002). Numerous studies in fragmented landscapes have shown that local genetic and species diversity are affected by the size and distance of a patch to the nearest similar habitat, because this affects the probability that seeds will reach the patch (Lienert, Fischer, Schneller, & Diemer, 2002; Vandepitte, Jacquemyn, Roldan-Ruiz, & Honnay, 2007).

Additionally, with increasing numbers of land-use types, genetic variation increased. An increase of land-use types is likely to be associated with greater spatial heterogeneity, smaller patch size and increased edge area; all of these may contribute to their being more or larger suitable sites, which in turn would enhance the possibilities for population persistence and gene exchange. Dispersal limitation was suggested to be the dominant force regulating the recovery of forest-herb diversity and thus for genetic variation in newly established populations (Vellend, 2003). In a study showing high levels of genetic variation in fragmented populations of the self-pollinated herb *Anthyllis vulneraria*, it was concluded that this could only have occurred through historically high rates of seed dispersal (Honnay et al., 2006). *G. urbanum* has hairy fruits with a long hook, which enables epizoochorous adhesive dispersal (Fischer, Poschlod, & Beinlich, 1996; Gorb & Gorb, 2002), and Tackenberg, Römermann, Thompson, and Poschlod (2006) showed that these typically have a moderately high attachment potential. Several studies assessed colonization of forest patches by forest herbs, including *G. urbanum*. Whereas *Geum* was often found in secondary forests and thus rated as a good colonizer (Butaye, Jacquemyn, & Hermy, 2001), low colonization ability has also been reported (Grashof-Bokdam & Geertsema, 1998). These inconsistencies may indicate differences of landscape structure or composition. In less forested landscapes, rare long-distance dispersal events may play a more important role in the colonization process (Verheyen, Guntenspergen, Biesbrouck, & Hermy, 2003). However, the traffic networks within landscapes act as barriers to migration and seed dispersal as genetic variation tended to decrease with increasing area of roads. Roads have a direct traffic impact on the movement pattern of animals (Mader, 1984; Trombulak & Frissell, 2000) and thus an indirect effect on the seed dispersal potential of *G. urbanum*.

Overall, these results provide evidence that complex landscapes with a high density and connectivity of uncultivated, perennial habitats may enhance genetic diversity and species diversity of all habitat types, including semi-natural habitats and arable fields. Therefore, management effort should be focused on reducing land-use intensity and increasing habitat connectivity in order to enhance diversity in European agricultural landscapes. However, habitat and landscape conditions may have not only a direct effect on species richness patterns but also long-term effects by selecting for particular combinations of species traits and migration potential. The degree to which habitat fragmentation and deterioration affect the persistence of spatially structured populations depends not only on the present but also on the historical landscape conditions and on the rate of landscape change. Therefore, long-term studies are of high importance to understand ecological processes. Both studies about temporal changes and spatial determinants of species diversity and genetic variation could show that the incorporation of spatially and temporally explicit historical attributes into landscape ecology studies can greatly expand our knowledge about the factors driving biotic patterns in fragmented agricultural landscapes.

Acknowledgements We are thankful to E.G. Mahn, R. Schubert and W Westhus who provided primary data of their historical relevés. We thank H. Auge, I. Kühn and C. Dormann for discussion of the statistical analyses. In addition, we are grateful to P. Edwards for his contributions. This study was financially supported by the Energy, Environment and Sustainable Development Program (FP5) of the European Commission (contract number EVK2-CT-2000-00082).

References

Aerts, R., & Berendse, F. (1988). The effect of increased nutrient availability on vegetation dynamics in wet heathlands. *Vegetatio*, *76*, 63–69.

Arens, P., Durka, W., Wernke-Lenting, J. H., & Smulders, M. J. M. (2004). Isolation and characterization of microsatellite loci in *Geum urbanum* (Rosaceae) and their transferability within the genus Geum. *Molecular Ecology Notes*, *4*, 209–212.

Baessler, C., & Klotz, S. (2006). Effects of changes in agricultural land-use on landscape structure and arable weed vegetation over the last 50 years. *Agriculture, Ecosystems & Environment*, *115*, 43–50.

Bailey, S. (2007). Increasing connectivity in fragmented landscapes: An investigation of evidence for biodiversity gain in woodlands. *Forest Ecology and Management*, *238*, 7–23.

Billeter, R. C., Liira, J., Bugter, R., Arens, P., Augenstein, I., Aviron, S., et al. (2008). Indicators for biodiversity in agricultural landscapes: A pan-European study. *Journal of Applied Ecology*, *45*, 141–150.

Brown, J. H., & Kodric-Brown, A. (1977). Turnover rates in insular biogeography: Effect of immigration on extinction. *Ecology*, *58*, 445–449.

Bruland, G. L., Osborn, T. Z., Reddy, K. R., Grunwald, S., Newman, S., & DeBusk, W. F. (2007). Recent changes in soil total phosphorus in the Everglades: Water conservation area 3. *Environmental Monitoring and Assessment, 129*, 379–395.

Bruun, H. H. (2000). Deficit in community species richness as explained by area and isolation of sites. *Diversity and Distribution, 6*, 129–135.

Butaye, J., Jacquemyn, H., & Hermy, M. (2001). Differential colonization causing non-random forest plant community structure in a fragmented agricultural landscape. *Ecography, 24*, 369–380.

Carpenter, S. R., Caraco, N. F., Correl, D. L., Howarth, R. W., Sharpley, A. N., & Smith, V. H. (1998). Nonpoint pollution of surface waters with phosphorus and nitrogen. *Ecological Applications, 8*, 559–568.

Chapin, III, F. S., Walker, B. H., Hobbs, R. J., Hooper, D. U., Lawton, J. H., Sala, O. E., et al. (1997). Biotic control over the functioning of ecosystems. *Science, 277*, 500–504.

Colijn, F., Hesse, K. J., Ladwig, N., & Tillmann, U. (2002). Effects of large-scale uncontrolled fertilisation process along the continental coastal North Sea. *Hydrobiologia, 484*, 133–148.

Cousins, S. A. O., & Aggemyr, E. (2008). The influence of field shape, area and surrounding landscape on plant species richness in grazed ex-fields. *Biological Conservation, 141*, 126–135.

Cousins, S. A. O., & Eriksson, O. (2001). Plant species occurences in a rural hemiboreal landscape: Effects of remnant habitats, site history, topography and soil. *Ecography, 24*, 461–469.

Crawley, M. J. (2007). *The R Book.* West Sussex PO19 8SQ, England: John Wiley and Sons Ltd.

de Blois, S., Domon, G., & Bouchard, A. (2001). Environmental, historical, and contextual determinants of vegetation cover: A landscape perspective. *Landscape Ecology, 16*, 421–436.

Debinski, D. M., & Holt, R. D. (2000). A survey and overview of habitat fragmentation experiments. *Conservation Biology, 14*, 342–355.

Dieringer, D., & Schlötterer, C. (2003). Microsatellite analyser (MSA): A platform independent analysis tool for large microsatellite data sets. *Molecular Ecology Notes, 3*, 167–169.

Dormann, C. F., Schweiger, O., Augenstein, I., Bailey, D., Billeter, R. C., De Blust, G., et al. (2007). Effects of landscape structure and land-use intensity on similarity of plant and animal communities. *Global Ecology and Biogeography, 16*, 774–787.

Duelli, P. (1997). Biodiversity evaluation in agricultural landscapes: An approach at two different scales. *Agriculture, Ecosystems & Environment, 62*, 81–91.

Dunning, J. B., Danielson, B. J., & Pulliam, H. R. (1992). Ecological processes that affect populations in complex landscapes. *OIKOS, 65*, 169–175.

Dupré, C., & Ehrlén, J. (2002). Habitat configuration, species traits and plant distribution. *Journal of Ecology, 90*, 796–805.

Durka, W., & Ackermann, W. (1993). SORT – Ein Computerprogramm zur Bearbeitung von floristischen und faunistischen Artentabellen. *Natur und Landschaft, 68*, 16–21.

Endels, P., Jacquemyn, H., Brys, R., Hermy, M., & De Blust, G. (2002). Temporal changes (1986–1999) in populations of primrose (*Primula vulgaris* Huds.) in an agricultural landscape and implications for conservation. *Biological Conservation, 105*, 11–25.

Eriksson, O. (2000). Functional roles of remnant plant populations in communities and ecosystems. *Global Ecology & Biogeography, 9*, 443–449.

Fahrig, L. (2003). Effects of habitat fragmentation on biodiversity. *Annual Review of Ecology Evolution and Systematics, 34*, 487–515.

Fischer, S. F., Poschlod, P., & Beinlich, B. (1996). Experimental studies on the dispersal of plants and animals on sheep in calcareous grasslands. *Journal of Applied Ecology, 33*, 1206–1222.

Fischer, M., & Stöcklin, J. (1997). Local extinctions of plants in remnants of extensively used calcereous grasslands 1950–1985. *Conservation Biology, 11*, 727–737.

Frankham, R., Ballou, J. D., & Briscoe, D. A. (2002). *Introduction to conservation genetics.* Cambridge: Cambridge University Press.

Gabriel, D., Thies, C., & Tscharntke, T. (2005). Local diversity of 'arable weeds' increases with landscape complexity. *Perspectives in Plant Ecology, Evolution and Systematics, 7*, 85–93.

Gorb, E., & Gorb, S. (2002). Contact separation force of the fruit burrs in four plant species adapted to dispersal by mechanical interlocking. *Plant Physiology and Biochemistry, 40*, 373–381.

Grashof-Bokdam, C. J., & Geertsema, W. (1998). The effect of isolation and history on colonization patterns of plant species in secondary woodland. *Journal of Biogeography, 25*, 837–846.

Green, R. E., Cornell, S. J., Scharlemann, J. P. W., & Balmford, A. (2005). Farming and the fate of wild nature. *Science, 307*, 550–555.

Gustavsson, E., Lennartsson, T., & Emanuelsson, M. (2007). Land use more than 200 years ago explains current grassland plant diversity in a Swedish agricultural landscape. *Biological Conservation, 138*, 47–59.

Haeupler, H. (2002). Die Biotope Deutschlands. In S. Klotz, I. Kühn, & W. Durka (Eds.), *BIOLFLOR – Eine Datenbank mit biologisch-ökologischen Merkmalen zur Flora von Deutschland* (pp. 247–272). Bonn – Bad Godesberg: Bundesamt für Naturschutz.

Hanski, I. (1998). Metapopulation dynamics. *Nature, 396*, 41–49.

Hanski, I., & Ovaskainen, O. (2003). Metapopulation theory for fragmented landscapes. *Theoretical Population Biology, 64*, 119–127.

Helm, A., Hanski, I., & Pärtel, M. (2006). Slow response of plant species richness to habitat loss and fragmentation. *Ecology Letters, 9*, 72–77.

Herzog, F., Steiner, B., Bailey, D., Baudry, J., Billeter, R. C., Bukácek, R., et al. (2006). Assessing the intensity or temperate European agriculture at the landscape scale. *European Journal of Agronomy, 24*, 165–181.

Hietala-Koivu, R. (2002). Landscape and modernizing agriculture: A case study of three areas in Finland in 1954–1998. *Agriculture, Ecosystems & Environment, 91*, 273–281.

Hill, M. O., Roy, D. B., & Thompson, K. (2002). Hemeroby, urbanity and ruderality: Bioindicators of disturbance and human impact. *Journal of Applied Ecology, 39*, 708–720.

Hilli, M., Kuitunen, M. T., & Suhonen, J. (2007). The effect of land use change on the vascular plant species turnover in boreal lakes. *Biodiversity and Conservation, 16*, 3951–3962.

Honnay, O., Coart, E., Butaye, J., Adriaens, D., Van Glabeke, S., & Roldan-Ruiz, I. (2006). Low impact of present and historical landscape configuration on the genetics of fragmented *Anthyllis vulneraria* populations. *Biological Conservation, 127*, 411–419.

Honnay, O., Hermy, M., & Coppin, P. (1999). Impact of habitat quality on forest plant species colonization. *Forest Ecology and Management, 115*, 157–170.

Honnay, O., Jacquemyn, H., Bossuyt, B., & Hermy, M. (2005). Forest fragmentation effects on patch occupancy and population viability of herbaceous plant species. *New Phytologist, 166*, 723–736.

Ihse, M. (1995). Swedish agricultural landscapes – patterns and changes during last 50 years, studied by aerial photos. *Landscape and Urban Planning, 31*, 21–37.

Johansson, L. J., Hall, K., Prentice, H. C., Ihse, M., Reitalu, T., Sykes, M. T., et al. (2008). Semi-natural grassland continuity, long-term land-use change and plant species richness in an agricultural landscape on Öland, Sweden. *Landscape and Urban Planning, 84*, 200–211.

Khan, F. A., & Ansari, A. A. (2005). Eutrophication: An ecological vision. *The Botanical Review, 71*, 449–482.

Kiesel, G. (1980). Vegetationskundliche Untersuchungen im Raum Harzgerode und der Umgebung von Greifenhagen. Diplomthesis University Halle-Wittenberg..

Kleijn, D., Baquero, R. A., Clough, Y., Días, M., De Esteban, J., Fernándes, F., et al. (2006). Mixed biodiversity benefits of agri-environment schemes in five European countries. *Ecology Letters, 9*, 243–254.

Klotz, S., Kühn, I., & Durka, W. (2002). *BIOLFLOR – Eine Datenbank mit biologisch-ökologischen Merkmalen zur Flora von Deutschland*. Bonn – Bad Godesberg: Bundesamt für Naturschutz.

Leimu, R., Mutikainen, P., Koricheva, J., & Fischer, M. (2006). How general are positive relationships between plant population size, fitness and genetic variation? *Journal of Ecology, 94*, 942–952.

Lennon, J. J., Kunin, W. E., & Hartley, S. (2002). Fractal species distributions do not produce power-law species-area relationships. *OIKOS, 97*, 378 386.

Levin, S. A. (1992). The problem of pattern and scale in ecology. *Ecology, 73*, 1943–1967.

Lienert, J., Fischer, M., Schneller, J., & Diemer, M. (2002). Isozyme variability of the wetland specialist *Swertia perennis* (Gentianaceae) in relation to habitat size, isolation, and plant fitness. *American Journal of Botany, 89*, 801–811.

Liira, J., Schmidt, T., Aavik, T., Arens, P., Augenstein, I., Bailey, D., et al. (2008). Plant functional group composition and large-scale species richness in European agricultural landscapes. *Journal of Vegetation Science, 19*, 3–14.

Lindborg, R. (2007). Evaluating the distribution of plant life-history traits in relation to current and historical landscape configurations. *Journal of Ecology, 95*, 555–564.

Lindborg, R., & Eriksson, O. (2004). Historical landscape connectivity affects present plant species diversity. *Ecology, 85*, 1840–1845.

Linhart, Y. B., & Premoli, A. C. (1994). Genetic variation in central and disjunct populations of *Lilium parryi*. *Canadian Journal of Botany, 72*, 79–85.

Lönn, M., & Prentice, H. C. (2002). Gene diversity and demographic turnover in central and peripheral populations of the perennial herb *Gypsophila fastigiata*. *OIKOS, 99*, 489–498.

Loreau, M., Naeem, S., Inchausti, P., Bengtsson, J., Grime, J. P., Hector, A., et al. (2001). Biodiversity and ecosystem functioning: Current knowledge and future challenges. *Science, 294*, 804–808.

Lososová, Z., Chytrý, M., & Kühn, I. (2008). Plant attributes determining the regional abundance of weeds on central European arable land. *Journal of Biogeography, 35*, 177–187.

Luoto, M., Rekolainen, S., Aakkula, J., & Pykälä, J. (2003). Loss of plant species richness and habitat connectivity in grasslands associated with agricultural change in Finland. *Ambio, 32*, 447–452.

Mader, H. J. (1984). Animal habitat isolation by roads and agricultural fields. *Biological Conservation, 29*, 81–96.

Mahn, E. G., & Schubert, R. (1961). Vegetationskundliche Untersuchungen in der mitteldeutschen Ackerlandschaft. IV. Die Pflanzengesellschaften der Umgebung von Greifenhagen (Mansfelder Bergland). *Wiss. Z. Univ. Halle, Math. -Nat. Reihe, X*, 179–246.

Mahn, E. G., & Schubert, R. (1962). Vegetationskundliche Untersuchungen in der mitteldeutschen Ackerlandschaft. VI. Die Pflanzengesellschaften nördlich von Wanzleben (Magdeburger Börde). *Wiss. Z. Univ. Halle, Math. -Nat. Reihe, XI*, 765–816.

McCollin, D., Moore, L., & Sparks, T. (2000). The flora of a cultural landscape: Environmental determinants of change revealed using archival sources. *Biological Conservation, 92*, 249–263.

McGarigal, K., Cushman, S. A., Neel, M. C., & Ene, E. (2002). FRAGSTATS: Spatial pattern analysis program for categorical maps, computer software program produced by the authors at the University of Massachusetts, Amherst, from http://www.umass.edu/landeco/research/fragstats/fragstats.html.

McIntyre, S., Lavorel, S., Landsberg, J., & Forbes, T. D. A. (1999). Disturbance response in vegetation – towards a global perspective on functional traits. *Journal of Vegetation Science, 10*, 621–630.

Moilanen, A., & Hanski, I. (1998). Metapopulation dynamics: Effects of habitat quality and landscape structure. *Ecology, 79*, 2503–2515..

Muller, P., Gusewell, S., & Edwards, P. J. (2003). Impacts of soil conditions and agricultural land use on plant species richness of Alpine pastures in the south of Glarus. *Botanica Helvetica, 113*, 15–36.

Naveh, Z., & Lieberman, A. S. (1984). *Landscape ecology: Theory and application*. New York: Springer.

Nei, M. (1986). Definition and estimation of fixation indices. *Evolution, 40*, 643–645.

Nei, M. (1987). *Molecular evolutionary genetics*. New York: Columbia University Press.

Pannell, J. R., & Barrett, S. C. H. (1998). Baker's law revisited: Reproductive assurance in a metapopulation. *Evolution, 52*, 657–668.

Petit, S., Griffiths, L., Smart, S. S., Smith, G. M., Stuart, R. C., & Wright, S. M. (2004). Effects of area and isolation of woodland patches on herbaceous plant species richness across Great Britain. *Landscape Ecology, 19*, 463–471.

Piessens, K., Honnay, O., Nackaerts, K., & Hermy, M. (2004). Plant species richness and composition of heathland relics in north-western Belgium: Evidence for a rescue-effect? *Journal of Biogeography, 31*, 1683–1692.

Pigott, C. D. (1971). Analysis of the response of *Urtica dioica* to phosphate. *New Phytologist, 70*, 953–966.

Pinheiro, J. C., & Bates, D. M. (2000). *Mixed-effects models in S and S-PLUS*. New York: Springer.

Piorr, H.-P. (2003). Environmental policy, agri-environmental indicators and landscape indicators. *Agriculture, Ecosystems & Environment, 98*, 17–33.

Prentice, H. C., Lönn, M., Rosquist, G., Ihse, M., & Kindström, M. (2006). Gene diversity in a fragmented population of *Briza media:* Grassland continuity in a landscape context. *Journal of Ecology, 94*, 87–97.

Pysek, P., & Leps, J. (1991). Response of a weed community to nitrogen fertilization: A multivariate analysis. *Journal of Vegetation Science, 2*, 237–244.

Robinson, R. A., & Sutherland, W. J. (2002). Post-war changes in arable farming and biodiversity in Great Britain. *Journal of Applied Ecology, 39*, 157–176.

Rosenzweig, M. L. (1995). *Species diversity in space and time*. Cambridge: Cambridge University Press.

Rothmaler, W. (2002). *Exkursionsflora von Deutschland, Gefäßpflanzen: Kritischer Band*. Jena: Gustav Fischer Verlag.

Sala, O. E., Chapin, III, F. S., Armesto, J. J., Berlow, E., Bloomfield, J., Dirzo, R., et al. (2000). Global biodiversity scenarios for the year 2100. *Science, 287*, 1770–1774.

Schubert, R., & Mahn, E. G. (1959). Vegetationskundliche Untersuchungen in der mitteldeutschen Ackerlandschaft, I. Die Pflanzengesellschaften der Gemarkung Friedeburg (Saale). *Wiss. Z. Univ. Halle, Math. –Nat. Reihe, VIII*, 965–1012.

Schweiger, O., Maelfait, J.-P., van Wingerden, W., Hendrickx, F., Billeter, R. C., Speelmans, M., et al. (2005). Quantifying the impact of environmental factors on arthropod communities in agricultural landscapes across organizational levels and spatial scales. *Journal of Applied Ecology, 42*, 1129–1139.

Slatkin, M., & Voelm, L. (1991). F_{ST} in a hierarchical island model. *Genetics, 127*, 627–629.

Slivka, J. (1970). A one-sided nonparametric multiple comparison control percentile test: Treatments versus control. *Biometrika, 57*, 431–438.

Smith, R. S., Shiel, R. S., Bardgett, R. D., Millward, D., Corkhill, P., Rolph, G., et al. (2003). Soil microbial community, fertility, vegetation and diversity as targets in the restoration management of a meadow grassland. *Journal of Applied Ecology, 40*, 51–64.

Sosnoskie, L. M., Luschei, E. C., & Fanning, M. A. (2007). Field margin weed-species diversity in relation to landscape attributes and adjacent land-use. *Weed Science, 55*, 129–136.

Soule, M. E., & Yang, S. Y. (1973). Genetic-variation in side-blotched lizards on islands in Gulf of California. *Evolution, 27*, 593–600.

Staatliche Zentralverwaltung für Statistik. (1980). Statistisches Jahrbuch der DDR 1980. *25. Jg*. Berlin: Staatsverlag der DDR.

Staatliche Zentralverwaltung für Statistik. (1987). Statistisches Jahrbuch der DDR 1987. *32. Jg*. Berlin: Staatsverlag der DDR

Staatliche Zentralverwaltung für Statistik. (1960). Statistisches Jahrbuch der DDR 1959. *5 Jg*. Berlin: Deutscher Zentralverlag.

Statistisches Bundesamt. (1993). Sonderreihe mit Beiträgen für das Gebiet der ehemaligen DDR. Ausgewählte Zahlen zur Agrarwirtschaft 1949–1989. Heft 8. Wiesbaden.

Statistisches Bundesamt. (2000). Düngemittelversorgung. *Produzierendes Gewerbe. R. 8.2.*

Statistisches Landesamt Sachsen-Anhalt. (2000). Statistisches Jahrbuch des Landes Sachsen-Anhalt 2000/ 2001.

Statistisches Landesamt Sachsen-Anhalt. (2003). Statistisches Jahrbuch des Landes Sachsen-Anhalt 2003, Teil 2. Kreisübersichten

Stoate, C., Boatman, N. D., Borralho, R. J., Rio Carvalho, C., de Snoo, G. R., & Eden, P. (2001). Ecological impacts of arable intensification in Europe. *Journal of Environmental Management, 63*, 337–365.

Tackenberg, O., Römermann, C., Thompson, K., & Poschlod, P. (2006). What does diaspore morphology tell us about external animal dispersal? Evidence from standardized experiments measuring seed retention on animal-coats. *Basic and Applied Ecology, 7*, 45–58.

Taylor, K. (1997). *Geum urbanum* L. *Journal of Ecology, 85*, 705–720.

Tilman, D., Fargione, J., Wolff, B., D'Antonio, C., Dobson, A., Howarth, R., et al. (2001). Forecasting agriculturally driven global environmental change. *Science, 292*, 281–284.

Tilman, D., May, R. M., Lehman, C. L., & Nowak, M. A. (1994). Habitat destruction and the extinction debt. *Nature, 371*, 65–66.

Tilman, D., Reich, P. B., & Knops, J. M. H. (2006). Biodiversity and ecosystem stability in a decade-long grassland experiment. *Nature, 441*, 629–632.

Trombulak, S. C., & Frissell, C. A. (2000). Review of ecological effects of roads on terrestrial and aquatic communities. *Conservation Biology, 14*, 18–30.

Vandepitte, K., Jacquemyn, H., Roldan-Ruiz, I., & Honnay, O. (2007). Landscape genetics of the self-compatible forest herb *Geum urbanum*: Effects of habitat age, fragmentation and local environment. *Molecular Ecology, 16*, 4171–4179.

Vellend, M. (2003). Habitat loss inhibits recovery of plant diversity as forests regrow. *Ecology, 84*, 1158–1164.

Vellend, M. (2005). Species diversity and genetic diversity: Parallel processes and correlated patterns. *The American Naturalist, 166*, 199–215.

Vellend, M., & Geber, M. A. (2005). Connections between species diversity and genetic diversity. *Ecology Letters, 8*, 767–781.

Venables, W. N., Smith, D. M., & R Development Core Team. (2006) *An Introduction to R – Notes on R: A Programming Environment for Data Analysis and Graphics.*.

Verheyen, K., Guntenspergen, G. R., Biesbrouck, B., & Hermy, M. (2003). An integrated analysis of the effects of past land use on forest herb colonization at the landscape scale. *Journal of Ecology, 91*, 731–742.

Wagner, H. H., & Edwards, P. J. (2001). Quantifying habitat specificity to assess the contribution of a patch to species richness at a landscape scale. *Landscape Ecology, 16*, 121–131.

Walker, K. J., Critchley, C. N. R., Sherwood, A. J., Large, R., Nuttal, R., Hulmes, S., et al. (2007). The conservation of arable plants on cereal field margins: An assessment o new agri-environment scheme options in England, UK. *Biological Conservation, 136*, 260–270.

Warren, J., Lawson, C., & Belcher, K. (2008). *The agri-environment*. New York: Cambridge University Press.

Westhus, W. (1980). Die Pflanzengesellschaften der Umgebung von Friedeburg (Kr. Hettstedt) und Wanzleben während des Zeitraumes 1978/1979 und ihr Vergleich mit Untersuchungsergebnissen von 1958/1959 bzw. 1961/1962. Diplomathesis, Martin-Luther-University Halle-Wittenberg.

Wiens, J. A., Stenseth, N. C., van Horne, B., & Ims, R. A. (1993). Ecological mechanisms and landscape ecology. *OIKOS, 66*, 369–380.

Wilson, J. B., & Chiarucci, A. (2000). Do plant communities exist? Evidence from scaling-up local species-area relations to the regional level. *Journal of Vegetation Science, 11*, 773–775.

Wright, S. (1943). Isolation by distance. *Genetics, 28*, 114–138.

Young, A., Boyle, T., & Brown, T. (1996). The population genetic consequences of habitat fragmentation for plants. *Trends in Ecology & Evolution, 11*, 413–418.

Chapter 21

Integration of Long-Term Environmental Data by the Example of the UNECE Heavy Metals in Mosses Survey in Germany: Application of a WebGIS-Based Metadata System

Gunther Schmidt, Christian Aden, Lukas Kleppin, Roland Pesch, and Winfried Schröder

Abstract In Germany, environmental monitoring data are collected at different locations and in different environmental components by several federal monitoring networks. According to the scientific and legal demands, these data should be made available by integrating them into a web-based geographical information system (WebGIS) and described by harmonised metadata. This enables a spatial and temporal assessment of long-term environmental impacts induced by climate change, for example, as being one issue of the holistic approach, achievable through research strategies of the ILTER (International Long-Term Ecological Research). This chapter on hand describes a WebGIS that was implemented by using Open Source software and filled with geodata and metadata on relevant long-term environmental networks in Germany (LTER-D). The latter were gathered by using an online questionnaire being answered for a total of nine different monitoring networks describing atmosphere, soils, groundwater, biota and marine ecosystems at around 300 monitoring sites. The questionnaire and the WebGIS itself were implemented by using Open Source software, solely. Feasibility and surplus of the WebGIS were demonstrated by investigating the spatiotemporal trends of the metal bioaccumulation in Germany's ecoregions. The technological framework implemented for gathering and linking of environmental data from different monitoring networks proved to be an adequate and effective tool for an integrated investigation of long-term environmental changes. Hence, efforts should be improved to promote long-term ecosystem research in the LTER-D community and to accelerate integration of monitoring data and results into European and global long-term monitoring programmes.

Keywords Data acquisition · Environmental monitoring · Moss monitoring · Online questionnaire · WebGIS

21.1 Background and Goal

In the 1970s, it became clear that the planning of environmental and economic development required information not only on economics and society, but also on environmental conditions. From an ecological point of view, each environmental component like soil, water, air, animals and plants has to be examined at each of several monitoring sites to allow a holistic view at the state of the environment (Ellenberg, Fränzle, & Müller, 1978). This holds also true for monitoring the transfer of materials, energy and information between the different media. Based on this, environmental monitoring in Germany has to be realised by integrating three different levels of investigation: (1) ecosystem research (local but comprehensive monitoring on ecosystem structures, relations and fluxes), (2) ecological environmental monitoring (nationwide, but general monitoring on environmental conditions) and (3) environmental specimen bank (retrospective research). This monitoring system should also work as an early warning system to detect environmental changes (SRU, 1991). The federal nature protection law (§ 12 BNatSchG), the environmental monitoring concept of the Federal Ministry for the Environment,

G. Schmidt (✉)
Institute of Landscape Ecology, University of Vechta, PO Box 1553, 49364 Vechta, Germany
e-mail: gschmidt@iuw.uni-vechta.de

F. Müller et al. (eds.), *Long-Term Ecological Research*,
DOI 10.1007/978-90-481-8782-9_21,

Nature Conservation and Nuclear Safety (BMU, 2000) as well as the preamble of the administrative agreement between the German government and the federal states on the exchange of environmental data specify the following targets that should be complied with when carrying out environmental monitoring: The monitoring should be coordinated and based on harmonised or standardised methods (Keune & Mandry, 1996) so that the data can be compared and used for statistical analysis and modelling. Monitoring should be complete with regard to the data needed for an ecological description and interpretation of the environmental status. Each of the ecoregions (Bailey, 1995, pp. 24–26) of Germany should be represented without gaps by an adequate number of monitoring sites. This ecoregional representativeness of monitoring networks is of crucial relevance for the validity of the sample data coming out of it (Cao, Williams, & Larsen, 2002, Tirler, Donega, Voto, & Kahr, 2003). The monitoring data should allow for spatial extrapolation in order to bridge geographical gaps and for supporting long-term research on environmental changes. The flow of data should be efficient and the data should be available for scientists, especially for statistical testing of hypotheses and modelling data. The latter aspect also implies important technical issues, because of the enormous amount of information and data collected. For example, environmental monitoring networks require information exchange, which has to be supported by an adequate and efficient information platform that handles documentation and exchange of metadata (site descriptions, quality control data), measuring data and geodata. A cheap system which is compatible with all common operating systems would be best to publish data quickly without having special technical knowledge at the user's site and where user management ensures data access and quality.

Corresponding to this, the chapter on hand tackles the establishment of a web-based geo-information system (WebGIS), which was developed for efficient data retrieval and exchange of relevant environmental data collected in the Long-Term Ecological Research Network, Germany (LTER-D). The WebGIS 'LTER-D' should be used to gather data from different measurement networks and research projects in Germany dealing with environmental issues. The aim is to analyse long-term ecological impacts like global warming or forest dieback with integrated data from different environmental monitoring sites measured in different periods of time and located all over Germany. Facilitating the access to data describing these long-term processes is one major goal in this context. As an example of use, the UNECE moss monitoring is introduced, which was established in Germany in 1990 and which was supported in the latest campaign (2005) by a WebGIS data retrieval system (WebGIS 'MossMet') (Pesch, Schmidt et al., 2007). This was to simplify and harmonise sample documentation by metadata as well as to facilitate analyses on both metadata and measuring data.

21.2 ILTER, ALTER-Net, LTER-D

Changes in our environment are taking place in periods of time that can not be sufficiently covered by distinct research projects which usually last only a couple of years. Projects are limited in time and spatial scale and therefore mostly depict only snapshots of decade-long processes and reactions of ecosystems caused by changing environmental conditions. With these limitations it is difficult to distinguish between short-term fluctuations and long-term and global trends. A holistic approach, achievable through research strategies of the ILTER (International Long-Term Ecological Research),[1] is necessary to characterise the state and the change of ecosystems, the emergence of environmental impacts and the change of ecosystem processes in future times. These environmental changes may be induced by climate change, causing floods, desertification or glacial withdrawal. Other long-term environmental changes are eutrophication of waters, soil acidification, forest dieback, soil degradation or regression of biodiversity. The ILTER Network was founded in 1993 and provides a platform for world-wide communication in terms of long-term and large-scale environmental research. Worldwide, in 2006, 32 countries are members of the ILTER Network running a national LTER Network. The vision of the ILTER Network is to contribute to the advancement of the global environment that in turn, advances the health, prosperity, welfare and security of humanity. The mission of the ILTER Network is to develop and effectively deliver sound scientific information and predictive understanding of ecological processes associated with large temporal and spatial scales to the scientific community, policy makers and society in

[1] http://www.ilternet.edu

general. This information is needed to better conserve, protect and manage ecosystems at local, regional and global scales; their biodiversity and the services that they provide.

The research project ALTER-Net[2] (A Long-term Biodiversity, Ecosystem and Awareness Research Network) sponsored by the EU from 2004 to 2009 should develop a European long-term inter-disciplinary research facility for research on the complex relationship between ecosystems, biodiversity and society and to support policy assessment and development on the conservation and sustainable use of biodiversity in the European Union. An information platform will be established, which provides data for research and reports on biodiversity-related issues. Twenty-four partner institutes are involved in the ALTER-Net project; all of them have in-depth experience in biodiversity research, monitoring and/or communication. There are six overarching objectives which are grouped into 'integration goals' and 'research goals' and are considered to achieve long-lasting integration of the project partners. The 'integration goals' cover problems like knowledge management and network development, whereas the 'research goals' include scientific analysis in terms of risk assessment, indicator definition or cause–effect determination.

The European activities in long-term ecosystem research are going to be bundled at the home of LTER-Europe.[3] The aim of the site is to provide information about national LTER networks in Europe and to act as a focal point for the development of a pan-European LTER network to provide an integrated system for detecting and understanding environmental change at a range of scales. In 2010, eighteen European countries currently have formal LTER networks (i.e. formally accepted into ILTER). A further one has substantially developed networks and is working towards formal ILTER acceptance. Three other countries are in the early stages of network development. Apart from general information or links to national networks and contact addresses no further information, not to mention free data sets, are – at this stage – available.

The German network LTER-D attempts to provide a basis for communication, documentation and cooperation of scientists in the long-term, system-oriented and inter-disciplinary monitoring of the environment in Germany. The German LTER network intends to become an overarching and integrating structure for system-oriented ecological research. It will enhance the efficiency of the German research community in that field, which is highly important for environmental protection and a sustainable future. 'Gains in efficiency should arise from synergy effects brought about by intensive collaboration. Networking of the German LTER research community is urgently needed, in order to catch up with the international LTER process, as most European nations have already proceeded much further.'[4]

The national LTER networks integrated in ILTER are far from a harmonised and linked information community. They are rather autonomic platforms dealing with national projects on long-term ecological research. Neither the content management nor the technical framework for knowledge transfer and data exchange was coordinated. Especially small countries like Costa Rica[5] (forest dieback) or Taiwan[6] (biodiversity, climate monitoring) having relatively small research budgets are playing a pioneering role in information processing and exchange and even offer metadata and measurement data from various monitoring networks. Sophisticated national LTER networks were established as well in China and Austria.

The Chinese LTER network CERN[7] was founded in 1988 and coordinates activities gained at five disciplinary centres (soils, atmosphere, water, biology, marine ecosystems). The mission of CERN is to promote ecosystem conservation and improvement, environmental quality enhancement and agricultural development and to advance the studies in ecology and related inter-disciplines. Its mandate includes monitoring, research and demonstration on typical ecosystems in China. Apart from data management and processing, CERN aims at covering four objectives: prognoses on development of resources and ecology, administration of scientific research, publication of integrative environmental status reports and information platform for

[2] http://www.alter-net.info

[3] http://www.lter-europe.ceh.ac.uk

[4] Program of LTER-D (Friedrichstadt, 24.03.05): http://www.lter-d.de

[5] http://crlter.ots.ac.cr

[6] http://lter.npust.edu.tw/tern

[7] http://www.cern.ac.cn

decision-making processes in terms of economic and ecological issues. CERN serves as an important facility to control desertification, soil erosion, salinisation and eutrophication.

The LTER-Austria network[8] was founded in 2002, emerging from a compilation of research projects and institutions dealing with ecosystem research by the Austrian Network for Environmental Research (ÖNUF). For a long-term ecological research two ('High Alps' and 'Eisenwurzen') so-called multifunctional research platforms (MFRPs) were established, which are designated to serve as regional reference systems. These MFRPs were built on already existing research facilities, institutions and networks and gain added value by integrating single research results under a common roof of coordination. With the help of the MFRPs, environmental changes caused by socio-economic development in land use and global changes should be documented to detect, for example, polluters and to reveal dependent cause–effect relationships on a regional level. Data management is promoted by using an integrative information system called 'MORIS' (Monitoring and Research Information System). MORIS serves as an information platform based on a relational database system where metadata as well as measuring data from different environmental monitoring networks are administrated. MORIS allows hierarchical data attribution and offers different tools for data query and selection, for data import and export as well as for data documentation by photographs or field protocols. The integration and linkage of geodata and their spatial visualisation and analysis is planned but not yet realised.

In this context, the Chair of Landscape Ecology at the University of Vechta developed the technical framework to build an information platform where spatially referenced information on relevant research programmes in LTER-D is compiled (Schmidt & Loesewitz, 2005). This information includes metadata on monitoring networks, geodata on different ecological features as well as access to measurement data derived at these sites. Metadata were collected using an electronic questionnaire (Section 21.3.1). The Metadata Information System is enhanced by functionalities offered by a WebGIS. This includes logical and spatial queries in the information pool that should be extended with measurement data. The goal is to gather specific data according to relevant topics from long-term monitoring, in particular observation areas, and to analyse them across both the levels of ecological hierarchy (Allen & Starr, 1982) and the according scientific disciplines.

21.3 Metadata Survey and WebGIS Application

21.3.1 Metadata Inquiry

For collecting metadata, a web-based interface was implemented that allows standardised data survey and efficient data processing. These metadata include information on measured parameters, intervals and methods of measuring. The metadata were linked directly to the spatial information of the monitoring sites by simultaneous databank use and access. The metadata retrieval was realised using the Open Source software 'PHP Surveyor' contributed by SourceForge.net. This software may easily be installed and used even by non-computer scientists (Fig. 21.1). The PHP Surveyor is based on the script language PHP and the database system MySQL. A graphical user interface (GUI) allows easy and semi-automatic design of online questionnaires and offers sophisticated user management and administration. Lists of interviewees may be uploaded in csv format; data export is possible via SQL files and csv files.

For compiling relevant parameters the items of the questionnaire described below were adjusted to several data collection systems and concepts like GMES (Global System for Monitoring of Environment and Security),[9] GEOSS (Global Earth Observation System of Systems)[10] and MUDAB[11] (Marine Umweltdatenbank) and GOOS[12] (Global Ocean Observing System) for aquatic ecosystems and the so-called core data set for the integrated environmental monitoring in Germany (Mattern et al.,

[8] http://www.umweltbundesamt.at/umweltschutz/oekosystem/lter_allgemein/lter_national

[9] http://www.esa.int/esaLP/LPgmes.html

[10] http://www.epa.gov/geoss

[11] http://www.bsh.de/de/Meeresdaten/Umweltschutz/MUDAB-Datenbank/index.jsp

[12] http://ioc.unesco.org/goos

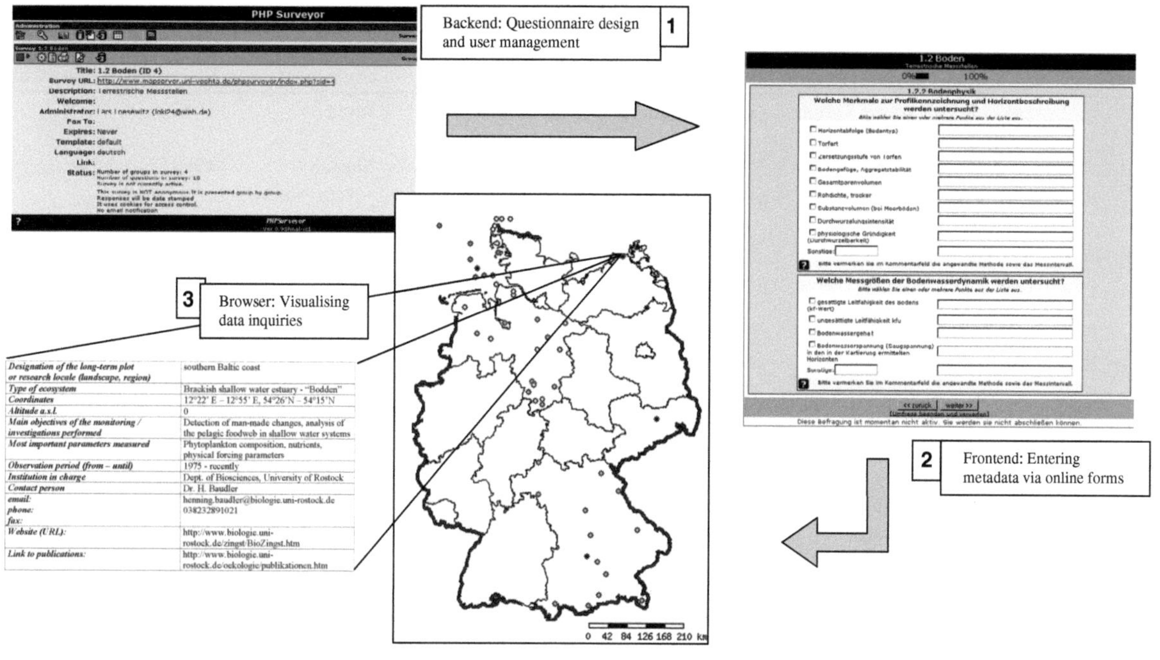

Fig. 21.1 Metadata inquiry by PHP Surveyor (online questionnaire)

2005; Schröder, Pesch, & Schmidt, 2006). For a harmonised data retrieval, there was also a cross-check with already established international environmental information systems, which are united under the label of ILTER, especially the Chinese Ecosystem Research Network (CERN) and the UK Environmental Change Network (ECN).[13]

The questionnaire is designed hierarchically. On the upper level, terrestrial and aquatic monitoring networks were distinguished. The terrestrial sites were subdivided into four components (atmosphere, soils, ground water and biota), whereas the aquatic monitoring sites were divided up into three parts (marine ecosystems, running water and standing water). From that level, different categories for each environmental medium were assigned to query-specific items like parameter, measuring method and interval as well as general information on the responsible institution. Those items which are part of the so-called core data set (Mattern et al., 2005) are attributed to enhanced queries in the databank. These parameters allow, in this sense, treatment of one of the ten general cause-relationship-hypotheses in ecosystem research (e.g. accumulation of toxic substances in terrestrial ecosystems) by providing related (meta-) data sets documented in the questionnaire (Section 21.3.3). For each member of LTER-D who consents into joining the inquiry, an individual access to the PHP Surveyor was created where the described items could be answered.

21.3.2 Inquiry Results

A total of 41 LTER-D member institutions were interested in participating in the online metadata survey. For these institutions online accounts were set up for the PHP surveyor to document the respective environmental monitoring network. At the end a total of 14 answered questionnaires were retrieved describing atmosphere, soils, groundwater, biota and marine ecosystems at nine different monitoring networks with around 300 monitoring sites (Table 21.1). Most of the monitoring sites are located in the North Sea and in forests of Lower Saxony and Bavaria. All questionnaires were exported in the PostgreSQL geodata base in SQL format and linked with the corresponding geodata sets locating the monitoring sites for building up the WebGIS 'LTER-D' (Section 21.3.3). Enhanced query tools enable to extract information

13 http://www.ecn.ac.uk

Table 21.1 Institutions and respective monitoring networks documented in the metadata survey by PHP surveyor

Institution	Federal state	Monitoring network	No. of sites	Environmental component	Start date	End date
AFS	Lower Saxony	Revitalisation of the river Ise	56	Biota	1987	–
Alfred-Wegener-Inst	Bremen	German bight	4	Marine ecosystems (benthos)	1969	–
Alfred-Wegener-Inst.	Bremen	Sylt tidal basin	101	Marine ecosystems (benthos)	2004	–
BayCEER	Bavaria	Lehstenbach	3	Atmosphere	1994	–
BayCEER	Bavaria	Lehstenbach	1	Soils	1992	–
BayCEER	Bavaria	Lehstenbach	14	Groundwater	1988	–
BFA FI /ISH	Hamburg	Wadden Sea	70	Marine ecosystems (fish)	1974	–
GSF	Bavaria	Klostergut Scheyern	1	Soils	1990	–
GSF	Bavaria	Klostergut Scheyern	1	Biota	1990	–
NFV	Lower Saxony	ICP forest (level II)	20	Soils	1992	–
NFV	Lower Saxony	ICP forest (level II)	20	Soils	1989	–
SAW	Saxony	Landscape monitoring	5	Soils	2000	2007
UFT	Bremen	Rumble dumpsite Siedenburg	1	Soils	1979	–
UFT	Bremen	Rumble dumpsite Siedenburg	1	Biota	1979	–

on environmental measurements concerning long-term monitoring issues. Due to the fact that there was less information on long-term monitoring networks gathered by the online inquiry the WebGIS 'LTER-D' was enriched by adding additional data on the German UNECE moss monitoring survey (Section 21.4). This was to demonstrate what considerable benefits could be derived from information integrated in the WebGIS 'LTER-D' in terms of long-term ecosystem research in LTER-D.

21.3.3 WebGIS LTER-D

The use of web-based information systems in natural sciences and engineering is increasing rapidly. The number of applications is enormous due to the heterogeneous requirements of science and economy. Systems designed both to enable retrieving and visualising experimental data from research projects and supporting land-use planning and engineering are two major examples of application. In addition, those systems are appropriate vehicles for publishing and illustrating research results and to accomplish legal report requirements (e.g. in context of the EU-Water Framework Directive). The access to and the documentation of environmental data should be as easy as possible. Therefore, it is important to provide a user-friendly interface that could be used without specialised GIS-Software or a deeper understanding in information processing. To control and manage data access, a user administration tool was implemented that enabled information exchange on different levels. It is possible to map the monitoring sites as well as to retrieve metadata and measurement data by performing spatial and logical queries. The establishment of such a flexible and multifunctional system fulfiling the above-mentioned requirements is realised economically and technically best by using platform-independent Open Source software components (Schmidt & Loesewitz, 2005).

21.3.3.1 Open Source

The WebGIS LTER-D was developed with Open Source software, which reduces the costs for the implementation of the system. Only the hardware environment has to be purchased. On the other hand, the use of Open Source allows (in contrast to proprietary software) the adjustment of the software to individual needs and purposes. Additionally, commercial software is often distributed by licence models charging yearly licence fees. Modification of the software is forbidden and mostly impossible due to hidden source code. Nowadays, the use of free software has become widely accepted in all fields of information technologies (Spath & Günther, 2005). Examples are operating systems (Linux systems) or business software (Open Office) and GIS-software (GRASS-GIS). Open Source software is labelled under the terms of the Open Source Initiative (OSI)[14] (Williams, 2002) that defines access, distribution and modification of the software. Examples for licence models of the OSI are the GNU Public License or the Lesser General Public License and BSD License.

21.3.3.2 Standards and Components

The Open Geospatial Consortium (OGC) is an international organisation comprised of business companies, universities and public facilities and defines open standards for interfaces to process various types of geodata on the Internet. The standards and specifications (e.g. for geospatial data or the catalogue service implementation) are supposed to provide access to complex spatial information and to ensure interoperability between map services located anywhere on the earth (Mitchell, 2005). There are also standards for the management of metadata given by the international standardisation organisation (ISO) like ISO 19115. Standards for metadata are essential for WebGIS platforms to allow clients the use of catalogue services for geodata. The metadata describe the quality, spatial expansion, contents and other important characteristics of the geodata layers (Kresse & Fadaie, 2004). Projects like INSPIRE (Infrastructure for Spatial Information in Europe)[15] or Geoportal.Bund[16] are using these ISO and OGC standards for publishing and sharing European and German data sets as well.

For the establishment of the WebGIS, a server programme is required in order to provide the functionality of a 'spatial' communication. Clients send requests to a server to provide them with environmental data having a spatial reference. For this purpose, a combination of the Apache HTTP-Server[17] with the UMN Mapserver[18] working on the operating system Debian Sarge[19] represents an adequate and efficient solution. The UMN Mapserver is capable of reading several data sources like raster and vector data stored in files, databases or integrated into a GIS. The UMN Mapserver meets the standards of the OGC. With the help of the Open Source database management system PostgreSQL[20] in conjunction with the spatial extension PostGIS[21] and the use of the Geometry Engine Open Source (GEOS), a spatial database backend was built up and combined with metadata or other related data. The spatial extension consists of a library, which allows the storage of vector-based geodata (point, line and polygon) and is able to work with different spatial frames of reference within the databases of PostgreSQL. Only with this extension, the OGC compatibility will be ensured (Brinkhoff, 2005). Within this backend, simple GIS functionalities such as buffering and intersecting are also provided. More specialised applications could be implemented with a powerful GIS-backend like GRASS (Neteler & Mitasova, 2004).

The system architecture of the LTER-D WebGIS is set up by different software components as illustrated in Fig. 21.2. The main task of the Apache HTTP-server is to enable communication between web clients (frontend) and the server system (backend) via HTTP (Hypertext Transfer Protocol). Based on client requests, the UMN Mapserver generates raster maps of the chosen geodata layers. A common geo-projection for each data set is not necessary; the mapserver handles this issue via defined projection codes. Spatial queries in the generated maps are possible by clicking directly into the mapframe. Doing this, related map attributes stored in the relational geodata base will be displayed in a new window. The visualisation of the geodata and the interactions between the

[14] http://www.opensource.org/licenses
[15] http://inspire.jrc.ec.europa.eu
[16] http://www.geoportal.bund.de
[17] http://www.apache.org
[18] http://mapserver.org/
[19] http://www.debian.org/
[20] http://www.postgresql.org/
[21] http://postgis.refractions.net/

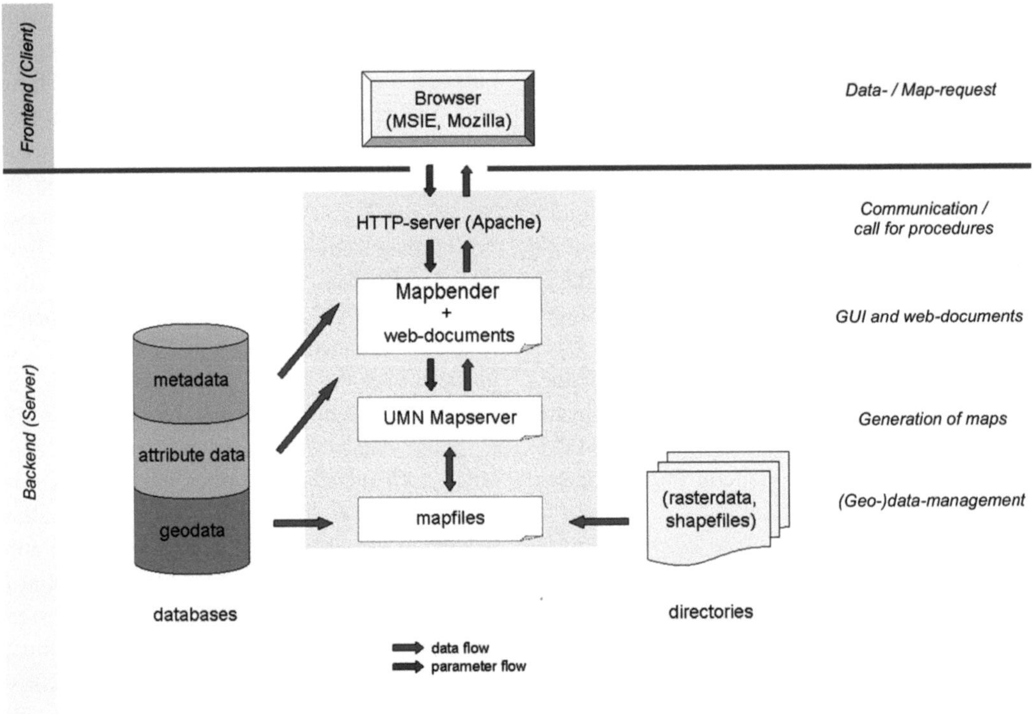

Fig. 21.2 Software architecture WebGIS LTER-D

client and the data are realised by the Mapbender software (Fig. 21.3). Mapbender is an Open Source WebGIS client suite developed by CCGIS[22] and is used for the browser-based visualisation of geodata as well as for the administration of web mapping services. All basic tools for interacting with map contents are implemented in Mapbender. Basic tools are used for navigation (zoom, pan, distance measuring etc.) and queries in metadata and geodata as well. Additionally, the software enables users to build up a spatial data infrastructure (SDI) with servers localised anywhere in the world. Considering OGC standards[23], the SDI is an interoperable system that can be joined with other SDIs.

21.3.3.3 Query Tools

One major objective of the WebGIS 'LTER-D' is the map-based access to metadata and measuring data from relevant environmental monitoring networks operated by different institutions all over Germany. It is possible to receive information by clicking into the mapframe as well as by performing logical queries. The latter is realised by using different query templates as illustrated in Fig. 21.4. Queries regarding interesting institutions, environmental media or hypotheses on related cause–effect relationships are possible. The result of the query is a list of all institutions observing this medium or providing related metadata on the respective hypothesis. After choosing the interesting institution, the respective questionnaire will be shown (Fig. 21.4). The query of metadata about institutions and monitoring networks is realised technically by using the structured query language (SQL). To increase the performance of the queries, all questionnaires are assigned to the respective environmental medium and the relating institution. The linkage between questions and hypotheses was done manually according to the 'core data set' of Mattern et al. (2005). Besides metadata and measuring data compiled by PHP surveyor questionnaires, we integrated additional surface maps and other geodata that may be useful for enhanced spatial analyses. This includes

22 http://www.ccgis.de/

23 http://www.opengeospatial.org/standards

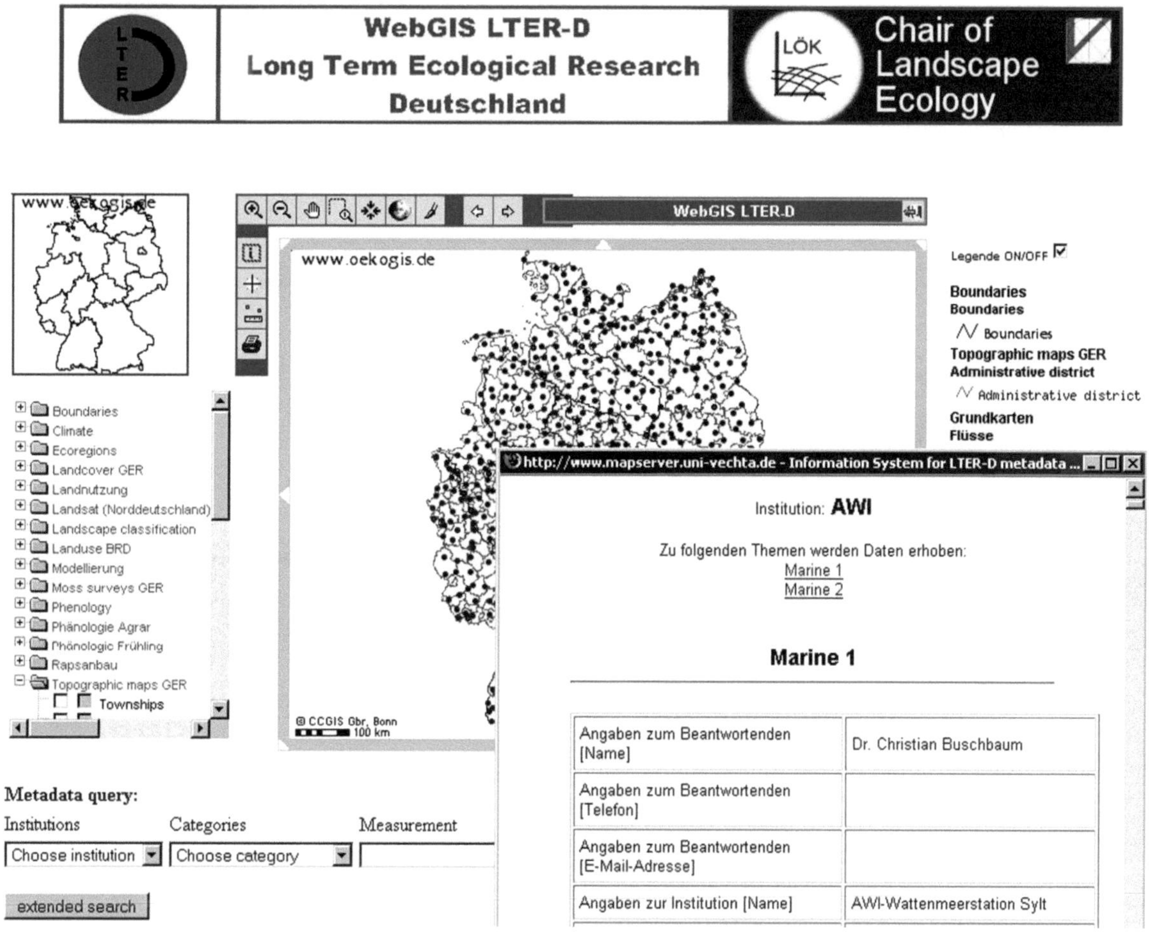

Fig. 21.3 GUI WebGIS LTER-D (*left*) and according metadata template for the one monitoring site (*right*)

administrative boundaries, topographical maps showing river networks, urbanised areas and roads as well as elevation maps and more. Geodata comprising ecological features like climate data, maps of land use, phenological data, soil maps and ecoregions are included as well.

As described in Section 21.3.2, the response to the online inquiry was unsatisfactory. Only a total of 14 monitoring networks were described by metadata (Table 21.1). To demonstrate the practicability of the WebGIS, data from the UNECE moss monitoring survey were integrated in the WebGIS 'LTER-D' and analysed, exemplarily. The moss monitoring serves as an example of a nationwide well-harmonised environmental monitoring network that could be used for answering long-term environmental issues in terms of trends in accumulation of heavy metals (Section 21.4).

21.4 Example of Application: Moss Monitoring Germany

21.4.1 Background

The UNECE moss surveys of 1990, 1995, 2000 and 2005 enabled to map spatiotemporal trends of the metal concentrations in mosses throughout Europe (Rühling, 1994; Rühling & Steinnes, 1998; Harmens et al., 2004, 2007a, 2007b). The goal of these national surveys was to assess the atmospheric deposition expressed as bioaccumulation rates of metals and, from 2005 on, also nitrogen (N) in mosses at minimum distances to emission sources like motorways (min. 300 m), settlements (min. 300 m) or industries (min. 1,000 m). The moss method thereby has

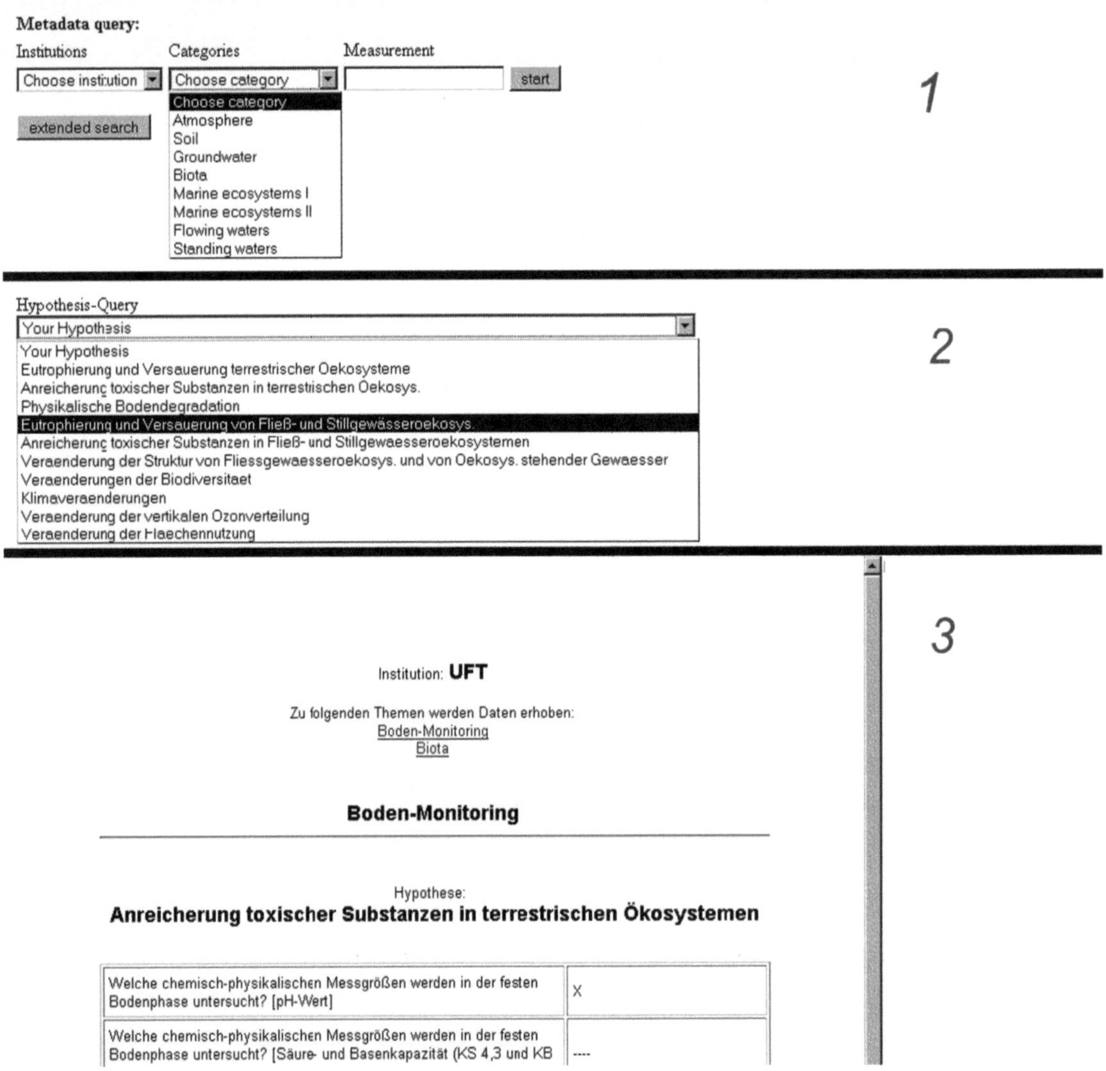

Fig. 21.4 Query tools of LTER-D WebGIS. Query by institutions, categories of environmental media and keywords (1), query by hypothesis (2) and query results by answered questionnaire (3)

proven useful to monitor spatiotemporal trends of the atmospheric bioaccumulation of metals (in mosses) (Markert, Breure, & Zechmeister, 2003) throughout Europe: In Germany, the spatial patterns of the metal bioaccumulation mirror the location of large industrial and urbanised areas in each of the four surveys (Pesch, Schröder, et al., 2007). Hot spots were detected predominantly in the urbanised and industrialised Ruhr Area, the densely populated Rhein-Main region and in the historically grown industrial regions of former East Germany (e.g. Halle-Leipzig region). As in all other participating countries, the experimental survey design followed the UNECE (2001) monitoring manual. The manual provides information on the moss species to collect and critical distances to keep from trees and emission sources and, thus, helps to assure the comparability of the measurement data.

The information collected in the moss surveys consists of (1) measurement data on arsenic (As), cadmium (Cd), chromium (Cr), copper (Cu), iron (Fe), mercury (Hg), nickel (Ni), lead (Pb), antimony (Sb), vanadium (V), titanium (Ti), zinc (Zn) and nitrogen (N) accumulation in the mosses and (2) site-specific metadata to characterise the sampling sites with regard to vegetation, land use, elevation and the distance of the sites to trees and emission sources (e.g.

roads, motorways, human settlements or industries). In previous studies, these metadata were used to assess factors influencing the metal bioaccumulation (Schröder et al., 2008). It could be shown that the metal loads in the mosses correlate significantly with factors related to the emission situation around the monitoring sites. Nevertheless, uncertainties remain when trying to relate the measured metal concentrations to deposition rates since moss-specific criteria, the canopy drip effect, the sea-spray effect, precipitation and elevation show significant associations as well. Furthermore, it has to be considered that the metal concentrations do not only reflect atmospheric depositions but also geogenic background concentrations. Besides predicting or measuring the exposition (predicted environmental concentrations, PEC) expressed as deposition loads, it is still an open issue to define critical concentrations below them these depositions have no ecotoxicological effects (predicted no effect concentrations, PNEC) (Fränzle, Straskraba, & Jorgensen, 1995).

In a further study, Schröder and Pesch (2004) introduced an ordinal-scaled index, which was computed by percentile statistics ranging from 1 for low to 10 for high metal accumulation. The indices were derived for both monitoring sites and surface grids and correspond to the average rank of the element loads of chosen metal elements. Two such indices were calculated: The MMI 1990–2000 aggregates the element loads of Cr, Cu, Fe, Ni, Pb, Ti, V and Zn. These elements were analysed in moss samples collected across the entire German territory in each of the three campaigns 1990, 1995 and 2000. The MMI 1990–2000 therefore allows detecting spatiotemporal trends of the metal bioaccumulation since 1990. The MMI 1995–2000 aggregates the same eight elements and additionally As, Cd, Sb and Hg. In 1990, the latter four metals were either not analysed at all or just in parts of the country. In future works, the two indices will be updated using the measurement data from 2005.

21.4.2 Spatiotemporal Trends of the Metal Bioaccumulation in Germany's Ecoregions

The geostatistical surface estimations and both MMI were intersected with a map on terrestrial ecoregions depicting the ecological coverage of Germany. The map was derived from available data on ecological landscape characteristics using the decision tree algorithm called Classification and Regression Trees (CART) (Schröder, Schmidt, & Hornsmann, 2006). The data consisted of raster maps on the potential natural vegetation (PNV, Federal Office for Nature Protection), elevation above sea level (UNEP), soil texture (digital soil map of Germany 1:1,000,000) as well as on monthly averages on temperature, evaporation and precipitation (1961–1990) and global radiation (1981–1990) provided by the German Meteorological Service. The PNV can be defined as the vegetation that would establish without human interference under present climate and soil conditions and may be interpreted as an integral indicator for the ecological conditions in terrestrial ecosystems (Schröder, Schmidt, & Hornsmann, 2006). To derive ecological landscape units for Germany by CART, the PNV was set as the target variable, whereas the above-mentioned maps on elevation, soil texture and climate were chosen as the independent variables. The decision tree was at first grown to 73 endnodes and then pruned back automatically to 21 endnodes. Since each of the endnodes was the result of a certain sequence of decision rules, they were applied on the available raster data to calculate the ecoregions for Germany.

By intersection of the ecoregions with the moss raster data, which were calculated by means of geostatistics from the site-specific survey data, the metal bioaccumulation could be regionalised for areas that are quantitatively described with regard to abiotic landscape characteristics such as soil texture, climate and elevation above sea level. This enables detection of temporal trends of the metal bioaccumulation for landscape units that are distinctly characterised with regard to their physical, geographical characteristics. As can be seen in Fig. 21.5, a continuous decrease of the MMI 1990–2000 can be observed for the two ecoregions 26 and 43 from 1990 to 1995 to 2000. The same temporal development holds true regarding almost all of the element-specific raster data available for all 21 ecoregions in all three campaigns 1990, 1995 and 2000, although there are exceptions. Significant increases of the metal bioaccumulation can be observed for Fe in three ecoregions, for Cr in 15 ecoregions and for Zn in 20 ecoregions. The same can be stated for the temporal development of Zn from 1990 to 2000 in ecoregions 20, 30, 42 and 43. Regarding the highest spatial differentiation of ecological landscape units, similar tendencies can be observed.

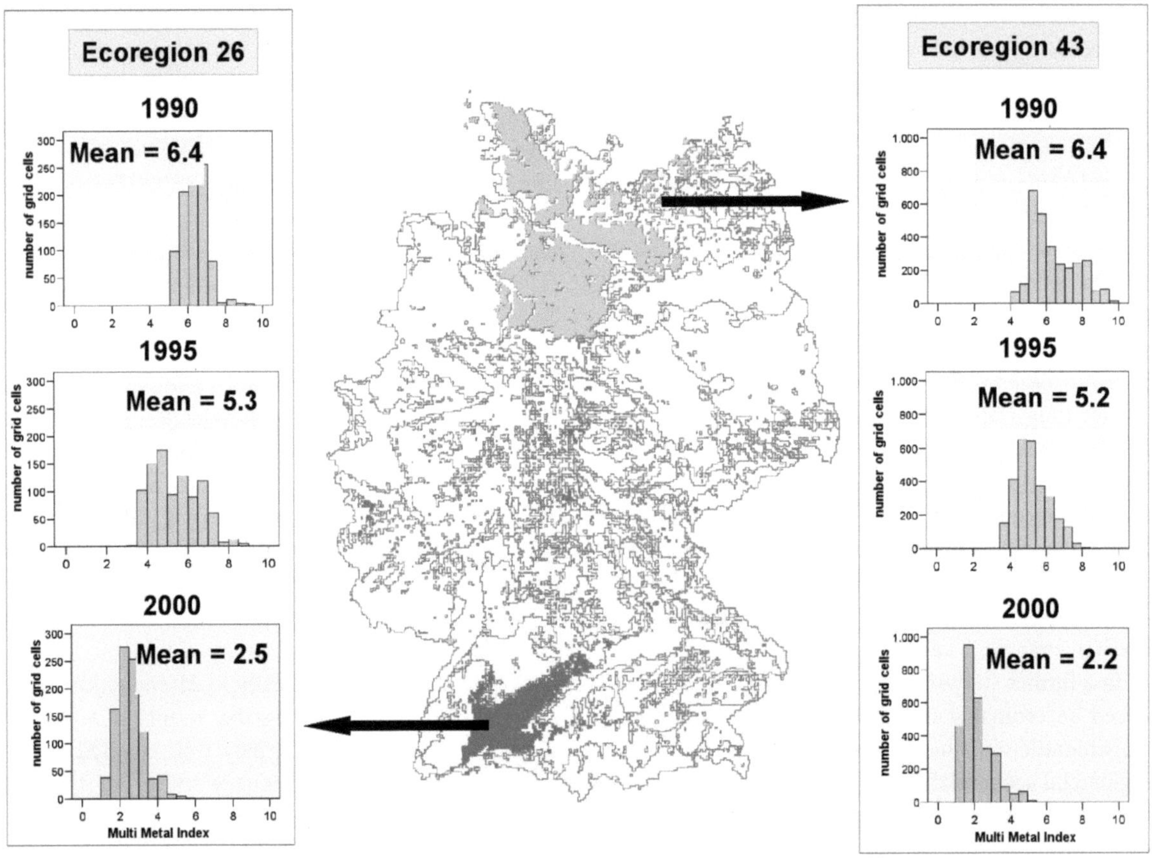

Fig. 21.5 Temporal development of the MMI 1990–2000 in ecoregions 26 and 43 from 1990 to 1995 to 2000

21.4.3 The WebGIS MossMet

The WebGIS developed for LTER-D (Section 21.3.3) was enhanced for optimising data handling in the moss survey 2005. This WebGIS called 'MossMet' was implemented with the help of Open Source components considering the described technical standards (Pesch, Schmidt, et al., 2007). By this, the metadata can be integrated within the information system via Internet by the moss samplers, directly. The WebGIS 'MossMet' comprehensively documents the metadata, the measurement values and statistically derived metal bioaccumulation indices regionalised for ecoregions depicting the landscape coverage of Germany. In the German moss survey 2005, the WebGIS 'MossMet' was applied, routinely. The WebGIS 'MossMet' allows digitising new sampling sites as well as to analyse and query existing metadata. It is furthermore possible to visualise all monitoring data from the surveys and to relate them with geo-information on land cover, traffic or other environmental monitoring networks. The metadata survey was improved by developing integrated query templates, which made the use of the PHP surveyor dispensable.

21.5 Discussion

Environmental monitoring should provide information about environmental conditions and changes for resource managers and policy makers and should be open to the scientific community as well as to the public. To achieve a holistic view, data from several research and monitoring programmes should be compiled and integrated into one comprehensive information platform according to the concept of Ellenberg et al. (1978), the recommendations of SRU (1991) and the requirements defined by BMU (2000). This should enable a pan-media (air, water, soils and biota), nationwide view of long-term environmental changes in Germany. First efforts in metadata compilation have been realised (Klitzing, 2002; Knetsch & Schröder, 2002; Schröder, Schmidt, & Pesch, 2003; Schröder, Pesch, & Schmidt, 2004; Schröder, Pesch, et al., 2006), but due to political, technical or financial restrictions the implementation of such a global information system for environmental monitoring purposes is still difficult. Examples for a successful integrative analysis of environmental data cover a wide range of applications from terrestrial to marine monitoring issues (Buchwald, Gigante, Pöhlker, & Schmidt, 2006; Daschkeit et al., 2002; Graef, Schmidt, Schröder, & Stachow, 2005; Schlüter, Schröder, & Vetter, 2004; Schröder & Schmidt, 2005). German geodata infrastructures like Portal-U[24] and Geoportal.Bund or the European INSPIRE directive are initiated to define standards for geodata exchange and provide a WebGIS platform where relevant data should be shared. Up to now one main problem of those platforms is the lack of appropriate data sets that should be uploaded or integrated by interested users or institutions. Missing metadata and geodata input is as well one crucial problem when building up the WebGIS 'LTER-D'. The technical infrastructure of the WebGIS 'LTER-D' has been implemented, and yet the interoperability to Portal-U and Geoportal.Bund could be ensured by using the same OGC and ISO standards. But due to the sluggish feedback – only 14 monitoring entered the metadata survey of LTER-D – the contents in the system are still rare. That is why the surplus of the WebGIS technology was demonstrated by the example of the measurement data on metal accumulation in mosses and the sampling site describing metadata collected in Germany 1990, 1995 and 2000. The WebGIS 'MossMet' was used successfully in the survey 2005/2006. Referring to user comments on the system's applicability, we conclude that the WebGIS 'MossMet' effectively supports the compilation, quality control and integrated assessment of metadata and measurement data. For other environmental issues, the WebGIS technology was successfully applied too.

[24] http://www.portalu.de

For the monitoring of genetically modified organisms (GMO), a WebGIS 'GMO monitoring' was created that should help in the surveillance on possible ecological impacts of GMO cultivation according to § 16 GenTG (Aden, Schmidt, & Schröder, 2007). A WebGIS on the spatial distribution of GMO fields in Germany is in preparation (Kleppin, Aden, Schmidt, & Schröder, 2009).

21.6 Conclusions and Outlook

Web-based geographical information systems help us to manage, to publish and to share environmental data efficiently and economically. Open Source software like the UMN mapserver and GRASS allow the implementation of such systems without intense costs, high performance and functionality. The access to sensitive data sets can be controlled by user management and the client needs no sophisticated knowledge or equipment. Projected or current studies can be conducted and documented promptly and interactively by using online input masks or questionnaires. It could be demonstrated that the implementation of a web-based information platform is an adequate tool for the integration of relevant data for long-term ecosystem research. Metadata from different monitoring networks covering terrestrial as well as marine ecosystems were integrated in one joint platform by using Open Source software components, for metadata query as well as for geodata linkage and visualisation. After creating an efficient and user-friendly technical framework for LTER-D, efforts should be improved to use this vehicle for data compilation, to promote long-term ecosystem research in the LTER-D community and to accelerate integration of monitoring data and results into European and global long-term monitoring programmes.

References

Aden, C., Schmidt, G., & Schröder, W. (2007). Ein webbasiertes Geografisches Informationssystem für das Monitoring gentechnisch veränderter Organismen. In B. Breckling, M. Dolek, A. Lang, H. Reuter & R. Verhoeven (Eds.), *Naturschutz und Biologische Vielfalt 49: GVO-Monitoring vor der Umsetzung* (pp. 97–112). Bonn: Bundesamt für Naturschutz.

Allen, J. M., & Starr, T. B. (1982). *Hierarchy: Perspectives for ecological complexity*. Chicago: Chicago University Press.

Bailey, R. G. (1995). *Ecosystem geography*. New York, Berlin, Heidelberg: Springer.

BMU (Bundesministerium für Umwelt, Naturschutz und Reaktorsicherheit). (2000). *Umweltbeobachtung. Stand und Entwicklungsmöglichkeiten*. Bonn: BMU.

Brinkhoff, T. (2005). *Geodatenbanksysteme in Theorie und Praxis*. Heidelberg: Wichmann.

Buchwald, R., Gigante, D., Pöhlker, J., & Schmidt, G. (2006). Loss of biodiversity by eutrophication – a long-year-study in deciduous forests of the Weser-Ems-region (NW Germany). *Verhandlungen der Gesellschaft für Ökologie, 36*, 182.

Cao, Y., Williams, D. D., & Larsen, D. P. (2002). Comparison of ecological communities. The problem of sample representativeness. *Ecological Monographs, 72*(1), 41–56.

Daschkeit, A., Fränzle, O., Sterr, H., Schottes, P., Schröder, W., & Bartels, F. (2002). Klimafolgen für Sylt. Integrative Analyse und Bewertung. In A. Daschkeit & P. Schottes (Eds.), *Sylt. Klimafolgen für Mensch und Küste* (pp. 279–321). Berlin: Springer.

Ellenberg, H., Fränzle, O., & Müller, P. (1978). *Ökosystemforschung im Hinblick auf Umwelt- und Entwicklungsplanung*. Berlin, Bonn: Umweltbundesamt.

Fränzle, O., Straskraba, M., & Jorgensen, S.E. (1995). Ecology and ecotoxicology. *Ullmann's Encyclopedia of Industrial Chemistry, 7*, 19–154.

Graef, F., Schmidt, G., Schröder, W., & Stachow, U. (2005). Determining ecoregions for environmental and GMO monitoring networks. *Environmental Monitoring and Assessment, 108*, 189–203.

Harmens, H., Buse, A., Büker, P., Norris, D., Mills, G., Williams, B., et al. (2004). Heavy metal concentration in European mosses: 2000/2001 survey. *Journal of Atmospheric Chemistry, 49*, 425–436.

Harmens, H., Norrisa, D. A., Koerbera, G. R., Buse, A., Steinnes, E., & Rühling, A. (2007a). Temporal trends in the concentration of arsenic, chromium, copper, iron, nickel, vanadium and zinc in mosses across Europe between 1990 and 2000. *Atmospheric Environment, 41*, 6673–6687.

Harmens, H., Norrisa, D. A., Koerbera, G. R., Buse, A., Steinnes, E., & Rühling, A. (2007b). Temporal trends (1990–2000) in the concentration of cadmium, lead and mercury in mosses across Europe. *Environmental Pollution, 151*, 368–376.

Keune, H., & Mandry, P. (1996). Harmonization of environmental data. The requirements for developing a consistent view of the environment world-wide. In W. Schröder, O. Fränzle, H. Keune, & P. Mandry (Eds.), *Global monitoring of terrestrial ecosystems* (pp. 51–56). Berlin: Ernst & Sohn.

Kleppin, L., Aden, C., Schmidt, G., & Schröder, W. (2009): WebGIS als Instrument für Planung und Monitoring des Anbaus von Bt-Mais. In: J. Strobl, Th. Blaschke, Th., & G. Griesebener (Eds.), *Angewandte Geoinformatik 2009* (pp. 588–597). Heidelberg: Wichmann.

Klitzing, von F. (2002). Umweltbeobachtung des Bundes, Integration der Beobachtungsprogramme anderer Ressorts. *UBA-Texte, 65/02*, 1–66.

Knetsch, G., & Schröder, W. (2002). *Use of GIS and statistics for an environmental monitoring system in Germany*. Paper presented at the 16th international conference EnviroInfo, Vienna, Austria.

Kresse, W., & Fadaie, K. (2004). *ISO standards for geographic information*. New York, Berlin, Heidelberg: Springer.

Markert, B., Breure, A., & Zechmeister, H. G. (2003). *Bioindicators, biomonitors*. Amsterdam: Elsevier.

Mattern, K., Hain, B., & Schönthaler, K. (2005). Integrated environmental monitoring – an ecosystem based approach. In: German MAB national Committee (Ed.), *Full of life – UNESCO biosphere reserves, model regions for sustainable development* (pp. 122–128). New York, Berlin, Heidelberg: Springer.

Mitchell, T. (2005). *Web mapping illustrated*. Sebastopol, Cambridge: OReilly.

Neteler, M., & Mitasova, H. (2004). *Open source GIS: A GRASS GIS approach*. Boston: Kluwer Academic Publishers.

Pesch, R., Schmidt, G., Schröder, W., Aden, C., Kleppin, L., & Holy, M. (2007). Development, implementation and application of the WebGIS MossMet. In K. Tochtermann & A. Scharl (Eds.), *The Geospatial Web. How geo-browsers, social software and the Web 2.0 are shaping the network society* (pp. 191–200). London: Springer.

Pesch, R., Schröder, W., Matter, Y., Holy, M., Goeritz, A., & Genßler, L. (2007). *Moos-Monitoring 2005/2006: Schwermetalle IV und Gesamtstickstoff*. Berlin: Umweltbundesamt.

Rühling, A. (1994). *Atmospheric heavy metal deposition in Europe – estimation based on moss analysis*. Copenhagen, Denmark: Nordic Council of Ministers.

Rühling, A., & Steinnes, E. (1998). *Atmospheric heavy metal deposition in Europe 1995–1996*. Copenhagen, Denmark: Nordic Council of Ministers.

Schlüter, M., Schröder, W., & Vetter, L. (2004). Marine-Geo-Information-System for visualization and typology of marine geodata (Mar_GIS). *Geotechnologien Science Report, 4*, 77–78.

Schmidt, G., & Loesewitz, L. (2005). *A WebGIS data retrieval system for use in environmental monitoring*. Paper presented at the 19th international conference EnviroInfo, Brno, Czech Republic.

Schröder, W., & Pesch, R. (2004). Metal accumulation in mosses. Spatial analysis and indicator building by means of GIS, geostatistics and cluster techniques. *Environmental Monitoring and Assessment, 98*, 131–156.

Schröder, W., & Schmidt, G. (2005). Metadatenbanken und GIS als technische Unterstützung der stofflichen Expositions- und Wirkungsanalyse in der Umweltbeobachtung. In S. Fränzle, B. Markert, & S. Wünschmann, S. (Eds.), *Lehrbuch der Technischen Umweltchemie* (pp. 211–227). Landsberg: ecomed.

Schröder, W., Pesch, R., & Schmidt, G. (2004). Soil monitoring in Germany. Spatial representativity and methodical comparability. *Journal of Soils and Sediments, 4*(1), 49–58.

Schröder, W., Pesch, R., & Schmidt, G. (2006). Identifying and closing gaps in environmental monitoring by means of metadata, ecoregionalisation and geostatistics. The UNESCO biosphere reserve Rhön (Germany) as an example. *Environmental Monitoring and Assessment, 114*, 461–488.

Schröder, W., Schmidt, G., & Pesch, R. (2003). Harmonization of environmental monitoring. Tools for examination of methodical comparability and spatial representativity. *Gate to Environmental and Health Sciences, 7*, 1–13.

Schröder, W., Schmidt, G., & Hornsmann, I. (2006). Eine multivariat-statistische abgeleitete ökologische Raumgliederung für Deutschland zur Analyse und Bewertung der Messnetze der Umweltbeobachtung. In O. Fränzle, F. Müller, & W. Schröder (Eds.), *Handbuch der Umweltwissenschaften. Grundlagen und Anwendungen der Ökosystemforschung* (pp. 1–100). Landsberg: ecomed.

Schröder, W., Pesch, R., Englert, C., Harmens, H., Suchara, I., Zechmeister, H., et al. (2008). Metal accumulation in mosses across national boundaries: Uncovering and ranking causes of spatial variation. *Environmental Pollution, 151*, (377–388).

Spath, D., & Günther, J. (2005). *Open Source Software – Strukturwandel oder Strohfeuer? - Eine empirische Studie zu Trends und Entwicklungen zum Einsatz von Open Source Software in der öffentlichen Verwaltung und IT-Unternehmen in Deutschland.* Retrieved June 8, 2007, from http://www.iao.fraunhofer.de/d/oss_studie.pdf.

SRU (Der Rat von Sachverständigen für Umweltfragen). (1991). *Allgemeine ökologische Umweltbeobachtung (Sondergutachten)*. (Stuttgart).

Tirler, W., Donega, M., Voto, G., & Kahr, G. (2003). Quick evaluation of long term monitoring samples and its representativeness. *Organohalogen Compounds, 60*, 509–512.

UNECE (United Nations Economic Commission for Europe). (2001). *Monitoring of Atmospheric Heavy Metal Deposition in Europe Using Bryophytes. Experimental Protocol 2000/2001 Survey*. Retrieved April 6, 2007, from http://icpvegetation.ceh.ac.uk.

Williams, S. (2002). *Free as in freedom. Richard Stallman's crusade for free software*. Sebastopol, Cambridge: OReilly.

Part VI
Concepts and Results: Presenting and Interpreting Long-Term Ecological Processes: Linking Research and Applications

Chapter 22

Monitoring of Ecosystems: Two Different Approaches – Long-Term Observation Versus Success Control

Vera Luthardt

Abstract In the broad field of monitoring, this chapter highlights two specific approaches: long-term ecosystem observation and success control. Different target settings in nature protection determine very strictly the concrete monitoring programmes as demonstrated by two topical examples. As the first example, the 'Long-term observation of ecosystems in the biosphere reserves of Brandenburg' is operating for more than 10 years now. It comprises a broad variety of areas while observing all components relevant to the respective ecosystem. The fundamental approach of this concept is introduced. Criteria used for the definition of ecosystem types in combination with utilization forms and for the selection of observation areas are described. A closer look at the aims and programmes of fens exemplifies the approach of the project. In order to outline the differences between long-term observations and success control, the second example depicts a success control programme for the rewetting of peatlands in forests.

Keywords Biosphere reserves · Monitoring of mires · Nature conservation management · Observation goals · Success control programme for rewetting peatlands

22.1 Ecosystems and Different Approaches to Monitor Them

The terms 'ecosystem' and 'monitoring' are often used in a different way. For this reason relevant terms will be characterized in the following.

V. Luthardt (✉)
Department of Landscape Management and Nature Conservation, Hochschule für nachhaltige Entwicklung (FH) Eberswalde, University of Applied Sciences Eberswalde, Friedrich-Ebertstr. 28, 16225 Eberswalde, Germany
e-mail: vluthardt@hnee.de

An ecosystem is determined as '... an ecological system composed of living organisms (plants, animals, and microbes) and their nonliving environment (climate and soil in the case of terrestrial ecosystems; aqueous environment and substrate in aquatic ecosystems). To be an ecosystem, these components must be spatially arranged and have the appropriate interactions that lead to the capture and storage of energy as biomass, a trophic structure, a circulation of nutrients, and change over time (ecological succession). Ecosystems are characterized by five major attributes: structure, function, complexity, interaction of the components, and change over time' (Kimmins, 1997, p. 525).

Also the term 'monitoring' is used with very different meanings. Hurford (In Hurford & Schneider, 2006, p. 3) defines the term very strictly as '... intermittent surveillance carried out in order to ascertain the extent of compliance with a predetermined standard or the degree of variation from an expected norm'. However, this narrow definition does not always correlate with its daily practical use. For this reason 'monitoring' is employed here in the wider sense of environmental observation.

As a matter of fact not only the terms but also the requirements of observation programmes are often mixed up that may lead to dissatisfaction of all participants. Thus, one of the most important demands for the development of monitoring programmes is the very clear and strict definition of the concrete aim and needs of the observation programme.

In environmental and nature conservation a broad variety of approaches and observation goals have been developed (see, e.g., Artiola, Pepper, & Brusseau, 2004; Hurford & Schneider, 2006; Schönthaler et al., 2003). Not only sectoral issues, such as the

F. Müller et al. (eds.), *Long-Term Ecological Research*,
DOI 10.1007/978-90-481-8782-9_22,

Fig. 22.1 General aims of ecosystem monitoring as an integrated approach

General aims of the monitoring of ecosystems

1. Documentation of the ecosystem development (fluctuations, dynamics, succession stages of all parts of the ecosystem and their interactions)
⇒ Receipt of comparative data of close to nature ecosystems
⇒ New knowledge about medium- and long-term effects of different utilization forms and intensities on the ecosystems
2. Conclusions from the assessment of the actual state of areas under the viewpoint of nature conservation
3. Basic data lines for the evaluation, improvement and verification of ecological models
4. Formulation of new strategies for future use and management of ecosystems
5. Design of an „early warning system“
6. Input for public relations about the development of ecosystems.

Table 22.1 Comparison of the two monitoring approaches: long-term observation and success control

	Long-term observation (= surveillance[a])	Success control (= experimental management[a])
Description	Value-free acquisition of data about sectoral or ecosystem development	Survey of short- and medium-term effects referring to the achievement of a measure's goal
Background	Without pressure to succeed	Clear assessment of the success
Demands	Standardized and clearly assessed methodology, spatially and temporally repeatable, organized database management and continuous evaluation	Fast and (semi)quantitative, comprehensible measurement of the direct effect of the conservation action
Measured indicators	For *all compartments* of the sector or ecosystem	For the *specific object, function or part* which is changed by the measure
Time intervals	e.g. 3–6–9–12–15... years	1–2–3–(5)–(10)... years
Responsible clients	Ministry for Research; Federal offices for the protection of nature and environment; Research institutes	Client of the concrete management measure

[a]Terms of Hurford and Schneider (2006)

development of red-list species or changes in natural replenishment, but also general goals can be addressed (Fig. 22.1).

The main goal of a long-term ecosystem observation is the recording, documentation and evaluation of the local development of different habitats. For this purpose its holistic concept integrates all relevant ecosystem components and functions. But in recent years a new task arose with growing importance: the evaluation, improvement and verification of ecological models. Model concepts can be designed, tested and eventually adapted on the basis of monitored data, allowing a more precise and secure prediction of future developments. The more data are collected throughout a long period of time, the better this mission can be fulfilled. Consequently data collection and modelling should be coordinated with one another as soon and as far as possible.

Completely different needs on ecosystem monitoring are demanded by practical nature protection management. In this case a fast control of management measures is of main priority. As new knowledge is quickly needed for future projects, results of the management effect on the conditions of an ecosystem should be gained as soon as possible. Although the financial resources for such missions are usually strongly limited, the expectations on the information content of such success control activities are often highly exaggerated.

The several possible monitoring objectives result in different data collection requirements. As an example for different goals and demands for monitoring approaches you can see the comparison in Table 22.1.

22.2 Two Specific Examples

In the following section we describe the two unlike approaches to survey ecosystems as introduced in the previous chapter.

Each type will be discussed on behalf of one practical example from north-east Germany: first the 'Long-term observation of ecosystems in the biosphere reserves of Brandenburg' and second 'Monitoring the success of rewetting of peatlands in the forests of Brandenburg'.

22.2.1 Long-Term Observation of Ecosystems – A Task of Biosphere Reserves

One of the tasks of biosphere reserves (BR), as described by the MAB committee of UNESCO in 1995 (MAB Deutsches Nationalkomitee, 1996), is the design and implementation of ecological monitoring with standardized methods. This requirement is formulated especially for biosphere reserves because they provide an insight into the development of ecosystems with relatively little direct human impact as well as of ecosystems shaped by human land use. The implementation of such a monitoring programme forms one functional criterion among other that is decisive for the evaluation of biosphere reserves applied by UNESCO every 10 years.

In response to this MAB requirement, the Brandenburg federal environmental agency has funded the design and installation of a monitoring plan for the observation of ecosystems in three UNESCO biosphere reserves in Brandenburg – 'Schorfheide', 'Spreewald' and 'River Landscape Elbe'. As a result environmental monitoring has been conducted for 10 years now (Luthardt, Vahrson, & Dreger, 1999; Luthardt et al., 2005). Detailed information can be found at the home page: http://lanuweb.fh-eberswalde.de/oeub/

The monitoring programme has been developed as illustrated in Fig. 22.2. Different categories of ecosystems were identified as follows: forests, lakes, rivers, fens and peatlands, grassland and fields. These were combined with the most important forms of land use to *ecosystem-use-types* which reflect the diversity of nature–human impact combinations in the biosphere reserves.

Table 22.2 shows the criteria used for the classification of the ecosystem types and related ecosystem-use-types.

After classification of present inventory as described above, selection of particular ecosystems for the project was limited on behalf of the following factors (Vahrson, Luthardt, & Dreger, 2000).

In the first step four criteria were used for a decision referring to the ecosystem-use-type:

- dominance in the particular biosphere reserve,
- nationwide scarcity in Germany,
- close-to-nature state (with a high share of the area),
- special management.

Second another four factors limited the choice of observation sites additionally:

- conformance to the observation focus prescribed by MAB committee (Table 22.3),
- goals defined by the BR administration referring to the development of land use (e.g. the effect of organic farming in comparison to conventional land use in the Schorfheide and Spreewald BR),
- prescriptions on the distribution of tasks between the BR,
- available budget.

The ultimate choice of observation sites has been made in a close cooperation between the federal environmental agency of Brandenburg and the respective BR administration. As a result the number of current monitoring areas is documented in Table 22.4.

For every ecosystem category a set of variables related to the monitoring objectives were identified. Abiotic and biotic components that are considered relevant to the ecosystem are represented by parameters with a high indicator level for the present ecosystem conditions. The data are collected by using standardized methods. Highest importance is given to the precise documentation of the applied methods and of the spatial location of sample points via Differential Global Positioning System (DGPS).

The measurements are repeated in time intervals which range from 3 years for more dynamic parameters to 6 and 12 years for parameters considered more stable. For this a complex time schedule of measurements has been developed (see Luthardt et al., 2006). The first results confirm the impression that our monitoring strategy is going to satisfy the demands of feasibility, repeatability and affordability to achieve the main goals in a good manner. The methodology can be transferred to other large protected areas.

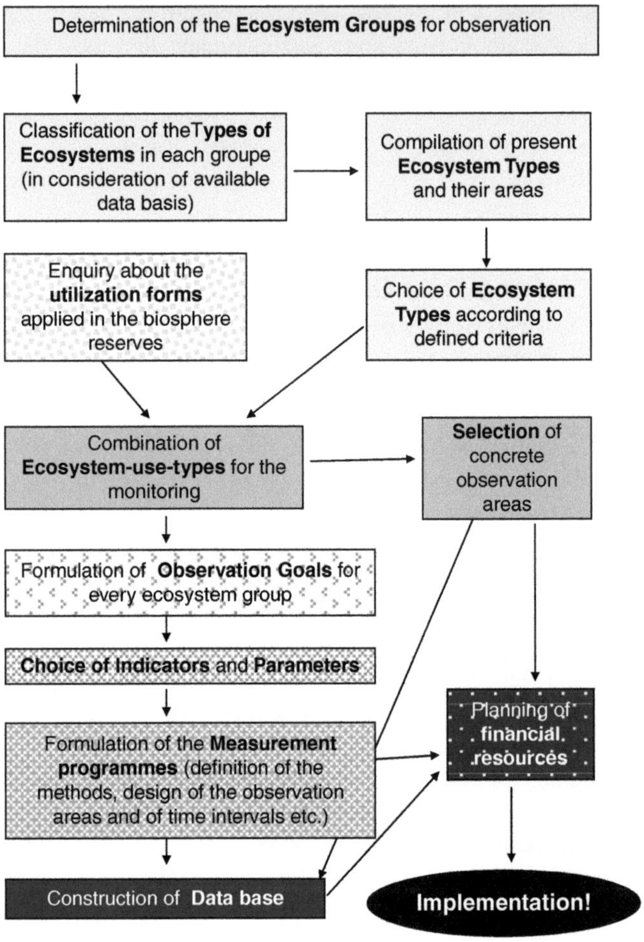

Fig. 22.2 General steps to create the programme for the long-term observation of ecosystems in the biosphere reserves of Brandenburg

A closer look at the concrete monitoring programme for the ecosystem type *semi-natural fens* will exemplify the project. Here, following trends are examined under the conditions of climatic changes, alterations in the hydrologic balance in the catchments areas, changes of the input of material by air and rainfall and in specific cases with the influence of measures like rewetting or forest transformation:

1. changes in the water budget,
2. development of the soil status (trophic conditions, acid/base relation, degradation level),
3. changes in the plant community, vegetation structure and growth of biomass,
4. changes to the habitat function for paludose plant and animal species,
5. sequences of succession stages,
6. fluctuations of vegetation units.

The monitored variables, which are chosen to meet the observation goals, are shown in Table 22.5.

Every 3 years we report on the ecosystem's development, including evaluation and prognosis (Fig. 22.3). We are presently in the third and fourth measurement cycle, so that we are able to make initial statements about trends now.

The results are of high importance for the future management of the biosphere reserves, especially for the control of land use intensity. Due to the detailed

Table 22.2 Criteria for the classification of ecosystem types and related ecosystem-use-types applied in the long-term ecosystem observation in the biosphere reserves of Brandenburg

Ecosystem group	Criteria for classification and inventory of the ecosystem types	Selected utilization forms and specific management measures
Forests	Configuration of features: • vegetation structure (like tree species, layers) • growth relevant ecological factors (like soil fertility and humidity, climatic factors) • important processes (like net primary production, competition)	• forests in strictly protected zones • forests in nature conservation areas with management • forests with specific management (like forest transformation, understory)
Fens and peatland	(a) Natural fens; combination of • ecological type • plant formation and • dominant plant species (b) Drained peatland, combination of • plant formation • peat depth • degradation level of the soil	(a) Natural fens • surrounded by forests • surrounded by fields • surrounded by pastures (b) Drained peatland • without use • with mowing • with grazing • with rewetting
Lakes	Combination of • ecological and hydrological type (primary and actual) • size, depth, maturity, stratification • dominant vegetation of the water and the bank	• lakes in strictly protected zones • lakes in nature conservation areas with different intensities of use like fishing, bathing
Rivers and ditches	Combination of • ecological type (crenal/rhithral/potamal) • structure • bottom soil • catchment area • dynamics	• natural streams • important draining lines • backwater • canals • ditches in dikelands, strictly protected zones and nature conservation areas with different intensities of agricultural use • streams close to fish ponds
Mineral grassland	Combination of • biotope • substratum • relief • hydromorphical type	• meadow • pasture of cows • pasture of sheep • fallow land
Agricultural land	Combination of • soil substrate • soil type • relief • hydromorphical type	• conventional production – only plant production/with animal husbandry • organic farming – only plant production/with animal husbandry • fallow land
Kettle holes	• dynamics of the water level • habitat quality • shadowing	No separate selection, combination with the selected field areas

documentation of the development of specific ecosystem types the outcomes substantiate the effects of nature conservation against the background of global climatic changes. The results can be directly processed in the revision of the environmental management plan of the biosphere reserves.

At the moment we are working on the optimization of semi-automatic analyses and on modelling techniques for the prediction of climatic changes in the BR.

22.2.2 Success Control – Indispensable Part of Nature Conservation Management

One of the most critical limiting factors for implementing management activities in nature conservation are the financial resources. They have a significant effect on the success of conservation strategies. Gaining experience of how ecosystem management

Table 22.3 Overview of ecosystems with a high priority for monitoring in the biosphere reserves of Brandenburg: Schorfheide-Chorin (SC), Spreewald (SW) and River Landscape Elbe (FE) (summary from AG BR, 1995, supplemented in 2003)

Type of ecosystem	SC BR	SW BR	FE BR
Rivers and banks	–	X	X
Lakes, oxbow lakes	X	–	X
Bogs and transitional mires	X	–	–
Fens and swamps	–	X	–
Peatlands in different stages of regeneration or degeneration	X	–	–
Fields/fallow land	X	–	–
Meadows, pastures and humid grassland	–	X	X
Sand dunes, dry grassland	–	–	X
Horticulture	–	X	–
Alder Forests	–	X	–
Beech–pine forests/pure pine forests	X	–	–
Pine forests on sand dunes	–	–	X
Riparian forests	–	–	X

Table 22.4 Number of monitoring areas in the three biosphere reserves of Brandenburg

	SC BR	SW BR	FE BR
Total area of the BR [km^2]	1291	484	535
Ecosystem group			
Woodland and forests	21	8	8
Lakes	30/4[a]	0	0/4[b]
Rivers (sections) and ditches	0	13	13
Fens and peatland	7	4	2
Mineral grassland	4	5	5
Fields (agricultural land)	5	2	0
Sum	**71**	**32**	**32**

[a]Kettle holes
[b]Special pools of water

Table 22.5 Measurement schedule for the ecosystem group 'Fens': parameters, their purpose and time intervals of investigation

Parameter	Goal	Interval
Monitoring area		
Photo documentation	V, Su, FV, (WB)	6a
Soil		
Topsoil horizons	Soi, (WB)	6a
'Einheitswasserzahl' of the topsoil layers of OHDE	Soi, (WB)	6a
pH	Soi, (HF)	6a
$CaCO_3$	Soi, (HF)	6a
C total, C org, N total, C/N-ratio	Soi, (HF)	6a
Ground water level	Soi, (HF)	1a
Vegetation		
Plant species – diversity (list of all species in the whole area)	HF, (WB), (Soi), (Su), (FV)	6a
Diversity of the vegetation (transect)	V, Su, FV, (WB), (Soi), (HF)	6a
Plant-association (Vegetation analysis of Braun-Blanquet)	V, Su, FV, (WB), (Soi), (HF)	6a
Zoning of vegetation units	V, Su, FV, (WB), (Soi), (HF)	12a
Fauna		
Species diversity of dragonflies (transects)	HF, (WB)	3a
Species diversity of amphibians (list of all species and reproduction activities)	HF, (WB)	3a
Species diversity of butterflies (transects)	HF, (Su)	3a
Catchment area		
Biotope mapping	(WB)	6a

6a = every sixth year; Abbreviation of goal in brackets = indirect parameter (1 = WB – changes in the *w*ater *b*udget, 2 = Soi – development of the *soi*l status, 3 = V – changes in the plant community, *v*egetation structure and growth of biomass, 4 = HF – changes to the *h*abitat *f*unction for paludose plant and animal species, 5 = Su – sequences of *su*ccession stages of the vegetation, 6 = FV – *f*luctuations of *v*egetation units)

achieves conservation goals so that we can develop more successful plans for the next project forms an urgent task.

In contrast to the previous example of long-term observation we want to introduce now a programme applying 'success control': the government project 'Mire conservation in the forests of Brandenburg'. This project is financed by the Federal Department of Forestry since 2004. Applicants are private forest owners, forestry agencies, etc. They are responsible for the success control of their individual sub-project. In order to come up with all expectations the method has to be simpler than in the previous example, easy to describe and to understand and should cause as low costs as possible.

Here monitoring should aim at documenting the direct effect of the measure ('Implementation monitoring', Block, Franklin, Ward, Ganey, & White, 2001) as well as reflecting the effect on, e.g., umbrellas species ('Effectiveness monitoring', dito). The goal definition has to focus on the restoration of ecosystem functions – as already recommended

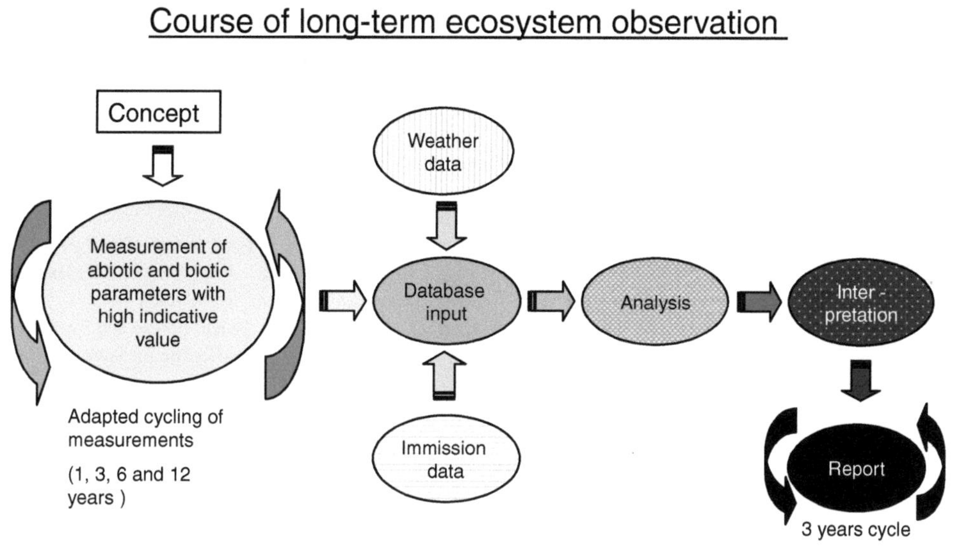

Fig. 22.3 The continuous process of long-term ecosystem observation applied in the monitoring programme in the biosphere reserves of Brandenburg

by Stanturf, Schoenholtz, Schweitzer, and Shepard (2001).

So the most important goals for the restoration of fens surrounded by forests in East Germany influenced by former land use, melioration and climatic changes are as follows:

1. maximum increase of the water level in peatland,
2. improvement of the wetland character to the point of peat growth,
3. resettlement/expansion of specific paludose plant and animal species.

The success of management measures is indicated by a significant improvement of peatland conditions in regard to the three goals described above.

Hereby a qualitative comparison of the conditions before and after the conservation management is preferred over quantifications (such as recommended by Matthews & Endress, 2008).

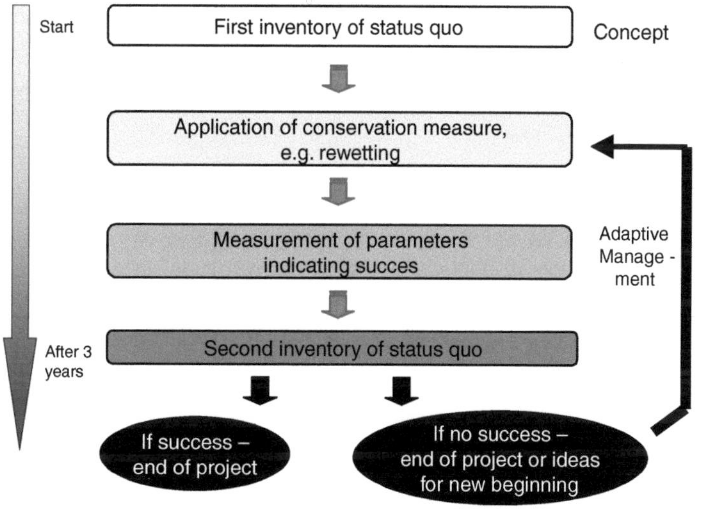

Fig. 22.4 The chronological course of success control for the government project 'Mire conservation in the forests of Brandenburg'

Table 22.6 Indicators and methodology of the success control programme for rewetting forest peatlands

Goal	Indicator	Method	Interval
Increase of water level	1. Relief 2. Formation of the lagg 3. Visible fluctuation of the water level per year 4. Actual state of the drainage system/barrage	Mapping	Twice: before and 3 years after the measure
-	Water level over the year	Gauge data	Each month after the beginning of the measure for 3 years
Improvement of the wetland character	1. Dominant vegetation types 2. Vitality and growth form of woody plants	Mapping	Twice: before and 3 years after the measure
New peat accumulation	1. Presence of peat forming plants like mosses, sedges, reed 2. Vitality of the moss cover	Mapping	Twice: before and 3 years after the measure
Resettlement/expansion of specific paludose plants – and animal species	1. Dominant vegetation types 2. Occurrence of paludose plant groups like mosses, sedges, reed 3. Spatial expansion of both in the area	Mapping	Twice: before and 3 years after the measure

The chronological course of the success control programme is described in Fig. 22.4.

The methodology basically consists of two procedures:

1. Mapping of the area with a special standardized method before and 3 years after the measure;
2. Measurement of the water level with one water gauge (vegetation usually reacts with delay) continuously since the beginning of the measure until 3 years after.

In order to evaluate whether the goals are achieved, following specific indicators were employed, summarized in Table 22.6.

We chose especially parameters which indicate that essential functions and values will readjust over time (like demanded by Short, Burdick, Short, Davis, & Morgan, 2000; Ruiz-Jaen & Aide, 2005). One of the most characteristic functions of fens is the peat-forming process and the habitat suitability for paludose species. The prerequisite for this is a high water level throughout the year.

The mapping method is well described, simply structured and has a standardized form. As long as the peatland area does not exceed 25 ha mapping lasts approximately 1 day. Every employee, who is educated in biotope mapping, can learn this method in a short time.

The time expenditure on gauge measurements is higher. But according to our experience foresters or private owners are often willing to take care of this job in combination with their field work. Further detailed information about the method can be found at the homepage: www.dss-wamos.de.

So the effect of success control is fast results for further practical implementation with relatively low costs and effort at the same time. It produces new knowledge about the effects caused by concrete conservation actions. *However, it usually delivers only little information applicable to a broader ecological context or to modelling purposes!*

22.3 Conclusions

The concept of any form of monitoring programme should be strictly aligned with the pursued goal. The definition of goals is the first step to be taken, followed by the selection of indicators, parameters and investigation areas. Also the evaluation of data is subject to answer goal-specific questions. It is useless to overstuff monitoring programmes with parameters and analysis options. Otherwise, the result will be a 'data cemetery', meaning that many data are left unexploited or in the other case excessive requests for additional assessments cannot be fulfilled. The general differentiation between long-term observation for the clarification of nature scientific problems and success control as a tool for fast assessment and improvement of nature protection measures is especially important.

The same applies for restoration purposes. New knowledge for restoration ecology forming a branch of fundamental science can be collected only by long-term observation due to the broad basis of ecological parameter (Morris, Alonso, Jefferson, & Kirby, 2006; Aronson & Vallejo, 2006). In contrast, ecological restoration is the process of assisting the recovery of an ecosystem that has been degraded, damaged or destroyed (Harris & Van Diggelen, 2006). As a matter of fact short-term success control is not capable of observing long-term maturation of the ecosystem. Nevertheless, the indicators of the initial stages prove the successful or unsuccessful initiation of this process caused by the restoration management.

Long-term observation programmes increasingly gain significance as a data source and evaluation basis for prediction models. From the very beginning of the establishment of new programmes or during the appraisal of existing ones, a direct collaboration with modellers is recommended, concerning the selection of sample plots, number of repetitions, parameters, appropriate database design and evaluation conditions. This forms an important prerequisite to fully utilize the collected monitoring data.

Acknowledgements I sincerely thank Hamish Kimmins for critical discussions and review of the chapter. Also thanks to Katrin Haggenmueller for the creation of various illustrations and the English revision.

References

AG BR (Ständige Arbeitsgruppe der Biosphärenreservate in Deutschland) (1995). *Biosphärenreservate in Deutschland. Leitlinien für Schutz, Pflege und Entwicklung*. Berlin, Heidelberg, New York: Springer.

Aronson, J., & Vallejo, R. (2006). Challenges for the practice of ecological restoration. In J. v. Andel & J. Aronson (Eds.), *Restoration ecology: The new frontier* (pp. 234–247). Oxford: Blackwell Science).

Artiola, J., Pepper, L., & Brusseau, M. (2004). *Environmental monitoring and characterization*. Netherlands: Elsevier.

Block, W. M., Franklin, A. B., Ward, J. P., Ganey, J. L., & White, G. C. (2001). Design and implementation of monitoring studies to evaluate the success of ecological restoration on wildlife. *Restoration Ecology, 9*, 293–303.

Harris, J., & Van Diggelen, R. (2006). Ecological restoration as a project for global society. In J. v. Andel & J. Aronson, J. (Eds.), *Restoration ecology: The new frontier* (pp. 3–15). Oxford: Blackwell Science.

Hurford, C., & Schneider, M. (Eds.). (2006). *Monitoring nature conservation in cultural habitats*. Netherland: Springer.

Kimmins, J. P. (1997). *Forest ecology. A foundation for sustainable management*. Canada: Prentice Hall.

Luthardt, V., Brauner, O., Dreger, F., Friedrich, S., Garbe, H., Hirsch, A.-K., et al. (2006). *Methodenkatalog zum Monitoring – Programm der Ökosystemaren Umweltbeobachtung in den Biosphärenreservaten Brandenburgs*. Eberswalde: University of Applied Sciences Eberswalde.

Luthardt, V., Brauner, O., Witt, B., Friedrich, S., Zeidler, M., Hofmann, G., et al. (2005). *Lebensräume im Wandel – Bericht zur ökosystemaren Umweltbeobachtung (ÖUB) in den Biosphärenreservaten Brandenburgs. Fachbeiträge des LUA Brandenburg, 94*. Potsdam: Ministry of environment.

Luthardt, V., Vahrson, W. G., & Dreger, F. (1999). Konzept und Aufbau der Ökosystemaren Umweltbeobachtung für die Biosphärenreservate Brandenburgs. *Natur und Landschaft, 74*, 135–143.

MAB Deutsches Nationalkomitee. (1996). *Kriterien für Anerkennung und Überprüfung von Biosphärenreservaten der UNESCO in Deutschland*. Bonn: Federal Agency for nature conservation.

Matthews, J. W., & Endress, A. G. (2008). Performance criteria, compliance success, and vegetation development in compensatory mitigation wetlands. *Environmental Management, 41*, 130–141.

Morris, R., Alonso, A., Jefferson, R., & Kirby, K. (2006). The creation of compensatory habitat: Can it secure sustainable development? *Journal for Nature Conservation, 14*, 106–116.

Ruiz-Jaen, M., & Aide, T. (2005). Restoration success: How is it being measured? *Restoration Ecology, 13*, 569–577.

Schönthaler, K., Meyer, U., Pokorny, D., Reichenbach, M., Windhorst, W., & Schuller, D. (2003). *Ökosystemare Umweltbeobachtung – vom Konzept zur Umsetzung*. Berlin: Erich Schmidt Verlag.

Short, F., Burdick, D., Short, C., Davis, R., & Morgan, P. (2000). Developing success criteria for restored eelgrass, salt marsh and mud flat habitats. *Ecological Engineering, 15*, 239–252.

Stanturf, J., Schoenholtz, S., Schweitzer, C., & Shepard, J. (2001). Achieving restoration success: Myths in bottomland hardwood forests. *Restoration Ecology, 9*, 189–200.

Vahrson, W.-G., Luthardt, V., & Dreger, F. (2000). Flächenauswahl und Ökosystemares Monitoring in den Biosphärenreservaten Schorfheide-Chorin und Spreewald. *Umweltwissenschaften und Schadstoff- Forschung, 12*, 362–372.

Chapter 23

National Parks as Model Regions for Interdisciplinary Long-Term Ecological Research: The Bavarian Forest and Šumavá National Parks Underway to Transboundary Ecosystem Research

Marco Heurich, Burkhard Beudert, Heinrich Rall, and Zdenka Křenová

Abstract National parks are protected areas that have been excluded from human intervention and exploitation in order to safeguard the species inventory and natural processes in a way as 'true to nature as possible'. As permanently protected ecosystems in a process of near-natural development, national parks serve as extremely attractive control areas for ecosystem research and, especially, for scientific, long-term monitoring. In the midst of Europe, in a landscape that has been utilised for millennia, the existence of extensive protected areas can provide answers to an abundance of basic questions that cover an enormously wide variety of themes.

National park research has several functions such as to develop the scientific foundation for the implementation of national park goals; monitor the efficiency of national park management; research the undisturbed development of the biocenoses; study the socio-economic and socio-ecological relationships between the national park, its visitors, and its periphery; and determine anthropogenic influences and their effects on the ecological communities.

Based on these basic considerations, a research concept for the Bavarian Forest was developed, and the practical implementation of the individual monitoring elements is presented in this chapter. In the next step, the mass reproduction of the spruce bark beetle (*Ips typographus* L.) is used as an example of how a long-term monitoring concept can help to explore and explain the causes of processes, their effects on the ecosystems, and their human dimensions.

M. Heurich (✉)
Bavarian Forest National Park, Freyunger Straße 2, 94481 Grafenau, Germany
e-mail: marco.heurich@npv-bw.bayern.de

Keywords Long-term monitoring · National park · Spruce bark beetle · Disturbance · Cross-border co-operation · Research concept · Ecosystem-based research · Forest development · Element fluxes · Human dimension · Species diversity · Global change

23.1 Research – A National Park-Specific Challenge

23.1.1 Introduction

There are only few examples in Central Europe of how nature is able to develop when allowed to do so without human intervention. In view of the global ecological crisis, it has become even more important to observe and understand natural processes and the interactions between natural ecosystems.

National parks are extensive, protected areas that have been largely excluded from human intervention and exploitation in order to safeguard the species inventory and natural processes in a way as 'true to nature as possible'. Thus, they serve a central role as an ecological baseline. When compared to managed systems, they can help to improve our understanding of the human impact on ecosystems (Sinclair, 1998).

Consequently, according to the definition of the International Union for the Conservation of Nature (IUCN), research is one of the most important purposes of national parks. This is confirmed in the action plan for protected areas in Europe (IUCN, 1994). Correspondingly, scientific ecological observation is

F. Müller et al. (eds.), *Long-Term Ecological Research*,
DOI 10.1007/978-90-481-8782-9_23, © Springer Science+Business Media B.V. 2010

described as a purpose of the national parks in German and Bavarian environmental law. Moreover, the German national parks are designated 'Natura 2000' areas according to the FFH-Directive of the European Union. Because of this status, monitoring of the species inventory and ecosystem quality is required.

Research in national parks must, however, adhere to certain rules so that neither the gathering of data nor the removal of material will affect or impair the natural course of development and, therefore, the quality of the national park as a protected area.

23.1.2 The Role of National Park Research

As permanently protected ecosystems in a process of near-natural development, national parks serve as extremely attractive 'control' areas for ecosystem research and, especially, for scientific, long-term monitoring. In the midst of Europe, in a landscape that has been utilised for millennia, the existence of extensive protected areas of the IUCN category II (national parks) can provide answers to basic questions that cover an enormously wide variety of topics. Examples are the practicability of renaturation concepts and the development of 'secondary' wilderness, potential problems with the willingness of local citizens to accept land-use restrictions, regional economic effects of an attractive tourist industry, and monitoring of the efficiency of visitor guidance and educational programs.

The diversity of functions includes the following:

1. *Development of scientific principles for the implementation of national park goals:* For the realisation of long-term planning goals and for high efficiency within the national park management, a thorough examination of the living and inanimate natural resources within the national park are absolutely necessary (inventory). Fundamental research should include information on the natural landscape, aspects of the natural and cultural history of the area, settlement characteristics and infrastructure, and the former and present forms of land use.
2. *Efficiency control of the national park management:* The continuous control of the efficiency of measures implemented to achieve the national park goals is of fundamental importance for an effective national park management. To assure their optimisation in the sense of adaptive management, these measures require constant feedback through scientific observation. The results of this research serve to formulate recommendations for further support measures and improved realisation of the nature protection goals. This research must also consider the area surrounding the parks, since there is a broad interaction between the development of the national park and its environs.
3. *Research and documentation of the development of natural communities in the absence of human intervention:* A focal point of the scientific observation is the research and documentation of the essential natural development of communities in the absence of human intervention. The spatial and temporal dynamics of natural communities and species should be documented by means of long-term monitoring programs (permanent monitoring areas, inventories) within representative components of the landscape. More specific aspects can be investigated through supplemental project-related research.
4. *Documentation of anthropogenic influences and their effects on the biocenoses:* There are no longer any natural landscapes on earth that are completely devoid of human influence. Compared to sylviculturally managed forests, however, the forests in a national park are able to develop without intervention for purposes of human utilisation. Therefore, for comparative research, in which anthropogenic influences and their effects on biocenoses are to be documented, the forests of the national park can be regarded as control areas.
5. *Documentation of the interactions between socio-economic and socio-ecological systems:* Research must also aim to identify the interactions between the national park, national park surroundings, national park visitors, and the inhabitants of the national park region. One important aspect in regard to the socio-economic influences is the significance of the national park for the economic development of the region. Also decisive for implementation of the nature protection goals are the socio-ecological effects; that is, to what degree visitors, recreational users, and developments in the surrounding area will influence the natural development of processes in the national park.

6. *Production and interpretation of research results for nature protection, education, natural forest management, science, and public relations:* Based on the scientific observations achieved in the national park, it is possible to derive knowledge on interactions between humans and nature in non-utilised ecosystems. Such knowledge may also be valuable in regard to economically used areas outside of the national park, and might help to describe the interactions between man and nature in general. The understanding of natural processes that occur in the national park forests will primarily serve the goals of nature protection, science, and natural forest management. Research also supports the educational and public relations missions of the national park. The results of this research must be prepared in a manner so that it is available on short notice and in a generally understandable form.
7. *Data documentation and management:* Data documentation and management, which must accompany this research, are absolutely essential for the success of long-term monitoring programs. Data quality management entails procedures to insure the integrity, security, and the availability of data over years and even decades. This is assured by documentation of the associated metadata, which specify purpose, scope, and quality of the data.

23.2 Practical Implementation of a Research Concept: An Example from the Bavarian Forest National Park

23.2.1 Study Area

The Bavarian Forest National Park is located in southeastern Germany along the border to the Czech Republic and encompasses an area of 24,369 ha (Fig. 23.1). Together with the bordering Bohemian Forest, this area, also known as the Inner Bavarian Forest, comprises one of the largest contiguous forested areas in Central Europe. The national park ranges from an elevation of 600 m above sea level (a.s.l.) with its lowest valley (Kolbersbach), to the ridges, with the highest elevation at 1,453 m a.s.l. (Grosser Rachel). The border between Germany and the Czech Republic is identical with the ridgeline of the Bohemian Forest, which is also the watershed between the Danube and Elbe Rivers. The interaction between various site factors, such as climate and soils, are clearly reflected in the study area, especially in the altitude-dependent zonation of the forests (Elling, Bauer, Klemm, & Koch, 1987).

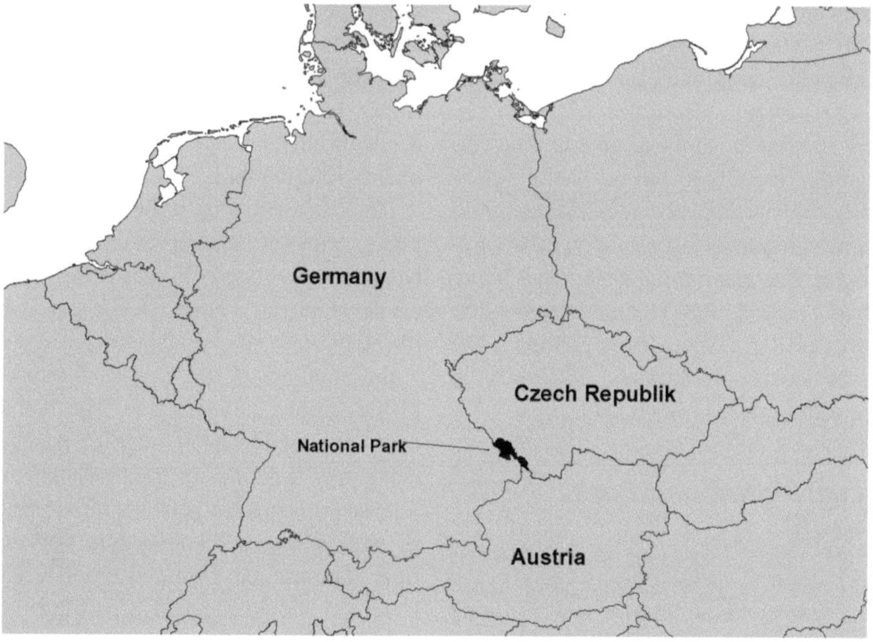

Fig. 23.1 Location of the National park in Central Europe

With annual mean temperatures of 2.0–5.0°C, the high altitude and peak regions above 1,150 m a.s.l. are the coldest locations in the national park. Annual precipitation ranges between 830 and 2,280 mm, whereby in the peak regions, additional precipitation in the form of fog can add up one third to the total. The dominant forest community is the montane spruce forest (Calamagrostio villosae Piceetum barbilophozietosum), which, in its natural state, is almost exclusively characterised by Norway spruce (*Picea abies* L.). It is interspersed with Mountain ash (*Sorbus aucuparia* L.) and Sycamore maple (*Acer pseudoplatanus* L.).

The intermediate slope areas, between 700 and 1,150 m a.s.l. and with mean annual temperatures between 4.4 and 7.2°C, are the climatically most favourable locations within the national park. Annual precipitation amounts to between 830 and 1,820 mm. Because of the relatively favourable climatic conditions, this is the zone of the montane mixed forest, with European beech (*Fagus sylvatica* L.), silver fir (*Abies alba* MILL.), and Norway spruce as the main tree species. These forests can be divided into two types: the Broad-buckler-fern-fir-beech forest (Luzulo luzuloides-Fagetum) on poor soils and the Woodruff-fir-beech forest (Galion odorati-Fagetum) on richer soils.

At altitudes between 600 and 800 m a.s.l. are the 'valley bottoms'. Topographically, these are flat areas and troughs in which cold air descending from the higher elevations becomes trapped. Typical for these locations are frequent, early and late frosts. In connection with the spacious occurrence of wet soils, this determines the lower limit of distribution for beech. The natural forest association of the valley bottoms is the spruce wetland forest (Calamagrostio villosae Piceetum bazzanietosum), in which the eponymous spruce is the dominant tree species with some Silver birch (*Betula pendula* ROTH.) and Downy birch (*Betula pubescens* EHRH.). Annual precipitation in the valley bottoms is between 1,030 and 1,630 mm; the mean annual air temperature is 3.5–6.5°C.

23.2.2 A Model for National Park Research

Ever since the founding of the first German National Park in the Bavarian Forest in 1970, nearly two dozen protected areas with similar goals have been established within the German-speaking regions of Europe alone. Therefore, it is necessary for these parks to develop area-specific identities based on their respective fields of activity. This would help to avoid expensive, parallel research projects. The research in the Inner Bavarian Forest National Park is concentrated on a defined, central theme: the long-term development of near-natural forests after abandonment of economic use and the effects on biotic as well as on abiotic components of the ecosystems. This goal has been further developed under the overall concept 'control factors of natural forest development'. This offers possibilities and stimulating prospects for use in practical protected area management, the educational obligation, and in basic ecosystem research. The spruce bark beetle infestation, in particular, which has been ongoing since the mid-1980s, has significantly affected the forests of the national park. For this reason, its causes and effects on the ecosystem will be described in detail in the following sections.

23.2.3 Thematic Long-Term Timeline

National park research is based on three principles: inventory of the biogeographic constellation (site characteristics, species, and species communities), monitoring (repeated inventory and time-appropriate measurement of significant ecosystem components and their influencing factors), and project-related research (specific aspects, processes, and functional relationships). In combination, the specific contributions rendered through these three principles provide optimal quality information.

The 'thematic long-term timeline' (Fig. 23.2) comprises a bundle of representative monitoring units based on the concept 'control factors of natural forest development'. This concept appears as a recurrent theme throughout national park research and stipulates a minimum amount of long-term monitoring for each of the various subject areas. The individual, thematically assigned projects must be designed to connect along this timeline. In this way, the model requires intense co-operation between monitoring and research as well as close co-ordination between researchers from the national park and other research institutes. Research should proceed according to the following motto: 'as much monitoring as necessary, as much project-related research as possible!'

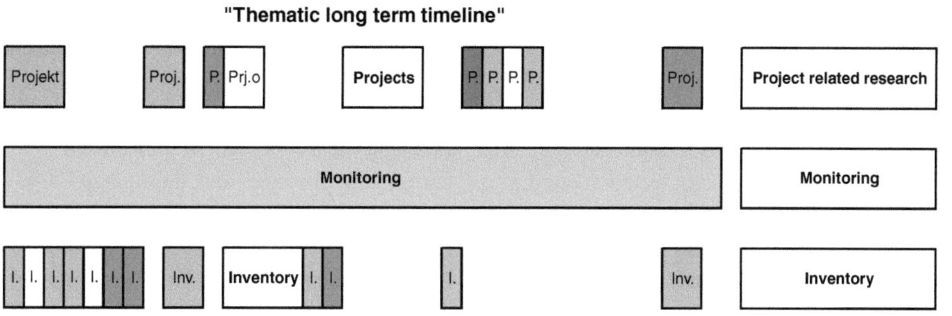

Fig. 23.2 Research concept of the 'thematic long-term timeline'

23.2.4 Elements of the Thematic Long-Term Timeline

23.2.4.1 Inventory

An inventory involves an initial recording of relevant basic data, in so far as it is meaningful for the interpretation of the situation at the outset and is unexpendable for the monitoring concept. In the early years of the national park, basic mapping surveys were performed (geology, soils, plant societies, species composition, and distribution of the higher plants and vertebrate fauna). Forest and land-use history were also researched. The Bavarian Land Survey Office developed a terrain model and carried out a surface water survey (Elling et al., 1987; Petermann & Seibert, 1979; Hauner, 1980; Strobl & Haug, 1993). Now, even 30 years after the establishment of the national park, there are still conspicuous gaps in knowledge concerning lower plant species, soil organisms, and the invertebrate fauna.

23.2.4.2 Monitoring

Monitoring is an important pillar of long-term environmental observation as required by the IUCN (1994). Ideally it is based on inventories and is often realised by repeated inventories. The topics (species, processes, site conditions) to be monitored are derived from inventories, whereby the time intervals between follow-up surveys strongly depend on the expected rate of change. In accordance with the concept, relevant long-term themes have been selected and combined to form a central theme.

Meteorology

Because of the elevation range (between 600 and 1,453 m a.s.l.) and the particularities of the landscape, several weather stations are needed to adequately characterise the climate conditions of the area. Therefore, four permanent stations have been established to record the weather conditions in the three ecological altitudinal zones of the area (Elling et al., 1987, see Section 23.2.1): Klingenbrunn Railway Station (750 m a.s.l.; since 1976) and Taferlruck station (770 m a.s.l.; since 1981) in the valley bottoms, Waldhäuser station (940 m a.s.l.; since 1972) on the middle slope, and Waldschmidhaus (1,360 m a.s.l.; since 1998) at high altitude. The extensive measuring program proceeds according to the World Meteorological Organisation (1983) and is illustrated in Table 23.1. Beginning in 1974, four phenological garden plots have been established at the different ecological altitudinal zones according to international standards, whereby the slopes were divided into upper and lower slopes. Their purpose is to document the onset of phenological stages of woody species typical to the

Table 23.1 Parameters of weather observation

WMIT	Mean wind speed
WMAX	Maximum wind speed
WR	Wind direction
TMIT	Mean air temperature
TMIN	Minimum air temperature
TMAX	Maximum air temperature
RF	Relative humidity
ERDMIN	Minimum temperature above soil
GS	Global radiation
SD	Sunshine duration
NS	Precipitation

region in response to the trend of the driving weather conditions (Schnelle & Volkert, 1957).

In addition to these permanent stations, project-specific weather stations have also been established at five geobotanic long-term observation sites, in the Grosse Ohe catchment area (three weather stations), and along a line transect covering the entire range of elevations. Air temperature and humidity are measured at these stations (30 data recorders and precipitation gauges). The data are used for the interpretation of an extended set of biological data collected at these plots.

Water and Element Fluxes

The 19 km^2 catchment area of the Grosse Ohe lies completely within the 'nature zone' of the national park. Intensive, long-term scientific investigations on the interactions between climate, atmospheric deposition, and forests have been carried out here since 1978 as part of the collaborative project 'water cycling and element budgets in the near-natural catchment Grosse Ohe'. Originally, the aim of this collaborative project was to investigate the changes in the water cycle following the transition from managed to near-natural forests. The basic program includes monitoring of precipitation (eight stations) and water storage in snow cover (12 stations) in the catchment area and stream discharge and weather observations at Taferlruck gauge and meteorological station. The die-back of silver fir and the dramatic decline in fish populations, which were traced to increasing loads of acidifying and eutrophicating air pollutants, was the motive to introduce wet and dry deposition measurements in the 1980s, which then were intensified in two completely forested and unmanaged sub-catchments. Since 1987, the Bavarian Water Management Agency (LFW, 2004) has monitored groundwater in the high elevation catchment area Markungsgraben (890–1,355 m a.s.l.) with regard to groundwater protection from input via deposition and seepage water.

In the adjacent sub-catchment of the Forellenbach (0.7 km^2), which predominantly covers the lower and middle slopes, an intensive and interdisciplinary ecosystem monitoring program was launched in 1990. It is part of the 'International Co-operative Programme on Integrated Monitoring of Air Pollution Effects on Ecosystems' (ICP IM) within the framework of the UN/ECE 'Convention of Long-Range Transboundary Air Pollution' (CLRTAP). This programme is meant to document the state of ecosystems and changes caused by anthropogenic impacts, such as atmospheric pollutants and climate change. These objectives require long-term observation of physico-chemical parameters, and of biotic components of the ecosystem that indicate environmental changes. The Federal Environmental Agency (UBA) and the national park administration are carrying out the extended measuring program shown in Fig. 23.3 (Beudert & Breit, 2004; FEI 2004).

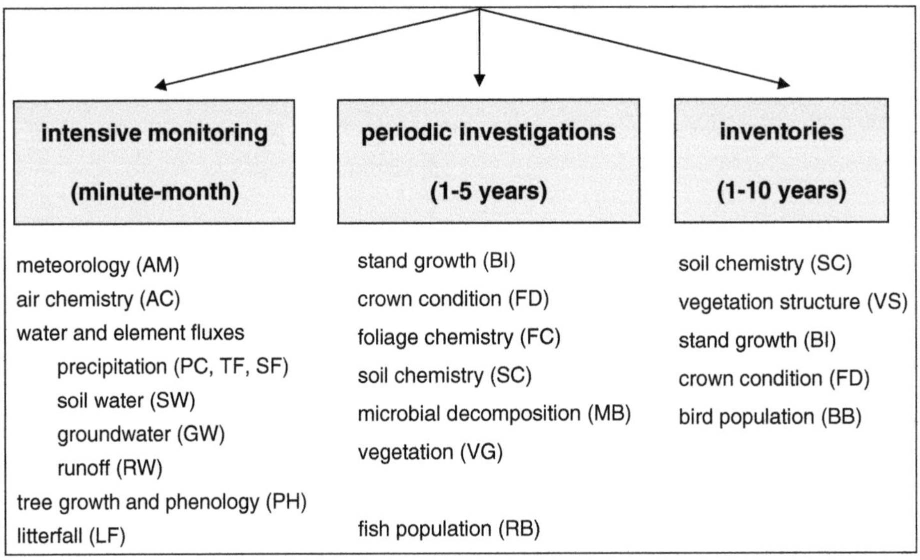

Fig. 23.3 Sub-programs performed in the Forellenbach area

Forest Development

1. *Remote sensing:* Forest development is being documented through aerial photographic series, which have been produced annually since 1988. An aerial photographic survey from 1980 provides reference material. Colour infrared images (CIR) at a scale of between 1:10,000 and 1:15,000 are used. They are archived at the national park administration. Starting in the year 2000, the images were scanned and are now available as digital orthophotos. Since 2004, the aerial photographs have been produced with a digital camera (DMC from ZIImaging). Now, the entire data flow is in digital format and analyses can be easily performed using colour infrared and true colour images as well (Heurich & Rall, 2006).
2. *Forest inventories:* The condition of the forest, on a landscape level, is determined by means of a permanent inventory sampling procedure (control sampling method) on permanently marked plots. This procedure is based on a 200 × 200 m grid, which covers the entire national park. Each intersecting point of the grid is the centre of a sample plot. There is a total of 5,841 sample points inside the national park. The sample areas are in the shape of concentric circles of up to 500 m^2. The recording parameters are presented in Table 23.2. Due to the considerable resource requirements, sampling is performed in 10-year intervals (1971, 1981, 1991, and 2002). To document the course of regeneration of tree species after the bark beetle calamity in the high altitudes, additional inventories are carried out in the high altitudes of the Rachel–Lusen region. Since 1996, these regeneration inventories have been carried out every 2 years (Rall, 1995; Heurich & Neufanger, 2005).
3. *Permanent geobotanical monitoring areas:* These are permanent sampling plots with an area of more than 1,000 m^2, in which vegetation and structural characteristics are recorded. Structural characteristics include living trees and shrubs, their regeneration stages, and deadwood. Small structural elements, such as upended root plates, larger rock outcroppings, and tree stumps were also recorded (Fischer, Abs, & Lenz, 1990; Jehl, 1995, Heurich & Jehl, 2001).

Table 23.2 Parameters of permanent forest inventory

Plot information	Tree information	Additional ecological meaningful information
Forest compartment	Position	Dead wood
Number of inventory map	Tree species	Ant colonies
Coordinate in the sampling grid	Diameter at breast height	Root plates
Slope	Height	Woodpecker hole
Elevation	Age	Special sites
Soils	Vertical layer	Herb/moos layer
Exposition	Bark peeling by red deer	Shrub layer
–	Browsing of terminal shoot	–

For the selection of long-term sample plots, special emphasis was placed on the following aspects:

1. Forest development in near-virgin forest stands in very old forest reserves (11 sample plots)
2. Long-term plant-sociological monitoring in the most important forest types of the Bavarian Forest National Park (four sample plots)
3. Forest development on cleared and not cleared windthrow areas covering all of the altitudinal zones of the Bavarian Forest National Park (six sample plots)
4. Forest development after bark beetle infestation in high elevation spruce forests of the national park (28 sample plots).

Species Diversity

Changes in bird species composition and population densities in relation to the increasing supply of deadwood and old trees in the Bavarian Forest National Park have been studied in a long-term monitoring area on the Grosser Spitzberg since 1986. The study area comprises 100 ha, in which a territorial and grid mapping survey of the bird population is carried out with at least twelve survey dates per year.

A new monitoring program to document the species diversity in the national park started in 2005. Four transects were set up, ranging from the valley bottoms to the high altitude zones. Circular sampling plots are situated every 100 m along the transects. This results in nearly 300 sampling plots distributed across the entire altitudinal range (600–1,400 m a.s.l.), covering the natural and structural diversity of the forest stands within the national park. In addition, two transects were placed in the Rachel–Lusen region, where large areas have been affected by spruce bark beetle infestation, and two were set up in the Falkenstein–Rachel region, where countermeasures against the spreading of bark beetle for a 30-year transition period have been agreed upon. The precise measurement of the sample plots makes it possible to perform accurate follow-up investigations. In each of the study plots, the following species groups are being surveyed: vascular plants, ferns, mosses, lichens, lignicolous fungi, xylobiontic beetles, carabid beetles, true bugs, neuropterans, sawflies, hoverflies, aculeata (stinging wasps), ants, centipedes, millipedes, isopods, pseudoscorpions, Opiliones (harvestmen), spiders, molluscs, breeding bird species, small mammals, and bats. In the beginning, the forest structure of transect plots was determined by means of terrestrial surveys, aerial laser scanning images (LiDAR), and CIR-aerial photo analysis (Bässler & Hahn, 2005).

23.2.4.3 Project-Related Research

Despite the great relevance of inventories and monitoring programs in ecosystem research, additional investigations on special aspects and processes are needed to promote our understanding of ecosystem functions. This is the field of short-term projects and supplemental approaches, which in turn profit from long time series of the underlying conditions.

23.3 The Spruce Bark Beetle Infestation as an Example for the Realisation of the Research Concept

The large, eight-toothed spruce bark beetle (*Ips typographus* L.) is dependant on Norway spruce (*Picea abies* L.) for its reproduction. In spring, when air temperatures rise above 16.5–20°C, usually in April or May, the beetles emerge from their winter quarters. They fly to suitable trees and bore into the bark, where after laying eggs, the larvae will develop. Within several weeks, the feeding damage caused by the larvae results in the death of the tree. Normally, spruces are able to defend themselves against the insects by the secretion of toxic resins. If the trees are weakened, however (snow damage, windthrow, or drought), even small numbers of beetles are able to bed beneath the bark. The time required for development from egg to imago generally depends on temperature and can be as little as 6 weeks. In favourable years, up to three generations of beetles can develop. During exponential population growth, the beetles are even able to kill healthy trees.

In commercial forests, sylvicultural methods, such as beetle traps and decortication, are used to combat the spruce bark beetle. In the national park, however, the beetle population is allowed to develop freely. This was the prerequisite for the massive reproduction of the spruce bark beetle. The causes and effects on the ecosystem are described below. It must be kept in mind that the present epidemic is not unique in space and time. Bark beetle attacks have been historically documented for large areas of the Eastern Bavarian mountain ranges for the years following 1870 and 1929 (Elling et al., 1987; Rall, 1995). Also, outbreaks of bark beetles (different beetle species and different conifer species) simultaneously occurred in the northern hemisphere in the 1990s (Christiansen & Bakke, 1988; Ehnström, Annila, Austarå, Harding, & Ottosson, 1998; Polevoi, Scherbakov, Humala, & Naldeev, 2006).

23.3.1 Development of the Deadwood Areas

In 1983 and 1984 strong storms knocked forest stands totalling up to 173 ha to the ground. Approximately half of the windthrow was located within the 'nature zone' of the national park. Here, the windthrown trees were left untreated in the forest. Due to the large breeding area that had become available, the beetle population increased substantially, thereby leading to the infestation of healthy trees as well. Gradually, the infestation decreased until 1992. This was followed suddenly by a period in which the annual increase in deadwood-covered areas rose sharply. The largest increase was in 1996 with a total of 827 ha of newly infested areas. Between 1997 and 2000, the infestation rate became more stable with 400–600 ha a year. In 2001, it decreased to 55 ha. Between then and 2006, the annual increase of newly infested areas rose almost exponentially to 411 ha. In the meantime, the extent

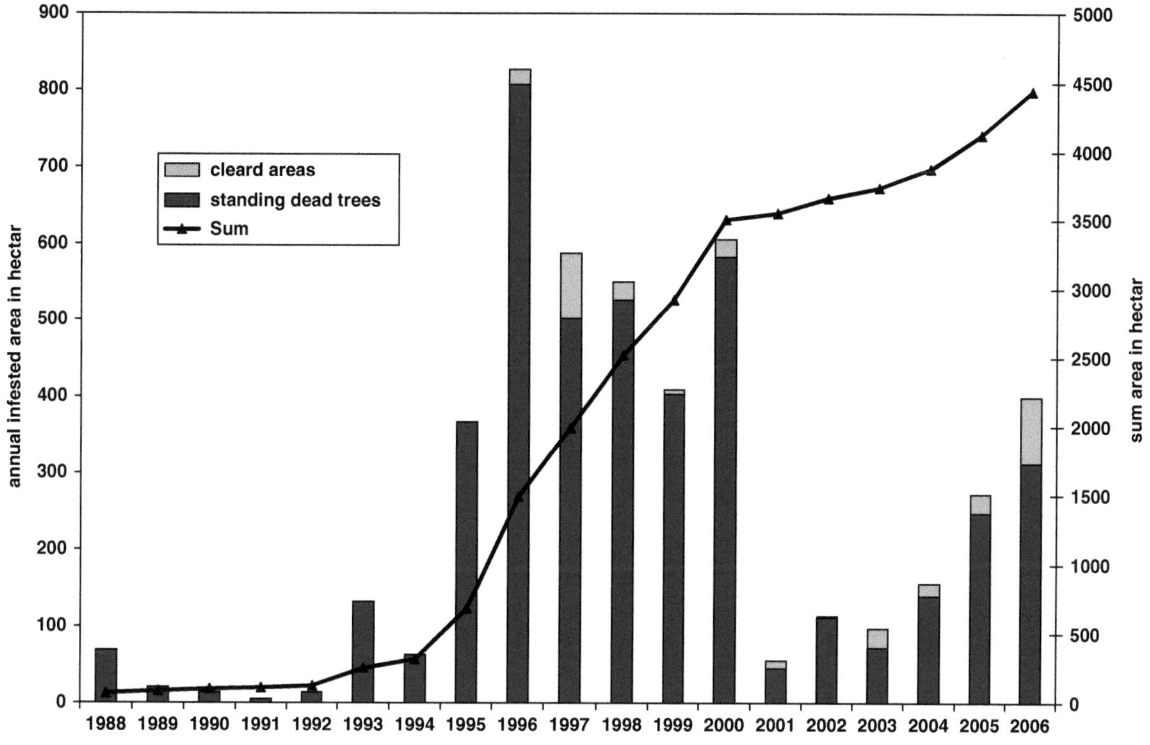

Fig. 23.4 Dead spruce stands in the Rachel–Lusen area of the national park. *Left ordinate*: per annum, *second ordinate*: cumulative

of the total affected area has increased to 4,766 ha. On 324 of these hectares, the spruce bark beetle has been combated in order to protect adjacent, privately owned forests from economic damage (Nüßlein et al., 1999; Heurich & Rall, 2006) (Fig. 23.4).

23.3.2 Causes of the Bark Beetle Outbreak

Since the beginning of forest management in the 19th century, the proportion of spruce has increased to more than 70% at the cost of the other tree species that formerly made up the forest (European beech, Silver fir) (Plochmann, 1961). Since the founding of the national park, living tree biomass has increased by 40% due to reduced harvesting. These changes improved the quantity and quality of bark beetle habitat. The recently observed climate changes, as described below, further help to explain this phenomenon. They show contrasting effects on the vitality of spruce and the dynamics of bark beetle population:

1. The winter storms of 1990, followed by some moderate storms and snow damage (1999, 2000, and 2006), were responsible for high numbers of fallen and damaged trees, which served as initial breeding habitats for the bark beetle epidemic.
2. The strong regional warming in April and May during the last decades resulted in an earlier swarming time of bark beetles (Fig. 23.5).
3. The strong warming during the entire summer season extended the breeding season of the bark beetle and favoured reproductive success, while mortality due to wet and cold weather conditions decreased (Fig. 23.5).
4. Extraordinarily warm and dry summers (1992, 1995, 1998, and 2003) have weakened the vitality of the spruce.
5. The number of years with flowering and fruiting of spruce increased significantly compared to the 1953–1988 period. This has probably been supported by recovery from airborne acidification. But the high carbon costs of reproduction may have weakened the power of resistance against pathogens.

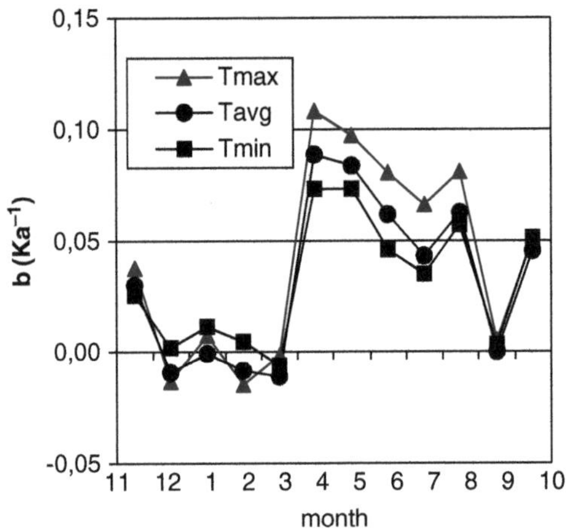

Fig. 23.5 Monthly trend functions (K a^{-1}) of mean air temperature and mean temperature extremes at Waldhäuser meteo station

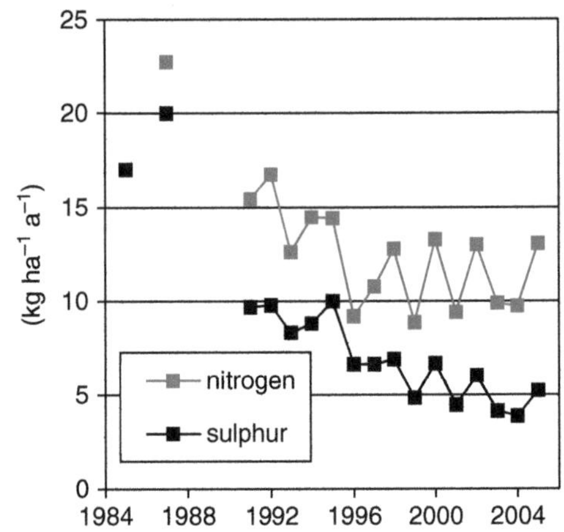

Fig. 23.6 Total deposition of inorganic nitrogen and sulphur (SO_4–S) on beech stand B1 Schachtenau

6. The deposition of air pollutants in the region's forests decreased significantly during the last 15 years (Fig. 23.6). Consequently, the yellowing of spruce needles that was attributed to acidifying gases has disappeared to a large extent. Therefore, one may argue that air pollutants have become less important in impairing tree vitality.
7. Results of project-related research have demonstrated that, according to current knowledge, antagonists, intraspecific competition, and endogenous factors have no significant influence on the population dynamics of the spruce bark beetle.

In summary, weather conditions and climate trends that favour the growth of the bark beetle population are harmful to the population of spruce, which in turn, is adapted to cool and moist conditions.

23.3.3 The Influence of the Spruce Bark Beetle on the Ecosystem

23.3.3.1 Element Fluxes

Stand Level

The effects of bark beetle infestation can be studied in spruce stand F1 (about 110 years old). In 1995, the timber volume over bark was 998 m^3 ha^{-1} and the volume growth increment was 16 m^3 ha a^{-1} (1990–1995). Beech in the third tree layer ($n = 240$) contributed about 28 m^3 ha^{-1} to the total stock volume.

Nitrogen in the stand biomass amounted to 1,303 kg N ha^{-1}, but 89% of the total nitrogen pool is stored in the soil (10,524 kg N ha^{-1}), predominantly in the upper mineral soil layers (0–40 cm depth). The potential net mineralisation of organic N was estimated to about 200 kg N ha^{-1} a^{-1}.

From June 1996 to autumn 1997, more than 99% of the spruce died after a bark beetle attack. By 1999, most of the dead, standing trees had already fallen or broken.

This insect pest has caused dramatic changes in the stand's structure, the micro-climatic conditions, and the relationships between producers and decomposers. Nutrient losses into the groundwater have been induced, and the ecosystem's nutrient pools have been affected.

Changes in the Balance of Nitrogen

Within 2 months after the bark beetle attack, N concentrations began to increase in the organic layer. Between 1997 and 2000, maximum concentrations in seepage water were 21, 43, and 28 mg l^{-1} at 0, 40, and 100 cm depth, respectively.

Between 1992 and 1996, there was an output of 5 ± 2 kg N ha^{-1} a^{-1} through seepage water, while net

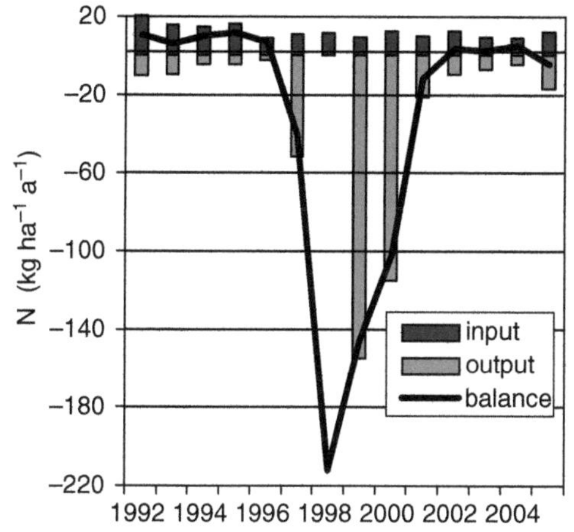

Fig. 23.7 *N* mass balance for the spruce plot F1

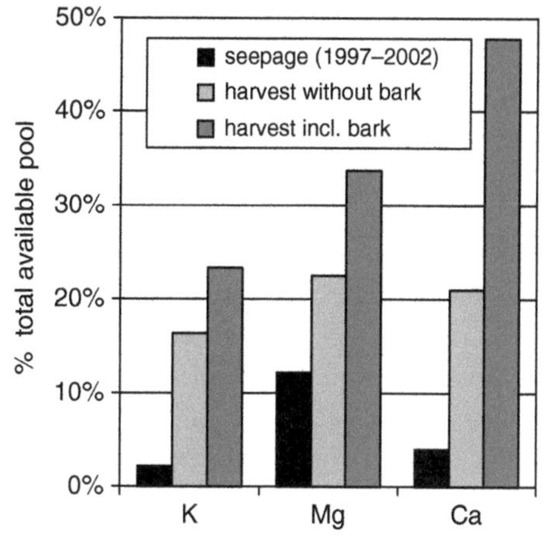

Fig. 23.8 Bark beetle-induced losses of base cations via seepage water (% of available pool) in comparison with their export via hypothetical harvest at spruce plot F1

retention in this spruce ecosystem was 10 kg ha^{-1} a^{-1} (Fig. 23.7). After the bark beetle attack in 1996, N losses increased rapidly to more than 200 kg $ha^{-1}a^{-1}$ in 1998 and then decreased until 2002 (9 kg ha^{-1} a^{-1}). Within these 6 years of excess mineralisation, N losses added up to more than 500 kg ha^{-1}. In contrast, the breakdown of canopy structures minimised interception gains of aerosols and thus the deposition of N to 11 ± 2 kg N ha^{-1} a^{-1}, which equals the rate of wet deposition. These additional losses, exceeding the previous 'normal' export of nitrate, summed up to 494 kg N ha^{-1} (1997–2002), which amounts to about 38% of the N pool in stand biomass and to about 4% of the ecosystem's total reserves.

Changes of Nutritional Base Cations

This 6-year period of excess mineralisation also demonstrated a dramatic increase in losses of nutritional base cations. Mean output rates were 6.4 (K^+), 13.6 (Ca^{2+}) and 7.3 kg ha^{-1} a^{-1}(Mg^{2+}), which is two times (Ca^{2+}) and three times (K^+, Mg^{2+}) higher than it was before the die-back. In order to demonstrate the relevance of these quantities for the nutrition of the following tree generation, the difference in nutrient losses after (1997–2002) and before (1992–1996) the die-back was related to pools available to plants in this system, i.e., the total contents in stand biomass and exchangeable contents in the fine earth fraction of the soil. As shown in Fig. 23.8, K^+ and Ca^{2+} losses were small (2 and 4%, respectively), while Mg^{2+} losses amounted to 12% of the system's reserves, of which the soil comprises only one third.

However, these losses via seepage appear small in comparison to the nutrient export by a hypothetical harvest: harvesting of timber would reduce the pool available to plants by only 16% (K^+), 22% (Mg^{2+}), and 21% (Ca^{2+}). With timber over bark, 23% (K^+), 33% (Mg^{2+}), and 48% (Ca^{2+}) would be removed. Thus, the seepage losses presented above probably do not affect the nutrient supply of the regenerating stand. And, without removal by harvest, additional nutrients will be available from slowly decomposing stand biomass. In managed forests, at sites with similar or even poorer fertility, nutrient losses via seepage of a similar order have to be kept in mind in addition to the removal by harvest, for example, after large-scale storm damage.

Catchment Level

Groundwater and runoff reflect the effects and interactions of changes in deposition load and of altered biochemical processes triggered by the spruce die-back.

Between 1992 and 1998, mean annual runoff of the Forellenbach amounted to 59 (± 3) % of the annual precipitation (1,560 mm). Since 1999, when dead spruce stands accounted for more than 25% of the area, runoff has increased to 68 (± 4) % of the 1,660 mm annual precipitation. This is due to reduced evapotranspiration losses from these sites, estimated at 630 (± 0.07) mm (1992–1998) and, subsequently, at 530 (± 0.13) mm by the difference between precipitation and runoff. Additionally, maximum flood discharge and mean groundwater table levels have increased slightly (Klöcking et al., 2005).

Until 1997, NO_3^- concentrations in runoff water and in groundwater showed a slight decrease, at least partly as a response to decreasing N deposition. Parallel to the shift in runoff ratio, NO_3^- concentrations in runoff and in two out of three groundwater wells increased to 7–8 mg l^{-1}, indicating the import of highly concentrated soil water from disturbed spruce sites. The lower elevation groundwater catchment is apparently the most affected, since NO_3^- concentrations at this hillside have the highest median (20 mg l^{-1}) and the highest maximum value (28 mg l^{-1}). To date, the threshold concentration for drinking water of 50 mg l^{-1}nitrate has not been reached. These findings are of particular importance, because the adjoining municipalities acquire most of their drinking water supply from inside the national park area.

The most important improvement of runoff water chemistry is the reduction of episodic acidification during high flood periods, especially during snow melt. In the beginning of the 1990s, SO_4^{2-} concentrations increased by about 2.5 mg l^{-1} as discharge increased by one order of magnitude. Since 2000, the slope of the concentration–discharge relation has fallen to below 1 mg l^{-1}, indicating the reduction of sulphate pools in the mineral soils, which had been built up during the decades of high immission loads (Fries & Beudert, 2007).

23.3.3.2 Forest Development

Impact on Forest Structures

Windthrows and bark beetle infestation have caused drastic changes in this formerly largely managed forest area. While the wood biomass within the park had steadily increased between 1979 and 1991 from 301 to 416 m^3 ha^{-1}, this was followed by a drastic reduction to 302 m^3 ha^{-1} by the year 2002. The distribution of stem biomass (58% living, 29% dead standing, and 13% dead lying) is now (since 2002) more similar to the distribution in virgin forests. In 1991, 20 years after the establishment of the national park, dead trees accounted for less than 10% of total biomass (456 m^3 ha^{-1}).

Furthermore, the tree species composition, especially in living stands above age 100, which had declined from 6,405 to 3,680 ha, has changed significantly. While spruce decreased from 71.5 to 44.6% of the total tree number, beech increased from 22.9 to 45.3%, and fir increased from 4.3 to 8.1%. In the deadwood areas, spruce is the most common species, comprising 75% of the total. The pioneer tree species, especially Mountain ash, make up 14%. Beech comprise 9%; fir and the valuable hardwood species are less important; they make up only 1.2% and 5%, respectively.

Regeneration of the Mountain Spruce Forest

By 2005, about 91% of the natural spruce stands at high elevations had died. Surveys on these sites showed that the remaining spruce stands consist of young trees (<55 years). However, many young spruce trees also died after the bark beetle attacks, particularly at stand edges. Few old spruce trees survived as single trees or in small clusters. Although the number of living trees remaining is relatively small, theoretical calculations indicated that their seeds could reach at least 66% of the study area.

Because local people voiced their fear that there might be no sufficient natural regeneration, inventories were started in 1996. These inventories showed a successful regeneration of tree species far beyond expectation (Fig. 23.9). In 1991, an average of only 978 trees taller than 20 cm were found per hectare.

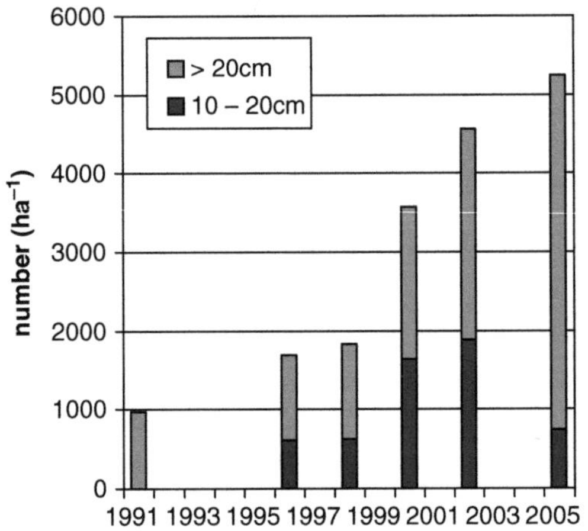

Fig. 23.9 Number of young trees (ha^{-1}) in high-elevation spruce stands

By 2005, this number had significantly increased to 4,502 trees per hectare. If trees between 10 and 20 cm in height are included, the total number of plants per hectare equalled 5,240. Although the major share of the younger trees is still too small to be proclaimed a successful regeneration, the regeneration density corresponds to the target numbers for tree density in managed lowland forests. In the light of these facts, it would appear that the regeneration of high elevation spruce stands is favoured by natural disturbances, such as windthrow and insect pests.

Change in Plant Species Communities

The die-back of single spruce and spruce stands has changed the environmental conditions and, thus, the relative competitive power of the species in the understorey. Analysis of the 2002 forest inventory revealed that the coverage of shrubs, grasses, and herbaceous plants in bark beetle affected plots had increased by 20 and 6% on slope and high-elevation sites, respectively, while the coverage on valley sites had decreased by 11%. On slope sites, all groups of species had benefited from the increase in available light after the spruce die-off. The only species suffering a net decrease in coverage were mosses. This was due to losses in the valleys (−12%) and on higher elevation sites (−15%). These specialists for low radiant intensities under dense spruce cover cannot successfully compete with species groups that are better adapted to light. Especially dwarf shrubs and grasses were the beneficiaries of the altered habitat conditions.

From 1990 to 2000, inventory plots in the Forellenbach area, which were not yet influenced by bark beetle infestation, showed no significant changes in numbers, abundance, and coverage of non-woody species. In contrast, abundance and coverage of *Vaccinium myrtillus, Rubus idaeus, Dryopteris carthusiana, Calamagrostis villosa,* and*Epilobium angustifolium* have increased significantly on heavily affected plots. Despite the fact that pioneer species could profit from spruce die-back, the number of species characteristic of old forest stands has not been diminished.

Despite the very high mineralisation rates at these sites, species with high nitrogen demands could not reach any level of importance. This is demonstrated by the decreasing ELLENBERG N values. Apparently, plant species that are already established possess greater competitive strength. They are able to convert the increased amounts of light and nutrients into additional biomass and reproduce mainly by vegetative means.

23.3.3.3 Species Diversity

Alterations in the Forest Ecosystems

The massive changes in the forest vegetation caused by the spruce bark beetle naturally result in changes in habitat conditions for many species. Important characteristics of the new habitat conditions are the abundance of deadwood, the reduced shading by trees, and the related increase in biomass production on the forest floor. As a result, the microclimate of the area becomes altered: temperatures, especially at ground level, increase; precipitation and radiation increase, relative humidity decreases.

Vascular plants react to such changes with increased coverage and species richness, since species of the forest and species of clearings are then able to exist side by side. The presence of blossoms increases; especially berry-bearing plants are observed to blossom and fruit more intensively.

Small mammals also profit from the dying of the old forest stands and react with increased reproduction rates and population growth. An exemplary investigation detected average densities of 26 animals per hectare in closed stands and 70 animals per hectare in stands of deadwood in the same year. It is noteworthy that, in contrast to cleared areas, the original small mammal population did not collapse; instead, forest species remained dominant. The dominant species of these areas were *Clethrionomys glareolus*, *Apodemus flavicollis*, and *Sorex araneus.* In grassy areas, as opposed to closed forest, *Microtus agrestis* was also found. The occurrence of additional species was restricted to individuals of *Muscardinus avellanarius*, *Glis glis,* and *Sorex minutus* (Huber, 2000).

In a monitoring area within the transition zone between slope and high altitude, the total number of bird species increased from 48 to 73. The population density also increased over the years. This was interrupted only by the population collapse immediately after the die-off of the old stands, which is due to the fact that deadwood stands without ground vegetation are unattractive for many bird species. Among the losers of the changes in habitat conditions are the species of the tree crowns and seed eaters, such as *Loxia curvirostra, Regulus regulus, Parus ater,* and *Parus cristatus*, as well as *Spinus spinus* and *Pyrrhula pyrrhula.* Among the winners are species of the park-like landscapes and the forest steppe, such as *Prunella modularis, Parus major, Sylvia* sp., *Phylloscopus* sp., *Turdus merula,* and *Phoenicurus phoenicurus* (Scherzinger, 1999).

The results of entomological studies indicate that the open areas become habitats for a large number of thermophilic insect species. Relative species richness, as well as the highest number of specialised species (saproxylic beetles, true bugs, Diptera, Hymenoptera, Lepidoptera, Heteroptera), were found in these disturbed habitats; all of these species are missing in closed forest systems. Comparative studies in the adjacent Šumavá National Park revealed that spruce forest climax communities provide habitats for many relic and endemic species of soil Collembola, which can also be found in dead forests but are almost absent on clearings.

Up to now, we have only some episodic knowledge on the influence of the bark beetle on biodiversity, but the recently started monitoring will give us new insights into this exciting field. In conclusion, bark beetles are considered as pest species in managed forests. From an ecological point of view, however, *Ips typographus* (L.) is a keystone species for biodiversity. It assumes this role by killing trees and creating gaps in forests and, thus, structural diversity.

Alterations in Aquatic Ecosystems

In response to hydrochemical recovery, the size of the brown trout population has increased markedly since 2000. In addition, fish biomass has become at least twice as high as before (Fig. 23.10). The biogeochemical effects of spruce die-back seem to interact positively with recovery from airborne acidification: input of litter, increased radiation and water temperature, increased NO_3^-, and (to a smaller extent) base cation concentrations have all helped to significantly increase the food supply for fish.

The recovery of the fish populations has, in turn, had a positive effect on the otter. While in the 1970s and 1980s, otters could only be detected sporadically in the lower elevations of the national park, recent surveys indicate that, since 2004, otters have come to settle all streams up to an altitude of 1,000 m a.s.l. The importance of the recovering fish stock for the otter population is being confirmed through nutrient analysis. While in the 2004 surveys, otter scat contained 91.5% fish remains and only 8.5% remains of amphibians and small mammals, the percentage of fish remains in the 1970s was only less than 50% (Hodl-Rohn & Becker, 1978).

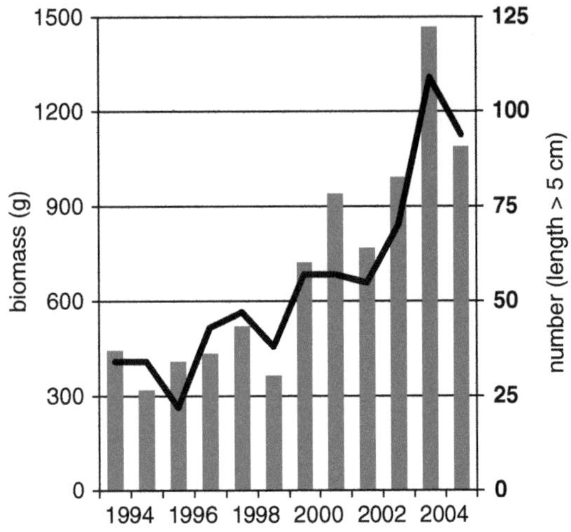

Fig. 23.10 Size and biomass (grams, fresh weight) of brown trout in the Forellenbach

23.3.3.4 Human Dimensions

In the cultural landscape of Central Europe, developments in the forests that are not human induced and that are supported by the willingness to forgo intervention after such events as windthrow, snow damage, and bark beetle calamities lead to completely unaccustomed images and viewpoints. The large area covered with deadwood often comes as a shock to local people and tourists alike. Depending on their personal perception, understanding of nature, life experiences, and character, people who are confronted with this scene will react differently. They may reflect on the experience within their own minds or more openly express their feelings and opinions. In the Bavarian Forest, people have even formed supportive societies to organise public activities on occasion, in which they express their rejection or support for the national park principle: ‘let nature be’. The existence of national parks, which in Germany are exclusively an institution of the federal state, is dependent on the political decision-making process of a democratically elected government. A large protected area, such as the Bavarian Forest National Park, which according to the IUCN guidelines (1994) also has the goal to protect natural processes, is especially dependent in the long-term on the ability of the regional population, as well as tourists, to develop a basic understanding of the natural process of forest regeneration and to accept the idea as a majority. Acceptance at the local level is also promoted when the protected area is attractive for tourists and correspondingly serves to strengthen the regional economy. Therefore, in the sense of a management-directed research approach, socio-economic studies and acceptance analyses have been carried out every several years since the 1970s. The respective results of these resident and visitor monitoring programs serve especially to test and, if necessary, adapt public relations activities, information, and education-related work. Especially noteworthy are the investigations of Kleinhenz (1982) on the importance of the national park for the tourism industry and the studies of Rentsch (1986) on acceptance, in which it was found that the acceptance of the national park increases with increasing distance from it. After the greatest increase in the spruce bark beetle infestation two acceptance analyses among tourists were performed (Suda & Pauli, 1998; Suda, Beck, & Zormaier, 2000; Suda & Feicht, 2002). Both came to the conclusion that the occurrence of deadwood over larger areas would not discourage guests from a future visit to the national park. On the contrary, of 600 visitors surveyed in 2001, 73% completely agreed with the national park statement: 'Let nature be' and 16% agreed somewhat. Only 6% of the questioned vacationers fluctuated between agreement and disagreement; 3% tended more towards non-acceptance, and only 1% could not accept the statement at all. Large changes in the natural characteristics and infrastructure of the national park made it necessary to perform a new survey of the regional economy and acceptance situation. After expansion of the national park (1997) and opening of the 'House of the Wilderness' (Haus zur Wildnis) in August 2006, the University of Munich began an investigation of the regional economic effects of the national park in the Bavarian–Czech border area.

This produced interesting results in regard to opinions on the mass reproduction of the spruce bark beetle: while 32.5% of the local tourism businesses completely agreed with the statement 'the bark beetle development in the national park is detrimental to tourism', only 3.3% of the visitors to the region were of the same opinion. This reveals the high degree of scepticism harboured by the local population towards the national park. The study found that, from an economic view, the advantages to the region can more than be compensated for by the costs of the establishment of the park. Government expenditures for the national park amount to more than 12 million € per year. The national park administration directly employs 200 people and provides indirectly for the equivalent of 939 additional full-time jobs in the tourism industry. Each Euro invested in the national park by the government, is more than doubled by the money spent privately by the visitors. The amount of money spent by visitors exceeds government funding by a factor of 1.31 (Job et al., 2008). These figures illustrate how well economy and ecology can be harmonised.

Since the results of ecosystem dynamics do not always correspond to the human ideal of an intact natural system, in which the individual components exist in a harmonic state of equilibrium, we must learn to accept the unaccustomed, natural changes that occur within ecosystems. Apparently, this learning process has progressed more quickly among the national park's visitors than it has among the local population. Knowledge of the great economic significance of the national park will help to increase local acceptance of the protected area.

23.4 Cross-Border Co-operation

The Šumavá National Park (Bohemian Forest) encompasses an area of 69,030 ha and borders directly on the Bavarian Forest National Park. After man and nature in this area had been separated by the Iron Curtain for decades, it has finally become possible to join together to protect the threatened natural resources.

The years following the opening of the borders showed the development of a close co-operation between the two protected areas, which with a total of 93,280 ha, comprise the largest stringently protected forest area in Central Europe. The border fence, which had separated the regions, was removed, and gradually, natural development was able to proceed without interference. The relationships between the people on both sides of the border intensified as well. The two national park administrations played an important role in the advancement of this process. For example, the possibility for many people to visit the protected area from both sides of the border and to cross the border as well was established. In addition, information signs, displays, and brochures in both parks are bilingual. Many environmental education programmes are designed to reach citizens on both sides of the border.

Collaboration between the national park administrations is supported by a common memorandum that was signed by the respective ministers of the two nations in 1999. At the core of this memorandum is the statement that the two national parks are to be dedicated to the realisation of the goals of national park criteria according to the IUCN. This means that, in large portions of the national parks, natural environmental forces and the uninhibited dynamics of natural communities will be permitted to develop undisturbed by human intervention.

To achieve these goals, intense co-operation in the fields of research, environmental protection, management, public relations, and education is required. For this purpose, the memorandum designated individual task groups, each of which includes German and Czech representatives. The first step involved an exchange of information and data and the opportunity to become acquainted with one another in advance of the regular meetings. An important goal is the compilation of cross-border information on endangered species, which are used as a basis for the common institution of necessary protective measures. One of the main problems in this process was the language barrier, since only few task group members were able to speak the language of the other country. For this reason, the meetings were initially conducted in the presence of an interpreter. However, this soon proved to be tedious and non-conducive towards the goal of true and close collaboration. The use of English as a working language has quickly helped to allay these problems, especially in regard to research.

A good example for cross-border collaboration is the designation of 'natural zones', in which natural processes are allowed to take their course over large areas. The results of the research in both national parks are prepared for presentation to the general public. The experiences from the park on the German side of the border, which was described in this chapter and which is 21 years older, were applied for public relations work in the Šumavá National Park. As a result, it was possible to gradually increase the size of the 'natural zone'. The results of research on the influence of the spruce bark beetle on forest development and the water balance played a significant role in this respect.

Close collaboration was especially important in regard to the bark beetle problem. This involved co-ordination for the designation of the central zones in both national parks. After the storm catastrophe 'Kyrill' in 2007, the two administrations worked in close co-operation. Cross-border agreements were drawn up to determine the areas in which the windthrow was not to be processed. In this way, each administration was able to count on the support of the other. As a result, they were able to argue for the designation of the largest 'natural zones' possible.

There are, however, some problems with the collaboration. These are often due to the specific inventory systems, which are based on traditionally different and, sometimes, specific national methods. Forest inventories, for example, are based on different standards, which are difficult to reconcile. By working together on projects supported by the European Union, some of these problems could be resolved. Data compilation and evaluation could be harmonised. Instead of performing forest inventories using each nation's conventional methods, a uniform remote sensing method was employed in the territories of both countries. In addition, a common survey of habitat types, according to criteria of the European Union, was harmonised for use on both sides of the border.

In the course of the past several years, consciousness for the shared responsibility of the two national park administrations for this unique Central European forest ecosystem has increased. With the introduction of the newly coined term 'Greater Bohemian Forest Ecosystem', it has been made clear that this protected area can only be preserved through international co-operation and the inclusion of the people who live in the area.

References

Bässler, C., & Hahn, C. (2005). Vorstellung des Langzeitprojektes "Bioindikation und Prognosen zur Auswirkung des Klimawandels auf den Nationalpark Bayerischer Wald": Methodik der Arealkartierung – Einrichtung eines Transekts entlang des vertikalen Temperaturgradienten. *Mycologia Bavarica, 8*, 35–42.

Beudert, B., & Breit, W. (2004). *Zwölf Jahre Integrated-Monitoring-Programm an der Meßstelle Forellenbach im Nationalpark Bayerischer Wald.* Grafenau: Nationalparkverwaltung Bayerischer Wald.

Christiansen, E., & Bakke, A. (1988). The spruce bark beetle of Eurasia. In A. A. Berryman (Ed.), *Dynamics of forest insect populations* (pp. 479–503). New York: Plenum Press.

Ehnström, B., Annila, E., Austarå, Ø, Harding, S., & Ottosson, J. G. (1998). *Insect pests in the nordic countries 1982–1986. Rapport fra skogforskningen, Supplement 2.* Ås: Norsk institutt for skogforskning.

Elling, W., Bauer, E., Klemm, G., & Koch, H. (1987). Klima und Böden – Waldstandorte. *Wissenschaftliche Schriftenreihe des Nationalparks Bayerischer Wald, 1*, 1–150.

FEI. (Ed.). (2004). *Manual for Integrated Monitoring, 2004.* Retrieved from http://www.environment.fi/default.asp?contentid=17110&lan=en.

Fischer, A., Abs., C., & Lenz, F. (1990). Natürliche Entwicklung von Waldbeständen nach Sturmwurf. Ansätze einer Urwaldforschung in der Bundesrepublik. *Forstwissenschaftliches Centralblatt, 109*, 309–326.

Fries, H., & Beudert, B. (2007). Report on national ICP INTEGRATED MONITORING activities in Germany. In Kleemola, S. & M. Forsius (Eds.), 16th annual report 2007, *International Cooperative Programme on Integrated Monitoring of Air Pollution Effects on Ecosystems.* The Finish Environment 26/2007, S. 68–81, ISBN 978-952-11-2758-8, Helsinki.

Hauner, U. (1980). Eiszeitliche Formen und Ablagerungen. *Wissenschaftliche Schriftenreihe des Nationalparks Bayerischer Wald, 6*, 1–198.

Heurich, M., & Jehl, H. (2001). Waldentwicklung im Bergwald nach Windwurf und Borkenkäferbefall. *Nationalpark Bayerischer Wald: Wissenschaftliche Reihe, 14*, 1–182.

Heurich, M., & Neufanger, M. (2005). Die Wälder des Nationalparks Bayerischer Wald. *Nationalpark Bayerischer Wald: Wissenschaftliche Reihe, 16*, 1–176.

Heurich, M., & Rall, H. (2006). Hochlageninventur 2005 und Luftbildauswertung 2003 bis 2005. Ergebnisse der Untersuchungen zur Waldentwicklung im Nationalpark Bayerischer Wald. *Berichte aus dem Nationalpark, 3/2006*, 1–21.

Hodl-Rohn, I., & Becker, B. (1978). Fischotter. *Wissenschaftliche Schriftenreihe des Nationalparks Bayerischer Wald, 3*, 1–60.

Job, H., Mayer, M., Woltering, M., Müller, M., Harrer B., & Metzler, D. (2008). Der Nationalpark Bayerischer Wald als regionaler Wirtschaftsfaktor. *Berichte aus dem Nationalpark, 1/2008*, 1–20.

Kleinhenz, G. (1982). *Die fremdenverkehrswirtschaftliche Bedeutung des Nationalparks Bayerischer Wald.* Grafenau: Verein der Freunde des Nationalparks Bayerischer Wald e.V.

Klöcking, B., Schwarze, R., Beudert, B., Suckow, F., Lasch, P., Badeck, F., et al. (2005). Effects of bark beetle infestation on the water and element budgets of two catchments in the Bavarian Forest National Park (in German). *Water cycling and element budgets in the near-natural Große Ohe Catchment* (Vol. 8, 169p). ISSN 0937-0056, Grafenau, Germany.

LfW. (2004). *15 Jahre integriertes Messnetz Stoffeintrag- Grundwasser, Methoden und Ergebnisse, Informationsbericht Nr. 2/2002.* München: Bayerisches Landesamt für Wasserwirtschaft.

Nüßlein, S., Faißt, G., Weißbacher, A., Moritz, K., Zimmermann, L., Bittersohl, J., et al. (1999). *Zur Waldentwicklung im Nationalpark Bayerischer Wald 1999.* Freising: Bayerische Landesanstalt für Wald und Forstwirtschaft.

Petermann, R., & Seibert, P. (1979). Pflanzengesellschaften des Nationalparks Bayerischer Wald. *Wissenschaftliche Schriftenreihe des Nationalparks Bayerischer Wald, 4*, 1–142.

Plochmann, R. (1961). 150 Jahre Waldbau im Staatswaldgebiet zwischen Osser und Dreisessel. *Beihefte zum Forstwissenschaftlichen Centralblatt, 13*, 1–130.

Polevoi, A. V., Scherbakov, A. N., Humala, A. E., & Naldeev, D. F. (2006). *Ips typographus* outbreak as a consequence of extensive windfall in the Vodlozersky national park. [Вспышка короеда-типографа (Ips typographus L.) как одно из последствий массового ветровала в национальном парке "Водлозерский" // Водлозерские чтения: Естественнонаучные и гуманитарные основы природоохранной, научной и просветительской деятельности на охраняемых природных территориях Русского Севера. Материалы науч.-практич. конф., посвященной 15-летию Национального парка "Водлозерский"] In Anonymous, Proceedings of the 15 years anniversary conference of the Vodlozersky national park [Материалы науч.-практич. конф., посвященной 15-летию Национального парка "Водлозерский"] (pp. 96–102). Petrozavodsk.

Rall, H. (1995). Die Wälder im Nationalpark Bayerischer Wald: Von forstwirtschaftlicher Prägung zur natürlichen Entwicklung. In: Nationalparkverwaltung Bayerischer Wald (Ed.), *Nationalpark Bayerischer Wald – 25 Jahre auf dem Weg zum Naturwald* (pp. 9–58). Grafenau: Nationalparkverwaltung Bayerischer Wald.

Rentsch, G. (1986). Die Akzeptanz des Nationalparks Bayerischer Wald durch die lokale Bevölkerung. Diplomarbeit, Technische Universität München.

Schnelle, F., & Volkert, E. (1957). Vorschläge zur Errichtung Internationaler Phänologischer Gärten eines Grundnetzes für internationale phänologische Beobachtungen. *Meteorologische Rundschau, 10*, 130–133.

Sinclair, A. R. E. (1998). Natural regulation of ecosystems in protected areas as ecological baselines. *Wildlife Society Bulletin, 26*, 399–409.

Strobl, R., & Haug, M. (1993). Eine Landschaft wird Nationalpark. Die Geschichte des Waldes und seiner Besiedlung. *Wissenschaftliche Schriftenreihe des Nationalparks Bayerischer Wald, 11*, 1–180.

Suda, M., & Feicht, E. (2002). *Wahrnehmung, Bewertung und Konsequenzen großflächig abgestorbener Bäume im Bereich des Nationalparks Bayerischer Wald aus der Sicht von Touristen. Projektbericht am Lehrstuhl für Forstpolitik und Forstgeschichte der TU München.* München: Technische Universität München.

Suda, M., & Pauli, B. (1998). *Wahrnehmung und Bewertung großflächig abgestorbener Waldbestände im Nationalpark*

Bayerischer Wald. Projektbericht am Lehrstuhl für Forstpolitik und Forstgeschichte der TU München. München: Technische Universität München.

Suda, B., Beck, R., & Zormaier, F. (2000). Making forest restoration politically feasible. In: Hasenauer, H. (Ed.), *Forest ecosystem restoration – Proceedings of the International Conference held in Vienna, Austria 10–12 April 2000* (pp. 264–270). Vienna: Universität für Bodenkultur.

WMO. (1983). *Guide to meteorological instruments and methods of observation* (4th ed.). Geneva: Secretariat of the World Meteorological Organization.

Chapter 24

Turning Long-Term Monitoring into Policy – Using the National Park Schleswig-Holstein Wadden Sea as an Example

Britta Diederichs

Abstract Environmental monitoring in its broadest sense aims at determining the quality and extent of human influence on the environment and to record long-term changes (Kellermann & Riethmüller, 1998). It can provide a basis for political actions; however, the influence of monitoring on political decisions has been controversial (see, e.g., de Jong, 2006).

Using the protection of the National Park Schleswig-Holstein Wadden Sea as a best practice example, this chapter shows the strength of long-term monitoring data as a solid basis for conservation concepts. Along with an efficient organisational structure – i.e. in the frame of the trilateral cooperation between the Netherlands, Denmark and Germany for the protection of the Wadden Sea as a whole – monitoring is able to influence or even induce political decisions. A number of achievements of the trilateral cooperation are presented. Future challenges, which will arise from the implementation of EU directives and international conventions like the Convention on Biological Diversity (CBD), are discussed with a focus on the role of monitoring data within such processes.

Keywords National Park · Monitoring · Data management · TMAP · Schleswig-Holstein

B. Diederichs (✉)
Schleswig-Holstein Agency for Coastal Defence, National Park and Marine Conservation, Dep. 3: National Park and Marine Conservation, Schlossgarten 1, 25832 Tönning, Germany
e-mail: britta.diederichs@lkn.landsh.de

24.1 Monitoring and Research – A Decisive Prerequisite for the Protection of the Wadden Sea

Due to its particular natural qualities, the Wadden Sea of Schleswig-Holstein was declared a National Park in October 1985. The National Park Act passed by the state parliament aims at protecting natural processes. The National Park Act states that 'Natural processes shall be permitted to take place as unimpeded as possible. The National Park shall be conserved as a habitat for the plant and animal species that occur in it naturally, as well as for the relationships occurring between such species and their habitats. Nature in its entirety, including its natural development, and including all plants, animals and ecosystems, has a value in and of itself and must be protected as such.' However, the Wadden Sea is also an area where people live, work and recreate. The mainland areas and islands have been inhabited for centuries, and humans have been working as farmers, fishermen and lately as hosts for tourists. The National Park Act therefore also mandates that 'Unreasonable impairments of the interests and traditional uses of the local population shall be avoided. All usage interests shall be fairly balanced with the protection purpose in general, and shall be fairly balanced in individual cases.'

The development of sound concepts for nature conservation and for environmentally friendly uses in the National Park in Schleswig-Holstein is therefore necessary to enable this balance of interest. Up to now these concepts have been developed under the auspices of the National Park Administration and are based on the results from monitoring and research programmes. The 'Integrated Report on

F. Müller et al. (eds.), *Long-Term Ecological Research*,
DOI 10.1007/978-90-481-8782-9_24,

Ecosystem Research Schleswig-Holstein Wadden Sea' summarises the results from a large-scale research project from 1989 to 1995 and still serves in addition to ongoing monitoring programmes and specific research projects as a basis for conservation concepts and thus management of the protected area.

The need for monitoring and research data as a basis for nature conservation concepts does not only exist in Schleswig-Holstein but in the whole Wadden Sea area. As early as in the beginning of the 1970s environmental scientists stated that the ecosystem of the Wadden Sea cannot be divided according to national borders. The Wadden Sea is, from an ecological point of view, one coherent system. Since 1978, the Netherlands, Denmark and Germany have been working together on the protection and conservation of the Wadden Sea, covering management, monitoring and research, as well as political matters. In 1994 the Trilateral Monitoring and Assessment Program (TMAP) was established 'to continuously evaluate the ecological state of the Wadden Sea as a whole, in order to be able to decide on relevant trilateral policy measures' and 'to assess the status of implementation of the ecological targets' (TMAG, 2000).

These ecological targets cover typical Wadden Sea habitats, species, water and sediments, as well as landscape and culture. The trilateral Wadden Sea Plan (Cwss, 1997) states not only these targets but also common principles and work programmes and provides a strategic focus for the management of the area. This politically adopted target concept is a worldwide unique approach to strategic management of an international shared ecosystem (Moser & Brown, 2007).

24.2 Conservation of Birds as an Example for the Trilateral Procedure

The Wadden Sea ranks among the most important wetlands to migratory waterbirds in the world. Its vast area of intertidal mudflats hosts numerous bird species breeding in the tundra from arctic Canada in the west to northern Siberia in the east. They use the Wadden Sea either as a stop-over site between the arctic breeding areas and the wintering areas in Africa or stay in the area to winter or to moult (Koffijberg et al., 2003). The safeguarding of high-tide roosts is one of the most important tools for the conservation and protection of birds in the Wadden Sea and has thus been recorded as a target in the Wadden Sea Plan. To achieve this target, trilateral projects and actions are defined in the Wadden Sea Plan: 'An inventory of all important and potential roosting sites along the coastline of each country, in conjunction with an evaluation of available knowledge on the necessity for undisturbed roosting sites, in order to investigate the possibilities for creating undisturbed roosting sites.' For this purpose long-term monitoring data from trilaterally coordinated waterbird counts from the Joint Monitoring and Assessment Program (TMAP) from the period 1990 to 2000 were analysed for species which highly depend on high-tide roosts (Koffijberg et al., 2003). The report showed that apart from natural factors, the level of anthropogenic disturbance is one of the most important factors determining numbers observed at high-tide roosts. These disturbances put an extra constraint on the birds' narrow energetic balance and tight time schedule for migration. Case studies in several parts of the Wadden Sea point out that recreational activities are the most important observed sources of disturbance. Although more than 80% of the high-tide roosts are located within protected areas, the report shows that disturbance of roosting birds occurs in all parts of the Wadden Sea.

Conflicts between waterbirds and recreational activities around the birds' roosting sites are primarily expected in May and in July–October due to the fact that both peak in these periods. In order to reduce this conflict Koffijberg et al. (2003) recommended that a spatial and temporal zoning of recreational activities as well as a convincing visitor information system should be further developed.

The above can serve as an example of how monitoring and research results can lead from the formulation of a status quo (Wadden Sea as an important area) over the identification of conflicts (recreational activities disturbing migratory birds at high-tide roost sites at a certain time) to concrete recommendations. It shows that measures (implementation of protected areas and a visitor information system) and an assessment of the outcome are necessary to adjust the actions for protection.

24.3 Structure of the Trilateral Cooperation on the Protection of the Wadden Sea

The Trilateral Governmental Conference (TGC) is the highest decision-making body in the framework of the cooperation. Ministerial conferences are held every 3–4 years. At the conferences the work of the last years is reviewed, discussed and agreements for the next period are adopted. The Quality Status Report (QSR) provides an important basis for decisions at the TGC. It gives a comprehensive overview of the ecosystem, recent developments, the impact of human activities, the implementation status of the ecological targets formulated in the Wadden Sea Plan (Cwss, 1997) and the effects of the policies implemented by the three Wadden Sea countries. Between the conferences the work programme based on the Wadden Sea Plan (Cwss, 1997) and the conference decisions are organised, conducted and assessed by several hierarchically organised working groups of the cooperation (Fig. 24.1). The Schleswig-Holstein National Park Administration and the Ministry for Agriculture, the Environment and Rural Areas (MLUR) are members of several trilateral groups and are responsible for implementing the agreed actions for the protection of the Wadden Sea in Schleswig-Holstein.

As shown by an external evaluation report on the Trilateral Wadden Sea Cooperation this way of cooperation and its organisational structure have been very effective in meeting its original objective of a comprehensive protection of the Wadden Sea (Cwss, 1982; Moser & Brown, 2007). The cooperation is perceived (internationally as well as among its key stakeholders) as an extremely successful, pioneering and world-class model for the protection and management of a transboundary ecological system of international importance. Nevertheless, after 30 years of cooperation the long-term vision, mission and strategy for the cooperation, including programmatic and institutional development, have to be refreshed.

24.4 Present Achievements and Existing Threats

24.4.1 Milestones of the Trilateral Cooperation

Today, almost the entire Wadden Sea area is comprehensively protected under national legislation and the EU Natura 2000 network (Birds and Habitats Directives), in addition to its international designations as Ramsar Site, PSSA (Particularly Sensitive Sea Area)

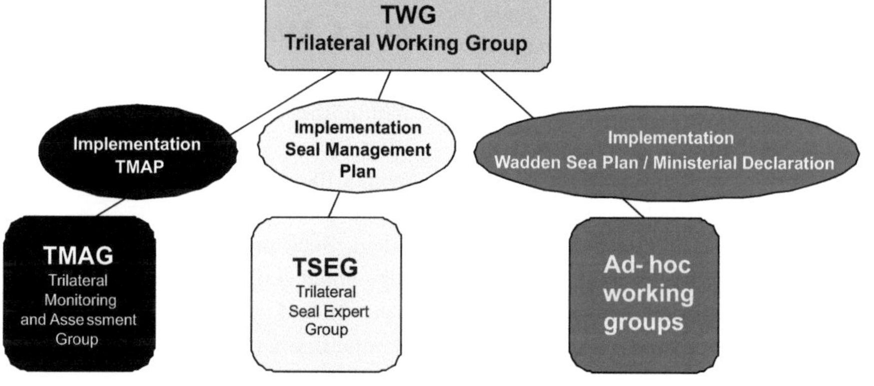

Fig. 24.1 Status quo of the organisational structure of the trilateral cooperation (Source: Reprinted with permission from Harald Marencic, Common Wadden Sea Secretariat (CWSS), http://www.waddensea-secretariat.org)

and UNESCO Biosphere Reserve (except Denmark). The nomination for the Dutch–German Wadden Sea as a Natural World Heritage Site has been submitted to UNESCO in January 2008 (Cwss, 2008). The current level of nature protection of the Wadden Sea could not have been achieved without the work conducted within the trilateral working groups. Some of the benefits resulting from achievements in protection and the established organisational structures are mentioned below.

24.4.2 Benefit for the Environment (Animals, Plants, Ecosystems)

Owing to the protection and the accompanying management activities a lot has been achieved for the ecosystem as a whole and its inhabitants. Although economic activities like commercial fishing and grazing of salt marshes are allowed up to a certain degree in huge parts of the Wadden Sea, in other parts natural processes are almost undisturbed. Within the zero use zone (3% of the National Park area) in the vicinity of the island of Sylt all resource use (including fishing) has been fully prohibited. Further offshore a cetacean protection area (28%) was designated because of the importance as a nursery area for harbour porpoises. Within zone I, which includes the ecologically most valuable areas (about 36% of the National Park area), strict regulations apply including extensive restrictions to public admittance. In general, any activity which could cause destruction, damage or change to the protected area or any part thereof or that could lead to lasting disturbance is prohibited. This means that, e.g., hunting and cockle fishery are completely prohibited within the National Park. The same holds for the erection of wind turbines. All of these management activities aim at protecting not only species and habitats but natural processes in general. Benefits can also be seen on a species level though. Prohibition of hunting in the Wadden Sea area (in the Netherlands in 1962, in Lower Saxony in 1971, in Schleswig-Holstein in 1973 and in Denmark in 1977), e.g., was one precondition for the harbour seal population to increase. While in the 1970s less than 5,000 individuals were counted within the Wadden Sea, the population reached a maximum (since the official counting start) of about 20,250 individuals in 2008 (www.waddensea-secretariat.org). Another example of a species, which benefits from management activities in the Wadden Sea area, is the barnacle goose. While the Russian-Baltic barnacle goose population plummeted to a minimum of 10,000 individuals in the early 1950s due to hunting pressure in Northern Europe and Siberia (Ganter et al., 1999), the trilateral counts (results from 1992 to 2000) show an increase to about 250,000 individuals resting solely in the Wadden Sea area (Blew & Südbeck, 2005), with a current further increase. Flensted (2008) suggests that this population increase is partly due to a better protection of this species and their roosting sites, e.g. better feeding opportunities and fewer disturbances also in the Wadden Sea.

24.4.3 Benefit for Tourists and the Economy of the Region

The 14 million day trippers and 17 million overnight stays recorded at the North Sea coast of Schleswig-Holstein every year demonstrate the importance of tourism for the National Park region. The Wadden Sea has achieved a good reputation as a tourist destination and the National Park is increasingly used as a brand for advertising within the tourist sector. A huge variety of guided tours not only improves the service to visitors but also strengthens the economy of the region. During the last few years specially certificated 'National Park Partners' took over an important role as multiplicators for the idea of the National Park. In this way, nature protection and the strengthening of the local economy can be furthered.

24.4.4 Benefit for Future Generations

Within the trilateral framework several Interreg or other EU co-financed projects were implemented and reports like the Quality Status Report (2004) and the Policy Assessment Report (Cwss, 2006) were published. Such projects improve and spread the knowledge about ecological, economic and cultural connections of and in the Wadden Sea region. Initiatives like the International Wadden Sea School transfer this knowledge to the youth. This helps to preserve this unique ecosystem for future generations.

24.4.5 *Benefit for a Better Exchange of Views*

Within the trilateral cooperation the Wadden Sea Forum was established to compile a broad range of stakeholders' knowledge. Coastal protection and nature protection work closer together than in former times. Nature conservation NGOs are irreplaceable partners on the way to an adequately protected Wadden Sea. Parallel to the bird management the grown cooperation of stakeholders becomes apparent in various ways, e.g. in November 2008 at an international workshop on goose management. The workshop was attended by some 30 representatives from science, agriculture, nature conservation societies, hunters and responsible authorities to discuss best practice solutions for handling geese grazing on farmland. We learned that problems like these can only be addressed adequately through the cooperation of all partners.

Networks like LTER-D function as important platforms bringing together scientists and administrators as well as governmental and non-governmental authorities. These transboundary networks will gain more importance in the future as they will further improve the understanding of observed changes within the Wadden Sea which is an important prerequisite for the protection of this area.

24.4.6 *Threats*

Human activities and external threats like oil exploitation, military tests, fishery, hazardous substances and nutrients, sea level rise and several planned construction activities like power cables for wind farms, river deepening and flood barrages could still pose a threat to this valuable area in Schleswig-Holstein.

In some cases monitoring data can clearly show changes in the population size of a specific species. In order to gain more knowledge on the causal factors which govern population trends in the Wadden Sea area, a strong cooperation of national park agencies with universities and other research institutes is urgently needed.

24.5 Topical Challenges

24.5.1 *EU Directives (HD, WFD) and Adaption of the Monitoring and Assessment Programme*

The implementation of the Water Framework Directive (WFD), becoming operative in 2000, had not only consequences for the European environmental policy but also direct consequences for monitoring and management in the Wadden Sea. Overall the directive aims at protecting all waters (including surface and ground waters) in a holistic way and at achieving a good ecological status by the year 2015. For that matter monitoring is an important instrument (as is also the case within the trilateral cooperation for the protection of the Wadden Sea) and is therefore mandatory. Monitoring programmes are required to establish a coherent and comprehensive overview of the status of water bodies (surveillance monitoring). If the monitoring results reveal moderate or even worse, water quality measures have to be taken for improvement. The success of the measures has to be controlled by a so-called operational monitoring. This is a good example of how monitoring is used as an instrument for nature policy at the EU level.

The same is true for the Habitats Directive (HD) and the Birds Directive (BD). For both directives long-term monitoring data provided a fundamental basis for the designation of protected areas like the 'Ramsar Site Schleswig-Holstein Wadden Sea and adjoining coastal areas' as well as the HD site at the Schleswig-Holstein North Sea coast 'National Park Schleswig-Holstein Wadden Sea and adjoining coastal areas'. Moreover monitoring is required to determine whether a 'Favourable Conservation Status' is achieved or management measures have to be carried out.

Following the EU directives, the establishment of a sufficient monitoring and assessment programme in the Schleswig-Holstein part of the Wadden Sea requires much less effort than in other regions due to monitoring and reporting already conducted at a high quality level not only in Schleswig-Holstein but in a cooperative manner within the whole Wadden Sea area (Fels, 2001; TMAG, 2001). As part of the trilateral cooperation 25 parameters (like nutrients, contamination in blue mussels, salt marshes, breeding and migratory birds, land use, eelgrass and weather) are monitored (TMAP Common Package) and assessed, spanning a broad range from physiological processes over population development to changes in landscape and morphology, to get a holistic view of the status of the Wadden Sea. Analyses showed that the existing TMAP Common Package covers the monitoring requirements under the EU directives to a large extent, e.g. information on the main relevant quality elements of the WFD, or information on Annex I species of the Birds Directive, but can offer only limited information on subtidal habitats and fish communities (TMAG, 1997, 2001). For instance, new monitoring strategies have to be developed for subtidal habitats. Also, 'breeding success' is proposed to be included in the revised TMAP as a new parameter to assess the target of a 'natural breeding success'. This 'early warning' aspect will complement the existing bird monitoring programme and will thus allow for a better and integrated assessment of the status of breeding bird populations. But despite this the existing monitoring programmes already comply with the requirements of the EU directives and other international agreements to a large extent.

Quantitative aspects for the assessment required by the WFD do not exist in the Wadden Sea Plan so far. Therefore several trilateral ad hoc groups started their work in 2005 in order to develop quantitative assessment tools. These assessment tools rely on sound long-term monitoring data. Two examples of assessment tools developed by the trilateral working groups are presented below.

As eelgrass is a biological quality component of the WFD as well as an integral component of habitat type mudflats of the HD, great efforts have been made to elaborate a first classification of the actual water quality for the Schleswig-Holstein Wadden Sea on the basis of existing long-term data on the development of eelgrass (Figs. 24.2 and 24.3) (Schanz & Reise, 2007).

The trilateral bird expert group developed an alert system for bird population changes. This helps to detect threats as early as possible. The first evaluation

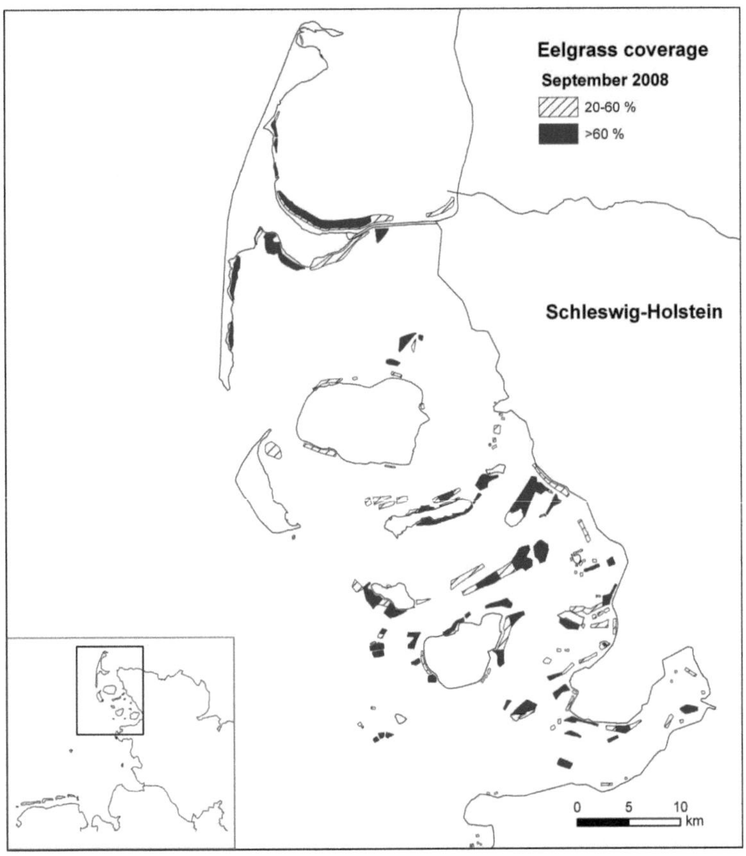

Fig. 24.2 Eelgrass distribution in the North Frisian Wadden Sea (Source: Reise, 2008, map source: LKN – National Park Administration, edited by Diederichs) – a basis for assessment in terms of WFD and HD

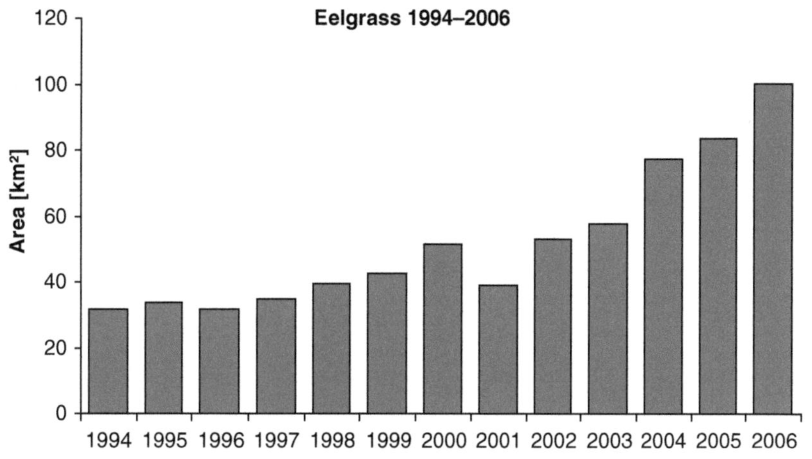

Fig. 24.3 Area [km^2] of seagrass beds (>20% coverage) recorded by aerial surveys in the North Frisian Wadden Sea in August 1994–2006 (Source: Reise & Kohlus, 2007, edited by Diederichs)

of migratory bird data gathered in the period of 1980–2000 in the framework of TMAP alerted the scientists. Of the 34 species, for which the Wadden Sea is an essential stepping stone on their migration route, about 40% showed significant declining trends. After analysing the newest data a slight improvement of the situation was observed. But negative trends were still detected for a few species, amongst which are those feeding on mussels such as the oystercatcher (Reineking & Südbeck, 2007). Alarmed by these results the National Park Administration of the Lower Saxony Wadden Sea, the Common Wadden Sea Secretariat and the Institute of Avian Research 'Vogelwarte Helgoland', Wilhelmshaven organised an international workshop in Wilhelmshaven, Germany, in August 2006 to discuss causes and consequences of seriously declining trends in migratory waterbirds in the Wadden Sea as well as to formulate future ecological research needs and necessary management measures (Reineking & Südbeck, 2007). This example shows both the need for long-term monitoring data and for new analysis and assessment methods to judge population trends. As shown in Section 24.2 observations are the first step on the way to concrete management measures.

The development of such assessment tools is a big challenge which is complicated by the fact that an intercalibration with the management of freshwater areas and their typical methods needs to be done. The revision of the TMAP and the Wadden Sea Plan which has already begun is thus necessary and will result in a revised TMAP taking into account the requirements not only of the EU directives (Birds, Habitats, Water Framework) but also of the Trilateral Cooperation (Wadden Sea Plan) and other international obligations (e.g. OSPAR, CBD, CMS).

Despite these difficulties, the implementation of the EU directives also has potential and offers advantages. Schuchardt and Scholle (2007) underline the potential of precise WFD targets and their surveillance as an important benefit for a holistic ecological assessment of the Wadden Sea in the context of the Quality Status Report. Using 29 ecological targets for 12 issues the assessment within the scope of the QSR 2004 already gives a holistic view of the ecological status of the Wadden Sea. In some cases target evaluation was not possible due to the lack of proper data and quantitative criteria though. These shortcomings have to be amended in the TMAP revision.

24.5.2 Data Handling Requirements by EU Directives – A Modern Data Information System – Essential for Effective Data Transfer

Not only has the implementation of the WFD and other EU directives like INSPIRE (Infrastructure for Spatial Information in the European Community, EU directive 2007/2/EC) raised the standard for monitoring and assessment, but also new needs for data handling have been added. Granting an efficient data and information transfer between the participant stakeholders of the WFD is one of the greatest challenges. Moreover the development of a digital reporting system corresponding to EU guidelines is required to

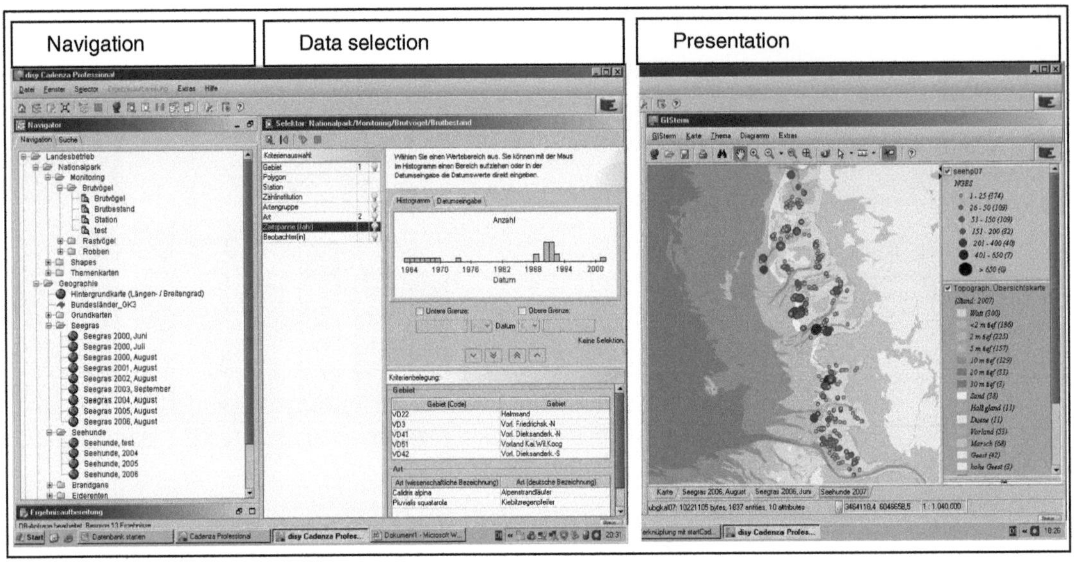

Fig. 24.4 Screenshot of the National Park database – front end disy Cadenza. *Left:* Navigation of available data. *Middle:* Criteria-based data selection (selection of time, location, etc.). *Right:* Presentation of selected data in form of maps, charts or diagrams

allow data access to decision-makers as well as to the public. Only in this way a close cooperation with local stakeholders and participation of organisations is guaranteed.

To fulfil the reporting requirements the National Park Administration decided to implement the reporting software Cadenza in 2006. This software provides a wide range of functionalities including interactive design features for table assessments, diagrams and map presentations (Fig. 24.4). The system allows experts to prepare information and assessments by grouping them into tasks and providing decision-makers, specialists and occasional users with relevant information via intranet and Internet. A test phase showed that this tool is able to fulfil the above-mentioned challenges. We suppose that a successful implementation of Cadenza will make a considerable contribution to the political decision-making process based on monitoring data.

24.5.3 EU Directives and Management Difficulties – Process Protection Versus Species Protection – How to Find Priorities?

At the 1992 Earth Summit in Rio de Janeiro world leaders agreed on a comprehensive strategy for 'sustainable development'. The goal is to meet our needs while ensuring that we leave a healthy and viable world for future generations. One of the key agreements adopted at Rio was the Convention on Biological Diversity (CBD).

The CBD obliges the member states to develop national strategies, plans and programmes for the maintenance and sustainable use of biological diversity or to adjust existing strategies to conserve the biological diversity. In Germany it is expected that the network of Natura 2000 sites – of which the Wadden Sea Area is part – will contribute the biggest share to the country's obligations for the maintenance of biological diversity.

Though in general the EU Habitats Directive, the National Park law and the Convention on Biological Diversity aim at the same – i.e. biological diversity – the implementation and management of protected sites in line with the above-mentioned directives and law causes priority conflicts. Management in National Parks in Germany follows the principle 'let nature on its own devices'. Human impacts have to be minimised. To maintain a favourable conservation status of a single species as required under Natura 2000 this might not always be enough and actions then have to be taken. In some cases endangered species might be protected without disturbing natural processes (e.g. by identifying protection zones for rare bird species breeding on beaches). In other cases other solutions have to be found though. For instance, a reduced number and area of sandbanks due to sea level rise might in the future cause problems for birds breeding in this habitat (e.g. Little tern). The creation of artificial sandbanks might be beneficial for the protection of those species. This action would be in contradiction to the hands-off principle though. In this case two goals, namely species and process protection compete and managers have to decide on the best solution in each case.

We should keep in mind though that in general 'Ecosystem functioning and resilience depends on a dynamic relationship within species, among species and between species and their abiotic environment as well as physical and chemical interactions within the environment. The conservation of these interactions and processes is of greater significance for the long-term maintenance of biological diversity than simple protection of species' (Malawi Principle 5, UNEP, 1998).

24.6 Conclusions

The protection of the Wadden Sea on an international cross-border level shows that several steps have to be taken to save or even improve the status of this unique ecosystem. Monitoring and research along with an efficient data handling system are needed to enhance the knowledge about species, habitats and processes. Assessment based on monitoring data and research leads to recommendations for a better management. Agreements at the Trilateral Governmental Conferences and good public relations assure the implementation of these recommendations and the public acceptance. The chapter shows that managing the transboundary Wadden Sea area adequately also means to face new challenges like implementing the EU directives and a functioning data handling system on the national and international level.

The evaluation of the Trilateral Cooperation in 2007 has shown that the existing organisational structures are very effective in meeting the Trilateral Cooperation's original objective from 1982 of a comprehensive protection of the Wadden Sea. Many aspects of its work like the politically adopted targets and the Wadden Sea Plan are world-class in quality (Moser & Brown, 2007). Nevertheless, after 30 years Moser and Brown (2007) recommend some changes and adaptions, e.g. revision of the definition and refreshment of the long-term vision, mission and strategy of the cooperation, including both programmatic and institutional development. If the implementation of these recommendations succeeds the cooperation will continue to be one of the leading global models for transboundary protected area management. However, in order to improve the ecological status of the Wadden Sea a lot still has to be done. The summary of the target evaluation of the QSR (2005) shows that only 4 out of 29 targets have been achieved yet. Another eight targets have not been met, but show a positive development. Partners and networks are urgently needed to reach these targets.

Acknowledgments Thanks to the CWSS for their excellent homepage and especially Harald Marencic for providing Fig. 24.1. Jörn Kohlus supplied eelgrass data from the Trilateral Monitoring and Assessment Program (TMAP). Karsten Laursen, Dörte Poszig, Dr. Klaus Koßmagk-Stephan and Kai Eskildsen made helpful suggestions to improve the chapter.

References

Blew, J., & Südbeck, P. (Eds.). (2005). *Migratory waterbirds in the Wadden Sea 1980–2000. Wadden Sea Ecosystem No. 20.* Wilhelmshaven: Common Wadden Sea Secretariat, Trilateral Monitoring and Assessment Group, Joint Monitoring Group of Migratory Birds in the Wadden Sea.

Common Wadden Sea Secretariat. (Ed.). (1982). *Copenhagen Declaration – Trilateral Wadden Sea Plan. Ministerial Declaration of the Eighth Trilateral Governmental Conference on the Protection of the Wadden Sea.* Wilhelmshaven: Common Wadden Sea Secretariat.

Common Wadden Sea Secretariat. (Ed.). (1997). *State Declaration – Trilateral Wadden Sea Plan. Ministerial Declaration of the Eighth Trilateral Governmental Conference on the Protection of the Wadden Sea.* Wilhelmshaven: Common Wadden Sea Secretariat.

Common Wadden Sea Secretariat. (Ed.). (2006). *Schiermonnigkoog Declaration. Ministerial Declaration of the Tenth Trilateral Governmental Conference on the Protection of the Wadden Sea.* Wilhelmshaven: Common Wadden Sea Secretariat.

Common Wadden Sea Secretariat. (Ed.). (2008). *Nomination of the Dutch-German Wadden Sea as World Heritage Site – Volume one.* Wilhelmshaven: Common Wadden Sea Secretariat.

De Jong, F. (2006). *Marine eutrophication in perspective – On the relevance of ecology for environmental policy.* Berlin: Springer.

Essink, K., Dettmann, C., Farke, H., Laursen, K., Lüerßen, G., Marencic, H., et al. (Eds.). (2005). *Wadden sea quality status report 2004. Wadden Sea Ecosystem No. 19.* Wilhelmshaven: Common Wadden Sea Secretariat.

Fels, P. (2001). *Implementation of the EU habitat directive for monitoring and reporting in the Wadden sea area. Report.* Utrecht: Universiteit Utrecht.

Flensted, K. (2008). *Goose management – opportunities and conflicts. Statement at the Workshop Goose Management in Ribe, 19–20 – November 2008. Workshop Report.* Retrieved from http://www.waddensea-secretariat.org/news/news/goose-management-2008.html.

Ganter, B., Larsson, K., Syroechkovsky, E.V., Litvin, K. E., Leito, A., & Madsen, J. (1999). Barnacle goose *Branta leucopsis*: Russian and Baltic populations. In J. Madsen, T. Fox, & J. Cracknell (Eds.), *Goose populations of the western palearctic. Wetlands International Publ no. 48* (pp. 270–283). Rönde, Denmark: National Environmental Research Institute.

Kellermann, A., & Riethmüller, R. (1998). Umweltbeobachtung im Wattenmeer. In NPA & UBA (Eds.), *Umweltatlas Wattenmeer. Bd. 1. Nordfriesisches und Dithmarscher Wattenmeer* (pp. 1–8). Stuttgart: Ulmer.

Koffijberg, K., Blew, J., Eskildsen, K., Günther, K., Koks, B., Laursen, K., et al. (2003). *High tide roosts in the Wadden Sea: A review of bird distribution, protection regimes and potential sources of anthropogenic disturbance. A report of the Wadden Sea Plan Project 34. Wadden Sea Ecosystems No. 16.* Wilhelmshaven: Common Wadden Sea Secretariat, Trilateral Monitoring and Assessment Group, JMMB.

Moser, M., & Brown, A. (2007). *Trilateral Wadden Sea cooperation external evaluation report.* Retrieved from http://www.waddensea-secretariat.org/news/Evaluation/Wadden%20Sea%20Evaluation%20Report%20Final070621.pdf.

Reineking, B., & Südbeck, P. (Eds.). (2007). *Seriously declining trends in migratory waterbirds: causes-concerns-consequences. Proceedings of the International Workshop on 31 August 2006 in Wilhelmshaven, Germany. Wadden Sea Ecosystem No. 23.* Wilhelmshaven: Common Wadden Sea Secretariat, Wadden Sea National Park of Lower Saxony, Institute of Avian Research, Joint Monitoring Group of Migratory Birds in the Wadden Sea.

Reise, K., & J. Kohlus (2007). Seagrass recovery in the Northern Wadden Sea? *Helgoland Marine Research, 62*, 77–84.

Schanz, A., & Reise, K. (2007). *Referenz und Klassifizierungsansatz für die Makrophytenvegetation des Nordfriesischen Wattenmeeres, gemäß der Vorgaben der EG-Wasserrahmenrichtlinie (EG-WRRL). Report im Auftrag des Landesamtes für Natur und Umwelt des Landes Schleswig-Holstein, Flintbek 08/2006.* List: Alfred-Wegener Institut für Polarforschung, Wattenmeerstation Sylt.

Schuchardt, B., & Scholle, J. (2007). *Referenzbedingungen der biologischen Qualitätskomponenten nach EU-Wasserrahmenrichtlinie in den Küsten- und Übergangsgewässern in NL, GER und DK – Stand der Entwicklung und Vergleich mit den Zielen des trilateralen Wattenmeerplans. Bericht im Auftrag des Common Wadden Sea Secretariat.* Bremen: Bioconsult Schuchardt & Scholle GbR.

TMAG. (1997). *Implementation of the Trilateral Monitoring and Assessment Program (TMAP). Final Report December 1997.* Wilhelmshaven: Common Wadden Sea Secretariat.

TMAG. (2000). *The Trilateral Monitoring and Assessment Program (Updated TMAP Manual from June 2000).* Wilhelmshaven: Common Wadden Sea Secretariat.

TMAG. (2001). *TMAP Evaluation Report. Final draft, February 2001.* Wilhelmshaven: Common Wadden Sea Secretariat.

UNEP. (1998). *Convention on biological diversity & report of the workshop on the ecosystem approach.* (UNEP/CBD/COP/4/Inf.9). Retrieved from http://www.cbd.int/doc/meetings/cop/cop-04/information/cop-04-inf-09-en.pdf.

Chapter 25

Design and Importance of Multi-tiered Ecological Monitoring Networks

K. Bruce Jones, Heye Bogena, Harry Vereecken, and Jake F. Weltzin

Abstract Multi-scaled ecological monitoring networks offer significant potential to address a wide range of challenging environmental problems. Knowledge gained through these networks will be critical in understanding, detecting, and forecasting ecological changes that affect important ecological services upon which society depends. The networks will provide information necessary for societies to adapt to broad-scale changes such as those associated with land use, demographic, and climate change. Several new multi-tiered monitoring programs are being developed to evaluate ecological changes and associated drivers of change at a range of spatial and temporal scales. Additionally, existing ecological monitoring programs, such as the Long-Term Ecological Research (LTER) program, are attempting to improve their capacities to extrapolate results to larger spatial extents by developing a standard set of measures or indicators and by facilitating cooperation among scientists within and among the various monitoring networks. Despite these attempts, several issues remain in integrating existing monitoring programs. We discuss these issues and review existing programs within a multi-tiered monitoring framework that explicitly incorporates citizen-based monitoring. Direct involvement of the public is seen as a critical element in expanding and maintaining existing and new ecological monitoring networks. We provide examples of two emerging ecological monitoring networks, the Terrestrial Observatory Network (TERENO) and the USA National Phenology Network (USA-NPN), to convey some of the complexities and challenges confronting the design and implementation of multi-tiered ecological monitoring networks.

Keywords Ecological monitoring · Multi-tiered monitoring networks · Monitoring design · Ecological indicators · Monitoring networks

25.1 Introduction

The 20th century has witnessed an unprecedented influence of humankind on fundamental environmental and physical processes that sustain life on earth (EEA, 2004; McCarthy, Canziani, Leary, Dokken, & White, 2001). Through modifications of the earth's surface and global biogeochemical cycles, humankind has increased the pace and amount of productive ecosystems lost to desertification processes, reduced availability and reliability of clean and abundant sources of water and food, increased risks to natural hazards such as floods and fires, and increased disease and exposure to harmful chemicals through environmental modifications and release of harmful chemicals into the environment (Houghton, 1994; Ojima, Galvin, & Turner, 1994; Bellamy et al., 2005; Young & Harris, 2005). We are also starting to observe broad-scale ecological impacts of global climate change, including changes in the timing of ecological processes (Westerling, Hidalgo, Cayan, & Swetnam, 2006) and decline of widely distributed species such as the polar bear (Stirling & Derocher, 2007). The specific human influences and drivers of these observed changes are complex and multi-scaled in nature (Millennium Ecosystem Assessment, 2005).

K.B. Jones (✉)
U.S. Geological Survey, 12201 Sunrise Valley Drive, Reston, VA, USA
e-mail: kbjones@usgs.gov

F. Müller et al. (eds.), *Long-Term Ecological Research*,
DOI 10.1007/978-90-481-8782-9_25,

There has been an increased recognition of the role that multi-tiered, long-term ecological monitoring programs can play in understanding complex relationships between humans and environment and in developing policies and strategies to reduce impacts and secure a more sustainable future (Parr, Sier, Battarbee, Mackay, & Burgess, 2003). Multi-tiered monitoring is needed to provide early warning and forecasting of potentially irreversible conditions and trends in processes that maintain important ecological services upon which humankind depends (Clark et al., 2001). As such, it is critical to understand scaling functions and complex feedback loops and time lags, and how these functions and relationships vary within and among different biophysical and socio-ecological systems (Carpenter et al., 2006). Multi-tiered ecological research and monitoring frameworks also are needed to determine relationships between natural and anthropogenic factors, and status and trends in fundamentally important ecological processes (McDonald et al., 2002). Finally, long-term ecological monitoring is needed to evaluate alternative future management scenarios (Baker et al., 2004) and to monitor ecological responses to different environmental management prescriptions and policies within an adaptive management framework (Christensen et al., 1996; Bormann, Haynes, & Martin, 2007).

Comprehensive, multi-tiered monitoring programs provide critical information needed in broad-scale, periodic environmental assessments, such as European environmental assessments (EEA, 2007a), the Australian Report of the Environment (Beeton et al., 2006), the US Environmental Protection Agency's Report on the Environment (EPA, 2003), the H. John Heinz Center III State of the Nation's Ecosystems Report (Heinz, 2008), and the Millennium Ecosystem Assessment (Millennium Ecosystem Assessment, 2005). However, most of these assessments have concluded that current monitoring data and networks are insufficient to address current and emerging environmental problems.

Currently, we lack coordinated, multi-tiered ecological monitoring networks on national scales, let alone continental or global scales, to understand complex ecological processes and upon which to base environmental management and policy decisions. Moreover, we lack coordinated monitoring programs that provide integrated data across multiple ecological systems at appropriate temporal and spatial scales. Existing research and monitoring networks tend to address specific issues, and as a result, represent different geographies, temporal and spatial scales, and attributes of the environment (CENR, 1997). Moreover, a critical need exists to synthesize biophysical, ecological, social, and economic information to increase our understanding of the significance of interactions among processes that affect ecological systems. New developments in science and technology provide new opportunities for collecting and organizing data that could greatly expand our monitoring capabilities.

In this chapter, we discuss existing and emerging ecological monitoring programs within the context of a multi-tiered monitoring framework. We describe the importance of each tier in the framework and synergy gained through explicit linkages between tiers. We also describe and discuss some of the challenges and opportunities related to implementing and sustaining long-term ecological monitoring, including ecological indicators. Finally, we describe in more detail two contrasting networks, the multi-tiered network TERENO and the four-tiered network USA-NPN.

25.2 Multi-tiered Ecological Monitoring

Ecological research and monitoring occur at scales ranging from a few meters to vast areas across the globe. Most in situ research and monitoring are limited to specific areas for specific purposes, e.g., to understand biogeochemical fluxes within a small watershed (Lins, 1994). Ecological research is often limited in scope and spatial extent, especially when it involves experimental manipulation. In situ ecological monitoring beyond specific geographic areas is limited by costs and consistency. However, many of our most pressing questions and environmental challenges can only be solved by understanding processes and interrelationships across many temporal and spatial scales. As such, many recent reviews of existing monitoring programs have stressed the need for multi-tiered monitoring.

The Committee on the Environment and Natural Resources has suggested implementation of a three-tiered approach to ecological research and monitoring, including a tier consisting of intensive research sites, a tier consisting of a large number of widely distributed in situ monitoring sites, and a tier consisting of spatially extensive remote sensing data (Fig. 25.1, CENR,

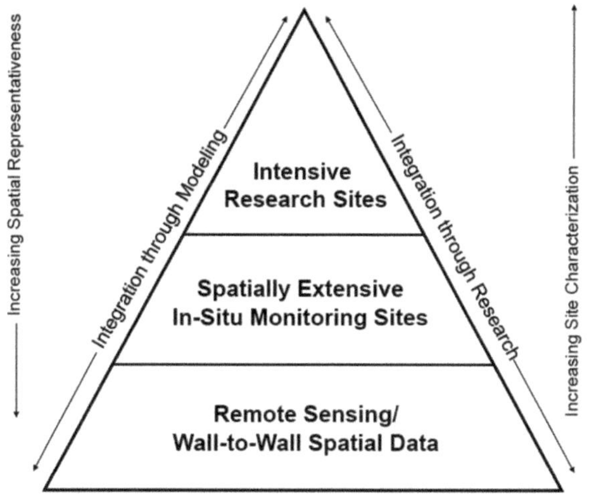

Fig. 25.1 The Committee on Natural Resources and the Environment (CENR) multi-tiered monitoring framework (CENR, 1997). The intensive research site tier includes programs such as the Long-Term Ecological Research (LTER) program (Kratz et al., 2003), the spatially extensive in situ monitoring tier programs such as the Forest Inventory and Analysis (FIA) program (Bechtold et al., 2005), and the remote sensing tier programs such as the National Land Cover Database (Homer et al., 2004)

1997). Successful implementation of this kind of monitoring system requires implementation of a core set of similar or comparable measurements within each tier, and measurements that can be linked quantitatively across tiers (CENR, 1997, see later discussion). The USA National Phenology Network (USA-NPN), a program launched in 2006, has added a fourth tier into the multi-tiered network: a citizen-based program intended to extend the spatial coverage of the broadly distributed in situ monitoring network (Betancourt et al., 2007). Similarly, a citizen-based monitoring approach has been the foundation of the European Phenology Network (van Vliet et al., 2003).

Multi-tiered research and monitoring programs potentially provide benefits above and beyond benefits derived from individual programs or tiers, including

- An understanding of important synchronies among ecological characteristics and processes across space and time, and analysis of changes in the timing and synchrony of important biological, ecological, and hydrologic relationships;
- Cross-scale analysis of ecologically important attributes and processes and their relationships;
- Evaluation of cascading effects of natural and anthropogenic drivers and stressors across scales, as well as the magnitude of ecological change and lag times;
- Detection and evaluation of ecological thresholds and tipping points (for forecasting);
- Detection of surprises in ecological processes and how they cascade across spatial and temporal scales;
- Early warning of ecological process changes that affect important ecological services;
- How scaling functions and importance of variables in predicting ecological conditions and responses vary within and among biophysical settings;

Below we discuss each of the four monitoring tiers and give some examples of programs that represent each tier. Characterization of existing monitoring programs into tiers is imperfect but serves the purpose of describing each tier.

25.2.1 Description of Tiers

25.2.1.1 Intensive Research Sites

Intensive research sites are the foundation for ecological research in multi-tiered monitoring networks. Research is conducted at scales suitable for intensive data collection on ecological processes (ranging from a few square meters to the size of small watersheds), including fluxes and flows of biota, energy, water, and materials and nutrients. Often, researchers are involved with experimental manipulations or gradient analysis to test hypotheses regarding ecological state and process variables, and indicators (Gosz, 1992; Peterson & Waring, 1994; Müller, 1998; Dunne, Saleska, Fischer, & Harte, 2004). Many of these sites have relatively long-term data sets which permit research on lag times, community interactions, and cyclical phenomena in ecosystems. They also permit analysis of thresholds or tipping points and horizontal landscape interactions (e.g., surface flows and fluxes of materials, nutrients, energy, and biota). Intensive research sites also permit analysis of response of ecosystems to human intervention and management. Finally, they provide valuable information on species traits and functions and on changes in soil, vegetation, and atmospheric interactions that cannot be measured over more extensive monitoring networks or via remote sensing. However, recent advances in remote sensing and mapping of detailed biophysical data now make it possible to extrapolate some functions and processes over broader areas (Asner, Scurlock, & Hicke, 2003).

Most intensive research sites are not part of a monitoring network per se. Typically, they are associated with federal science agencies, local conservation programs, or universities and only have loose affiliations to broader programs. However, several important intensive research site networks provide important biophysical and ecological data over diverse landscapes. Fluxnet is a global network of some 200 sites (42 countries) that are instrumented with sensors on towers that measure fluxes of gases such as carbon and nitrogen, as well as energy and water vapor (Falge et al., 2002). Ameriflux is a regional network (Western Hemisphere but mostly in the USA) of about 100 sites that measure gas exchange, as well as expanded studies and monitoring of vegetation and canopy attributes (Hargrove, Hoffman, & Law, 2003).

Perhaps the best-known intensive research site network is the Long-Term Ecological Research (LTER) program (Kratz, Deegan, Harmon, & Lauenroth, 2003). The US LTER Network is funded primarily by the US National Science Foundation, and is comprised of 26 sites that represent different biophysical settings and ecoregions (including two urban regions in the states of Arizona and Maryland). The network was established in 1980, although many of the sites added at that time already had previous research and monitoring activity, which provides an opportunity to extend analyses back several decades. The variety and number of biophysical and ecological attributes and processes measured at these sites far exceeds those measured by any other intensive research network, although there is considerable variability in what is measured at LTER sites. Despite being investigator driven, the network has attained some measure of standardization and data networking that has resulted in an understanding of changes in ecosystem characteristics and processes at regional scales (Rastetter et al., 2003). Extrapolation of LTER research results to broader scales has been facilitated by development of spatially extensive biophysical data and standardization of measurements. Within Europe, several countries are developing LTER networks, including Germany (LTER-D, http://www.lter-d.ufz.de/). Additionally, an international LTER program (ILTER, http://www.ilternet.edu/) consisting of multi-country networks has been established in an attempt to improve collaboration, data exchange, and broader-scale experimental designs.

The National Ecological Observatory Network (NEON, 2006) is a continental-scale research platform for discovering and understanding the impacts of climate change, land-use change, and invasive species on ecology (Senkowsky, 2003; Keller, Schimmel, Hargrove, & Hoffman, 2008). It will consist of distributed sensor networks and experiments, linked by advanced cyberinfrastructure to record and archive ecological data over long time periods. Using standardized protocols, NEON will gather essential data for developing the scientific understanding and theory required to forecast and resolve ecological challenges within the USA. It differs from the LTER program in that it is being designed as a nationally linked network with common measurement capacities and research themes. Using a statistical analysis of ecoclimatic state variables and wind vectors, NEON has divided the USA into 20 ecoclimate domains. Each domain will host one fully instrumented NEON Core Site located in a wildland area. Every NEON Core Site will include a standard set of instruments to collect biological, biophysical, biogeochemical, and land-use and land-management data. NEON proposes to implement a rapid deployment system to study sudden events on the landscape, such as wildfires, natural catastrophes, disease outbreaks, or the emergence of an invasive species. Finally, each domain will deploy a set of moveable sites to provide the flexibility to evaluate specific environmental gradients within each ecoclimatic domain, and potentially, across the entire network. NEON is in an early stage of development and implementation.

Although these research networks have produced valuable information on ecological processes, it is difficult to extrapolate results to other biophysical settings. This is because (1) there are a small number of sites in each network, (2) there is seldom a core set of ecological measurements across the networks (because they are largely investigator driven), and (3) the sites are typically selected opportunistically (rather than based on a network sampling design). Linkage and/or incorporation of these networks into a multi-tiered network will depend on development and adoption of a nested set of ecological measures and standards that can be functionally related to each other (see later discussion on TERENO).

25.2.1.2 Spatially Extensive Science Monitoring Networks

Spatially extensive in situ monitoring networks can contribute substantially to an understanding of site and landscape-scale ecological changes over broad geographic areas, including regional, national, and continental scales (McDonald et al., 2002; Bechtold and Patterson, 2005; Paulsen, Stoddard, Holdsworth, Mayio, & Tarquinio, 2006). These networks can be used to develop ecological measures and models and collect data on ecological and biophysical indicators resulting from intensive site studies and broad-scale gradient studies (Ator, Olsen, Pitchford, & Denver, 2003). Several networks co-locate stressors and ecological measurements and indicators in an attempt to relate observations of conditions and trends to the types and magnitude of stressors. Spatially extensive networks are also critically important in developing and validating ecological, habitat, and biophysical models within and among different biophysical settings (Boyer et al., 2006; Donohue, McGarrigle, & Mills, 2006; Phillips, Anderson, & Schapire, 2006; Scott, Harvey, Alexander, & Schwarz, 2007), and in facilitating broad-scale environmental assessments (Hamilton, Miller, & Myers, 2004; Gilliom et al., 2006; Paulsen et al., 2006). When the in situ site network is spatially extensive and collects a standard set of biophysical measurements, it is possible to develop statistically based, multi-scaled models covering broad geographic areas (Jones et al., 2001; Donohue et al., 2006). Moreover, it is possible to explain how environmental stressors (representing different scales) change in their importance in explaining ecological conditions within and among specific geographic areas across broad regions (O'Connor et al., 1996; Jones et al., 2006).

Examples of spatial extensive science networks that collect ecological data (indicators) include the US National Water-Quality Assessment Program (NAWQA, Mueller & Sparh, 2006), the Environmental Monitoring and Assessment Program (EMAP, McDonald et al., 2002), the US National Wadeable Stream Survey (Paulsen et al., 2006), the Forest Inventory and Analysis Program (FIA, Bechtold et al., 2005), the Countryside Survey (Haines-Young et al., 2006), the National Inventory of Landscapes in Sweden (NILS, Esseen, Glimskar, Stahl, & Sundquist, 2006), and the Irish Aquatic Monitoring Survey (Donohue et al., 2006). All of these networks use either random or stratified random survey designs. Stratified random designs use biophysical and ecological classifications as their primary strata. Additionally, several spatially extensive networks provide biophysical and hydrologic information important in assessing broad-scale ecological and environmental conditions. For example, the USGS stream gage network consists of more than 7300 sites used to measure stream and river flow across the USA (Hester, Carsell, & Ford, 2006). The network provides real-time and archived flow information (early 1900s to the present) that contributes to flood prediction and warning (O'Connor & Costa, 2004) and relationships between climate variability and change and water availability (Cayan, Redmond, & Riddle, 1999). Additionally, because of its long history, the network permits an assessment of how land-use and land cover changes have affected hydrological and ecological processes (Jennings & Jarnagin, 2002). Finally, spatially extensive climate and weather networks provide valuable information to interpret vegetation condition and changes over broad regions (Schwartz, Reed, & White, 2002).

A primary limitation of these networks is that data collection is limited to specific ecological themes or ecosystems rather than multiple ecosystems. For example, FIA focuses on forests, EMAP on surface water, USGS-NAWQA on biologic parameters, surface- and groundwater quality, and the USGS-stream gage network on streamflow. During its early days of development, EMAP attempted to sample many US ecosystems (Messer, Linthurst, & Overton, 1991), but logistical issues and costs forced the program to focus on surface waters, estuaries, and landscape composition and patterns (McDonald et al., 2002). Additionally, because of costs, many of these programs have had to scale back on the spatial extent of their sampling networks (Hester et al., 2006) and the number of sites resampled. This dramatically decreases the ability to detect changes and trends in ecological conditions over broad areas, which is the primary stated goal of monitoring programs. Significant effort is needed to relate sampling and/or experimental designs of monitoring networks and to standardize data collection for a common set of ecological metrics and indicators (Larsen et al., 2007).

25.2.1.3 Spatially Extensive Citizen Networks

Citizen monitoring networks have gained great popularity and momentum over recent years. In Australia, the Waterwatch program has established more than 7,000 sites through collaboration with states, catchment care groups, and local communities (Commonwealth of Australia, 2003). NatureWatch is a citizen-based program in Canada that collects data on both plants and animals (NatureWatch, 2007). The European Phenology Network (van Vliet et al., 2003) consists of 12 citizen-based networks in Europe that collect data on both plant and animal phenology. The GLOBE program (http://www.globe.gov) fosters global and regional environmental awareness through citizen science and education programs.

Most citizen monitoring networks involve relatively simple measurements and observations on plants and animals (e.g., date of first leaf production, emergence of a species from hibernation, first appearance of a migratory bird). Although simple, these data can provide valuable clues on how species and ecosystems are responding to major environmental stressors, including climate change and urbanization (Harvey, 2006). They also provide three valuable services that may help sustain and expand monitoring networks over time. First, they provide education and awareness of environmental issues to people from all walks of life. Second, they often involve participation by children and therefore might help reverse or slow the trend of children being disconnected from nature (Louv, 2005). Finally, when people are involved in volunteer programs, and can see how their involvement benefits our understanding of and solutions to problems, they develop pride and ownership in the programs. This leads to a greater likelihood of sustainability and growth of these types of programs. Two challenges for citizen-based monitoring are (1) implementing measurements and indicators that are not too complex, yet ecologically meaningful, and (2) developing ways to improve consistency and quality across the network.

The US Breeding Bird Survey (BBS) is an example of a program that uses volunteers, but that has maintained fairly rigorous standards for data collection through training and quality control methods (Peterjohn & Sauer, 1999). Collected on more than 4,000 road-based transects (39.4 km), with some transects dating back 40 years, this network has been used extensively to assess bird population trends (Peterjohn & Sauer, 1999; Sauer, Fallon, & Johnson, 2003), and to link observed changes to environmental stressors, including forest fragmentation (Boulinier et al., 2001) and housing density (Pidgeon et al., 2007).

A primary issue with the BBS network is the representativeness of the road surveys. Roads create edges and this may result in increased observations of edge species. Representativeness of samples is a key issue for many spatially extensive monitoring networks since samples are often selected in non-random ways (Larsen et al., 2007).

Of all the tiers in the four-tiered monitoring framework, the citizen-based network may be the most critical to develop and maintain (Harvey, 2006). These networks are cost-effective, take great advantage of the Internet, and increase education of environmental problems and ownership in the process and results. However, priority must be given to link and integrate existing citizen monitoring programs into multi-tiered ecological monitoring frameworks, and where possible should maintain a strong scientific basis in design and implementation. Successful integration of citizen scientists into a nationwide, multi-tiered ecological monitoring program is embodied in the structure of the USA National Phenology Network (see later discussion on the USA-NPN).

25.2.1.4 Remote Sensing Monitoring

Over the last two decades, we have seen great advances in and deployment of sensors that measure the detailed features of the earth (GEOSS, 2005). These include a wide range of biophysical features, including temperature, moisture, and characteristics of soil, geology, landform, elevation, water, and vegetation, including phenology, structure, and chemical composition. Moreover, advances in computing and digital communication have dramatically increased our ability to process, store, and analyze remote sensing imagery.

Remote sensing data have facilitated land surface and ecological assessments over broad areas. Data from the Advanced Very High-Resolution Radiometer (AVHRR) and the Moderate-resolution Imaging Spectroradiometer (MODIS) provide daily to monthly measures of important vegetation indices at 1 km and 250 m spatial resolutions, respectively. Monthly composites for AVHRR (8 km resolution) are available for the entire globe dating back to 1981 (http://glcf.umiacs.umd.edu/data/gimms/). These data have been used extensively in combination with spatially extensive biophysical data (e.g., digital elevation models, soils maps, spatially interpolated climate data) to extrapolate results of intensive site studies to broad geographical areas, including nitrogen mineralization (Fan, Randolph, & Ehman, 1998), forest transpiration and photosynthesis (Anselmi, Chiesi, Giannini, Manes, & Maselli, 2004), global leaf area (Asner et al., 2003), and regional net primary productivity and water yield (Ollinger, Aber, & Federer, 1998). They also have been used to forecast crop yield and productivity (Reynolds et al., 2000; Kastens, Price, Kastens, & Martinko, 2001), fuels and fire regimes (Rollins, Keane, & Parsons, 2004), and global terrestrial primary production (Running et al., 2004). They have been used to evaluate trends in vegetation indices (Paruelo and Lauenroth, 1995; Schwartz et al., 2002; Lupo, Linderman, Vanacker, Bartholome, & Lambin, 2007) and to relate vegetation trends to broad-scale disturbance and environmental management (Wylie et al., 2008). Additionally, global scale land cover data have been developed from these data (Loveland et al., 2000) and from the Medium Resolution Imaging Spectrometer (MRIS, 300 m resolution, Defourny et al., 2006; Clevers et al., 2007). A primary challenge with these data is spatial resolution. Many features and processes associated with land-surface change go undetected by these sensors, hence the need to include finer scale remote sensing and in situ monitoring networks (see later discussion on TERENO).

Moderate-resolution imagery such as that from Landsat (30 m) and the Advanced Spaceborne Thermal Emission and Reflection Radiometer (ASTER, 15–90 m) has been used extensively to monitor earth-surface conditions and change (Stefanov, Christensen, & Ramsey, 2001; Loveland et al., 2002; Wickham, Riitters, Wade, Coan, & Homer, 2007; Jones et al., 2008). Additionally, Landsat data have been used to characterize land cover status and change at continental scales (Homer, Huang, Yang, Wylie, & Coan, 2004; Australian Government, 2005; Haines-Young and Weber, 2006), and are used in many of the existing multi-scale monitoring networks described above. Land cover data generated from Landsat imagery also have been used in combination with other spatial coverages or grids of biophysical data to produce continental-scale biophysical classifications (Jongman et al., 2006). Seasonal and annual analysis of vegetation characteristics is possible, but because of costs in acquiring and processing these data, such analyses have been limited to relatively small geographic areas (Bohlman, Adams, Smith, & Peterson, 1998). However, increased availability of multi-date remote sensing data via web portals at little or no cost may increase the feasibility of conducting these types of analyses.

High-resolution spatial data (1–10 m) generated from satellites such as IKONOS and Quickbird permit analysis of detailed land-surface characteristics, including water clarity, impervious surface, and submerged aquatic vegetation (Sawaya, Olmanson, Heinert, Brezonik, & Bauer, 2003). High-resolution aerial photography also provides detailed information on fine-scale landscape features such as hedgerows and fencerows (Bartuszevige, Gorchov, & Raab, 2006). Moreover, airborne instruments such as the light detecting and ranging (LIDAR) instrument provide detailed information on topography and vegetation structure and composition (Anderson et al., 2006; Streutker & Glenn, 2006).

Hyperspectral data from sensors on satellites (e.g., Hyperion) and aircraft (e.g., airborne visible and infrared imaging spectrometer or AVIRIS) have provided detailed information on vegetation composition and chemistry (Martin et al., 1998; Ramsey, Rangoonwala, & Ehrlich, 2005). They also have been used in combination with radar data to evaluate vegetation structure and density (Treuhaft, Asner, Law, & Van Tuyl, 2002). Radar has also been used to track movement patterns in birds (Diehl, Larkin, & Black, 2003). Unmanned air vehicles (UAVs) offer significant potential to detect fine-scale surface characteristics important to ecological monitoring (Binenko, Donchenko, Andreev, & Ivanov, 2001).

The primary limitations in using high-resolution spectral or spatial data in ecological monitoring programs are cost (acquisition and analysis) and availability. As a result, most programs have used or propose to use these data in selected geographic areas (e.g., as part of random sample or along environmental gradients) nested under broader-scale remote sensing data (e.g., AVHRR, MODIS, Landsat, or ASTER data). A nested design for use of sensors is highlighted later in Section 25.3.1.

25.2.2 Ecological Indicators

Development and application of indicators are essential components of multi-tiered ecological monitoring and assessment programs (Hunsaker, Carpenter, & Messer, 1990; Walker & Reuter, 1996; Kurtz, Jackson, & Fisher, 2001; McDonald et al., 2002; Jorgensen, Xu, Salas, & Marques, 2005; Müller & Lenz, 2006). Moreover, comprehensive environmental assessments such as the Millennium Ecosystem Assessment (Carpenter et al., 2006), the Australia State of the Environment Report (Beeton et al., 2006), the European Environmental Assessment (EEA, 2007b), and the US State of the Nation's Ecosystems Report (Heinz, 2008), depend on indicators to draw conclusions about status and trends in environmental resources.

Existing monitoring programs use a wide variety of ecological indicators. The type and number of indicators used depend on the purpose and scale of the monitoring (Heinz, 2006). For example, the FIA monitoring program collects data on forest characteristics and structure (Bechtold & Patterson, 2005), the National Wadeable Streams Survey focuses on aquatic ecosystem indicators (Paulsen et al., 2006), and the BBS on breeding bird abundance and distribution (Peterjohn & Sauer, 1999). Differences in sampling designs and indicators make it difficult to aggregate or compare the results from these programs (Heinz, 2006). Moreover, some programs, such as the National Wadeable Stream Survey and NAWQA, collect data on similar ecosystems (e.g., streams), but differences in sampling methodologies make it difficult to compare or combine the results from these programs (Whitacre, Roper, & Kershner, 2007). Emerging monitoring programs such as the Global Earth Observation System of Systems (GEOSS), the USA-NPN, NEON, and TERENO are attempting to standardize measurement approaches, and implement a core set of indicators to facilitate multi-scale ecological analysis and interpretation.

Spatially extensive monitoring programs have focused on ecological indicators that measure status and trends in ecosystem characteristics and structure (Kurtz et al., 2001). Existing programs have been challenged with implementing indicators that capture ecological processes and functions (Kurtz et al., 2001; Groffman et al., 2006; Carpenter et al., 2006), especially those that relate to fluxes and flows across landscapes (Voinov, Costanza, Boumans, Maxwell, & Voinov, 2004; Peters et al., 2006; Urban & Keitt, 2001; Reiners & Driese, 2001). Moreover, programs have been challenged with identifying tipping points or ecological thresholds that might indicate the potential for rapid changes in ecological conditions (Groffman et al., 2006).

Development of in situ-automated monitors offers significant potential to measure fundamental ecological processes across spatially extensive monitoring networks (Hart & Martinez, 2006). Automated sensor networks have been established to monitor seismic activities across North American and the globe (Ekström, Humphreys, & Levander, 1998; Fischer & van der Hilst, 1999). They have been used as part of ecological forecasting networks, including outbreak and fate of harmful algal blooms (Valette-Silver & Scavia, 2003). Many of the emerging monitoring networks described in this chapter (e.g., GEOSS, NEON, and TERENO) propose to deploy an array of automated, in situ sensors, and these sensors should contribute substantially to our understanding of ecological functions and processes.

Advances in remote sensing technology and increasing availability of relatively fine-scale spatial databases over extensive areas offer significant potential to scale process-related results from intensive research site networks to broader geographic areas (Rastetter, Aber, Peters, Ojima, & Burke, 2003; Anselmi et al., 2004).

Power laws and allometric measurements offer significant potential to link and interpret ecological measurements and processes across spatial and temporal scales (Schneider, 2001; Belgrano & Brown, 2002; Brown et al., 2002; Kerr, Kharouba, & Currie, 2007; Li, Gorshkov, & Makarieva, 2007). However, the primary limitation of this approach is the degree to which power and algometric scaling functions exist for the range of ecological processes important to monitoring programs.

An important issue related to multi-scaled monitoring networks is the application of indicators that assess responses of ecological systems to environmental management and policies. Many governments and agencies want to know whether their policies and programs are having a positive impact on the environment. Most multi-scaled monitoring programs use ecological indicators to assess status, changes, and trends in ecological conditions rather than ecological responses to environmental policies and management practices.

However, some existing monitoring programs measure ecological condition and stressors known to affect ecological condition in an attempt to determine probable cause. For example, EMAP and the National Wadeable Streams Survey co-locate measurements of chemicals and physical habitat disturbance with in situ ecological monitoring sites to establish potential relationships between biological and ecological indicators and stressors (Kurtz et al., 2001; Paulsen et al., 2006, respectively). Moreover, these programs measure remote sensing-based indicators of land use and landscape composition and pattern that have been quantitatively linked to stream ecological conditions (Donohue et al., 2006; Jones et al., 2006).

Despite efforts to link indicators of ecological condition and natural and anthropogenic stressors, we lack the information needed to evaluate the effectiveness of policy and environmental management options. The adaptive management concept acknowledges ecosystem complexity and uncertainties in cause-and-effect relationships between environmental management and ecological conditions, and uses monitoring and ecological indicators to adjust environmental policies and management prescriptions (Bormann et al., 2007). The success of this type of approach depends on defining clear environmental management objectives and maintaining commitment to long-term ecological monitoring (Dale & Beyeler, 2001; Bormann et al., 2007). Emerging programs such as NEON, TERENO, and the USA-NPN (see later discussion) are implementing designs and measurements that might help establish linkages between natural and anthropogenic drivers and stressors and ecological processes and conditions.

25.3 Example Multi-tiered Monitoring Networks

Comprehensive, multi-tiered ecological research and monitoring capacities are key objectives of emerging programs including NEON (Senkowsky, 2003; Keller et al., 2008) and GEOSS (GEOSS, 2005). GEOSS is attempting to increase global- and continental-scale continuity among different earth observation programs, including in situ- and remote sensing-based observation systems (GEOSS, 2005). The goal is to enhance the capacity to forecast and assess key environmental attributes and processes upon which humankind depend. Through an integrated earth observation system GEOSS hopes to (1) reduce loss of life and property from natural and human-induced disasters, (2) understand environmental factors affecting human health and well-being, (3) improve management of energy resources, (4) understand, assess, predict, mitigate, and facilitate adaption to climate variability and change, (5) improve water resource management through better understanding of the water cycle, (6) improve weather information, forecasting, and warning, (7) improve the management and protection of terrestrial, coastal, and marine ecosystems, and (8) support sustainable agriculture and combat desertification. GEOSS has spawned observation networks focused on specific environmental themes, including the Biodiversity Observation Network (GEO BON, Scholes et al., 2008). Additionally, programs like the Consortium of Universities for Advancement for Hydrological Science (CUASHI, Loescher et al., 2007), the Water and Environmental Research Systems Network (WATERS, Montgomery et al., 2007), and ILTER are attempting to improve networking of existing monitoring sites and networks through standardization of metrics and collaboration of investigators across multiple institutions. Many of these programs are attempting to relate ecological and other characteristics and processes to environmental attributes and processes valued by societies across a range of spatial and temporal scales.

Several existing programs have implemented multi-tiered monitoring, including but not limited to the Countryside Survey (Barr, Bunce, & Heal, 1994; Haines-Young et al., 2006), EMAP (McDonald et al., 2002), NAWQA (Hamilton et al., 2004), and the Canadian Ecological Integrity Program (EMAN, 2007). Moreover, multi-tiered monitoring approaches have been implemented to forecast environmental and ecological conditions (Vallette-Silver & Scavia, 2003; Running et al., 2004). For example, the National Oceanic and Atmospheric Administration (NOAA) has implemented an ecological forecasting system that combines in situ monitors, process models, and satellite imagery to forecast distributions of harmful algal blooms, stinging nettles, and movement and persistence of spilled oil and potentially harmful chemicals (Valette-Silver & Scavia, 2003). The goal of the forecasting system is to provide early warning to the public and environmental managers to reduce human health and environmental risks.

All of these programs have implemented spatially extensive science monitoring tiers as noted above. However, only the USA-NPN explicitly includes all four monitoring tiers in its fundamental network design (Betancourt et al., 2007). In the following section, we highlight two emerging ecological monitoring networks, TERENO and USA-NPN, to provide detailed examples of the development of multi-tiered programs.

25.3.1 TERENO – A Multi-tiered Terrestrial Monitoring Network

The Terrestrial Observatory Network (TERENO) is a proposed network of terrestrial observatories. Although biosphere–atmosphere systems are extremely complex, the terrestrial component in most process-based climate and biosphere models is typically represented in a conceptual and often rudimentary way. Remedying this deficiency is therefore one of the most important challenges in environmental and terrestrial research, and we suggest that terrestrial observatories could be an important step toward a new quality in environmental and terrestrial research. For the first phase, three terrestrial observatories in Germany have been identified: the Lower Rhine Basin, the metropolitan area Leipzig-Halle, and the Northern pre-Alps including the long-term research stations Hoeglwald and Scheyern. A fourth observatory is planned in the northeast German lowlands. TERENO started in 2008 and is funded by the Helmholtz Association (http://www.tereno.net). Each observatory has a number of intensive measurement sites located along a specific set of environmental gradients. TERENO is an example of a hierarchically organized monitoring network, similar in design to NEON.

25.3.1.1 Implementation of TERENO

A terrestrial system in this context is defined as a system consisting of the subsurface environment, the land surface including the biosphere (organised in ecosystems), the lower atmosphere, and the anthroposphere. These systems are organized along a hierarchy of spatial scales and structures ranging from the local scale to the regional scale (Fig. 25.2). Hydrological units will be used as the basic scaling units, because catchments are bordered by surface-water divides representing laterally well-defined system boundaries.

Each observatory within the TERENO framework will consist of real-time measurement platforms to allow terrestrial systems directly influenced by human activities to be observed and to enable the development of forecasts important to ecological changes. Furthermore, scientific experiments will be carried out across the nested hierarchy of scales ranging from the local scale (small test sites) to large catchments (Fig. 25.2).

The establishment of terrestrial observatories within the TERENO network will contribute to

- studying the impact of land-use changes, climate change, socio-economic developments, and human interventions on the evolution of terrestrial systems and their functioning and analyzing the interactions and feedback between the soil-vegetation and atmosphere compartments in these systems across a variety of scales;
- developing methods for upscaling parameters, fluxes, and state variables that describe processes controlling matter fluxes in soil–plant–atmosphere systems representative of the functioning of the systems at the selected scales based on theoretical and modeling concepts;
- providing high-quality data to validate existing and newly developed model concepts (e.g., inverse modeling, stochastic data fusion approaches) and upscaling theories to estimate effective parameters, fluxes, and state variables at various scales;
- bridging the gaps among scales, measurements, and modeling in hydrological and terrestrial sciences and developing and improving decision support systems for environmental management from the perspective of sustainable development;
- improving continuously integrated models that predict the evolution of: "man-made" terrestrial systems;
- promoting and supporting the development and use of early-warning systems (flooding, freshwater quality, etc.);
- bringing together different disciplines to enable the analysis of interactions between natural patterns and processes of landscapes and anthropogenic patterns and processes on different scales.

25.3.1.2 Criteria Applied in TERENO

The terrestrial observatories in Germany have a spatial scale ranging between 2,000 and 20,000 km^2.

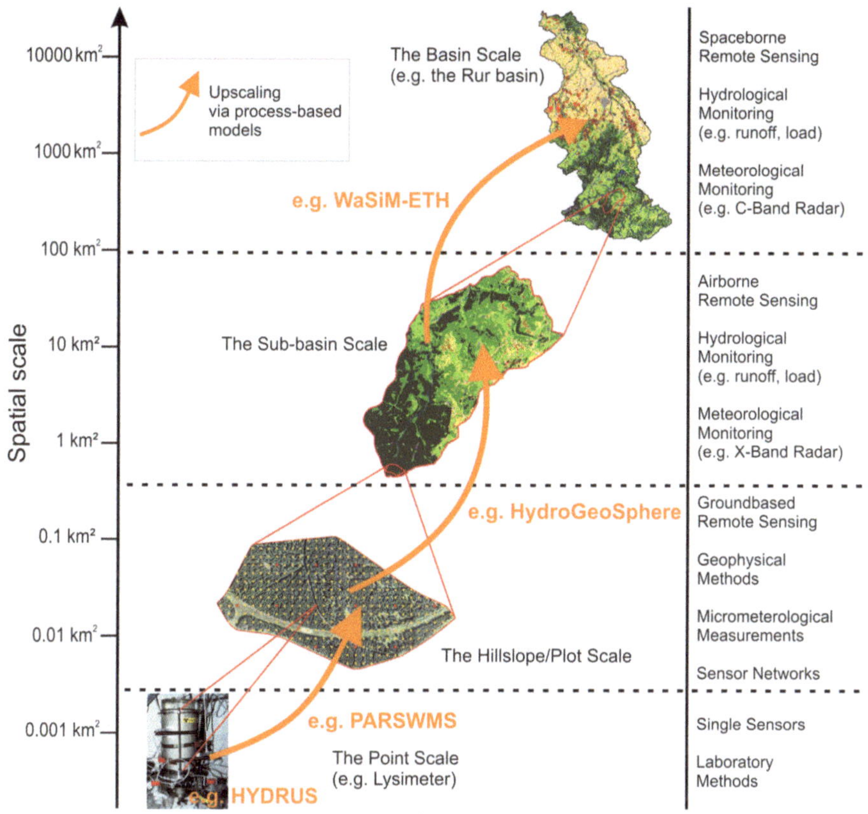

Fig. 25.2 Nested hierarchy of sensors and measurements as proposed by the TERENO multi-tiered monitoring network, including the point, plot, sub-basin, and basin scales. See the following references for details: HYDRUS (Šimůnek et al., 2008), PARSWMS (Hardelauf et al., 2007), HydroGeoSphere (Jones, Sudicky & McLaren, 2008), WASIM-ETH (Kunstmann et al., 2006)

These scales are appropriate to characterize terrestrial-atmospheric feedback and human-use gradients. Since hydrologic processes exert a fundamental control on aquatic and terrestrial metabolism and nutrient cycling, catchments represent an ideal fundamental unit of these observatories. A distinct variety in climate, land use, demography, etc., will be present over the whole catchment area in order to design sampling strategies to capture a range of natural and human use environmental gradients. By combining different observatories within a network, even broader scale relationships between anthropogenic and natural factors can be established.

Near-natural zones (e.g., national parks) will serve as reference sites in the TERENO network. Comparative monitoring strategies that take into account the surrounding landscapes will enable the analysis of long-term environmental changes due to human impact. Each observatory will incorporate prospective changes (e.g., open cast mining, ecosystem restoration, deforestation, land-use changes, economic and demographical characteristics, etc.) in order to analyze the effects of regional changes and disturbances.

Installations for the measurement of environmental and socio-economic quantities, influencing variables and indicators should be available for subsequent integration into the TERENO network. Within each terrestrial observatory, small-scale research facilities and test areas have to be in place in order to perform detailed process studies. By using hierarchical scales, this detailed information can be transferred to the regional scale.

Individual observatories within the TERENO network will be equipped with research instruments designed for long-term measurements. Important parameters, fluxes and state variables (PFS) for the compartments of the atmosphere, hydrosphere, soil, hydrogeology, and biosphere, as well as for socio-economic and demographic problems have to be monitored simultaneously. Different strategies will be considered depending on the time scale of expected changes. Rapid, short-term variations will need to be monitored by fixed installed instruments networked via state-of-the-art communications in order to enable near-real-time measurements and to simplify data management, e.g., smart sensor networks. Smart sensor networks will consist of a multitude of small sensors nodes that are able to observe phenomena, e.g., temperature or soil moisture fields, with high temporal and spatial resolution. Slower variations of system states might best be observed by a system of mobile and flexible sensor networks that will operate periodically on a regular or event-driven basis thus allowing a more efficient and wider-area analysis (e.g., similar to the NEON design). The monitoring and scientific programme has to be accompanied by high-performance systems for data storage and processing.

25.3.1.3 Terrestrial Observatory Design

Each observatory within the network will consist of all test facilities/areas listed in Table 25.1 in a nested configuration. This nested design is important to overcome the distinct gap between the process scale (lysimeter, test plot) and the field and catchment scale, and to foster integration of biodiversity and socio-economic aspects. The nested concept can be illustrated for soil moisture. Soil moisture plays a key role in partitioning water and energy fluxes, providing moisture to the atmosphere for precipitation, and controlling the pattern of groundwater recharge. Large-scale soil moisture variability is driven by space–time precipitation and radiation pattern. At local scales, atmospheric processes (wind, solar radiation, and temperature), land cover, soil conditions, and topography act to redistribute soil moisture.

Intensive soil moisture monitoring sites will be collocated with climate towers for coordinated observation. Measurements taken from soil water content sensors can be made at several depths, ranging from the ground surface to the water table. Wireless sensor network technology has the potential to reveal dynamic changes in soil moisture with a high spatial–temporal resolution from the plot to sub-catchment scale (Bogena, Huisman, Oberdörster, & Vereecken, 2007). Ground penetrating radar techniques should be applied for soil water content determination at the intermediate scale (Huisman, Sperl, Bouten, & Verstraten, 2001). Radar-based airborne and satellite platforms are able to measure surface-near soil moisture at the sub-catchment to regional scale (Krajewski et al., 2006). New generations of radiometer and radar remote sensing instruments will become available within the next years. Ideally, these instruments will

Table 25.1 Research facilities and test areas classified according to their scale and their characteristic

Research facility/test areas	Scale [m^2]	Characteristics
Local scale	10^0–10^1	• Ideally defined system boundaries • Boundary conditions can be set • Very high information density • Very well suited for process studies
Plot-field scale	10^1–10^2	• Well-defined system boundaries • Boundary conditions can partly be set • High information density • Well suited for process studies • Low biodiversity
Field/slope	10^2–10^5	• Partly defined system boundaries • Boundary conditions can partly be set • High to medium information density • Suited for process studies • Low to medium biodiversity
Sub-catchment	10^5–10^6	• Well-defined system boundaries • Boundary conditions cannot be set • Medium information density • Well suited for catchment-scaled process studies • Medium biodiversity • Low socio-economic relevance
Catchment	10^6–10^9	• Moderately defined system boundaries • Boundary conditions cannot be set • Low information density • Suited for catchment-scaled process studies • Medium to high biodiversity (not necessarily) • High socio-economic relevance
Basin/region	$> 10^9$	• System boundaries are often poorly defined • Boundary conditions cannot be set • Very low information density • Poorly suited for process studies • Very high biodiversity • Very high socio-economic relevance

provide either coarse spatial resolution data with high repeat cycles, or high-resolution data with low repeat cycles (e.g., ESA-SMOS: L-Band, 3-day revisit, 40 km resolution, launch 2008; RADARSAT-2: C-Band, 23-day revisit, 3–100 m resolution, launched 2006). However, the received signal is strongly influenced by the vegetation and surface structure, and the sampling depth is restricted to the uppermost soil (2–5 cm) (Walker, Houser, & Willgoose, 2004). Consequently, direct measurements are still indispensable in areas with significant vegetation and litter cover.

25.3.1.4 Scale Transfer

In order to enable the transfer of detailed information from local measurements to the regional scale or policy scale the instrumentation of each observatory follows a hierarchical design (Fig. 25.2). The scaling will be accomplished via process-based models. For example, multi-step outflow experiments in the laboratory on lysimeter (e.g., Weihermüller, Huisman, Graf, Herbst, & Vereecken, 2007), taken from representative locations within each observatory, will be used to determine important soil parameters inversely via a process based soil hydraulic model (e.g., HYDRUS, Šimůnek, VanGenuchten, & Šejna, 2008). In return these parameters can be used to parameterise field scale models like PARSWMS (Hardelauf et al., 2007) or HydroGeoSphere (Jones, Sudicky & McLaren, 2008). Important fluxes of energy, water and matter (e.g., evapotranspiration, infiltration, groundwater flow, runoff discharge etc.) will be measured in selected test sites in order to further parameterise and validate these models.

In a next step the model-based process representations will be scaled up to the regional or policy scale using a catchment model like WASIM-ETH (e.g., Kunstmann, Heckl, & Rimmer, 2006), which will reproduce management activities such as reservoir use and groundwater withdrawal. Additionally, a catchment model is able to assimilate remote sensing data (e.g., land-use classification, precipitation radar, soil moisture). This scaling procedure will help bridge the gap between monitoring, models, and management.

In order to monitor the processes controlling matter fluxes in soil–plant–atmosphere systems, hydrological monitoring can be extended by adding other relevant compartments (e.g., soil, vegetation, and atmosphere) and by a socioeconomic analysis. This may include the identification of socioeconomic forces that drive the intensification and/or de-intensification processes (e.g., demographic and technological change) and the feedback to the ecological system (e.g., on the state of soil, water and air quality) as well as the analysis of ecological impacts of land-use changes across and within sectors (e.g., from agricultural to industrial types or from chemical to energy production).

25.3.2 USA-NPN – A Four-Tiered Biological Monitoring Network

Phenology is the study of periodic plant and animal life cycle events (e.g., timing of leafing and flowering of plants, development of agricultural crops, emergence of insects, or migrations of animals) that are influenced by environmental changes. Phenological events integrate across driving variables (e.g., warming, precipitation), and over time reflect environmental variability and change. Phenological data have a number of applications, including conservation (e.g., species interactions), ecosystem dynamics (e.g., water and carbon exchange), agriculture (e.g., timing of production), and can be coupled with process models to assist natural resource management as well as predict health risks and hazards such as fire.

The USA National Phenology Network (USA-NPN; http://www.usanpn.org) is an emerging partnership of federal agencies, the academic community, and the general public to monitor and understand the influence of phenology on national resources, health, and production (Betancourt et al., 2005, 2007). The mission of USA-NPN is to establish a spatially dense and evenly distributed network of phenological observation sites across the country, and to implement a simple and effective means to input, report, and utilize coordinated observations of the activity of plants and animals, including the resources to provide information for a wide range of decisions made routinely by private citizens, industry, government agencies, and the nation as a whole.

USA-NPN can benefit from and contribute to existing local, state, and federal organizations that are concerned about nature and climate. The USA-NPN seeks to (1) coordinate research activities with existing networks to advance phenological science and develop mechanistic phenological models to support improvement of climate and ecosystem models; (2) maximize the representation of phenological monitoring sites at the national and regional scale to enable biological baseline characterization and trend detection; and (3) provide data and information to policy makers to support land-management decisions related to agriculture, forestry, and wildlife conservation. Principles for collaboration among network participants include mutually beneficial activities, shared vision on science and education, minimization of the demand on the capacities of partners, feedback to improve collaboration; and transparent data and information sharing policies.

25.3.2.1 Implementation of a Multi-tiered Monitoring Structure

USA-NPN includes four observation components or tiers (Fig. 25.1, CENR, 1997). Each tier represents different levels of spatial coverage and related environmental information (as described in Section 25.2.1). First, USA-NPN will interface with existing networks of local intensive sites focused on detailed environmental measurements and process studies, such as the AmeriFlux network of ˜100 research sites, and the NSF-supported National Ecological Observatory Network (NEON) and the Long-Term Environmental Research (LTER) Network.

The next conceptual tier includes spatially extensive environmental networks focused on standardized observations distributed across the country (similar to the US cooperative observer weather station network), including National Park Service Inventory and Monitoring sites, National Wildlife Refuge System sites, Organization of Biological Field Stations sites, botanical gardens and arboreta, etc.

The USA-NPN citizen volunteer programs will leverage off of existing volunteer and education networks; this tier will allow all interested members of the public to participate in phenological observation while enabling the use of phenology as an integral part of environmental education. USA-NPN will develop and distribute standardized educational/outreach materials, training modules and other volunteer training materials that can be coupled with education and outreach programs being conducted by collaborating networks and organizations.

Finally, the USA-NPN will support the development and application of remote sensing products for land surface phenology to extend surface observations across landscapes, regions or the nation as a whole. In addition, ground observations will be collected in such a manner that they can be used to validate or interpret remotely sensed images. Scaling algorithms will serve as a tool to interpret satellite observations collected over the last several decades, thus providing a longer time series of phenology measures from which to establish short- and medium-term trends.

This multi-tiered approach is designed to create a national-scale observatory, where observations of phenology and phenological processes are collected using standardized protocols at ~40,000 locations across the nation. The intensive and extensive observation tiers provide a strong scientific basis for the monitoring network, and facilitate integration of localized process studies with spatially extensive observations. However, they represent a limited number (i.e., several thousand) of observation locations. Thus, inclusion of volunteers, who with training can collect high-quality observations of phenology, is required to create the high-density observatory network required to understand climate change impacts across the diversity of ecosystem types represented in the nation. Finally, inclusion of remote sensing for land surface phenology creates a continuous distribution of phenological observations (e.g., at the scale of 30 m based on Landsat imagery) that can be coupled with ground observations to understand phenology at landscape to national scales.

25.3.2.2 Criteria and Design for USA-NPN Observatories

In contrast to TERENO, which has a focus on hydrological processes on a catchment scale, USA-NPN is focused primarily on biological activity of plants and animals, on scales from individuals to populations and communities, and – based on patterns of vegetation activity (determined using spectral reflectance in the form of vegetation indices) – land surface phenology at landscape to regional scales. Thus, the spatial scale of observations can range from <1 m^2 to 100,000 km^2. The sheer number of in situ observations (targeted at 40,000 for the nation) will thus include a broad range of climates, land use/land cover types, along a continuum of natural to human-modified environmental.

Observations can be combined within or across climatic regions or use types to investigate broad-scale relationships between climate variation and biological activity.

In contrast to TERENO or the US-based National Ecological Observatory Network (NEON), USA-NPN has minimal needs for infrastructure because of the strong volunteer component for the project (Fig. 4.1). The defining characteristic of a typical, non-network USA-NPN volunteer observation station revolves around the individual observer, who registers a site and associated individual organism(s) for monitoring. Observers can collect data for either plants or animals following national, standardized, and vetted monitoring protocols. The USA-NPN plant phenology monitoring program consists of 211 plants with national or regional importance. Plant species are distributed across the USA to create a spatial network of observations with sufficient overlap to allow inter-correlation of species responses across the entire nation. The USA-NPN is in the process of developing an animal phenology monitoring program, which will focus on mammals, birds, invertebrates, amphibians, reptiles, and fishes. Finally, the network is developing an approach to coordinate and standardize remote sensing for assessment of land-surface phenology (e.g., across remote sensing platforms) among the various agencies and institutions charged with understanding phenology at broad spatial scales.

25.4 Concluding Remarks

Multi-tiered ecological monitoring programs offer significant promise to help understand and address some of the most pressing environmental issues faced by society. Several new programs, including GEOSS, NEON, and the USA-NPN, and efforts to expand and link existing programs (e.g., ILTER) offer significant potential to address these needs. However, several important issues must be resolved in order for these networks to work in harmony. This includes developing and deploying a set of standard measurements or indicators that can be scaled and compared across multiple monitoring tiers and programs. Suites of indicators and models need to be developed and deployed that work in harmony to detect changes in ecological synchronies and relationships that affect important ecological functions across scales. Moreover, research is needed to establish thresholds or tipping points to detect major phase shifts in important ecological processes and associated structure. Existing and emerging multi-tiered monitoring programs must incorporate a greater number of biological indicators than currently exist. We know very little about status and trends in individual species, biological communities, and overall biological diversity. Moreover, little is known about the contribution of biodiversity in maintaining ecological resilience. Multi-tiered monitoring programs offer great potential to address these questions. Significant advances are needed in developing demographic, social, economic, cultural, and political indicators and models to explain and forecast ecological conditions across multiple scales. Multi-tiered monitoring networks offer significant potential to improve our understanding of these relationships.

Differences in monitoring designs and spatial and temporal coverage present significant challenges in integrating existing monitoring networks. Most networks were established to address specific questions for specific ecological systems, making their integration difficult. However, programs like GEOSS are attempting to improve linkages among different monitoring networks through refinement of measurement and sampling protocols, and through optimization of models. Emerging programs such as NEON and TERENO are incorporating cross-scale and multiple network linkages into their initial designs. These as well as other multi-scaled monitoring programs offer significant potential to understand and address a wide range of today's most pressing environmental problems.

Similar to the USA-NPN, multi-tiered monitoring programs should incorporate where possible a spatially extensive citizen science tier into their design. Many agency-based monitoring programs have strong outreach and education programs, but do not use citizens as part of their data collection network. Conversely, many citizen science monitoring programs are independent of agency-based monitoring programs. Given the costs of spatially extensive science networks, this may be the only way to achieve a dense network of in situ monitoring sites across broad geographic areas. Advances in web-based applications and the Internet provide a great opportunity to expand such networks. Moreover, direct involvement by the public will help maintain interest in and funding for long-term ecological monitoring.

References

Anderson, J., Martin, M. E., Smith, M.-L., Dubayah, R. O., Hofton, M. A., Hyde, P., et al. (2006). The use of waveform lidar to measure northern temperate mixed conifer and deciduous forest structure in New Hampshire. *Remote Sensing of Environment, 105*, 248–261.

Anselmi, S., Chiesi, M., Giannini, M. Manes, F., & Maselli, F. (2004). Estimation of Mediterranean forest transpiration and photosynthesis through the use of an ecosystem simulation model driven by remotely sensed data. *Global Ecology and Biogeography, 13*, 371–380.

Asner, G., Scurlock, J. O., & Hicke, J. A. (2003). Global synthesis of leaf area index observations: Implications of ecological and remote sensing studies. *Global Ecology and Biogeography, 12*, 191–205.

Ator, S. W., Olsen, A. R., Pitchford, A. M., & Denver, J. M. (2003). Application of a multipurpose unequal probability stream survey in the Mid-Atlantic Coastal Plain. *Journal American Water Resources Association, 39*, 873–885 .

Australian Government. (2005). *Greenhouse gas emissions from land use change in Australia: Results of the National Carbon Accounting System 1988–2003.* Canberra: Australian Greenhouse Office.

Baker, J. P., Hulse, D. W., Gregory, S. V., White, D., VanSickle, J., Berger, P. A., et al. (2004). Alternative futures for the Willamette River Basin, Oregon. *Ecological Applications, 14*, 313–324.

Barr, C. J., Bunce, R. G. H., & Heal, O. W. (1994). Countryside Survey 1990: A measure of change. *Journal of the Royal Agricultural Society of England, 155*, 48–58.

Bartuszevige, A. M., Gorchov, D. L., & Raab, L. (2006). The relative importance of landscape and community features in the invasion of an exotic shrub in a fragmented landscape. *Ecography, 29*, 213–222.

Bechtold, W. A., & Patterson, P. L. (Eds.). (2005). *The enhanced forest inventory and analysis program – national sampling design and estimation procedures. General Technical Report SRS-80.* Asheville, NC: U.S. Department of Agriculture, Forest Service, Southern Research Station.

Beeton, R. J. S., Buckley, K. I., Jones, G. J., Morgan, D., Reichelt, R. E., & Dennis, T. (2006). *Australia state of the environment 2006. Independent Report to the Commonwealth Minister for the Environment and Heritage.* Canberra: Department of the Environment and Heritage.

Belgrano, A., & Brown, J. H. (2002). Oceans under the macroscope. *Nature, 419*, 128–129.

Bellamy, P. H., Loveland, P. J., Bradley, R. I., Lark, R. M., & Kirk, G. J. D. (2005). Carbon losses from all soils across England and Wales 1978–2003. *Nature, 437*, 245–248.

Betancourt, J. L., Schwartz, M. D., Breshears, D. D., Brewer, C. A., Frazer, G., Gross, J. E., et al. (2007). Evolving plans for the USA National Phenology Network. *Earth Observing System Transactions, American Geophysical Union, 88*, 211.

Betancourt, J. L., Schwartz, M. D., Breshears, D. D., Cayan, D. R., Dettinger, M. D., Inouye, D. W., et al. (2005). Implementing a US National Phenology Network. *Earth Observing System Transactions, American Geophysical Union, 86*, 539–542.

Binenko V. I., Donchenko, V. K., Andreev V. L., & Ivanov R. V. (2001). Some results and perspectives of using unmanned air vehicles for ecological monitoring tasks. *Ecological Chemistry, 10*, 21–31.

Bogena, H. R., Huisman, J. A., Oberdörster, C., & Vereecken, H. (2007). Evaluation of a low-cost soil water content sensor for wireless network applications. *Journal of Hydrology, 344*, 32–42.

Bohlman, S. A., Adams, J. B., Smith, M. O., & Peterson, D. L. (1998). Seasonal foliage changes in the eastern Amazon Basin detected from Landsat Thematic Mapper satellite images. *Biotropica, 30*, 376–391.

Bormann, B. T., Haynes, R. W., & Martin, J. R. (2007). Adaptive management of forest ecosystems: Did some rubber hit the road? *BioScience, 57*, 186–191.

Boulinier, T., Nichols, J. D., Hines, J. E., Sauer, J. R., Flather, C. H., & Pollock, K. H. (2001). Forest fragmentation and bird community dynamics: Inference at regional scales. *Ecology, 82*, 1159–1169.

Boyer, E. W., Alexander, R. B., Parton, W. J., Li, C., Butterbach-Bahl, K., Donner, S. D., et al. (2006). Modelling denitrification in terrestrial and aquatic ecosystems at regional scales. *Ecological Indicators, 16*, 2123–2142.

Brown, J. H., Gupta, V. K., Li, B.-L., Milne, B. T., Restrepo, C., & West, G. B. (2002). The fractal nature of nature: Power laws, ecological complexity and biodiversity. *The Philosophical Transactions of the Royal Society, 357*, 619–626.

Carpenter, S. R., DeFries, R., Dietz, T., Mooney, H. A., Polasky, S., Reid, W. V., et al. (2006). Millennium ecosystem assessment: Research needs. *Science, 314*, 257–258.

Cayan, D. R., Redmond, K. T., & Riddle, L. G. (1999). ENSO and hydrologic extremes in the Western United States. *Journal of Climate, 12*, 2881–2893.

CENR. (1997) *Integrating the Nation's environmental monitoring and research networks and program: A proposed framework.* Washington, DC: Committee on Environment and Natural Resources, Subcommittee on Ecological Systems.

Christensen, N. L., Bartuska, A. M., Brown, J. H., Carpenter, S., D'Antonio, C., Francis, R., et al. (1996). The report of the Ecological Society of America Committee on the scientific basis for ecosystem management. *Ecological Applications, 6*, 665–691.

Clark, J. S., Carpenter, S. R., Barber, M. S., Collins, S., Dobson, J., Foley, A., et al. (2001). Ecological forecasts: An emerging imperative. *Science, 293*, 657–660.

Clevers, J. G. P. W., Schaepman, M. E., Mücher, C. A., DeWit, A. J. W., Zurita-Milla, R., & Bartholomeus, H. M. (2007). Using MERIS on Envisat for land cover mapping in the Netherlands. *International Journal of Remote Sensing, 28*, 637–652.

Commonwealth of Australia. (2003). *Waterwatch Australia national technical manual. Module 1.* Canberra: Environment Australia.

Dale, V. H., & Beyeler, S. C. (2001). Challenges in the development and use of ecological indicators. *Ecological Indicators, 1*, 3–10.

Defourny, P., Vancutsem, C., Bicheron, P., Brockmann, C., Nino, F., Schouten, L., et al. (2006). *GLOBCOVER: A 300 m global land cover product for 2005 using ENVISAT MERIS time series. ISPRS Commission VII Mid-term Symposium "Remote Sensing: From Pixels to Processes", Enschede, the Netherlands, 8–11 May 2006*. Retrieved from http://www.itc.nl/isprsc7/symposium/proceedings/TS04_1.pdf.

Diehl, R. H., Larkin, R. P., & Black, J. E. (2003). Radar observations of bird migration over the Great Lakes. *The Auk, 120*, 278–290.

Donohue, I., McGarrigle, M. L., & Mills, P. (2006). Linking catchment characteristics and water chemistry with the ecological status of Irish rives. *Water Research, 40*, 91–98.

Dunne, J. A., Saleska, S. R., Fischer, M. L., & Harte, J. (2004). Integrating experimental and gradient methods in ecological climate change research. *Ecology, 85*, 904–916.

EEA (2004). *Impacts of Europe's changing climate – an indicator-based assessment. EEA Report, No. 2/2004*. Copenhagen European Environment Agency.

EEA. (2007a). *Europe's environment: The fourth assessment. European Environment Agency, TH-78-07-127-EN-C*. Copenhagen European Environment Agency.

EEA. (2007b). *Halting the loss of biodiversity by 2010: Proposal for a first set of indicators to monitor progress in Europe. EEA Technical report No 11/2007*. (Copenhagen European Environment Agency).

Ekström, G., Humphreys, G., & Levander, A. (1998). USArray – Probing the continent. *IRIS Newsletter, 16*, 4–6.

EMAN. (2007). *Ecological monitoring and assessment network, environment Canada*. Retrieved from http://www.emanrese.ca/eman/.

EPA. (2003). *Draft report on the environment 2003. U.S. EPA, Report Number 260-R-02-006*. Washington, DC: U.S. Environmental Protection Agency.

Esseen, P. A., Glimskar, A., Stahl, G., & Sundquist, S. (2006). *Faltinstruktion for nationaell inventering av landskapet i Sverige, NILS*. Umea: Department of Forest Resource Management, Swedish University of Agriculture.

Falge, E., Baldocchi, D., Tenhunen, J., Aubinet, M., Bakwin, P., Berbigier, P., et al. (2002). Seasonality of ecosystem respiration and gross primary production as derived from FLUXNET measurements. *Agricultural and Forest Meteorology, 113*, 53–74.

Fan, W., Randolph, J. C., & Ehman, J. L. (1998). Regional estimation of nitrogen mineralization in forest ecosystems using geographic information systems. *Ecological Applications, 8*, 734–747.

Fischer, K. M., & van der Hilst, R. D. (1999). Geophysics enhanced: A seismic look under the continents. *Science, 285*, 5432–1365.

GEOSS. (2005). *The Global Earth Observation System of Systems (GEOSS) 10-Year Implementation Plan*. Retrieved from http://earthobservations.org/.

Gilliom, R. J., Barbash, J. E., Crawford, C. G., Hamilton, P. A., Martin, J. D., Nakagaki, N., et al. (2006). *The Quality of Our Nation's Waters: Pesticides in the Nation's Streams and Ground Water, 1992–2001. U.S. Geological Survey Circular 1291*. Reston, Virginia U.S. Geological Survey.

Gosz, J. R. (1992). Gradient analysis of ecological change in time and space: Implications for forest management. *Ecological Applications, 2*, 248–261.

Groffman, P. M., Baron, J. S., Blett, T., Gold, A. J., Goodman, I., Gunderson, L. H., et al. (2006). Ecological thresholds: The key to successful environmental management or an important concept with no practical application. *Ecosystems, 9*, 1–13.

Haines-Young, R., & Weber, J.-L. (2006). *Land Accounts for Europe 1990–2000: Towards integrated land and ecosystem accounting*. Copenhagen: European Environment Agency.

Haines-Young, R. H., Barr, C. J., Black, H. I. J, Briggs, D. J, Bunce, R. G. H., Clarke, R. T., et al. (2006). *Accounting for nature: Assessing habitats in the UK countryside*. Norwich: Queen's Printer and Controller of HMSO.

Hamilton, P. A., Miller, T. L., & Myers, D. N. (2004). *Water quality in the Nation's streams and aquifers: Overview of selected findings 1991–2001. U.S. Geological Survey Circular 1265*. Reston, Virginia: U.S. Geological Survey.

Hardelauf, H., Javaux, M., Herbst, M., Gottschalk, S., Kasteel, R., Vanderborght, J., et al. (2007). PARSWMS: A parallelized model for simulating three-dimensional water flow and solute transport in variably saturated soils. *Vadose Zone Journal, 6*, 255–259.

Hargrove, W. W., Hoffman, F. M., & Law, B. E. (2003). New analysis reveals representativeness of AmeriFlux network. *Earth Observing System Transactions, American Geophysical Union, 84*, 529.

Hart, J. K., & Martinez, K. (2006). Environmental sensor networks: A revolution in the earth system science? *Earth-Science Reviews, 78*, 177–191.

Harvey, C. (2006). Monitoring change: Citizen science and international environmental treaty-making. *Papers on International Environmental Negotiation: Ensuring a Sustainable Future, 15*, 77–90.

Heinz (The H. John Heinz Center). (2006). *Filling the gaps: Priority data needs and key management challenges for National Reporting on Ecosystem Condition*. Washington, DC: The H. John Heinz III Center for Science, Economics, and the Environment.

Heinz (The H. John Heinz Center). (2008). *The State of the Nation's Ecosystems 2008: Measuring the land, waters, and living resources of the United States*. Chicago: Island Press.

Hester, G., Carsell, K., & Ford, D. (2006). *Benefits of USGS stream gauging program*. Denver, Colorado National Hydrologic Warning Council.

Homer, C., Huang, C., Yang, L., Wylie, B., & Coan, M. (2004) Development of a 2001 national land-cover database for the United States. *Photogrammetric Engineering and Remote Sensing*, 70, 829–840.

Houghton, R. A. (1994). The worldwide extent of land-use change. *BioScience, 4*, 305–313.

Huisman, J. A., Sperl, C., Bouten, W., & Verstraten, J. M. (2001). Soil water content measurements at different scales: Accuracy of time domain reflectometry and ground-penetrating radar. *Journal of Hydrology, 245*, 48–58.

Hunsaker, C., Carpenter, D., & Messer, J. (1990). Ecological indicators for regional monitoring. *Bulletin of the Ecological Society of America, 71*, 165–172.

Jennings, D. B., & Jarnagin, S. T. (2002). Impervious surfaces and stream flow discharge: A historical perspective in a Mid-Atlantic subwatershed. *Landscape Ecology, 17*, 471–489.

Jones, J. P., Sudicky, E. A., & McLaren, R. G. (2008) Application of a fully-integrated surface-subsurface flow model at the watershed-scale: A case study. *Water Resources Research, 44*, DOI:10.1029/2006WR005603.

Jones, K. B., Nash, M. S., Wade, T. G., Walker, J., Neale, A. C., Müller, F., et al. (2008). Cross-European landscape analyses: Illustrative examples using existing spatial data. In I. Petrosillo, F. Müller, K. B. Jones, G. Zurlini, K. Krauze, S. Victorov, B.-L. Li, W. G. & Kepner (Eds.), *Use of landscape sciences for the assessment of environment security* (pp. 258–316). Dordrecht: Springer.

Jones, K. B., Neale, A. C., Nash, M. S., VanRemotel, R. D., Wickham, J. D., Riitters, K. H., et al., (2001). Predicting nutrient and sediment loadings to streams from landscape metrics: A multiple watershed study from the United States Mid-Atlantic Region. *Landscape Ecology, 16*, 301–312.

Jones, K. B., Neale, A. C., Wade, T. G., Cross, C. L., Wickham, J. D., Nash, M. S., et al. (2006). Multiscale relationships between landscape characteristics and nitrogen concentrations in streams. In J. Wu, K. B. Jones, H. Li, & O. L. Loucks (Eds.), *Scaling and uncertainty analysis in ecology: Methods and applications* (pp. 205–224). Berlin: Springer.

Jongman, R. H. G., Bunce, R. G. H., Metzger, M. J., Mucher, C. A., Howard, D. C., & Mateus, V. L. (2006). Objectives and applications of a statistical environmental stratification of Europe. *Landscape Ecology, 21*, 409–419.

Jorgensen, S. E., Xu, F. L., Salas, F., & Marques, J. C. (2005). Applications of indicators for the assessment of ecosystem health. In S. E. Jorgensen, R. Costanza, & F.-L. Xu (Eds.), *Handbook of ecological indicators for assessment of ecosystem health* (pp. 5–66). Boca Raton: CRC Press.

Kastens, J. H., Price, K. P., Kastens, D. L., & Martinko, E. A. (2001). *Forecasting Pre-harvest Crop Yields using Time Series Analysis of AVHRR NDVI Composite Imagery. Proceedings, Annual Convention of the American Society of Photogrammetric Engineering and Remote Sensing 2000, April 23–27.* St. Louis: American Society of Photogrammetric Engineering and Remote Sensing.

Keller, M., Schimmel, D. S., Hargrove, W. W., & Hoffman, F. M. (2008). A continental strategy for the National Ecological Observatory Network. *Frontiers in Ecology and the Environment,* 6, 282–284.

Kerr, J. T., Kharouba, H. M., & Currie, D. J. (2007). The macroecological contribution to global change. *Science, 316*, 1581–1584.

Krajewski, W. F., Anderson, M. C., Eichinger, W. E., Entekhabi, D., Hornbuckle, B. K., Houser, P. R., et al. (2006). A remote sensing observatory for hydrologic sciences: A genesis for scaling to continental hydrology. *Water Resources Research, 42*, DOI:10.1029/ 2005WR004435.

Kratz, T. K., Deegan, L. A., Harmon, M. E., & Lauenroth, W. K. (2003). Ecological variability in space and time: Insights gained from the US LTER program. *BioScience, 53*, 57–67.

Kunstmann, H., Heckl, A., & Rimmer, A. (2006). Physically based distributed hydrological modelling of the Upper Jordan catchment and investigation of effective model equations. *Advanced Geoscience, 9*, 123–130.

Kurtz, J. C., Jackson, L. E., & Fisher, W. S. (2001). Strategies for evaluating indicators based on guidelines from the Environmental Protection Agency's Office of Research and Development. *Ecological Indicators, 1*, 49–60.

Larsen, D. P., Olsen, A. R., Lanigan, S. H., Moyer, C., Jones, K. K., & Kincaid, T. M. (2007). Sound survey designs can facilitate integrating stream monitoring data across multiple programs. *Journal of the American Water Resources Association, 43*, 384–397.

Li, B.-L., Gorshkov, V. G., & Makarieva, A. M. (2007). Allometric scaling as an indicator of ecosystem state: A new approach. In I. Petrosillo, F. Müller, K. B. Jones, G. Zurlini, K. Krauze, S. Victorov, B.-L. Li, & W. G. Kepner (Eds.), *Use of landscape sciences for the assessment of environment security* (pp. 107–117). Dordrecht: Springer.

Lins, H. F. (1994). Recent developments in water, energy, and biogeochemical budgets research. *Earth Observing System Transactions, American Geophysical Union, 75*, 433–439.

Loescher, H. W., Jacobs, J. M., Wendroth, O., Robinson, D. A., Poulos, G. S., McGuire, K., et al. (2007). Enhancing water cycle measurements for future hydrologic research. *American Meteorological Society, 88*, 669–676.

Louv, R. (2005). *Last child in the woods: Saving our children from nature-deficit disorder*. Chapel Hill, NC: Algonquin Books of Chapel Hill.

Loveland, T. R., Reed, B. C., Brown, J. F., Ohlen, D. O., Zhu, Z., Yang, L., et al. (2000). Development of a global land cover characteristics database and IGBP discover from 1 Km AVHRR data. *International Journal of Remote Sensing, 21*, 1303–1330.

Loveland, T. R., Sohl, T. L., Stehman, S. V., Gallant, A. L., Sayler, K. L., & Napton, D. E. (2002). A strategy for estimating the rates of recent United States land-cover changes. *Photogrammetric Engineering and Remote Sensing, 68*, 1091–1099.

Lupo, F., Linderman, M., Vanacker, V., Bartholome, E., & Lambin, E. F. (2007). Categorization of land-cover change processes based on phenological indicators extracted from time series of vegetation index data. *International Journal of Remote Sensing, 11–12*, 2469–2483.

Martin, M. E., Newman, S. D., Aber, J. D., & Congalton, R. G. (1998). Determining forest species composition using high spectral resolution remote sensing data. *Remote Sensing and the Environment, 65*, 249–254.

McCarthy, J. J., Canziani, O. F., Leary, N. A., Dokken, D. J., & White, K. S. (2001): *Climate Change (2001) Impacts, Adaptation, and Vulnerability, IPCC Third Assessment Report.* Cambridge: Cambridge University Press.

McDonald, M., Blair, R., Dlugosz, J., Hale, S., Hedtke, S., Heggem, D., et al. (2002). EPA's Environmental Monitoring and Assessment Program (EMAP) in the 21st Century. *Hydrological Science and Technology, 18*, 133–143.

Messer, J.J., Linthurst, R. A., & Overton, W. S. (1991). An EPA program for monitoring ecological status and trends. *Environmental Management, 17*, 67–78.

Millennium Ecosystem Assessment. (2005). *Ecosystems and human well-being: Synthesis*. Washington, DC: Island Press.

Montgomery, J. L., Harmon, T., Haas, C. N., Hooper, R., Clesceri, N. L., Graham, W., et al. (2007). The Waters Network: An integrated environmental observatory network for water research. *Environmental Science and Technology, 41*, 6642–6647.

Mueller, D. K., & Sparh, N. E. (2006). *Nutrients in streams and rivers across the Nation — 1992–2001. U.S. Geological Survey Scientific Investigations Report 2006–5107*. Reston: U.S. Geological Survey.

Müller, F. (1998). Gradients in ecological systems. *Ecological Modelling, 108*, 3–21.

Müller, F., & Lenz, R. (2006). Ecological indicators: Theoretical fundamentals of consistent applications in environmental management. *Ecological Indicators, 6*, 1–5.

NatureWatch. (2007). Retrieved from http://www. naturewatch.ca/english/.

NEON. (2006). *Integrated science and education plan for the National Ecological Observatory Network*. Retrieved from http://www.neoninc.org/documents/ ISEP_2006Oct23.pdf.

O'Connor, J. E., & Costa, J. E. (2004). Spatial distribution of the largest rainfall-runoff floods from basins between 2.6 and 26,000 km^2 in the United States and Puerto Rico. *Water Resources Research, 40*, DOI:10.1029/2003WR002247.

O'Connor, R. J., Jones, M. T., White, D., Hunsaker, C., Loveland, T., Jones, B., et al. (1996). Spatial partitioning of environmental correlates of avian biodiversity in the conterminous United States. *Biodiversity Letters, 3*, 97–110.

Ojima, D. S., Galvin, K. A., & Turner, B. L., II. (1994). The global impact of land-use change. *BioScience, 44*, 300–304.

Ollinger, S. V., Aber, J. D., & Federer, C. A. (1998). Estimating regional forest productivity and water yield using an ecosystem model linked to GIS. *Landscape Ecology, 13*, 323–334.

Parr, T. W., Sier, A. R. J., Battarbee, R. W., Mackay, A., & Burgess, J. (2003). Detecting environmental change: science and society – perspectives on long-term research and monitoring in the 21st Century. *The Science of the Total Environment, 310*, 1–8.

Paruelo, J. M., & Lauenroth, W. K. (1995). Regional patterns of Normalized Difference Vegetation Index in North American shrublands and grasslands. *Ecology, 76*, 1888–1898.

Paulsen, S., Stoddard, J., Holdsworth, S., Mayio, A., & Tarquinio, E. (2006). *Wadeable streams assessment: A collaborative assessment of the Nation's streams*. EPA 841-B-06–002. (Washington, DC: EPA Office of Research and Development.

Peterjohn, B. G., & Sauer, J. R. (1999). Population status of North American grassland birds from the North American Breeding Bird Survey, 1966–1996. *Studies in Avian Biology, 19*, 27–44.

Peters, D. P. C., Gosz, J. R., Pockman, W. T., Small, E. E., Parmenter, R. R., Collins, S., et al. (2006). Integrating patch and boundary dynamics to understand and predict biotic transitions at multiple scales. *Landscape Ecology, 21*, 19–33.

Peterson, D. L., & Waring, R. H. (1994). Overview of the Oregon Transect Ecosystem Research-Project. *Ecological Applications, 4*, 211–225.

Phillips, S. J., Anderson, R. P., & Schapire, R. E. (2006). Maximum entropy modeling of species geographic distributions. *Ecological Modelling, 190*, 231–259.

Pidgeon, A. M., Radeloff, V. C., Flather, C. H., Lepczyk, C. A., Clayton, M. K., Hawbaker, T. J., et al. (2007). Associations of forest bird species richness with housing and landscape patterns across the USA. *Ecological Applications, 17*, 1989–2010.

Ramsey, III, E., Rangoonwala, A., & Ehrlich, R. (2005). Mapping the invasive species, Chinese Tallow with EO1 satellite Hyperion hyperspectral image data and relating tallow percent occurrences to a classified Landsat Thematic Mapper landcover map. *International Journal of Remote Sensing, 26*, 1637–1657.

Rastetter, E. B., Aber, J. D., Peters, D. P. C., Ojima, D. S., & Burke, I. C. (2003). Using mechanistic models to scale ecological processes across space and time. *BioScience, 53*, 68–76.

Reiners, W. A., & Driese, K. L. (2001). The propagation of ecological influences through heterogeneous environmental space. *BioScience, 51*, 939–950.

Reynolds, C. A., Yitayew, M., Slack, D. C., Hutchinson, C. F., Huete, A., & Petersen, M. S. (2000). Estimating crop yields and production by integrating the FAO Crop Specific Water Balance model with real-time satellite data and ground-based ancillary data. *International Journal of Remote Sensing, 21*, 3487–3508.

Rollins, M. G., Keane, R. E., & Parsons, R. A. (2004). Mapping fuels and fire regimes using remote sensing, ecosystem simulation, and gradient modelling. *Ecological Applications, 14*, 75–95.

Running, S. W., Nemani, R. R., Heinsch, F. A., Zhao, M., Reeves, M., & Hashimoto, H. (2004). A continuous satellite-derived measure of global terrestrial primary production. *BioScience, 54*, 547–560.

Sauer, J. R., Fallon, J. E., & Johnson, R. (2003). Use of North American Breeding Bird Survey data to estimate population change for bird conservation regions. *Journal of Wildlife Management, 67*, 372–389.

Sawaya, K. E., Olmanson, L. G., Heinert, N. J., Brezonik, P. L., & Bauer, M. E. (2003). Extending satellite remote sensing to local scales: Land and water resource monitoring using high-resolution imagery. *Remote Sensing of Environment, 88*, 144–156.

Schneider, D. C. (2001). The rise of the concept of scale in ecology. *BioScience, 51*, 545–553.

Scholes, R. J., Mace, G. M., Turner, W., Geller, G. N., Jürgens, N., Larigauderie, A., et al. (2008). Toward a global biodiversity observing system. *Science, 22*, 1044–1045.

Schwartz, M. D., Reed, B. C., & White, M. A. (2002). Assessing satellite-derived start-of-season measures in the conterminous USA. *International Journal of Climatology, 22*, 1793–1805.

Scott, D., Harvey, D., Alexander, R., & Schwarz, G. (2007). Dominance of organic nitrogen from headwater streams to large rivers across the conterminous United States. *Global Biogeochemical Cycles, 21*, GB1003/1-GB1003/8, DOI:10.1029/2006GB002730.

Senkowsky, S. (2003). NEON: Planning for a new frontier in biology. *BioScience, 53*, 456–461.

Šimůnek, J., VanGenuchten, M. T., & Šejna, M. (2008). Development and applications of the HYDRUS and STANMOD software packages and related codes. *Vadose Zone Journal, 7*, 587–600.

Stefanov, W. L., Christensen, P. R., & Ramsey, M. S. (2001). Monitoring urban land cover change – An expert system approach to land cover classification of semiarid to arid urban centres. *Remote Sensing of Environment, 77*, 173–185.

Stirling, K., & Derocher, A. E. (2007). Melting under pressure: The real scoop on climate warming and polar bears. *The Wildlife Professional, Fall 2007*, 24.

Streutker, D. R., & Glenn, N. F. (2006). LiDAR measure of sagebrush steppe vegetation heights. *Remote Sensing of Environment, 102*, 135–145.

Treuhaft, R. N., Asner, G. P., Law, B. E., Van Tuyl, S. (2002). Forest leaf area density profiles from the quantitative fusion of radar and hyperspectral data. *Journal of Geophysical Research-Atmospheres, 107*, 4568–4578.

Urban, D., & Keitt, T. (2001). Landscape connectivity: A graph-theoretic perspective. *Ecology, 82*, 1205–1218.

Valette-Silver, N. J., & Scavia, D. (2003). Introduction to ecological forecasting: New tools for coastal and marine ecosystems management. In: Valette-Silver, N. J. & Scavia, D. (Eds.), *Ecological forecasting: New tools for coastal and ecosystem management, NOAA Technical Memorandum NOS NCCOS 1* (pp. 1–4). Silver Spring: National Oceanic and Atmospheric Administration Data Centre.

Van Vliet, A. J. H., Braun, P., Bruegger, R., Bruns, E., Clevers, J., Estreguil, C., et al. (2003). *The European Phenology Network; nature's calendar on the move*. Wageningen: Wageningen University.

Voinov, A., Costanza, R., Boumans, R. M. J., Maxwell, T., & Voinov, H. (2004). Patuxent landscape model: Integrated modeling of a watershed. (In R. Costanza, & A. Voinov (Eds.), *Landscape simulation modelling* (pp. 197–232). New York: Springer-Verlag.

Walker J. P., Houser, P. R., & Willgoose, G. R. (2004). Active microwave remote sensing for soil moisture measurement: A field evaluation using ERS-2. *Hydrologic Processes, 1811*, 1975–1997.

Walker, J., & Reuter, D. G. (1996). *Indicators of catchment health: A technical perspective.* Melbourne: CSIRO Publishing.

Weihermüller, L., Huisman, J. A., Graf, A., Herbst, M., & Vereecken, H. (2007). Multistep outflow experiments for the simultaneous determination of soil physical and CO_2 production parameters. *Geophysical Research Abstracts, 9*, 1742.

Westerling, A. L., Hidalgo, H. G., Cayan, D. R., & Swetnam, T. W. (2006). Warming and earlier spring increases western U.S. forest Wildlife Activity. *Science, 313*, 940–943.

Whitacre, H. W., Roper, B. B., & Kershner, J. L. (2007). A comparison of protocols and observer precision for measuring physical attributes of streams in Oregon and Idaho. *Journal of the American Water Resources Association, 43*, 1–15.

Wickham, J. D., Riitters, K. H., Wade, T. G., Coan, M., & Homer, C. (2007). The effect of Appalachian mountaintop mining on interior forest. *Landscape Ecology, 22*, 179–187.

Wylie, B. K., Zhang, L., Bliss, N., Jia, L., Tieszen, L. L., & Jolly, W. M. (2008). Integrating modelling and remote sensing to identify ecosystem performance anomalies in the boreal forest, Yukon River Basin, Alaska. *International Journal of Digital Earth, 1*, 196–220.

Young, S. S., & Harris, R. (2005). Changing patterns of global-scale vegetation photosynthesis, 1982–1999. *International Journal of Remote Sensing, 26*, 4537–4563.

Part VII
Future Demands and Challenges

Chapter 26

Conceptualising Long-Term Socio-ecological Research (LTSER): Integrating the Social Dimension

Simron J. Singh, Helmut Haberl, Veronika Gaube, Clemens M. Grünbühel, Petru Lisivieveci, Julia Lutz, Robin Matthews, Michael Mirtl, Angheluta Vadineanu, and Martin Wildenberg

Abstract In order to support the emerging network of long-term ecological research (LTER) sites across Europe, the European Union has launched ALTER-Net, a network aiming at lasting integration of long-term socio-economic, ecological and biodiversity research. Due to its high population density and long history of human habitation, however, Europe's ecosystems are generally intensively used. Social and natural drivers are so inextricably intertwined that the notion of 'socio-ecological' systems is appropriate. Traditional natural science-based approaches are insufficient to understand these integrated systems, as they cannot adequately capture their relevant socio-economic dimensions. This is particularly relevant because the EU launched ALTER-Net has an explicit aim to support sustainability, a goal that requires integration of socio-economic and ecological dimensions. As such, LTER is challenged to significantly expand its focus from ecological to socio-ecological systems, thus transforming itself from LTER to long-term socio-ecological research or LTSER. In order to support this transformation, this chapter explores several approaches for conceptualising socio-economic dimensions of LTSER. It discusses how the socio-economic metabolism approach can be combined with theories of complex adaptive systems to generate heuristic models of society–nature interaction which can then be used to integrate concepts from the social sciences. In particular, the chapter discusses possible contributions from the fields of ecological anthropology and ecological economics and shows how participatory approaches can be integrated with innovative agent-based modelling concepts to arrive at an integrated representation of socio-ecological systems that can help to support local communities to move towards sustainability.

Keywords Agent-based modelling · Complex adaptive systems · Participation · Long-term ecological research (LTER) · Long-term socio-ecological research (LTSER) · Society–nature interaction · Socio-economic metabolism

26.1 Introduction

Long-term ecological research (LTER) has gained much significance in the last decades due to the recognition that several relevant questions can only be answered by monitoring and analysing changes in patterns and processes in ecosystems over long time scales (Hobbie, Carpenter, Grimm, Gosz, & Seastedt, 2003; NRC, 2004) rather than through short-term studies (Dearing et al., 2006). Collecting evidence of the impacts of climate change on ecosystems, for instance, requires a long-term approach, not only because many of the variables are changing slowly but also because spatial and temporal variability of some of these variables (e.g. seasonal temperatures) make it difficult to discriminate the signal of climate change from the background noise (Greenland & Kittel, 2002). Moreover, ecosystems involve numerous and complex interactions between physical, chemical and biological components, and this complexity may generate dynamics endogenously, thus sometimes masking the

S.J. Singh (✉)
Institute of Social Ecology, Faculty of Interdisciplinary Studies, Klagenfurt University, Schottenfeldgasse 29, 1070 Vienna, Austria
e-mail: simron.singh@uni-klu.ac.at

F. Müller et al. (eds.), *Long-Term Ecological Research*,
DOI 10.1007/978-90-481-8782-9_26, © Springer Science+Business Media B.V. 2010

effects of changes in exogenous drivers. A large number of parameters and factors must be monitored – that is, measured consistently over long periods of time – in order to be able to reliably detect changes in the functioning of ecosystems and their components. Therefore, most classic LTER sites are small, often comprising only a few hectares. Many LTER sites, above all in the US LTER network, are deliberately selected to represent ecosystems with little current direct human influence in order to facilitate the detection of signals of global environmental change (Haberl et al., 2006).

There is increasing evidence, however, that classical disciplinary ecosystem research is not sufficient to guide action to conserve valuable ecosystems (Delbaere, 2005; Vadineanu, 1998, 2001). In Romania, for example, six decades of ecosystem research from 1900 to 1960 in the Lower Danube Wetlands System did not succeed in protecting the region from adverse management and development policies. Despite the fact that two LTER sites are located in this region, neo-classical economists, together with those from the field of agriculture, civil engineering and water management, went as far as to regard the wetlands as wastelands. In their opinion, the only 'economically viable option' was to utilise the region in a mono-functional fashion (crop, wood or fish) by establishing management systems subsidised by high amounts of external energy and material inputs (Vadineanu, Adamescu, Cazacu, Bodescu, & Danielescu, 2001). Progress could only be made when, from 1970s onwards, ecosystem research was complemented with an understanding of the major anthropogenic drivers. The results helped for understanding structural and functional changes driven by a set of major anthropogenic drivers and pressures and further for adopting first decisions regarding conservation of Danube Delta and Small Islands of Braila. An economic valuation of the gains from multi-functional farming and other ecosystem services helped to convince relevant policy makers to adopt new, bio-economically oriented management strategies extended for the entire Lower Danube Wetlands (~10,000 km^2) that aim to integrate economic and ecological goals (Vadineanu, Adamescu, Vadineanu, Christofor, & Negrei, 2003). It is increasingly being realised that the implementation of some top-down decisions and management plans cannot be implemented without social contextualisation and considerations.

The example of the Lower Danube Wetland System shows that interdisciplinary approaches, in this case a combination of economic and ecological expertise, are required in exploring viable management strategies oriented towards sustainability and nature conservation. The lesson for LTER is that, in order to generate knowledge which would be useful to resolve society's problems related to sustainability, it needs to extend its focus beyond classical ecological research. Traditional ecological approaches often seem to regard human activities as disturbances to otherwise properly functioning ecosystems (Haberl et al., 2006). If LTER is to contribute in finding solutions to sustainability problems it must go beyond a focus on patterns and processes in ecosystems and their alteration due to changes in global environmental conditions. It has to include an analysis of socio-economic activities that actively change and use ecosystems and of the socio-economic significance of these ecological changes. In order to be able to do so, LTER would have to include approaches from the social sciences within its framework (Redman, 1999; Redman, Grove, & Kuby, 2004). This would require a considerable shift in thinking within the LTER community that is far-reaching enough to warrant a re-labelling of the enterprise to 'long-term socio-ecological research', abbreviated LTSER (Haberl et al., 2006).

Such an LTSER approach is consistent with the emerging agenda of sustainability science (Clark, Crutzen, & Schellnhuber, 2004; Kates et al., 2001; Parris & Kates, 2003) that emphasises the sustainable use of natural resources to meet the needs of the present as well as those of future generations. Most ecological problems are due to the ways society interacts with nature. Investigations into the interactions between natural ecosystems and human activities would require not only approaches from the social sciences to be taken into account but also an up-scaling of present LTER sites to regions in which substantial human populations reside (Wilbanks & Kates, 1999). Obviously, in doing so, scientists are confronted with a complex interplay of various ecosystems and societal dynamics; that is, the focus would change from ecosystems to socio-ecological systems. Although this entails a considerable increase in complexity of the endeavour, it has important benefits. Such integrated LTSER platforms (Haberl et al., 2006) can be of much higher utility than LTER sites in supporting local populations in finding solutions to pressing sustainability problems, thus providing them decision support for viable future options. Furthermore, an understanding of society–nature interaction at local scales would provide reasonable estimates of the level of impact that local activities have on ecosystems compared to impacts caused by processes on larger scales, such as global environmental change.

Such considerations, together with the recognition that European ecosystems are typically used and transformed by humans to a much larger extent than in the USA (due to Europe's much higher population density and long history of human habitation), have motivated a strong orientation towards including socio-economic components in the emerging European long-term ecological research networks. In particular, within its sixth Framework Programme (FP6), the EU commission has set up a Network of Excellence to foster integration of socio-economic and ecological expertise in long-term ecological research. This network, called ALTER-Net (Delbaere, 2005, http://www.alter-net.info), acknowledges the need to integrate socio-economic knowledge in several of its work packages, but a proper understanding of how to tackle this task, both conceptually and methodologically, is still to emerge. In this chapter we will discuss concepts for including social sciences into LTER, thus transforming it to LTSER. However, we are cautious not to propose a full-scale social science programme, but to offer a selection of promising social science concepts that could readily be integrated and are beneficial to LTER research and conservation goals. The chapter is organised as follows: First, we discuss general concepts of socio-ecological systems. Then we elaborate on some of the general areas where social science could contribute in this regard together with a description of a few social science methods that could prove useful. We conclude by discussing an example from the emerging Austrian LTSER platform in which some of these approaches are currently being tested.

26.2 Conceptualising Socio-ecological Systems

26.2.1 Sustainability Science

Ever since the publication of the Brundtland Commission report, 'sustainable development' has been high on the political and scientific agenda. While such a vague definition of sustainable development allows for a broad political consensus, it drove scientists on a vigorous path to operationalise the concept over the next decades. The massive amount of scientific research that followed soon came to warrant the label of 'sustainability science' (Kates et al., 2001, p. 641) offering a variety of interpretations and approaches to sustainability by the natural and social sciences alike. It is increasingly acknowledged that environmental problems owe much to the way humans interact with their natural environment. This concept is challenging, as it requires interdisciplinary cooperation across the social/natural sciences divide, and of course LTSER has to be interdisciplinary as well, if it should contribute to monitoring progress towards sustainability. To be successful in this context, it is necessary to observe societies, natural systems, and their interaction over time, asking the following questions: (1) Which changes do socio-economic activities cause in natural systems? (2) Which socio-economic forces drive these changes, and what can we do to change them? (3) How do changes in natural systems impact on society? (4) How, if at all, can society cope with the changes it has set in motion? (Haberl, Fischer-Kowalski, Krausmann, Weisz, & Winiwarter, 2004).

Natural systems undergo significant changes as a matter of course. For example, temperature, precipitation, sea level, atmospheric chemistry and biodiversity have fluctuated drastically in the last thousands and millions of years, driven by both endogenous (e.g. geological or biotic processes) and exogenous (e.g. meteorite impacts) phenomena (Schellnhuber, 1999; Schlesinger, 1997). Such natural systems may be seen as self-organising dynamic systems which may be near equilibrium for limited periods, but may as well rapidly fluctuate between different states to maintain stability (Holling, 1986; Scheffer, Carpenter, Foley, Folke, & Walker, 2001). Thus any discussion of sustainability needs to recognise this intrinsic potential for change and that maintenance of any equilibrium over long time spans is unrealistic.

If equilibrium is unattainable, what else could sustainability mean? Sustainability has been defined as meeting the needs of the present without compromising the ability of future generations to meet their needs (WCED, 1987). A more equitable distribution of resources between regions and within nations is also often regarded as one of its cornerstones (UNEP, 2002). Sustainability defined this way is therefore anthropocentric, as it demands that human-induced changes in ecosystems must not affect society's survival or well-being, thus 'creating and maintaining our options for prosperous social and economic development' (Folke et al., 2002, p. 3). Some authors stress the need to 'expect the unexpected' (Holling, 1986) and improve society's ability to cope with uncertainty and surprise, defined as a situation in which perceived reality departs qualitatively from expectation (Berkes and Folke, 1998, p. 6). This has led to thinking about sustainability as 'the capacity to create, test, and maintain adaptive capability' (Holling, 2001, p. 390) which is related to the resilience of social–ecological systems (Carpenter, Walker, Anderies, & Abel, 2001).

An influential definition of sustainability is that societies should live 'within the regenerative capacity of the biosphere' (Wackernagel et al., 2002, p. 9266). This focus on the biophysical foundations of sustainability makes it obvious that a transition towards sustainability would not just require minor changes in current trends, but a radical reorientation. Even today, when only about one fourth or maybe one third of humanity lives in relative prosperity, humanity consumes each year an amount of natural resources which would take the biosphere 1.2 years to regenerate (Wackernagel et al., 2002). Hence, analyses of the biophysical dimension of society–nature interaction (Wackernagel, 1999) are of high importance for sustainability science and also for LTSER.

26.2.2 Conceptualising Long-Term Socio-ecological Research

There has been a long-standing debate on whether or not to view the natural world (dominated by biophysical realities such as matter and energy) as distinct from the human world (conceived as a system of recursive human communication and culture) (Croll & Parkin, 1992; Descola & Pálsson, 1996; Teich, Porter, & Gustafsson, 1997). This separation of nature and society has its roots in 'Cartesian dualism' manifested in the 'great divide' between the natural and social sciences. The question on how to view the world – natural and cultural realms as two entities or the latter embedded in the former – remains unresolved and is not within the scope of this chapter to discuss. However, the challenge of sustainability requires pragmatic approaches to overcome disciplinary boundaries between the natural and social sciences. Several approaches to conceptualise socio-ecological systems (also sometimes referred to as 'human–environment systems' or 'socio-environmental systems') – that is, systems that include natural as well as social components and drivers – are currently being explored, among others the DPSIR approach (EEA, 1995; Holten-Andersen, Paalby, Christensen, Wier, & Andersen, 1995, Haberl et al., 2009), the ecosystem service approach (Millennium Ecosystem Assessment, 2005) or CBD's Ecosystem Approach (Allen, Bandurski, & King, 1993). We here focus on the socio-economic metabolism approach that has recently gained recognition within emerging scientific fields such as industrial ecology, ecological economics, human ecology, land-change science and many other fields that study society–nature interactions. This approach is based on the idea of viewing 'society as a hybrid of the realm of culture, of meaning, of communication, and of the natural world' (Fischer-Kowalski & Weisz, 1999). In other words, society is seen to be composed of a system of recurrent self-referential communication, and material components, namely, a defined human population as well as a physical infrastructure such as buildings, machines, artefacts in use and animal livestock.

In order to aid interdisciplinary efforts, therefore, a useful heuristic model of the study of socio-ecological systems can be constructed, as shown in Fig. 26.1, by drawing two overlapping spheres, one depicting the study of 'natural' or 'biophysical' processes and a second representing the study of 'cultural' or 'symbolic' processes, including symbolic communication in a semantic sense. LTSER then would be the integrative field dealing with the processes of coupled socio-ecological systems, thus transcending classical LTER that aims to focus on more or less undisturbed natural systems (small protected plots), and even 'extended LTER' that also deals with strongly human-modified ecosystems.

This heuristic model can be used as a mind map to identify and locate the contributions from different disciplines and to understand the interrelations between them. One research strategy that has been followed is the analysis of material and energy flows between the biophysical structures of society and the other components of the biophysical sphere of

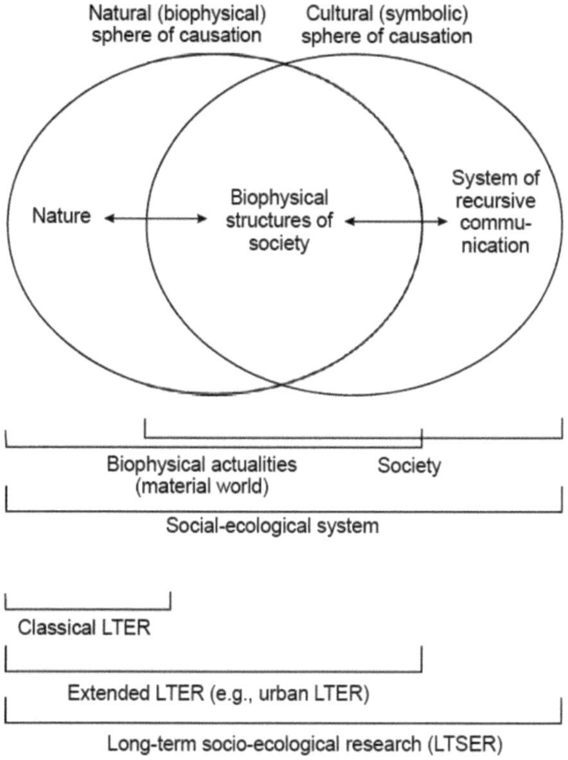

Fig. 26.1 A conceptual model of society–nature interaction (based on Boyden, 1992; Fischer-Kowalski & Haberl, 1997; Fischer-Kowalski et al., 1997; Haberl et al., 2004; Sieferle, 1997) used here as a basis for conceptualising long-term socio-ecological research (LTSER)

causation. This approach, often denoted as 'socio-economic metabolism' (e.g. Ayres & Simonis, 1994; Fischer-Kowalski, 1997b, Fischer-Kowalski et al., 1997; Matthews et al., 2000), regards society as a physical input–output system drawing material and energy from its environment, maintaining internal physical processes and dissipating wastes, emissions and low-quality energy to the environment. The central idea of the metabolism approach is to view societies as organising and maintaining flows of materials and energy for their production and reproduction. Such engagements with biophysical processes serve to produce and maintain not only a society's own biological existence but also that of their livestock and the whole range of artefacts such as buildings, infrastructure, machinery.

The concept of 'socio-economic metabolism' has already been described elsewhere in detail (e.g. Fischer-Kowalski & Weisz, 1999; Daniels & Moore, 2001). Current work in this context organises its accounts in a way that is compatible with established tools for societal self-observation, above all, social and economic statistics upon which practically all modelling in economics and the social sciences is based. These tools facilitate the analysis of mutual relations between symbolic (e.g. money flows) and biophysical aspects (e.g. material flows) of society. By means of this 'double compatibility' – towards ecological and socio-economic models and data – the socio-economic metabolism approach can establish a link between socio-economic variables and biophysical patterns and processes.

The analysis of material and energy flows related to economic activities alone, however, is not sufficient to capture society–nature interactions. One important aspect not adequately grasped by the metabolism approach is land use – one of the most important socio-economic driving forces of global change (Meyer & Turner, 1994). Land use can be conceptualised as 'colonisation of nature' (Fischer-Kowalski & Haberl, 1997; Haberl et al., 2001; Weisz et al., 2001), an approach that emphasises the fact that these human interventions into ecosystems are undertaken deliberately with the intention to modify natural systems according to society's needs and wants. Colonisation intensity in ecosystems can be analysed empirically by comparing currently prevailing ecosystem patterns and processes with those patterns and processes that would be expected without human intervention. An example of this approach is the calculation of the 'human appropriation of net primary production', or HANPP (Vitousek, Ehrlich, Ehrlich, & Matson, 1986) which is defined as the difference between NPP of potential (i.e. hypothetical, non human-modified) vegetation with the amount of NPP remaining in currently prevailing ecosystems after harvest, i.e. the amount of trophic energy diverted by humans from ecosystems (Haberl, 1997).

The notion of a 'MEFA framework – Material and Energy Flow Accounting' – has been proposed (Haberl et al., 2004; Krausmann, Haberl, Erb, & Wackernagel, 2004) to describe an integrated, consistent accounting framework consisting of data on socio-economic metabolism and on the colonisation of nature. Three parts of the MEFA framework have been elaborated in considerable detail: (1) Material flow accounting (MFA) has received most attention (e.g. Eurostat, 2001; Weisz et al., 2006). (2) Energy flow accounting (EFA) methods consistent with MFA have been proposed and applied (Haberl, 2001a, 2001b). (3) The Human Appropriation of Net Primary Production, or HANPP, proposed almost 20 years ago (Vitousek et al., 1986), has been further developed in a way that makes it consistent with material and energy flow accounting (Haberl et al., 2001). The MEFA framework is useful to analyse how we depend on and use the following three core functions of ecosystems for humans (Dunlap & Catton, 2002):

1. 'Resource supply': Land serves as a source of inputs for socio-economic metabolism by providing renewable and non-renewable resources (e.g. air, water, biomass, fossil fuels and minerals).
2. 'Waste absorption': The biosphere absorbs socio-economic outputs such as wastes or emissions.
3. 'Occupied space for human infrastructure': Humans occupy areas for housing, work space, infrastructure (including transportation), recreation, education and many other culturally important human activities.

Analyses of socio-economic metabolism and the colonisation of nature are important in an LTSER context above all because they can provide a link between natural-science based approaches (due to the obvious relevance of resource extraction, dissipation of wastes and land use for ecosystems) and approaches from the social sciences. Useful research questions can be derived by asking, for example, what cultural, economic or political conditions are most important in driving changes in socio-economic metabolism and land use and what role individual actors play.

26.2.3 Dynamics of Socio-ecological Systems

In a long-term perspective it is particularly important to pay attention to the temporal dynamics of socio-ecological systems. One body of theory that has recently gained attention derives from the recognition that society and nature co-evolve in a non-linear fashion (Abel, 1998; Norgaard, 1994; Weisz, 2002). This notion has led to an interest in the theory of complex adaptive systems as a means to understand socio-ecological systems. In fact, the theory of complex adaptive systems is more a collection of theories rather than a single theory, but these theories share some common characteristics in that they describe socio-ecological systems that are composed of hierarchical structures, whose dynamics are very sensitive to initial conditions, and which exhibit self-organisation, emergent phenomena and possibly sudden transitions from one stable equilibrium to another (Kay, Regier, Boyle, & Francis, 1999). These characteristics may make such systems unpredictable in some cases.

The term SOHO, which stands for 'Self-Organising Holarchic Open' systems,' has been proposed to describe ecosystems, based on the concept of a 'holon' (Kay et al., 1999). The term holon was devised by Koestler (1967) as part of an attempt to bridge the gap between individual behaviour at the micro-level and aggregate behaviour at the macro-level. A holon is a system that has a unique identity and is semi-autonomous, but in turn is composed of other subsystems (in themselves holons), and simultaneously forms part of a larger unit of organisation, with the overall system being referred to as a 'holarchy' or 'hierarchy of holons.' Such holons arise as a result of the system self-organising around an attractor to dissipate the flow of energy (high-quality energy) through itself (Odum & Pinkerton, 1955; Hall, 1995), and, as such, can be seen as non-equilibrium dissipative structures (Prigogine, 1976). As with the adaptive cycle concept, there is an appreciation that in SOHO systems, while their self-organising behaviour provides some ability to maintain themselves at an attractor despite changes in their environment, can also suffer catastrophic collapse or reorganisation into other attractors when certain variables within them reach specific thresholds, or if there is some kind of external perturbation. Which particular attractor domain a system finds itself in, therefore, also depends on its history.

Working independently, a number of groups have arrived at similar descriptions of the characteristics of socio-ecological systems, generally recognising that change is an intrinsic property of such systems, and that static equilibrium is seldom reached. For example, from a traditional ecological perspective on succession, Holling has developed the idea of an adaptive cycle (Holling, 1973, 1986). Initially this was a two-stage model in which the dynamics of biological communities consisted of an *r* phase or exploitation phase, in which rapid acquisition of resources is a successful strategy, and a *k* phase or consolidation phase leading to a stable climax state, in which conservation of accumulated capital is a successful strategy. This model was then extended to include an Ω phase or *creative destruction* phase, in which the *k* stage breaks down and an α phase or reorganisation phase in which new patterns are emerging (Gunderson & Holling, 2002).

Resilience was postulated to be a function of the potential and connectedness of the system and to be closely linked to the four phases just described, but to peak and to start declining before the peak of the *k* phase is reached, due to the increasingly fragile dependence of all the ecosystem components on one another, and increasing vulnerability to both internal and external shocks and stresses. Adaptive cycles were seen to operate at different hierarchical scales at different rates, with transfer of information between scales occurring via a limited number of variables (~5, i.e. 'the rule of hand'). From time to time, interactions between slower changing variables at one level and faster changing variables at another could trigger entry into the creative destruction and reorganisation phases of one or more of the cycles. These processes can also lead to a new stable system state – as opposed to the traditional view of having one climax state for one system – that may remain so for a while. The combined system of interacting adaptive cycles operating at different scales was termed 'panarchy' (Gunderson & Holling, 2002).

This metaphor was extended by conceptualising socio-ecological systems as being located on stability landscapes (in a topological rather than topographical sense) which contain basins of attraction representing a range of possible states with similar characteristics (Walker, Holling, Carpenter, & Kinzig, 2004). Socio-ecological systems were hypothesised to cycle within a particular basin of attraction, although external perturbations at critical times may, depending on circumstances, transform it into a neighbouring basin of attraction representing a significantly different type of system, particularly if it is close to a critical threshold (Walker & Meyers, 2004). Resilience was then defined as the amount of effort required to move from one basin of attraction into another. Basins of attraction and resilience are merely system characteristics, neither intrinsically good nor bad, and it is only when particular basins of attraction are considered more desirable than others that the concept of value enters. The notion of sustainability can then be thought of as the process of maintaining the system in a desirable basin. The adaptability of the system is the degree to which the components of the system can influence its internal dynamics and hence maintain its resilience (Walker et al., 2004).

While these approaches have been proposed by researchers mostly originating from the natural sciences and have then also been applied by interdisciplinary teams, many social scientists prefer the notion of 'transitions' between different, qualitatively states in socio-ecological systems, for example, between agrarian and industrial society (Fischer-Kowalski & Haberl, 2006). Transitions are also usually broken down into a formal sequence of phases. A common distinction is between a take-off phase in which the status quo is still in place, but there are various symptoms of its initial destabilisation, followed by an acceleration phase in which many rapid changes take place, and a subsequent stabilisation phase where changes are slowing down and the features of a new equilibrium begin to crystallise (Martens & Rotmans, 2002). A specific feature of the transition idea is that transitions take place between two qualitatively distinct states. No linear, incremental path leads from one state to the other, but rather a dynamic, possibly chaotic process of change. The transition notion allows qualitatively different states to be distinguished, in contrast to theories of growth or modernisation that assume certain homogeneity of the basic setting and gradual change over time. To some extent, socio-ecological transitions as understood in this tradition might be seen as flips between different states in a stability landscape, as suggested by the above-discussed model of Walker et al. (2004), but with the additional condition that these system changes are largely irreversible due to the legacies (Foster et al., 2003) emerging during the transition process.

Spontaneity and emergence are other important ingredients of the transition notion. It is neither possible for one state to be deliberately transformed into the other, nor for the process to be fully controlled – at least at present. Particularly if, as in sustainability research, the concept is applied to complex systems (such as societies or technology regimes), one is dealing with autocatalytic or autopoietic processes (Varela, Maturana, & Uribe, 1974) to which concepts of orderly governance, steering or management cannot be applied. It is commonly assumed, however, that there is increased potential for disturbance or of intervention into the system during the take-off phase of a transition, when the old interrelations are breaking apart but no clear directionality of change has yet been established (Rotmans, Kemp, & Van Asselt, 2001; Berkhout, Smith, & Stirling, 2003). There need not be a contradiction between this three-phase scheme (take off – acceleration – stabilisation) and the four-phase adaptive cycle (also called the 'lazy eight') scheme discussed above. However, there is an obvious similarity between the stabilisation phase and the highly connected, stable *k* phase and between the pioneer-dominated transient *r* phase and the acceleration phase. Whether or not it is helpful for the analysis of socio-ecological systems to break up the take-off phase in two phases, one of release (Ω) and one of reorganisation (α), may depend on the circumstances.

A socio-ecological transition that is currently being investigated in great detail due to its significance for current sustainability problems is the 'great transformation' (Polanyi, 1957) from agrarian to industrial society. Historical examples that have recently been investigated include the UK, the forerunner of industrialisation, and Austria (Krausmann & Haberl, 2002; Krausmann, 2004; Schandl & Schulz, 2002). These case studies suggest that industrialisation replaces one set of sustainability problems – above all those related to balancing land, labour, agricultural productivity and population (Boserup, 1965, 1981; Netting, 1993) – with new ones, above all those related directly or indirectly to the surging use of non-renewable resources, most prominently fossil fuels (Haberl et al., 2004; Hall, Cleveland, & Kaufmann, 1986). From the perspective of a member of industrial society – which most researchers are – this could be misinterpreted as a more or less completed process, encapsulated in the safe territory of the past. This point of view neglects, however, that about two-thirds, if not three quarters of the world population are currently in the midst of exactly this transition process, thus aggravating many of the global sustainability problems such as atmospheric change, climate change and rapid land-use change (Fischer-Kowalski & Haberl, 2006).

A theme common to all of these approaches is that because many aspects of the behaviour of socio-ecological systems are to some extent unpredictable due to cross-scale interactions, emergent properties and the existence of thresholds, the only realistic way to intervene into them is to follow a process of 'adaptive learning'. As societies and ecosystems co-evolve, decisions and actions made within society alter ecosystems and vice versa, and the perception of this co-evolutionary process in turn influences future decisions. This has implications for the role of researchers in the study of such systems – no longer are they external observers, but they have become intrinsic parts of the system, an approach sometimes referred to as 'post-normal' science (Funtowicz & Ravetz, 1993; Luks, 1996; Waltner-Toews, Kay, Neudoerffer, & Gitau, 2003; Waltner-Toews & Kay, 2005). The task of researchers is to understand the current structure and dynamic processes of socio-ecological systems, to identify possible attractors and the pathways between them and to inform other stakeholders within the system of possible trajectories. This results in a self-reflective research process, one that explicitly considers the perspectives of involved citizens, scientists and managers, and the dominant narratives of each of these groups. Cybernetics has termed such a process a second-order observation approach. In this kind of research process, the problem of communication between different actors cannot be managed easily (Waltner-Toews & Kay, 2005), but is discussed in more detail in Section 26.3.4 on participatory approaches below.

26.2.4 Structure of Socio-ecological Systems

As already mentioned, socio-ecological systems can be seen as 'open systems' operating far from equilibrium (Kay et al., 1999), with material, energy and information flowing both into and out of them. It is the way in which their internal social, economic and biophysical components are organised in relation to one another that determines how these flows are used and traded (Matthews & Selman, 2006). Networks of different types within socio-ecological systems contribute significantly to their functioning and hence to their resilience and adaptivity. For example, ecological networks determine the relationships between different species contained in the system (e.g. who eats whom or who lives on whom). Similarly, social networks play an important role in the way that information persists in the system, the rate at which it spreads through the system and the degree to which this information can be used. Some authors have developed models of social networks to extend the classical approach to innovation diffusion by distinguishing between the effects of initial information passed to a subset of farmers from institutions and subsequent discussion among farmers themselves (Deffuant et al., 2002).

The structures of such networks have a major influence on the way they function, and there has been significant interest in the theoretical aspects of network structure recently, particularly 'small-world' (Watts & Strogatz, 1998) and 'scale-free' (Barabási, Albert, & Jeong, 2000) networks, and how this structure influences network resilience (Barabási, Albert, Jeong, & Bianconi, 2000). There is a rich literature on modelling studies of the relationship between species diversity and stability in ecological food webs (McCann, 2000) although there is no clear consensus – in some cases there has been a positive correlation and in others a negative. For example, a case study found that resilience in ecological food webs (i.e. networks) increased with their connectivity, but was independent of species richness and omnivory (Dunne, Williams, & Martinez, 2002). These studies were on well-defined but static food webs (i.e. the networks were not growing), and it may be that part of the reason for these conflicting results is due to the dynamic nature of the relationships. Interestingly, the adaptive cycle discussed above hypothesises that resilience is positively correlated with capacity and connectedness in the *r* phase, but negatively correlated in the *K* phase. Certainly, the socio-ecological systems described by the adaptive cycle framework are dynamic and growing rather than static, so there is a clear need to investigate the relationships between these properties in that context. Barabási, Albert, and Jeong (2000) It was found that scale-free networks could be generated, first, if they were growing, and second, if new additions to the network were not attached randomly, but 'preferred' already highly connected ones (Barabási, Albert, Jeong, et al., 2000). It seems reasonable to assume that other network architectures can be generated by different assembly rules.

In relation to socio-economic networks, the role and development of inequality of the nodes is of particular interest. Tilly's concept of 'durable inequality' that explains the process by which networks are formed and grow, and the concomitant diffusion of new ideas resulting in rewards persistently accumulating to particular groups (e.g. wealthier households and men) is an area that needs to be explored (Tilly, 1988). Three different types of information processing in a socio-ecological system have been proposed which they have termed egalitarian, hierarchical and distributed, and which can be related to the structure of the social networks operating (van der Leeuw & Aschan-Leygonie, 2005). However, while much of the evolution of inequality, networks and information transfer has been theorised, little has been simulated, despite it being identified as an emerging research agenda. Modelling studies need to explore more sophisticated general concepts of socio-economic networks, including that of 'supracriticality' as a function of network size and connectivity (Kauffmann, 1995), and the idea that wide-ranging networks of many 'weak ties' facilitate transfer of information, the diffusion of innovation and ultimately promotes economic development, while limited networks of strong ties lead to economic stagnation (Granovetter, 1973, 1995).

26.3 Contribution to LTSER from the Social Sciences

26.3.1 *Coping with Diversity: A Plethora of Social Science Paradigms*

Classical LTER, if it considers social factors at all, is dominated by paradigms and concepts of society, or humans, which have been developed by natural scientists. LTSER implies the need not only to consider human interventions into ecosystems but to acknowledge that socio-economic dimensions of environmental change cannot be adequately understood without integrating researchers from the social sciences. This owes to the fact that changes in biodiversity and ecosystems are determined to a large extent by human activities, which cannot be grasped without this expertise (Delbaere, 2005).

One feature that characterises social sciences is the heterogeneity of paradigms and theories. Selecting an approach that is useful for LTSER requires that social scientists understand the questions asked, and problems faced, and paradigms applied in ecological research. The concepts selected must be compatible with natural scientific reasoning, they must explain social processes without denying biophysical processes, and they must open points of contact to which natural sciences can connect and relate. That is, neither must society be seen as 'just one additional component' or subsystem of ecosystems, nor is it useful to conceptualise humans as a mere disturbance to otherwise well-functioning ecosystems.

At the same time, social scientists have to accept that society is related to, even dependent on, ecosystems, a notion that seems hard to swallow for social scientists since the work of Max Weber and Emile Durkheim (Catton, 2002). The basic existence of society as an entity comprising biophysical structures, as discussed above, implies that it is dependent on, and in this sense part of and influenced by, natural processes (Dunlap & Catton, 1994). Society would not exist without human organisms and their capacity for reproduction and without being able to organise all the other biophysical flows (its socio-economic metabolism) required to maintaining its integrity. Society can, however, not be understood merely by observing its biophysical structures. In the social sciences, the most common usage of the term 'society' refers to a functional and delineated social unit that shares common cultural traits as well as to a politico-administrative unit in which decision making and executive powers are applied (Weisz et al., 2001).

This requires another, qualitatively different, system to be taken into account, namely the purely symbolic system of recursive communication (Luhmann, 1986, 1995). This system is immaterial, and it is autopoetic, that is, it is self-referential and is able to structure itself and create dynamics endogenously (Varela et al., 1974). We do not follow Luhmann, however, in denoting this system of recursive communication as society, because this would deprive society of all biophysical components – a notion of society obviously not very useful to understand society–nature interaction. Rather, we adopt the notion of society as a hybrid of this system of recursive communication and a part of the material world denoted here (Fig. 26.1) as 'biophysical structures of society' (Fischer-Kowalski & Weisz, 1999; Weisz et al., 2001). 'Culture' consists of material components or artefacts, such as religious objects, tools, machinery, infrastructure, and 'immaterial' or symbolic elements such as beliefs, arts, knowledge, languages.

Similarly, the notion of 'economy' gains a double meaning: in its symbolic representation, it may be thought of as the communicative processes involved in organising production and consumption – in industrial society the communicative subsystem delineated by the communicative code of money (almost the only set of issues addressed by professional economists today) – on a biophysical level it can be understood as the organisation of the biophysical flows required to sustain a society's biophysical structures, i.e. humans, livestock and artefacts. Obviously, then, human and animal labour and artefacts required for production ('capital') are important in this context.

A third important theme is the study of governance (Ostrom, 1996). In order to support a transition towards sustainability, LTSER explores decision-making processes at different scales to understand conflict as a basis for reconciling divergent goals amongst stakeholders (Adams, Brockington, Dyson, & Vira, 2003; Dietz, Ostrom, & Stern, 2003) and to reduce the vulnerability of people, places, and ecosystems (Turner et al., 2003). Within LTSER, good governance is understood as the combined effort of society to implement and enforce rules related to the provision of individual and collective goods and services to sustain local livelihoods without compromising ecosystem health. This requires understanding of how access to, use and exchange of resources are managed and negotiated in practice, questions obviously necessitating the inclusion of stakeholders in the research process.

This discussion reveals at least three entry points for social sciences that will be discussed in the remainder of this section: (a) the analysis of culture, both in its biophysical and its symbolic meaning. Here we can build on a tradition sometimes referred to as 'ecological anthropology' (e.g. Orlove, 1980) or 'cultural ecology' (e.g. Steward, 1955; White, 1959) that has yielded a rich body of studies of society–nature interaction (e.g. Boserup, 1965; Grünbühel, Haberl, Schandl, & Winiwarter, 2003; Netting, 1993; Singh, 2003). (b) The analysis of the interplay of the economy and ecosystems, a field of expertise today mostly referred to as 'ecological economics' (Costanza, Cumberland, Daly, Goodland, & Norgaard, 1998; Martinez-Alier, 1990). An aspect that may be particularly valuable at local and regional scales that play an important role in LTSER is the focus on the interrelations between time use, land use and income generation (Schandl & Grünbühel, 2005). (c) A focus on governance involving the participation of stakeholders and their influence in decision making (Hare & Pahl-Wostl, 2002; Pahl-Wostl, 2002; Kasemir, Jäger, Jaeger, & Gardner, 2003). A final subsection will discuss how these concepts can be integrated with approaches from the natural sciences in innovative modelling tools.

26.3.2 Building on the Rich Tradition of Ecological Anthropology

In social and cultural anthropology, stress on the ecological dimension in the study of social and cultural systems is relatively recent and came about as a consequence of increasing interest in ecosystem research within biological ecology. Ecological anthropology, or an ecosystem approach to anthropology, focuses on the study of complex relations between people and their environments. In other words, it is a study of the relations among population dynamics, social organisation, culture of human populations and the environment they inhabit from both synchronic and diachronic perspectives (Orlove, 1980). Practitioners of this discipline direct their attention to an understanding of the ways by which a particular population, intentionally or unintentionally, shapes its environment and is shaped by it (Barnard & Spencer, 1996). Ecological anthropology draws much from the systems theory approach, whereby human populations are regarded as one of the components of the ecosystem. In its most classic sense, ecological anthropology promulgated the study of energy flows between the human and the ecosystem, in particular the energetic return upon investment, and as such, interpreting cultural behaviours and subsistence patterns (Rappaport, 1967, 1971).

An important research agenda in treating human populations as part of ecosystems has been to understand human adaptability to various forms of environmental stress – physiological, cultural and behavioural. In doing so, there has been increasing interest in using ecological anthropology as a strategy for studying a wide range of human responses to environmental problems, to social constraints and to past solutions to environmental problems (Vayda & MacKay, 1977). Studies relate to the various ways populations have responded to environmental forces, such as high levels of frost in highland New Guinea (Waddell, 1975), storms (Bayliss-Smith, 1974), droughts (Lees, 1974; Reyna, 1975), famines (Krech, 1978; Orlove & Custtred, 1980) and earthquakes (Oliver-Smith, 1977).

Several studies in ecological anthropology emphasise the 'long-term' aspect by which mechanisms of change can be understood. This relates to the idea of 'transition' as discussed earlier in this chapter. Ecological anthropologists have carried out several studies to highlight the various drivers and initial conditions that may be attributed to changes in the socio-ecological system. Prominent among them are an examination of demographic variables and production systems (Boserup, 1965), the response of populations to sudden or prolonged environmental stress (see previous paragraph), the formation and consolidation of adaptive strategies (Bennett, 1976; Bettinger, 1978) and response to globalisation of economic and production processes (Friedman, 1974).

For LTSER, the field of ecological anthropology appears to be quite promising insofar as it allows a wide range of studies and methods to be harnessed, providing insights into the ways humans interact with their environment. In particular, we are able to gain from long-term studies relating to how populations have transformed their environment and how they have been transformed as a consequence. These relate to an understanding of socio-economic drivers of environmental change within the framework of LTSER. Ecological anthropology also provides insights into human decision making and the choices they possibly would make under given resource constraints and opportunities. At the same time, the global environmental change community can enhance its understanding on the ways human populations respond to various forms of environmental stress, which in turn may provide valuable insights into possible human responses to ongoing global environmental change.

26.3.3 Relating Time, Land and Income: Applying Concepts from Ecological Economics to the Local Level

Ecological economics studies the relations between the economy and the environment in order to address questions of sustainability using a variety of methods (Martinez-Alier, 1987). One specifically useful approach for LTSER is the analysis of the interrelations between time use, land use and income at the local and regional levels (e.g. Schandl & Grünbühel, 2005). In linking time use, land use and income generation we start with the so-called magic triangle of sustainability (Fischer-Kowalski, 1997a) on a local level. Looking at the 'magic triangle of sustainability' we identify three

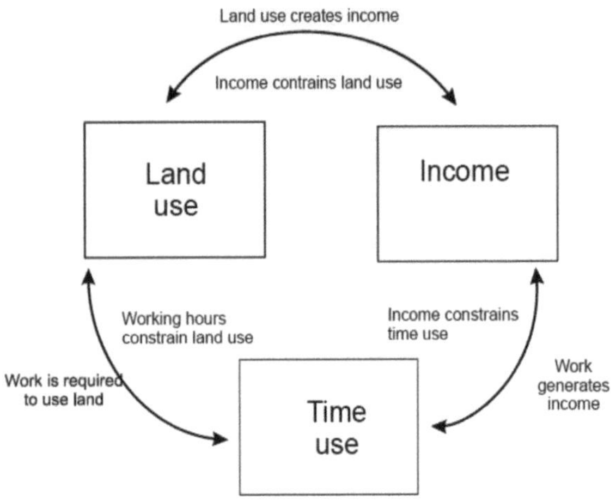

Fig. 26.2 Basic interrelations between time use, land use and income assumed in the internal structure of agricultural agents (= farms)

dimensions along which a social unit can be analysed: the economic, the social and the ecological dimension. On a local level the ecological dimension might be defined as the way land is used in a specific area. The social dimension might be defined as the way and quality of life of a specific social unit in a specific area. Our main indicator to describe the way/quality of life is the way time is used ('time use') by the members of the social unit. The economic dimension might be defined as the monetary income of a specific social unit (household, person, community) in a specific area. The three dimensions are highly interdependent (Fig. 26.2).

We start with the interplay between land use and time use, given the condition that the system is closed in regard to available working hours. For example, at the local level, a village would comprise of a fix amount of land ('total available land') and a certain amount of time ('total available time') depending on the number of inhabitants. Of the total available land, only part is available for income generation (part may be required, for example, for living space). Similarly, the 'disposable human time' is only a part of the total time available since every individual requires a certain amount of time for 'basic personal reproduction' (such as sleeping and eating) and 'extended personal reproduction' (such as education and leisure). In this way, both land and time are a biophysical constraint. A specific way of using land requires a specific amount of working hours and vice versa. As land use requires working hours, it constrains the time budget that can be used for other activities not related to land use (leisure time, reproduction, etc.).

Similarly, land and income constrain each other. This is to say that each square meter of land is able to generate a certain amount of income. In other words, income is constrained by the quality and quantity of land available. The income on the other side determines how the land is used: if there is a need for more income, it may be that more land, if available, will be brought under production or the use of existing land is intensified. Furthermore, the quality of land use relates to the means of production or available technology a social unit can afford (such as tillers and tractors).

Similarly, the relation between time use and income can also be determined. The disposable working time in a given society is determined by several factors such as the total number of inhabitants, life expectancy, dependency ratio and working age. This may vary by culture and region. The disposable working time has a direct bearing on the amount of money an individual can earn. Alternatively, the need for less income would mean less working hours required, and the need for a higher income may restrict time to be invested in other activities such as leisure and vacations.

All three dimensions have their inherent dynamics as they are subject to specific systems dynamics. This is to say, how time is used largely depends on the social/cultural system a social unit is part of (e.g. social values and norms, infrastructure). The income is highly dependent on the dynamics of economic systems (market, prices, etc.). Finally, land use is constrained by the specific features of the local ecosystem (e.g. rice does not grow in arid areas) as well as on global environmental dynamics. As research shows, these considerations are especially applicable in rural areas and farming systems. Several ongoing studies are now integrating land-time-income analysis within computer models such as agent-based modelling (Gaube, Adensam, Erb, & Haberl, 2005). It would have to be proved how and if they apply to industrial areas. LTSER sites can be used as test areas for undertaking land-time-income analysis so as to provide insights into these biophysical constraints and opportunities in relation to possible economic options and ecological impacts.

26.3.4 Participation and Decision Making

A major influence on long-term studies has been the development of decision-making models (or actor-based models). These models permit better analysis of the parameters of behaviour and variations of behaviour within the different human populations, including conflict and competition and an understanding of the processes which generate economic, political and social relations. Decision-making models may either be cognitive/naturalistic or microeconomic models. The former borrows much from cognitive anthropology that depicts actual psychological processes of decision making by locating alternatives and the procedures for choosing among them, while the latter resemble economic models of choice making under a set of resource constraints (Orlove, 1980).

It has repeatedly been argued in this chapter that sustainability research cannot solely focus on ecosystems, but has to analyse social parameters as well. However, in most cases analysing social parameters is not enough to enable us to set powerful interventions counteracting a decrease of biodiversity levels or the quality of ecosystems. What is needed is a specific method of analysis – an analysis that allows actors to participate throughout the whole research process, starting with defining the problem that should be analysed and ending with planning or initialising specific interventions. In other words, social actors or social systems should be able to learn throughout the research process or should at least be stimulated in some way. Participation is the key to achieve this (see for example Hare & Pahl-Wostl, 2002; Pahl-Wostl, 2002).

The term participation can be defined as awareness of and identification with the research conducted in a particular locale as well as active dialog with the researchers and stakeholders with respect to the process. Research should take place with the inclusion of the relevant stakeholders in a social system, integrating their interests and defining common research goals. The classical research approach starts out with a research question, which the researcher defines according to the current state of the art and its research demands. Participatory research means defining the research question and the scientific interest together with relevant stakeholders. In order to be able to do this, the area where the study is being conducted must be known to the researchers and the actors/stakeholders identified.

Stakeholders are actors that are directly involved in the problem to be investigated. After having defined the actors in the field, the group of stakeholders who will be engaged in the process may have to be narrowed down for practical reasons. While the selected group of stakeholders should bear some relevance to the problem, it is up to the researcher to lay down the criteria of selection as long as the criteria are transparent and potentially subject to criticism. In most cases, the selection criteria for stakeholders highly depend on the needs of the actors involved, the willingness to cooperate and the level of influence they wield in guiding social processes in the study area.

Research questions and goals are defined in cooperation with the stakeholder group in order to make practical use of the research results. Nevertheless, social goals and visions of the future must be translated into scientific categories and variables for their usefulness in the research process, i.e. they have to be scientifically operationalised. In return, scientific evidence has then to be transposed into socially relevant information in order to serve as a basis for decision making.

According to the intense cooperation with the actors, the dynamics, goals and directions participative research follows, strongly depend on the needs of the actors and on the inner dynamics of the system analysed. Thus, research design differs from classical research approaches as it has to be more flexible in various aspects (e.g. definitions of research goals, selections of actors involved, milestones planned, and methods applied). Ownership is crucial in participatory research. Research results can never be applied to local decision making if the actors are not aware of the research questions and problems addressed. The stakeholder group must (at least partly) share the motivations and interests of the researcher and appreciate the methods applied in the field. Local actors must support research activities and perceive them as a possible basis for taking decisions on their own future. Ideally, participative research contributes to democratisation and involves the citizen in public decisions of the community.

26.3.5 The Utility of Models for Integrating Socio-economic and Ecological Approaches

Because socio-ecological systems are complex, and because it is not possible to manipulate real systems experimentally, modelling can help in integrating these processes into a common framework in order to analyse the likely impacts of internal and external drivers, both solely and in conjunction with one another. Integrated models should focus not only on the economic consequences but also on the social and ecological outcomes. Modelling can be especially useful to integrate social science-based approaches with concepts from the natural sciences (van der Leeuw, 2004). Existing models are, however, largely unattractive in this respect, as most of them reflect theories and concepts developed within single scientific disciplines. Global integrated assessment models such as IMAGE, on the other hand, that were derived by coupling a large number of existing disciplinary-based 'sectoral' models (Alcamo, Kreileman, & Leemans, 1996) are of limited applicability at local scales such as those dealt with in LTSER. A new generation of models that can deal with local situations and aim to integrate biophysical (e.g. land use) issues with socio-economic factors is currently being explored by many groups of researchers around the world (e.g. McConnell, 2001; Janssen, 2004; Matthews, 2006).

In this context, the aim of modelling goes beyond prediction and is seen as a strategy for exploring how elements of socio-ecological systems traditionally described by different disciplines are related to one another and for learning how they might be combined to arrive at a useful representation of the overall situation. Modelling may be useful to understand how the situation might unfold, thus helping researchers and stakeholders to devise ways of achieving desirable outcomes (Fig. 26.3). This non-linear process raises interesting questions about reflexivity and its ability to introduce uncertainty in its own right; for example, a model that could predict the behaviour of the stock market would be used by investors to alter their behaviour, but if everyone were to do this, the predictions of the model would be falsified. Should, therefore, models of socio-ecological systems contain representations of themselves in order to realistically model the behaviour of the social processes they contain?

A modelling paradigm which is potentially suitable for linking the biophysical and socio-economic characteristics of a system is multi-agent simulation or agent-based modelling. Originating from the field of distributed artificial intelligence, these models consist of a number of 'intelligent' virtual agents which are sensitive to other agents and interact with both them and their environment and can change their actions as a result of events in the process of interaction. Agents typically have only partial knowledge of the system

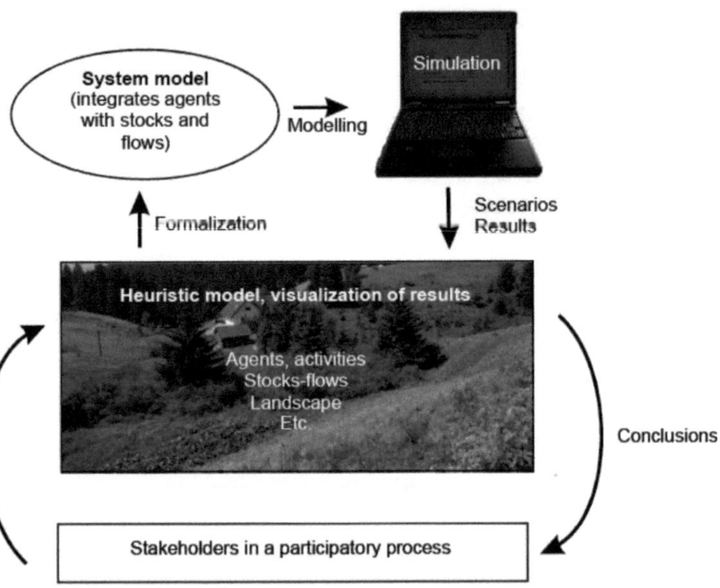

Fig. 26.3 Schematic representation of the use of models in a participatory process. Own figure, drawn based on a concept of Berger (2004)

as a whole, but a key characteristic is their ability to communicate and exchange information with each other. The behaviour of the whole system (i.e. the virtual community) depends on the aggregated individual behaviour of each agent. Agent-based modelling is dynamic as compared to traditional approaches such as linear programming and is particularly suitable for looking at processes over time (Axtell, Andrews, & Small, 2002; Janssen, 2004; McConnell, 2001; Pahl-Wostl, 2002).

- Reactive agents decide on actions directly on the basis of what they have sensed around them. For example, they may select a particular land use based on soil type.
- Deliberative agents are those that, for example, reflect upon alternative courses of action, and select one of them for execution, i.e., they plan. An example of this might be that, for a particular land unit, a number of land uses are possible, but one is selected on the basis of one or more criteria, such as relative return and/or labour availability.
- Adaptive agents change their behaviour in the light of changing circumstances, implying an element of learning – e.g. based on the performance in the past year of a particular land use, new or modified land uses are selected.
- Social agents communicate and cooperate with other agents, keeping histories of their interactions and updating their beliefs (using Bayesian updating) after observing their environment and the behaviour of other agents.

Depending on the type of agent, each will consist of one or more of the following main components – (a) a component to manage communication with other agents, (b) a second to maintain knowledge (i.e. the agent memory), (c) a third to make decisions based on its knowledge and perception of the outside world and (d) the fourth to perceive the outside world. Thus, triggers for action by the agent may originate from three sources – those as a result of its own internal state, those as a result of requests from other agents and those as a result of some condition within the environment.

Validation of such models is difficult. Validation of individual components against observed data is possible and necessary, although this does not test for any errors introduced through linking them at a higher level. Two other approaches of validation are as follows:

- Accuracy in simulating a historical situation, for example, changes in commodity prices or demographic changes over a particular time period. Such an attempt at validation was made by using a model to simulate various aspects of three widely contrasting socio-economic regions, the results of which reflected current development trends in two of the regions with reasonable accuracy (Van Keulen, 1993).
- Comparing simulated results with outcomes expected by 'experts'. Validation would be positive when, for example, patterns of knowledge transmission predicted by the model matched those observed by sociologists or development professionals.

However, as the purpose of the model is usually to explore options for effecting change in rural communities, rather than predicting them, it is perhaps more important that the structure of the model and the assumptions incorporated into it are transparent and therefore well-documented. Provided these are known, they can also provide a focus for debate, and sensitivity analysis can be carried out to determine their relative importance to the overall system.

26.4 Integrating Ecosystems with Participation and Time Use: A Case Study from the Austrian LTSER Platform 'Eisenwurzen'

In this section, we will present a case study from a recently concluded project in the Austrian LTSER platform 'Eisenwurzen', one of the 10 European LTSER platforms, where we implement several of the approaches discussed above. The name of the region, Eisenwurzen ('origin of the iron'), refers to its historical role as an important iron-supplying region of Europe: A few hundred years ago, about 15% of the total amount of iron produced in Europe came from that region (Gruber, 1998). The topography ranges from hilly to mountainous, and mostly belongs to the northern limestone Alps. Due to its mountainous

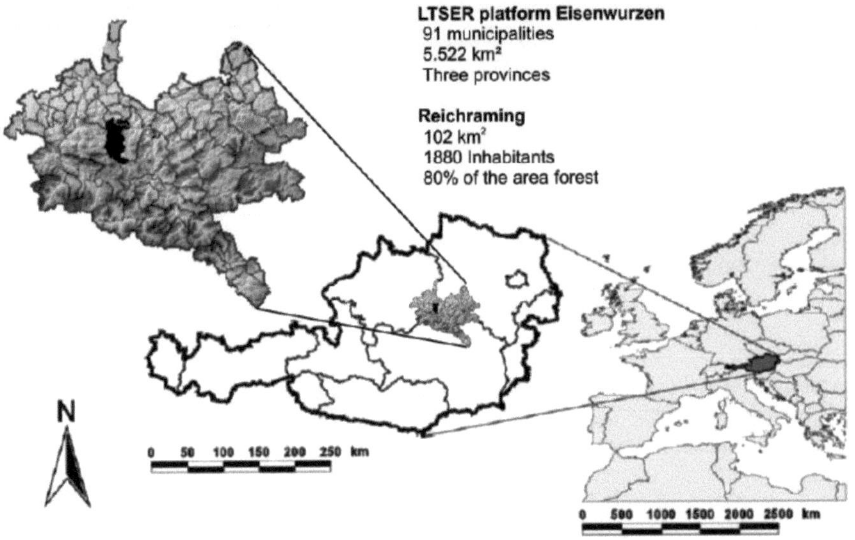

Fig. 26.4 Location of the study area. Reichraming (*marked black*) is one of the 91 municipalities belonging to the LTSER platform 'Eisenwurzen' which includes parts of the territory of three Austrian provinces, Oberösterreich ('Upper Austria'), Niederösterreich ('Lower Austria') and Steiermark ('Styria')

character, the region is hardly attractive for agriculture, in particular for intensive, high-yielding mechanised agriculture. A massive reforestation resulted which is today perceived as an important challenge for the continuation of human habitation in the region. It is estimated that forests presently cover at least 80% of the area of Reichraming. Agriculture is almost exclusively based on extensive cattle rearing and suffers from low incomes. Commuting to regional centres such as Steyr and Linz accounts for a significant proportion of gainful employment.

Within the larger region that is included in the LTSER platform is the municipality of Reichraming the area where the project is being implemented (see Fig. 26.4). The project 'LTSER Eisenwurzen' aims to develop an integrated model SERD (Simulation of Ecological Compatibility of Regional Development) to be able to simulate changes in income and workload of farmsteads as well as land use and material/substance flows (Gaube et al., 2009). The model will consist of two modules: (1) an agent-based actor's model and a social-economic and biophysical approach integrating stock and flow model (Fig. 26.5). Dynamics in the model will be driven by assumptions on changes in the external conditions, both socio-economic and political as well as environmental framework.

The agent-based model in SERD relates different groups of actors to each other depending on their impact on land use:

1. Primary actors: they are those land owners who affect land use change in the region directly. Important among them are the agrarian households who decide based on their knowledge, individual preferences and information about the environment and other agents how to use land. The decision-finding process of all farms is carried out along a 'sustainability triangle' in which the three core sustainability dimensions (social/ecological/economic) are dynamically interlinked with each other. The social dimension is represented by time invested by inhabitants on a farm, the economic dimension by the farm's income and the ecological dimension by land-use patterns (Fig. 26.2). Beside farmers, the municipality, the forestry agency of Austria, the National Park Kalkalpen and some of the households and enterprises are further land users of Reichraming and hence primary actors.
2. Secondary agents: these are groups of actors that have a direct impact on the primary agents as they set their framework conditions, e.g. landowners who do not use their land but lease it to others.

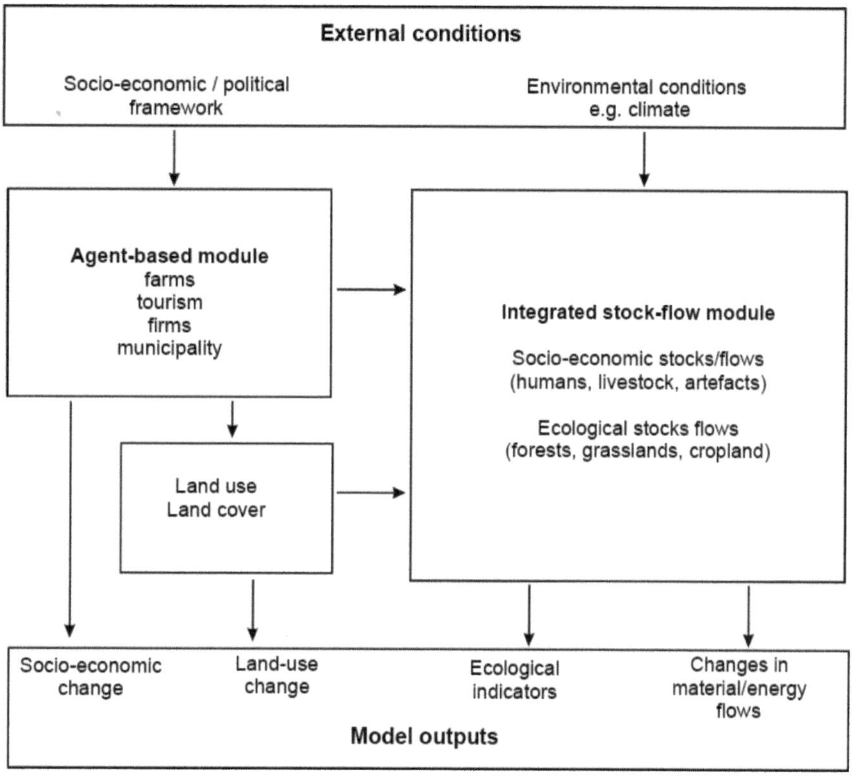

Fig. 26.5 Concept of the integrated system model SERD

3. Aggregated agents: finally, there are the aggregated agents such as the local government, associations and networks who have influence on the region.

An integral part of the project is the inclusion of the agents/stakeholders in the research process. Workshops will be organised where assumptions for the model will be discussed at all stages of the project. The actor model will be coupled with a social/economic/ecological material/substance flow model. Outputs of the agent-based model concerning changes in land-use and agricultural practices as well as in population and infrastructure are simultaneously inputs for the integrated stock-flow module of SERD which are further divided into two types of stocks and flows. On the one hand there is the socio-economic information about humans, livestock and artefacts, and on the other hand there are ecological stocks and flows from various land categories such as forests, grasslands and cropland. Substance flows, especially carbon and nitrogen flows are sensitive indicators for changes of relevant ecological processes and allow assessments on the ecological compatibility of changes in agricultural practices, land use, economic and social conditions.

The model outputs of SERD can be classified into four parts:

1. Assessment of social and economic trends,
2. Illustration of expectable land-use patterns in the future,
3. Changes of ecological indicators, and
4. Changes of local and regional substance flows.

SERD will allow the simulation of future scenarios, e.g. on the effects of improved collaboration between agriculture, tourism and the National Park, on the income and time use of farmsteads as well as land-use patterns and substance flows. A second application of SERD will be its ability to back-cast based on historical sources and data.

26.5 Conclusions

Integrating socio-economic dimensions into long-term ecological research is a challenging and ongoing endeavour. It requires fundamentally new, inter- and transdisciplinary scientific approaches. This is difficult due to the long-standing history of specialisation in natural sciences, social sciences, and the humanities (Huber, 1989, p. 67, own translation): 'The very idea to transcend the division between natural and social sciences (...) does not seem adequate. It is not possible to make the thick branches of a tree ungrown. (...) It would be no minor achievement if people on both sides would cease to believe to be closer to reality or to truth than the others. (...) We should strive (...) for a state of peaceful coexistence and, on the basis of this (...), create a controlled external trade that benefits both sides, hoping this could lead to some kind of co-evolution.'

Exploring integrated approaches to grapple with socio-ecological systems requires both conceptual, theoretical discussions such as those included in Sections 26.2 and 26.3 , and practical work in interdisciplinary case studies such as the Lower Danube Wetland System mentioned in the Introduction and the Austrian LTSER study discussed in Section 26.4. On a theoretical level, we suggest that the concepts of socio-ecological metabolism, transition concepts from the social sciences, and theories of complex adaptive systems can be combined to provide a good basis for integrating approaches from natural sciences, social sciences and the humanities.

On a practical level we suggest that concrete, interdisciplinary studies should be launched to be carried out at the emerging LTSER platforms with the explicit aim to test different approaches for inter- and transdisciplinary integration. In particular, we feel that the following considerations could be helpful in this context:

Transdisciplinarity – that is, the integration of stakeholders in the research process – is a useful tool to foster problem-oriented work. Challenging interdisciplinary teams to work in a problem-oriented way is useful to help in overcoming traditional boundaries between scientific disciplines. It helps in structuring research questions in novel ways, thus inspiring innovation in interdisciplinary projects. It also requires a high level of self-reflection of the researchers, as they themselves become part of the system to be analysed. On the other hand, high standards of scientific excellence in such projects are essential in order to prevent them from becoming purely consultancy work.

The use of models either formalised or heuristic is an important tool to foster interdisciplinary integration. Even though socio-ecological systems may be too complex to ever be adequately represented, let alone predicted, by formal models, the very process of constructing the model is of great help in fostering mutual learning in interdisciplinary teams (van der Leeuw, 2004). Moreover, these models can be used in transdisciplinary processes together with stakeholders and help to structure discussions on policy options to support sustainability.

One scientific paper, such as this one, cannot, of course, address all the challenges involved in designing LTSER. Other papers have focused on the situation in the United States (Redman et al., 2004) and on themes of LTSER (Haberl et al., 2006). Several groups within ALTER-Net are working on complementary projects tackling issues such as those related to scales and levels of socio-ecological systems (Dirnböck, personal communication) and on the utility of combining approaches from Political Ecology, Ecological Economics and Social Ecology in understanding drivers of biodiversity (Waetzold, personal communication). This multitude of approaches is certainly warranted, given the complexity of the problem at hand. We urgently require ever more cases where integrative approaches are tried and tested by interdisciplinary teams in order to foster mutual learning to address the sustainability issue.

Acknowledgements We acknowledge financial support by ALTER-Net, a Network of Excellence funded by the EU within its sixth Framework Programme, by the Austrian Federal Ministry of Education, Science and Culture within the projects 'partizipA' (Cultural Landscapes Research) and the project 'LTSER Eisenwurzen' (proVISION). This research also contributes to the Global Land Project (GLP). We thank H. Adensam, M. Fischer-Kowalski, K.H. Erb, S. Gingrich and F. Krausmann for useful discussions.

References

Abel, T. (1998). Complex adaptive systems, evolutionism, and ecology within anthropology: Interdisciplinary research for understanding cultural and ecological dynamics. *Georgia Journal of Ecological Anthropology, 2*, 6–29.

Adams, W. M., Brockington, D., Dyson, J., & Vira, B. (2003). Managing tragedies: Understanding conflict over common pool resources. *Science, 302*, 1915–1916.

Alcamo, J., Kreileman, E., & Leemans, R. (1996). Global models meet global policy: How can global and regional modellers connect with environmental policy makers? What

has hindered them? What has helped? *Global Environmental Change, 6*, 255–259.
Allen, T. H. F., Bandurski, B., & King, A. (1993). *The ecosystem approach: Theory and ecosystem integrity (Report to the Great Lakes Advisory Board).* Windsor, Ontario: International Joint Commission.
Axtell, R. L., Andrews, C. J., & Small, M. J. (2002). Agent-based modelling and industrial ecology. *Journal of Industrial Ecology, 5*, 10–13.
Ayres, R. U., & Simonis, U. E. (1994). *Industrial metabolism: Restructuring for sustainable development.* Tokyo, New York, Paris: United Nations University Press.
Barabási, A. L., Albert, R., & Jeong, H. (2000). Scale-free characteristics of random networks: The topology of the world-wide web. *Physica A: Statistical Mechanics and its Applications, 281*, 69–77.
Barabási, A.-L., Albert, R., Jeong, H., & Bianconi, G. (2000). Power-law distribution of the World Wide Web. *Science, 287*, 2115a.
Barnard, A., & Spencer, J. (1996). *Encyclopedia of social and cultural anthropology.* New York: Routledge.
Bayliss-Smith, T. (1974). Constrains on population growth: The case of the Polynesian outlier atolls in the precontact period. *Human Ecology, 2*, 259–295.
Bennett, J. W. (1976). *The Ecological transition: Cultural anthropology and human adaption.* New York: Pergamon Press.
Berkes, F., & Folke, C. (1998). Linking social and ecological systems for resilience and sustainability. In F. Berkes & C. Folke (Eds.), *Linking social and ecological systems. Management practices and social mechanisms for building resilience* (pp. 1–26). Cambridge: Cambridge University Press.
Berkhout, F., Smith, A., & Stirling, A. (2003). *Socio-technical regimes and transition contexts.* Brighton: University of Sussex.
Bettinger, R. L. (1978). Alternative adaptation strategies in the prehistoric Great Basin. *Journal of Anthropological Research, 34*, 27–46.
Boserup, E. (1965). *The conditions of agricultural growth – the economics of agrarian change under population pressure.* London: Earthscan.
Boserup, E. (1981). *Population and technological change – A study of long-term trends.* Chicago: The University of Chicago Press.
Carpenter, S., Walker, B., Anderies, J. M., & Abel, N. (2001). From metaphor to measurement: Resilience of what to what? *Ecosystems, 4*, 765–781.
Catton, W. R., Jr. (2002). Has the Durkheim legacy misled sociology? In R. E. Dunlap, F. H. Buttel, P. Dickens, & A. Gijswijt (Eds.), *Sociological theory and the environment. Classical foundations, contemporary insights* (pp. 90–115). Lanham, Boulder, New York, Oxford: Rowman & Littlefield.
Clark, W. C., Crutzen, P. J., & Schellnhuber, H.-J. (2004). Science for global sustainability. In H.-J. Schellnhuber, P. J. Crutzen, W. C. Clark, M. Claussen, & H. Held (Eds.), *Earth system analysis for sustainability. Report of the 91st Dahlem Workshop* (pp. 1–28). Cambridge, MA, London: MIT Press.
Costanza, R., Cumberland, J., Daly, H. E., Goodland, R., & Norgaard, R. B. (1998). *An introduction to ecological economics.* Boca Raton: CRC Press.
Croll, E., & Parkin, D. (1992). Anthropology, the environment and development. In E. Croll & D. Parkin (Eds.), *Bush base: Forest farm* (pp. 1–10). London: Routledge.
Daniels, P. L., & Moore, S. (2001). Approaches for quantifying the metabolism of physical economies, Part I: Methodological overview. *Journal of Industrial Ecology, 5*, 69–93.
Dearing, J. A., Graumlich, L. J., Grove, R., Grübler, A., Haberl, H., Hole, F., et al. (2006). *Integrating socio-environment interactions over centennial timescales: Needs and issues. Dahlem Workshop Report 96.* Cambridge, MA: The MIT Press.
Deffuant, G., Huet, S., Bousset, J. P., Henriot, J., Amon, G., & Weisbuch, G. (2002). Agent-based simulation of organic farming conversion in Allier department. In M. Janssen (Ed.), *Complexity and ecosystem management: The theory and practice of multi-agent systems* (pp. 158–187). Cheltenham, UK: Edward Elgar.
Delbaere, B. (2005). European policy review: Biodiversity research to support European policy. *Journal of Nature Conservation, 13*, 213–214.
Descola, P., & Pálsson, G. (1996). *Nature and society. Anthropological perspectives.* London, New York: Pergamon.
Dietz, T., Ostrom, E., & Stern, P. C. (2003). The struggle to govern the commons. *Science, 302*, 1907–1912.
Dunlap, R. E., & Catton, W. R., Jr. (1994). Struggling with human exemptionalism: The rise, decline and revitalizaton of environmental sociology. *The American Sociologist, 25*, 5–29.
Dunlap, R. E., & Catton, W. R. (2002). Which function(s) of the environment do we study? A comparison of environmental and natural resource sociology. *Society and Natural Resources, 15*, 239–249.
Dunne, J. A., Williams, R. J., Martinez, N. D. (2002). Network structure and biodiversity loss in food webs: Robustness increases with connectance. *Ecology Letters, 5*, 558–567.
EEA. (1995). *Europe's environment: The Dobris assessment.* Copenhagen: European Environment Agency.
Eurostat. (2001). *Economy-wide material flow accounts and derived indicators. A methodological guide.* Luxembourg: Office for Official Publications of the European Communities.
Fischer-Kowalski, M. (1997a). Dynamik und Selbststeuerung industrieller Gesellschaften. In M. Fischer-Kowalski, H. Haberl, W. Hüttler, H. Payer, H. Schandl, V. Winiwarter, & Zangerl-Weisz, H. (Eds.), *Gesellschaftlicher Stoffwechsel und Kolonisierung von Natur. Ein Versuch in Sozialer Ökologie* (pp. 203–221). Amsterdam: Gordon & Breach Fakultas.
Fischer-Kowalski, M. (1997b). Society's metabolism: On the childhood and adolescence of a rising conceptual star. In M. Redclift & G. R. Woodgate (Eds.), *The international handbook of environmental sociology* (pp. 119–137). Cheltenham, Northhampton: Edward Elgar.
Fischer-Kowalski, M., & Haberl, H. (1997). Tons, joules and money: Modes of production and their sustainability problems. *Society and Natural Resources, 10*, 61–85.
Fischer-Kowalski, M., & Haberl, H. (2006). *Global environmental change and socio-ecological transitions.* Cheltenham: Edward Elgar.

Fischer-Kowalski, M., Haberl, H., Hüttler, W., Payer, H., Schandl, H., Winiwarter, V., et al. (1997). *Gesellschaftlicher Stoffwechsel und Kolonisierung von Natur. Ein Versuch in Sozialer Ökologie.* Amsterdam: Gordon & Breach Fakultas.

Fischer-Kowalski, M., & Weisz, H. (1999). Society as hybrid between material and symbolic realms. Toward a Theoretical Framework of Society-Nature Interaction. *Advances in Human Ecology, 8*, 215–251.

Folke, C., Carpenter, S., Elmqvist, T., Gunderson, L. H., Holling, C. S., Walker, B., et al. (2002). Resilience and sustainable development: Building adaptive capacity in a world of transformations. *Ambio, 31*, 437–440.

Foster, D., Swanson, F., Aber, J., Burke, I., Brokaw, N., Tilman, D., et al. (2003). The importance of land-use legacies to ecology and conservation. *BioScience, 53*, 77–88.

Friedman, J. (1974). Marxism, structuralism and vulgar materialism. *Man, 9*, 444–469.

Funtowicz, S. O., & Ravetz, J. R. (1993). Science for the post-normal age. *Futures, 25*, 739–755.

Gaube, V., Adensam, H., Erb, K.-H., & Haberl, H. (2005, August). *Modelling the impacts of CAP 2006 on sustainable regional development considering land use change, social structure and economic performance. An Austrian case study.* Paper presented at the "45th Congress of the Regional Science Association", Amsterdam.

Gaube, V., Kaiser, C., Wildenberg, M., Adensam, H., Fleissner, P., Kobler, J., et al. (2009). Combining agent-based and stock-flow modelling approaches in a participative analysis of the integrated land system in Reichraming, Austria. *Landscape Ecology, 24*(9), 1149–1165.

Granovetter, M. (1973). The strength of weak ties. *American Journal of Sociology, 78*, 1360–1380.

Granovetter, M. (1995). *Getting a job: A study of contacts and careers.* Chicago: University of Chicago Press.

Greenland, D., & Kittel, T. G. F. (2002). Temporal variability of climate at the US Long-Term Ecological Research (LTER) sites. *Climate Research, 19*, 213–231.

Gruber, B. (1998). Die Rohstoffe der oberösterreichischen Eisenwurzen. In: R. Sandgruber (Ed.), *Land der Hämmer – Heimat Eisenwurzen. Die Oberösterreichische Landesausstellung 1998 und die Region Pyhrn-Eisenwurzen* (pp. 24–35). Salzburg: Residenz-Verlag.

Grünbühel, C. M., Haberl, H., Schandl, H., & Winiwarter, V. (2003). Socio-economic metabolism and colonization of natural processes in SangSaeng village: Material and energy flows, land use, and cultural change in Northeast Thailand. *Human Ecology, 31*, 53–87.

Gunderson, L., & Holling, C. S. (2002). *Panarchy. Understanding transformations in human and natural systems.* Washington, DC: Island Press.

Haberl, H. (1997). Human propriation of net primary production as an environmental indicator: Implications for sustainable development. *Ambio, 26*, 143–146.

Haberl, H. (2001a). The energetic metabolism of societies, Part I: Accounting concepts. *Journal of Industrial Ecology, 5*, 11–33.

Haberl, H. (2001b). The energetic metabolism of societies, Part II: Empirical examples. *Journal of Industrial Ecology, 5*, 71–88.

Haberl, H., Erb, K.-H., Krausmann, F., Loibl, W., Schulz, N. B., & Weisz, H. (2001). Changes in ecosystem processes induced by land use: Human appropriation of net primary production and its influence on standing crop in Austria. *Global Biogeochemical Cycles, 15*(4), 929–942..

Haberl, H., Fischer-Kowalski, M., Krausmann, F., Weisz, H., & Winiwarter, V. (2004). Progress towards sustainability? What the conceptual framework of material and energy flow accounting (MEFA) can offer. *Land Use Policy, 21*, 199–213.

Haberl, H., Winiwarter, V., Andersson, K., Ayres, R. U., Boone, C., Castillo, A., et al. (2006). From LTER to LTSER: Conceptualizing the socioeconomic dimension of long-term socioecological research. *Ecology and Society, 11*, 13. From URL: http://www.ecologyandsociety.org/vol11/iss2/art13/.

Haberl, H., Gaube, V., Díaz-Delgado, R., Krauze, K., Neuner, A., Peterseil, J., et al. (2009). Towards an integrated model of socioeconomic biodiversity drivers, pressures and impacts. A feasibility study based on three European long-term socio-ecological research platforms. *Ecological Economics, 68*(6), 1797–1812.

Hall, C. A. S. (1995). *Maximum Power. The Ideas and Applications of H.T. Odum.* Niwot, Colorado: The University Press of Colorado.

Hall, C. A. S., Cleveland, C. J., & Kaufmann, R. K. (1986). *Energy and resource quality, the ecology of the economic process.* New York: Wiley Interscience.

Hare, M., & Pahl-Wostl, C. (2002). Stakeholder categorisation in participatory integrated assessment processes. *Integrated Assessment, 3*, 50–62.

Hobbie, J. E., Carpenter, S. R., Grimm, N. B., Gosz, J. R., & Seastedt, T. R. (2003). The US long term ecological research program. *BioScience, 53*, 21–32.

Holling, C. S. (1973). Resilience and stability of ecological systems. *Annual Review of Ecological Systematics, 4*, 1–24.

Holling, C. S. (1986). The resilience of terrestrial ecosystems: Local surprise and global change. In W. C. Clark & R. E. Munn (Eds.), *Sustainable development of the biosphere* (pp. 292–320). Cambridge: Cambridge University Press.

Holling, C. S. (2001). Understanding the complexity of economic, ecological, and social systems. *Ecosystems, 4*, 390–405.

Holten-Andersen, J., Paalby, H., Christensen, N., Wier, M., & Andersen, F.M. (1995). *Recommendations on Strategies for Integrated Assessment of Broad Environmental Problems. Report Submitted to the European Environmental Agency (EEA) by the National Environmental Research Institute (NERI).* Copenhagen: National Environmental Research Institute.

Huber, J. (1989). Eine sozialwissenschaftliche Interpretation der Humanökologie. In B. Glaeser(Ed.), *Humanökologie, Grundlagen präventiver Umweltpolitik* (pp. 57–75). Opladen: Westdeutscher Verlag.

Janssen, A. M. (2004). Agent-based models. In J. Proops & P. Safonov (Eds.), *Modelling in ecological economics* (pp. 155–172). Cheltenham, UK, Northampton, MA, USA: Edward Elgar.

Kasemir, B., Jäger, J., Jaeger, C. C., & Gardner, M. T. (2003). *Public participation in sustainability science. A handbook.* Cambridge, UK: Cambridge University Press.

Kates, R. W., Clark, W. C., Corell, R., Hall, J. M., Jaeger, C. C., Lowe, I., et al. (2001). Sustainability science. *Science, 292*, 641–642.

Kauffmann, S. (1995). *At home in the universe: The search for laws of complexity.* London: Penguin Books.

Kay, J. J., Regier, H. A., Boyle, M., & Francis, G. (1999). An ecosystem approach for sustainability: Addressing the challenge of complexity. *Futures, 31*, 721–742.

Koestler, A. (1967). *The ghost in the machine.* London, UK: Arkana.

Krausmann, F. (2004). Milk, manure and muscular power. Livestock and the industrialization of agriculture. *Human Ecology, 32*, 735–773.

Krausmann, F., & Haberl, H. (2002). The process of industrialization from the perspective of energetic metabolism. Socioeconomic energy flows in Austria 1830–1995. *Ecological Economics, 41*, 177–201.

Krausmann, F., Haberl, H., Erb, K.-H., & Wackernagel, M. (2004). Resource flows and land use in Austria 1950–2000: Using the MEFA framework to monitor society-nature interaction for sustainability. *Land Use Policy, 21*, 215–230.

Krech, S. I. (1978). Disease, starvation and Northern Athapaskan social organisation. *American Ethnology, 5*, 710–732.

Lees, S. H. (1974). Hydraulic development as a process of response. *Human Ecology, 2*, 159–175.

Luhmann, N. (1986). *Ökologische Kommunikation: Kann die moderne Gesellschaft sich auf ökologische Gefährdungen einstellen*? Opladen: Westdeutscher Verlag.

Luhmann, N. (1995). *Social systems.* Stanford: Stanford University Press.

Luks, F. (1996). Post-Normal Science, Dematerialisierung und die Ökonomie – Über den (wirtschafts) wissenschaftlichen Umgang mit Umweltproblemen. In J. Köhn & M. J. Welfens (Eds.), *Neue Ansätze in der Umweltökonomie* (pp. 89–108). Marburg: Metropolis.

Martens, P., & Rotmans, J. (2002). *Transitions in a globalising world.* Lisse: Swets & Zeitlinger Publishers.

Martinez-Alier, J. (1987). *Ecological economics. Energy, environment and society.* Oxford: Blackwell.

Martinez-Alier, J. (Ed.) (1990). *Ecological economics. Energy, evironment and society.* Oxford, UK; Cambridge, USA: Blackwell.

Matthews, E., Amann, C., Fischer-Kowalski, M., Bringezu, S., Hüttler, W., Kleijn, R., et al. (2000). *The weight of nations: Material outflows from industrial economies.* Washington, DC: World Resources Institute.

Matthews, R. (2006). The People and Landscape Model (PALM): An agent-based spatial model of livelihood generation and resource flows in rural households and their environment. *Ecological Modelling, 194*, 329–343.

Matthews, R., & Selman, P. (2006). Landscape as a focus for integrating human and environmental processes. *Journal of Agricultural Economics, 57*, 199–212.

McCann, K. S. (2000). The diversity-stability debate. *Nature, 405*, 228–233.

McConnell, W. (2001). *Agent-based models of land-use and land-cover change.* Belgium: LUCC International Project Office).

Meyer, W. B., & Turner, B. L. I. (1994). *Changes in land use and land cover, a global perspective.* Cambridge: Cambridge University Press.

Millennium Ecosystem Assessment. (2005). *Ecosystems and human well-being.* Washington, DC: Island Press.

Netting, R. M. (1993). *Smallholders, householders. farm families and the ecology of intensive, sustainable agriculture.* Stanford: Stanford University Press.

Norgaard, R. B. (1994). The coevolution of economic and environmental systems and the emergence of unsustainability. In R. England (Ed.), *Evolutionary concepts in contemporary economics* (pp. 213–225). Ann Arbor: University of Michigan Press.

NRC. (2004). *NEON, addressing the nations environmental challenges.* Washington, DC: The National Academy Press.

Odum, H. T., & Pinkerton, R. C. (1955). Time's speed regulator: The Optimum efficiency for maximum power output in physical and biological systems. *American Scientist, 43*, 331–343.

Oliver-Smith, A. (1977). Traditional agriculture, central places, and postdisaster urban relocation in Peru. *American Ethnology, 4*, 102–116.

Orlove, B. S. (1980). Ecological anthropology. *Annual Review of Anthropology, 9*, 235–273.

Orlove, B. S., & Custtred, G. (1980). The alternative model of agrarian society in the Andes: Households, networks and corporate groups. In B. S. Orlove & G. Custtred (Eds.), *Land and power in Latinamerica: Agrerian economics and social processes in the Andes* (pp. 31–54). New York: Holmes & Meier.

Ostrom, E. (1996). *Governing the commons. The evolution of institutions for collective action.* Cambridge: Press Syndicate of the University of Cambridge.

Pahl-Wostl, C. (2002). Participative and stakeholder-based policy design, evaluation and modelling processes. *Integrated Assessment, 3*, 3–14.

Parris, T. M., & Kates, R. W. (2003). Characterizing and measuring sustainable development. *Annual Review of Environment and Resources, 28*, 559–586.

Polanyi, K. (1957). *The great transformation: The political and economic origins of our time.* Boston: Beacon Press.

Prigogine, I. (1976). Order through fluctuations: Self-organisation and social systems. In E. Jantsch & C. H. Waddington (Eds.), *Evolution and consciousness: Human systems in transition* (pp. 93–130). Reading, MA: Addison-Wesley.

Rappaport, R. A. (1967). *Pigs for the ancestors.* New Haven: Yale University Press.

Rappaport, R. A. (1971). The flow of energy in an agricultural society. *Scientific American, 225*, 116–136.

Redman, C. L. (1999). Human dimensions of ecosystem studies. *Ecosystems, 2*, 296–298.

Redman, C. L., Grove, J. M., & Kuby, L. H. (2004). Integrating social science into the long-term ecological research (LTER) network: Social dimensions of ecological change and ecological dimensions of social change. *Ecosystems, 7*, 161–171.

Reyna, S. P. (1975). Making do when the rain stops: Adjustment of the domestic structure to climate variation among the Barma. *American Ethnologist, 2*, 535–551.

Rotmans, J., Kemp, R., & Van Asselt, M. (2001). More evolution than revolution: Transition management in public policy. *Foresight, 3*, 15–31.

Schandl, H., & Grünbühel, C. M. (2005). Using land-time budgets to analyse farming systems and poverty alleviation policies in Lao PDR. *International Journal of Global Environmental Issues, 5*, 142–180.

Schandl, H., & Schulz, N. B. (2002). Changes in United Kingdoms' natural relations in terms of society's metabolism and land use from 1850 to the present day. *Ecological Economics, 41*, 203–221.

Scheffer, M., Carpenter, S., Foley, J. A., Folke, C., & Walker, B. (2001). Catastrophic shifts in ecosystems. *Nature, 413*, 591–596.

Schellnhuber, H. J. (1999). 'Earth system' analysis and the second Copernican revolution. *Nature, 402*, C19-C23.

Schlesinger, W. H. (1997). *Biogeochemistry – An analysis of global change.* San Diego: Academic Press.

Singh, S. J. (2003). *In the sea of influence: A world system perspective of the Nicobar Islands.* Lund: Lund University.

Steward, J. H. (1955). *Theory of culture change. The methodology of multilinear evolution.* Urbana: University of Illinois Press.

Teich, M., Porter, R., & Gustafsson, B. (1997). *Nature and society in historical context.* Cambridge: Cambridge University Press.

Tilly, C. (1988). *Durable inequality.* Berkeley: University of California Press.

Turner, B. L. I., Matson, P. A., McCarthy, J. J., Corell, R. W., Christensen, L., Eckley, N., et al. (2003). Illustrating the coupled human-environment system for vulnerability analysis: Three case studies. *Proceedings of the National Academy of Science, 100*, 8080–8085.

UNEP (2002). *Global Environment Outlook 3. Past, present and future perspectives. London: United Nations Environment Programme (UNEP).* London: Earthscan.

Vadineanu, A. (1998). *Sustainable development, theory and practice.* Bucarest: Bucarest University Press.

Vadineanu, A. (2001). *Sustainable development: Theory and practice regarding the transition of socio-economic systems towards sustainability.* Bucarest: UNESCO – CEPES.

Vadineanu, A., Adamescu, M. C., Cazacu, C., Bodescu, F., & Danielescu, S. (2001). *Past and future management of the Lower Danube Wetlands. Working paper of the Departement of System Ecology.* Bucarest: Bucarest University.

Vadineanu, A., Adamescu, M. C., Vadineanu, R. S., Christofor, S., & Negrei, C. (2003). Past and future management of the Lower Danube Wetland System: A bioeconomic appraisal. *The Journal of Interdisciplinary Economics, 14*, 415–447.

Van der Leeuw, S. E. (2004). Why Model? *Cybernetics and Systems, 35*, 117–128.

Van der Leeuw, S. E., & Aschan-Leygonie, C. (2005). A long-term perspective on resilience in socio-natural systems. In H. Liljenström, & U. Svedin (Eds.), *Micro, Meso, Macro. Addressing Complex Systems Couplings* (pp. 227–264). New Jersey: World Scientific.

Van Keulen, H. (1993). Options for agricultural development: A new quantitative approach. In F. W. T. PenningDeVries, P. Teng, & K. Metselaar (Eds.), *Systems approaches for agricultural development* (pp. 355–365). Dordrecht: Kluwer Academic Publishers.

Varela, F. G., Maturana, H. R., & Uribe, R. (1974). Autopoiesis: The organization of living systems, its characterization and a model. *Biosystems, 5*, 187–196.

Vayda, A. P., & MacKay, B. (1977). Problems in the identification of environmental problems. In T. P. Bayliss-Smith & R. G. A. Feachem (Eds.), *Subsistence and survival: Rural ecology in the Pacific* (pp. 476–498). New York, London: Academic.

Vitousek, P. M., Ehrlich, P. R., Ehrlich, A. H., & Matson, P. A. (1986). Human appropriation of the products of photosynthesis. *BioScience, 36*, 363–373.

Wackernagel, M. (1999). Ecological economics forum: Why sustainability analysis must include biophysical assessments. *Ecological Economics, 29*, 13–15.

Wackernagel, M., Schulz, N. B., Deumling, D., Linares, A. C., Jenkins, M., Kapos, V., et al. (2002). Tracking the ecological overshoot of the human economy. *Proceedings of the National Academy of Science, 99*, 9266–9271.

Waddell, E. (1975). How the Enga cope with frost: Responses to climatic perturbations in the Central Highlands of New Guinea. *Human Ecology, 3*, 249–73.

Walker, B., Holling, C. S., Carpenter, S. R., & Kinzig, A. (2004). Resilience, adaptability and transformability in social-ecological Systems. *Ecology and Society, 9(2)*, 5. Retrieved from URL: http://www.ecologyandsociety.org/vol9/iss2/art5.

Walker, B., & Meyers, J. A. (2004). Thresholds in Ecological and Social-Ecological Systems: A Developing Database. *Science and Society, 9(2)*, 3. Retrieved from URL: http://www.ecologyandsociety.org/vol9/iss2/art3.

Waltner-Toews, D., & Kay, J. J. (2005). The evolution of an ecosystem approach: The diamond schematic and an adaptive methodology for ecosystem sustainability and health. *Ecology and Society, 10*(1), 38. Retrieved from URL: http://www.ecologyandsociety.org/vol10/iss1/art38/..

Waltner-Toews, D., Kay, J. J., Neudoerffer, C., & Gitau, T. (2003). Perspective changes everything: Managing ecosystems from the inside out. *Frontiers in Ecology and Environment, 1*, 23–30.

Watts, D. J., & Strogatz, S. H. (1998). Collective dynamics of `small-world' networks. *Nature, 393*, 440–442.

WCED. (1987). *World commission on environment and development: Our common future.* New York: Oxford University Press.

Weisz, H. (2002). Gesellschaft-Natur Koevolution: Bedingungen der Möglichkeit nachhaltiger Entwicklung. Dissertation, Humboldt Universität, Berlin..

Weisz, H., Fischer-Kowalski, M., Grünbühel, C. M., Haberl, H., Krausmann, F., & Winiwarter, V. (2001). Global environmental change and historical transitions. *Innovation – The European Journal of Social Sciences, 14*, 117–142.

Weisz, H., Krausmann, F., Amann, C., Eisenmenger, N., Erb, K.-H., Hubacek, K., et al. (2006). The physical economy of the European Union: Cross-country comparison and determinants of material consumption. *Ecological Economics, 58*, 676–698.

White, L. A. (1959). *The evolution of culture.* New York: McGraw-Hill.

Wilbanks, T. J., & Kates, R. W. (1999). Global change in local places: How scale matters. *Climatic Change, 43*, 601–628.

Chapter 27

Integrating Social Sciences into Long-Term Ecological Research

Cornelia Ohl and Scott M. Swinton

Abstract In Europe and North America, long-term ecological research (LTER) networks are changing their treatment of human activity from exogenous 'disturbances' to endogenous behaviour. The engagement of social scientists in LTER networks currently takes forms ranging from nonexistent, to research in parallel with ecological research but with minimal interaction, to truly collaborative long-term socio-ecological research (LTSER). Successful collaboration of social and ecological scientists can be facilitated by a 'jazz band' approach that allows shifting multidisciplinary leadership along with disciplinary research solos. Socio-ecological simulation modelling can serve as a common tool for analysing complex dynamics of the interacting systems. The design criteria for an LTSER network should include socio-economic as well as ecological factors in order to ensure that findings can be extrapolated in both dimensions, an approach currently being followed in Europe. However, due to evolving societal needs, socio-ecological research should also be occurring outside the network of LTSER sites. New governmental initiatives on both sides of the Atlantic have the potential to enable more and better socio-ecological research than in the past.

Keywords Socio-ecological research · Shifting multidisciplinary leadership and disciplinary research solos · 'Jazz band' approach

C. Ohl (✉)
UFZ-Helmholtz Centre for Environmental Research, Permoserstraße 15, 04318 Leipzig, Germany;
German Emissions Trading Authority at the Federal Environment Agency, Bismarckplatz 1, 14193 Berlin, Germany
e-mail: Cornelia.Ohl@uba.de

27.1 Introduction

At local, regional and global levels, human activity affects numerous ecosystem functions that are crucial for human well-being (MASR, 2005). The emission of greenhouse gasses influences climatic conditions (Oberthür & Ott, 1999). The release of ozone depleting substances jeopardises essential protection mechanisms of the atmosphere (Benedick, 1991). Extraction, transport and transformation of natural resources change our landscape, influence biodiversity and redefine the state of ecosystems (Tilman, Polasky, & Lehman, 2005). The multiple feedbacks between society and nature are fast, drastic and large scale; they condense to three words: *Global ecosystem change.*

Global ecosystem change needs to be understood better if it is to be managed effectively. During recent decades, global ecosystem changes have been increasingly driven by human activity. Past human activity generated numerous benefits for economic and social development. Today, however, we humans increasingly fear such negative consequences as the exhaustion of resources and nature's diminishing capacity to absorb and to neutralise human pollution and waste. These troubling aspects of global ecosystem change put at risk both present well-being and opportunities for future generations. These troubling aspects also explain why global ecosystem change shows up in various forms on both national and international political agendas (e.g. COM, 2005; UNMP, 2005).

The long-term ecological research (LTER) networks were established to monitor, analyse and predict the local impacts of global ecological change. Up to now, the networks have focused on biological science. However, in order to better understand the dynamics

F. Müller et al. (eds.), *Long-Term Ecological Research*,
DOI 10.1007/978-90-481-8782-9_27, © Springer Science+Business Media B.V. 2010

of the interaction of social and natural systems, a knowledge base of social and behavioural science is needed that is both supplementary and complementary to standard ecological research.

The integration of social and behavioural research challenges existing ecological research networks, like LTER, by raising questions about how to integrate and compare research foci, how to monitor linked environmental and socio-economic processes and how to identify suitable research sites designed to integrate socio-economic dimensions into traditional ecological research. In this chapter, from a social science point of view, we focus on the process of transforming LTER to LTSER – long-term *socio-ecological* research – both the efforts that have been made and the challenges that remain to be faced. The chapter is organised as follows: Section 27.2 elaborates on how socio-economic research has related to ecological research up to now and what factors have shaped the status quo; Section 27.3 presents how socio-ecological research is evolving and the path ahead.

27.2 Linking *Ecological* Research with Human Behaviour

The publication in 1997 of Vitousek et al.'s *Science* article on 'human domination of the Earth's ecosystems' shifted the centre of debate about the human role from interloper to dominator of the planet's ecosystems. The role of humans as intentional and unintentional drivers of global ecosystem change has been explored much more thoroughly through the Millennium Ecosystem Assessment (Nelson et al., 2006; Kareiva, Watts, McDonald, and Boucher, 2007). Among direct drivers, the study identifies climate change, agricultural nutrient use, land conversion and the spreading of diseases and invasive species. But the MEA highlights the importance of indirect human drivers, notably demographic and economic growth as well as social, cultural and technological change.

The recognition of the overwhelming importance of the human footprint has led social scientists to begin exploring ways to manage that footprint intentionally. Agricultural ecosystems, which are already managed by humans, have been proposed as useful learning opportunities to see how human management is influenced by spatial heterogeneity and scale, dynamic uncertainties and incentives from markets and policy (Antle et al., 2001; Swinton, Lupi, Robertson, & Landis, 2006; Swinton, Lupi, Robertson, & Hamilton, 2007).

27.2.1 Overview of Literature on Site-Based Socio-ecological Research

In the LTER system, research into how ecological and social systems are linked has followed several different models. The earliest LTER research focused on pristine systems where the human footprint was apparently absent. But in the 1990s LTER research began to examine systems that were increasingly disturbed by humans. In order to understand human effects on water quality in the North Temperate Lakes and on land use in the southern Appalachian mountains of North America, two LTER sites added social science research activities in 1994 (Gragson & Grove, 2006). The social scientists at these sites undertook parallel social research to understand why and how human systems had evolved to affect water and land use as they had. The social scientists expanded the geographic scale of research from the ecological observation sites to the surrounding area where human activity could influence those or similar biomes (a lake watershed in one case, a mountainous region in the other). Without special funding, other LTER sites, such as the crop agroecology site at the Kellogg Biological Station, have subsequently followed this approach.

A second model for socio-ecological research was inaugurated with two new urban LTER sites in 1996. At these locations, social and ecological scientists collaborated in geographic sampling plans to understand social–ecological linkages explaining the evolution of vegetation in the slow-growing city of Baltimore, Maryland, and the rapidly expanding city of Phoenix, Arizona (Gragson & Grove, 2006). The collaborative research design was established at a spatial scale relevant to both human and ecological research from the outset. The social scientists involved were primarily sociologists and geographers, working closely with ecologists on jointly designed research.

A third model of socio-ecological research is emerging in response to competitive grant funding for interdisciplinary science. Such projects often have social scientists in the lead, with research scale and sampling designed around human drivers. Examples of this are the U.S. National Science Foundation's (NSF's) Dynamics of Coupled Natural and Human Systems and Human and Social Dynamics programmes of the early 2000s. These programmes have attracted a wide variety of social science disciplines. In early 2007, the former became the NSF's first permanent interdisciplinary research programme. The design of the research framework programmes (FPs) in Europe also shows a clear trend towards integration. From the fifth to the seventh FP the European Commission launched more and more topics calling for integrated research activities of the applicants (http://cordis.europa.eu/en/sitemap.htm#eu-research); a trend which is also observed in EU's member states. In 1999, for example, the German Federal Ministry of Education and Research (BMBF) established the Social–Ecological Research programme (SÖF) focusing on sustainability issues at the interface between nature and society (http://www.sozial-oekologische-forschung.org/en/724.php).

Much LTER social science research to date has explored relatively disciplinary questions, disciplinary at least from a social science perspective. At the North Temperate Lakes (NTL) site, sociologists Peter Nowak and William Freudenberg have led in applying models of socio-economic inequality to explaining disproportional incidence of lake water pollution affecting less prosperous populations (Nowak, Bowen, & Cabot, 2006). In parallel, resource economist Bill Provencher has explored how water clarity affects lake-front home values (Spalatro & Provencher, 2001). At the Coweeta LTER in the southern Appalachian Mountains, anthropologist Ted Gragson has explored the effects of historical land use legacies on current ecosystem functioning (Gragson & Bolstad, 2006), while forest economists David Wear and David Newman have modelled spatial land use change (Wear & Bolstad, 1998). At the KBS-LTER row crop agroecological site, agricultural economist Scott Swinton has modelled the economic role of natural enemies in regulating agricultural pests (Zhang & Swinton, 2009).

The urban socio-ecological research sites have established a more interdisciplinary modus operandi. Their work includes exploration of social class interactions with urban ecosystems and the use of geographical information (remote sensing of vegetative cover and government or marketing databases on socio-economic status) to uncover human influences on land cover (Grove et al., 2006; Grimm, Grove, Pickett, & Redman, 2000; Pickett et al., 2001). In looking at social structure effects on ecology, they have refined spatial resolution to the parcel level, allowing examination of how parcel history and property rights regimes affect local ecology.[1]

Focusing on Europe we find that ecosystem research is on the one hand rich and varied but on the other hand dispersed and disconnected (Parr, 2006). Required information and knowledge for the implementation of policy responses to global ecosystem change phenomena are therefore not easily provided. Against this background the emerging LTER system serves as an archetype for the development of pan-European research. The aim is to develop a common understanding of the main drivers and pressures for global ecosystem change, the development of common frameworks for monitoring activities as well as methods, tools and policies for improvement and cost-effective management of ecosystems.

The European LTER network started to grow only recently. In 2003 representatives of 23 countries started discussions on how to strengthen and better integrate LTER activities in Europe. At present, 15 European countries have formal LTER networks, four countries have substantially developed networks and a further four are in the early stages of network development (http://www.lter-europe.ceh.ac.uk/European_LTER.htm). In terms of socio-ecological research the most developed site in Europe is 'Eisenwurzen' in Austria. It is designed as a multifunctional research platform (Mirtl, 2004). The focus of research interest is the whole landscape as a scientific basis for sustainable regional development. The site includes diverse spectrum of land uses, including wilderness, pristine forest, national park, and lake, agricultural and urban uses. Elements of this approach take into account regional characteristics and user groups, general conditions posed by global and social changes and national economic areas (http://www.umweltbundesamt.at/en/umweltschutz/oekosystem/lter_allgemein/).

The most recently selected site is Leipzig-Halle. Like 'Eisenwurzen', the region covers a spectrum of cultural landscapes, including former mining areas, urban and agricultural land, and different administrative units with explicit gradients of precipitation,

[1] Grove, J. Morgan. Personal communication by email to Scott M. Swinton, October 6, 2007.

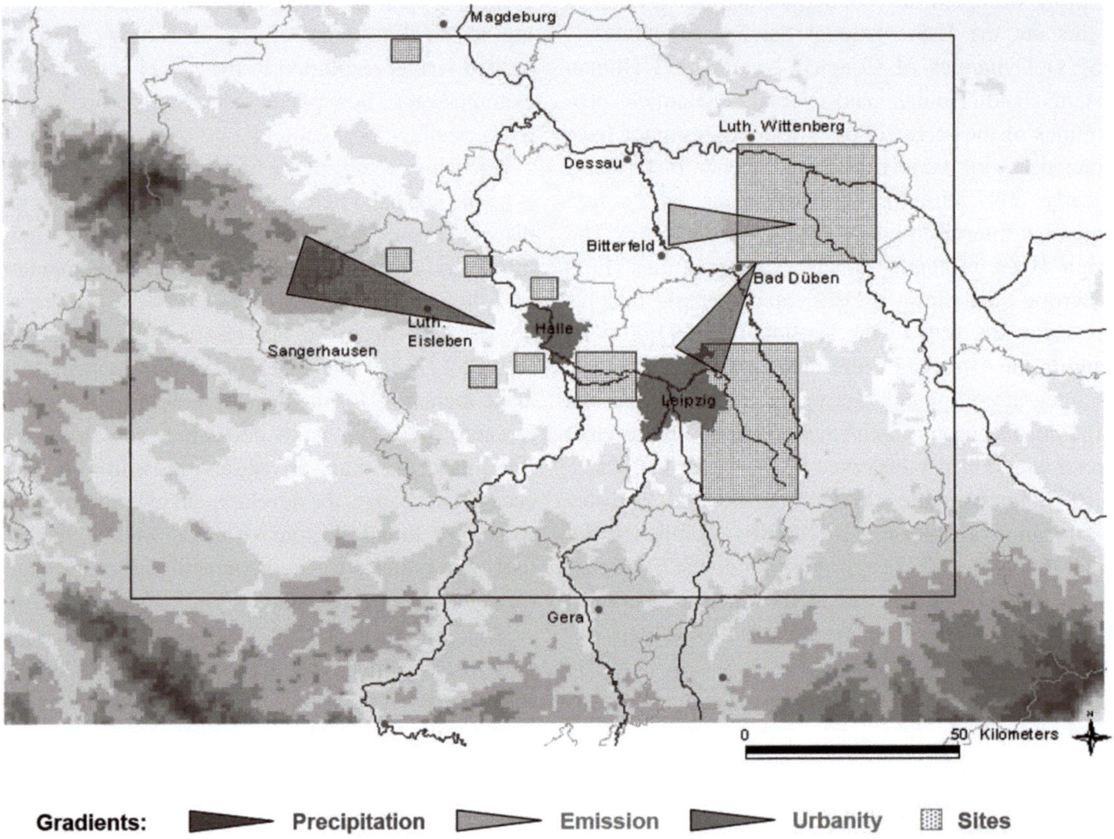

Fig. 27.1 LTSER site Leipzig-Halle, Germany (underway)

emissions and population density (see Fig. 27.1). The conceptual frame for the multidisciplinary approach, however, is still being developed. The idea is to concentrate research activities of natural and social scientists on a number of core themes, such as land use and institutional change. The research questions will be addressed via different so-called competence platforms, including monitoring and observation, experiments, modelling and concept formation, management support and policy design. A science–society interface will explicitly focus on collaboration of science and society to take up societal problems and needs, as well as the exchange of knowledge, ideas and experiences. It will organise networking on biodiversity issues and bring stakeholders and decision makers in contact with the scientists. See 'UFZ Biodiversity Research Strategy (2007)'.

In short, socio-ecological research in the European LTER network, like the network itself, is still emerging. National level efforts are being supplemented by funding from the European Commission for development of a pan-European LTER network. This period of rapid development of research infrastructure provides a rich opportunity to develop a balanced socio-ecological research approach.

Although the LTER system provides a valuable structure for understanding long-term ecological change, on neither side of the Atlantic has it yet met its potential for offering insights into how and why humans contribute to global ecosystem change. At present, there exist several sites where social science research is underway in parallel with ecological research, as well as a handful where truly collaborative socio-ecological research is happening. The system needs to adapt further if it is to make major strides towards understanding human behaviour with respect to ecosystems and inform insightful policy responses to undesired ecosystem changes. In short, the LTER framework needs to treat human behaviour and institutions as endogenous – rather than exogenous – elements of ecosystem change.

27.2.1.1 Impediments and Approaches to Integrative Socio-ecological Research

If more multidisciplinary integration remains to be done, it is because the task is not easy. Integrated models that treat both the human and the ecological dimension as interacting endogenous parameters are still the exception rather than the rule (e.g. Jentsch, Wittmer, Jax, Ring, & Henle, 2003; Perrings, 2002; Shogren, Parkhurst, & Settle, 2003). Two roadblocks for integrative research have been identified (Wätzold et al., 2006): On a practical level individual researchers fear a loss of in-depth, disciplinary knowledge in the course of integration (Roughgarden, 2001). Second, current institutional research structures and incentives that are organised around academic disciplines may be ill-equipped to address interdisciplinary research topics (Committee on Facilitating Interdisciplinary Research, 2004). One structural example of fiscally orphaned interdisciplinary research funding comes from the NSF Department of Environmental Biology, which overwhelmingly funds the U.S. LTER system but has no mandate to fund social science research. On a more abstract theoretical level, the methods and tools of single disciplines often ignore complexity relevant to other disciplines. Drechsler et al. (2007) surveyed 60 models related to biodiversity conservation that were randomly chosen from eight ecological and economic journals. They found that socio-economic models dealing with environmental behaviour of humans are usually formulated and solved in a static analytical setting that considers no uncertainty. On the other hand the ecological models are solved numerically or through simulations and deal with uncertainties and system dynamics explicitly.

Important mismatches exist between the scale of management and the scale(s) of the ecological processes being managed (Dirnböck et al., 2006 and the literature cited therein). Boundaries of the ecosystem may not coincide with political and administrative borders. And trade and communication are only two examples of societal activities that are widely independent of geographical scales (Haberl et al., 2006). Moreover the relevance of human interventions at particular scales (e.g. at upper policy levels) is sometimes disregarded by ecological scientists (Carpenter et al., 2006; Cumming, Cumming, & Redman, 2006). Ecological studies often focus on a well-defined study area and a time frame meeting the needs of the ecological problem only. Disregard for economic scale issues notoriously afflicted Costanza et al.'s (1997) estimates of the value of the world's ecosystem services (Bockstael, Freeman, Kopp, Portney, & Smith, 2000; Pearce, 1998). Reasonably, the report of the U.S. National Research Council on *Decision Making for the Environment* (2005) calls for fuller development of the social and behavioural science knowledge base for environmental decision making.

Integration is a means to improved knowledge generation and policy design; it is not an end in itself. Consequently the issue arises of precisely which disciplines are needed to conduct problem-oriented research, which methods and tools capture the most important aspects of the problem and which are most convenient to foster integrative research activities. Just as a botanist cannot ably handle all ecological science research topics, so a sociologist cannot ably handle all social science research topics. The selection of suitable disciplines and research methods needs to be driven by specific research problems and the resources available to address them. But there are at least two major approaches to aid in this process.

Computer simulation models can serve as useful integrating tools to analyse complex systems. They encourage an understanding of underlying mechanisms and the key factors and processes responsible for certain outcomes. Besides allowing prediction of future changes, simulation models can also be used for virtual experiments that would be impossible in nature but are relevant to support management decisions. Simulation experiments are especially suited to studying the effects of random processes over time to study uncertainty (King, Lybecker, Regmi, & Swinton, 1993). And in recent years, ecological-economic modelling has been developed to integrate ecological and economic knowledge and solve problems in nature conservation (Johst, Drechsler, & Wätzold, 2002; Drechsler et al., 2007; Tschirhart, 2004; Wätzold, Lienhoop, Drechsler, & Settele, 2008).

Multidisciplinary integration can also be achieved through careful team design. Based on the authors' experience, effective socio-ecological research teams can be developed according to what we term the 'jazz band' model. They are composed of members with strong disciplinary backgrounds who are individually capable of research leadership. Indeed, a key to success is recruitment of members possessing both strong disciplinary backgrounds and the interest and ability to listen and learn from scientists in other fields. Nourished by core funds, such teams can establish cross-disciplinary research priorities, whereby appropriate individual researchers take turns leading on priorities that fit their comparative advantage. Members take turns, alternately leading specific research efforts and supporting colleague research leaders by informing from a different disciplinary perspective. When successful, the jazz band approach resonates with both interdisciplinary harmonies and disciplinary solos. Jazz band research can operate at both the site level and across sites, but it is probably most practical at the site level.

An effective *modus operandi*, such as the jazz band model, is a necessary but not sufficient condition for effective, sustained multidisciplinary research. For enduring success, the integration of social sciences into multidisciplinary LTER research teams requires two added conditions: a way to attract top-level social scientists and the means to sustain their contributions. A system dominated by one community of scientists – ecologists, in this case – may have difficulty attracting top-level scientists from other communities. One successful model for attracting excellent social scientists to multidisciplinary teams with natural scientists has been competitive research funding where proposals are reviewed by leading scientists from all fields involved. The NSF Coupled Natural and Human Systems programme offers a successful model. A second sine qua non for sustained multidisciplinary integration is dedicated funding for that purpose. Disciplinary funding, such as that of the NSF Biological Sciences division that supports pilot social science activities in the U.S. LTER system, should not by itself be expected to sustain balanced long-term multidisciplinary efforts.

27.2.1.2 Conceptual Framework for Site-Based Research Treating Human Behaviour as Endogenous

Research on the human role in ecosystem change is making a much-needed transition to treating humans as endogenous to the system. Much early (and continuing) research in the LTER system has focused on how ecosystem structure and function responds to 'disturbances' caused by humans. Examples include changes in water-borne nutrient levels, wildlife habitat availability and ambient temperatures. However, in the years since Vitousek, Mooney, Lubchenco, and Melillo

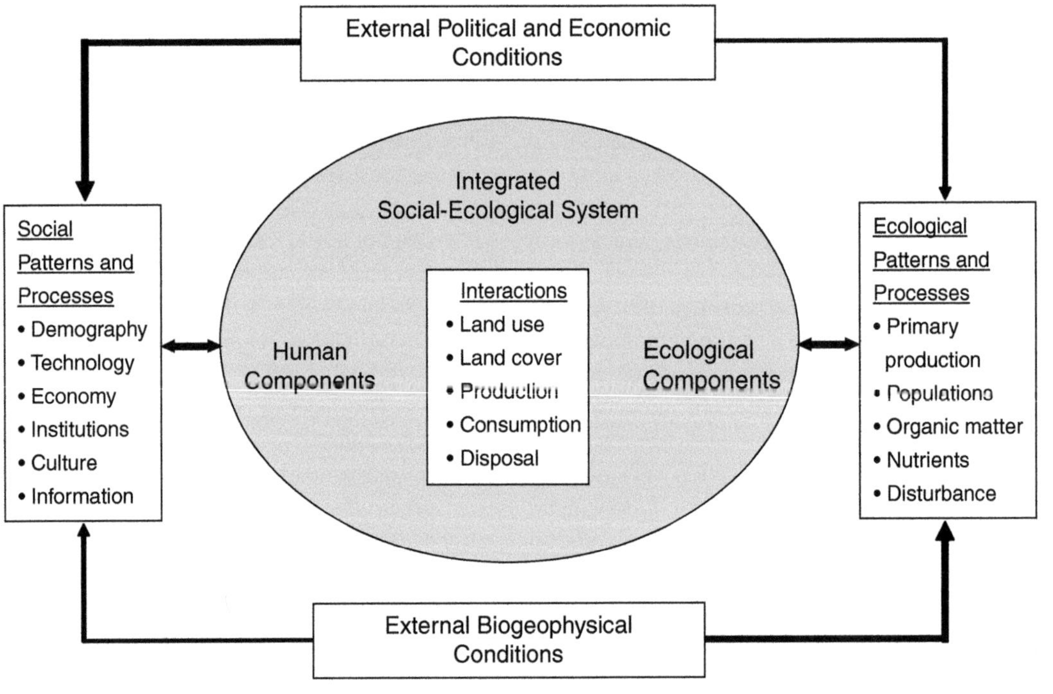

Fig. 27.2 Conceptual framework for socio-ecological systems from Redman et al. (2004). (Reprinted with kind permission from Springer Science+Business Media.)

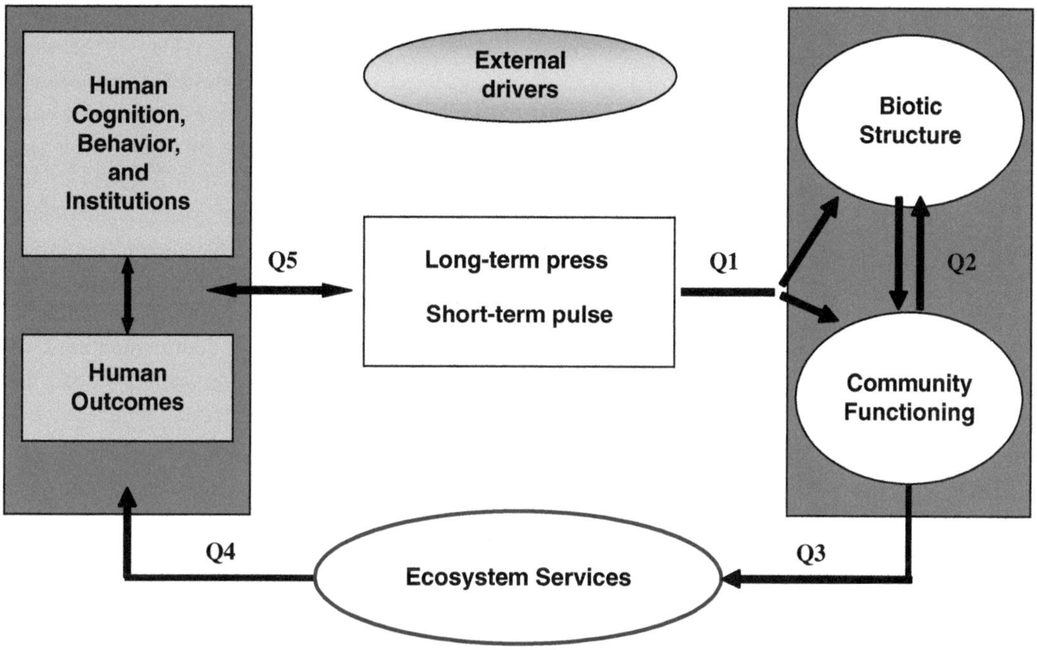

Fig. 27.3 Socio-ecological system diagram from *Integrated Science for Society and the Environment: A Strategic Research Initiative* (Collins et al., 2007)

(1997) call for studying the human role in ecosystem dynamics, alternative conceptual frameworks have begun to emerge that treat humans as endogenous. When presented in iconic form, these models typically illustrate the human environment on one side and ecosystems on the other side, and they focus on the nature of interaction.

Two conceptual frameworks for socio-ecological systems deserve mention, due to their influence on perceptions among LTER researchers in North America. Redman, Grove, and Kuby (2004) paint a picture with both exogenous political and economic conditions and exogenous biophysical conditions that affect both human and ecological systems (Fig. 27.2). Those forces act on patterns and processes that are on the one hand social and on the other hand ecological. The interaction of these sets of patterns and processes is the integrated social-ecological system, where interactions include land use, land cover, production and consumption and disposal activities. A very recent model, developed in a U.S. LTER network planning effort, presents a cyclical relationship that splits the interaction area into two parts, with one comprised of human actions towards ecosystems (long-term presses and sudden pulses) and the other comprised of services generated by ecosystem that are felt by humans (Fig. 27.3) (Collins et al., 2007). An ideal model would draw on both of these sources. The Redman et al. (2004) model explicitly recognises the important role of external forces, including biophysical ones like volcanoes and meteors, and socio-economic ones like human desires for wealth, fame and happiness. The Collins et al. (2007) model more explicitly details effects of one part of the system on another, rather than leaving that to a general interaction zone.

27.3 Progress Towards Long-Term Socio-ecological Research

27.3.1 Building ALTER Net, a European LTSER Research System to Link Ecological Research and Human Behaviour

'A Long-Term Biodiversity, Ecosystem and Awareness Research Network' (ALTER Net) is a 'Network of Excellence', organised as a partnership of 24 organisations from 17 European countries. It was created under the EU's sixth Framework Programme to develop integrated research capacity for long-term biodiversity and ecosystem research. Its goals are to integrate social and ecological research on the relationships between biodiversity, ecosystem services and societal responses at various levels and to design communication and outreach activities for transferring knowledge to and from scientists, policy makers and the public (Parr, 2006). An important means to reach these goals is the development of a network of research sites where problem-oriented research should be performed by natural and social scientists (Siepel & Furman, 2006).

ALTER Net has contributed to both the growth of nationally supported LTER Networks across Europe and the recent establishment of LTER Europe, which was officially founded in June 2007 (http://www.lter-europe.ceh.ac.uk/index.htm). One aim is to attract researchers from outside ALTER Net, mainly from the International LTER (ILTER) community. A second aim is the integration of the natural and social scientists at common spatial scales. This creates need for a new type of research sites – the LTSER sites.

In ecological research, the term 'site' is used to describe an area where natural processes are being studied in undisturbed or strictly controlled environments. In the past a host of ecological and taxonomical criteria have been applied to define and select the sites for ecological research. Recently a novel set of criteria was proposed to meet the needs of long-term *socio-ecological* research (Ohl, Krauze, & Grünbühel, 2007). These criteria were proposed as guidelines for ecological research sites that intend to expand their research scope as well as for the purposeful design of new LTSER sites.

The proposed criteria are subdivided into five *site* and three *pool* criteria. Site-based research should allow for a better understanding of local sociological, political psychological and economic processes, which directly or indirectly reshape dynamics of ecosystems. Site criteria focus on (1) economic diversity, (2) conservation and environmentally relevant policy, (3) local conflict, (4) demography and (5) land use and cover.[2]

Apart from selection of specific sites it is also crucial to focus on the characteristics of the whole pool of selected sites for purposes of cross-scale analysis, data aggregation and sharing. Proposed pool criteria focus on: (1) vulnerability of economic systems, (2) local trade relations and (3) differences in economic development. Both sets of criteria were derived on the basis of European Environment Agency's framework of 'driving forces – pressures – state – impact – response' (DPSIR) (EEA, 1999a, 1999b), which was adopted by ALTER Net for communicating knowledge on environmental questions such as those regarding biodiversity (Svarstad, Petersen, Rothman, Siepel, & Wätzoldt, 2008).

27.3.1.1 Site-Based Research

ALTER Net has developed the idea of long-term socio-ecological research (LTSER) sites to demonstrate their potential for interdisciplinary and policy-relevant research. In the past the LTER network used to limit the spatial extent of research sites to study ecological processes separately from human stimuli. In order to better understand human behaviour, ALTER Net (WP R1) has supplemented ecological criteria for research site selection by five explicitly human criteria:

- Criterion 1, *economic diversity*, is derived from the importance to ecosystem change of how people generate and spend their income. For example in the primary sector income is generated by fishing, forestry, hunting and agriculture, activities that impact ecosystems directly (e.g. by altering the stock of fish). Likewise, food and energy consumption have major effects on the use of renewable and non-renewable resources. Sites selected in accordance with the economic diversity criterion ensure that within the research sites a critical mass of people are found which derive income from different sectors of the economy.

[2] A criterion for the spatial extent of the sites to be selected - although important in ecological studies - was intentionally not included in the list. The size of the social system is determined by other criteria, such as income, markets and demography. Therefore the number and characteristics of ecological units (ecosystems, watersheds) should be adjusted to the meta-level aim of improving research collaboration between natural and social scientists to enhance our understanding of the interacting systems, society and nature.

- Criterion 2, *conservation and environmentally relevant policies*, influence human effects on ecosystems via laws and regulations (that restrict undesired behaviour) and taxes and subsidies (that stimulate desired behaviour). Long-term studies at the site level should provide insight into the success and failures of policy measures intentionally designed to manage the interaction of society and nature. They should also focus on the ecological implications of policy measures that have unintentional effects on the state of the ecosystem.
- Criterion 3, *conflicts and their resolution*, are at the root of various environmental change phenomena (e.g. climate change and biodiversity loss). A core focus of social science research at different LTSER sites should therefore be on local modes of conflict resolution and their abilities to resolve conflicts that affect ecosystems.
- Criterion 4, *demography*, focuses on more quantitative aspects of environmental change, specifically the demographic structure of the site population as well as human migration patterns. Demographic changes influence inter alia the type and intensity of urbanisation, the rates of resource extraction, consumption, waste production and pollution release.
- Criterion 5, *land use and cover*, reflects the interplay among the four aforementioned criteria. This criterion calls for including different land uses within each research site: natural/semi-natural, agricultural and urban or suburban land uses. This should allow for a differentiated assessment of the drivers of land use and cover change as well as of the impacts on the ecosystems and biodiversity in particular.

27.3.1.2 Pool-Based Research

A well-designed LTSER network should enable comparison of basic socio-ecological processes for identification of patterns of interaction as well as best practices adjusted for local context. Therefore the pool of selected sites should consist of a number of contrasting as well as similar sites. To find a set of such sites pool criterion (1), *vulnerability of economic systems*, explicitly calls for at least one site where the income of local people depends on a service provided by the ecosystem. Pool criterion (2), *local trade relations*, calls for areas with similar resource endowments but varying dependency on economic markets. And criterion (3) ensures that with regard to economic development, different sites are found along north–south and east–west gradients.

In a first step, the ALTER Net proposed ten sites to initiate long-term site and pool-based socio-ecological research. These are in Aberdeenshire in Scotland (UK), Nora in Sweden (S), Veluwe in the Netherlands (NL), Pilica catchments in Poland (PL), Pleine Fougères in France (F), Eisenwurzen in Austria (A), Balaton Lake and catchment in Hungary (H), Braila islands in Romania (ROM), Donana in Spain (ES) and Leipzig-Halle in Germany (D). The size of these sites, the number of administrative units and the main ecologically relevant types of land use and cover are shown in Table 27.1.

Table 27.1 Characteristics of LTSER sites in Europe

Name	Size (km^2)	Administrative unit(s)	Type
Nora (S)	1,200	2 communities (in 1 county)	Boreal forest
Aberdeenshire (UK)	6,317	1 county	Atlantic agricultural land
Veluwe (NL)	2,186	21 municipalities (in 1 province)	Atlantic lowland forest and agriculture
Pilica river catchment (PL)	3,921	23 districts (in 5 counties partial)	Continental agricultural land
Pleine Fougères (F)	90 (site)	Municipalities in 1 department	Atlantic agricultural land
Eisenwurzen (A)	5,522	25 municipalities (in 3 provinces)	Mountain forest and agriculture
Balaton (H)	6,367	4 counties (partial)	Continental lake and catchment
Braila (ROM)	2,439	3 counties (partial)	Inland delta of Donau
Doñana (ES)	1,100	1 municipality	Mediterranean delta
Leipzig-Halle (G)	25,000	3 federal states (partial)	Continental agricultural and urban land

Adapted from Siepel 2006.

Do these sites sufficiently consider the proposed socio-economic criteria and do those criteria capture the key individual and social behaviour needed to describe and explain the central interactions between society and nature? So far, this is an open question. Nonetheless, it is an important milestone to initiate site- and pool-based research that offers the possibility of analysing the co-evolution of society, biodiversity, ecosystems and management strategies in different regions at different spatial scales from an integrative, long-term point of view.

27.3.2 Socio-ecological Research Outside LTER Networks

While carefully designed, multidisciplinary LTSER site research holds great potential, there also exist important opportunities for socio-ecological research that do not readily coincide in scale with either LTSER sites or pooled sites. Important human activities coincide with institutions such as markets and governments. Geographic boundaries for these institutions pose both challenges and opportunities for socio-ecological research. Because these institutional boundaries often do not coincide with ecological boundaries, there may be inconsistencies. But there is also a wealth of natural experiments. For example, a national border like the one between Haiti and the Dominican Republic can offer an opportunity to understand how two governance regimes affect one eco-region. Larger human scales are also needed in order to understand human motives behind such major drivers of global ecosystem change as migration flows. Simply to observe that human populations are rising in migratory destination regions begs the question of why people are moving there.

Another reason for socio-ecological research outside of LTER networks is to compensate for the downside of long-term networks, which is their inflexibility. Once a long-term project is established, it can be difficult to adjust in response to changing environmental and policy needs. Inflexibility takes two forms: (1) the research site and (2) the researchers. Needs for both may change. Hence, research systems that incorporate the capability for flexible, competitive socio-ecological research will be better able to respond to inevitably changing needs – albeit without necessarily having all the strengths of long-term research.

27.3.3 The Path Ahead for Socio-ecological Research in Europe and North America

During the past decade, research into understanding human and social behaviour has made impressive advances in the context of existing LTER systems. Ecological researchers have shown not just openness but determination to incorporate the capability for high-quality social science research. That determination has given birth to the European ALTER Net and the growth of U.S. socio-ecological research at both the new urban LTER sites and at several existing LTER sites with human managed ecosystems. Despite these achievements, much social science research to date has been conducted in parallel with ecological research, but has fallen short of the multidisciplinary ideal of socio-ecological research.

Two new developments point towards a brightening future for genuine socio-ecological research. First, the European research network ALTER Net and the U.S. purpose-designed socio-ecological research sites are only beginning to reach the stage of development where their long-term nature is beginning to pay off. With attention to integrating teams successfully (whether by the simulation modelling or the jazz band research team approach or both), these new research institutions show great promise.

Second, in 2007, members of the U.S. LTER network developed a new plan for network-level research programme called 'Integrative Science for Society and Environment' (ISSE). The ISSE conceptual diagram (Fig. 27.3) explicitly calls for integration of ecological and social science. The planned programme would not only expand site-based and pool-based LTSER, it would also create a competitive research fund to support research on shifting topics by researchers who need not be members of the existing LTER network. Such competitive funding has the potential to attract new and excellent social science researchers into the existing LTER system. Of course, such integration will only be sustainable if funded as an explicitly multidisciplinary programme. A further approach is being explored in Europe. National governments provide basic support for the national networks (for example the German Research Foundation, DFG, supports parts of LTER Germany, the *DFG-Exploratorien*) and the research institutions take responsibility for running a L(S)TER site (as the Helmholtz Centre for

Environmental Research – UFZ does with Leipzig-Halle). But it is also expected that further research grants will be applied for on a competitive basis, e.g. under the EU's seventh Framework Programme. If the ISSE and EU initiatives can succeed in attracting fresh, able minds and addressing evolving socio-ecological research priorities, they will contribute further to strengthening the contribution of socio-ecological research to meeting the challenges of global ecosystem change.

Acknowledgements We gratefully acknowledge support from ALTER Net (A Long-Term Biodiversity, Ecosystem and Awareness Research Network) funded by the EU's sixth Framework Programme, as well as support from the National Science Foundation under Long-Term Ecological Rresearch (KBS-LTER) Grant No. 0423627. We thank Frank Wätzold, Irene Ring, Ted Gragson, Morgan Grove and Bill Provencher for helpful comments. We are also indebted to Prof. em. Wolfgang Haber and to an anonymous referee for stimulating comments.

References

Antle, J. M., Capalbo, S. M., Elliott, E. T., Hunt, H. W., Mooney S., & Paustian, K. H. (2001). Research needs for understanding and predicting the behavior of managed ecosystems: Lessons from the study of agroecosystems. *Ecosystems, 4,* 723–735.

Benedick, R. E. (1991). *Ozone diplomacy.* Harvard: University Press.

Bockstael, N. E., Freeman, A. M., Kopp, R. J., Portney, P. R., & Smith, V. K. (2000). On measuring economic values for nature. *Environmental Science & Technology, 34,* 1384.

Carpenter, S. R., DeFries, R., Dietz, T., Mooney, H. A., Polasky, S., Reid, W. V., et al. (2006). Millennium ecosystem assessment: Research needs. *Science, 314,* 257–258.

Collins, S. L., Swinton, S. M., Anderson, C. W., Benson, B. J., Brunt, J., Gragson, T., et al. (2007). *Integrated science for society and the environment: A strategic research initiative.* Publication #23 of the U.S. LTER Network, LTER Network Office. Albuquerque: University of New Mexico.

COM. (2005). *Draft declaration on guiding principles for sustainable development. Communication from the Commission of the European Communities to the Council and the European Parliament 218.* Retrieved from http://eur-lex.europa.eu/LexUriServ/site/en/com/2005/com2005_0218en01.pdf.

Committee on Facilitating Interdisciplinary Research, National Academy of Sciences, National Academy of Engineering, Institute of Medicine. (2004). *Facilitating interdisciplinary research.* Washington, DC: The National Academies Press.

Costanza, R., D'Arge, R., DeGroot, R., Farber, S., Grasso, M., Hannon, B., et al. (1997). The value of the world's ecosystem services and natural capital. *Nature 387,* 253–260.

Cumming, G. S., Cumming, D. H., & Redman, C. L. (2006). Scale mismatches in socio-ecological systems: Causes, consequences, and solutions. *Ecology and Society, 11,* 14.

Dirnböck, Th., Bezák, P., Dullinger, S., Haberl, H., Lotze-Campen, H., Mirtl, M., et al. (Eds.). (2006). *Scaling issues in long-term socio-ecological biodiversity research: A review of European cases. Social Ecology Working Paper 100.* Vienna: KlagenfurtUniversity.

Drechsler, M., Grimm, V., Mysiak, J., & Wätzold, F. (2007). Differences and similarities between ecological and economic models for biodiversity conservation. *Ecological Economics, 62,* 232–241.

Drechsler, M., Johst, K., Ohl, C., & Wätzold, W. (2007). Designing cost-effective payments for conservation measures to generate spatiotemporal habitat heterogeneity. *Conservation Biology, 21,* 1475–1486.

EEA. (1999a). *Environment in the European Union at the turn of the century. Environmental assessment report No 2.* Copenhagen: European Environment Agency.

EEA. (1999b). *Environmental indicators: Typology and overview. Technical report No 25.* Copenhagen: European Environment Agency.

Gragson, T. L., & Grove, M. (2006). Social science in the context of the Long Term Ecological Research Program. *Society and Natural Resources, 19*(1), 93–100.

Gragson, T. L., & Bolstad, P. V. (2006). Land use legacies and the future of Southern Appalachia. *Society and Natural Resources, 19*(1), 175–190.

Grimm, N., Grove, J. M., Pickett, S. T. A., & Redman, C. L., (2000). Integrated approaches to long-term studies of urban ecological systems. *Bioscience, 50,* 571–584.

Grove, J. M., Cadenasso, M. L., Burch, W. R., Pickett, S. T. A., Schwarz, K., Neil-Dunne, J., et al. (2006). Data and methods comparing social structure and vegetation structure of urban neighborhoods in Baltimore, Maryland. *Society and Natural Resources, 19*(1), 117–136.

Haberl, H., Winiwarter, V., Andersson, K., Ayres, R. U., Boone, C., Castillo, A., et al. (2006). From LTER to LTSER: Conceptualizing the socioeconomic dimension of long-term socioecological research. *Ecology and Society, 11*(2). Retrieved from URL: http://www.ecologyandsociety.org/vol11/iss2/art13/.

Jentsch, A., Wittmer, H., Jax, K., Ring, I., & Henle, K. (2003). Biodiversity. Emerging issues for linking natural and social sciences. *GAIA, 12,* 121–128.

Johst, K., Drechsler, M., & Wätzold, F. (2002). An ecological-economic modelling procedure to design effective and efficient compensation payments for the protection of species. *Ecological Economics, 41,* 37–49.

Kareiva, P., Watts, S., McDonald, R., & Boucher, T. (2007). Domesticated nature: Shaping landscapes and ecosystems for human welfare. *Science, 316,* 1866–1869.

King, R. P., Lybecker, D. W., Regmi, A., & Swinton, S. M. (1993). Bioeconomic models of crop production systems: Design, development, and use. *Review of Agricultural Economics, 15,* 393–405.

MASR. (2005). *Ecosystems and human well-being: Biodiversity synthesis* (Millennium Ecosystem Assessment Scientific Report). Washington, DC: Island Press.

Mirtl, M. (2004, July). *LTER in Austria: Implementing multifunctional research platforms (MFRPs) for LTER.*

Paper presented at the ILTERN Coordinating Committee Meeting 7–9 July, Manaus, Brazil. Retrieved from http://www.ilternet.edu/meetings/.

Nelson, G. C., Bennett, E., Berhe, A. A., Cassman, K., DeFries, R., Dietz, T., et al.. (2006). Anthropogenic drivers of ecosystem change: An overview. *Ecology and Society, 2*, 29. Retrieved from http://www.ecologyandsociety.org/vol11/iss2/ art29/.

Nowak, P., Bowen, S., & Cabot, P. E. (2006). Disproportionality as a framework for linking social and biophysical systems. *Society and Natural Resources, 19*, 153–173.

Oberthür, S., & Ott, H. E. (1999). *The Kyoto protocol*. Berlin: Springer.

Ohl, C., Krauze, K., & Grünbühel, C. (2007). Towards an understanding of long-term ecosystem dynamics by merging socio-economic and environmental research. Criteria for Long-Term Socio-Ecological Research Sites Selection. *Ecological Economics*, *63*, 383–391.

Parr, T. (2006). Research Integration for the CBD. *ALTERNews, 02*, Retrieved from www.alter-net.info.

Pearce, D. 1998. Auditing the Earth. *Environment, 40*, 23.

Perrings, C. (2002). Modelling sustainable ecological-economic development. *International Yearbook of Environmental and Resource Economics, 2001/2*, 179–201.

Pickett, S. T. A., Cadenasso, M. L., Grove, J. M., Nilon, C. H., Pouyat, R. V., Zipperer, W. C., et al. (2001). Urban ecological systems: Linking terrestrial ecological, physical, and socioeconomic components of metropolitan areas. *Annual Review of Ecology and Systematics, 32*, 127–157.

Redman, C. L., Grove, J. M., & Kuby, L. H. (2004). Integrating social science into the long-term ecological research (LTER) network: Social dimensions of ecological change and ecological dimensions of social change. *Ecosystems, 7*, 161–171.

Roughgarden, J. (2001). Guide to diplomatic relations with economists. *Bulletin of the Ecological Society of America, 82*, 85–88.

Shogren, J. F., Parkhurst, G. M., & Settle, C. (2003). Integrating economics and ecology to protect nature on private lands: Models, methods and mindsets. *Environmental Science and Policy, 6*, 233–242.

Siepel, H. (2006). *LTSER-sites in Europe. A preliminary selection*. Retrieved from www.alter-net.info.

Siepel, H., & Furman, E. (2006). Biodiversity at Stake: Do we need a multi-disciplinary approach? *ALTERNews, 02*. Retrieved from www.alter-net.info.

Spalatro, F., & Provencher, B. (2001). An analysis of minimum frontage zoning to preserve lakefront amenities. *Land Economics, 77*, 469–481.

Svarstad, H., Petersen, L. K., Rothman, D., Siepel, H., & Wätzoldt, F. (2008). Discursive biases of the environmental research framework DPSIR. *Land Use Policy, 25*, 116–125.

Swinton, S. M., Lupi, F., Robertson, G. P., & Hamilton, S. K. (2007). Ecosystem services and agriculture: Cultivating agricultural ecosystems for diverse benefits. *Ecological Economics, 64*, 245–252.

Swinton, S. M., Lupi, F., Robertson, G. P., & Landis, D. A. (2006). Ecosystem services from agriculture: Looking beyond the usual suspects. *American Journal of Agricultural Economics, 88*, 1160–1166.

Tilman, D., Polasky, S., & Lehman, C. (2005). Diversity, productivity and temporal stability in the economies of humans and nature. *Journal of Environmental Economics and Management, 49*, 405–426.

Tschirhart, J. (2004). A new adaptive system approach to predator-prey modeling. *Ecological Modelling, 176*, 255–276.

UFZ Biodiversity Research Strategy. (2007). *Biodiversity and ecosystem functions under global change*. Concept paper. Helmholtz Centre for Environmental Research – UFZ.

U.S. National Research Council. (2005). *Decision making for the environment*.

UNMP. (2005). *Investing in development: A practical plan to achieve the millennium development goals, overview* (United Nations Millennium Project). Washington, DC: Communications Development Inc.

Vitousek, P. M., Mooney, H. A., Lubchenco, J., & Melillo, J. M. (1997). Human domination of earth's ecosystems. *Science, 277*, 494–499.

Wätzold, F., Drechsler, M., Armstrong, C. W., Baumgärtner, S., Grimm, V., Huth, A., et al. (2006). Ecological-economic modelling for biodiversity management: Potential, pitfalls, and prospects. *Conservation Biology, 20*, 1043–1041.

Wätzold, F., Lienhoop, N., Drechsler, M., & Settele, J. (2008): Estimating optimal conservation in agricultural landscapes when costs and benefits of conservation measures are heterogeneous in space and over time, *Ecological Economics, 68*(1–2), 295–305.

Wear, D. N., & Bolstad, P. (1998). Land-use changes in Southern Appalachian landscapes: Spatial analysis and forecast evaluation. *Ecosystems, 1*, 575–594.

Zhang, W., & Swinton, S. M. (2009). Incorporating natural enemies in an economic threshold for dynamically optimal pest management. *Ecological Modelling, 220*, 1315–1324.

Chapter 28

Ecosystem Manipulation and Restoration on the Basis of Long-Term Conceptions

Oliver Dilly, Seth Nii-Annang, Joachim Schrautzer, Peter Schwartze, Vera Breuer, Eva-Maria Pfeiffer, Werner Gerwin, Wolfgang Schaaf, Dirk Freese, Maik Veste, and Reinhard F. Hüttl

Abstract Ecosystems are affected by anthopogenic activities at a global level and, thus, are manipulated world-wide. This chapter addresses the impacts of apparent and non-apparent manipulations and restoration by human activities in Europe with a focus on the temperate zone. Agricultural management practices induced evident site-specific modification of natural ecosystem structures and functions whereas forests and natural grasslands and also aquatic systems are considered as being less manipulated. Ecosystems such as mires, northern wetlands and the tundra, have received attention due to their vulnerability for conserving carbon and biodiversity and for identifying the role of non-apparent manipulations on ecosystem functioning. Drastic types of ecosystem manipulation include open-cast mining activities that occur worldwide and induce perturbation of large areas across landscapes. Such harsh human impacts create the need for remediation and restoration measures for mining regions that address classical food and fodder services and also nature conservation and novel social benefits. Recultivation therefore offers the opportunity to introduce new land-use types and to study processes of initial ecosystem development that are still poorly understood.

Keywords Ecosystem management practices · Human impact · Rehabilitation · Open-cast mining

O. Dilly (✉)
School of Integrated Climate System Sciences, Klima Campus, University of Hamburg, Grindelberg 5, 20144 Hamburg, Germany
e-mail: oliver.dilly@uni-hamburg.de

28.1 Introduction

Human activities have modified environmental conditions worldwide (Rockström et al., 2009). In recent decades land-use changes in combination with climate change has represented a major threat to the functioning of our ecosystems (IPCC, 2007). Accordingly, the European Commission and the European Environmental Agency have released a number of directives to preserve environmental resources such as soil (SEC, 2006) and biodiversity (COM, 2006; EEA, 2006).

In recent decades, great attention has been given to ecosystem research with reference to (i) forest decline during the 1970s and (ii) landscape structures and functions such as the Bornhöved Lakes district (Fränzle et al., 2008; Chapter 4), Solling beech forests (Ellenberg et al., 1986) and Norway spruce forests (Gollan & Heindl, 1998) and finally agricultural systems during the 1970s–1990s. These studies which started with long-term conceptions were funded for around one decade for broad time-series investigations and then slowed down. Ecosystem experiments such as the long-term agricultural field trials with the Rothamsted Park Grass Experiment and the Bad Lauchstädt Static Fertiliser Experiment were older and started in 1856 and 1902, respectively (Smith et al., 1997).

Both the "agricultural" and "ecosystem" research trails represent more or less manipulated treatments against each other and allowed for hypothesis testing by specific experimental and statistical design. Manipulation is used here as the modification of natural ecosystems by human activities to gain specific service. The combination of several disciplines

F. Müller et al. (eds.), *Long-Term Ecological Research*,
DOI 10.1007/978-90-481-8782-9_28, © Springer Science+Business Media B.V. 2010

enabled the in-depth understanding of ecosystem structures and processes under human impacted and complex landscapes (Fränzle et al., 2008). While agricultural experiments aimed at optimising food production with reference to both economic and ecological outcomes, a number of national and international long-term ecosystem research projects aimed at analysing the role of human impacts on biodiversity and biogeochemical cycling and their resilience to restoration and re-cultivation measures. Recent developments showed that the production of food is largely competing with the production of energy which requires new research on the impact of this type of land-use change.

Ecosystem manipulations such as open-cast mining activities represent drastic impacts for example when polluted and acidifying sediments are deposited at the surface after lignite mining. Mining activities have led to the destruction of mature ecosystems. However, the factors controlling the initial ecosystem development are still poorly understood. False chronosequences were therefore considered for ex-ante evaluation of the efficiency of re-cultivation and restoration practices for example, in terms of successful re-vegetation and the establishment of biodiversity and resilience of ecosystem structures and functions.

This chapter addresses both apparent and non-apparent (hidden) ecosystem manipulation and ecosystem threats such as degradation. It points mainly towards management practices for restoring degenerated fen grasslands and open-cast mining regions particularly with respect to new and multifunctional land-use types in Europe with a focus on the temperate zone.

28.2 Ecosystem Manipulation and Management

Ecosystem management and manipulation on the basis of ecosystem functioning is essential for human needs (Table 28.1). While selective harvesting represented the manipulation measure in old cultures, men bred and propagated selectively to profit at a higher degree from ecosystem services such as food, fibre and energy supply (Costanza et al., 1997).

Tillage practices aim at supporting germination and the rapid growth of the target vegetation. Furthermore, organic and mineral fertilisation and also irrigation practices have introduced nutrients limiting plant productivity. Nitrogen is typically applied to agricultural soils at high rates of up to some hundred kg N ha^{-1} $year^{-1}$ supporting plant assimilation activity and, thus, increasing biomass production and also yield quality and quantity. In addition, phosphorus and other nutrients, and also inorganic and organic pollutants including some pesticides have extensively impacted both managed and natural ecosystem structures and productivity.

The increase in carbon dioxide and nitrogen compounds in the atmosphere can be classified as a

Table 28.1 Key ecosystem manipulation types and the respective impact and ecosystem functions

Type of manipulation	Degree of manipulation	Impact	Ecosystem functions
Changing land use	Partial or complete	Increase of plant yield and change in biodiversity	Food and raw material
Fertilisation with phosphorus, potassium, zinc, iron, ...	Doses exceeding limitation	Increase of plant yield formation	Nutrient cycling
Harvesting	Up to 2 kg dry mass m^{-2} y^{-1}	Withdrawal of nutrients and information	Food and raw material
Liming	Addition of 400 g $CaCO_3$ m^{-2} y^{-1}	Stabilisation of soil pH value at 5–7	Nutrient cycling
Nitrogen fertilisation	Addition of 4 g to 80 g N m^{-2} y^{-1}, organic and inorganic	Increase of plant yield formation	Nutrient cycling
Monoculture	Complete	Specific target yield, decrease in biodiversity	High yields
Mix	Partial or complete	Targeted functions	Biodiversity and nutrient cycling
Sawing	0.1 to 100 g information input per m^2	Targeted stands	Food and raw material
Tillage and clearing	Low or deep, up to 60 cm ploughing and loosening	Soil aeration, soil clearance, reduce biodiversity	High yields

non-apparent side-effect of ecosystem manipulation favouring net primary productivity of plants and ecosystems. Both the non-apparent nitrogen input and in particular the modern nitrogen fertilisation practices, when combined with biomass removal for bioenergy production, were discussed to exacerbate soil C losses (Khan et al., 2007).

The role of reducing vegetation diversity, applying tillage practices and increasing fertilisation intensity were widespread ecosystem-related researchers' and practitioners' objectives during the last decades. This was done with the view of optimising both ecosystem productivity and minimising ecological side-effects. Ecosystem manipulation and management practices affected both short-term safety of the harvests and long-term environmental impacts and have contributed to extensive modifications of global biochemical cycling and biodiversity. Their interactions have become visible for biotic ecosystem components in agriculture and forestry (Dilly et al., 2008).

Recently, the ecological footprint which is an estimate of the area needed to support a population's lifestyle and includes the consumption of food, fuel, wood, and fibres and pollution by heavy metals and persistent organic pollutants (POPs) has been studied. In addition, carbon dioxide, nitrous oxides and methane are also considered as part of the footprint (EEA, 2007). The footprint shows that the non-apparent human impact is becoming more and more important at remote biomes on our planet and may have an impact on ecosystem structures and functions of similar magnitude to apparent drastic human manipulations.

28.3 Ecosystem Degradation and Climate Change

Global warming and regional climate change are considered to affect the functioning of ecosystems and may contribute to ecosystem degradation. In Europe, the most severe impacts on ecosystems are expected in southern and northern Europe.

In dry lands and dry regions, desertification is a major threat to the ecosystems and represents a significant concern for the growing human population. Inappropriate land-use and over-exploration of natural resources leads to land degradation (Breckle et al., 2001). The United Nations Conventions to Combat Desertification (UNCCD) defines desertification as the degradation of the land in arid, semi-arid and dry sub-humid areas caused by various factors, including climate changes and human activities. Dry land ecosystems and especially desert margins are very vulnerable to desertification. Aside of the human impact and land characteristics, the water regimes are controlling desertification. In Europe, desertification is considered as a serious threat in the Mediterranean ecosystems and it is believed it will accelerate through climate change in the next century. Since ecosystems differ with reference to water and energy use efficiency, manipulation practices need to balance water–nutrient interactions (Hatfield et al., 2001). Arid environmental conditions and ineffective irrigation systems may increase salinity problems in southern Europe and, thus, halophytes may increase the productivity in salt-affected ecosystems (Wucherer et al., 2005). Phytomelioration by planting on saline soils can lead to a more rapid closure of the vegetation cover and helps to minimise the widespread negative effects of salt-induced desertification. Studying natural halophytes and ecosystem processes are important for regions where salinity has reached a level that desalinisation techniques are too costly. Beside salinisation, increasing fire frequency represents a threat to Mediterranean ecosystems as restoration is particularly difficult after large-scale disturbance of ecosystems with high plant biodiversity (Aronson et al., 1993).

Drastic effects of climate warming are observed in the Arctic and Sub-Arctic ecosystems, which are remote from human settlements and represent sinks for exogenous inputs such as nutrients and pollutants from distant sources. In the near future they may be affected indirectly by human activities contributing to, for example, global warming, the melting of the permafrost and increasing UV-B radiation. Especially ecosystems with permafrost are sensitive to climate changes. In the Northern Hemisphere regions in which permafrost occurs occupy approximately 25% or 23 million km^2 of the total land area. The thickness of permafrost varies from less than one metre to more than 1,500 m. Contemporary permafrost ecosystems were formed during cold glacial periods, and have persisted through warmer interglacial periods including the last 10,000 years. The recent permafrost degradation is recorded in the thickening of the active layer, the lowering of the permafrost table, the reduction in the thickness and area coverage and also the complete disappearance of the permafrost. Permafrost degradation starts when the temperature remains higher than 0°C. The temperature in permafrost regions is increasing in most locations in the Arctic and Sub-Arctic with a typical increase of about 1–2°K in permafrost temperature (IPCC, 2007). The active layer depth is increasing in some locations and there are some locations in interior Alaska where the active layer doesn't re-freeze completely every year (Romanovsky, Burgess, Smith, Yoshikawa, & Brown, 2002). Long-term permafrost thawing has already started in some locations in natural undisturbed conditions reflecting ongoing climate warming and non-apparent human impact. The impact of temperature-induced change on permafrost ecosystems are the distinct shift of vegetation to the north with a higher extent of shrubs and trees (Wilmking et al., 2005), higher C-turnover in cryosols (Gundelwein et al., 2007) and increasing release of climate-relevant trace gases like CO_2 and CH_4 (Kutzbach et al., 2007, Wille et al., 2008). Thus, human activities have obviously manipulated ecosystem structures and processes.

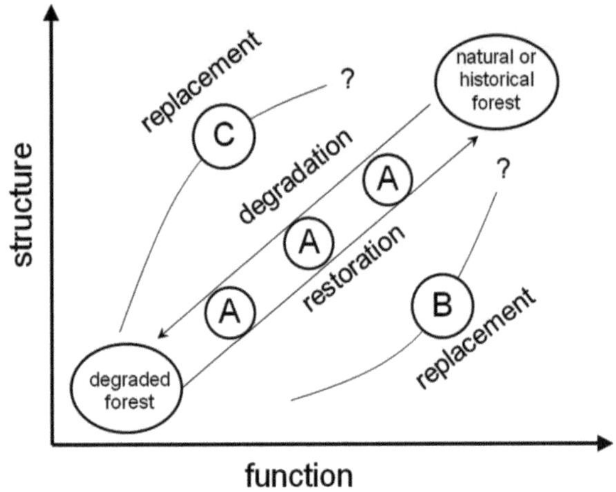

Fig. 28.1 Concepts of forest restoration in relation to function and structure of the forest. (**a**) Represents various degrees of degraded forest that can also be replaced by (**b**) and (**c**) forest type after Löf, Madsen, and Stanturf (2008)

28.4 Restoration

28.4.1 Definition and General Importance

Restoration is defined as repairing ecosystems with respect to health, integrity and sustainability either by recovering the historic structure or the biological function (Aronson et al., 1993). Restoration addresses (i) species, (ii) ecosystems and landscapes, and (iii) ecosystem services (Ehrenfeld, 2000; Zerbe & Wiegleb, 2008), which is demanding for ecosystems with biodiversity damages, large-scale disturbances and that are under the influence of climate change.

Restoration strategies with reference to long-term effects of climate change is a focus for the scientific community (Harris et al., 2006) using nature-based silviculture with conversion of evergreen Norway spruce stands to mixed forests with broadleaf beech and oak under European temperate conditions (Stanturf & Madsen, 2005) and testing tree species or genotypes adjusted to changing environmental conditions (Kriebitzsch et al., 2008). Species selection and the resulting ecosystem manipulation will have long-term effects on the development and functioning of the forests under changing environmental conditions. Thus, the restoration of natural forest is vital for the stabilisation of ecosystem structures and functions (Fig. 28.1) with respect to biomass production, energy fluxes and nutrient cycling as well as ecosystem diversity and resilience. Regulatory mechanisms in highly impacted ecosystems such as degraded wetlands and ecosystems destroyed by open-cast mining activities are still poorly understood.

28.4.2 Degenerated Wetlands and Plant Diversity

Natural fens represent ecosystems that sequester carbon and gain nutrients via ground and surface water. Fens are habitats for hygrophilous plants and animals since the water table fluctuates at the surface. In north-western Europe more than 90% of the pristine fens (Scheuchzerio-Caricetea, Phragmitetea) have been converted into intensively managed grasslands (Joosten & Couwenberg, 2001). The moderate drainage in combination with low fertilisation and one or two cuts annually resulted in the

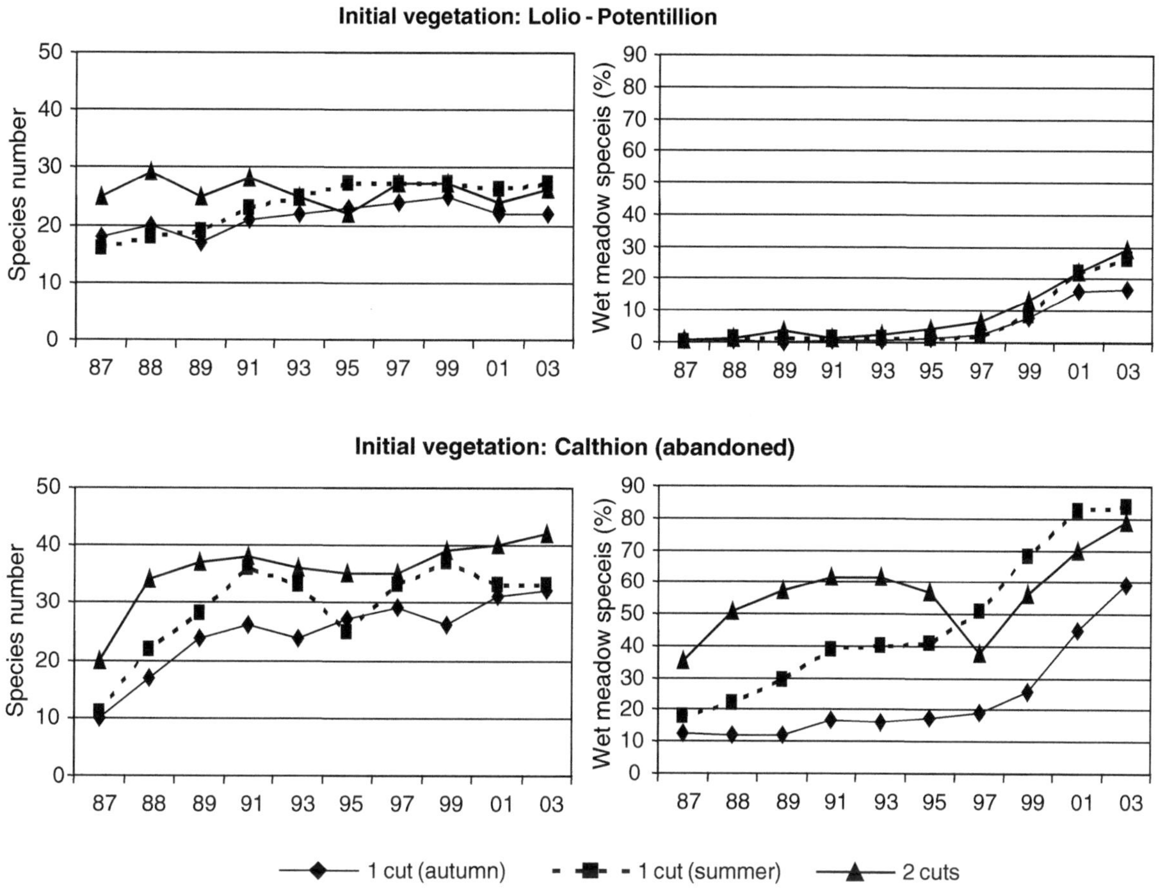

Fig. 28.2 Dynamics of species richness (vascular plants) and coverage (%) of target species (wet meadow species) in permanent plots with different initial vegetation. Plot size 16 m^2

development of species-rich wet meadows (Calthion) in which few plants characteristic for pristine fens could survive (Rosenthal et al., 1998). Strong drainage, high fertilisation and the change from mowing to grazing with high live stock densities have produced species-poor wet grasslands (Lolio-Potentillion). The abandonment of wet meadows owing to decreasing prices of agricultural products induced secondary succession with dominating tall species, which reduced light availability within the stands (Kotowski et al., 2001, Schrautzer & Jensen, 2006) and displaced light-demanding species of wet meadows. In addition, thick litter layers prevented species germination and establishment, especially if seeds are small (Jensen & Gutekunst, 2003).

Conservation projects have been implemented in Germany and north-western European countries since the 1980s to investigate mitigation strategies for reducing negative impacts of land-use changes, for example maintaining and restoring species-rich wet meadows by mowing or grazing without fertilisation. However, few studies achieved data sets longer than 10 years to identify significant secondary succession structures and processes. For example, results from North Rhine-Westphalia in Western Germany indicated that vegetation dynamics is controlled by the initial species composition. Species richness in previously abandoned wet meadows of the Calthion increased already within a few years independent of the mowing regime (Fig. 28.2). In contrast, species richness in previously exploited wet grasslands (Lolio-Potentillion) remained more or less constant suggesting that seed bank and external seed dispersal and also germination and establishment conditions at these sites are poor (Schopp-Guth, 1997). Low restoration potentials for intensively drained wet grasslands

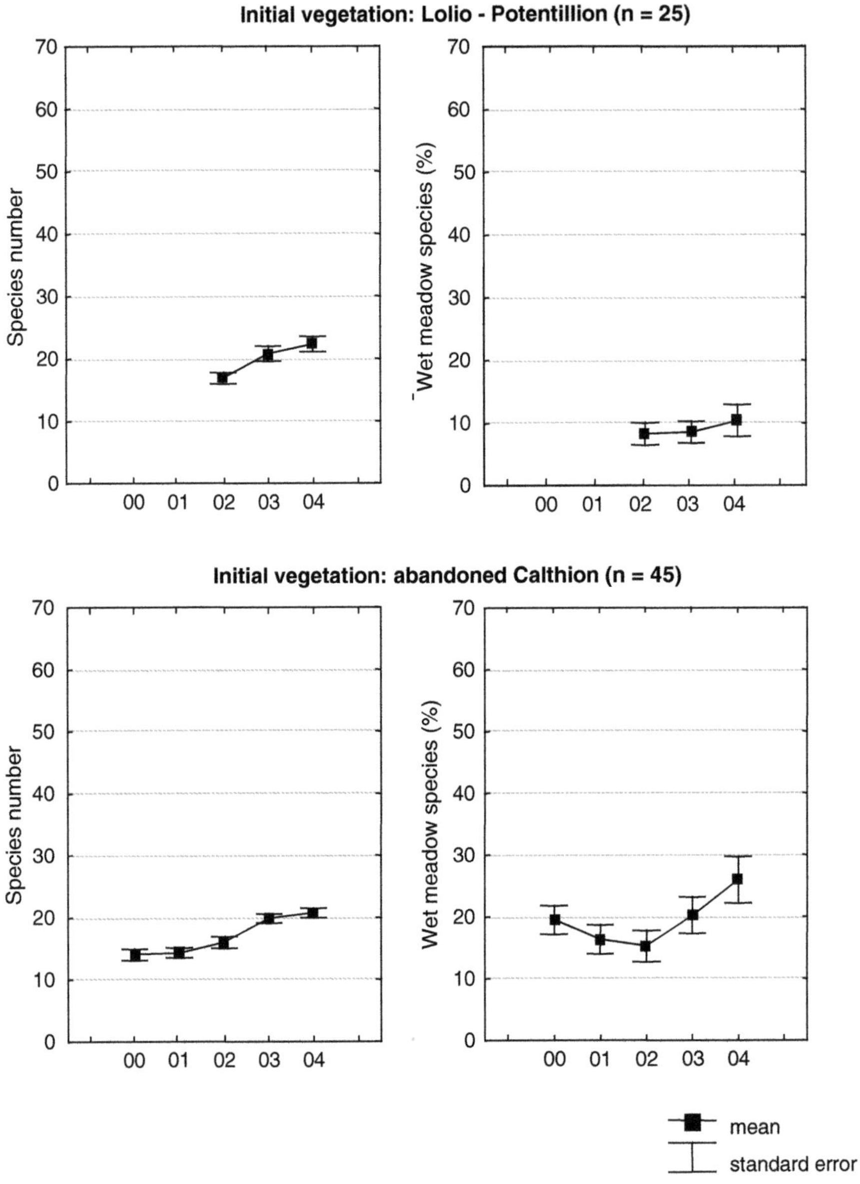

Fig. 28.3 Effects of large-scale grazing on species richness (vascular plants) and coverage (%) of target species in two fen areas of the Eider valley near Kiel. Plot size 25 m^2

were also detected by mowing without fertilisation (Rosenthal, 1992; Sach & Schrautzer, 1994). However, some target species of wet meadows that were scarcely present initially became abundant after 10 years (Fig. 28.2). Higher potentials for the re-establishment of species in abandoned wet meadows were explained by species-rich seed banks (Jensen, 1998) and better growth conditions due to higher water tables. The abundance of target species increased earlier with single mowing in summer or twice yearly in comparison to single autumn mowing.

Large-scale grazing systems are used as a new strategy for maintaining and restoring species-rich, semi-natural plant communities in European cultural landscapes (Finck et al., 2002). Two large pastures in the Eider valley in Northern Germany (Schrautzer et al., 2002) responded with reference to their land-use history (Fig. 28.3). Large-scale grazing was

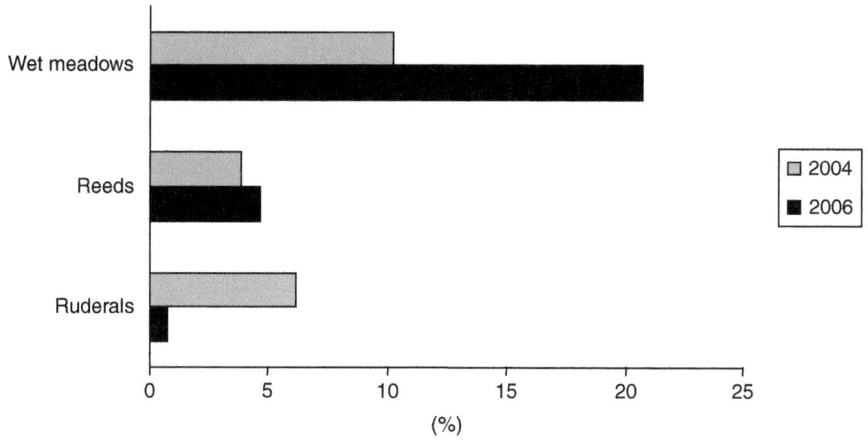

Fig. 28.4 Mean proportions of species groups related to total coverage of vascular plants within stands of the Lolio-Potentillion located in a rewetted fen area of the Eider valley near Kiel. Plot size 2 m^2, $n = 32$

shown to have a rapid effect on species richness and coverage of target species in the previously intensively used (initial vegetation: Lolio-Potentillion) and in the previously abandoned (initial vegetation: Calthion) area. The increase in species richness was attributed to the increase in grazing-induced soil gaps that favoured germination of new species (Schrautzer et al., 2004). However, such an increase in species richness may end after 6 years as observed in Danish wet grasslands (Hald & Vinther, 2000).

Active rewetting measures by ditch blocking and extensive grazing in the Eider valley led to the rapid occurrence of characteristic plant species groups in a previously intensively drained fen (Fig. 28.4). The water table increased approximately 20 cm over two years and reached levels that are typical for wet meadows of the Calthion. The establishment of wet meadow species and the decrease of ruderal species within three years suggested that nutrients decreased through the rewetting measures.

28.4.3 Open-Cast Mining Regions and Soil Recultivation

Open-cast mining activities induced landscape fragmentation and can be considered as one of the most severe anthropogenic disturbance (Walker & Willig, 1999). Both terrestrial and aquatic ecosystems are frequently affected, for example during lignite mining in Lusatia in eastern Germany (Hüttl & Weber, 2001). In terrestrial ecosystems, disturbances and destructions of soil and habitat functions and consequently the biocoenoses were observed. The surface water and groundwater are affected quantitatively and qualitatively (Grünewald, 2001).

The conveyer belt technology for lignite mining in Lusatia has enabled large-scale excavation of lignite seams since the 1920s. The annual lignite mining amounted to 195 million tons from 17 mines in the late 1980s and decreased to 60 million tons after the German unification. Presently, five open-cast mines are active in the Lusatian district and are negotiated to operate until 2030–2040. More than 78,000 ha are affected by mining activities in Lusatia (Drebenstedt, 1998). The conveyer belt technology mixed the upper quaternary and lower tertiary sediments inducing the oxidation of pyrite from the tertiary marine-brackish sediments thus leading to strong acidification and high salinity (Hüttl & Weber, 2001). Acid top soils were ameliorated with lime and fly ash derived from lignite power plants. Soil chemical processes in post-mining regions fluctuated largely during the initial phase when compared to the pre-mining situation and the site-specific geochemical composition largely controls the element fluxes of developing systems (Schaaf, 2001). In addition, geogenic- and lignite-derived carbon in these sandy nutrient poor soils may play a significant role in the water and nutrient supply of the developing ecosystems. These conditions have a large influence on both above and below ground biomass development and humus accumulation which are key components of the ecosystem nutrient cycle.

Table 28.2 Recultivated open cast sites for chronosequence studies

Name	Dumped in	Substrate	Melioration	Vegetation and stand age
Weissagker Berg	1991	lignite + pyrite containing sand	28 t CaO per ha at 60–80 cm soil depth in 1996	*Pinus sylvestris*, 8 years
Bärenbrück	1977	lignite + pyrite containing loamy sand	190 t CaO per ha at 40 cm soil depth in 1978	*Pinus nigra*, 22 years
Meuro	1970	lignite + pyrite containing loamy sand	160 t CaO per ha at 60 cm soil depth in 1971	*Pinus sylvestris*, 27 years
Domsdorf	1946	lignite + pyrite containing loamy sand	50 t CaO per ha at 30 cm soil depth in 1963	*Pinus sylvestris*, 40 years

Table 28.3 Soil chemical parameters of recultivated sites in Lusatia

Site	Age [Years]	Soil depth [cm]	pH [H_2O]	C_t	N_t [mg g^{-1}]	S_t	CEC [mmolc kg^{-1}]	Fe_o	Fe_d [mg g^{-1}]	Al_o	Al_d
Weissagker Berg[1]	16	20	3.7	14.4	0.6	4.4	32.8	3.1	3.9	0.4	0.4
–	–	100	2.8	16.4	0.4	5.6	22.0	4.1	4.4	0.6	0.6
Bärenbrück	30	20	4.0	53.8	1.0	9,8	97.2	7.5	8.9	1.5	1.0
–	–	100	2.5	62.0	1.2	13.6	72.1	7.0	6.5	1.9	1.5
Meuro	37	20	5.3	31.1	0.7	2.3	99.7	4.2	6.3	0.6	0.6
–	–	100	2.8	44.6	1.0	3.0	68.3	3.6	4.2	0.3	0.3
Domsdorf	61	20	5.5	52.5	1.1	1.5	203.8	7.1	10.9	0.6	0.7
–	–	100	2.9	58.0	1.3	2.3	73.5	7.1	9.4	0.4	0.4

[1]Data from Heinkele and Weiß (BTU Cottbus, 1999, pers. Comm.)
[2]Not detected nd. Oxalate extractable iron and aluminium Feo Alo. Dithionite extractable iron and aluminium, Fed Ald.

Since forests were predominant at 60% before mining, the post-mining regions have to be re-forested for legislative reasons; they were recultivated mainly with *Pinus sylvestris* L. (Table 28.2). Soil properties, soil chemistry and water fluxes in forest ecosystem compartments were analysed along "false time series" to identify the key processes and temporal trends in soil and ecosystem. For example, four sites with comparable lignite- and pyrite-containing mine substrates and planted with pine trees covered a period of 60 years (Table 28.2). The substrates were ameliorated with different amounts of fly ash depending on their estimated potential lime requirements. The incorporation depths represented the available technology at the time. Sites were fertilised with 100 kg nitrogen ha^{-1}, 100 kg phosphorus ha^{-1} and 80 kg potassium ha^{-1} before afforestation.

Because of fly ash amelioration the top soils showed high pH-values, whereas the subsoils were in the pH range of 2.5–2.9 (Table 28.3). High total soil carbon, nitrogen and sulphur contents in these young soils were derived from lignite fragments. The actual cation exchange capacity of the substrates was 10–30-fold higher than at natural sites in the region (Weisdorfer et al., 1998) and was correlated with the geogenic carbon contents (Schaaf et al., 2001). Iron and aluminium availability were high throughout the profiles. The soil solution is characterised by extremely low pH values of up to 2.2 and high electrical conductivity with maximal values of 13.5 mS cm^{-1} and also high ionic strength of maximal 0.55 mol L^{-1}. The indicator values increased generally with soil depth (Table 28.3). High electrical conductivity values occurred at the younger sites. Soil solution contained mainly calcium (up to 490 mg Ca L^{-1}), magnesium (up to 316 mg Mg L^{-1}), aluminium (up to 1,760 mg Al L^{-1}), iron (up to 580 mg Fe L^{-1}), and sulphate (up to 11,900 mg SO_4 l^{-1}). Application of fly ash increased the pH values by three units in the ameliorated horizons compared to the non-ameliorated subsoil. Sulphate dominated anions at the sites and with increasing soil depth. However, the cation composition varied with soil depth and age and was dominated by calcium and magnesium in the ameliorated horizons. Acidic metals like iron and aluminium increased with soil depth and decreasing pH values.

Recultivated landscapes develop on substrates that are not equivalent to "undisturbed" landscapes. Lignite

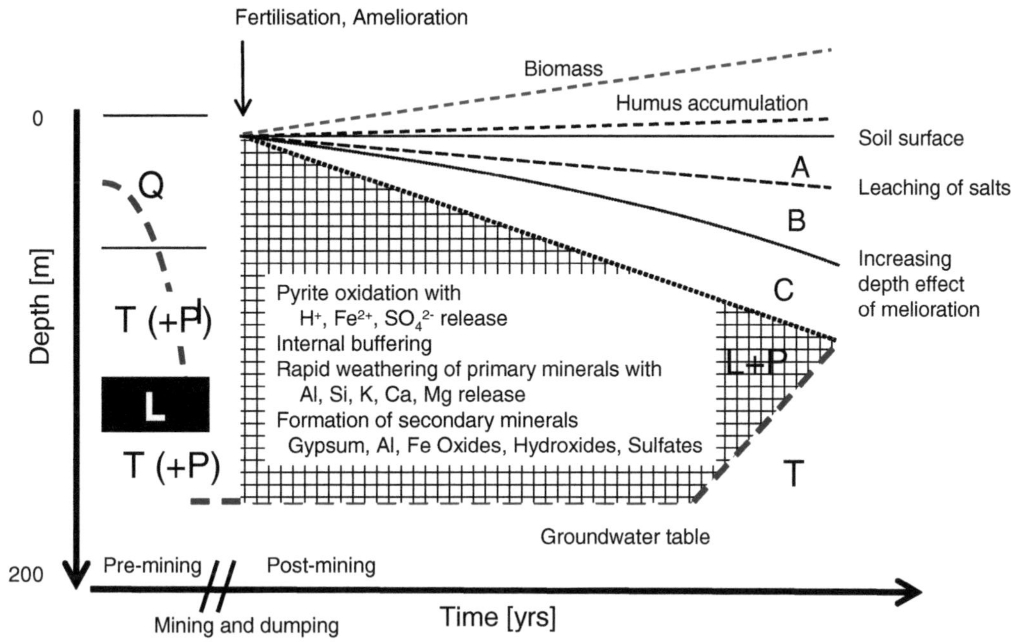

Fig. 28.5 Biogeochemical processes during ecosystem development at the post-mining sites, Abbreviations: Quaternary sediments Q. Tertiary sediments T. Lignite seam L. Leached zone free of easily soluble salts and labile phases A. Amelioration horizon B. Pyrite P oxidation zone C. Initial pyrite oxidation zone due to pre-mining groundwater pumping I

fragments and secondary minerals of post-mining soils control element cycling and nutritional status of the new forest ecosystems. The amelioration measures necessary for forest restoration additionally changed the prevailing condition. Thus, "strange" soil types and properties developed as already pointed out by Jenny (1941).

In addition, ecosystem processes are controlled by the artificial groundwater level that was lowered under the lignite seam for up to approximately 120 m below surface. The simultaneous aeration induced pyrite oxidation and soil acidification (Fig. 28.5). Feldspar weathering can contribute to buffer in-situ the acidity (Heinkele et al., 1999). Pyrite oxidation released acidity, sulphate and iron and thus favoured the weathering of primary minerals discharging aluminium, calcium, magnesium and potassium and consequently the precipitation of secondary salt like gypsum, oxides, hydroxides and also aluminium and iron sulphates. Pyrite-free and salt-poor layers developed rapidly by leaching of soluble and labile secondary salt and mineral phases like gypsum, anhydrite or magnesium sulphate heptahydrates (zone A; Schaaf, 2003). These processes are affected by the amelioration measures and recultivation practices. Large amounts of lignite ashes are incorporated in the topsoils (zone B) resulting in high calcium and magnesium availability. This introduces a large acid neutralisation capacity compared to the substrate-internal buffering and also enhances formation of gypsum and other sulphate salts like $MgSO_4$ that is leached rapidly from the profiles (Schaaf et al., 1999). The high soil pH value induced the precipitation of iron and aluminium oxides and hydroxides.

28.4.4 Development of Initial Ecosystems

Knowledge of the interactions of processes and structures during ecosystem development is an important requirement for the successful restoration and management of landscapes. However, initial ecosystem stages under temperate conditions in Central Europe are uncommon and often affected by human activities. New ecosystem structures with homogenous sediments and without plants are created by severe human impacts since mining operations induce the complete replacement of the natural landscapes (Hüttl &

Gerwin, 2004). "Point zero" of ecosystem development was analysed in for example the volcanic regions of Hawaii, Mount St. Helens and Surtsey Island and in the Arctic, Antarctic and Alps (Jones & DelMoral, 2005) and also at the Hydrohill artificial grassland catchment in China (Kendall et al., 2001).

Multi-disciplinary long-term studies have created a new body of knowledge regarding ecosystem structures and processes. Most research programs commenced with the analyses of functions in mature ecosystems with complex interrelationships (Likens & Bormann, 1995). The investigations of processes and balances of water and elements were done to draw conclusions on stability, elasticity, and resilience of mature ecosystem structures and processes.

Water catchments are usually the basis of such ecosystem studies (Wright et al., 1994). Ecosystem approaches allow the estimation of balances for landscape parts but suffer on natural heterogeneity and clear boundary definition (Table 28.4). The prerequisites are definite catchments for estimating balances from punctual measurements and for up-scaling.

Artificially constructed catchments delineate boundaries on and below the surface and also allow for manipulating inner structures and sediment composition. In contrast, natural watersheds have to be re-constructed based on measurements and models. Artificial catchments at a landscape scale have rarely been realised and, if realised, it addressed mainly hydrological issues and did not represent an interdisciplinary approach (Nicolau, 2002).

Young ecosystems may be characterised by highly variable processes and low degree of interacting structures. Abiotic processes seem pre-dominant initially and biotic processes become more important over time when many "hot spots" have been established. Thus, the initial phase of ecosystem development may be regarded as an unbalanced system. Basic assumptions on the behaviour and function of mature ecosystems may not be valid for young systems. Initial stages of ecosystems can be regarded as the "Geo-System", subsequent stages as the "Hydro-Geo-System" and advanced stages as the "Bio-Hydro-Geo-System" which corresponds to the term "ecosystem" (Fig. 28.6). Thresholds should distinguish the three phases. It may be hypothesised that structures and processes that represent drivers during the early phases also control the mature ecosystem.

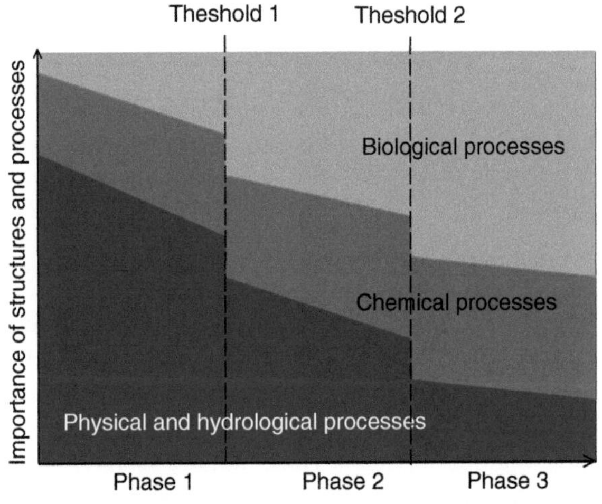

Fig. 28.6 Initial ecosystem genesis with reference to physical (hydrological), chemical and biological structures and processes

Table 28.4 Studies on ecosystem research

	Initial ecosystem	Small ecosystem catchment	Artificial ecosystem catchment	Artificial initial ecosystem catchment
	Island Surtsey, Island Mount St. Helens, U.S.A.	Hubbard Brook, U.S.A. Bornhöved Lakes District, Germany Solling, Germany	Oil Sands Environmental Research Network OSERN, Canada	Hühnerwasser' SFB TRR 38
Primary succession	+	–	–	+
Balance calculation	–	+	+	+
Surrounding conditions knowledge	–	–	+	+
Scale at landscape level	+	+	–	+
Long-term investigations	+	+	–	+

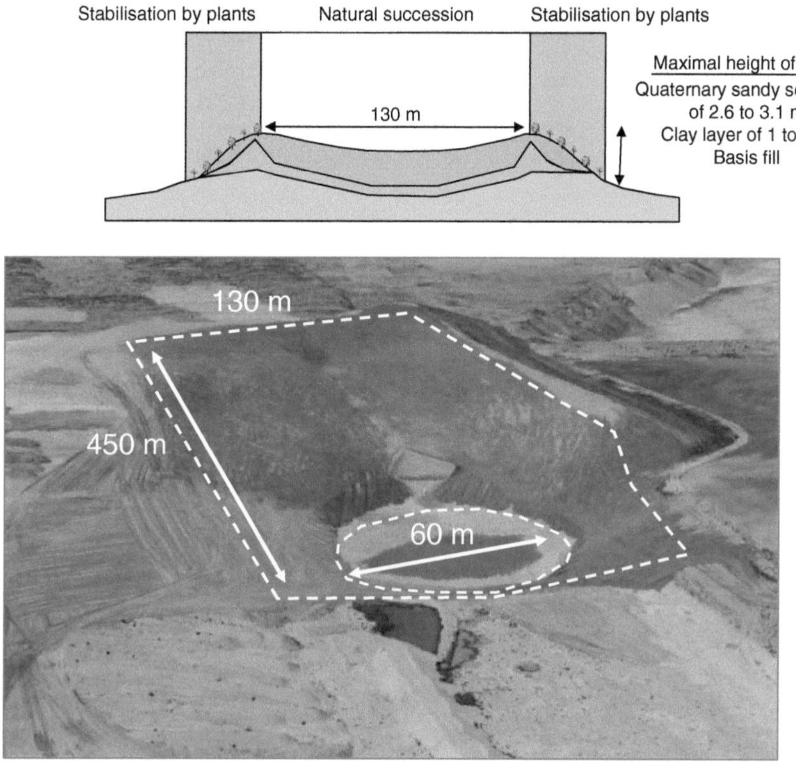

Fig. 28.7 Cross section (*above*) and aerial view (*below*) of the artificial catchment "Hühnerwasser" in the recultivation area of the open-cast lignite mine "Welzow-Süd" in eastern Germany

Recent scientific approaches have investigated initial artificial systems that have not been human impacted and managed. The post-mining landscapes provide opportunities for innovative research on ecosystem development. The mining technique allows for the construction of complete landscape sections with defined inner structures and clear boundaries as done with the "Hühnerwasser" (chicken creek) catchment in the Lusatia mining region (Table 28.4, Fig. 28.7). However, artificial catchments have specific conditions such as the sediment height over the impermeable layer and some early stage artefacts attributed to endogenous C and Fe oxidation controlled by mixing and loosening (aerating) of sediments during the catchment's construction (e.g., oxidation of Fe, S and organic matter) and also due to subsequent compaction processes. In addition, the local genetic resources are controlling factors. Therefore, false chronosequences on young sand dunes and glacial retreat areas can be considered to generalise conclusions from young ecosystems. As a key objective, the energetic metabolism, defined here as "energomics" derived from linking "energy" and "omics", can be a key issue to be tested with proceeding ecosystem development at different levels and in different discipline perspectives (Dilly et al., 2001; Dilly, 2006). For example, the soil carbon metabolism was found to become more efficient at the Rotmoosferner in southern Austria with advanced succession (Table 28.5).

28.5 New Ecosystem Management Types

Related to European policy-relevant perspectives on biomass and biodiversity issues (COM, 2006; EEA, 2006), it became clear that modern land use must integrate the production of food and fodder and also raw and energy materials in conjunction with biodiversity conservation and climate issues. Simultaneously, tourist interests need to be considered. Therefore, multifunctional land and ecosystem use became an important issue in Europe and helped to fulfil multiple short-term and long-term human needs particularly in densely populated regions. Along with policy-driven

Table 28.5 Energomics in topsoil during succession at the Rotmoosferner moraine (Insam and Haselwandter, 1989) $qCO_2\,[mg\ CO_2\text{-}C\ g^{-1}C_{mic}\ h^{-1}]$; $C_{org}\,[mg\ C\ g^{-1}\ soil]$ $qCO_2/C_{org}\,[mg\ CO_2\text{-}C\ g^{-1}\ C_{mic}\ h^{-1}/(g\ C\ g^{-1}\ soil)]$

Site	Age [years]	Metabolic quotient, $qCO_2\,[mg\ CO_2\text{-}C\ g^{-1}C_{mic}\ h^{-1}]$	C_{org} content $[mg\ C\ g^{-1}\ soil]$	qCO_2/C_{org} ratio $[mg\ CO_2\text{-}C\ g^{-1}\ C_{mic}\ h^{-1}\ (mg\ C\ g^{-1}\ soil)^{-1}]$
I	1	3.9	1.9	2.058
II	5	5.7	5.6	1.013
III	65	3.8	15.7	0.245
IV	135	3.0	22.2	0.133
V	1000	2.0	70.8	0.028

land-use changes, the importance of regionalised sustainability impact assessment with reference to environmental, social and economic issues (Fig. 28.8) is recognised (Helming et al., 2007) for which special attention has to be given to sensitive regions and their sustainability issues (Dilly, Camilleri, et al., 2008). Post-industrial regions, mountains, islands and coasts can be considered as sensitive areas due to cross-cutting land-use functions and specific behaviour of sustainability indicators which vary from the EU average and undergo frequently high temporal changes (Dilly, Camilleri, et al., 2008; Ling et al., 2007). For beneficial development in sensitive regions, stakeholder views are taken into account (Dilly et al., 2007).

Agroforestry systems have been recently considered as a new land-use type in temperate regions and in particular in post-industrial regions (Grünewald et al., 2007; Dupraz & Liagre, 2008). Agroforestry can be defined as a land-use type and technology where woody perennials such as trees, shrubs, palms and bamboos are cultivated in some spatial and temporal arrangement together with agricultural crops and animals for beneficial ecological and socio-economical

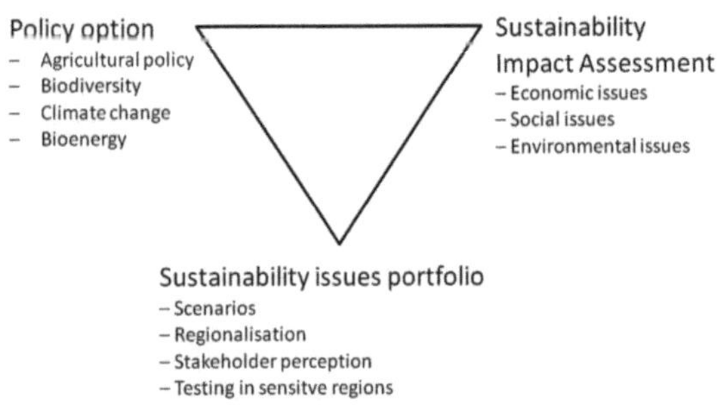

Fig. 28.8 Triangle between policy options, sustainability impact assessment and sustainability issue portfolio

interactions. This land-use type can increase the land surface with land equivalent ratio of can significantly increase the land surface with land equivalent ratios (Dupraz, personal Comm., 2008).

Agroforestry land-use systems may meet the dynamic productivity requirements with regard to the competition between food and bio-energy production and can contribute to biodiversity issues and the CO_2 mitigation activities with reference to adaptation to climate change and thus can be defined as multi-functional systems. Energy production from both agricultural and forest biomass provides energy security in regions with decentralised structures and creates jobs. However, positive CO_2 mitigation effects should not induce higher N_2O and CH_4 emissions and thus cumulative negative effects to the climate system. Agroforestry landscapes with tree and hedge structures can significantly favour habitat corridors for animals (Ewers & Didham, 2007).

The German Government adopted the Renewable Energy Act in 2000 to increase the share of renewable energy with biomass contributions in the electricity market from 5 to 10% by the year 2010. In 2006, 1.56 million ha representing 13% of the German arable land were used for growing renewable resources. Additional material came from 11.8 million ha of the German forests representing 33% of the total land area in Germany. Thus, the area for non-food crops and the competition between food, fodder and raw material is increasing. Additional biomass potentials are perceived at marginal areas, set-aside land without losing the European common agricultural policy payments and sites polluted with heavy metals and organic xenobiotics.

To reduce the risks of monocultures, the Commission adopted a comprehensive EU strategy specifically dedicated to soil protection (EEA,

Fig. 28.9 Alley-cropping land-use system (schematic above, modelled below) in the sub-continental post-mining region of Eastern Germany; visualisation by Lenné3D GmbH, Germany

2006). Some ecosystems and soil degradation phenomena may be accelerated by human manipulation and non-sustainable land-use practices. Also the production of bio-energy holds some risks concerning soil quality. The main harmful processes of soil degradation are erosion, compaction, loss of nutrients, lack of water, decrease of biodiversity and salinisation. Regional stakeholders are also aware of these problems (Dilly, Doerrie, et al., 2007).

Ecosystem management has already adapted to changing climatic conditions at regional scales in the sub-continental lower Lusatia region in Eastern Germany in order to secure the provision of crops and woody biomass in regions with significant drought. In Lusatia, black locust showed highest biomass yields at three and six years in comparison to willow and poplar (Grünewald et al., 2007). However, the high biomass production may impact negatively on seepage water at the landscape level (Kamm et al., 2006). Agroforestry is of interest with regard to the reduction of severe soil erosion and as a measure to adjust to climate change and to diversify large-scale monoculture farmland (Fig. 28.9). Furthermore, the ecological functions of agroforestry systems are related to soil protection through carbon sequestration, nitrogen and phosphorus cycling, and organic matter accumulation (Kamm et al., 2006) and also biodiversity preservation. The economic benefits rely on the potential of agroforestry systems to produce biomass for energetic transformation and chemical processing in bio-refineries. Climate change with increasing temperature has stimulated a discussion on the re-vitalisation of new land-use types that are typical for warmer and dryer conditions; for example viticulture that is presently concentrated in south-west Germany received attention from eastern and northern Europe (Dilly, Zeihser, Hüttl, Kendzia, Wüstenhagen, & Dähnert, 2007).

Besides diversification, efficient cycling of nutrients, elements and energy which depends largely on the structure of the biocoenoses with plants, microorganisms and animals and their respective ecophysiology driving the input, allocation and mineralisation is an important goal (Ehrenfeld, 2000; Nii-Annang, Grünewald, Freese, Hüttl, & Dilly, 2008). The energetic strategy of organisms is categorised as opportunistic and energetically inefficient r- and autochthonous and energetically efficient K-strategists (Dilly, 2005). The developing discipline that studies energetic metabolism and its role in ecosystem functioning and services can be termed as energomics (Dilly, 2006).

The net primary productivity and the partitioning of carbon may vary substantially between biome-characteristics of plants at the landscape level as shown under temperate conditions and observed at the Bornhöved Lakes District in northern Germany (Dilly et al., 2008). Alder trees fixed more carbon at the light-rich forest–lake interface than beech trees in forest stands and could consequently allocate more below-ground carbon. The increased ecosystem acquisition by alder and the increased soil carbon input was associated with higher biotic carbon in the soil organic carbon and reduced soil microbial energetic efficiency. This energetic allocation seems attributable to the vegetation's strategy to support beneficial microbial activities and the plant–soil interactions for acquiring nutrients limiting growth.

It is important to acknowledge that soil microbial biodiversity is generally high and, furthermore, small organisms are considered as more sensitive to environmental conditions and are generally less energetically efficient (Table 28.6). However, it is unclear to what extent the soil microbial diversity and the microbial energetic efficiency are relevant to ecosystem stability and resilience and also sustainable ecosystem functioning. The biotic web is generally regulated by the amount, composition and turnover time of biological substrates in soil and thus the free energy of formation for example –917 GJ mol^{-1} for glucose, −369 for acetate, −51 for methane and −79 for ammonium (Dilly, 2005). This energy derived mainly from plant inputs and the above-ground plant biodiversity is also considered to play an essential role in the establishment of stable ecosystems (Hector et al., 1999). Mature ecosystems were identified as more stable and more efficient in energy and nutrient use than young ecosystems and similarly non-manipulated ecosystems more levelled than manipulated ones. Modern techniques such as molecular tools (Dilly et al., 2004) and stable isotype analysis (Kendall et al., 2007; Zyakun & Dilly, 2005) will help to answer fundamental questions on long-term investigation sites.

As mentioned above, the water supply and the water use efficiency of the vegetation is crucial for ecosystem functioning. The water requirement for CO_2 fixation can vary by more than one order of magnitude (Table 28.7) with values between 167 and 1,500 L H_2O per kg C. Water use efficiency is dependent on species, human impact and diurnal and seasonal changes in environmental conditions such as temperature, water supply and shading. The plants with C4 metabolism are considered as being more efficient than C3 plants (Marschner, 1986). Plant growth is reduced at low water availability particularly in dry summers and on soils with low water-holding capacity (Grünewald et al., 2007).

It is generally agreed that temperature increase induced by climate change at constant annual precipitation rates will stress the vegetation and human

Table 28.6 Energy use efficiency of lower and higher organisms after Fritsche (1999)

	Cell diameter [μm]	Biomass-specific metabolic activity [$\mu l\ O_2 mg^{-1}$dry mass h^{-1}]
Bacteria	1	1,000
Fungi, yeast	10	100
Plants, animals	100	10

Table 28.7 Water use efficiency of plants with reference to manipulation measures

Plant	Manipulation	mmol CO_2 mol^{-1} H_2O	L H_2O kg^{-1} C	Source
Wheat	Low NPK fertilisation	9.0	167	Gorny and Garzynski (2002)
Wheat	High NPK fertilisation	8.5	176	Gorny and Garzynski (2002)
Douglas fir	None	8.1 ± 2.4	185	Ponton et al. (2006)
Aspen	None	5.4 ± 2.3	247	Ponton et al. (2006)
C4 plant	ND	5.0	300	Marschner (1986)
Alder	None. Peripheral leaves	3.5	429	Dilly et al. (2008)
Grasses	None	3.0	500	Dilly et al. (2008)
Grassland	None	2.6 ± 0.7	577	Ponton et al. (2006)
C3 plant	ND	2.5	600	Marschner (1986)
Alder	None. Inner crown	1.0	1,500	Dilly et al. (2008)

society worldwide. Increasing CO_2 concentration in itself will reduce drought stress by reducing transpiration (Kruijt et al., 2007). High biomass production for food, fibre and bio-energy requires an enormous amount of water particularly when not selecting water-use efficient species. Under dry climatic conditions like in Lusatia or in the lee of mountains, this may lead to very little formation of ground water. The increase in water use efficiency at different scales still requires research which should include modern stable isotopic characteristics (Marshall et al., 2007).

28.6 Conclusions and Future Demands

Natural and historical ecosystems have largely been manipulated which was shown in long-term agricultural and forest experiments. The combination of interdisciplinary studies gave extensive knowledge on biotic and abiotic structures and processes in manipulated and managed ecosystems. Even ecosystems that are remote from regions populated by humans can be considered as manipulated. Ecosystems with similar degrees of human impact showed similar adjustments in nutrient use efficiency in the above-ground biota and soil biota. The restoration of abandoned wet meadows can be more successful than that of previously intensively managed wet pastures. The adjustment of water tables was important and management practises such as large-scale grazing and haymaking with one early summer cut or two cuts per year could enhance plant diversity of degenerated fen grasslands. Harsh manipulations are present in industrial and post-industrial regions and at urban sites with for example heavy metals, pollutants and destruction.

Ecosystem manipulation and restoration measures should be addressed more carefully in the future, for example for mitigating and adapting to climate change. Ecosystem management practices should consider (i) biomass production optimisation by the adaptation of traditional and new ecosystems to climatic and edaphic conditions with for example agroforestry; (ii) planning tools at regional levels; (iii) exploring environmental benefits such as the reduction of soil erosion, the accumulation of organic matter, efficient water use, the preservation and increase of biodiversity and CO_2 sequestration at the ecosystem and landscape level. Long-term studies are generally required to test for example new established agroforestry systems, land-use changes and developed wetland restoration. False chronosequences may support the identification of fundamental processes with high dynamics in the initial ecosystem development induced by substrate composition, and intensive transformation processes, and changes of element pools. Such a chronosequence still showed large disturbances after several decades and it can therefore be expected that these effects will remain for centuries. Initial ecosystem development studies are important since knowledge on ecosystem potentials and the interacting abiotic and biotic components is scarce.

While earlier ecosystem research projects addressed environmental, social and economic issues separately, modern approaches combined these three pillars of sustainable development. This methodology supports the identification of cross-cutting environmental, social and economic issues and the translation of environmental findings towards the society and economic benefits and also to implement environmental technologies for socio-economic and environmental benefits (ETAP, 2007).

Acknowledgements This work was supported by the Integrated EU Project "Sustainability Impact Assessment: Tools for Environmental, Social and Economic Effects of Multifunctional Land Use in European Regions – SENSOR", the German Research Foundation (SFB/TRR 38), the Brandenburg Ministry for Science, Research and Culture and Vattenfall Europe Mining AG.

References

Aronson, J., Floret, C., LeFloćh, E., Ovalle, C., & Pontainer, R. (1993). Restoration and rehabilitation of degraded ecosystems in arid and semi-arid lands. I. A view from the south. *Restoration Ecology, 1*, 8–17.

Breckle, S. W., Veste, M., & Wucherer, W. (Eds.) (2001). *Sustainable land-use in deserts*. Heidelberg, Berlin, New York: Springer.

COM. (2006). *Halting the loss of biodiversity by 2010 – and beyond. Sustaining ecosystem services for human well-being. Communication from the Commission of the European Communities to the Council and the European Parliament 216.* Retrieved from http://europa.eu/scadplus/leg/en/lvb/l28176.htm.

Costanza, R., D'Arge, R., DeGroot, R., Farber, S., Grasso, M., Hannon, B., et al. (1997). The value of the world's ecosystem services and natural capital. *Nature, 387*, 253–260.

Dilly, O. (2005). Microbial energetics in soil. In: Buscot, F. & Varma, A. (Eds.), *Microorganisms in soils: Roles in genesis and functions* (pp. 123–138). Berlin: Springer.

Dilly, O. (2006). Evaluating soil quality in ecosystems based on modern respiratory approaches. In: Cenci, R. & Sena, F. (Eds.), *Biodiversity-bioindication to evaluate soil health* (pp. 59–64). Luxembourg: Office for Official Publications of the European Communities.

Dilly, O., Bloem, J., Vos, A., & Munch, J. (2004). Bacterial diversity during litter decomposition in agricultural soils. *Applied and Environmental Microbiology, 70*, 468–474.

Dilly, O., Camilleri, M., Dörrie, C., Formosa, S., Galea, R., Hallenbarter, D., et al. (2008). Key sustainability issues and the spatial classification of sensitive regions in Europe. In: Helming, K., Perez-Soba, M. & Tabbush, P. (Eds.), *Sustainability impact assessment of multifunctional land use* (pp. 471–494). Berlin: Springer.

Dilly, O., Doerrie, C., Schneider, B.-U., & Hüttl, R. F. (2007). Abschätzung der Folgen von Bioenergieförderung in der brandenburgischen Lausitz. *Forum der Forschung, 20*, 35–40.

Dilly, O., Eschenbach, C., Kutsch, W. L., Kappen, L., & Munch, J. C. (2008). Eco-physiological key processes in agricultural and forest ecosystems. In: Fränzle, O., Kappen, L., Blume, H. P., & Dierssen, K. (Eds.), *Ecosystem organization of a complex landscape: Long-term research in the Bornhöved Lake District, Germany. Ecological Studies 202* (pp. 61–81). Berlin: Springer.

Dilly, O., Winter, K., Lang, A., & Munch, J. C. (2001). Energetic eco-physiology of the soil microbiota in two landscapes of southern and northern Germany. *Journal of Plant Nutrition and Soil Science, 164*, 407–413.

Dilly, O., Zeihser U., Hüttl, R. F., Kendzia G., Wüstenhagen D., & Dähnert D. (2007). Perspektiven für einen modernen Weinbau in der Niederlausitz. *Forum der Forschung, 20*, 80–84.

Drebenstedt, C. (1998). Planungsgrundlagen der Wiedernutzbarmachung. In: Pflug, W.(Ed.), *Braunkohletagebau und Rekultivierung* (pp. 487–512). Berlin: Springer.

Dupraz, C., & Liagre, F. (2008). *Agroforesterie. Des arbres et des cultures*. Paris: Editions France Agricole.

EEA. (2006). *Progress towards halting the loss of biodiversity by 2010*. Copenhagen: European Environmental Agency.

EEA. (2007). *Europe's environment. The fourth report*. Luxembourg: Office for Official Publications of the European Communities.

Ehrenfeld, J. G. (2000). Defining the limits of restoration: The need for realistic goals. *Ecological Restoration, 8*, 2–9.

Ellenberg, H., Mayr, R., & Schauermann, J. (1986). *Ökosystemforschung – Ergebnisse des Solling-Projektes 1966–1986*. Stuttgart: Ulmer Verlag.

ETAP. (2007, December 01). *Environmental technologies action plan*. Retrieved from http://ec.europa.eu/environment/etap/index_en.htm.

Ewers, R. M., & Didham, R. K. (2007). The effect of fragment shape and species' sensitivity to habitat edges on animal population size. *Conservation Biology, 21*, 926–936.

Finck, P., Riecken, U., & Schröder, E. (2002). Pasture landscapes and nature conservation. New strategies for preservation of open landscapes in Europe. In: Redecker, B., Finck, P., Riecken, U., & Härdtle, W. (Eds.), *Pasture landscapes and nature conservation* (pp. 1–14). Berlin: Springer.

Fränzle, O., Kappen, L., Blume, H. P., & Dierssen, K. (Eds.). (2008). *Ecosystem organization of a complex landscape: Long-term research in the Bornhöved Lake District, Germany. Ecological Studies 202*. Berlin: Springer.

Fritsche, W. (1999). *Mikrobiologie*. Jena: Gustav Fischer Verlag.

Gollan, T., & Heindl, B. (1998). Bayreuther Institut für Terrestrische Ökosystemforschung. In: Fränzle, O., Müller, F., & Schröder, W. (Eds.), *Handbuch der Ökosystemforschung. Grundlagen und Anwendungen der Ökosystemforschung. V-4.6* (pp. 1–18). Landsberg: Ecomed.

Górny, A. G., & Garczynski, S. (2002). Genotypic and nutrition-dependent variation in water use efficiency and photosynthetic activity of leaves in winter wheat (*Triticum aestivum* L.). *Journal of Applied Genetics, 43*, 145–160.

Grünewald, U. (2001). Water resources management in river catchments influenced by lignite mining. *Ecological engineering, 17*, 143–152.

Grünewald, H., Brandt, B. K. V., Schneider, B. U., Bens, O., Kendzia G., & Huettl, R. F. (2007). Agroforestry systems for the production of woody biomass for energy transformation purposes. *Ecological Engineering, 29*, 319–328.

Gundelwein, A., Müller-Lupp, T., Sommerkorn, M., Haupt, E. T. K., Pfeiffer, E.-M. & Wiechmann, H. (2007). Carbon in tundra soils in the Lake Labaz region of arctic Siberia. *European Journal of Soil Science, 58*, 1164–1174.

Hald, A. B., & Vinther, E. (2000). Restoration of a species-rich fen-meadow after abandonment: Response of 64 plant species to management. *Applied Vegetation Science, 3*, 15–24.

Harris, J. A., Hobbs, R. J., Higgs, E., & Aronson, J. (2006). Ecological restoration and global climate change. *Restoration Ecology, 14*, 170–176.

Hatfield, J. L., Sauer, T. J., & Prueger, J. H. (2001). Managing soils to achieve greater water use efficiency: A review. *Agronomy Journal, 93*, 271–280.

Hector, A., Schmid, B., Beierkuhnlein, C., Caldeira, M. C., Diemer, M., Dimitrakopoulos, P. G., et al. (1999). Plant diversity and productivity experiments in European grasslands. *Science, 286*, 1123–1127.

Heinkele, T., Neumann, C., Rumpel, C., Strzyszcz, Z., Kögel-Knabner, I., & Hüttl, R. F. (1999). Zur Pedogenese pyrit- und kohlehaltiger Kippsubstrate im Lausitzer Braunkohlerevier. In: Hüttl, R. F., Klem, D., & Weber, E. (Eds.), *Rekultivierung von Bergbaufolgelandschaften* (pp. 25–44). Berlin: de Gruyter.

Helming, K., Sieber, S., Tscherning, K., König, B., Müller, K., Wiggering, H., et al. (2007). Land use functionality as a frame for impact assessment. In: Knierim, A., Nagel, U. J., & Schäfer, C. (Eds.), *Managing economic, social and biological transformations. Proceedings of the First Green Week Scientific Conference. 2007* (pp. 33–44). Weikersheim: Margraf Publishers.

Hüttl, R. F., & Gerwin, W. (2004). Entwicklung und Bewertung gestörter Kulturlandschaften. Fallbeispiel Niederlausitzer Bergbaufolgelandschaft. Cottbuser Schriften zur Ökosystemgenese und Landschaftsentwicklung, Cottbus, Band 2.

Hüttl, R. F., & Weber, E. (2001). Forest ecosystem development in post-mining landscapes: A case study of the Lusatian lignite district. *Naturwissenschaften, 88*, 322–329.

Intergovernmental Panel on Climate Change. (2007). The physical science basis. In S. Solomon et al. (Eds.), Contribution of working group I to the fourth assessment report of the Intergovernmental Panel on Climate Change. Cambridge: Cambridge University Press.

Jenny, H. (1941). *Factors of soil formation*. New York: McGraw-Hill.

Jensen, K. (1998). Species composition of soil seed bank and seed rain of abandoned wet meadows and their relation to aboveground vegetation. *Flora, 139*, 345–359.

Jensen, K., & Gutekunst, K. (2003). Effects of litter on establishment of grassland plant species: The role of seed size and successional status. *Basic and Applied Ecology, 4*, 579–587.

Jones, C. C., & DelMoral, R. (2005). Effects of microsite conditions on seedling establishment on the foreland of Coleman Glacier, Washington. *Journal of Vegetation Science, 16*, 293–300.

Joosten, H., & Couwenberg, J. (2001). Bilanzen zum Moorverlust – das Beispiel Europa. In: Succow, M., & Joosten, H. (Eds.), *Landschaftsökologische Moorkunde* (pp. 406–409). Stuttgart: Schweizerbart'sche Verlagsbuchhandlung.

Kamm, B., Schneider, B. U., Hüttl, R. F., Grünewald, H., Gusovius, H. J., Stollberg, C., et al. (2006). Lignocellulosic feedstock biorefinery – Combination of technologies of agroforestry and a biobased substance and energy economy. *Forum der Forschung, 19*, 53–62.

Kendall, C., Elliott, E. M., & Wankel, S. D. (2007). Tracing anthropogenic inputs of nitrogen to ecosystems. In: Michener, R. & Lajtha, K. (Eds.), *Stable isotopes in ecology and environmental science* (pp. 375–449). Oxford: Blackwell Publishing.

Kendall, C., McDonnell, J. J., & Gu, W. (2001). A look inside 'black box' hydrograph separation models: A study at the Hydrohill catchment. *Hydrological Processes, 15*, 1877–1902.

Khan, S. A., Mulvaney, R. L., Ellsworth, T. R., & Boast, C. W., (2007). The myth of nitrogen fertilization for soil carbon sequestration. *Journal of Environmental Quality, 36*, 1821–1832.

Kotowski, W., VanAndel, J., VanDiggelen, R., & Hogendorf, J. (2001). Responses of fen plant species to groundwater level and light intensity. *Plant Ecology, 155*, 147–156.

Kriebitzsch, W.-U., Beck, W., Schmitt, U., & Veste, M. (2008). Bedeutung trockener Sommer für Wachstumsfaktoren von verschiedenen Herkünften der Rotbuche (Fagus sylvatica L). *AFZ-Der Wald, 5*, 246–248.

Kruijt, B., Witte, J.-P. M., Jacobs C. M. J., & Kroon, T. (2007). Effects of rising atmospheric CO_2 on evapotranspiration and soil moisture: A practical approach for the Netherlands. *Journal of Hydrology, 349*, 257–267.

Kutzbach, L, Wille, C., & Pfeiffer, E.-M. (2007). The exchange of carbon dioxide between wet arctic tundra and the atmosphere at the Lena River Delta, Northern Siberia. *Biogeosciences, 4*, 869–890.

Likens, G. E., & Bormann, F. H. (1995). *Biogeochemistry of a forested ecosystem*. New York: Springer Verlag.

Ling, C., Handley, J., & Rodwell, J. (2007). Restructuring the post-industrial landscape: A multifunctional approach. *Landscape Research, 32*, 285–309.

Löf, M., Madsen, P., & Stanturf, J. (2008). Restaurering av sydsvensks lövskog – några tankar kring ett nytt skötselkoncept. *Svensk Botanisk Tidskrift, 102*, 43–51.

Marschner, H. (1986). *Mineral nutrition of higher plants*. San Diego: Academic Press.

Marshall, J. D., Brooks, J. R., & Lajtha, K. (2007). Source of variation of the stable isotopic composition in plants. In: Michener, R. & Lajtha, K. (Eds.), *Stable isotopes in ecology and environmental science* (pp. 1–21). Oxford: Blackwell Publishing.

Nii-Annang, S. A., Grünewald, H., Freese, D., Hüttl, R. F., & Dilly, O. (2009). Microbial activity and soil quality in alley cropping systems after 9 years of recultivation of quaternary deposits in eastern Germany. *Biology and Fertility of Soils*. DOI 10.1007/s00374-009-0360-4.

Nicolau, J. M. (2002). Runoff generation and routing on artificial slopes in a Mediterranean-continental environment: The Teruel coalfield, Spain. *Hydrological Processes, 16*, 631–647.

Rockström, J., Steffen, W., Noone, K., Persson, A., Chapin, F. S., Lambin, E. F. (2009). A safe operating space for humanity. *Nature, 461*, 472–475.

Romanovsky, V. E., Burgess, M., Smith, S., Yoshikawa, K., & Brown, J. (2002). Permafrost temperature records: Indicators of climate change. *Earth Observing System Transactions, American Geophysical Union, 83*, 589.

Rosenthal, G. (1992). *Erhaltung und Regeneration von Feuchtwiesen. – Dissertationes Botanicae 182*. Berlin: Cramer.

Rosenthal, G., Hildebrandt, J., Zöckler, C., Hengstenberg, M., Mossakowski, D., Lakomy, W., et al. (1998). *Feuchtgrünland in Norddeutschland. Ökologie, Zustand, Schutzkonzepte*. Münster: Bundesamt für Naturschutz.

Sach, W., & Schrautzer, J. (1994). Phytomasse- und Nährstoffdynamik sowie floristische Veränderungen von Knickfuchsschwanz-Flutrasen (Ranunculo-Alopecuretum geniculati Tx. 37) unter extensiver Nutzung. *Flora, 189*, 37–50.

Schaaf, W. (2001). What can element budgets of false-time series tell us about ecosystem development on post-lignite mining sites? *Ecological Engineering, 17*, 241–252.

Schaaf, W. (2003). Leaching induced changes in substrate and solution chemistry of mine soil microcosms. *Water, Air and Soil Pollution – Focus, 3*, 139–152.

Schaaf, W., Gast, M., Wilden, R., Scherzer, J., Blechschmidt, R., & Hüttl, R. F. (1999). Temporal and spatial development of soil solution chemistry and element budgets in different mine soils of the Lusatian lignite mining area. *Plant and Soil, 213*, 169–179.

Schaaf, W., Neumann, C., & Hüttl, R. F. (2001). Actual cation exchange capacity in lignite containing pyritic mine soils. *Journal of Plant Nutrition and Soil Science, 164*, 77–78.

Schopp-Guth, A. (1997). Die Zusammensetzung des Diasporenpotentials unter Niedermoorböden Nordostdeutschlands. *Zeitschrift für Ökologie und Naturschutz, 2*, 87–98.

Schrautzer, J., Irmler, U., Jensen, K. Nötzold, R., & Holsten, B. (2004). Auswirkungen großflächiger Beweidung auf die Lebensgemeinschaften eines nordwestdeutschen Flusstales. *Schriftenreihe für Landschaftspflege und Naturschutz, 78*, 39–62.

Schrautzer, J., & Jensen, K. (2006). Relationship between light availability and species richness during fen grassland succession. *Nordic Journal of Botany, 24*, 341–353.

Schrautzer, J., Jensen, K., Holsten, B., Irmler, U., Kieckbusch, J., Noell, C., et al. (2002). The Eidertal pasture landscape – Mire restoration and species conservation in a river valley of Schleswig-Holstein (northwest Germany). In: Redecker, B., Finck, P., Härdtle, W., Riecken, U., & Schröder, E. (Eds.), *Pasture landscapes and nature conservation* (pp. 227–237). Berlin: Springer.

SEC. (2006). *Thematic Strategy for Soil Protection. Communication from the Commission to the Council, the European Parliament, the European Economic and Social Committee and the Committee of the Regions*. Retrieved from http://eur-lex.europa.eu/smartapi/cgi/sga_doc?smartapi!celexplus!prod!DocNumber&lg=en&type_doc=COMfinal&an_doc=2006&nu_doc=231.

Smith, P., Smith, U. J., Powlson, D. S., McGill, W. B., Arah, J. R. M., Chertov, O. G., et al. (1997). A comparison of the performance of nine soil organic matter models using datasets from seven long-term experiments. *Geoderma, 81*, 153–222.

Stanturf, J., & Madsen, P. (Eds.). (2005). *Restoration of boreal and temperate forests*. Boca Raton: CRC Press.

Walker, L. R., & Willig, M. R. (1999). An introduction to terrestrial disturbance. In: Walker, L. R. (Ed.), *Ecosystems of disturbed grounds. Ecosystems of the world 16* (pp.1–16). Amsterdam: Elsevier.

Weisdorfer, M., Schaaf, W., Blechschmidt, R., Schütze, J., & Hüttl, R. F. (1998). Soil chemical response to drastic reductions in deposition and its effects on the element budgets of three Scots pine ecosystems with different pollution history in NE-Germany. In: Hüttl, R. F. & Bellmann, K. (Eds.), *Changes in atmospheric chemistry and effects on forest ecosystems. Nutrients in Ecosystems* (pp. 187–225, Vol. 3). Dordrecht: Kluwer Academic Publishers.

Wille, C., Kutzbach, L., Sachs, T., Wagner, D., & Pfeiffer, E.-M. (2008). Methane emission from Siberian arctic polygonal tundra: Eddy covariance measurements and modelling. *Global Change Biology, 14*, 1395–1408.

Wilmking, M., D'Arrigo, R., Jacoby, G., & Juday, G. (2005). Increased temperature sensitivity and divergent growth trends in circumpolar boreal forests. *Geophysical Research Letters, 32*(15), L15715.1–L15715.4.

Wright, R. F., Lotse, E., & Semb, A., (1994). Experimental acidification of Alpine catchments at Sogndal, Norway: Results after 8 years. *Water, Air & Soil Pollution, 72*, 297–315.

Wucherer, W., Veste, M., Herrera-Bonilla, O., & Breckle, S.-W. (2005). Halophytes as useful tools for rehabilitation of degraded lands and soil protection. *Proceedings of the First International Forum on Ecological Construction of the Western Beijing, Beijing* (pp. 87–94). Retrieved February 2008 from http://www.desertconsult.de/PDF/62Halopyhten%20as%20tools_Bejing2005.pdf.

Zerbe, S., & Wiegleb, G. (2008). *Renaturierung von Ökosystemen in Mitteleuropa*. Heidelberg: Spektrum Akademischer Verlag.

Zyakun, A., & Dilly, O. (2005). Respiratory quotient and priming effect in an arable soil induced by glucose. *Applied Biochemistry and Microbiology, 41*, 512–520.

Chapter 29

Exploratories for Large-Scale and Long-Term Functional Biodiversity Research

Markus Fischer, Elisabeth K.V. Kalko, K. Eduard Linsenmair, Simone Pfeiffer, Daniel Prati, Ernst-Detlef Schulze, and Wolfgang W. Weisser

Abstract Current changes in biodiversity and their functional consequences for ecosystem processes matter for both fundamental and applied reasons. In most places the most important anthropogenic determinant of biodiversity is land use. The effects of type and intensity of land use are modulated by climate and atmospheric change, nutrient deposition and pollution and by feedback effects of changed biological processes. However, it is not known whether the genetic and species diversity of different taxa responds to land-use change in similar ways. Moreover, consequences of changing diversity for ecosystem processes have almost exclusively been studied in model experiments of limited scope. Clearly, there is an urgent scientific and societal demand to investigate the relationships between land use, biodiversity and ecosystem processes in many replicate study sites in the context of actual landscapes. Furthermore, these studies need to be set up in long-term frameworks. Moreover, because monitoring and comparative observation cannot unravel causal mechanisms they need to be complemented by manipulative experiments. In the 'Exploratories for large-scale and long-term functional biodiversity research' (see www.biodiversity-exploratories.de), we provide a platform for such successful long-term biodiversity research. The biodiversity exploratories aim at contributing to a better understanding of causal relationships affecting diversity patterns and their change, developing applied measures in order to mitigate loss of diversity and functionality, integrating a strong research community to its full potential, training a new generation of biodiversity explorers, extending the integrated view of functional biodiversity research to society and stimulating long-term ecological research in Germany and globally. Our experience has several implications for long-term ecological research and the LTER network including the necessity of formulating common research questions, establishing a joint database, applying modern tools for meta-analysis or quantitative review and developing standardised experimental and measurement protocols for facilitating future data synthesis.

Keywords Community ecology · Ecosystem processes · Ecosystem services · Landscape ecology · Large-scale ecological research · Large-scale experiments · Population biology

29.1 Introduction

In this chapter, we shortly review links between changes in land use, biodiversity and ecosystem processes. In particular we outline that the studies of effects of land use on biodiversity are often limited to single taxa and that the causal study of the relationship between biodiversity and ecosystem processes has almost exclusively focused on model ecosystems. Based on these considerations we highlight the scientific and societal demand for research in this context. Then we introduce the 'Exploratories for large-scale and long-term functional biodiversity research' (www.biodiversity-exploratories.de), the research project which we established to provide

M. Fischer (✉)
Institute of Biochemistry and Biology, University of Potsdam, Maulbeerallee 1, D-14469 Potsdam, Germany; Institute of Plant Sciences, University of Bern, Altenbergrain 21, 3013 Bern, Switzerland
e-mail: markus.fischer@ips.unibe.ch

F. Müller et al. (eds.), *Long-Term Ecological Research*,
DOI 10.1007/978-90-481-8782-9_29, © Springer Science+Business Media B.V. 2010

a platform and critical mass for successful long-term biodiversity research. Finally we draw conclusions from the exploratories for the national and international Long-Term Ecological Research (LTER) initiatives.

29.2 Changes of Land Use and Its Intensity

In recent decades, socio-economic, agro-ecological, climatic and geophysical shifts caused enormous changes in land use and land-use intensity ranging from local and regional (Hietel, Waldhardt, & Otte, 2005) to national (Stöcklin, Bosshard, Klaus, Rudmann-Maurer, & Fischer, 2007), international and global scales (Houghton, 1994). These changes constitute crucial elements of the current global change (Millenium Ecosystem Assessment, 2005). Intensification of agriculture in Central Europe associated with the Common Agricultural Policy (CAP) of the European Union has been characterised by three main driving processes: the simplification and specialisation of agricultural landscapes leading to a decrease of permanent grasslands and semi-natural landscape elements, the transformation of less fertile areas into shrublands and early successional forests and an overall loss of landscape diversity and the increase in the use of entrants (herbicides, pesticides and fertilisers) per unit area (Shucksmith, Thomson, & Roberts, 2005). The spatial extent of agricultural land has been declining because high labour costs limit the competitiveness of agricultural products on the free market and because of recent changes in the CAP, such as a reduction in support for cereals and beef and the introduction of 'set-aside' payments to withdraw land from production (Shucksmith et al., 2005). Further changes in European agriculture are expected in the forthcoming years as CAP payments will become progressively decoupled from single-farm production. Although increasing emphasis on the production of biofuel is starting to change agricultural markets (Sims, Hastings, Schlamadinger, Taylor, & Smith, 2006) the trend of decreasing area of agricultural land is very likely to continue because biofuel production will only be attractive on the most productive land (Smeets, Faaij, & Lewandowski, 2004). Clearly, the decisions of farmers between types of land use and its intensity are mainly driven by product prices, the availability and costs of labour and machinery and to a large extent by the CAP as well to a lesser extent also by other measures of national or regional authorities. Similarly, type and intensity of forest use is affected by timber and firewood prices and national and regional regulations (Koskela, Ollikainen, & Pukkala, 2007).

In Germany, forests and grasslands are among the most important ecosystems that continue to be strongly affected by human requirements. Although the natural vegetation cover of Germany consists of deciduous forests, this type of ecosystem is now restricted to very small and highly fragmented areas whereas most of the previously forested area is replaced with agricultural land or forest plantations (Ellenberg, 1996). Furthermore, rather nutrient-poor grasslands that were formerly only used at low intensities by humans and formed the most species-rich species assemblages are now either threatened by complete abandonment or by intensification of land use (Poschlod, Bakker, & Kahmen, 2005).

Changes in land use from one type to another, say from forest to grassland or from grassland to field, are clearly defined. Moreover, it is common knowledge that the intensity of land use changes, e.g., from nutrient-poor semi-natural grasslands used at low intensity to highly fertilised and frequently mown grasslands or from hardly used natural beech stands to plantations of conifers (Poschlod et al., 2005). It is important to note, however, that beyond such coarse categories land-use intensity is not clearly defined at all. This is all the more disturbing as the term 'intensive land use' is very frequently used. Only a clear definition of the intensity of land use will allow the objective evaluation of effects of land-use intensity on biodiversity and ecosystem processes (Canals et al., 2007). It is therefore an urgent and interesting research question to find a scientifically sound and measurable definition. One approach could be measuring the amount, quality and frequency of input and output of material.

The biodiversity exploratories described below focus on forest and grassland habitats by choosing plots ranging from near-natural to intensively used land. In Germany, these landscape structures can be found in and around protected sites like National Parks and Biosphere Reserves and on cultivated public and private land.

29.3 Biodiversity

In this chapter and in line with the Convention on Biodiversity (CBD, 1992) the term biodiversity is meant to encompass all scales of biological organisation, spanning intra-specific genetic, morphological and demographic diversity, community diversity, the diversity of biological interactions between organisms and the diversity of ecosystems in the landscape (Wilson, 2001).

Usually, the diversity of plant and animal species is higher in more diverse landscapes (Clough et al., 2007; Hendricks et al., 2007; Rudmann-Maurer, Weyand, Fischer, & Stöcklin, 2008). However, it is much less well known how far biodiversity at lower levels is positively associated with each other. This is exemplified by the very scarcely studied relationship between species diversity and genetic diversity within species. This is of fundamental importance for evolutionary ecology and it matters for conservation, because a negative relationship would indicate a conflict between conservation measures promoting genetic diversity and those promoting species diversity, whereas a positive relationship would indicate synergies. A positive association between the two types of diversity has been predicted (van Valen, 1965; Vellend, 2003). However, whereas some studies showed a positive correlation between genetic variation as measured with molecular markers and species diversity, others did not (Vellend & Geber, 2005). Analogous studies using selectively non-neutral measures of genetic diversity, such as heritabilities and additive quantitative genetic variation, are still lacking. Moreover, studies considering the genetic variation of more than just one species of a community are still missing in this context.

More research efforts have been made to study the relationship between the species diversity of plants and animals. Clearly, plant species diversity also affects the diversity of higher trophic levels. Virtually all ecological models predict that an increase in resource diversity supports a more diverse array of consumers (Rosenzweig, 1995). Correspondingly, a positive relationship between plant and insect herbivore diversity has been found in many correlative studies of natural systems and in studies of agricultural monocultures and polycultures, although there are notable exceptions (e.g. Andow, 1991; Murdoch, Peterson, & Evans, 1972; Risch, Andow, & Altieri, 1983; Root, 1973). However, even though generally plant diversity appears to be positively related to animal diversity, there is large variation among different taxonomic groups in the exact pattern of this relationship, which is one of the reasons why the selection of biodiversity hotspots solely based on plant diversity can be misleading (Prendergast, Quinn, Lawton, Eversham, & Gibbons, 1993). In part this is due to species-specific effects such as the tenfold diversity of insect herbivores on oaks than on ash or lime (Brändle & Brandl, 2001). To date, the roles of direct effects of plant diversity on plant–herbivore interactions (Siemann, 1998; Siemann, Tilman, Haarstad, & Ritchie, 1998) and of indirect effects via plant productivity and herbivore density (Hunter & Price, 1992) for the different results are not very clear. In general, similar diversity considerations as for herbivory also apply to plant–pollinator interactions. Moreover, it is becoming increasingly clear that many flower visitation webs are characterised by asymmetric specialisation where specialised pollinators rather visit generalist plants and specialised plants are rather visited by generalist pollinators (Waser & Ollerton, 2006). The redundancy involved in these plant–pollinator webs implies that the local extinction of pollinators need not necessarily lead to a loss of the diversity of animal-pollinated plants (Memmott, Waser, & Price, 2004). Nevertheless, the degree of specialisation in plant–pollinator webs is higher than that in other plant–animal interactions such as seed dispersal webs (Blüthgen, Menzel, Hovestadt, Fiala, & Blüthgen, 2007). Finally, despite pioneering large-scale programs such as the UK soil biodiversity project (http://soilbio.nerc.ac.uk/), very little is known about the relationship between plant diversity and the diversity of other groups such as soil invertebrates and microbes at the plot level as well as at the landscape level.

Positive relationships between plant and animal diversity have also been proposed for vertebrates. For instance, small mammals such as voles and shrews (Soricidae) occupy central positions in many biotic and abiotic cycles within terrestrial ecosystems. In grasslands, voles prey selectively on grasses and herbs, collect and cash seeds and perturbate the ground introducing gap dynamics for plant succession (e.g. Seabloom & Reichman, 2001; Seabloom & Richards, 2003). Shrews prey heavily on invertebrates above and below ground (e.g, Churchfield, Hollier, & Brown, 1991; Churchfield & Rychlik, 2006). This may dilute predation pressure of herbivorous arthropods on plant resources and of carnivorous arthropods on other invertebrate taxa, ultimately leading to an increase in local plant and invertebrate diversity. However, despite of the profound impact small mammals exert on plants, almost nothing is known how the diversity of the one group influences the diversity of the other, the associated ecosystem processes and whether different land-use practices alter the functioning of these fundamental interactions.

29.4 The Effect of Land Use on Biodiversity

Changes in land use and its intensity are major determinants of biodiversity throughout the world (Millenium Ecosystem Assessment, 2005; Sala et al., 2000). In the cultural landscapes of Central Europe land use has clearly increased biodiversity during the last few centuries (Ellenberg, 1996). This is exemplified by the high number of highly diverse nutrient-poor grasslands, which are replacing less diverse former natural forests. Similarly, without land use biodiversity would be much lower in many areas of the world. Clearly, it is not correct to coarsely assign negative biodiversity effects to all kinds of land use. In principle changes in land use and land-use intensity allow to affect biodiversity and ecosystem processes both in positive and in negative ways.

Currently, however, biodiversity experiences a general and drastic decline especially at the local and regional scales, which in terrestrial habitats is mainly due to the ongoing changes in land use, and furthered by the direct effects of displacement of native species by invasive alien ones, of industrial emissions of nitrogen, organic compounds and heavy metals, and of climate change and by their indirect effects via changed land use (Millenium Ecosystem Assessment, 2005; Sala et al., 2000; Thuiller, 2007). Although the mostly negative relationship between biodiversity and current land-use change is rather well known for some taxa, in particular vertebrates, vascular plants and some invertebrate groups such as grasshoppers or butterflies, it is not known for many others. Because different taxa may respond differently to changes in land use, investigating the effects of land use on biodiversity of different taxa remains an important and urgent research demand. Understanding such differences is crucial as changes in land use also affect species interactions (Tylianakis, Tscharntke, Lewis, 2007), making it a challenge to disentangle the multiple dependencies between species in complex ecosystems.

The most obvious component contributing to land-use intensity in grasslands is fertilisation. The relationship between soil fertility and plant diversity has often been described by a hump-shaped curve with highest plant diversity at intermediate fertility levels (Grime, 1973; Mittelbach et al., 2001) and not, as one might expect alternatively, at the sites with most nutrients. A number of mechanisms have been discussed that lead to a decrease in diversity with increasing fertilisation (Rosenzweig, 1995; Suding et al., 2005). However, this relationship may depend on several ecological factors including human disturbance or seed limitation (Schmid, 2002). It has therefore been suggested that the hump-back curve constitutes an envelope function rather than an exact fit to the data (Schmid, 2002). Similarly, the effect of grazing intensity on plant diversity may vary depending on factors related to stand productivity, thus ranging from positive trends on fertile sites to negative trends on less fertile stands (Proulx & Mazumber, 1998). Because land use itself involves several forms of disturbance which can vary in intensity, the respective disturbances are likely to play crucial roles in the context of the land use – biodiversity relationships. According to the intermediate disturbance hypothesis (Connell, 1978; Grime, 1973), species diversity should be highest at intermediate levels of disturbance. Although this has been shown for a few taxa, it remains unclear how far this hypothesis can be generalised to other taxa and levels of biodiversity at various spatial and temporal scales (Hughes, Byrnes, Kirnbro, & Stachowicz, 2007).

In principle, similar considerations as for grasslands also apply to forest trees and understory plants. However, in managed forests the effect of land use on the diversity of plants is more difficult to evaluate because it strongly depends on the number and identity of the tree species that were planted (Hunter, 1999). As age-class forests are accompanied by succession in the associated plant communities, the diversity of these forests with regard to different age classes needs to be compared with unmanaged forests that are characterised by a more continuous age distribution of trees. Thus, the spatial arrangement of age classes in a managed forest provides different opportunities for species than the compact arrangement of a natural forest. An additional aspect of particular importance in Central European forests is the pressure of large herbivores such as deer and wild boar. Although hunting intensities vary considerably, herbivory by game is generally high. The degree to which this reduces or even hinders tree recruitment is a hotly debated question (Senn & Suter, 2003).

Land management has manifold effects on the diversity of higher trophic levels. Frequent mowing of grasslands or intense grazing leads to a decrease in the diversity of insect herbivores and parasitoids because there is insufficient time available for the species to complete their life cycle despite the presence of the host plant (Kruess & Tscharntke, 2002; Völkl, Zwölfer, Romstöck-Völkl, & Schmelzer, 1993). Ploughing leads to a strong reduction in soil animal diversity which recovers very slowly even if the above-ground plant diversity is already very high (Hendrix et al., 1986). Short-term fertilisation may lead to an increase in insect number and diversity independent of plant species diversity due to increased resource availability whereas after long-term fertilisation the detrimental effects on plant diversity also affect insect diversity (Perner et al., 2005; Siemann, 1998). Similar effects are likely for other animal groups. Predicting the consequences of land-use change for biodiversity involves disentangling direct effects of land use on the diversity of higher trophic levels from indirect effects mediated by plant diversity.

29.5 Ecosystem Services

Society is interested in the proper and continuous functioning of ecosystem processes and in particular in the services they provide. These include the pools and fluxes of water, carbon and nutrients, the maintenance of soil fertility, clean water and air, the provisioning of food and construction material and pollination and pest control (Christensen et al., 1996). The relationship between biodiversity and ecosystem functioning is a rapidly expanding research area in ecology and evolution (Hooper et al., 2005). To date, a number of model experiments have addressed the functional consequences of changes of plant diversity (Hector et al., 1999; Roscher et al., 2005; Tilman et al., 2001). It could be shown that ecosystem processes and services depend not only on the species composition of ecosystems but also, and mostly positively, on their species richness (Balvanera et al., 2006; Hooper et al., 2005; Naeem, Thompson, Lawler, Lawton, & Woodfin, 1994; Fig. 29.1). Moreover, higher plant species diversity matters for maintaining multiple ecosystem processes at the same time (Hector & Bagchi, 2007).

However, the model experiments are still mainly restricted to few types of communities and short time periods. Moreover, the functional consequences of changes in animal, fungal and microbial diversity have so far been mainly investigated in micro- and mesocosm studies and are hardly known for natural ecosystems. In the actual landscape context, positive

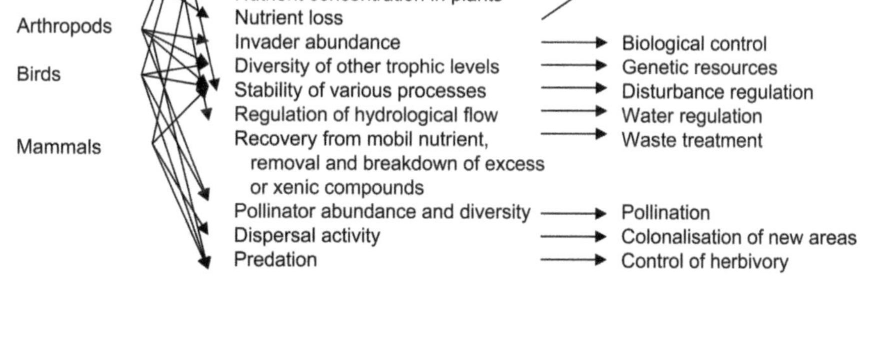

Fig. 29.1 Association of taxa with particular ecosystem processes and services (in part compiled from Schmid et al., 2001 and Hooper et al., 2005). Most of these associations will be investigated in the biodiversity exploratories in the landscape context

correlations between biodiversity and ecosystem processes have been reported for some taxa (Balvanera et al., 2006; Hooper et al., 2005). First experiments manipulating plant species diversity at several sites (Zeiter, Stampfli, & Newbery, 2006) indicate positive relationships of plant species diversity and production. However, replicated experiments manipulating the biodiversity also of other taxa at many sites in actual landscapes are missing completely.

29.6 Feedback Loops Between Changes in Land Use, Biodiversity and Ecosystem Processes

Causal relationships between land use, biodiversity and ecosystem processes are by no means one-way. Rather, there are many feedback loops. Typically, changes in land use change biodiversity, which in turn affects ecosystem processes and services. These changes affect the benefits associated with land use and its intensity and subsequently affect decisions on future land use (Schröter et al., 2005). For example, partial logging of a forest feeds back on its future use if concomitant changes in microclimate change growth rates and sapling recruitment. Such feedback loops between society, land use, biodiversity and ecosystem processes are illustrated by the arrows in Fig. 29.2.

Most countries have signed and ratified the Convention on Biodiversity (CBD, 1992) and therefore agreed that the preservation of biodiversity and its values has to become an integral part of the policy not only on conservation, agriculture and forestry, urban and regional planning but also across other political sectors (Stöcklin et al., 2007). This illustrates that the question of how to best preserve and sustainably use biodiversity leads to an interdisciplinary research demand both of fundamental biological relevance and of immediate applied and societal relevance. This research demand comprises considering the feedback loops between society, land use, biodiversity and ecosystem processes outlined in Fig. 29.2.

29.7 Current Biodiversity Research

Many of the research questions outlined above have already been addressed. However, the overwhelming majority of studies are either conducted by biodiversity researchers considering only one or very few taxa without the inclusion of ecosystem processes or by ecosystem researchers ignoring diversity issues.

Germany provides a good example of a country with profound knowledge of biodiversity and a highly proficient community of scientists addressing biodiversity and ecosystem research questions (Beck, 2004). However, most recent biodiversity data of Germany are being obtained in the framework of biodiversity monitoring studies carried out by the different federal states of Germany, or in the wake of environmental impact assessments in the process of urban and regional planning activities. Unfortunately, these data

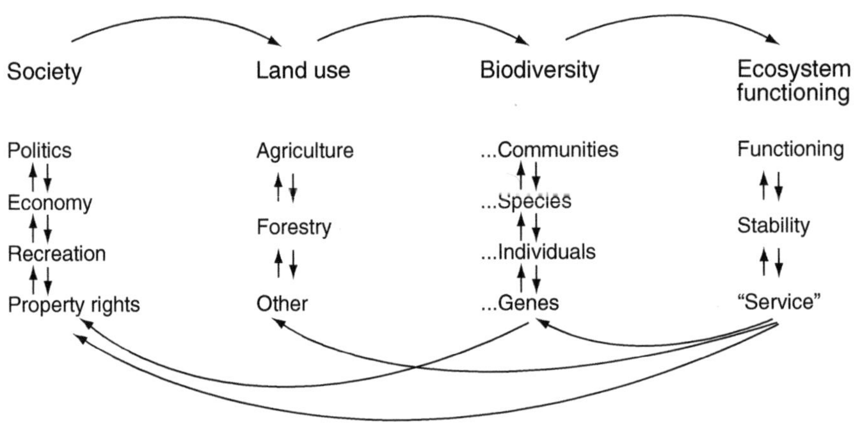

Fig. 29.2 Illustration of the concept underlying the biodiversity exploratories. Societal forces affect type and intensity of land use. This influences biodiversity at several levels of biological organisation, which in turn feeds back on ecosystem processes and the services they provide. *Arrows* indicate causal relationships and illustrate various feedback loops of the functioning of ecosystems on biodiversity and on decision-making by land users and society

are not centrally accessible. In addition, there are no uniform standards and in particular no enforcement of existing standards and no standardised quality control such that most data are not very reliable. Finally, while these data, if made centrally available, could provide a very good inventory of German biodiversity for particular taxa and for quantifying the effect of land use on biodiversity, they do not allow to study the consequences of biodiversity change for ecosystem functioning. As far as biodiversity research funded by German research funding organisation is concerned, current research activities on relationships between land use change, biodiversity change, and ecosystem processes are highly scattered across the country and mostly restricted to short-term studies at small scales.

29.8 Requirements for Functional Biodiversity Research

Addressing the comprehensive research objective of understanding the relationships between land use, biodiversity and ecosystem processes requires an integrative and interdisciplinary research approach in which different plant, animal, fungal and microbial taxa are studied together with a thorough assessment of the relevant ecosystem functions and processes (Figs. 29.1 and 29.2). This needs not only to be done in model systems such as microcosms and experimental gardens but to be realised in many replicate study plots in the landscape. Therefore, to adequately address the interrelationships of diversity and ecosystem processes it is pivotal to measure them at the same temporal and spatial scale with a sufficient number of geo-referenced replicate study plots. Furthermore, because different taxa may respond differently to variation in land use, it is crucial to study all taxa on the same plots. This will allow identification of possible conflicts in land-use management and provide baseline data for the protection of particular taxa. Moreover, as it is of utmost importance to unravel causal relationships and mechanisms rather than simple correlations, manipulative experiments must be performed at many replicate plots at the landscape level. Furthermore, relationships between land use, biodiversity and ecosystem processes may well differ between different landscapes depending on climatic, geological, topographic or anthropogenic differences. Therefore, it is indispensible to study these relationships in different landscapes. Clearly, the large research scope described in this paragraph makes clear that such research can only succeed if a critical mass of researchers of many biodiversity and ecosystem research disciplines, landscape ecology, forestry and agricultural science collaborates closely.

To date, most research covers only time periods of a few years that are too short to permit a clear distinction between different types of temporal pattern responses such as linear trends of increase or decrease, periodicity, threshold behaviour, or non-linear or chaotic trajectories, and time lags of response and feedback (Hansell, Craine, & Byers, 1997). Clearly, a comprehensive understanding of the amount and functional consequences of biodiversity changes requires a long-term perspective and joint consideration of climate and land-use changes (Hobbs, Yates, & Mooney, 2007).

29.9 Exploratories for Functional Biodiversity Research: General Considerations

The research demand outlined above forms the basis of the 'Exploratories for large-scale and long-term functional biodiversity research', a joint project and research platform funded by the Deutsche Forschungsgemeinschaft DFG (www.biodiversity-exploratories.de). The name 'exploratories' was selected instead of 'observatories' to emphasise the important role of manipulative experiments, which – in contrast to also important descriptive monitoring – allow identifying causal relationships. In the following we introduce the research topics addressed by the exploratories and outline the design and implementation of the project, which is currently in its initial phase 2006–2011.

The most important research objectives of the project focus on

- the effect of land use on biodiversity
- the relationships between genetic diversity, species diversity of many taxa and the diversity of biotic interactions and how these are affected by land use
- the functional effects of biodiversity on ecosystem processes in real landscapes

Fig. 29.3 Situation of the three biodiversity exploratories 'Schorfheide-Chorin', 'Hainich-Dün' and 'Schwäbische Alb' in Germany. For more detailed maps see Fig. 29.6

The three exploratories for functional biodiversity research (Fig. 29.3) are located in the biosphere reserve Schorfheide-Chorin in the State of Brandenburg (northeast Germany), the National Park Hainich and its surroundings in the State of Thuringia (central Germany) and the designated biosphere area Schwäbische Alb in the State of Baden-Württemberg (southwest Germany). The three landscapes selected for the exploratories represent different climatic, geological and topographical situations (Table 29.1) which are representative for large parts of Germany.

Of course, it is desirable to complement them with further landscapes in future exploratories, for instance in coastal, urban, post-industrial or alpine landscapes of Germany, which would also cover further climatic conditions (Fig. 29.4). Moreover, further biodiversity exploratories would also be very desirable in further landscapes abroad.

All three exploratories share a number of important features. They are situated in landscapes comprising grassland and forests at similar elevation and on similar bedrock but are used at very different intensities. Moreover, the various land-use types and intensities in the exploratories reflect conflicting land-use interests in Germany and Central Europe, with a strong pressure on intensification or abandonment of grasslands, and various forms of forest management. In addition, for each exploratory, background information is available on habitat types, vegetation, grassland and forest management and land-use history, and many taxa have already been or continue to be monitored for conservation purposes.

The forest plots either form part of large homogenous old-growth beech forests or are situated in patchy and diverse mixed forest stands dominated by beech, oak and pine and differing in land-use intensity. Intensities of grassland use decline from fertilised meadows, over pastures used by sheep, horse and cattle to at least recently unmanaged dry grasslands (Table 29.1). Many of the habitat types represented in the exploratories are of high conservation value for Germany and Central Europe.

The study of whole landscapes requires an intense collaboration between research institutions, land owners, land users and local authorities. Large proportions of the exploratories are protected as national park, nature reserve or protected zone of a biosphere reserve. Therefore, specific permits for entrance and work need to be obtained from regional environmental offices and land owners. In return, data on biodiversity and ecosystem processes obtained in the exploratories can be used for improved planning of conservation management and sustainable land use. These mutual benefits between stakeholders such as land owners and authorities of agriculture, forest and conservation on the one hand and researchers on the other are an important characteristic of the exploratories. To firmly establish the corresponding links between stakeholders and researchers from the very beginning, the research proposal was developed in collaboration with local authorities in all three exploratories.

Depending on state policies and land-use history of Brandenburg, Thuringia and Baden-Wuerttemberg, the degree of landscape fragmentation and property rights of suitable forest and grassland areas vary considerably. Thus, the number of involved public and private land owners differs strongly between the three exploratories and ranges from almost exclusive state forest in the Hainich exploratory to more than 80

Table 29.1 Characteristics of the three biodiversity exploratories

Exploratory characteristics	Schorfheide-Chorin	Hainich-Dün	Schwäbische Alb
General information			
Region	Northeastern Germany	Central Germany	Southwestern Germany
Federal State	Brandenburg	Thuringia	Baden-Württemberg
Protection status of parts of the exploratory	Biosphere reserve	National park and surrounding area	Biosphere reserve
Total exploratory size	~1,300 km^2	~1,500 km^2	~700 km^2
Topography	Low land	Hilly land	Low mountain
Altitude a.s.l.	3–140 m	300–400 m	720–840 m
Inhabitants per km^2	23	116	258
Geology and hydrology	Young glacial landscape, ground moraine, terminal moraine, lakes, moors, fens and mires	Limestone, few ponds, little streams with wetlands in the east, river 'Werra' in the west, few depressions	Calcareous bedrock karst phenomena such as huge cavities, dolines and springs
Climate			
Annual mean temperature	8–8.5°C	6.5–7.5°C	6–7°C
Annual mean precipitation	500–600 mm	750–800 mm	700–1,000 mm
Main soil types	Cambisol and Histosol	Luvisol and Stagnosol	Leptosol and Cambisol
Gradient of land-use intensity			
Forest	Old-growth unmanaged beech forest; age-class pure or mixed beech forest; age-class pine forest	Old-growth unmanaged beech forest; selection-cutting beech forest; age-class pure beech forests (>70% beech); age-class spruce forest	Unmanaged beech forest; age-class pure or mixed beech forest; age-class spruce forest
Grassland	Fertilised meadows; low-intensity cattle pastures; fallow land; unfertilised grassland	Fertilised meadows; pastures of different intensities; late mown pastures (sheep, cattle)	Fertilised meadows; low-intensity sheep and cattle pastures; fallow land; unfertilised dry grassland

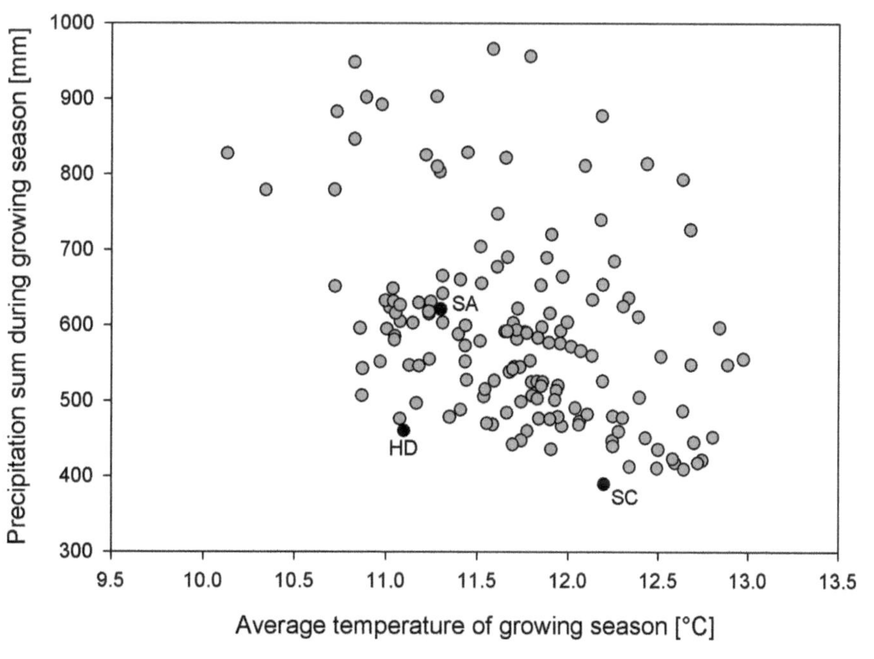

Fig. 29.4 The climatic space spanned by precipitation and mean temperature during the growing season in Germany. Growing season comprises all months with an average temperature of at least 3.0°C. The graph shows the position of each German 0.25° × 0.25° quadrat with forest. The *coloured symbols* denote the quadrats with the three exploratories Schorfheide-Chorin (SC), Hainich (H) and Schwäbische Alb (SA). Data according to Mitchell and Jones (2005) and Loveland et al. (2005)

different grassland owners in the Schwäbische Alb exploratory. Project presentations on the goals, practical perspectives and requirements of the exploratories for local populations and authorities turned out to be a very useful information tool. The positive feedback to these presentations by land owners and authorities very much facilitated plot selection and implementation.

29.10 Exploratories for Functional Biodiversity Research: Study Plots and Initial Measurements

Each exploratory comprises ~1000 km^2 with at least 1,000 study plots selected from the intersection points of a 100 × 100 m grid (Fig. 29.5). These study plots are called Grid Plots. The total area of each exploratory was pre-selected to cover sufficient areal proportions of land with different intensities of grassland and forest use. Half of the Grid Plots (500) are situated in forests and the other half (500) in grasslands. The Grid Plots have been selected from the total grid with the goal to permit future stratified sampling according to soil type and land-use intensities (Fig. 29.6).

For each of the 3,000 Grid Plots, soil samples were taken and are being analysed for a number of physical and chemical parameters. These include a designation of soil horizons, determination of humus and soil types, estimates of soil texture, measurements of total soil thickness and analysis of organic and anorganic soil carbon stocks. For all forest Grid Plots comprehensive forest inventories were obtained in circles of 12.5 m radius comprising detailed information on stand structure, the species, size and distribution of established and young trees and the amount and position of coarse woody debris. Moreover, vegetation records were taken on 400 m^2 in forest Grid Plots and on 16 m^2 in grassland Grid Plots using cover estimates. In addition, current land use and land-use history are being explored. More detailed information on land use, land-use history, and the surrounding landscape matrix is being obtained by structured interviews of land owners and tenants.

Out of the 1,000 Grid Plots per exploratory, a subsample of 100 plots, 50 in forests and 50 in grasslands, were selected for more detailed observations

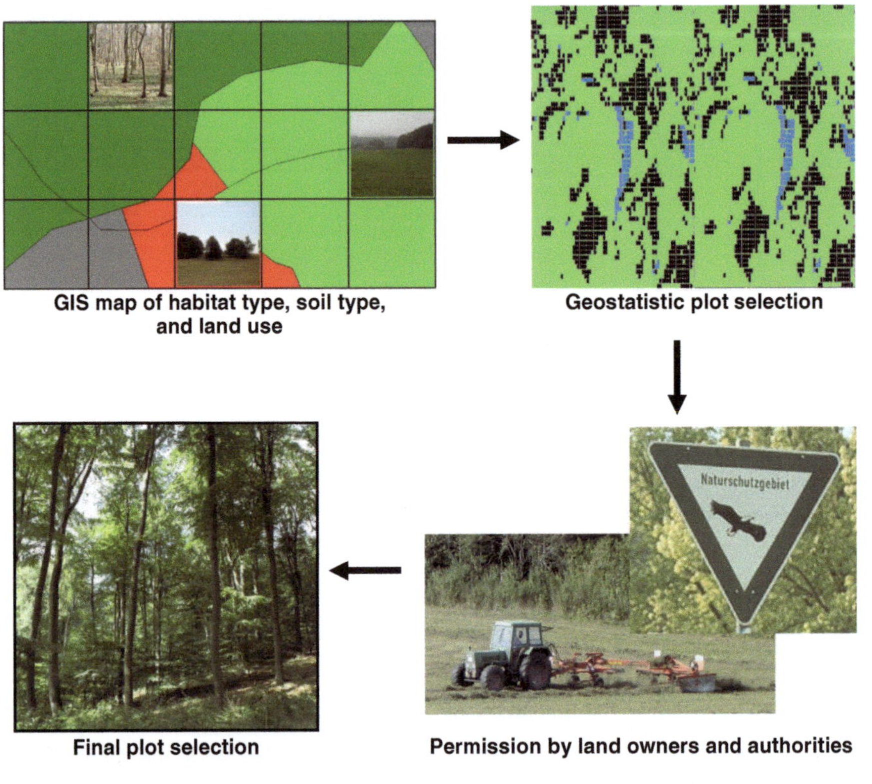

Fig. 29.5 Selection process of study plots within the biodiversity exploratories. First, information on types of habitat, soil and land use was compiled in GIS. Then, appropriate grid points on a 100 m grid overlaying the landscape of the exploratory were identified. Third, land users and authorities were asked for research permissions. In each of the three exploratories 1,000 plots were selected, 500 in grasslands and 500 in forests (Fig. 29.6)

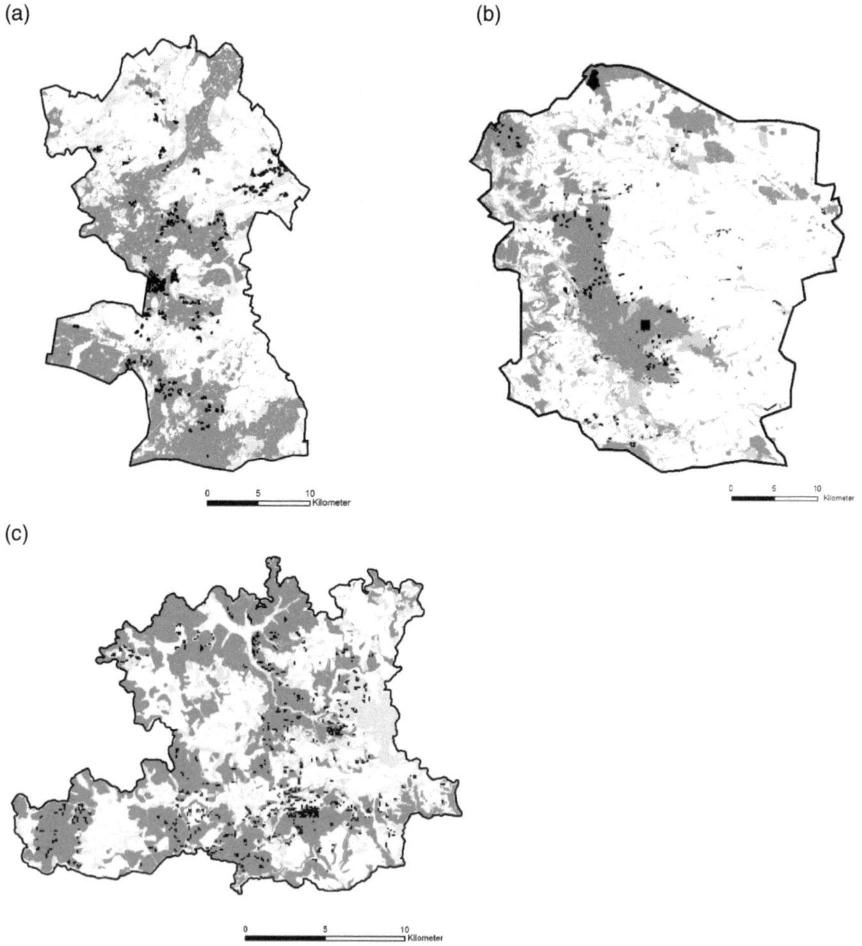

Fig. 29.6 Maps of the three biodiversity exploratories (**a**) 'Schorfheide-Chorin', (**b**) 'Hainich-Dün' and (**c**) 'Schwäbische Alb' in Germany. Maps show the outline of each exploratory, its forest (*dark grey*) and grassland areas (*light grey*) and the 1,000 study plots (*black dots*)

such as repeated vegetation surveys, assessments of bird, mammal and arthropod diversity, recordings of climate data and for experiments such as seed addition, topsoil disturbance and exclusion of certain taxa. These plots are called Experimental Plots. The main criterion for their selection was that they represent forests and grasslands of different land-use intensity while other potentially confounding variables such as soil type or situation in the topographic relief do not differ between plots. Further criteria were within-plot habitat homogeneity across a sufficiently large area to allow for enough space for the planned monitoring and experimental activities of 1 ha in forests and 0.25 ha in grassland. To avoid potential bias in plot selection due to environmental spatial autocorrelation, Experimental Plots of all land-use types are spread over the whole area of the exploratories and Experimental Plots of similar land use are separated by at least 300 m. Each Experimental Plot is being equipped with data loggers recording temperatures and moisture and humidity of soil and air. This equipment is operational since fall 2008. Out of the 100 Experimental Plots, 18 were selected for particularly detailed research, which can only be done with minimum replication.

29.11 Exploratories for Functional Biodiversity Research: Initial and Further Contributing Projects

The initial projects underway since July 2006 are addressing species diversity of selected taxa (plants, fungi, birds, bats and other mammals, selected insect taxa and microorganisms), genetic diversity of selected plant and microorganismal taxa and the proteomic diversity of soil water samples, which contain proteins of all kinds of taxa, including many of those living above ground (Schulze et al., 2005). The initial studies also focus on selected relationships between land-use intensity, diversity and associated ecosystem processes including biomass and carbon pools and fluxes, pollination services, predation, seed dispersal and community stability, which are assessed by monitoring and experimentation. The initial projects serve for assessing important biodiversity and ecosystem data and as examples for research projects exploring relationships between land use, biodiversity and ecosystem process in an integrated way.

A call of the Deutsche Forschungsgemeinschaft DFG for further projects contributing to the exploratories in December 2006 resulted in 171 project outlines and in almost half as many full research proposals. These proposals were reviewed in fall 2007 and the successful projects are addressing the diversity of further taxa, ecosystem processes and experiments since February 2008 (see www.biodiversity-exploratories.de for a full list of projects). Another call for further projects was issued in early 2008 and more projects started in early 2009. A further call is issued in early 2010.

29.12 Role and Significance of the Project

By funding the exploratories, the DFG is setting a signal for opening large-scale and long-term perspectives that are so urgently needed in ecology. The exploratories will pioneer the simultaneous assessments of the diversity of different plant, animal, fungal and microbial taxa combined with experimental manipulations and modelling approaches at the landscape level to unravel causal relationships and feedback loops between biodiversity and ecosystem processes. As these studies are done in well-replicated study plots representing gradients of land-use intensity they will allow us to understand biodiversity and ecosystem processes at the relevant scales and to draw meaningful conclusions for land management and conservation. The combination of a grid-based survey on large contiguous land areas with stratified Experimental Plots allows for combining research on a geographically large scale with investigations of processes which can only be conducted on a small number of intensively studied plots.

The exploratories are not meant to be mere research projects analysing land use, biodiversity and ecosystem processes across several spatial and temporal scales. They will also contribute to merging a broad-ranging and currently rather scattered scientific community of biodiversity and ecosystem researchers. This will be achieved by integrating biodiversity research at all levels of biological organisation, ranging from the molecular, individual and population level to the community level with ecosystem research in many replicate study plots in the landscape. The exploratories also integrate organismic and molecular research to overcome the unnecessary and unproductive polarisation between the two. The integrative function of the exploratories is furthered by supporting measures including conferences, workshops and courses for graduate students.

29.13 Conclusions from the Exploratories for the LTER Initiative

The national and international LTER initiatives aim at bringing together researchers and insights of existing and starting ecological research sites with a long-term perspective, originally only in the USA, but now also in Germany and globally (Kim, 2006). We all agree that the necessity and value of long-term studies in ecology cannot be overestimated. However, in Europe, mainly due to historical and financial reasons, the individual research sites in LTER constitute a patchwork of sites regarding the scope, variety of ecological questions, taxa and ecosystem measures that are addressed as well as in the statistical design of the projects including the generally small number of replicate study plots. In part these drawbacks can be overcome by joint analyses across LTER sites, which is one of the major advantages of creating an LTER network. Examples could be analyses of the relationship between diversity patterns of various taxa or associations between land use, plant diversity and plant biomass production because these variables are measured in many sites. Clearly, if measured over a long time, this can also serve for correlative studies of effects of global change. However, such joint analyses require a number of prerequisites. First, researchers of the different disciplines need to talk to and understand each other in order to find common grounds for true interdisciplinary synthesis (Fischer, Schreier, & Larigauderie, 1997). Second, common research questions need to be developed that are of fundamental ecological importance which cannot be addressed in single sites but require cross-site comparison of several LTER (and other) sites. Third, the analysis of these questions needs to be put into practice by developing common databases and by applying meta-analysis techniques or related approaches. Fourth, to increase the value of cross-LTER-site comparisons the independent environmental axis along which the LTER sites are arranged needs to be identified and quantified, e.g. by means of multivariate analyses of environmental and land-use data. Once such an axis has been identified, its relationship with variables describing ecosystem processes and biodiversity can be examined. Fifth, a standardised management, experimental and measurement protocol needs to be developed to increase the value of the LTER-site network by facilitating future data synthesis. Sixth, the across-site analysis would greatly benefit from a computer infrastructure linking the sites and providing automatic collection of standard variables measured in all sites, as planned in the NEON (National Ecological Observatory Network) initiative in the USA (Schimel, Hargrove, Hoffman, & MacMahon, 2007).

29.14 Concluding Remarks

We develop the biodiversity exploratories as an exemplary research platform for all scientists studying the feedback loops between biodiversity patterns, its consequences for ecosystem functioning and ecosystem services and land use. By combining monitoring of biodiversity and ecosystem measurements with experimental approaches, the exploratories will overcome the separation of disciplines that has hindered a comprehensive understanding of biodiversity and its functional consequences in the past. Both the spatial scale with sites across Germany and the temporal scale with a designated duration of more than 10 years assure that the results of the exploratories will uncover reliable results on the effects of land-use gradients and climate change on diversity patterns and ecosystem functioning in a real landscape. Much of the local and global change is still to come!

Acknowledgements We thank the Deutsche Forschungsgemeinschaft DFG for funding (SPP 1374), our institutions for administrating the exploratories, the participants of the exploratory project for their excellent and dedicated work, the authorities of the biosphere reserve Schorfheide-Chorin and the National Park Hainich and the START team of the biosphere area Schwäbische Alb for support, Sonja Gockel, Konstans Wells and Andreas Hemp for maps, Annett Börner and Ilka Mai for graphs and editorial help and Bernhard Schmid and Mark van Kleunen for comments on the chapter.

References

Andow, D. A. (1991). Vegetational diversity and arthropod population response. *Annual Review of Entomology*, *36*, 561–586.

Balvanera, P., Pfisterer, A. B., Buchmann, N., He, J.-S., Nakashizuka, T., Raffaelli, D., et al. (2006). Quantifying the evidence for biodiversity effects on ecosystem functioning and services. *Ecology Letters*, *9*, 1146–115.

Beck, E. (2004). German Research Programs, related to the understanding and conservation of biodiversity as an example of the impact of the Convention of Rio on an industrial nation. *Lyonia*, *6*, 7–18.

Blüthgen, N., Menzel, F., Hovestadt, T., Fiala, B., & Blüthgen, N. (2007). Specialization, constraints, and conflicting interests in mutualistic networks. *Current Biology*, *17*, 341–346.

Brändle, M., & Brandl, R. (2001). Species richness of insects and mites on trees: Expanding Southwood. *Journal of Animal Ecology*, *70*, 491–504.

Canals, L. M. I., Bauer, C., Depestele, J., Dubreuil, A., Knuchel, R. F., Gaillard, G., et al. (2007). Key elements in a framework for land use impact assessment within LCA. *International Journal of Life Cycle Assessment*, *12*, 5–15.

CBD. (1992). Convention on biological diversity (with annexes). Concluded at Rio de Janeiro on 5 June 1992. *United Nations Treaty Series*, *1760*, 142–302.

Christensen, N. L., Bartuska, A. M., Brown, J. H., Carpenter, S., D'Antonio, C., Francis, R., et al. (1996). The report of the Ecological Society of America Committee on the scientific basis for ecosystem management. *Ecological Applications*, *6*, 665–691.

Churchfield, S., Hollier, J., & Brown, V. K. (1991). The effects of small mammal predators on grassland invertebrates, investigated by field exclosure experiment. *Oikos*, *60*, 83–290.

Churchfield, S., & Rychlik, L. (2006). Diets and coexistence in Neomys and Sorex shrews in Bialowieza forest, eastern Poland. *Journal of Zoology*, *269*, 381–390.

Clough, Y., Holzschuh, A., Gabriel, D., Purtauf, T., Kleijn, D., Kruess, A., et al. (2007). Alpha and beta diversity of arthropods and plants in organically and conventionally managed wheat fields. *Journal of Applied Ecology*, *44*, 804–812.

Connell, J. H. (1978). Diversity in tropical rain forests and coral reefs – high diversity of trees and corals is maintained only in a non-equilibrium state. *Science, 199*, 1302–1310.

Ellenberg H. (1996). *Vegetation Mitteleuropas mit den Alpen in ökologischer, dynamischer und historischer Sicht.* Stuttgart: Ulmer.

Fischer, M., Schreier, E., & Larigauderie, A. (1997). Interdisziplinäre Forschung im "Integrierten Projekt Biodiversität" des Schweizerischen Nationalfonds: Ziele und Strukturen. *Zeitschrift für Ökologie und Naturschutz, 6*, 247–252.

Grime, J. P. (1973). Competitive exclusion in herbaceous vegetation. *Nature, 242*, 344–347.

Hansell, R. I. C., Craine, I. T., & Byers, R. E. (1997). Predicting change in non-linear systems. *Environmental Monitoring and Assessment, 46*, 175–190.

Hector, A., & Bagchi, R. (2007). Biodiversity and ecosystem multifunctionality. *Nature, 448*, 188–190.

Hector, A., Schmid, B., Beierkuhnlein, C., Caldeira, M. C., Diemer, M., Dimitrakopoulos, P. G., et al. (1999). Plant diversity and productivity experiments in European grasslands. *Science, 286*, 1123–1127.

Hendricks, F., Maelfait, J. P., Van Wingerden, W., Schweiger, O., Speelmans, M., Aviron, S., et al. (2007). How landscape structure, land-use intensity and habitat diversity affect components of total arthropod diversity in agricultural landscapes. *Journal of Applied Ecology, 44*, 340–351.

Hendrix, P. F., Parmelee, R. W., Crossley, Jr., D. A., Coleman, D. C., Odum, E. P., & Groffman, P. M. (1986). Detritus food webs in conventional and no-tillage agroecosystems. *BioScience, 36*, 374–380.

Hietel, E., Waldhardt, R., & Otte, A. (2005). Linking socio-economic factors, environment and land cover in the German Highlands, 1945–1999. *Journal of Environmental Management, 75*, 133–143.

Hobbs, R. J., Yates, S., & Mooney, H. A. (2007). Long-term data reveal complex dynamics in grassland in relation to climate and disturbance. *Ecological Monographs, 77*, 545–568.

Hooper, D. U., Chapin, F. S., Ewel, J. J., Hector, A., Inchausti, P., Lavorel, S., et al. (2005). Effects of biodiversity on ecosystem functioning: A consensus of current knowledge. *Ecological Monographs, 75*, 3–35.

Houghton, R. A. (1994). The worldwide extent of land-use change. *BioScience, 44*, 305–313.

Hughes, A. R., Byrnes, J. E., Kirnbro, D. L., & Stachowicz, J. J. (2007). Reciprocal relationships and potential feedbacks between biodiversity and disturbance. *Ecology Letters, 10*, 849–864.

Hunter, M. D., & Price, P. W. (1992). Playing chutes and ladders: Heterogeneity and the relative roles of bottom-up and top-down forces in natural communities. *Ecology, 73*, 724–732.

Hunter, M. L. (1999). *Maintaining biodiversity in forest ecosystems* Cambridge: Cambridge University Press.

Kim, E.-S. (2006). Development, potentials, and challenges of the International Long-Term Ecological Research (ILTER) Network. *Ecological Research, 21*, 788–793.

Koskela, E., Ollikainen, M., & Pukkala, T. (2007). Biodiversity conservation in commercial boreal forestry: The optimal rotation age and retention tree volume. *Forest Science, 53*, 443–452.

Kruess, A., & Tscharntke T. (2002). Species richness and parasitism in a fragmented landscape: Experiments and field studies with insects on *Vicia sepium. Oecologia, 122*, 129–137.

Loveland, T. R., Reed, B. C., Brown, J. F., Ohlen, D. O., Zhu, Z., Yang, L., et al. (2005). Development of a global land cover characteristics database and IGBP DISCover from 1 km AVHRR data. *International Journal of Remote Sensing, 21*, 1303–1330.

Memmott, J., Waser, N.M., & Price, M. V. (2004). Tolerance of pollination networks to species extinctions. *Proceedings of the Royal Society of London, Series B-Biological Sciences, 271*, 2605–2611.

Millennium Ecosystem Assessment. (2005). *Ecosystems and human well-being.* Washington, DC: Island Press.

Mitchell, T. D., & Jones, P. D. (2005). An improved method of constructing a database of monthly climate observations and associated high-resolution grids. *International Journal of Climatology, 25*, 693–712.

Mittelbach, G. G., Steiner, C. F., Scheiner, S. M., Gross, K. L., Reynolds, H. L., Waide, R. B., et al. (2001). What is the observed relationship between species richness and productivity? *Ecology, 82*, 2381–2396.

Murdoch, W. W., Peterson, C. H., & Evans, F. C. (1972). Diversity and pattern in plants and insects. *Ecology, 53*, 819–829.

Naeem, S., Thompson, L. J., Lawler, S. P. Lawton, J. H., & Woodfin, R. M. (1994). Declining biodiversity can alter the performance of ecosystems. *Nature, 368*, 734–737.

Perner, J., Wytrykush, C., Kahmen, A., Buchmann, N., Egerer, I., Creutzburg, S., et al. (2005). Effects of plant diversity, plant productivity and habitat parameters on arthropod abundance in montane European grasslands. *Ecography, 28*, 1–14.

Poschlod, P., Bakker, J. P., & Kahmen, S. (2005). Changing land use and its impact on biodiversity. *Basic and Applied Ecology, 6*, 93–98.

Prendergast, J. R., Quinn, R. M., Lawton, J. H., Eversham, B. C., & Gibbons, D. W. (1993). Rare species, the coincidence of diversity hotspots and conservation strategies. *Nature, 365*, 335–337.

Proulx, M., & Mazumber, A. (1998). Reversal of grazing impact on plant species richness in nutrient-poor vs. nutrient-rich ecosystems. *Ecology, 79*, 2581–2592.

Risch, S. J., Andow, D., & Altieri, M. A. (1983). Agroecosystem diversity and pest-control – data, tentative conclusions, and new research directions. *Environmental Entomology, 12*, 625–629.

Root, R. B. (1973). Organization of a plant-arthropod association in simple and diverse habitats – fauna of Collards (*Brassica oleracea*). *Ecological Monographs, 43*, 95–120.

Roscher, C., Temperton, V. M., Scherer-Lorenzen, M., Schmitz, M., Schumacher, J., Schmid, B., et al. (2005). Overyielding in experimental grassland communities – irrespective of species pool or spatial scale. *Ecology Letters, 8*, 419–429.

Rosenzweig, M. L. (1995). *Species diversity in space and time.* Cambridge: Cambridge University Press.

Rudmann-Maurer, K., Weyand, A., Fischer, M., & Stöcklin, J. (2008). The role of landuse and natural determinants for grassland vegetation composition in the Swiss Alps. *Basic and Applied Ecology, 9*, 494–503..

Sala, O. E., Chapin, F. S., Armesto, J. J., Berlow, E., Bloomfield, J., Dirzo, R., et al. (2000). Biodiversity – Global biodiversity scenarios for the year 2100. *Science, 287*, 1770–1774.

Schimel, D., Hargrove, W., Hoffman, F., & MacMahon, J. (2007). NEON: A hierarchically designed national ecological network. *Frontiers in Ecology and the Environment, 5*, 59.

Schmid, B. (2002). The species richness-productivity controversy. *Trends in Ecology & Evolution, 17*, 113–114.

Schröter, D., Cramer, W., Leemans, R., Prentice, I. C., Araujo, M. B., Arnell, N. W., et al. (2005). Ecosystem service supply and vulnerability to global change in Europe. *Science, 310*, 1333–1337.

Schulze, W. X., Gleixner, G., Kaiser, K., Guggenberger, G., Mann, M., & Schulze, E.-D. (2005). A proteomic fingerprint of dissolved organic carbon and of soil particles. *Oecologia, 142*, 335–343.

Seabloom, E. W., & Reichman, O. J. (2001). Simulation models of the interactions between herbivore foraging strategies, social behaviour, and plant community dynamics. *American Naturalist, 157*, 76–96.

Seabloom, E. W., & Richards, S. A. (2003). Multiple stable equilibria in grasslands mediated by herbivore population dynamics and foraging behaviour. *Ecology, 84*, 2891–2904.

Senn, J., & Suter, W. (2003). Ungulate browsing on silver fir (*Abies alba*) in the Swiss Alps: Beliefs in search of supporting data. *Forest Ecology and Management, 181*, 151–164.

Shucksmith, M., Thomson, K. J., & Roberts, D. (Eds.) (2005). *The CAP and the regions: The territorial impact of the common agricultural policy.* Wallingford: CABI Publishing.

Siemann, E. (1998). Experimental tests of effects of plant productivity and diversity on grassland arthropod diversity. *Ecology, 79*, 2057–2070.

Siemann, E., Tilman, D., Haarstad, J., & Ritchie, M. (1998). Experimental tests of the dependence of arthropod diversity on plant diversity. *American Naturalist, 152*, 738–750.

Sims, R. E. H., Hastings, A., Schlamadinger, B., Taylor, G., & Smith, P. (2006). Energy crops: Current status and future prospects. *Global Change Biology, 12*, 2054–2076.

Smeets, E. M. W., Faaij, A. P. C., & Lewandowski, I. M. (2004). *A quickscan of global bioenergy potentials to 2050 – an analysis of the regional availability of biomass resources for export in relation to the underlying factors.* Utrecht: Utrecht University.

Stöcklin, J., Bosshard, A., Klaus, G., Rudmann-Maurer, K., & Fischer, M. (2007). *Landnutzung und biologische Vielfalt in den Alpen – Fakten, Perspektiven, Empfehlungen.* Zürich: vdf Verlag.

Suding, K. N., Collins, S. L., Gough, L., Clark, C., Cleland, E. E., Gross, K. L., et al. (2005). Functional- and abundance-based mechanisms explain diversity loss due to N fertilization. *Proceedings of the National Academy of Sciences of the United States of America, 102*, 4387–4392.

Thuiller, W. (2007). Biodiversity – Climate change and the ecologist. *Nature, 448*, 550–552.

Tilman, D., Reich, P. B., Knops, J., Wedin, D., Mielke, T., & Lehman, C. (2001). Diversity and productivity in a long-term grassland experiment. *Science, 294*, 843–845.

Tylianakis, J. M., Tscharntke, T., Lewis, O. T. (2007). Habitat modification alters the structure of tropical host-parasitoid food webs. *Nature, 445*, 202–205.

Van Valen, L. (1965). Morphological variation and width of ecological niche. *American Naturalist, 99*, 377–390.

Vellend, M. (2003). Island biogeography of genes and species. *American Naturalist, 162*, 358–365.

Vellend, M., & Geber, M. A. (2005). Connections between species diversity and genetic diversity. *Ecology Letters, 8*, 767–781.

Völkl, W., Zwölfer, H., Romstöck-Völkl, M., & Schmelzer, C. (1993). Habitat management in calcareous grasslands – effects on the insect community developing in flower heads of Cynarea. *Journal of Applied Ecology, 30*, 307–315.

Waser, N. M., & Ollerton, J. (2006). *Plant-pollinator interactions: From specialization to generalization.* Chicago: University of Chicago Press.

Wilson, E. O. (2001). *The diversity of life.* London: Penguin Press Science.

Zeiter, M., Stampfli, A., & Newbery, D. M. (2006). Recruitment limitation constrains local species richness and productivity in dry grassland. *Ecology, 87*, 942–951.

Part VIII
Conclusions

Chapter 30

On the Way to an Integrative Long-Term Ecosystem Research – Milestones, Challenges, and some Conclusions

Felix Müller, Cornelia Baessler, Mark Frenzel, Stefan Klotz, and Hendrik Schubert

Long-Term Ecosystem Research has been successfully carried out for some decades now, but networking, tuning, and harmonization at least in Europe have just started. Several developmental targets have been obtained and very interesting results have been achieved. Regrettably, these milestones have been reached in a few regional cases only, and, i.e., due to the relatively short-term existence of long-term environmental investigations on the global scale, many urgent questions – with existential problems of mankind among them – could not be answered up to now. In spite of the comprehensive demands for future research including the global scale, some of the key queries may be discussed constructively. Therefore, as an attempt to summarize the complex information of the previous papers, we might refer to the questions which have been asked in the introduction of the book.

In Part II of the book *The Significance of Ecological Long-Term Processes* has been discussed and the focal questions have addressed (i) the basic processes suitable for long-term-research, (ii) the typical questions arising from these LTER objects, and (iii) the motivations to investigate these interrelationships.

The targets of long-term environmental research of course are widespread, but initially this branch of science is based on the demand of better understanding ecological systems and their behavior. While we know a lot about annual cycles and change processes within a small number of years, there are big gaps of knowledge if we spread the viewpoint toward longer periods. In this situation we can easily observe that such long-term processes have already been affected and altered by extreme environmental changes. As demonstrated by Müller et al. we need long-term research to understand ecosystem complexity, ecosystem self-organization, the role of hierarchies and scales, and the dynamics of large-scale variables which could function as strong constraints in the future. Theoretical knowledge and environmental policy have meanwhile accepted the inevitability of change, but to find optimal adaptation strategies, we do not know enough about the mechanisms of long-term dynamics, disturbances, rare events, and collapse in ecosystems and human–environmental systems. Furthermore, as Fath and Müller have worked out, even the constructive and complexifying branch of ecosystem development (e.g., orientor theory, succession theory) needs empirical tests which should be focal parts of LTER activities. Special attention should be paid to slow processes, episodic phenomena, rare events, and processes with high variability and complex interactions. Also the regularity hypothesis of breakdown dynamics, which has been depicted by the adaptive cycle of the resilience alliance, should be investigated with more emphasis: During the temporal long-term development of all systems (ecosystems as well as social and economic systems) complexity rises steadily; therefore also the support costs of the multiple individual interactions rise and the overall system has to allocate increasing portions of the captured energy into the maintenance demands to make secure the long-term existence of the entity. Due to the flow arrangements of mature ecosystems, the system's reserves for adaptation decrease and therefore the risk of collapse due to a reduced adaptability increases with what we would call developmental stage. Maybe, under several aspects, the attractiveness of such mature states in environmental

F. Müller (✉)
Ecology Centre, University of Kiel, Olshausenstraße 75, D-24118 Kiel, Germany
e-mail: fmueller@ecology.uni-kiel.de

F. Müller et al. (eds.), *Long-Term Ecological Research*,
DOI 10.1007/978-90-481-8782-9_30, © Springer Science+Business Media B.V. 2010

Table 30.1 The mission of the US-LTER Network encompasses six goals (from Chapter 5)

Understanding: To understand a diverse array of ecosystems at multiple spatial and temporal scales
Synthesis: To create general knowledge through long-term, interdisciplinary research, synthesis of information, and development of theory
Information: To inform the US-LTER and broader scientific community by creating well-designed and well-documented databases
Legacies: To create a legacy of well-designed and well-documented long-term observations, experiments, and archives of samples and specimens for future generations
Education: To promote training, teaching, and learning about long-term ecological research and the Earth's ecosystems and to educate a new generation of scientists
Outreach: To reach out to the broader scientific community, natural resource managers, policymakers, and the general public by providing decision support, information, recommendations, and the knowledge and capability to address complex environmental challenges

management has to be reflected critically. Additionally Schimming et al. have demonstrated the high value of long-term data sets and the high potentials for symbiotic collaboration between long-term research and monitoring. Thus, added values from LTER can be produced by jointly improving networks, indicators, and methodologies to optimize the early warning potential of monitoring schemes. Vice versa the monitoring data can be excellent sources for the test of hypotheses related to long-term ecological processes. These ideas have been explicated with reference to water quality and forest integrity, and of course also the consequences of climate change, the loss of ecosystem services, and the analysis of human–environmental interactions should be objects of monitoring and research.

In Part III *International Experience on Long-Term Ecological Research* is documented. While Gnauck et al. and Hostert et al. demonstrate the high potential of mathematical, statistical, remote sensing, and modeling techniques, Gosz et al. have provided a report about 28 years of LTER in the United States, describing the development of the program, its contents, questions, and outreach (Table 30.1) as well as the experience made throughout all of that time. Besides the broad experience and knowledge that has been accumulated in US-LTER, the value of organization and cooperation becomes extremely obvious by looking at this national example. But it also became obvious that long-term research on a national scale does not allow answering the full set of open questions asked in the first section. Therefore broadening the regional coverage by installation of an international network was a logical consequence. As part of the global LTER network (ILTER) the European LTER network unites 18 formal national networks and comprises a high number of existing ecological research sites. Mirtl demonstrates that long-term research across larger spatial scales allows for the detection and quantification of processes of ecosystems and socio-ecological systems that determine the sustainable provision of ecosystem services. As multifunctional network LTER-Europe has the potential to optimize a distributed research infrastructure and to stimulate the development of research projects meeting societal needs.

Part IV has put emphasis on the presentation and interpretation of research *Concepts and Results concerning Long-Term Ecological Processes*. The focal questions were related to the basic results of available long-term research projects, the focal processes that have been observed in different ecosystem and landscape types, and the question 'How can those results be used to link research and application?'

In Part IV, the focus is on the results of *Aquatic Long-Term Ecosystem Research*. In the chapters of van Beusekom et al., Stockmann et al., Schubert et al., and Köhler and Poschlod et al. the recent development of exemplary marine systems, brackish ecosystems, and different freshwater systems has been described on the base of long-term data sets. One result of these contributions is a list of main anthropogenic pressures which are provoking changes in the states of the ecosystems. While the modifications of water temperatures, i.e., in winter have been leading to changes of the plankton and benthos communities in most of the systems, the situation concerning the consequences of eutrophication is quite different. In the Wadden Sea reduced nutrient inputs are already changing the ecological situation, e.g., to be indicated by a decreasing number of phytoplankton blooms. The brackish systems do not show a similar behavior: Increasing eutrophication is still a major pressure for the investigated ecosystems, but the authors expect a future nutrient reduction also in the Baltic Sea. Lakes and rivers still are strongly influenced by high nutrient loads, and only in some cases a change toward mesotrophic conditions might have started. The data also show that lotic systems have been impacted by consequences of acidification processes. The development of all aquatic systems can be characterized by a change in species composition, whereby a 'loss of the extremes' in river flora can be marked, while invasive species have played a major role especially in marine ecosystems. The future trends might be dominated by climate change processes; not only the rise of water temperatures but also the sea level rise might play a major role in the future.

Similar tendencies can be seen in *Terrestrial Long-Term Ecosystem Research*, which are described in Part V. Koehler and Melecis describe successional traits of soil fauna over 20-year time slots. Biogeochemical long-term data sets are explained and interpreted with reference to ground water and stream water quality in small catchments (Lischeid et al.). Schindler et al. have demonstrated the consequences of different agricultural land use regimes for the water and nitrate behavior in agricultural landscapes. The flows and budgets of nitrogen and sulfur have been observed in the forest ecosystems of the Bornhöved Lakes Region since 1989. Tavares et al. have shown some general biogeochemical developmental tendencies for that period. Baessler et al. demonstrated the influence of regional and historical processes, as well as local and present environmental factors on local plant community and population structure. The authors disentangled landscape-specific patterns from general relationships consistent across landscapes as they extend both a species diversity and a landscape genetic approach to three agricultural landscapes in central Germany. Ecosystem research concepts can also be applied on broader scales. Respective approaches have been presented by Syrbe et al., who show results and study designs of landscape change investigations. A modern tool set for such landscape investigations is presented by Hostert et al. who have been using remote sensing methods in LTER investigations to investigate ecological patterns and to provide spatial explicit information. Enhancing the extent of investigations, Schmidt et al. have presented GIS applications with reference to the long-term development of airborne heavy metal inputs in Germany.

The chapters of Part VI are related to the *Linkage between Research and Applications*. The case studies from Biosphere Reserves and National Parks (Luthardt, Heurich et al., Diederichs) show an optimistic picture about the potential interrelations between research, monitoring, and environmental policy. This attitude is supported by Jones et al., who report about modern networking designs for long-term ecosystem research and application.

Part VII has been established to discuss *Future Demands and Challenges of Long-Term Ecosystem Research*. In four chapters ideas and conceptions for future research and application of long-term investigations are outlined and designed: Singh et al. as well as Ohl and Swinton argue for a better linkage between regional socio-economic investigations and long-term ecological research. They illustrate their requirements with materials about the actual foundation of the Long-Term Socio-Ecological Research Program. Other challenges are described by Dilly et al.: From their perspective long-term ecological research results directly flow into the development of principles and management strategies of restoration measures. In this case, long-term observations are necessary to control the potential success of restoration activities as well as artificial ecosystem creations. Finally, Fischer et al. have reported their program for a long-term research project focusing on the development of biodiversity items in different investigation areas which will be a guiding component of future LTER research in Germany.

Summarizing the strategic backbones of the interesting chapters ahead, it can be stated that long-term ecological research has been developed in different regional situations, representing an acknowledged and highly supported science which is reflected in founding measures and programs on the EU and the national level. Of course, all of these stages have to continue working on the long-term research questions, thus developing toward more mature stages in all regional and national divisions. Within the respective developmental process, we see one focal necessity for long-term research: That is *Integration*, and we understand this requirement as an emerging system of several linked components:

- *The integration of different temporal and spatial scales* can be understood as the basic motivation of all LTER activities. In this context we should not be exclusive (e.g., considering *only* temporal periods between decades and centuries) but be aware that also short-term processes influence long-term behavior: In many cases long-term dynamics are created due to the emergent connectivity of short-term processes. For example, contrasting the generation time of microbes and trees it is evident that long-term research might cover completely different temporal scales. All discussed developments are constrained by those structures and processes which have turnover and change rates longer than 100 years.

- *The integration of structural and functional items in ecosystems* and their linkage in self-organized entities: Ecosystem research is based on the attempt to link structural and functional items that are compositions of system components on the one hand and the flows between these structural items on the other. If we take a hierarchical view on this duality, it fades, because the only remaining difference will be the frequency of change. Therefore the questions how these different frequencies interact and how their development is organized will remain a focal objective of long-term ecosystem research.
- *The integration of different disciplinary process classes in ecosystems*: The subsequent requirement of course has to be understood as a cross-disciplinary thesis. Ecosystem integration needs disciplinary competence for all components, but most important is the motivation of the disciplinary actors to cooperate.
- *The integration of different processes between ecosystems*: If the target of LTER is a hierarchical analysis of processes on different temporal scales it has to be acknowledged that they generally are related to different spatial scales as well. There is a functional continuum from small and fast subsystems to large and slow items. Understanding this structure requires to take into account patches, ecosystems, landscapes, and landscape complexes in order to understand ecological organization.
- *The integration of emergent properties that are created in ecosystem development* consequently must be the focal target of theoretical long-term process analysis. There are several future tasks and challenges arising from this idea, such as

  • Testing *orientor theory* and the respective hypotheses on the base of long-term data sets;
  • Relativization of the *stability* concept, as long-term analyses demonstrate that stability per se does not exist, although it is still used as a political target;
  • Quantifying *resilience*, which has become more and more prominent, taking the place of a relatively 'tolerant' stability concept; in spite of many theoretical and conceptual discussions in the literature, data-based interpretations are extremely rare;
  • Assessing ecosystem *adaptability* which is demanded in an increasing degree due to the practical demands of global change impacts.

- *The integration of natural sciences and social sciences* seems to be a central target of LTER activities, and the LTSER concepts and the increasing numbers of LTSER platforms under construction in Europe underline the importance to integrate human and environmental systems.
- *The development of an integrated methodology of long-term ecological research* should be another future task. Due to the time-related starting point, temporal elements should be investigated with a much higher degree, indicators of temporal characteristics should be derived, and the hierarchical nature of interacting timescales should be investigated with much more emphasis than before, leading to an ecological theory of time which should be based on well-elaborated statistical, simulation, and data handling tools.
- *The integration of organizational units* is a focal activity of all national and international units of LTER and ILTER. This consists predominantly of capacity and network-building and the development of 'networking goods and services.' The whole process has a strong social component which makes intensive and meaningful communication inevitable. Put into practice this means that people of the different levels of LTER and between the levels have to meet regularly to be motivated by participation. Besides theoretical umbrellas like a common research strategy on the level of ILTER or LTER-Europe the practical issues and benefits are most important. One of these is, e.g., the commitment of national networks, institutions, and persons to the criteria necessary for networking in LTER. Especially important is and will be the

framework for information management as the crucial basis for standards in storing data and working with distributed databases. Moreover tangible products like the meta-data database available for LTER-Europe play an essential role for the identity and the self-organization of the LTER community. The main outcome of this network-building should be adjusted basic programs and rules to enable comparisons and research projects on biogeographical scales rather than on local or national scales.

- *The integration of ecosystem science and monitoring*: For the scientific mainstream innovation is the focal target. Therefore 'monitoring' is a word which makes several scientists shiver, because many (small-scale) repetitions are necessary to find a (broad-scale) proposition. But the related fear seems to be unauthorized because in spite of the potentially minor methodological thrill, long-term series provide exciting results, which are not only very interesting but also very important for an overall understanding of ecosystem function. Therefore, scientists and practitioners should interact much better to demonstrate that all parties can benefit from joint monitoring schemes.
- *The integration of theory and application*, as stated in the title of this book, must be one significant target of the future development of long-term ecological science. Only if this component is fulfilled can we guarantee an increasing provision of new scientific knowledge for the improvement of environmental practice and policy. Therefore new concepts of long-term ecosystem management should be developing hand in hand with the development of respective research strategies.

Reading the list above it becomes obvious that the success of our recent attempts of integration is crucial for the output of all the long-term research activities. Missing integration, on what of the above points ever, may reduce the output of decades of individual effort to monitoring lists sleeping for further decades in drawers. On the other hand, as demonstrated by the examples presented here, many of the integration tasks are already well developed and showing fruits encouraging further efforts.

Index

F. Müller et al. (eds.), *Long-Term Ecological Research*,
DOI 10.1007/978-90-481-8782-9, © Springer Science+Business Media B.V. 2010

FSC
www.fsc.org
MIX
Papier aus verantwortungsvollen Quellen
Paper from responsible sources
FSC® C105338

If you have any concerns about our products,
you can contact us on
ProductSafety@springernature.com

In case Publisher is established outside the EU,
the EU authorized representative is:
Springer Nature Customer Service Center GmbH
Europaplatz 3, 69115 Heidelberg, Germany

Printed by Libri Plureos GmbH
in Hamburg, Germany